MEDICAL AND VETERINARY ENTOMOLOGY

Second Edition

Dedication

To my father and mother whose unobtrusive sacrifices made my University training possible

2nd edition
This edition is dedicated to the late John R. Linley (1938–1994), my one-time colleague and friend of 35 years, whose untimely death has cut short a most productive career at its peak.

MEDICAL AND VETERINARY ENTOMOLOGY

Second Edition

Edited by

D.S. Kettle

Emeritus Professor
Department of Entomology
University of Queensland
Australia

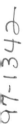

CAB INTERNATIONAL

CAB INTERNATIONAL　　　　Tel: +44 (0)1491 832111
Wallingford　　　　　　　　　Telex: 847964 (COMAGG G)
Oxon OX10 8DE　　　　　　　E-mail: cabi@cabi.org
UK　　　　　　　　　　　　　Fax:+44 (0)1491 833508

A catalogue record for this book is available from the British Library.

ISBN 0 85198 968 3 Hbk
ISBN 0 85198 969 1 Pbk

Typeset in 10/12 Plantin by Colset Pte Ltd, Singapore.
Printed and bound in the UK at the University Press, Cambridge

Contents

Acknowledgements   vii

Part I: General Introduction to Medical and Veterinary Entomology

   1.   Introduction   3

   2.   Classification of Arthropoda and the Medical Importance of
       Groups of Minor Significance   13

   3.   Classification and Structure of the Diptera   41

   4.   Mouthparts of Insects of Medical and Veterinary Importance and
       Host Finding   55

   5.   Internal Structure and Function of Insects   73

   6.   Species Complexes, and Variation in *Aedes aegypti* and *Aedes*
       albopictus   88

Part II: Insects and Acarines of Medical and Veterinary Importance
(A) Diptera

   7.   Culicidae (Mosquitoes)   109

   8.   Ceratopogonidae (Biting Midges)   152

   9.   Psychodidae – Phlebotominae (Sandflies)   177

  10.   Simuliidae (Blackflies)   192

  11.   Tabanidae (Horseflies, Deer Flies, Clegs)   211

  12.   Glossinidae (Tsetse Flies)   225

  13.   Muscidae and Fanniidae (Houseflies, Stableflies)   248

14. Calliphoridae, Sarcophagidae (Blowflies) and Myiasis   268

15. Oestridae (Gad Flies, Warble Flies and Stomach Bots)   292

16. Hippoboscidae (Keds, Louse Flies)   315

(B) Other Insects

17. Siphonaptera (Fleas)   323

18. Blood-sucking Hemiptera (Bugs)   344

19. Phthiraptera (Lice)   361

(C) Acari

20. Acari – Astigmata and Oribatida (Mange Mites, Beetle Mites)   383

21. Acari – Prostigmata and Mesostigmata (Chiggers, Blood-sucking
Mites)   414

22. Ixodida – Argasidae (Soft Ticks)   440

23. Ixodida – Ixodidae (Hard Ticks)   458

**Part III: Diseases of which the Pathogens are Transmitted by
Insects or Acarines**

24. Arboviruses   489

25. Typhus and Other Rickettsial Diseases   517

26. Relapsing Fevers, Borrelioses, Plague and Tularaemia   537

27. Malaria (*Plasmodium*) and Other Haemosporina (Sporozoa)   558

28. Babesiosis and Theileriosis   591

29. Trypanosomiases and Leishmaniases   612

30. Lymphatic Filariasis (*Wuchereria bancrofti, Brugia malayi, B. timori*) 638

31. Human Onchocerciasis (*Onchocerca volvulus*)   656

32. Other Helminths Transmitted by Insects   671

Index   691

Acknowledgements

The first edition was based on courses given to undergraduate and postgraduate students at the Universities of Edinburgh and Queensland. It was displayed as a poster presentation at the International Congress of Entomology in Hamburg and Tropical Medicine and Malaria in Calgary in 1984, where I listened with some concern, lest the newly published book contained any major mistake(s). For the second edition I decided it was preferable to go to the Congresses before attempting the revision. As a result of attending the Entomology Congress in Beijing, the Tropical Medicine and Malaria Congress in Thailand and the Arbovirus Research Symposium in Australia, I made contact with workers in various disciplines in many parts of the world. I am deeply grateful to these busy, working scientists who took the trouble to answer my queries, often embarrassing me with the generosity of their responses. They are listed alphabetically under their countries of residence in the next paragraph. My sincere apologies to anyone who has been inadvertently omitted from the list.

Australia: Joan Bryan, Peter Cranston, Craig Eisemann, Des Foley, Bruce Halliday, David Kemp, Tom McRae, Don Sands, Allan Saul, Margaret Schneider; *Austria*: Erich Kutzer; *Canada*: Jim Sutcliffe, *China*: Cheng-yi Li, Hou-yong Wu, Lu Bao Lin, Shan-qing Wang, Tongyan Zhao; *France*: P. Houvette; *Germany*: R. Garms; *India*: V.P. Sharma, Aruna Srivastava; *Indonesia*: Ruben Dharmawan; *Italy*: Mario Coluzzi; *Ivory Coast*: Christian Back; *Japan*: Akira Igarashi, Takahiro Uchida, Kimito Uchikawa; *New Zealand*: Bob Pilgrim, Hamish Spencer; *South Africa*: Richard Hunt; *Taiwan*: Jih-Ching Lien; *Thailand*: Supat Sucharit; *UK*: Peter Billingsley, Nick Burgess, Alan Clements, Chris Curtis, Brian Laurence, Mike Lehane, Steven Lindsay, Wallace Peters, David Tarry, Nigel Wyatt; *USA*: Arshad Ali, Dov Borovsky, Lane Foil, Neal Haskell, Nancy Hinkle, Jerome Hogsette, Daniel Leprince, the late John Linley, Stephen Murphree, James Oliver, Kenneth Pruess, Michael Rust, Andrew Spielman, Ronald Ward, Richard Wilkerson.

I also want to record my thanks to Professor Gordon Gordh for his support and encouragement; to Greg Daniels, who was always willing to resolve my computing problems; to Tina Mayes who prepared more than 60 illustrations for this edition and Roy Mayes who carefully read the final draft commenting on punctuation and inconsistency for which I am grateful; to my wife, Babs, for support and, when sought, ready comment. I am indebted to Mary O'Sullivan and the staff of the Biological Sciences Library at the University of Queensland for their cooperation and a willingness to search out publications unavailable in Brisbane; to Brian Laurence who generously offered illustrations he had drawn but not published. I appreciate the ready cooperation of Ward Cooper of Chapman & Hall; A. Danbi of the Onchocerciasis Control Program Liaison Office, Geneva for Fig. 31.1; David Thomson of the Office of Publications, WHO, Geneva for allowing the use of the Dengue distribution map (Fig. 24.1) which, at that time, had not been published. Lastly but not least it is a pleasure to record my thanks to Tim Hardwick, who was responsible for the first edition being published by Croom Helm and, with his competent supporting staff produced this edition for CABI.

I

GENERAL INTRODUCTION TO MEDICAL AND VETERINARY ENTOMOLOGY

Introduction 1

The science of Entomology should, strictly speaking, be restricted to the six-legged animals or Insecta, but the applied entomologist is expected to cover a wider field. The agricultural entomologist is often required to advise on soil nematodes and to distinguish between damage done by insects and that due to other pathogens such as viruses. The medical and veterinary entomologist is expected to deal with the Acari, i.e. ticks and mites, and to be reasonably informed on the other terrestrial Arthropoda, e.g. spiders, scorpions, etc. In this book I shall use the term 'medical entomologist' in place of the more accurate but cumbersome 'medical and veterinary entomologist', and references to 'medical entomology' should be taken to include 'veterinary entomology'. Similarly, when used in a general sense, the terms 'entomology' and 'insects' should be taken to include the Acari as well as the Insecta.

Medical entomology is concerned with the role of insects in the causation of disease in animals and humans. This concern is paramount and hence the centre of interest of the medical entomologist must be disease incidence and disease control, not the insect and insect control. Insect control is one means of disease control and, in a particular setting, may not be the most appropriate method. Where disease incidence is low and the insect widespread, it may be more realistic to treat cases as they arise rather than attempt to control the vector. The medical entomologist who forgets this primary involvement with disease and focuses attention solely on the insect does a disservice to the profession. The individual may be a better entomologist but not necessarily a good medical entomologist.

Animal and human disease arise basically from two causes: the presence of an introduced agent or pathogen upsetting the normal functioning of the organism, or a breakdown of the organism's integrating system leading to the development of organic disease. Medical entomology is obviously only concerned with diseases caused by pathogens. Sometimes the insect itself may be the pathogen as in scabies, a skin disease due to the presence of the mite, *Sarcoptes scabiei*, and pediculosis due to infestation with the human body louse, *Pediculus humanus*, or head louse, *P. capitis*. More commonly the role of the insect is as a vector of the pathogen from one host to another. Insects function as vectors in one of two ways, either mechanically or biologically. In mechanical transmission the insect acquires the pathogen from one source and deposits it in other locations, where it may infect a new host. The role of houseflies in the transmission of enteric diseases is mechanical. Houseflies are attracted equally to faeces and food on both of which they feed. Consequently organisms picked up on and in the body of a housefly, when it is feeding on faeces, are carried away and may be deposited on human food, when the fly feeds there. The housefly functions in a similar manner to a pathologist's platinum loop, which is used to lift a sample from faeces and apply it to a suitable agar plate for incubation and isolation of pathogens. The important feature to be recognized is that there are other routes for the spread of enteric diseases, e.g. by faecal contamination of drinking water or by human carriers of the pathogen handling food for human consumption.

In biological transmission the only natural route for the pathogen to take from host to host is through an insect. Thus, while it is possible to transmit malaria by blood transfusion, in nature the only way the malarial organism is passed from person to person is through the bite of the *Anopheles* mosquito. The role of the insect is very important in biological transmission and insect control is a major weapon in the armoury of disease control. There is a fundamental difference in the effect of insect control on disease incidence depending on whether transmission is mechanical or biological. Elimination of the insect vector of a biologically transmitted disease will eliminate the disease whereas eradication of a mechanical vector will only reduce the incidence, but not eliminate the disease. Elimination of *Anopheles* mosquitoes eradicates malaria but removal of houseflies only reduces the incidence of enteric disease; the pathogens will continue to spread through other routes.

Another major difference between biological and mechanical transmission involves the onset and duration of the vector's infectivity. Infectivity of a mechanical vector declines sharply with time and by 24 hours is, to all intents and purposes, nil (see Chapter 11). Biologically transmitted organisms undergo a cycle of development in the vector. Consequently there is a period after ingestion of the pathogen when the infected insect is not infective. When the pathogen's cycle is complete the insect is infective and usually remains so for the rest of its life. Mechanical or biological transmission is a character of the pathogen not of the insect vector. For example, although horseflies (Diptera, Tabanidae) are mechanical vectors of *Trypanosoma evansi*, the causative organism of surra in horses and camels, they are biological vectors of *Trypanosoma theileri* a benign parasite of cattle (see Chapter 29).

The greater part of the medical entomologist's work is concerned with diseases in which insects are biological vectors. In the simplest situation there are three different organisms involved: the pathogen, the vertebrate host and the insect vector. Such is the case with malaria involving the malarial parasite *Plasmodium*, people and the *Anopheles* mosquito. The parasite cycles between people and the mosquito. Theoretically there are three ways in which the disease may be controlled: (i) by breaking the human/mosquito contact and preventing transmission; (ii) by eradicating the *Anopheles* mosquito; and (iii) by chemotherapy of the population to eliminate the source of infection of the mosquito. In practice all three approaches are attempted with the emphasis being determined by local circumstances. Eradication of the mosquito is not necessarily required. It is enough to reduce the *Anopheles* population to a level below which no transmission occurs. Transmission is a matter of probability and the epidemiology of malaria is a quantitative relationship between three different organisms: humans, *Plasmodium* and *Anopheles*, each with their own ecologies (see Chapter 27).

In medical entomology many disease situations are complicated by the existence of a fourth component, which, from the human point of view, is often referred to as the reservoir host. It is the main population of the pathogen and the source of human infections. In such diseases there are two cycles: (i) reservoir host, pathogen, vector; and (ii) people, pathogen, vector. The pathogen is the same in both cases but the vector is likely to be different. For example, sylvan yellow fever is circulated among primates in the Bwamba Forest, Uganda, through *Aedes africanus* but the virus is transmitted to humans by *Aedes simpsoni* which feeds on both humans and monkeys and bites the latter when they invade banana plantations. From the pathogen's point of view the significant cycle is that among the reservoir host and human involvement is an accident of little quantitative importance. From the human standpoint the important feature is the relationship between the two cycles.

Certain consequences follow from this dual cycle. Firstly, treatment of human cases may make little or no difference to the incidence of the disease which is maintained by 'spill-over' from the reservoir host cycle. This is the case with flea-borne (murine) typhus caused by *Rickettsia mooseri* and transmitted by the tropical rat flea *Xenopsylla cheopis*. Reduction of the disease incidence depends either on dealing with the reservoir cycle or ensuring the separation of the two cycles. In the case of murine typhus, elimination requires reducing the close contact between rats and people. Adequate rat-proofing of human dwellings and storage of food in rat-proof containers will effectively eliminate the risk of murine typhus to people but leave the reservoir cycle in the rat unaffected. Dealing with the reservoir cycle is often more a theoretical than a practical possibility. For example, more than 200 species and subspecies of rodents are capable of harbouring *Yersinia pestis*, the causative organism of human plague. Clearly, in this case, elimination of the reservoir cycle is impractical and ecologically undesirable.

It should be evident now that the role of the medical entomologist is as a member of a team and, to be effective, it is necessary to understand the main features of the biology of both pathogen and host. Only then will the medical entomologist be in a position to appreciate those aspects of the ecology and biology of the vector, which are of the greatest importance in the epidemiology of an insect-borne disease. Thus to the medical entomologist the study of tsetse flies, *Glossina* spp., is inextricably bound up with the control of trypanosomiasis in humans and domestic animals. This approach is not to be interpreted too narrowly and limited solely to those aspects of tsetse biology which are of obvious epidemiological significance but wider investigations must be justified ultimately in their contribution to our understanding of the epidemiology of trypanosomiasis. This is not to deny the value of using medically important insects for the study of fundamental biological problems. The yellow fever mosquito, *Aedes aegypti*, has been colonized for many years and has proved a suitable subject for many biological investigations but it would be simplistic to regard all these investigations as studies in medical entomology. Such studies are clearly of interest to the medical entomologist but so are other biological investigations of fundamental phenomena which are undertaken on organisms of no direct importance to the medical entomologist.

In accordance with this understanding of medical entomology the chapters of this book are arranged in three parts. The first deals with basic introductory entomology, the second with the recognition, biology and bionomics (ecology) of insects of medical importance, and the third with pathogens and diseases, emphasizing those aspects which are relevant to an understanding of the insects' role in the epidemiology of disease. Part III, restricted to pathogens of which insects are biological vectors, will be highly selective. It will deal cursorily, if at all, with diagnosis, pathology, clinical symptoms and structure and identification of pathogens. Insect-borne pathogens are to be found in a wide range of life forms: Viruses, Rickettsiales, Bacteria, Spirochaetales, Protozoa, Cestoda and Nematoda.

Generalizations and Quantitative Information

It is necessary in a book attempting to deal with a topic as broad as medical and veterinary entomology, to have recourse to generalizations. The reader should appreciate that, when such all-embracing statements are found in the text, they are not infallible. They are considered to have a high probability of being correct but there is also a finite probability of the existence of exceptions. Nevertheless, if this limitation is borne in mind, generalizations can be a valuable aid to learning. The place for the consideration of exceptions is in the advanced text or monograph. Every teacher will have learned with dismay that however briefly in the course of a lecture an exception is mentioned, there is a perverse streak in the learning processes of students whereby the exception will be the one firm fact that will be retained.

For example, it would not be unreasonable to make the general statement that all species of *Anopheles* mosquitoes are biologically capable of transmitting malaria. There are exceptions to this generalization. In the laboratory, some species of *Anopheles* are difficult to infect with certain strains of plasmodia to which they are not exposed in nature. However, the generalization that all species of *Anopheles* are biologically competent to transmit malaria has value but it must not be extrapolated and misinterpreted as indicating that all species of *Anopheles* are of equal importance in a field situation. This is quite untrue. The significant difference between species of *Anopheles*, which are major vectors of malaria in the field, compared with those which are unimportant, does not lie in their susceptibility to infection but in their behaviour, in particular the frequency with which they feed on the human population.

The medical and veterinary entomologist is concerned with the epidemiology of human and animal disease. The qualitative aspects of disease transmission form the framework within which epidemiology seeks to quantify the relationships between the various components (pathogen, vector, host) of the system. For this reason a certain amount of numerical information will be included in the text. As with generalizations, such numerical data should not be regarded as having the precision of physical constants but as indicators of the order of magnitude of the factor concerned. This information should provide the reader with a numerical framework within which he or she can begin to quantify a problem. It is important, for example, to know the approximate fecundity of a vector, the frequency with which it oviposits and the length of its life cycle. Many of these quantities are variable being dependent on environmental conditions, particularly temperature, and therefore the figures quoted will refer to optimal conditions. The numerical data presented in the text can prove useful, provided that they are regarded as approximations.

Nomenclature

The classification of living organisms is built on the concept of a species. Species are the building blocks out of which the edifice displaying relationships within the animal kingdom is constructed. A precise definition of a species is difficult, if not impossible. Definitions range from the valuable, but largely theoretical, 'a species is a population of animals which interbreed, producing fertile offspring', to the more practical, but decidedly vague, definition that, 'a good species is one recognized by a good taxonomist'. Interspecific sterility alone will not allow the coexistence (sympatry) of closely related (sibling) species without the development of an isolating specific-mate recognition system (Paterson, 1993). Notwithstanding the difficulty of defining the term 'species' it remains the basic unit of classification.

Each species has a name based on the binominal system devised by the Swedish naturalist, Linnaeus, in the eighteenth century. The name consists of two components, the generic and specific. The specific name, once published, is inviolable but the genus to which the species is referred may change with increased knowledge of the group. The full name of a species includes the author and the year in which the description was first published, e.g. *Musca domestica* Linnaeus, 1785, the common housefly. When, at a later date, a species is placed in a different genus from that to which it was referred by the author, then the author's name is placed in parentheses, e.g. *Aedes aegypti* (Linnaeus, 1762). When Linnaeus described *aegypti* he placed it in the genus *Culex* which he had created in 1758. The genus *Aedes* was not established by Meigen until 1818. The reader should note that the generic name always begins with an upper case letter while the specific name starts with a lower case letter, although botanical taxonomists use an upper case letter when a species has been named after a particular person.

Two sources of error can arise within this system. A species may be described more than once or several species may be confused and regarded as a single species. The first error is, in theory, the easier to deal with. Two or more names for the same species are synonyms and the earliest description has priority. For example the yellow fever mosquito was named *Culex aegypti* by Linnaeus in 1762, *Culex argenteus* by Poiret in 1787, and *Culex fasciatus* by Fabricius in 1805. These three names are synonyms and the earliest description, that of Linnaeus, has priority and the specific name of the mosquito is *aegypti*. Synonyms confound the reader but do not invalidate observations on the several 'species' since the observations are equally applicable to the single species.

In a large genus it is convenient to establish subgenera and the subgenus is then shown in parentheses after the generic name, e.g. *Aedes* (*Stegomyia*) *aegypti* (Linnaeus, 1762). *Stegomyia* was established by Theobald in 1901 as a genus distinct from *Aedes* but later workers include it as a subgenus of *Aedes*. The decision as to whether a group of species is best regarded as constituting a genus or included as a subgenus within an existing genus, is largely a matter of judgement by the individual taxonomist, who will need to possess a wide range of knowledge of the whole group.

The second error is more important. It involves homonyms in which two or more species have been confused and referred to by the same name. This often invalidates all previous biological data because it is uncertain as to which of the two or more entities particular observations apply. For example, the common European malarial mosquito *Anopheles maculipennis* Meigen, 1818, has proved to be a complex of at least six species (see Chapter 6), including a revised description of *An maculipennis* Meigen, 1818; *An sacharovi* Favre, 1903; *An labranchiae* Falleroni, 1926; and *An melanoon* Hackett, 1934. It is usual for the full name of the species to be given only in taxonomic papers. Most journals require the author's name, but not the year, to be given when a species is first cited in a scientific paper. After this the author's name is omitted and the generic name abbreviated to a single capital letter. Where such abbreviation might cause confusion, as for example when *A* could refer to *Aedes* or *Anopheles*, mosquito taxonomists use two-letter abbreviations for the genera of mosquitoes and *Aedes* and *Anopheles* are abbreviated to *Ae* and *An*, respectively (Reinert, 1975).

Sometimes a trinominal system is used with the third name denoting a subspecies, e.g. *Anopheles melanoon melanoon* Hackett, 1934, and *Anopheles melanoon subalpinus* Hackett and Lewis, 1935. *An m. melanoon* is widely distributed in southern Europe extending from Spain in the west to Iran in the east while *An m. subalpinus* has been recorded only from Albania and Turkey. It is desirable to keep the use of subspecific names to a minimum and to avoid the naming of numerous 'varieties'. Species complexes, such as that of *An maculipennis*, appear now to be common in widespread species and considerably complicate the work of the medical entomologist (see Chapter 6).

Species are placed in a genus (plural genera), genera are associated into tribes, tribes into subfamilies and so on into increasingly larger aggregations designated families, superfamilies and orders. Since in many cases the words for these different levels may use the same stem, it is important to appreciate the significance of the endings which, fortunately, are standardized.

Grade	*Ending*	*Example*
Order	– ptera (commonest ending)	Diptera
Superfamily	– oidea	Muscoidea
Family	– idae	Muscidae, Culicidae
Subfamily	– inae	Muscinae, Culicinae
Tribe	– ini	Culicini

Zoogeographical Regions

As a result of the past history of the continents, especially their degree of isolation from other land masses, and the geographical origin of the various animal groups, the faunas of different parts of the world are distinctive. It is useful to be able to describe the distribution of an insect or a disease by reference to zoogeographical regions rather than to national boundaries and names, which are constantly changing. For this purpose six main regions are recognized. They are:

Palaearctic: Europe, Africa north of the Tropic of Cancer but including the Sahara; China north of 30°N; Asiatic former USSR; Korea and Japan

Nearctic: USA, Canada, Greenland, Alaska and North Mexico

Afrotropical: the whole of Africa south of the Tropic of Cancer but excluding the Sahara

Oriental: India, Pakistan, south-east Asia, China south of 30°N, Malaysia and Indonesia

Australian: Australia, New Guinea, New Zealand and the Pacific Islands

Neotropical: Southern Mexico, Central America and South America.

There are common elements in the faunas of the Nearctic and Palaearctic regions and they are referred to collectively as the Holarctic region. The Afrotropical region was originally designated the Ethiopian region, but since Ethiopia forms only a very small part of the region, Crosskey and White (1977) proposed the more appropriate term, Afrotropical region.

Introduction to the Literature

The intention of this book is to provide the reader with sufficient background to be able to delve into the literature with some confidence. The text attempts to provide an outline sketch of the vectors, their biologies and the diseases they transmit.

References cited in the text are listed alphabetically at the end of each chapter. They do not represent an exhaustive coverage of the literature nor are the references cited necessarily the most important papers published. Very often they have been selected because they illustrate a particular point which has been made in the chapter. Their role is to provide the student with an entry into the literature. Each paper cited will, in turn, give references to earlier work and enable the reader to explore the subject as fully as necessary. The problem will be to keep abreast of new developments. Medical and veterinary entomologists are favourably placed for keeping up to date through the monthly *Review of Medical Entomology*, formerly the *Review of Applied Entomology B – Medical and Veterinary*. Two journals are devoted to the discipline – the *Journal of Medical Entomology*, founded by the late J. Linsley Gressit of the Bishop Museum, Hawaii, and now produced by the Entomological Society of America. A newer publication is *Medical and Veterinary Entomology* published by the Royal Entomological Society in London.

An abstracting journal of interest to the medical entomologist is the *Tropical Diseases Bulletin*. This publication covers the whole range of tropical diseases, whether or not they involve arthropods, and deals with aspects of the disease, e.g. pathology and medication, which are remote from medical entomology. Nevertheless the bulletin has a section devoted to medical entomology and is a useful source especially when a wider understanding of a particular disease is required. The *Annual Review of Entomology* is another publication in which articles of interest to the medical and veterinary entomologist appear. However, medical and veterinary entomology is only one aspect of entomology covered by the review.

Relevant recent publications are *Insects and Acarines of Medical Importance*, Lane and Crosskey (1993), *A Colour Atlas of Medical Entomology*, Burgess and Cowan (1993), *The Arthropods of Humans and Domestic Animals, a Guide to Preliminary Identification*, Walker (1994), *Biology of Blood-sucking Insects*, Lehane (1991) and *A Colour Atlas of Arthropods in Clinical Medicine*, Peters (1992). Checklists of preferred names have been produced by Pittaway (1991) for *Arthropods of Medical and Veterinary Importance* and by Wood (1992) for *Insects of Economic Importance*.

Special Programme for Research and Training in Tropical Disease (TDR)

The greatest contribution that medical and veterinary entomology can make to human development is in the tropics and subtropics where arthropod-borne diseases are most abundant (Anon., 1989). The tropics is also the region where the greatest number of developing countries are to be found. There has been a drive by the United Nations, independent foundations and grant-giving bodies, to improve human welfare in newly independent countries by eliminating or reducing substantially diseases affecting humans and their animals.

The United Nations Development Programme with the World Bank and the World Health Organization has formulated a 'Special Programme for Research and Training in Tropical Disease' (TDR). This programme, which commenced in 1975, listed seven major human diseases, of which five: malaria, filariasis, African trypanosomiasis, Chagas' disease and leishmaniasis, involve medical entomology. It was estimated (Anon., 1982) that in 1981, 1800 million people in 107 countries were exposed to malaria (see Chapter 27) and 150 million new cases occurred each year. Several hundred million people were affected by onchocerciasis and lymphatic filariasis (see Chapters 30, 31). African trypanosomiasis threatened 45 million people in 38 countries, and Chagas' disease 65 million people in Latin America with 24 million persons chronically infected (see Chapter 29). Leishmaniasis (see Chapter 29) claimed 400,000 new cases a year.

Following a review in 1992–1993 new targets and management structure for TDR came into effect on 1 January 1994 . Research on the diseases listed above will continue but the existing disease-specific research and development steering committees will be phased out and new steering committees established with the aim of achieving greater flexibility and interchange across as well as within diseases. Research and development activities will fall into three areas: strategic research, product research and development, and applied field research (Anon., 1993).

AIDS and Tropical Diseases

In the 1980s a new factor affecting human health has been the emergence of AIDS caused by infection with the Human Immunodeficiency Virus (HIV). Infection is widespread in the developing countries of the Afrotropical region and is spreading at a disturbing rate in South America and Asia, where in Thailand HIV infects 1% of the population (Anon., 1992a). This has two consequences for the control of vector-borne diseases. HIV infection leads to the expression of latent infections, e.g. visceral leishmaniasis (Peters *et al.*, 1990), and has a direct and indirect effect upon the health budget. In the Afrotropical region AIDS is affecting trained staff in their economically productive years leading to increased absenteeism, a loss of specialized personnel and a declining economy (Anon., 1992a). The control of vector-borne diseases now competes with the care of AIDS patients for the meagre health budget.

It is no surprise therefore that simple methods of control have been enthusiastically embraced by developing countries. The use of impregnated bednets to combat malaria and filariasis has been vigorously adopted. At the XIIIth International Congress of Tropical Medicine and Malaria in 1992 there were no fewer than 38 presentations on the use of impregnated bednets – 16 were poster displays (Anon., 1992b) and 22 were communications made at the two 'round-table' sessions (Curtis, 1993). In the new structure of TDR there will be an applied field research task force on 'Bednets'.

Entomophobia

In addition to the harm that insects can do to humans and their livestock by biting and transmitting pathogens there is the very real but irrational response that susceptible individuals can make to the presence of insects. It is a response out of all proportion to the real or imagined danger involved. It is often induced by spiders, bees or wasps. Such reactions appear laughable to an unaffected onlooker but are very real to the person reacting, who may know in hindsight that the response is unwarranted. These fears can be overcome by suitable desensitizing treatment (Hardy, 1988). Treatment is more difficult where the insects are imaginary. The insects claimed to be present are usually mites, bedbugs, fleas or lice and psychiatric counselling is required to eradicate such imaginary infestations (Leclercq and Musalek, 1992).

References

Anon. (1982) Special programme for research and training in tropical diseases. *TDR Newsletter* 18, 5–8.

Anon. (1989) *Geographical Distribution of Arthropod-borne Diseases and their Principal Vectors*. WHO Division of Vector Biology and Control. WHO/VBC/89.967, Geneva, Switzerland.

Anon. (1992a) *The Hidden Costs of AIDS. The Challenge of HIV to Development.* The Panos Institute, London, UK.

Anon. (1992b) *Abstracts of the XIIIth International Congress for Tropical Medicine and Malaria* 2, 261–264.

Anon. (1993) TDR undergoes major reorganisation. *TDR News* 43, 1–12.

Burgess, N.R.H. and Cowan, G.O. (1993) *A Colour Atlas of Medical Entomology.* Chapman & Hall, London.

Crosskey, R.W. and White, G.B. (1977). The Afrotropical region a recommended term in zoogeography. *Journal of Natural History* 11, 541–544.

Curtis, C.F. (1993) Workshop on bednets at the International Congress of Tropical Medicine. *Japanese Journal of Sanitary Zoology* 44, 65–68.

Hardy, T.N. (1988) Entomophobia: The case for Miss Muffet. *Bulletin of the Entomological Society of America* 34, 64–69.

Lane, R.P. and Crosskey, R.W. (eds) (1993) *Insects and Acarines of Medical Importance.* Chapman & Hall, London.

Leclercq, M. and Musalek, M. (1992) Délires d'infestation: entomophobie, acarophobie, dermatophobie parasitaire. Psychopathologie et thérapeutique. *Revue Medicale de Liège* 47, 305–313.

Lehane, M.J. (1991) *Biology of Blood-sucking Insects.* Harper Collins Academic, London.

Paterson, H.E.H. (1993) *Evolution and the Recognition Concept of Species – Collected Writings.* Johns Hopkins University Press, Baltimore.

Peters B.S., Fish, D., Golden, R., Evans, D.A., Bryceson, A.D.M. and Pinching, A. (1990) Visceral leishmaniasis in HIV infection and AIDS: clinical features and response to therapy. *Quarterly Journal of Medicine* 77, 1101–1111.

Peters, W. (1992) *A Colour Atlas of Arthropods in Clinical Medicine,* Wolfe Publishing Ltd, London, UK.

Pittaway, A.R. (1991) *Arthropods of Medical and Veterinary Importance. A Checklist of Preferred Names and Allied Terms.* CAB International, Wallingford, Oxon, UK.

Reinert, J.F. (1975) Mosquito generic and subgeneric abbreviations (Diptera: Culicidae). *Mosquito Systematics* 7, 105–110.

Walker, A. (1994) *The Arthropods of Humans and Domestic Animals, a Guide to Preliminary Identification.* Chapman & Hall, London.

Wood, A. (1992) *Insects of Economic Importance. A Checklist of Preferred Names.* CAB International, Wallingford, Oxon, UK.

Further Reading

Annual Review of Entomology, Annual Reviews Inc., Palo Alto, California, USA.

Journal of Medical Entomology, Entomological Society of America, Lanham, Maryland, USA.

Medical and Veterinary Entomology, Royal Entomological Society, London. UK.

Review of Medical and Veterinary Entomology, CAB International, Wallingford, UK.

Tropical Diseases Bulletin, CAB International, Wallingford, UK.

Classification of Arthropoda and the Medical Importance of Groups of Minor Significance

2

The largest phylum in the animal kingdom is the Arthropoda, which contains about 80% of the known species of animals. They are bilaterally symmetrical, segmented animals with jointed legs. Each segment consists of a dorsal sclerotized plate, the tergum, and a similar ventral plate, the sternum, the two being joined together laterally by membranous pleura (singular pleuron). The terga and sterna of successive segments are separated by intersegmental membranes which, together with the pleura, provide flexibility. The primitive arthropod was probably a worm-like creature with some cephalization at the anterior end, and metameric segmentation, i.e. each segment being similar to the one before and behind. Each segment bore a pair of appendages, and those on segments incorporated into the head formed the mouthparts. The exoskeleton provides a limit to growth and periodically arthropods have to develop a new skin under the existing one and then cast the old skin. The process is known as ecdysis or moulting and, in the Insecta, the interval between ecdyses is an instar. Internally arthropods have the typical invertebrate arrangement of ventral nerve cord and dorsal heart, the reverse of the vertebrate arrangement.

Arthropod Venoms

Before considering the classification of the Arthropoda it will be useful to examine the part played by arthropods in causing disease. Their main economic importance is as vectors of pathogens to humans and domestic animals. The vectors are species of insects and acarines, but a larger range of arthropods can cause severe reactions and even death by their stings or bites. Arthropod venoms were reviewed in a substantial work under the editorship of Bettini (1978). It includes six chapters on the venoms of spiders; six on Hymenoptera (ants, bees and wasps); four on scorpions; one each on centipedes, millipedes and ticks; one on other arachnids; and four on different insect orders, but not including the Diptera. It was originally

13

planned to include a chapter on the Diptera but this was abandoned as the information available was 'so scanty'.

Classification of Arthropoda

The Arthropoda may be divided into those with antennae, the Antennata, and
those lacking antennae but possessing chelicerae, the Chelicerata. The chelicerae
are mouthparts (see Chapter 20). Two classes of Arthropoda are especially important to the medical entomologist – the Insecta in the Antennata and the Arachnida
in the Chelicerata. Three other classes are of minor significance.

Class Crustacea

Strictly speaking the Crustacea are outside the scope of this book, but they are
common and of minor medical importance. With few exceptions the Crustacea are
aquatic, and the majority are marine. Their bodies are organized into a
cephalothorax and a posterior abdomen. They possess two pairs of antennae, which
may be reduced, and at least five pairs of legs. This class includes the crabs,
lobsters, shrimps and water fleas. They are medically important as intermediate
hosts of helminths, e.g. the river crab (*Eriocheir japonicus*) is an intermediate host
of the lung fluke *Paragonimus ringeri* and copepods of the genus *Cyclops* are
intermediate hosts of the guinea worm *Dracunculus medinensis*. Human beings
become infected by eating inadequately cooked river crabs or by drinking water
containing infected *Cyclops*. Predacious *Mesocyclops* are being used to control
container-breeding mosquitoes (Brown *et al.*, 1992).

Class Chilopoda (centipedes) (Lewis, 1981)

Chilopods are long, soft-bodied, terrestrial arthropods which are dorsoventrally
flattened (Fig. 2.1). They have one pair of antennae on the head and three pairs
of appendages associated with the mouth. Behind the head the body is metamerically segmented and composed of at least 17 segments, each of which, with the
exception of the last segment, bears a pair of legs. The first pair of legs act as powerful jaws which are pierced by a duct through which the secretion of the venom
glands is injected into the victim. The bite is painful and the effects may be long-
lasting, from some hours to three months, but only very rarely is the bite of a
chilopod fatal to humans. About 3000 species have been described. The soil-
dwelling, worm-like chilopods have up to 181 pairs of legs while the surface-
dwelling forms have 15 to 21 pairs. The larger, up to 26 cm long, brightly coloured,
tropical scolopendromorphs are the ones most likely to be encountered and may
attack humans when they are disturbed as they shelter in dark, humid places.

Class Diplopoda (millipedes) (Hopkin and Read, 1992)

Millipedes are metamerically segmented, long-lived (up to 11 years), terrestrial
arthropods which feed on decaying vegetable matter (Fig. 2.2). They have one pair

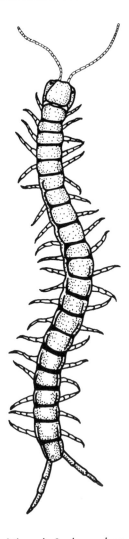

Fig. 2.1. Dorsal view of a chilopod, *Scolopendron morsitans*.
Redrawn from Lewis (1981).

of antennae and two pairs of appendages associated with the mouth. Each
apparent segment bears two pairs of legs and two pairs of spiracles, the respiratory
openings. Millipedes do not possess biting jaws of the centipede type but produce
defensive secretions from segmental glands of which there are not more than one
pair per segment. Secretions of some species act as anti-feedants, sedatives and
toxins to predators, and others species secrete a range of organic compounds
including benzoquinones. A few species secrete cyanide, enabling *Pachydesmus
crassicutis* to repel fire ants and *Apheloria corrugata* to discourage predacious safari
(doryline) ants. These secretions may cause considerable discomfort to humans if
they come into contact with sensitive skin or, if they get into the eyes, blindness,

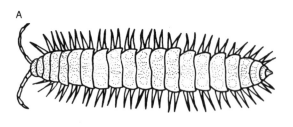

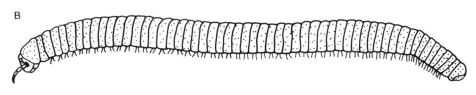

Fig. 2.2. **(A)** Dorsal view of a flat-back diplopod, *Polydesmus* sp. **(B)** Lateral view of a cylindrical diplopod, *Cylindroiulus* sp. Redrawn from Hopkin and Read (1992).

but they are not lethal. Subcycindrical millipedes may be up to 30 cm in length and flat-backed millipedes to 13 cm.

Class Arachnida (Grassé, 1949; Savory, 1977)

The Arachnida are carnivorous, terrestrial, chelicerate arthropods which have no antennae. They have one pair of chelicerae used in feeding, a pair of pedipalps whose function varies from order to order and four pairs of walking legs. The Arachnida is a large class containing nine orders of which three are of medical importance.

Order Scorpiones (scorpions) (Fig. 2.3)

Present-day knowledge of scorpions has been consolidated in *The Biology of Scorpions* edited by Polis (1990) and summarized by Cloudsley-Thompson (1992). About 1400 species of scorpions have been described throughout the world. They are large (up to 23 cm long), long lived (up to 25 years) arachnids with powerful chelate pedipalps whose abdomens end in a large globular sting terminating in a large curved spine. The pectines on the ventral surface are mechanoreceptors and contact chemoreceptors used by the male during mating to select a suitable place on which to deposit a spermatophore as the couple *promenade à deux*. The male then pulls the female over the spermatophore which the female detects with her pectines and spreads her genital opercula to take in its contents.

Scorpions can be roughly divided into a group that lives on or in the ground and a group that lives in vegetation. The latter hide away during the day under bark, in tree holes, among epiphytes, while ground dwelling scorpions live in burrows, under rocks and other surface debris. Burrows vary in depth from 15 to 100 cm and, more rarely, may extend to more that 200 cm in depth. Some scorpions

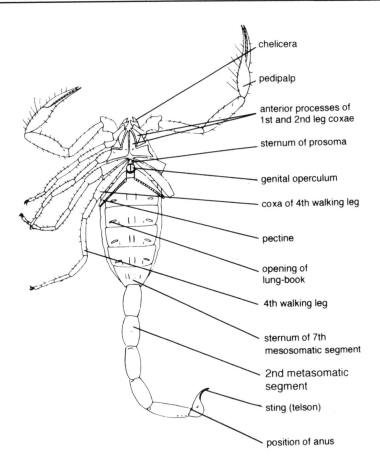

chelicera

pedipalp

anterior processes of
1st and 2nd leg coxae

sternum of prosoma

genital operculum

coxa of 4th walking leg

pectine

opening of
lung-book

4th walking leg

sternum of 7th
mesosomatic segment

2nd metasomatic
segment

sting (telson)

position of anus

Fig. 2.3. Ventral view of a scorpion. Source: Snow, K.R. (1970) *The Arachnids*, Routledge & Kegan Paul, London.

are ambush predators which remain motionless near the entrance to their burrows and detect prey by contact receptors on the pedipalps. Other scorpions are active predators and can orientate to and capture prey up to 15 cm away. The vibrations set up by the approaching prey are detected by the tarsal hairs and the slit sensillum on the two terminal segments of the legs. When large prey is seized in the pedipalps the sting is rapidly brought over the top of the body and thrust into the victim. Through the sting a venom is forcibly injected into the prey by muscular action. When, in the laboratory, ejection is artificially stimulated the venom may be propelled tens of centimetres. Small prey are devoured without being stung.

The effect of the toxin on mammals depends on the species of scorpion and is independent of size. In general scorpions that have massive claws are relatively harmless, while more dangerous species have slender chelae. The large *Hadrurus arizonensis* is comparatively harmless, while the small *Centruroides sculpturatus* is deadly. There is variation within a genus. Thus *C. sculpturatus* and *C. limpidus* are dangerous while *C. pantherinus* and *C. vitatus* are not (Minton, 1974). Most of the dangerous species belong to the family Buthidae which contains more than 500 species. Death is due to cardiac failure or to paralysis of the respiratory muscles and occurs within a few hours. Treatment may be provided by specially prepared sera. It has been estimated that over 5000 people die each year from scorpion stings (Cloudsley-Thompson, 1992). The venom of a single species of scorpion may contain one toxin which subdues insects, another that is effective against crustaceans and a third that deters mammalian predators (Simard and Watt, 1990).

Scorpion venom has been likened to Cobra venom because of the similarity of the victim's response but the two venoms are distinct. Scorpion venoms are homologous for amino acid sequences as are the venoms of elapid snakes but the sequences in the two groups are quite different. They also act differently on the nervous system. Elapid toxins produce an antidepolarizing block of the end plate, while scorpion venoms depolarize different target cells (Watt *et al.*, 1974). The result may be the same but the method of bringing it about is distinct. Scorpions will not attack but will defend themselves when apparently threatened. Most stings occur because scorpions have taken shelter in shoes or clothing or because

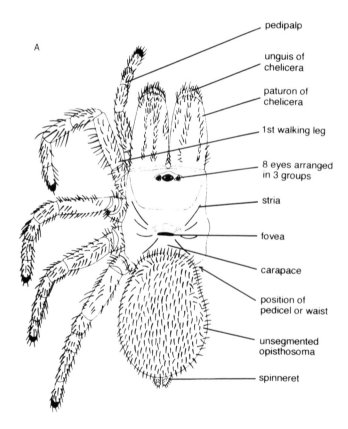

Fig. 2.4. (and opposite) Dorsal (**A**) and ventral (**B**) views of a mygalomorph spider. Source: Snow, K.R. (1970) *The Arachnids*, Routledge & Kegan Paul, London.

inquisitive humans too casually explore under stones or into holes in the ground. About 30 of the world's most dangerous species of scorpions have been reviewed by Keegan (1980).

Order Araneae (spiders) (Fig. 2.4)

More than 34,000 species of spiders have been described (Coddington and Levi, 1991). They are characterized by having a uniform prosoma (anterior portion of the body) joined by a narrow pedicel to an unsegmented opisthosoma (hind portion). The pedipalps are tactile, leg-like structures, shorter than the ambulatory legs. In the male they are modified as intromittent organs and male spiders are readily recognized by the terminal swelling on the pedipalp. The chelicerae are two-segmented but not chelate. The distal segment is sharply pointed and bears at its tip the opening of the poison duct. The poison gland may be contained in the basal segment of the chelicera or, more usually, occupies the anterior part of the prosoma. In the majority of spiders (suborder Araneamorphae) the chelicerae are fixed vertically with the terminal segment, the fang (unguis), being concealed

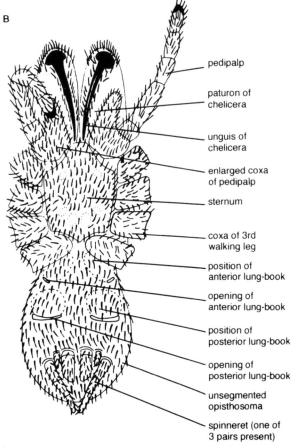

B

pedipalp

paturon of chelicera

unguis of chelicera

enlarged coxa of pedipalp

sternum

coxa of 3rd walking leg

position of anterior lung-book

opening of anterior lung-book

position of posterior lung-book

opening of posterior lung-book

unsegmented opisthosoma

spinneret (one of 3 pairs present)

Fig. 2.4 *contd*

in a groove in the larger basal segment. In action the fangs converge towards the midline horizontally. In the Mygalomorphae the chelicerae are directed forwards horizontally and in action the parallel fangs strike vertically downwards. Spider venom is adapted to the prey species on which the spider feeds, mostly invertebrates, but the mygalomorph spider, *Selonocosmia javanensis*, attacks and kills birds.

Most spiders are harmless being either unable to penetrate the human skin or having ineffective venom. Some of the larger species may cause a temporary local reaction and spiders of at least four genera can seriously affect humans (Coddington and Levi, 1991). The Australian funnel web spider, *Atrax robustus*, a mygalomorph, has caused the deaths of adult humans (Southcott, 1978) and all species of the genus should be regarded as dangerous. They are long-lived spiders (up to 20 years). The females form permanent nests in burrows and are only encountered by exploring humans while the males wander more widely at night in search of mates and are more likely to be encountered by passing humans (Main, 1985). An effective antivenom has been produced against *A. robustus* (Sutherland, 1980).

Species of *Latrodectus*, an araneomorph genus, produce a neurotoxic venom and cause a severe human reaction but rarely death in adults of good health. The *L. mactans* complex is a group of black spiders with red or yellow patches on the opisthosoma. They are widely distributed throughout the warmer parts of the world. *L. mactans* is the black widow spider of the USA; *L. tridecimguttatus* occurs in southern Europe; *L. hasselti* in Australia, *L. indistinctus* in southern Africa (Newlands and Atkinson, 1988) and three species, including *L. mactans*, in Brazil (Lucas, 1988). An effective antivenom has been available to combat bites of *L. hasselti* since 1956.

Araneomorph spiders of the genus *Loxosceles* produce a cytotoxic venom which causes tissue destruction. The initial bite may be painless and pass unnoticed, but locally extensive, necrotic patches develop around the location of the bite of *Loxosceles* in southern Africa (Newlands and Atkinson, 1988) and Brazil (Lucas, 1988). Over a 20-year period in Chile *Loxosceles laeta* caused five deaths among 133 cases (Schenone *et al.*, 1975). The brown recluse spider, *Loxosceles reclusa*, causes necrotic damage to humans in the USA (Wong *et al.*, 1987). In Brazil cases of necrosis formerly attributed to *Lycosa* spp. are now referred to *Loxosceles* and production of a lycosid antivenom discontinued. Lycosid bites are the commonest spider bites in Brazil but usually do not need medication. Spiders of the genus *Phoneutria* are aggressive and their neurotoxic venom is potentially fatal (Lucas, 1988). In southern Africa the venom of the genus *Sicarius* is cytotoxic and regarded as potentially lethal to humans (Newlands and Atkinson, 1988).

Order Acari (mites and ticks) (Fig. 2.5)

The Acari are sometimes treated as a subclass of the Arachnida. They are small arthropods 0.2–1.5 mm long, varying widely in form but in which the prosoma

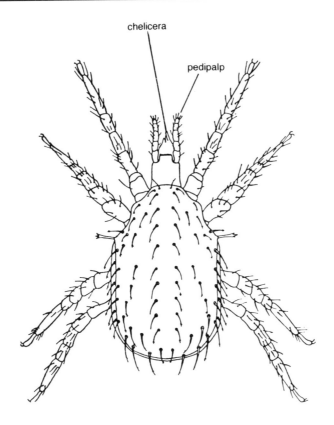

chelicera

pedipalp

Fig. 2.5. Dorsal view of an acarine, *Zercoseius ometes*. Reproduced with permission from Savoury, T.H. (1977) *Arachnida*, Academic Press, London.

and opisthosoma are broadly fused and abdominal segmentation is inconspicuous or absent. The pedipalps are short sensory structures associated with the chelicerae in a discrete gnathosoma. Acarines are extremely important agents of disease and vectors of pathogens, and will be considered in Chapters 20–23.

Orders of minor interest (Fig. 2.6)

The solifugids (sun or camel spiders) are an ancient order of Arachnida, being known as fossils from the Carboniferous period. They are large, hairy, nocturnal, carnivorous arachnids of desert areas in the tropics and subtropics. About 800 species have been described. Solifugids are easily recognized by their possession of large, powerful, chelate chelicerae with which they seize their prey (Fig. 2.6D). They will attack any suitable prey including small vertebrates and each other. However, they have no poison glands and rely solely on the crushing power of the chelicerae.

The Uropygi (whip-scorpions) and Amblypygi (whip-spiders) are flat bodied, mainly tropical arachnids, which are nocturnal predators with powerful pedipalps,

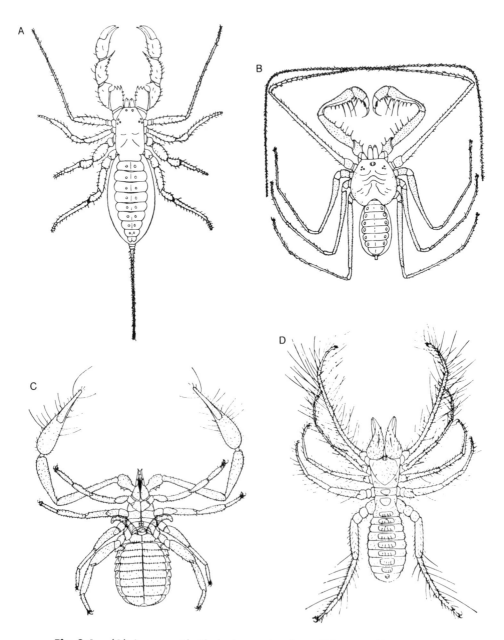

Fig. 2.6. (**A**) A uropygid, *Thelyphonus insularis*. (**B**) An amblypigid, *Stegophrynus dammermani*. (**C**) A pseudoscorpion, *Chelifer cancroides*. (**D**) A solifugid, *Galeodes arabs*. Reproduced with permission from Savoury, T.H. (1977) *Arachnida*, Academic Press, London.

but lacking poison glands. In both orders only the posterior three pairs of legs are used for walking and those of the first pair are tactile and, in action, are stretched out in front of the animal. In the amblypygids the first pair of legs is excessively long, more than twice the length of the walking legs (Fig. 2.6B). In the uropygids the abdomen terminates in a segmented flagellum, as long as the rest of the abdomen (Fig. 2.6A).

The pseudoscorpions are a widely distributed order of small arachnids (<8 mm) with large chelate pedipalps, superficially resembling small scorpions but they lack a sting (Fig. 2.6C). They are common in soil and decaying vegetation, and a few species are to be found in food stores and among books, presumably feeding on book-lice (psocids).

Superclass Hexapoda (Chapman, 1982; Davies, 1988; Naumann, 1991)

The Hexapoda are characterized, as the name implies, by having three pairs of walking legs. They also have three pairs of mouthparts, and the genital area placed posteriorly. This superclass contains the Insecta and three other groups of small hexapods. The latter are entognathous and prognathous, that is, the mouthparts and preoral cavity are enclosed and forwardly directed. Of the three groups only the Collembola (Springtails) are common, and are likely to be encountered when examining soil and material from humid habitats.

Class Insecta (Insects) (Fig. 2.7)

Insects are ectognathous and usually hypognathous, i.e. the mouthparts are exposed and directed ventrally. The body of an insect is organized into three regions – head, thorax and abdomen. The head bears a pair of sensory antennae, large compound eyes and three pairs of mouthparts (mandibles, 1st maxillae and labium or fused 2nd maxillae).

Structure of insects

Head (Fig. 2.8)

The insect head is a rigid capsule formed from a series of sclerotized plates. The anterior wall of the capsule is formed by the frons, which articulates dorsally with the vertex and ventrally with the clypeus, The dorsal and lateral walls are formed by the vertex and genae (singular gena) respectively. The incomplete posterior wall is the occiput, which articulates with the neck and surrounds the occipital foramen through which the nerve cord, gut and aorta pass.

The antennae are inserted between the eyes and, in action, are directed forwards. Their structure is very variable, ranging from being many-segmented, long and slender in cockroaches (Fig. 2.12) to being three-segmented, short and stout in fleas (Fig. 2.18). The compound eyes are placed dorsolaterally and an insect is said to be dichoptic when the eyes are separated and holoptic when they are contiguous in the mid-dorsal line. Many insects have ocelli (singular ocellus) on the

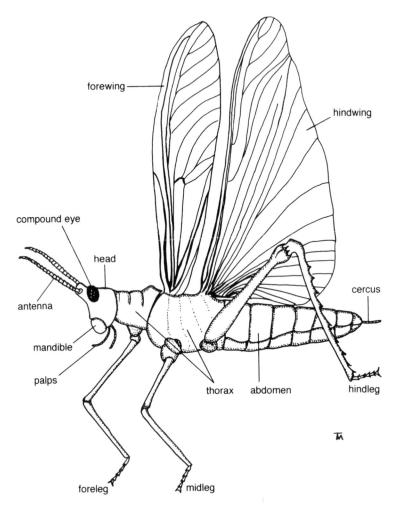

Fig. 2.7. Lateral view of a generalized insect.

head. They are dark, hemispherical structures projecting above the general level of the head surface. Typically there are three ocelli forming an inverted triangle anterodorsally, but in Diptera they are placed more dorsally and are located on the vertex.

The labrum or anterior lip, which is broadly joined to the clypeus, forms the anterior wall of the preoral cavity, exterior to the mouth. The other limits to the cavity are provided by the mouthparts, the mandibles and maxillae laterally, and the labium posteriorly. The maxillae and labium may carry sensory palps. The mouthparts of medically important insects are described in Chapter 4.

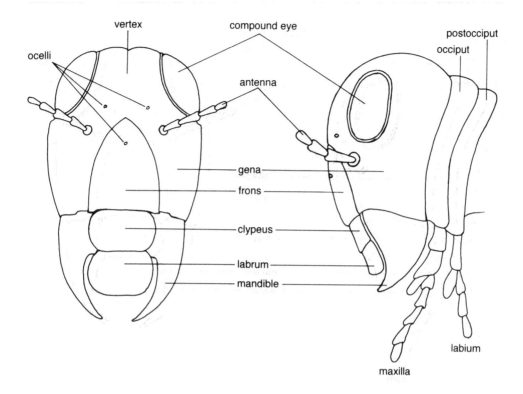

Fig. 2.8. Anterior (left) and lateral (right) views of a generalized insect head.

Thorax (Fig. 2.7)

The thorax is composed of three segments, named anteroposteriorly, prothorax, mesothorax and metathorax. In a typical adult insect each segment bears a pair of legs and, in the Pterygota, the mesothorax and metathorax a pair of wings. When both pairs of wings are fully developed the meso- and metathoraces are fused to form the pterothorax, which is very much larger than the prothorax. In the Diptera, there is only one pair of functional wings, the mesothoracic, consequently the mesothorax is highly developed and the prothorax and metathorax correspondingly reduced (see Chapter 3). The lateral walls of the thorax are pierced by the mesothoracic and metathoracic spiracles, openings of the respiratory system.

The legs are made up of a number of segments which provide flexibility. There are two small segments at the base, the coxa which articulates with the thorax and the trochanter (Fig. 2.9). They are followed successively by two longer segments, the femur and tibia. The leg terminates in the tarsus, composed of one to five short segments or tarsomeres. The basal segment may be referred to as the metatarsus or basitarsus. The terminal tarsal segment bears the pretarsus, which in Diptera consists typically of a pair of claws and pad-like pulvilli (singular pulvillus) and a median empodium which may be setaceous (hair-like) or pulvilliform (Fig. 2.10).

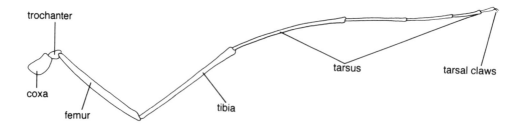

Fig. 2.9. Segments of a typical insect leg.

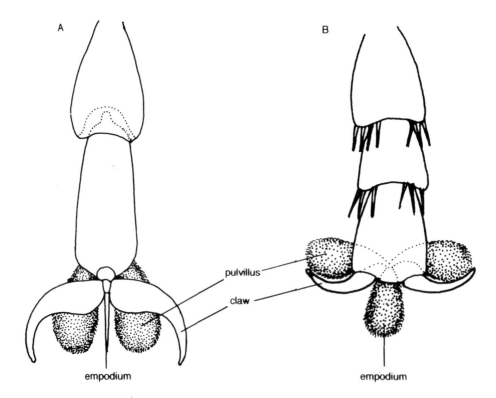

Fig. 2.10. Tarsal claws and associated structures.(**A**) *Lucilia cuprina* with large seta, which overlies the empodium, omitted. (**B**) A tabanid.

The typical insect wing is a thin, transparent membrane, composed of closely adherent upper and lower layers which enclose a series of longitudinal strengthening tubes, the veins. These are connected by cross-veins which are few in higher insects such as the Diptera. Each vein contains a trachea and an extension of the blood-containing haemocoele. The arrangement of veins in the wing is characteristic of the insect and referred to as its wing venation. It affords a very useful character for identification and it is often possible to decide the family or genus of an insect solely on its wing venation.

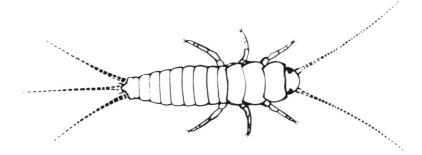

Fig. 2.11. Dorsal view of a thysanuran.

Abdomen

The abdomen is primitively 11-segmented, and only in the Apterygota are there appendages on the pregenital segments. In primitive insects the terminal abdominal segment may bear a pair of long, lateral cerci (singular cercus) and, less frequently, a median caudal filament or appendix dorsalis. Laterally the abdomen may bear up to eight pairs of spiracles. The gonopore opens on segment 9 in the male and behind segment 8 in the female. In the Pterygota highly modified appendages are associated with the gonopore and are involved in copulation and insemination. These terminalia, particularly those of males, provide useful characters for species identification, but their treatment is highly specialized and outside the scope of this book.

In the Diptera the first abdominal segment becomes progressively reduced and the 10th and 11th are fused to form the proctiger which bears the reduced cerci and the anal opening. In the higher Diptera the abdominal segments are telescoped so that only a small number are normally visible externally.

Subclass Apterygota (Fig. 2.11)

These primitive, wingless insects possess an appendix dorsalis. They continue to moult throughout their lives. Some of the Thysanura are minor household pests and include the silverfish, *Lepisma saccharina*, and the firebrat, *Lepismodes inquilinus*.

Subclass Pterygota

The Pterygota are winged or secondarily wingless insects in which an appendix dorsalis is rare and moulting ceases at sexual maturity. There are two distinct forms of development. In exopterygote (hemimetabolic) development the wing rudiments are present externally in the nymphal stages and increase in size at each moult, becoming functional only in the adult and in the winged preimaginal stage of the

Ephemeroptera (mayflies). Except in aquatic forms the nymphs are similar to the adults both in appearance and mode of life. In endopterygote (holometabolic) insects wings develop internally, and the larval stages are quite unlike the adult in appearance and lead an entirely different mode of life. Transition from larva to adult requires extensive reorganization and is effected in a non-feeding and usually non-motile stage, the pupa. This strategy allows the separate adaptation of larva and adult to their respective habitats and has been most successful as 86% of all insects are endopterygotes.

The type of life cycle has practical implications. There is usually only one control strategy possible against exopterygotes but there is a choice of two against endopterygotes when actions may be directed against the adult or the immature stages. Thus any action taken against exopterygote pests such as lice or cockroaches will operate equally on the immature stages and adults which share the same habitat, whereas with houseflies and mosquitoes there is a choice of actions. Control measures can be directed against the adults as, for example, when deposits of residual insecticides are applied to the inner surfaces of houses or animal shelters or, on the other hand, measures can be directed against the immature stages by eliminating breeding sites or by rendering them unsuitable. The strategy adopted will depend upon circumstances with the general rule being to attack the pest where it is more concentrated and accessible. Where breeding sites are abundant the only practical method may be to attack the adults. When breeding sites are limited or the adult insect is the pest it will be preferable to direct control measures against the immature stages and cut the pest off at its source.

Two main divisions are recognized within the Pterygota. Two ancient orders are often included in a single division, the Palaeoptera, while the bulk of insect species are in the Neoptera. Members of the Palaeoptera are characterized by an inability to fold the wings against the body when the insect is at rest. The wings are either held vertically above the body or horizontally at the sides of the body. The two orders involved are the Odonata (dragonflies, damselflies) and the Ephemeroptera (mayflies). Neither order is of medical importance but their immature stages are often abundant in aquatic habitats and will be found when searching for the larvae of mosquitoes or blackflies. The Odonata are predacious in all stages and both adults and nymphs may play a minor role as predators of mosquitoes.

Eleven of the 25 orders of extant Neoptera are included in the Endopterygota and four exopterygote orders in the Paraneoptera (= Hemipteroid assemblage). The remaining ten exopterygote orders are often referred to as the Blattoid–Orthopteroid group although Kristensen (1991) treats them as orders of *incertae sedis*. The Blattoid–Orthopteroid group includes many well-known and easily recognized insects such as locusts, crickets, stick insects, earwigs, preying mantids and termites. It includes only one order of medical importance: the Blattodea (cockroaches), which is frequently included with the praying mantids in a single order, the Dictyoptera (Davies, 1988) which may also include the termites (Kristensen, 1991). Members of the Blattoid–Orthopteroid group have generalized mandibulate (chewing) mouthparts and the forewings are usually thickened to provide protection to the folded hind wings. The Hemipteroid group is characterized by having specialized mouthparts which are often suctorial. Two of the four orders in the Hemipteroid group, the Phthiraptera (lice) and the Hemiptera (bugs) include families of medical and veterinary importance. The other two orders in this group are the plant-feeding thrips (Thysanoptera) and the scavenging psocids (Psocoptera). The two orders of major medical importance in the Endopterygota are the Siphonaptera (fleas) and the Diptera (mosquitoes, flies). The Endopterygota also includes three of the best-known orders of insects, the Coleoptera (beetles), the Lepidoptera (butterflies and moths) and the Hymenoptera (ants, bees and wasps). A very small number of these three orders have some medical and veterinary importance.

Order Blattodea (cockroaches) (Fig. 2.12)

Cockroaches are dorsoventrally flattened, exopterygote, terrestrial insects with long antennae and wings, when present, folded flat over the body. The mouthparts are mandibulate and the legs cursorial, i.e. adapted for running, at which cockroaches are very adept. The hardened forewings, called tegmina (singular tegmen), overlap and cover much of the dorsal surface of the body, protecting the more delicate hind wings, which are the effective flying organs. The abdomen ends in paired, jointed cerci. The eggs are enclosed in a hardened, purse-like case, the ootheca. Internally the nervous system has discrete segmental ganglia and there are numerous Malpighian tubes. More than 3500 species have been described of which a small number have become domestic pests. As synanthropic insects, i.e. adapted to living closely with people, they have been carried around the world in the baggage of humans. The main pest species are the large *Periplaneta americana* and *P. australasiae* and the smaller *Blatella germanica* and *Blatta orientalis*.

Cockroaches are active, nocturnal insects, whose main period of activity is after 'lights out'. The full extent of an infestation is only appreciated when the householder returns to the kitchen some time after all the lights have been put out. The sight can be most displeasing. Apart from their presence being aesthetically unacceptable they produce a characteristic offensive odour. They are scavengers which are attracted to any organic material which may serve as food. They are equally at home in the sewers and in the kitchen. The sewers of tropical cities often support unbelievable populations of cockroaches which on warm nights emerge and are attracted to lights around and inside houses.

Cockroaches will feed on human food, excreta, sputa and, when food is scarce, the bindings of books and even paper. This wide-ranging feeding habit makes cockroaches potential mechanical vectors of pathogens but their precise role in any particular situation has to be assessed individually. The relationship of cockroaches with pathogenic organisms and other life forms has been comprehensively reviewed by Roth and Willis (1957, 1960). When cockroaches feed on infected material their legs and mouthparts become contaminated and they can introduce the infection into the human environment. In addition, while feeding they defecate, and pathogens can remain fully viable after passage through the cockroach gut. In some situations cockroaches may be more important than houseflies as mechanical vectors of human disease.

At 27°C *B. orientalis* passes through seven to ten instars over a period of six

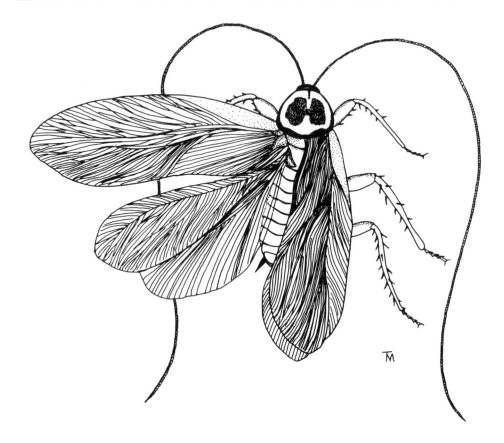

Fig. 2.12. Dorsal view of a cockroach, *Periplaneta australasiae*.

to seven months before becoming adult. Mated females live for three months and produce equal numbers of both sexes while unmated females live longer (4.5 months) and produce only females (Short and Edwards, 1991). *Blatta orientalis* and *Blattella germanica* are spreading in the United Kingdom facilitated by their ability to survive in outdoor habitats (Alexander *et al.*, 1991). This reinforces Schal and Hamilton's (1990) plea for the development of urban integrated pest control programmes for cockroaches based on a greater understanding of their ecologies, leading to a lesser dependence on insecticides.

Order Phthiraptera (lice)

Lice are small, dorsoventrally flattened, exopterygote, wingless, obligatory ectoparasites of birds and mammals. Two different forms of lice have evolved – the Mallophaga retain the primitive insectan mandibulate mouthparts and feed on epidermal structures of birds and mammals, while the Anoplura have evolved specialized mouthparts for blood-feeding and are found only on mammals.

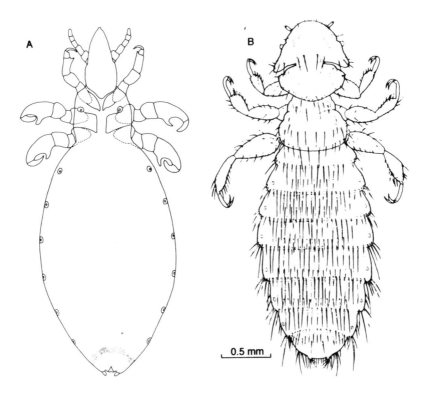

Fig. 2.13. (**A**) Dorsal view of an anopluran, *Linognathus vituli*, the long-nosed cattle louse (Phthiraptera, Anoplura). (**B**) Dorsal view of *Paraheterodoxus insignis* (Phthiraptera, Mallophaga).

Suborder Anoplura (= Siphunculata) (sucking lice) (Fig. 2.13A)

Sucking lice have relatively long, narrow heads with retracted mouthparts which are not discernible externally (see Chapter 4). The three thoracic segments are fused to form a single structure. The most important siphunculate louse is *Pediculus humanus*, a parasite of humans. It has had a major impact on human history and social development by being the biological vector of epidemic typhus and epidemic relapsing fever (see Chapters 25 and 26).

Suborder Mallophaga (chewing lice) (Fig. 2.13B)

The heads of chewing lice are broad to accommodate mandibulate mouthparts and their associated muscles. These mouthparts are obvious externally. The prothorax is free from the mesothorax, which may be fused to, or separate from, the metathorax. The chewing lice are mainly ectoparasites of birds but some species occur on mammals.

The Phthiraptera are considered further in Chapter 19.

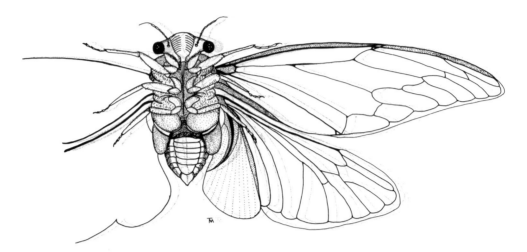

Fig. 2.14. Ventral view of an hemipteran, the double drummer, *Thopha saccata.*

Order Hemiptera (bugs) (Fig. 2.14)

The Hemiptera are exopterygote insects with highly specialized mouthparts produced into a ventrally reflected proboscis. Most Hemiptera feed on the fluid contents of plants either by tapping the phloem or by piercing the cells of the mesophyll. This habit of piercing plants makes them ideal vectors of plant pathogens and they are vectors of viruses which cause disease among crops and are therefore of considerable importance to agriculturalists and horticulturalists. Some Hemiptera are predacious and a few, including the Cimicidae (bedbugs) and some Reduviidae (assassin bugs), feed on blood (see Chapter 18). Some reduviids are vectors of Chagas' disease caused by *Trypanosoma cruzi* (Chapter 29).

Order Coleoptera (beetles) (Fig. 2.15)

Beetles are endopterygote, mandibulate insects with thickened forewings known as elytra (singular elytron), which meet edge to edge in the mid-dorsal line. The Coleoptera is the largest order of the Insecta, containing 40% of the known species of insects, but it is only of very minor importance in medical entomology. When some beetles are disturbed they produce a vesicating fluid. Members of the Meloidae are known as blister-beetles, of which the best-known member is the southern European species *Lytta vesicatoria* popularly known as Spanish-fly (Fig. 2.15A). The active agent in the vesicating secretions is cantharidin, a highly toxic material, which at one time was used medicinally as a counter irritant (Reynolds, 1982) and illegally as an aphrodisiac.

Worldwide there are some 600 species of *Paederus* sensu lato (Fig. 2.15A), a genus of rove beetles (Staphylinidae). Adults and larvae are beneficial predators which frequent mostly moist habitats. Adults are active in daylight and attracted to light at night which brings them into close contact with humans, especially in

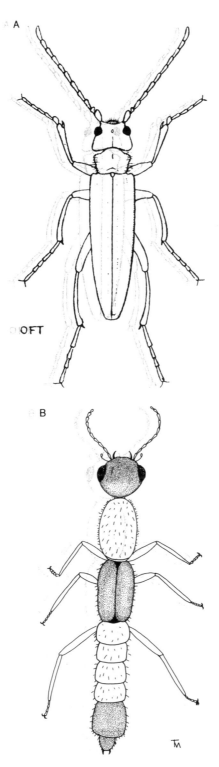

Fig. 2.15. Coleoptera: dorsal views of *Lytta vesicatoria* (**A**), and *Paederus cruenticollis* (**B**). Source: *L. vesicatoria* from Patton, W.S. (1931) *Insects, Ticks and Venomous Animals. Part II, Public Health*, H.R. Grubb, Croydon.

the tropics. They secrete a vesicating fluid when crushed or physically disturbed as happens when *Paederus* lands on bare skin and is injudiciously brushed off or when it becomes trapped between clothing and skin. There is no immediate response to the fluid but in a day or two's time an angry weal appears, followed by the formation of blisters. The condition is known as dermatitis linearis on the skin and conjunctivitis in the eye. The erythema may persist for months and the conjunctivitis can produce temporary blindness. The toxin, pederin, is the most complex non-proteinaceous insect secretion known. Experimentally it has suppressed cancerous tumours in mice and rats and has stimulated regeneration of damaged tissues. It has been used in Chinese medicine for more than 1000 years (Frank and Kanamitsu, 1987). These conditions have been recorded in the southern hemisphere and Oriental region. In Australia the species usually involved is *P. cruenticollis* (McKeown, 1951), and in Uganda *P. sabaeus*, which does not secrete haemolymph but releases it when the cuticle is broken (McCrae and Visser, 1975). Individual mature larvae of the dermestid beetle, *Trogoderma inclusum*, are covered with more than 3000 barbed, spear-headed hastisetae and 2000 slender spicisetae capable of penetrating the human skin. Their effect appears to be purely mechanical and involves no toxin (Okumura, 1967).

Order Hymenoptera (ants, bees, wasps) (Fig. 2.16)

The Hymenoptera are endopterygote, mandibulate insects in which the forewings are larger than the hind wings and both fore- and hind wings are coupled mechanically to function as a single entity. Many Hymenoptera have a complex social organization. The female ovipositor is often modified into a sting which is used to immobilize prey, as in the hunting wasps, or in defence of the colony, as in ants and bees. Although a wasp or bee sting is painful the main danger lies in an allergic response, which may lead to anaphylactic shock and rapid collapse of the person stung. In this way wasp and bee stings caused 61 deaths in England and Wales during the period 1959 to 1971 (Somerville *et al.*, 1975).

The biochemical, pharmacological and behavioural aspects of the venoms of Hymenoptera have been reviewed by Piek (1986). Two conditions are recognized by Schmidt (1986) – venom hypersensitivity (allergy) and a more subtle psychological aspect of insect venom hypersensitivity. The latter is the more difficult condition to treat. The fear engendered by stinging insects is out of all proportion to the risks involved. In the USA 2 out of every 100,000 deaths were due to hymenopteran stings which compares favourably with 5 deaths from lightning strike. Leclercq (1986) estimated the annual mortality from hymenopteran stings throughout the world to be one to six deaths per 10 million people. Death occurs in sensitized individuals within an hour of being stung from respiratory failure brought on by anaphylactic shock. Individuals can be desensitized with graduated injections of venom and venom sac proteins (Schmidt, 1986).

Less commonly wasps can cause damage by biting and Braverman *et al.* (1991) report that over a period of ten years *Vespula germanica* has sporadically caused damage to cows' teats in Israel in the summer months.

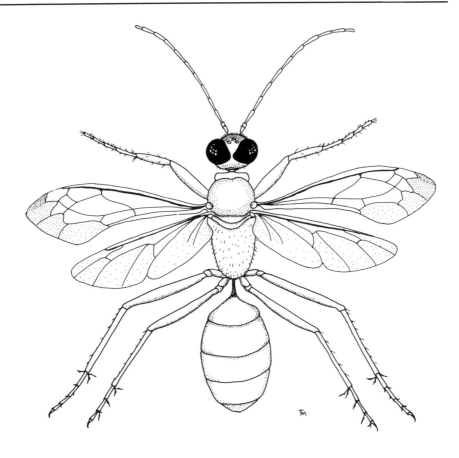

Fig. 2.16. Dorsal view of an hymenopteran (Sphecidae).

Order Lepidoptera (butterflies, moths) (Fig. 2.17)

Lepidoptera are endopterygote insects whose body, legs and wings are covered with detachable scales, which are morphologically flattened hairs. The scales produce colourful patterns which make members of this order so attractive and the object of collectors. The larval stage is a caterpillar with chewing mandibulate mouthparts, while the adult has a coiled proboscis for feeding on nectar. In West Africa a very few moths, mainly species of *Acyophora* (Noctuidae), have been found to feed on secretions of the eyes of cattle (Buttiker and Nicolet, 1975; Nicolet and Buttiker, 1975). Most of the 17 species of noctuid moths of the largely Asian genus *Calyptra* are fruit piercing (Bänziger, 1983). One species, *C. eustrigata*, actually pierces the skin of mammals and feeds on blood. It is nocturnal, feeding between 20.00 and 02.00 h with the actual feeding process taking 12 to 30 min. *C. eustrigata* has been observed feeding on a range of large mammals including elephant, rhino, tapir and Artiodactyla (Bänziger, 1975).

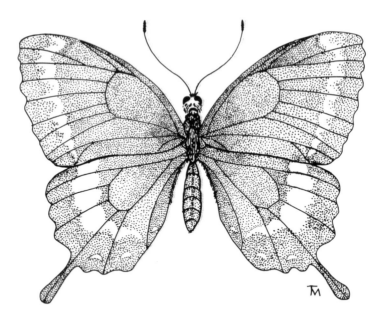

Fig. 2.17. Dorsal view of a lepidopteran, *Papilio canopus*.

Many caterpillars are covered with long urticating hairs (setae) which may or may not have a poison gland at the base. Last instar *Euproctis similis* larvae are estimated to have 2 million urticating setae per caterpillar. It was responsible for an outbreak of dermatitis in Shanghai in 1971 (Wirtz, 1984) and in Australia *Euproctis edwardsi*, with its vast numbers of fine irritating hairlets, is regarded as the most important urticating caterpillar (Southcott, 1988). In the USA the toxic setae of the gypsy moth, *Lymantria dispar*, inject histamine. In 1981 these moths defoliated nearly 5 million hectares of forest with an estimated 50,000 caterpillars per tree. They caused widespread dermatitis in the locality from direct contact with caterpillars or with windblown setae or contaminated clothing. The dermatitis was short lived, lasting a few days to two weeks. In Venezuela caterpillars of the saturnid moth, *Lonomia archilous*, are capable of injecting a powerful anticoagulant which can result in severe bleeding (Wirtz, 1984). In addition allergic responses can be induced among sensitive individuals by exposure to the scales and hairs of Lepidoptera. A few caterpillars secrete noxious fluids, e.g. formic acid by *Dicranura vinula* (Richards and Davies, 1977) and *Heterocampa mantea* (Kearby, 1975).

Order Siphonaptera (fleas) (Fig. 2.18)

Fleas are laterally flattened, wingless, endopterygote, blood-sucking ectoparasites of birds and mammals. The larva is apodous (legless) while in the adult the hind legs are saltatorial, i.e. adapted for leaping, and the source of the familiar escape response of adult fleas. Fleas are important vectors of disease, of which the major one is plague (Chapter 26) and are dealt with in Chapter 17.

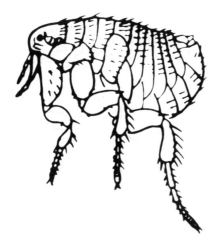

Fig. 2.18. Lateral view of a flea (Siphonaptera).

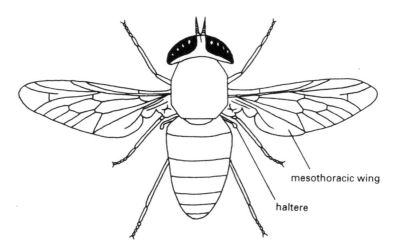

Fig. 2.19. Dorsal view of a dipteran.

Order Diptera (two-winged flies) (Fig. 2.19)

Diptera are endopterygote insects with only one pair of functional wings, the mesothoracic pair, the metathoracic pair being modified into halteres, which play an important role in flight. Halteres are stalked, knobbed structures which function as alternating gyroscopes providing, in most Diptera, essential sensory information for the stabilization of flight. Some Diptera, e.g. horseflies (Tabanidae), can still fly properly when the halteres have been surgically removed, but most Diptera, e.g. the blowfly, *Calliphora*, cannot (Wigglesworth, 1972). Dipterous larvae are

apodous; those of the higher Diptera have very reduced heads and some are known
as maggots. From a medical point of view the Diptera is the most important order
of insects and its classification will be considered in detail in the next chapter.

References

Alexander, J.B., Newton, J. and Crowe, G.A. (1991) Distribution of Oriental and German
    cockroaches, *Blatta orientalis* and *Blattella germanica* (Dictyoptera) in the United
    Kingdom. *Medical and Veterinary Entomology* 5, 395–402.
Bänziger, H. (1975) Skin-piercing blood-sucking moths. 1. Ecological and ethological studies
    on *Calpe eustrigata* (Lepidoptera: Noctuidae). *Acta Tropica* 32, 125–144.
Bänziger H. (1983) A taxonomic revision of the fruit-piercing and blood-sucking moth genus
    Calyptra Ochsenheimer [= *Calpe* Treitschke] (Lepidoptera: Noctuidae). *Entomologica
    Scandinavica* 14, 467–491.
Bettini, S. (ed.) (1978) *Arthropod Venoms.* Springer-Verlag, Berlin.
Braverman, Y., Marcusfeld, O., Adler H. and Yakobson, B. (1991) Yellow jacket wasps can
    damage cows' teats by biting. *Medical and Veterinary Entomology* 5, 129–130.
Brown, M.D., Mottram, P., Fanning, I.D. and Kay B.H. (1992) The peridomestic
    container-breeding mosquito fauna of Darnley Is. (Torres Strait) (Diptera: Culicidae),
    and the potential for its control by predacious *Mesocyclops* copepods. *Journal of the
    Australian Entomological Society* 31, 305–310.
Buttiker, W. and Nicolet, J. (1975) Observations complémentaires sur les lépidoptères
    ophthalmotropes en Afrique occidentale. *Revue d'Elevage et de Medécine Vétérinaire Pays
    Tropicaux* 28, 319–329.
Chapman, R.F. (1982) *The Insects: Structure and Function.* Hodder & Stoughton, London.
Cloudsley-Thompson, J.L. (1992) Scorpions. *Biologist* 39, 206–210.
Coddington, J.A. and Levi, H.W. (1991) Systematics and evolution of spiders (Araneae).
    Annual Review of Ecology and Systematics 22, 565–592.
Davies, R.G. (1988) *Outlines of Entomology.* Chapman & Hall, London.
Frank, J.H. and Kanamitsu, K. (1987) *Paederus,* sensu lato (Coleoptera: Staphylinidae):
    natural history and medical importance. *Journal of Medical Entomology* 24, 155–191.
Grassé, Pierre-P. (1949) *Traité de Zoologie, Anatomie, Systématique, Biologie. VI. Onycho-
    phores, Tardigrades, Arthropodes – Trilobitomorphes Chélicérates.* Masson et Cie, Paris.
Hopkin, S.P. and Read, H.J. (1992) *The Biology of Millipedes.* Oxford University Press,
    Oxford, UK.
Kearby, W.H. (1975) Variable oakleaf caterpillar larvae secrete formic acid that causes skin
    lesions (Lepidoptera: Notodontidae). *Journal of the Kansas Entomological Society* 48,
    280–282.
Keegan, H.L. (1980) *Scorpions of Medical Importance.* University Press of Mississippi,
    Jackson, Mississippi.
Kristensen, N.P. (1991) Phylogeny of extant arthropods. In: Naumann, I.D. (ed.) *The Insects
    of Australia – a Textbook for Students and Research Workers.* Melbourne University
    Press, Melbourne, pp. 125–140.
Leclercq, M. (1986) Piqûres d'insectes et d'arachnides physiopathologie et immunologie
    thérapeutique. Mise au point et observations inédites. *Revue Médicale de Liège* 49,
    545–565.
Lewis, J.G.E. (1981) *The Biology of Centipedes.* Cambridge University Press, Cambridge,
    UK.

Lucas, S. (1988) Spiders in Brazil. *Toxicon* 26, 759–772.

Main, B.Y. (1985) Mygalomorphae. *Zoological Catalogue of Australia* 3, 1–48.

McCrae, A.W.R. and Visser, S.A. (1975) *Paederus* (Coleoptera: Staphylinidae) in Uganda. I: Outbreaks, clinical effects, extraction and bioassay of the vesicating toxin. *Annals of Tropical Medicine and Parasitology* 69, 109–120.

McKeown, K.C. (1951) Dermatitis apparently caused by a staphylinid beetle in Australia. *Medical Journal of Australia* 2, 772–773.

Minton, S.A. (1974) *Venom Diseases.* Charles C. Thomas, Springfield, Illinois.

Naumann, I.D. (ed.) (1991) *The Insects of Australia – a Textbook for Students and Research Workers.* Melbourne University Press, Melbourne.

Newlands, G. and Atkinson, P. (1988) Review of southern Africa spiders of medical importance, with notes on the signs and symptoms of envenomation. *South African Medical Journal* 73, 235–239.

Nicolet, J. and Buttiker, W. (1975) Observations sur la kératoconjonctivitis infectieuse du bovin en Côte d'Ivoire. 2. Etude sur la rôle vecteur des lépidoptères ophthalmotropes. *Revue d'Elevage et de Medécine Vétérinaire Pays Tropicaux* 28, 125–132.

Okumura, G.T. (1967) A report of canthariasis and allergy caused by *Trogoderma* (Coleoptera: Dermestidae) *California Vector Views* 14, 19–22.

Piek, T. (1986) *Venoms of the Hymenoptera. Biochemical, Pharmacological and Behavioural Aspects.* Academic Press, Orlando, USA.

Polis, G.A. (ed.) (1990) *The Biology of Scorpions.* Stanford University Press, Stanford, USA.

Reynolds, J. (ed.) (1982) *Martindale the Extra Pharmacopoeia,* 28th edn. The Pharmaceutical Press, London, p. 1689.

Richards, O.W. and Davies, R.G. (1977) *Imms' General Textbook of Entomology.* Chapman & Hall, London.

Roth, L.M. and Willis, E.R. (1957) The medical and veterinary importance of cockroaches. *Smithsonian Miscellaneous Collections* 134(10), 1–147.

Roth, L.M. and Willis, E.R. (1960) The biotic associations of cockroaches. *Smithsonian Miscellaneous Collections* 141, 1–470.

Savory, T.H. (1977) *Arachnida.* Academic Press, London.

Schal, C. and Hamilton, R.L. (1990) Integrated suppression of synanthropic cockroaches. *Annual Review of Entomology* 35, 521–551.

Schenone, H., Rubio, S., Villarroel, F. and Rojas, A. (1975) Epidemiologia y curso clinico del loxoscelismo. Estudio de 133 casos causados por la mordedura de la araña de los rincones (*Loxosceles laeta*). *Boletin Chileno de Parasitologia* 30, 6–17.

Schmidt, J.O. (1986) Allergy to Hymenoptera venoms. In Piek, T. (ed.) *Venoms of the Hymenoptera. Biochemical, Pharmacological and Behavioural Aspects.* Academic Press, Orlando, USA. pp. 509–546.

Short, J.E. and Edwards, J.P. (1991) Reproductive and developmental biology of the oriental cockroach *Blatta orientalis* (Dictyoptera). *Medical and Veterinary Entomology* 5, 385–394.

Simard J.M. and Watt D.D. (1990) Venoms and toxins. In Polis, G.A. (ed.) *The Biology of Scorpions.* Stanford University Press, Stanford, USA, pp. 414–444.

Somerville, R., Till, D., Leclercq, M and Lecomte, J. (1975) Les morts par piqûre d'hymenoptères Aculéates en Angleterre et au Pays de Galles (Statistiques pour la periode 1959-1971). *Revue Médicale de Liège* 30, 76–78.

Southcott, R.V. (1978) *Harmful Arachnids.* Mimeographed by author, 2 Taylors Road, Mitcham, South Australia 5052.

Southcott, R.V. (1988) Some harmful Australian insects. *The Medical Journal of Australia* 149, 656–662.

Sutherland, S.K. (1980) The biochemistry and actions of some Australian venoms with notes on first aid. *Chemistry in Australia* 47, 351–356.

Watt, D.D., Babin, D.R. and Mlejnek, R.V. (1974) The protein neurotoxins in scorpion and elapid snake venoms. *Journal of Agricultural and Food Chemistry* 22, 43–51.

Wigglesworth, V.B. (1972) *The Principles of Insect Physiology*. Chapman & Hall, London.

Wirtz, R.A. (1984) Allergic and toxic reactions to non-stinging arthropods. *Annual Review of Entomology* 29, 47–69.

Wong, R.C., Hughes, S.E. and Voorhees, J.J. (1987) Spider bites. *Archives of Dermatology* 123, 98–104.

Classification and Structure of the Diptera

3

Before considering the classification of the Diptera it will be necessary to examine briefly some features of their structure. The antennae are typically many-segmented, long and slender in the Nematocera (Fig. 3.1) and three-segmented and short in the Cyclorrhapha. In the latter the characteristic antenna has two, short, basal segments, and a long, third segment bearing a dorsal arista (Fig. 3.2). In most Cyclorrhapha the antennae are surrounded dorsally and laterally by an inverted U-shaped frontal or ptilinal suture.

Structure of the Diptera (Colless and McAlpine, 1991; Crosskey, 1993)

Thoracic sclerites and chaetotaxy

The greater part of the dorsal surface of the thorax is formed by the scutum, which is usually divided into two by a transverse suture and a small, posterior scutellum (Fig. 3.3). The chaetotaxy, i.e. the arrangement of bristles and setae, includes the acrostichals, a row of bristles either side of the mid-dorsal line of the scutum; and a more-or-less parallel row of dorsocentral bristles between the acrostichals and more laterally placed bristles. The scutellum carries a row of bristles along its free margin.

The lateral walls of the thorax are formed from sclerites and their chaetotaxy provides useful characters for identification, particularly in the Cyclorrhapha. The notopleuron (*np*) is a triangular sclerite dorsolaterally placed, adjoining the transverse suture. The anepisternum (*an*) (= mesopleuron) lies below the noto-pleuron and posterior to the mesothoracic spiracle. It abuts on the ventral katepisternum (*kat*) (= sternopleuron). Posterior to these two sclerites there is an upper anepimeron (*ane*) (= pteropleuron) and a lower meron (= hypo-pleuron). The anepimeron lies below the insertion of the wing. The upper posterior corner of the meron adjoins the metathoracic spiracle.

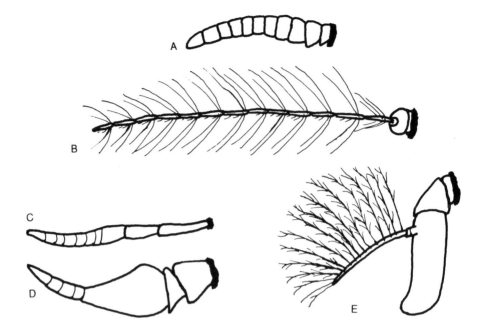

Fig. 3.1.   Antennae of Diptera. Nematocera: (**A**) blackfly (Simuliidae), (**B**) mosquito (Culicidae). Brachycera, Orthorrhapha, Tabanidae: (**C**) *Chrysops*, (**D**) *Tabanus*. Brachycera, Cyclorrhapha: (**E**) *Glossina* (tsetse fly).
Courtesy of Dr B.R. Laurence.

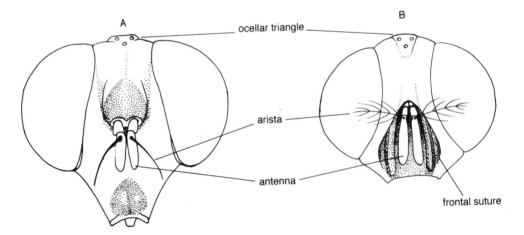

Fig. 3.2.   Heads of an aschizan (**A**), and a schizophoran (**B**).

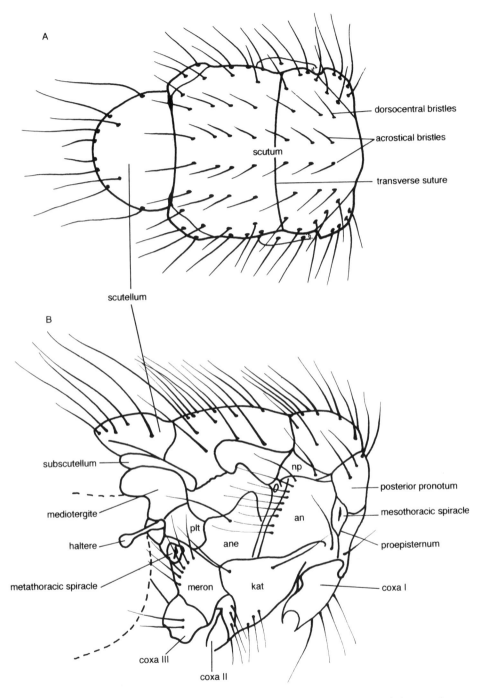

Fig. 3.3. Thoracic structure and chaetotaxy of a cyclorrhaphan. (**A**) Dorsal view. (**B**) Lateral view. Abbreviations: *an*, anepisternum; *ane*, anepimeron; *kat*, katepisternum; *np*, notopleuron; *plt*, pleurotergite.

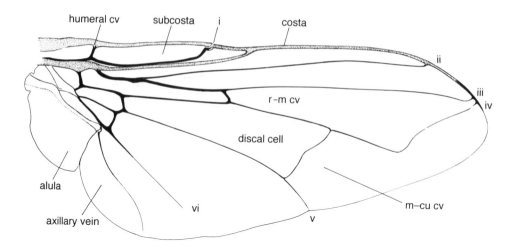

Fig. 3.4. Wing of *Musca domestica* (Cyclorrhapha, Muscidae). cv, Cross vein.

Wing structure and venation

The wing may have up to three basal lobes posteriorly – the alula and two squamae (singular squama). The alula has the same appearance as the wing membrane but is separated from it by a deep incision. The squamae or calypters are thicker and more opaque than the wing membrane. One is attached to the base of the wing and the other to the thorax, and when the latter is well developed it overlies the haltere, which is not visible from above.

Two systems of classifying wing veins are in general use. The Comstock–Needham system is widely used by entomologists, but many dipterists use a numerical system. The two systems will be compared using the wing venations of a muscid fly and a mosquito (Figs 3.4 and 3.5A). In both systems the vein supporting the anterior margin of the wing is the costa, and it usually ends towards the apex of the wing. Posterior to the costa there is a shorter, longitudinal vein, the subcosta, which is unbranched, but may be two-branched in other Diptera.

Proceeding posteriorly the longitudinal veins are successively the radius, media, cubitus and anal. The radius forks near the base of the wing, giving rise to an unbranched vein (R_1 or vein i) and the radial sector (Rs), which may have four branches (R_2 to R_5). Branches of veins are referred to by an initial followed by a number with the veins being numbered anteroposteriorly. In the muscid, vein ii ($R_2 + R_3$) is unbranched, but in the mosquito vein ii is branched, i.e. R_2 and R_3 are separate, and are referred to, in the numerical system, as the anterior and posterior branches of vein ii. This applies to other branched numbered veins. Vein iii (R_4 and $R_5 = R_{4+5}$) is unbranched in both the mosquito and muscid. The media vein may be four-branched (M_1 to M_4). In the muscid vein iv (M_{1+2}) is unbranched, but in the mosquito there is an anterior branch to vein iv (M_1) and a posterior branch (M_2). Vein v (M_{3+4}) is unbranched in muscids, and with anterior and posterior branches in the mosquito. Vein vi is formed from the cubitus and anal, and there may be an axillary vein.

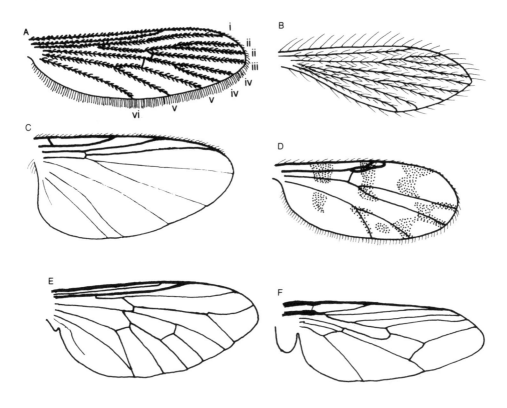

Fig. 3.5. Wings of blood-sucking Diptera (not to scale). (**A**) Mosquito (Culicidae). (**B**) Sandfly (Phlebotominae). (**C**) Blackfly (Simuliidae). (**D**) Biting midge (Ceratopogonidae). (**E**) Horsefly (Tabanidae). (**F**) Tsetse fly (Glossinidae). Courtesy of Dr B.R. Laurence.

In the Diptera there are relatively few cross-veins, but the following are usually present: a humeral cross-vein, running between the costa and subcosta at the base of the wing; an r–m cross-vein between veins iii and iv in the middle of the wing; and an m-cu cross-vein between veins iv and v. Areas of the wing separated by veins are referred to as cells, and named after the vein forming the anterior border of the cell. Cells do not have to be bound on all borders by veins, and are often open at the wing margin.

Classification of the Diptera (Colless and McAlpine, 1991; Crosskey, 1993)

The Diptera are divided into two suborders, the Nematocera and Brachycera. The Nematocera are usually small, delicate, gnat-like flies, with long filamentous antennae composed of many similar, freely articulated segments (more than six). The single pair of palps are three- to five-jointed and usually pendulous but are porrect,

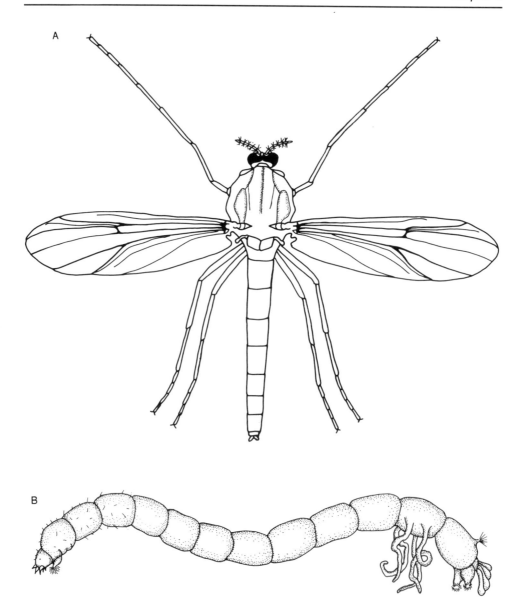

Fig. 3.6. Chironomidae. (**A**) Adult female. (**B**) Larva of *Chironomus tentans*. Larva redrawn from Johansen, O.A. (1934) *Aquatic Diptera*. Cornell University Agriculture Experiment Station, Memoir 164.

i.e. stiff and forwardly directed, in mosquitoes. The larval stage has a well-developed head. The Brachycera are small to large, stout-bodied flies with short antennae, often composed of three segments and never more than six, freely articulated segments (Fig. 3.1). The palps are one- or two-jointed and porrect. The larval stage has a reduced head.

Suborder Nematocera

Five families in the Nematocera are of medical and veterinary importance. They are the Culicidae (mosquitoes), Ceratopogonidae (biting midges), Chironomidae (midges), Simuliidae (blackflies) and Psychodidae (mothflies and sandflies). The characters by which the adults of these families may be recognized and the diseases and pathogens with which they are associated are given briefly below.

Family Chironomidae (midges) (Fig. 3.6)

A large cosmopolitan family with probably more than 2500 species in North America (Ali, 1991) and 10,000 species worldwide (Baur *et al.*, 1983). They are mostly small, delicate flies. Larger species may be confused with mosquitoes (but lack the forwardly directed proboscis) and smaller ones with biting midges (from which they can be separated by vein iv being unbranched and the mouthparts non-piercing). The immature stages are aquatic and adults may emerge in vast numbers. The males form large swarms, the source of the popular term 'dancing midges'. Adult midges can occur in plague-like numbers and the nuisance value of vast numbers being attracted to light together with the severe allergic responses produced in some people have resulted in control measures being taken against some species (Ali, 1995; Cranston, 1995).

Chironomids are unusual insects in that many species contain low molecular weight haemoglobins. Baur and his co-workers (Baur *et al.*, 1983) have shown chironomid haemoglobins to be potent allergens in humans. They tested 14 chironomid species from Africa, Australia and Europe belonging to five different genera and found that in spite of considerable variation in the electrophoretic patterns of their larval proteins, there was considerable immunological cross-reactivity mainly, or exclusively, due to common antigenic determinants of their haemoglobins. This cross-reactivity between the antigens of different chironomid species from different parts of the world means that sensitized persons may be at risk anywhere in the world where chironomids are a nuisance (Baur *et al.*, 1983).

Some of the species involved are *Cladotanytarsus lewisi* along the Nile in the Sudan (Cranston *et al.*, 1981), *Chironomus thummi thummi* in Europe (Wirtz, 1984) and *Kieferulus* (formerly *Cateronica*) *longilobus*, a salt water breeding chironomid in the Indo-Pacific region (Cranston *et al.*, 1990). Ali (1991) has reviewed nuisance species mainly in the USA and points out that in the world less than 100 species have been reported as nuisances.

Family Culicidae (mosquitoes) (see Chapter 7)

Adult culicids have scales on their wings and body and a characteristic wing vena-tion, in which the longitudinal veins (i to iv) run more or less parallel to the long

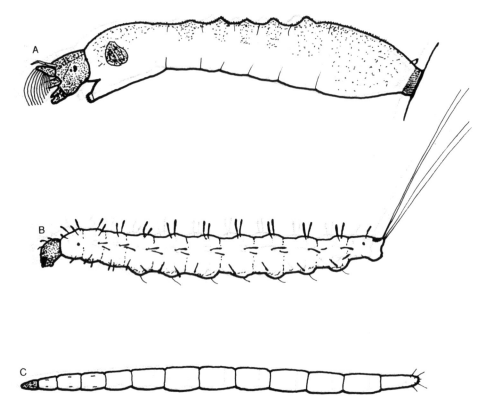

Fig. 3.7. Larvae of some blood-sucking Nematocera. (**A**) *Simulium* (Simuliidae). (**B**) *Phlebotomus* (Phlebotominae). (**C**) *Culicoides* (Ceratopogonidae). Courtesy of Dr B.R. Laurence.

axis of the wing and end at the wing tip. Veins ii and iv are forked while veins i and iii are unbranched (Fig. 3.5A). There is a long, forwardly directed proboscis, which is longer than the head and thorax combined and the palps are porrect. The immature stages of the Culicidae are aquatic (Fig.3.8A, C). Mosquitoes are vectors of pathogenic protozoans, viruses and nematodes which cause such diseases as malaria, yellow fever and lymphatic filariasis to humans and related diseases in domestic animals.

Family Ceratopogonidae (biting midges) (see Chapter 8)

Ceratopogonids are small, compact flies with short legs, a short vertical proboscis and pendulous palps (Fig. 8.1). The wing venation is reduced with only two veins (iv and v) in the posterior half of the wing (Fig. 3.5D). They are both two-branched. The immature stages of the important genus *Culicoides* are found mostly in the water/soil interface (Figs 3.7C and 3.8B). Members of this family are important pests of people and vectors of pathogens to livestock, including bluetongue virus to sheep, nodule-forming nematodes to cattle, and minor pathogens to humans.

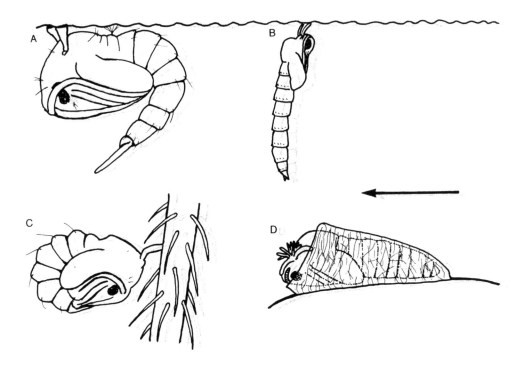

Fig. 3.8. Aquatic pupae of blood-sucking Nematocera. (**A**) Most mosquitoes (Culicidae). (**B**) *Culicoides* (Ceratopogonidae). (**C**) *Mansonia* (Culicidae), which attached to a plant. (**D**) *Simulium* (Simuliidae). Courtesy of Dr B.R. Laurence.

Family Psychodidae (sandflies, mothflies) (see Chapter 9)

These small flies have their bodies and wings covered with hairs but no scales. The wing venation is characterized by having veins ii and v branched, and iii and iv unbranched (Fig. 3.5B). The proboscis is short and directed vertically downwards and the palps are pendulous (Fig. 9.1). The true sandflies (Phlebotominae) are distinct from biting midges which in some countries are referred to as 'sandflies'. Their immature stages are terrestrial (Fig. 3.7B). Phlebotomines are vectors of several diseases of which the most important is leishmaniasis, caused by infection with the protozoan *Leishmania*. The mothflies are of no medical or veterinary importance.

Family Simuliidae (blackflies) (see Chapter 10)

Simuliids are small, hump-backed flies with a short vertical proboscis and pendulous palps. The antennae have 11 segments but appear shorter due to the individual segments being globular (Fig. 3.1A). The wing has strongly developed anterior veins and weakly developed posterior veins (iv, v and vi) (Fig. 3.5C). The immature stages of the Simuliidae are aquatic (Figs 3.7A and 3.8D). The adults are important vectors of filarial nematode worms, including *Onchocerca volvulus*, the cause of river blindness in humans.

Suborder Brachycera

The classification used here recognizes two divisions within the Brachycera, the Orthorrhapha and Cyclorrhapha. Within the Cyclorrhapha Colless and McAlpine (1991) recognize two series – Aschiza and Schizophora. The schizophoran families of medical importance, with the exception of the Chloropidae, are in the Muscoidea. Crosskey (1993) designates three infraorders within the Brachycera, one of which, the Muscomorpha equates to the Cyclorrhapha. He subdivides the Cyclorrhapha into two sections – Aschiza and Schizophora – and the Schizophora into two subsections – the Acalyptratae and Calyptratae. With the exception of the Chloropidae, the schizophoran families of medical importance are in the Calyptratae in which the medically important families are placed in three different superfamilies.

Division Orthorrhapha

The Orthorrhapha are large, stout-bodied flies, the adults and larvae of which are largely predacious. The structure of the antennae is variable, the palps are one- or two-segmented and the abdomen has seven visible segments. The head of the larva is retractable into the thorax and the pupa is free of the larval exuviae, i.e. it is obtect. Only one of the many families in this division, the Tabanidae, is of medical importance.

Family Tabanidae (horseflies, marchflies) (see Chapter 11)

Tabanids are stout-bodied flies with antennae, which are stiff and forwardly directed (Fig. 3.1C, D), and eyes, which in life, are often coloured. The wing venation of a tabanid is characteristic but not exclusive to the family. All the wing veins are well developed and there is an enclosed discal cell in the distal half of the wing (Fig. 3.5E). The squamae are large, obscuring the halteres from dorsal view. The empodium is padlike and hence the last tarsal segment of the leg carries three pads and two claws (Fig. 2.10). They are vectors of loiasis to humans and surra to livestock.

Division Cyclorrhapha

These flies have three-segmented antennae with the third segment bearing an arista on its dorsal surface (Fig. 3.2). The abdomen rarely has as many as seven segments visible. The larvae are acephalic, i.e. the head is vestigial, and they live in organic matter. The pupa is enclosed in the last larval skin or puparium, i.e. the pupa is coarctate. The breeding habits and immature stages of the cyclorrhaphan Diptera have been reviewed by Ferrar (1987). The Aschiza do not have a ptilinum (see below) and therefore lack a frontal suture on the head (Fig. 3.2). The

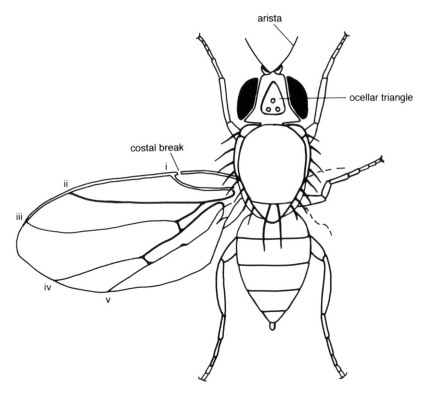

Fig. 3.9. A chloropid.

Schizophora have both a ptilinum and a frontal suture. When the fly emerges from the pupa the suture is open. This allows the flap, carrying the antennae, to hinge on its lower edge, permitting eversion of a balloon-like structure, the ptilinum, from the head. The larvae of Schizophora often burrow into the soil before pupariation to reduce predation and parasitization of the immobile puparium. The newly-emerged fly is able to force its way to the surface of the soil by alternately inflating and retracting the ptilinum. While it is being retracted the fly climbs up into the space vacated. Using this structure houseflies can reach the surface from considerable depths. Once the fly reaches the surface the ptilinum is withdrawn and the suture closed.

Family Chloropidae (eyeflies)

Eyeflies are small, very shiny flies about 2 mm long. Their antennae are short, with the third segment being nearly globular and the arista either bare or with very short branches. The ocellar triangle is very large (Fig. 3.9). The thorax has no distinct transverse suture and the squamae are small. The wings have no markings, the subcosta is rudimentary and vein vi is absent.

The adults of some species are attracted to body secretions and feed on sweat, at sores, and are particularly attracted to eyes, especially eyes which have a copious discharge. They scrape the conjunctival surface of the eye with the spiny tips of the pseudotracheal rings on the labella (see Chapter 4). This irritation increases the flow of secretion, attracting more eyeflies. By moving from one individual to another eyeflies can act as mechanical vectors of yaws (Nichols, 1936), and conjunctivitis, and in Colorado have been associated with vesicular stomatitis, a disease of livestock (Francy et al., 1988). In the Oriental region *Siphunculina funicola* is the main species visiting humans. In Sri Lanka *S. funicola* appears to have been replaced by the closely related *S. ceylonica*, whose relationship with humans has not been studied (Kanmiya, 1989). In the southern states of the USA the species involved are *Hippelates pusio* in the east and *H. collusor* in the west (Rogoff, 1978).

H. pusio oviposits in freshly cultivated or otherwise disturbed soil (Burgess, 1951; Dow and Hutson, 1958) where the larvae widely disperse in the soil and under favourable conditions peak emergence of adults will occur three weeks after oviposition. Other species breed in plant or vegetable debris.

Superfamily Muscoidea (Colless and McAlpine, 1991)

These are moderately large flies in which the thorax has a distinct transverse suture and the squamae are usually well developed. The commonest form of mouthparts in the Muscoidea is the lapping type, adapted for feeding on surface films and secretions. A small number of species have evolved piercing mouthparts adapted for feeding on blood, and a small number have vestigial, i.e. non-functional, mouthparts.

Family Glossinidae (tsetse flies) (see Chapter 12)

Tsetse flies are medium to large, brown, blood-sucking flies, which at the present day are confined to Africa south of the Sahara. They have a long, forwardly directed, piercing proboscis. The wing venation is unique being characterized by a 'hatchet-shaped' cell, more or less in the centre of the wing (Fig. 3.5F). The cell is limited by veins iv and v anteriorly and posteriorly, respectively, and cross-veins proximally and distally. Tsetse flies are viviparous, depositing fully-developed, non-feeding larvae, which burrow into the substrate and pupariate. Tsetse flies are biological vectors of pathogenic trypanosomes, which cause sleeping sickness in the human population and trypanosomiasis in animals.

Family Muscidae (houseflies, stableflies) and Fanniidae (lesser houseflies) (see Chapter 13)

Houseflies are medium sized, rather dull-coloured flies usually without bristles on the anepimeron and meron. Their mouthparts are either lapping or piercing. Vein iv is either parallel to vein iii or gently curved towards it in the distal half of the wing, and vein vi does not reach the wing margin (Fig. 3.4). Muscids breed commonly in material of vegetable origin, grass cuttings, straw, and the dung of herbivores. Houseflies are important mechanical vectors of intestinal pathogens, such as those which cause typhoid, and amoebic and bacillary dysenteries. Blood-sucking stableflies can act as mechanical vectors of surra in camels and horses.

Family Calliphoridae (blowflies) (see Chapter 14)

Calliphorids are medium to large flies often metallic in colour, dark green or blue. Their mouthparts are of the lapping type. The arista is plumose to beyond halfway. Vein iv is quite sharply bent towards vein iii and the two veins are very close together when they meet the wing margin. Both the meron and anepimeron bear bristles, and there are two stout bristles on the notopleuron. Calliphorid larvae feed on organic material of animal origin and serve a useful purpose in breaking down carrion, but some feed on living animal flesh causing a condition known as myiasis.

Family Sarcophagidae (grey fleshflies) (see Chapter 14)

Sarcophagids are similar in appearance to calliphorids but their bodies are grey with black longitudinal stripes on the thorax. In addition they have three or more bristles on the notopleuron and usually only the basal half of the arista is plumose. A few species of sarcophagids are associated with myiasis, but most breed in carrion.

Families Oestridae, Cuterebridae and Gasterophilidae (see Chapter 15)

These families were, at one time, included in a single family, an enlarged 'Oestridae', on the grounds that they were all large bee-like flies with vestigial mouthparts and usually densely clothed with bushy hair. The adults do not feed and therefore are relatively short-lived and uncommon. The larvae are endoparasites of vertebrates, causing myiasis in livestock.

Families Hippoboscidae (keds), Streblidae and Nycteribiidae
(see Chapter 16)

Members of the these families are leathery, dorsoventrally flattened, blood-sucking ectoparasites of birds and mammals. In adaptation to this mode of life the legs bear well-developed claws, and all nycteribiids and many other species are wingless. The females are viviparous with the larva being retained in the body of the female until it is fully developed. The larva is immobile and pupariates where it has been deposited. At one time it was considered that the female deposited a puparium and these three families were grouped together as the Pupipara. The main species of economic importance is the sheep ked, *Melophagus ovinus*.

References

Ali, A. (1991) Perspectives on management of pestiferous Chironomidae (Diptera), an emerging global problem. *Journal of the American Mosquito Control Association* 7, 260–281.

Ali, A. (1995) Nuisance, economic impact and possibilities for control. In: Armitage, P., Cranston, P.S. and Pinder, L.C.V. (eds) *The Chironomidae: the Biology and Ecology of Non-biting Midges.* Chapman & Hall, London, pp. 339–364.

Baur, X., Dewair, M., Haegele, K., Prelicz, H., Scholl, A. and Tichy, H. (1983) Common antigenic determinants of haemoglobin as basis of immunological cross reactivity between chironomid species (Diptera, Chironomidae): Studies with human and animal sera. *Clinical and Experimental Immunology* 54, 599–607.

Burgess, R.W. (1951) The life history and breeding habits of the eye-gnat *Hippelates pusio* Loew in the Coachella Valley, Riverside County, California. *American Journal of Hygiene* 53, 164–177.

Colless D.H. and McAlpine, D.K. (1991) Diptera (flies). In: Naumann, I.D. (ed.) *The Insects of Australia: a Textbook for Students and Research Workers.* Melbourne University Press, Melbourne, pp. 717–786.

Cranston, P.S. (1995) Medical significance. In: Armitage, P., Cranston, P.S. and Pinder, L.C.V. (eds) *The Chironomidae: the Biology and Ecology of Non-biting Midges.* Chapman & Hall, London, pp. 365–384.

Cranston, P.S., Gad-El-Rab, M.O. and Kay, A.B. (1981) Chironomid midges as a cause of allergy in the Sudan. *Transactions of the Royal Society of Tropical Medicine and Hygiene* 75, 1–4.

Cranston, P.S., Webb, C.J. and Martin J. (1990) The saline nuisance chironomid *Carteronica longilobus* (Diptera: Chironomidae): a systematic reappraisal. *Systematic Entomology* 15, 401–432.

Crosskey, R.W. (1993) Introduction to the Diptera. In: Lane, R.P. and Crosskey, R.W. (eds) *Medical Insects and Arachnids.* Chapman & Hall, London, pp. 51–77.

Dow, R.P. and Hutson, G.A. (1958). The measurement of adult populations of the eye gnat *Hippelates pusio. Annals of the Entomological Society of America* 5, 351–360.

Ferrar, P. (1987) *A Guide to the Breeding Habits and Immature Stages of Diptera Cyclorrhapha.* Entomonograph 8 in two parts. E.J. Brill, Scandinavian Science Press, Copenhagen.

Francy, D.B., Moore, C.G., Smith, G.C., Jakob, W.L., Taylor, S.A. and Calisher, C.H. (1988) Epizootic vesicular stomatitis in Colorado 1982: isolation of virus from insects collected along the northern Colorado Rocky Mountain Front Range. *Journal of Medical Entomology* 25, 343–347.

Kanmiya, K. (1989) Study on the eyeflies, *Siphunculina* Rondani from the Oriental region and Far East (Diptera: Chloropidae). *Japanese Journal of Sanitary Zoology* 40 (Supplement), 65–86.

Nicholls, L. (1936) *Framboesia tropica* – a short review of a colonial report concerning statistics and *Hippelates flavipes. Annals of Tropical Medicine and Parasitology* 30, 331–335.

Rogoff, W.M. (1978) *Methods for Collecting Eye Gnats (Diptera: Chloropidae).* United States Department of Agriculture, Science and Education Administration, ARM-W2.

Wirtz, R.A. (1984) Allergic and toxic reactions to non-stinging arthropods. *Annual Review of Entomology* 29, 47–69.

Mouthparts of Insects of Medical and Veterinary Importance and Host Finding

4

The mouthparts of medically important insects deserve special consideration because they are the main, but not the only, route whereby pathogens are transmitted from host to host. Even when no pathogen is involved, the mouthparts are the structures which pierce the skin and cause irritation. It is preferable to deal with the mouthparts on a comparative basis rather than to describe them separately for each group of insects.

Basic Chewing Mouthparts – Mallophaga (see Fig. 4.1)

The mouthparts of a primitive insect are chewing structures capable of dealing with a wide range of potential food material. Three pairs of appendages are associated with the mouth. They are the mandibles, first maxillae, and the labium, which is formed from the fused second maxillae. Both the maxillae and the labium may bear sensory palps. The mouthparts have evolved from appendages, and are therefore external to the mouth and border the preoral cavity. In hypognathous insects, the preoral cavity is bordered anteriorly by the labrum–epipharynx (referred to hereafter as the labrum), composed of an inner sensory surface, the epipharynx, and an external strongly-sclerotized labrum or upper lip. The lateral walls of the preoral cavity are formed by the mandibles and maxillae, and the posterior wall by the labium. The hypopharynx is an unpaired, median, tongue-like structure, located internally to the labium, and associated with the opening of the salivary duct.

At the start of feeding, potential food is palpated by the palps which have sensilla concentrated at their free ends. If the appropriate sensory information is received, the food is crushed by the mandibles and the exuding fluid passes over the sensory areas of the labium and maxillae. If the fluid contains the appropriate stimulating substances the maxillae and labium participate in feeding. Both structures have cutting blades which finely divide the food and during this process saliva is poured out and mixes with the food.

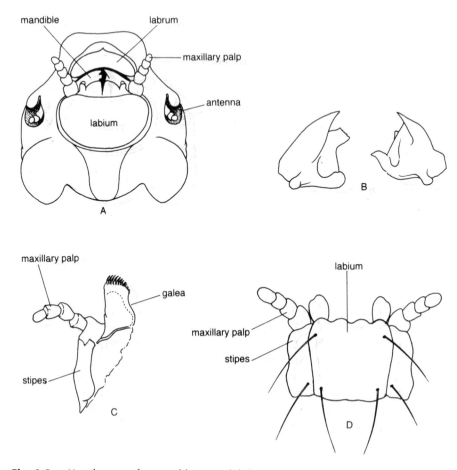

Fig. 4.1. Mouthparts of an amblyceran (Phthiraptera, Mallophaga). (**A**) Ventral view of head. (**B**) Mandibles. (**C**) Maxilla. (**D**) Labium and maxillae. Source: redrawn from Snodgrass, R.E. (1943) *Smithsonian Miscellaneous Collections* 104 (7) by permission of the Smithsonian Institution Press, Smithsonian Institution, Washington, D.C.

Chewing mouthparts of this type are found in cockroaches, beetles and locusts. They are modified in the Mallophaga with the labium being reduced to a simple broad plate to which the maxillae are attached laterally. There are no palps in the suborder Ischnocera, and only a pair of maxillary palps in the other suborder, the Amblycera.

Blood-sucking Mouthparts – Nematocera (see Fig. 4.2)

In blood-sucking Nematocera the mouthparts have to perform two functions, to pierce the skin and imbibe blood. The mouthparts are essentially the same in all blood-sucking Nematocera (Culicidae, Ceratopogonidae, Simuliidae and Phlebo-

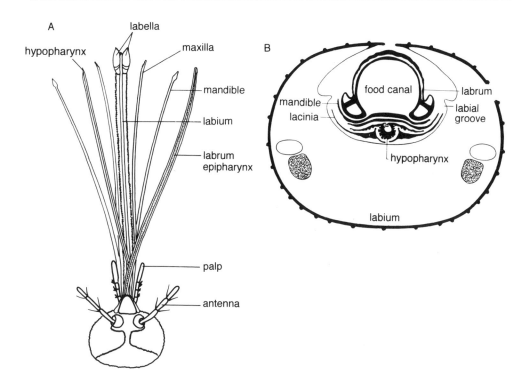

Fig. 4.2. (**A**) Mouthparts of a female mosquito (antennae cut short). (**B**) Transverse section towards the apex of proboscis of *Aedes aegypti*.
Source: (**A**) redrawn from Patton, W.S. and Evans, A.M. (1929) *Insects, Ticks, Mites and Venomous Animals. Part I: Medical.* H.R. Grubb, Croydon; B from *Insects of Australia* (1970), Melbourne University Press.

tominae), although they are greatly elongated in the Culicidae. They contain the same elements as in the Mallophaga. Only the maxillary palps are present. They have a sensory function in finding a host and Kline and Choate (1990) have shown a correlation between palpal structure and the type of host.

The primitive mouthparts have been greatly modified for feeding on blood. The labium forms a protective sheath for the effective structures and it ends in two sensory lobes, the labella (singular, labellum). The main cutting function is undertaken by the mandibles and maxillae (laciniae) which are slender structures, finely toothed at the distal ends. The labrum is curled inwards at the edges to form an almost complete tube, and the gap is closed by the mandibles to form the food canal. Blood is sucked up by two muscular pumps, which are separated from each other and the midgut by sphincter muscles. The cibarial pump operates at the base of the food channel and the pharyngeal pump between the cibarium and the midgut.

When a mosquito feeds, the labella test the surface of the skin and select a suitable location. The labrum, mandibles and maxillae (lacinia in Fig. 4.2B) are closely associated to form the fascicle, which operates as a single structure. The

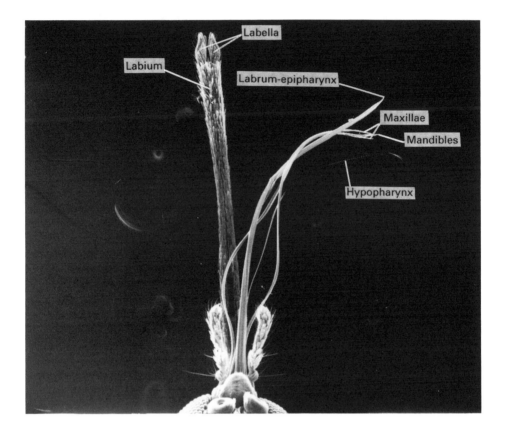

Plate 4.1. Scanning electron micrograph of mouthparts of a mosquito.
Magnification ×203. Courtesy of T. McRae.

tips of the labrum and the hypopharynx are also toothed. The fascicle moves up
and down, supported by the labella, and penetrates the skin. The fascicle is then
inserted and the flexible labium, which remains outside the host, becomes bowed
posteriorly. The fascicle is either inserted into a capillary or it is moved about to
lacerate capillaries and to facilitate the formation of a pool of blood (Gordon and
Lumsden, 1939).

Saliva is poured into the wound through a duct which runs the entire length
of the hypopharynx, opening at its tip. The function of the saliva is to bring about
the release of histamine and the consequent dilation of the capillaries, thus ensuring a good flow of blood. The saliva may or may not contain an anticoagulant.

The saliva of *Anopheles maculipennis* does, and that of *An claviger* does not (Yorke and Macfie, 1924). An anticoagulant is not essential for blood-feeding.

The irritation felt by sensitized individuals to mosquito bites is a reaction to injected saliva, which acts as an antigen. The response in an individual to being bitten by a species over a long period goes through four stages. There is an initial delayed skin reaction, later the reaction is both immediate and delayed, then with more exposure reaction is immediate and finally the individual becomes unresponsive, i.e. he or she has become immune to the bites of that particular species (Wikel, 1982).

Most Nematocera are pool feeders but the long proboscis of mosquitoes enables them to use either pool feeding or capillary feeding. In the blood-sucking Nematocera only the female takes blood. The mandibles and maxillae are reduced or absent in the male, which feeds only on nectar. This implies that only the female can be a vector of disease. Some pathogens, such as viruses and the sporozoites of the malaria parasite, pass down the salivary duct and into the host with the saliva. Infective larvae of filarial worms are too large to use that route. They enter the blood space in the labium and escape by rupturing it near the labella while the insect is feeding. The filarial larvae then enter the host through the feeding site.

Species of blood-sucking Nematocera differ in the degree to which they scrape away the surface of the skin or pierce it more cleanly. This has an effect on their vectorial status. The microfilariae of *Onchocerca volvulus* are in the skin and are acquired by *Simulium damnosum* as it scrapes through the skin. *Culicoides austeni* picks up microfilariae of *Mansonella perstans* from circulating human blood, but not those of *M. streptocerca*, which are in the skin. On the other hand, *Culicoides grahamii* more easily acquires *M. streptocerca* than *M. perstans*.

Blood-sucking and Lapping Mouthparts of Tabanids (see Fig. 4.3)

The mouthparts of tabanids combine the blood-sucking mouthparts of the Nematocera with the lapping mouthparts of the Cyclorrhapha. The fascicle lacks the delicacy which this structure shows in the Nematocera. The mandibles are flat, broad, saw-like blades, the maxillae are narrow, toothed files and the food canal is formed from a stout labrum and a narrow hypopharynx. The fascicle is accommodated in the labial gutter, a groove in the anterior side of the labium. The short, but stout-bodied labium bears terminally a pair of large, fleshy, inflatable labella. The detailed structure of the labella will be given when considering the lapping mouthparts of the Cyclorrhapha.

As in the Nematocera, only the female tabanid is haematophagous, i.e. blood-feeding, while the male feeds only on nectar. When the female tabanid feeds, the labella are retracted to expose the fascicle which pierces the skin and lacerates the tissues for pool-feeding. During this process the mandibles move with a scissor-like action and the maxillae move forwards and backwards (Dickerson and Lavoipierre, 1959b). Insertion of the large fascicle is usually painfully obvious. When feeding ceases the fascicle is withdrawn and as the labella come together they trap a film of blood. This is of great importance in the mechanical transmission of diseases

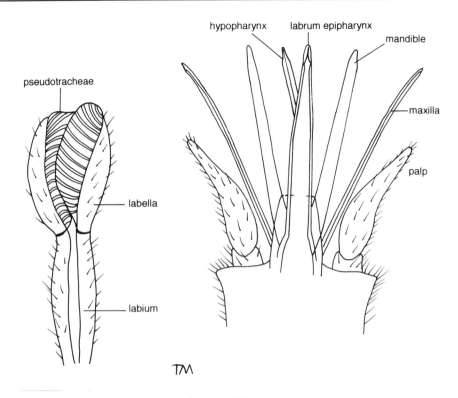

Fig. 4.3. Mouthparts of a female tabanid.

because the enclosed film is protected from drying and pathogens such as *Trypanosoma evansi* may survive for an hour or more. Foil *et al.* (1987) have calculated the volume of the enclosed film as about 10 nl. This is adequate for tabanids to be mechanical vectors of many pathogens (Foil, 1989). Tabanids are also biological vectors of the filaroid worm *Loa loa* in which the infective stage of the nematode enters the labium from which it escapes while the tabanid is feeding.

Can HIV be transmitted mechanically by biting flies?

This question is often asked of medical entomologists. Can a biting fly act as a 'flying syringe'? Foil *et al.* (1988) have attempted to answer this question using the best available data and conclude that between 83,000 and 167,000 tabanids would have to initiate feeding on an AIDS patient with a viraemia of one to two infected lymphocytes per millilitre of blood and then feed immediately upon a susceptible individual before one of the flies would be expected to transmit an infection!

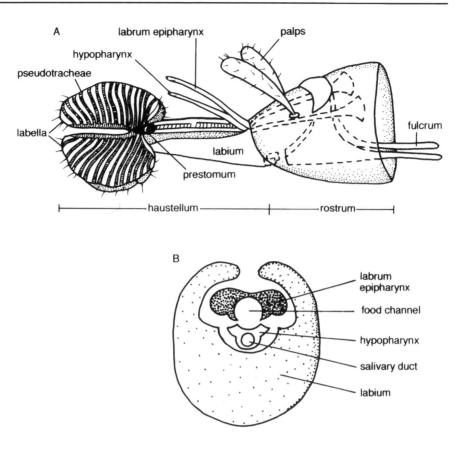

Fig. 4.4. (**A**) Mouthparts of *Musca domestica*. (**B**) Transverse section of labium. Redrawn from West, L.S. (1951) *The Housefly*, Comstock Publishing Company, New York.

Lapping Mouthparts of Cyclorrhapha (see Figs 4.4 and 4.5)

Most cyclorrhaphous flies have lapping mouthparts, designed for feeding on liquids, particularly when present in thin films. Under such conditions a food canal, which functions like a drinking straw, is ineffective as it cannot be immersed in a thin film. What is required is an absorbent structure and this has been developed from the labella, which have been greatly enlarged and modified for this function. At rest the inner surfaces of the labella are in close contact and kept moist by secretions from the labial salivary gland. These inner surfaces are covered with, more or less, parallel rows of pseudotracheae, which converge on the prestomum or opening of the food canal.

Pseudotracheae are incomplete tubes, the side walls of which are closely opposed, effectively completing the tube. The tubes are supported by numerous interrupted rings of chitin. It is the presence of these rings which causes the tubes to be named pseudotracheae by analogy with the tubular vertebrate trachea which

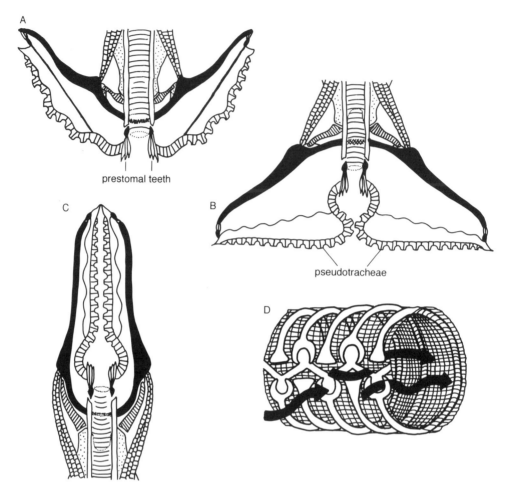

Fig. 4.5. Labella of *Calliphora*. (**A**) Direct feeding position. (**B**) Filtering position. (**C**) Resting position. (**D**) Schematic view of entry of material into the pseudotracheae via the interbifid gap and the longitudinal slit joining the interbifid spaces. Source: (**A**), (**B**) and (**C**) redrawn from Graham-Smith, G.S. (1930) *Parasitology*, Cambridge University Press. Source of (**D**) unknown.

is supported by cartilaginous rings. Each chitin ring has one end simple, and the other end bifurcate, and they are arranged on the pseudotracheae so that simple and bifurcate ends alternate. The membrane supported by the rings is complete everywhere except in the interbifid space, the gap between the arms of each bifurcation. Fluid flows through the interbifid gap by capillary action and is drawn to the prestomum by the sucking action of the cibarial pump acting via the food canal.

The interbifid space acts as a filter, limiting the size of particle which may enter. In *Musca domestica* the diameter of a pseudotrachea ranges from 8 to 16 µm and the interbifid space from 3 to 4 µm (Patton and Cragg, 1913). This clearly

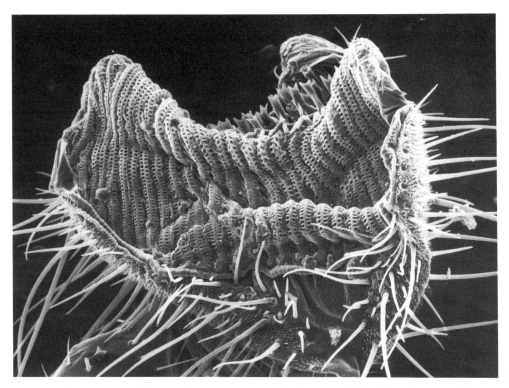

Plate 4.2. Scanning electron micrograph of everted labellum (position shown diagrammatically in Fig. 4.5A) of female *Lucilia cuprina*. Note parallel rows of pseudotracheae, and upwardly directed prestomal teeth. Magnification ×170. Courtesy of T. McRae.

influences the size of pathogen which may be ingested. However, the labella can be retracted and particles enter the food canal directly. The labella are broadly joined to the body of the labium, known as the haustellum, which bears the labial gutter anteriorly. The mandibles and maxillae have been lost, and only the labrum and the hypopharynx are housed in the labial gutter. The haustellum is a rather fleshy structure, supported posteriorly by the prementum, a chitinized plate, and anteriorly by the sclerotized floor of the labial gutter. The cuticle joining these two strengthening structures is thin and flexible.

The labium is joined proximally to a specialized part of the head capsule, the rostrum, which can be lowered or retracted by inflating or deflating air sacs in the head. When the rostrum is retracted and the haustellum raised to a horizontal position the mouthparts cannot be seen from dorsal view and are inconspicuous in profile. In action the rostrum is lowered by inflation of the air sacs and the haustellum directed downwards by muscular action. The labella are expanded by blood pressure and rotated to expose their inner surfaces which form a more or less flat, horizontal surface with the pseudotracheae on the underside. When the pseudotracheal surface is placed on films or liquids, fluid will flow into the pseudotracheae and on to the prestomum, where it will be sucked up the food canal

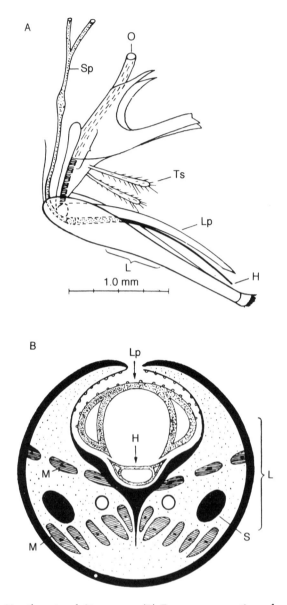

Fig. 4.6. (**A**) Mouthparts of *Stomoxys*. (**B**) Transverse section of proboscis. Abbreviations: *H*, hypopharynx; *L*, labium; *Lp*, labrum-epipharynx; *M*, muscles; *O*, oesophagus; *S*, tendon to labella; *Sp*, salivary gland; *Ts*, palp. Source: From Zumpt, F. (1973) *Stomoxyine Biting Flies of the World*. Gustav Fischer Verlag, Stuttgart.

by the cibarial pump and onwards to the gut. At rest the labella collapse as the blood escapes into the circulation, the rostrum is retracted and the haustellum raised to a horizontal position.

The minute teeth around the prestomum are used to scrape at material for direct ingestion into the food canal and are capable of damaging surface tissues of the host (Kovacs et al., 1990). In some muscid flies these prestomal teeth are enlarged and used to scrape away scabs from wounds and clots from milk, exposing the underlying fluid on which the muscid feeds. Patton and Cragg (1913) have described a series of Indian muscids of the genus *Philaematomyia* (now included in different subgenera of *Musca*), in which the teeth get larger and the scraping gets deeper until some can bore through unbroken skin by rapidly applying and withdrawing prestomal teeth. However, the evolution of the mouthparts of truly blood-sucking Muscoidea, e.g. *Glossina*, has followed a somewhat different route.

The Mouthparts of Blood-sucking Muscoidea (see Figs 4.6 and 4.7)

The ancestors of the blood-sucking Cyclorrhapha probably had lapping mouthparts and they had certainly lost the mandibles and maxillae. These are the skin-piercing tools of the blood-sucking Nematocera and a substitute had to be developed. The labella have been adapted for this purpose by being reduced in size and the pseudotracheae, if they were present originally, have been replaced by sharp teeth.

In action the labella are pressed on the skin, pulled apart by muscular action and, when the muscles relax, they recoil from the pressure of displaced blood. To function effectively the labella need to be rigidly supported in order to maintain adequate pressure on the host's skin. Rigidity is produced by reducing the zone of thin cuticle between prementum and labial gutter, thus strengthening the haustellum. However, a rigid haustellum cannot bend, as the labium does in the blood-sucking Nematocera when the fascicle is inserted into the host. It is necessary now for both the labium and the enclosed food canal to be inserted. This has been achieved by lengthening the haustellum and concentrating the labellar muscles at its base. The muscles are connected to the labella by long tendons which run throughout the space within the haustellum. To accommodate the concentrated muscles a noticeable swelling or bulb is developed at the base of the haustellum. The rostrum is reduced and the long, rigid haustellum cannot be concealed, being always visible from dorsal and/or lateral views. Mouthparts of this type are found in the Stomoxinae, *Glossina*, and the Hippoboscidae.

In feeding, the haustellum is turned vertically downwards and the sensory receptors of the labella are used to select the site for penetration. When the labella are pulled apart and recoiled, the teeth on the labella rasp away the skin. This action is repeated in rapid sequence and the haustellum pierces the skin. Saliva passes down the duct in the hypopharynx and blood is sucked up the food canal by the cibarial pump. In the Stomoxinae the mouthparts form a stout, robust structure (Fig. 4.6). In *Glossina* they have been refined and form a slender, relatively delicate, but highly efficient structure (Fig. 4.7). At rest the forwardly directed haustellum is protected by the palps, which flank it on either side. The main support of the haustellum is provided by the sclerotized labial gutter. The prementum is much reduced. The labrum is a delicate structure, held in the labial gutter by

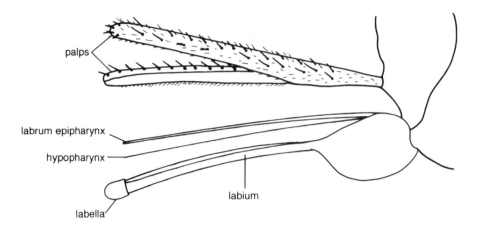

Fig. 4.7. Lateral view of mouthparts of *Glossina*.

teeth on its outer surface. It forms an almost complete tube and the hypopharynx is little more than a sclerotized salivary duct. The labella are narrow structures, each bearing three finely toothed rasps, additional larger teeth and sensilla (Jobling, 1933).

The haustellum is so long that, in feeding, the tsetse fly has to rear up on its hind legs in order to get the haustellum vertical. It should be noted that there is no connection between the salivary duct and the food canal except at the tip of the haustellum where they both open. This is important in the development of trypanosomes. In the hippoboscids the bulb of the haustellum is concealed in a recess in the head, and the thin proboscis is protected by the palps. The labrum, labial gutter and prementum are all sclerotized structures, but the hypopharynx is delicate.

Mouthparts of Fleas (see Fig. 4.8)

There have been different views regarding the homologies of the structures composing the mouthparts of fleas. The present view is that fleas have lost the mandibles but retained both labial and maxillary palps. The maxillae are represented by two elements, the lacinia and the stipes. The labrum is well developed and forms the food canal. The hypopharynx is very short, and an alternative route has to be found to convey the saliva to the distal end of the mouthparts. This is provided by the laciniae, which are grooved along their length, and the two grooves are closely opposed to form a tube. The labial palps form a protective sheath around the fascicle of laciniae and labrum. When the flea feeds, the mouthparts are steadied by the broad, triangular stipites (plural of stipes), which are directed laterally. The toothed laciniae pierce the skin, and as the fascicle is inserted the labial palps separate. Saliva is poured in and blood sucked up the food canal. Fleas

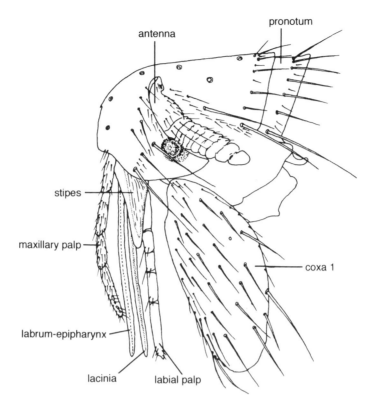

Fig. 4.8. Head and prothorax of a flea.

infected with plague bacilli become blocked and blood cannot be forced into the midgut by the cibarial pump. When the muscles of the pump relax, blood and part of the bacillary clot flow back into the host and infect it with plague bacilli.

Mouthparts of Blood-sucking Hemiptera (see Fig. 4.9)

The proboscis of an hemipteran consists of a jointed labium (rostrum) which encloses a fascicle formed from the closely opposed maxillae and mandibles. There are no palps, the hypopharynx is greatly reduced, and the labrum is short and entirely external to the labium and functional mouthparts. The mandibles, which form the inner core of the fascicle, bear two longitudinal grooves, which are closely opposed. One of these forms the food canal, and the other conveys saliva. When the hemipteran is feeding the fascicle penetrates the skin, and the jointed, labial sheath is bowed posteriorly in a similar manner to the labium of mosquitoes, except that in mosquitoes the labium is unjointed. In *Cimex and Rhodnius* the fascicle enters a blood vessel of suitable calibre, and they are capillary feeders (Dickerson and Lavoipierre, 1959a).

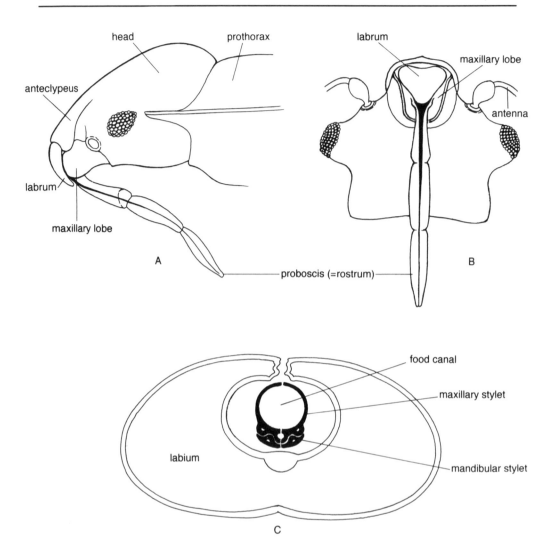

Fig. 4.9. Mouthparts of *Cimex lectularius*: (**A**) lateral view of head; (**B**) ventral view of head; (**C**) transverse section of proboscis (rostrum). Source: redrawn from Snodgrass, R.E. (1943) *Smithsonian Miscellaneous Collections* 104 (7) by permission of the Smithsonian Institution Press, Smithsonian Institution, Washington, D.C.

Mouthparts of Anoplura (see Fig. 4.10)

The mouthparts of blood-sucking lice are aberrant and impossible to homologize with those of other insects. They consist of three stylets, housed in the stylet sac, a diverticulum from the floor of the cibarium. The anterior opening of the foregut, the prestomum, is located at the anterior extremity of the head. It is sometimes referred to as the haustellum. Lined internally with small teeth the prestomum

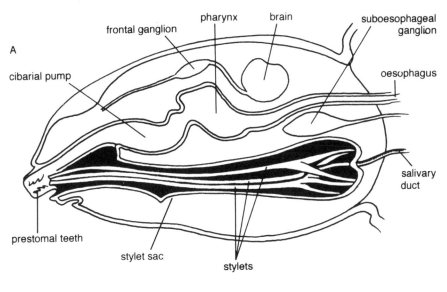

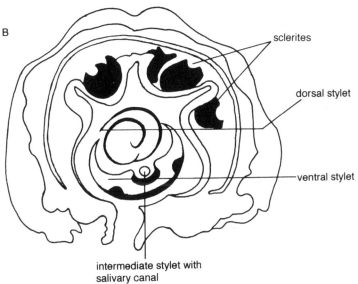

Fig. 4.10. (**A**) Longitudinal section through the head of an anopluran louse (Phthiraptera, Anoplura); (**B**) transverse section through head and stylet sac. Source: redrawn from Snodgrass, R.E. (1943) *Smithsonian Miscellaneous Collections* 104 (7) by permission of the Smithsonian Institution Press, Smithsonian Institution, Washington, D.C.

is everted in feeding, and the teeth firmly secure the louse to the skin of its host. The opening of the stylet sac is then brought up to the prestomum and the stylets pierce the skin. Blood is sucked up by the cibarial pump (Buxton, 1950).

Host Finding and Blood-feeding

Sutcliffe (1987) recognizes three phases in the search for a host by blood-sucking Diptera. They are the appetitive searching phase driven by endogenous rhythms and hunger, followed by activation and orientation when the insect encounters chemical stimuli (kairomones) indicating the presence of a host and attraction when the insect having located a host, alights to feed. Appetitive behaviour occurs at the time of day when the insect is normally active. Simuliids, tabanids, tsetse flies and the blood-sucking muscids are diurnal while most of the blood-sucking Nematocera are nocturnal or crepuscular.

Most blood-sucking Diptera are activated by carbon dioxide which acts both as an activating agent and as an attractant to mosquitoes (Gillies, 1980), and carbon dioxide emitting traps have been widely used in the study of mosquitoes – CO_2 emission is a feature of encephalitis virus surveillance (EVS) traps used for routine catches of mosquitoes. *An gambiae* s.s. is an exception being less activated by CO_2 and not orientating to it (Takken *et al.*, 1993). Hall *et al.* (1984) found that octenol (1-octen-3-ol) was a potent activator, bringing about upwind flight in *Glossina pallidipes* and *G. morsitans morsitans*. It increased the catch of tabanids threefold in canopy traps (French and Kline, 1989) and was as effective as CO_2 in increasing trap collections of *Anopheles* and *Aedes* but not of *Culex* (Kline, 1990). Octenol and CO_2 were equally effective in catching *Aedes taeniorhynchus* and *Culicoides furens* and together they acted synergistically producing significantly greater catches (Takken and Kline, 1989; Kline, 1990).

It has been estimated (Vale, 1977) that the maximum distance from which tsetse flies can be activated by a host animal is about 90 m. The final stage of landing and feeding is more complex and visual cues of colour and shape as well as odours emanating from the host play a role but none are so effective as the intact host (Bowen, 1991). Odours released from the host expand by molecular diffusion and turbulence to form an odour plume (Murlis *et al.*, 1992). Free flying wild tsetse flies orientate upwind in a host odour plume and change direction on entering or leaving a plume (Gibson and Brady, 1988). A comprehensive account of the physiology and behaviour of host finding is given by Lehane (1991).

The number of labial chemo- and mechanosensilla varies with the range of food on which an insect feeds. They are few (< 150) in obligate blood-feeders, e.g. *Glossina*, more numerous (*c.* 600) in blood and nectar feeders, e.g. simuliids, and most abundant (> 1600) in omnivores, e.g. the black blowfly, *Phormia regina* (McIver, 1987). Anopheline mosquitoes and *Lutzomyia* sandflies engorge as readily on plasma as on whole blood. The louse, *Pediculus humanus*, and the rat flea, *Xenopsylla cheopis*, ingest about half as much plasma as blood. Culicinine mosquitoes, simuliids, tsetse flies and tabanids require the cellular fraction or ATP (adenosine triphosphate) to be present for engorgement (Galun, 1987).

The labial chemoreceptors of *Glossina* are of two types and so distributed that one type (LR7), which is sensitive to ATP (Mitchell, 1976), is exposed before piercing and the other (LR5) after penetration (Rice *et al.*, 1973a). Another class of chemoreceptors is present in the labrocibarial region and monitors incoming material before swallowing (Rice *et al.*, 1973b). Feeding of *Glossina austeni, G. palpalis palpalis* and *G. tachinoides* is stimulated by the presence of ATP and less by ADP and AMP (Galun and Margalit, 1969; Galun, 1988; Galun and Kabayo, 1988). *Aedes aegypti* shows a similar feeding response to adenosine nucleotides (Galun *et al.*, 1963) as does the stable fly, *Stomoxys calcitrans* (Ascoli-Christensen *et al.*, 1991), while the rat flea, *X. cheopis*, responds only to ATP (Galun, 1966).

References

Allan, S.A., Day, J.F. and Edman, J.D. (1987) Visual ecology of biting flies. *Annual Review of Entomology* 32, 297–316.

Ascoli-Christensen, A., Sutcliffe, J.F. and Albert, P.J. (1991) Purinoreceptors in blood feeding behaviour in the stable fly, *Stomoxys calcitrans. Physiological Entomology* 16, 145–152.

Bowen, M.F. (1991) The sensory physiology of host-seeking behaviour in mosquitoes. *Annual Review of Entomology* 36, 139–158.

Buxton, P.A. (1950) *The Louse.* Edward Arnold, London.

Dickerson, G. and Lavoipierre, M.M.J. (1959a) Studies on the methods of feeding of bloodsucking arthropods. ll. The method of feeding adopted by the bed-bug (*Cimex lectularius*) when obtaining a blood-meal from the mammalian host. *Annals of Tropical Medicine and Parasitology* 53, 347–357.

Dickerson, G. and Lavoipierre, M.M.J. (1959b) Studies on the methods of feeding of blood sucking arthropods. lll. The method by which *Haematopota pluvialis* (Diptera: Tabanidae) obtains its blood-meal from the mammalian host. *Annals of Tropical Medicine and Parasitology* 53, 465–472.

Foil, L.D. (1989) Tabanids as vectors of disease agents. *Parasitology Today* 5, 88–96.

Foil, L.D., Adams, W.V., McManus, J.M. and Issel, C.J. (1987) Bloodmeal residues on mouthparts of *Tabanus fuscicostatus* (Diptera: Tabanidae) and the potential for mechanical transmission of pathogens. *Journal of Medical Entomology* 24, 613–616.

Foil, L.D., Seger, C.L., French, D.D., Issel, C.J., McManus, J.M., Ohrberg, C.L. and Ramsey, R.T. (1988) Mechanical transmission of bovine leukemia virus by horse flies (Diptera: Tabanidae). *Journal of Medical Entomology* 25, 374–376.

French, F.E. and Kline, D.L. (1989) 1-Octen-3-ol, an effective attractant for Tabanidae (Diptera). *Journal of Medical Entomology* 26, 459–461.

Galun, R. (1966) Feeding stimulants of the rat flea *Xenopsylla cheopis. Life Sciences* 5, 1335–1342.

Galun, R. (1987) Regulation of blood gorging. *Insect Science and its Application* 8, 623–625.

Galun, R. (1988) Recognition of very low concentrations of ATP by *Glossina tachinoides* Westwood. *Experentia* 44, 800.

Galun, R. and Kabayo, P. (1988) Gorging response of *Glossina palpalis palpalis* to ATP analogues. *Physiological Entomology* 13, 419–423.

Galun, R. and Margalit, J. (1969) Adenosine nucleotides as feeding stimulants of the tsetse fly *Glossina austeni* Newstead. *Nature, London* 222, 583–584.

Galun, R., Avi-Dor, Y. and Bar-Zeev, M. (1963) Feeding response in *Aedes aegypti*: stimulation by adenosine triphosphate. *Science* 142, 1674–1675.

Gibson, G.A. and Brady, J. (1988) Flight behaviour of tsetse flies in host odour plumes: the initial response to leaving or entering odour. *Physiological Entomology* 13, 29–42.

Gillies, M.T. (1980) The role of carbon dioxide in host-finding by mosquitoes (Diptera: Culicidae): a review. *Bulletin of Entomological Research* 70, 525–532.

Gordon, R.M. and Lumsden, W.H.R. (1939) A study of the behaviour of the mouthparts of mosquitoes when taking up blood from living tissue; together with some observations on the ingestion of microfilariae. *Annals of Tropical Medicine and Parasitology* 33, 259–278.

Hall, D.R., Beevor, P.S., Cork, A., Nesbitt, B.F. and Vale, G.A. (1984) 1-Octen-3-ol a potent olfactory stimulant and attractant for tsetse isolated from cattle odours. *Insect Science and its Applications* 5, 335–339.

Jobling, B. (1933) A revision of the structure of the head, mouthparts and salivary glands of *Glossina palpalis* Rob. Desv. *Parasitology* 24, 449–490.

Kline D.L. (1990) Semiochemicals for mosquito and biting midge surveillance and control. *Abstracts 2nd International Congress of Dipterology*, Bratislava, 27 August–1 September 1990, p. 112.

Kline, D.L. and Choate, P.M. (1990) The relationship between palpal morphology and host-seeking behaviour in adult mosquitoes. *Abstracts of the Second International Congress of Dipterology*, Bratislava, 27 August–1 September 1990, p. 113.

Kovacs, F., Medveczky, I., Papp, L. and Gondar, E. (1990) Role of prestomal teeth in feeding of the house fly, *Musca domestica* (Diptera: Muscidae). *Medical and Veterinary Entomology* 4, 331–335.

Lehane, M.J. (1991) *Biology of Blood-sucking Insects*. Harper Collins Academic, London.

McIver, S.B. (1987) Sensilla of haematophagous insects sensitive to vertebrate host-associated stimuli. *Insect Science and its Applications* 8, 627–635.

Mitchell, B.K. (1976) ATP reception by the tsetse fly, *Glossina morsitans* West. *Experentia* 32, 192–194.

Murlis, J., Elkington, J.S. and Cardé, R.T. (1992) Odour plumes and how insects use them. *Annual Review of Entomology* 37, 505–532.

Patton, W.S. and Cragg, F.W. (1913) *A Textbook of Medical Entomology*. Christian Literature Society for India, London.

Rice, M.J., Galun, R. and Margalit, J. (1973a) Mouthpart sensilla of the tsetse fly and their function. ll. Labial sensilla. *Annals of Tropical Medicine and Parasitology* 67, 101–107.

Rice, M.J., Galun, R. and Margalit, J. (1973b) Mouthpart sensilla of the tsetse fly and their function. lll. Labrocibarial sensilla. *Annals of Tropical Medicine and Parasitology* 67, 109–116.

Sutcliffe, J.F. (1987) Distance orientation of biting flies to their hosts. *Insect Science and its Applications* 8, 611–616.

Takken, W. and Kline, D.L. (1989) Carbon dioxide and 1-octen-3-ol, as mosquito attractants. *Journal of the American Mosquito Control Association* 5, 311–315.

Takken, W., Kemme, J. and van Essen, P.H. (1993) The influence of CO_2 and human steroids on the flight behaviour of *An gambiae s.s.* and *An stephensi*. In: Uren, M.F. and Kay, B.H. (eds) *Proceedings 6th Symposium on Arbovirus Research in Australia*, 7–11 December 1992, Brisbane, pp. 40–41.

Vale, G.A. (1977) The flight of tsetse flies (Diptera: Glossinidae) to and from a stationary ox. *Bulletin of Entomological Research* 67, 297–303.

Wikel, S.K. (1982) Immune responses to arthropods and their products. *Annual Review of Entomology* 27, 21–48.

Yorke, W. and Macfie, J.W.S. (1924) The action of the salivary secretion of mosquitoes and of *Glossina tachinoides* on human blood. *Annals of Tropical Medicine and Parasitology* 18, 103–108.

Internal Structure and Function of Insects

5

This chapter presents a general account of the internal structure and physiological organization of insects, emphasizing those features in which insects differ markedly from vertebrates, and which are relevant to medical entomology. A broad, but now somewhat dated, coverage of insect physiology is available in Wigglesworth (1972). Lehane (1991) covers adult blood-sucking insects and Clements (1992) all stages of mosquitoes.

Cuticle (see Fig. 5.1)

Many of the features peculiar to insects have their origin in the cuticle, which acts as both a limiting membrane or skin, and as a skeleton for the attachment of muscles. The cuticle is composed of a single layer of cells, the epidermis, which rests upon a basement membrane, and three outer non-living layers, the endocuticle, exocuticle and epicuticle. The cuticle of ticks (Ixodida) has a similar structure and function (Hackman, 1982). The physiology of the epidermis has been reviewed by Binnington and Retnakaran (1991).

Epicuticle

The epicuticle is about 1 μm in thickness and composed of several layers. Its innermost layer is a refractile, inelastic, amber-coloured membrane, the cuticulin layer, largely composed of lipoprotein. To allow for growth and stretching this layer is often wrinkled. Outside the epicuticle is a very thin, wax layer (0.25 μm), which is covered by a cement layer. The function of the cement is to protect the vital wax layer, which provides essential waterproofing to enable insects to survive desiccation in unsaturated atmospheres. Waterproofing is independent of the thickness of the cuticle, being solely dependent upon the existence of the wax layer, of which the most important component is an orientated monolayer closely adherent to the cuticulin layer (Beament, 1945).

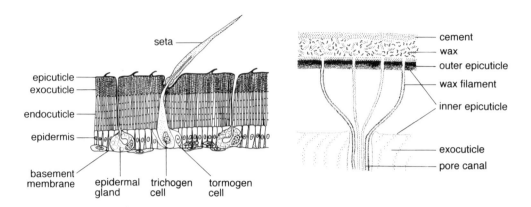

Fig. 5.1. Section of typical insect cuticle (left) and detail of the epicuticle (right). Source: Left, from Davies, R.G. (1988) *Outlines of Entomology*, Chapman & Hall, London. Right, from Gullan, P.J. and Cranston, P.S. (1994) *The Insects, an Ouline of Entomology*, Chapman & Hall, London.

The arrangement has applied implications. Firstly, the wax has a particular melting point, and there exists a critical temperature, about 5 to 10°C below the melting point, at which the monolayer becomes disorganized and water is rapidly lost from the organism. This temperature may be higher than the insect's thermal death point. Secondly, the wax layer can be removed by abrasive materials greatly increasing the rate of water loss. Abrasive dusts have been applied to clothing for the control of sucking lice. Thirdly, the wax layer is a barrier to the inward passage of foreign materials. Water-soluble materials are excluded, but lipid-soluble materials may dissolve in the wax and enter the insect. It is notable that the synthetic contact insecticides, e.g. DDT, are lipid-soluble and owe their activity on contact to this property.

Exocuticle

The exocuticle is a thick, rigid, amber-coloured layer which forms the main skeletal component. It is composed of chitin and protein. Chitin is a nitrogenous polysaccharide, composed of long chains of acetylated glucosamine residues. The rigidity is provided by the tanned protein, sclerotin. The exocuticle is absent from areas such as the integmental membranes, where the body is flexible.

Endocuticle

The endocuticle is a thick, elastic layer composed of chitin and protein, which has not been tanned. It provides flexible support. In some areas requiring great elasticity, as in the cibarial pump of Diptera (Rice, 1970), the rubber-like protein, resilin, is present in the endocuticle.

Pore canals

The endocuticle and exocuticle are penetrated by pore canals, protoplasmic extensions of the epidermal cells. There may be 200 pore canals per epidermal cell and it is likely that they penetrate the cuticulin layer and are involved in secreting the wax layer. The endocuticle and exocuticle are deposited by the pore canals while the cement layer is secreted by dermal glands which penetrate the cuticle.

Water absorption

The structure of the cuticle makes it more permeable to the passage of water into the body than out (Wigglesworth, 1972). The pupating larva of the oriental rat flea, *Xenopsylla cheopis*, can take up moisture from air at a relative humidity of 45% and may increase its weight by as much as 29% before pupation (Edney, 1947). Ixodid ticks have a similar capacity to absorb water from unsaturated air. This ability is important to the survival of starving or non-feeding stages.

Growth

Since parts of the cuticle are rigid, e.g. the head capsule, there arises a stage in the growth of an individual when a new cuticle is required. The epidermis separates itself from the existing cuticle, a process known as apolysis, and lays down the cuticulin layer of the epicuticle, followed by the exocuticle, which, at this stage, is as flexible as endocuticle, as it is not sclerotized until after the moult. The endocuticle is not secreted until after ecdysis. The cuticulin layer is, of course, wrinkled to provide room for expansion when the old cuticle is cast. The moulting fluid, containing appropriate enzymes, is secreted into the space between the old and new cuticles and completely digests the old endocuticle, the products of which are absorbed. This has the effect of reducing the ecdysial suture to epicuticle as no exocuticle is present under the suture. Shortly before ecdysis the wax layer is secreted and the cement layer shortly after the moult is completed. When an insect completes virtually all the moult except for the final discarding of the old cuticle the hidden new stage is referred to as pharate. Thus the mature flea pupa contains a pharate adult awaiting the stimulus of a host's arrival to cast its pupal cuticle and become a free adult.

The newly moulted insect is usually very pale because sclerotization has not begun. It immediately expands to a new size by increasing its volume by swallowing air or water. The newly emerged adult blowfly *Calliphora* increases its volume by 128% by swallowing air (Fig. 5.2). Formation of exocuticle and the laying down of the new endocuticle proceed rapidly. Newly emerged adults, which remain soft and pliable, are referred to as teneral.

Mosquito larvae have four instars in which only the head capsule is sclerotized. The abdomen and thorax continue to grow steadily throughout larval life but the head increases dramatically at each moult. The change shown in Table 5.1 was observed in *An sergenti* within a minute of the old skin being cast (Kettle, 1948).

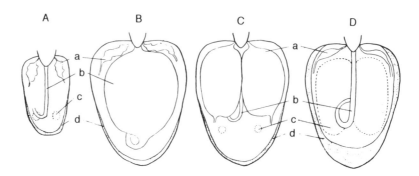

Fig. 5.2. Changes in the abdominal air sacs of *Lucilia* at various stages of adult development: (**A**) fly just emerged; (**B**) after five minutes – gut filled with air, abdomen distended; (**C**) after 10 hours – gut collapsed, air sacs distended; (**D**) fully fed for six days – ovaries and fat body fill abdomen, air sacs collapsed again. a, air sacs; b, gut; c, ovaries; d, fat body. Source: from Evans, A.C. (1935), *Bulletin of Entomological Research* 26, 115.

Table 5.1. Change observed in *Anopheles sergenti* at ecdysis.

	Head		Thorax	
	Breadth	Length	Breadth	Length
Before ecdysis	128 µm	116 µm	184 µm	265 µm
After ecdysis	230 µm	219 µm	196 µm	276 µm
% increase	80	89	4	6

The increase in volume was achieved by swallowing water and pumping blood anteriorly to expand the head capsule, which increased in volume sixfold. This stepwise increase of larval head size has a useful application. Most keys to the identification of *Anopheles* larvae apply to the 4th or final instar larvae. The 4th instar can be recognized by the size of the head capsule (Fig. 7.9).

The cuticle cast at ecdysis is called an exuviae (unchanged in the plural). Exuviae are taxonomically important because they retain the external features of the stage. It is therefore possible to associate a particular adult insect with its own larval and pupal exuviae.

Digestive Tract and Digestion (see Fig. 5.3)

The gut of an insect is composed of three main regions – foregut, midgut and hindgut. It is unusual in that the foregut and hindgut develop from ectodermal invaginations, and consequently are lined with a thin layer of cuticle. At ecdysis these cuticular linings are shed with the exuviae. In vertebrates the proctodeal and stomodeal invaginations are very short. Only the midgut of an insect is comparable to the vertebrate digestive tract.

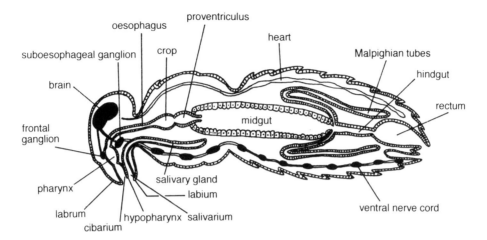

Fig. 5.3. Diagrammatic longitudinal section of a generalized insect.

The mouth leads into the buccal cavity and thence to the pharynx, which may be developed as a muscular pump moving food down the narrow oesophagus to the storage organ or crop. In Diptera the crop is a blind diverticulum from the oesophagus. In the Cyclorrhapha the crop functions as a store passing food to the midgut as required. In mosquitoes, and probably in other blood-sucking Nematocera, blood is stored in the midgut and sugar solutions in the crop. In the tabanid *Haematopota*, blood passes to the stomach and the overflow to the crop. The crop is purely storage and secretes no enzymes. Any digestion that occurs in the crop is due to the action of enzymes contained in the saliva.

The foregut ends in the proventriculus, which, in most medically important insects, has become a valve preventing regurgitation of food from the midgut. It is a very prominent structure in fleas being lined with backwardly directed spines said to be involved in rupturing red blood cells. Other blood-feeding insects manage without these spines. The spines are very important medically because it is in the interstices of the proventricular spines that plague bacilli multiply.

The midgut of many insects is lined with a non-adherent, thin, transparent tube, the peritrophic membrane. Its functions are to protect the midgut from abrasion by food particles and to prevent or restrict the passage of parasites and pathogens (Spence, 1991). In haematophagous Nematocera, e.g. adult mosquitoes, phlebotomine sandflies (Dolmatova, 1942) and simuliids (Lewis, 1953) a short-lived peritrophic membrane is secreted around the blood meal by the midgut epithelium. In dipteran larvae and most insects the peritrophic membrane is secreted at the anterior end of the midgut and moulded by an invagination of the oesophagus. It then forms a single continuous membrane which may extend into the hindgut (Fig. 5.4). In the context of vaccinating mammalian hosts against parasitic arthropods the peritrophic membrane may either restrict penetration of ingested immune components or serve as a target for immunological attack (Eisemann and Binnington, 1994). In *Glossina* it influences the development of trypanosomes. It is present in most adult Diptera, the larvae of mosquitoes,

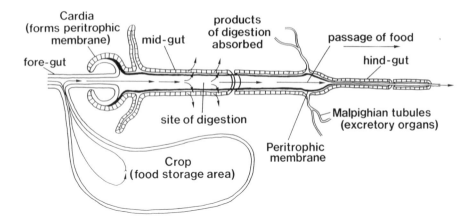

Fig. 5.4. Diagram of the digestive system of a dipteran. Courtesy of Dr C.H. Eisemann.

simuliids and fleas, but absent in adult fleas, and adult and nymphal Phthiraptera (Peters, 1992).

The ultrastructure of the midgut, the main digestive organ, of the most important blood-sucking insects has been reviewed by Billingsley (1990). In fleas and mosquitoes it forms a capacious stomach for the reception and digestion of blood. The oocysts of the malaria parasite develop on the outer side of the midgut in *Anopheles* mosquitoes. In blood-sucking bugs the anterior part of the midgut is storage and is separated from the posterior digestive part by a sphincter. In the Cyclorrhapha food is stored in the crop and the midgut is a relatively narrow tube with functional divisions. The anterior division is concentrative, removing excess water, the middle division digestive and the hind division absorptive. In most nematoceran larvae the midgut bears caeca anteriorly which are considered to be absorptive and secretory. In mosquitoes the ultrastructure of two of the four types of cells in the caeca are appropriate to ion and water transport (Clements, 1992). The midgut is separated by the pyloric sphincter from the hindgut. The midgut enzymes are adapted to the diet of the insect. Thus the midgut of blood-feeding *Glossina* secretes a very active protease but little carbohydrase and that of *Calliphora*, which feeds on sweet substances, produces little protease and abundant carbohydrases. *Chrysops*, which feeds on both nectar and blood, secretes an active protease and an active carbohydrase.

The main function of the hindgut is the absorption of water from the faeces and urine. Urine passes into the hindgut from the Malpighian tubules, which open just behind the midgut. When water is freely available as in aquatic insects and liquid-feeding adults, faeces and urine are watery, but most insects need to conserve water. Desiccation is an ever present threat. Water is absorbed from the hindgut especially by the rectal glands, and faeces are excreted as dry pellets and nitrogenous waste as uric acid. Water economy is practised in the excretion of nitrogenous waste, which passes, in solution, into the lumen of the Malpighian tubes as the relatively soluble potassium acid urate, where it is converted in the

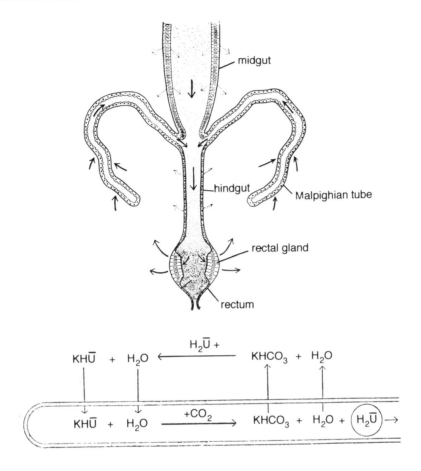

Fig. 5.5. Above, diagram of water-circulation in alimentary and excretory systems of an insect. Below, diagram showing possible mechanism by which uric acid is precipitated in the lower segment of the Malpighian tubules in *Rhodnius*. Source: from Wigglesworth, V.B. (1933) *Quarterly Journal of Microscopical Science* 75, 132.

presence of carbon dioxide to potassium bicarbonate and relatively insoluble uric acid. This enables a small amount of water to circulate repeatedly from hindgut to Malpighian tubes via the body cavity, and bring about the elimination of large amounts of nitrogenous waste (Fig. 5.5).

Within the Diptera, Bursell (1975) recognizes three groups on the basis of which substance(s) they are able to oxidize at a rate adequate to support flight. In species in which only the female is haematophagous, e.g. *Tabanus*, or in which neither sex is blood-sucking, e.g. *Musca*, energy for flight is provided by pyruvate oxidation. In facultative blood-sucking species, such as those in the Stomoxinae, pyruvate and proline are the substrates. In obligatory haematophagous Diptera, e.g. *Glossina*, energy is derived from the oxidation of proline.

Symbiotic Organisms in Blood-sucking Insects

Insects have different vitamin requirements from those of vertebrates. They do not require ascorbic acid (vitamin C) but do require a source of sterols and some specifically need cholesterol. All insects require certain fractions of the B group of vitamins and these are often provided by microorganisms living in the food source or in the gut of the insect. Blood is an incomplete food for insect growth and sexual maturity, and insects that feed in all stages of their life only on blood, have special arrangements for the cultivation of symbiotic microorganisms. In the blood-sucking reduviid bugs the organisms are free in the lumen of the midgut diverticula; in the Hippoboscidae the organisms (endocytobionts) inhabit cells in the wall of the midgut diverticula. In the Anoplura and *Glossina* a special zone of the midgut, the 'mycetome', more correctly referred to as the 'bacteriome' (Gassner, 1989), is composed of cells laden with microorganisms. In the bedbug *Cimex*, the 'mycetome' is remote from the gut, being located elsewhere in the body cavity.

Holometabous insects whose larval stage is free-living, e.g. mosquitoes, fleas, tabanids, *Stomoxys*, do not have 'mycetomes'. Presumably enough essential materials are acquired by their larvae to meet the needs of the adults. If the symbionts are removed from nymphs of *Pediculus* they die in a few days and female *Pediculus*, deprived of symbionts, are sterile. In adult *Glossina* symbionts affect reproduction but not longevity. In the *Culex pipiens* complex the microorganism (*Wolbachia pipientis*) is transmitted in the maternal cytoplasm and involved in reproductive cytoplasmic incompatibility (Gassner, 1989).

Respiration (see Fig. 5.6)

In insects air is carried direct to the tissues by special tubes, the tracheae. These are ectodermal invaginations and are consequently lined with thin cuticle which is thrown into folds which follow a spiral course round the trachea. These spiral thickenings resist compression and enable the tracheae to recoil quickly from deformation. The tracheae open to the exterior at the spiracles, of which there are typically two pairs on the thorax and eight pairs on the abdomen (Nikam and Khole, 1989). The tracheae anastomose, usually forming paired longitudinal trunks which run the length of the body and extend into the head. The tracheae branch among the tissues with the branches becoming narrower in diameter. The finest branches, the tracheoles, lack obvious spiral thickenings and are closely applied to the tissues and may even penetrate individual cells.

The fine endings of the tracheoles commonly contain fluid. When the tissue is active, as for example muscle following contraction, the liquid is withdrawn, presumably as a result of increasing osmotic pressure caused by metabolites in the tissue, and air temporarily extends to the termination of the tracheole. Oxygen reaches the tissues by diffusion along the tracheae, and in larger and/or very active insects diffusion is supplemented by active ventilation of the tracheae. Carbon dioxide, the final product of metabolism, may diffuse into the atmosphere along the tracheae but it is quite likely to diffuse through the general body cuticle rather as it does through the skin of a frog.

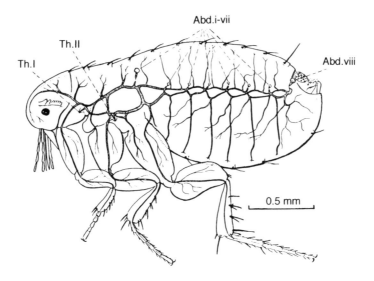

Fig. 5.6. Tracheal system of *Xenopsylla* (left side only). *Th.I, Th.II,* thoracic spiracles. *Abd. i-viii,* abdominal spiracles. Source: from Wigglesworth, V.B. (1935) *Proceedings of the Royal Society* (B) 118, 398.

Open spiracles represent a major source of water loss and elaborate arrangements are made to restrict this. The spiracles of most terrestrial insects have a closing mechanism, which limits water loss. It operates in response to the accumulation of carbon dioxide and not to a deficiency of oxygen. When the flea, *Xenopsylla cheopis,* is placed in an atmosphere of 5% carbon dioxide the spiracles remain open, and it loses water at twice the rate it does in air. It is of interest to note that in vertebrates respiratory movements are also controlled by the concentration of carbon dioxide in the circulating blood and not by a lack of oxygen.

In many insects the tracheal system has large thin-walled dilations, the air sacs. These aid in ventilating the tracheal system and are common in active insects but they also have the function of keeping the external shape of an insect constant in spite of extensive internal changes, e.g. egg maturation. The newly emerged adult *Lucilia* inflates itself to its full size by swallowing air. Initially this air is taken into the gut but 10 h later the gut has returned to its normal size and the air sacs are inflated. The volume of the air sacs is adjusted to keep the fly's total volume constant when food is imbibed or eggs matured. In the gravid female the ovaries almost fill the abdomen and the air sacs are insignificant (Fig. 5.2). Possession of a constant external shape must simplify flight by eliminating one variable, namely aerodynamic shape.

The respiration of aquatic larvae will be considered when dealing with individual families.

Insects are less dependent on a continuous supply of oxygen than vertebrates. Certain important cells in the vertebrate central nervous system are irreversibly damaged by being deprived of oxygen. Insects lack these cells and readily survive

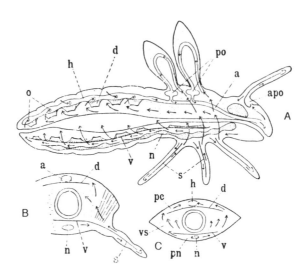

Fig. 5.7. Diagram of fully developed circulatory system: (**A**) lateral view; (**B**) transverse section of thorax; (**C**) transverse section of abdomen. Arrows show course of circulation. *a*, aorta; *apo*, accessory pulsatile organ; *d*, dorsal diaphragm with aliform muscles; *h*, heart; *n*, nerve cord; *o*, ostia; *pn*, perineural sinus; *po*, mesothoracic and metathoracic pulsatile organs; *s*, septa dividing appendages; *v*, ventral diaphragm; *vs*, visceral sinus. Source: from Wigglesworth, V.B. (1972) *Principles of Insect Physiology*, Chapman & Hall, London.

periods of oxygen deprivation. Insect tissues are able to build up an oxygen debt by respiring anaerobically. This debt is discharged when oxygen becomes available later. When the cockroach, *Blatta orientalis*, is deprived of oxygen for half an hour it sustains an oxygen debt which requires one and a half to two hours to be repaid and the extra oxygen consumed is equivalent to that expected to be used in half an hour (Davis and Slatter, 1926).

Circulatory System (see Fig. 5.7)

Insect blood or haemolymph is clear and contains no respiratory pigment. The carriage of oxygen to the tissues is the function of the tracheal system. The function of the circulatory system is to transport food to the tissues, to remove the products of metabolism, and to convey hormones around the body. The haemolymph is circulated by a dorsal heart, which is a longitudinal contractile tube. Haemolymph enters the heart via openings, the ostia, of which there is one pair in each abdominal segment. There are no veins obviously recognizable as such. The coelom which forms the body cavity in vertebrates is greatly reduced and its place has been taken by the haemocoele, which can be regarded as a greatly enlarged venous system. Haemolymph passes from the arteries to the haemocoele and from there by various routes to the dorsal pericardial sinus where it enters

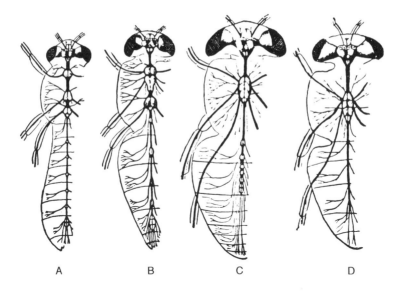

Fig. 5.8. Successive stages in the coalescence of thoracic and of abdominal ganglia in Diptera: (**A**) *Chironomus* (Nematocera); (**B**) *Empis* (Orthorrhapha); (**C**) *Tabanus* (Orthorrhapha); (**D**) *Sarcophaga* (Cyclorrhapha).

the heart through the ostia. In the heart the haemolymph is propelled forward by a steady contractile wave, which passes along it from the posterior end. In most insects there are supplementary pulsating organs which drive the haemolymph through the appendages. In mosquitoes there are pulsating organs which drive haemolymph into the antennae (Clements, 1992).

Nervous System (see Figs 5.3 and 5.8)

Primitively the nervous system of an insect consists of a ring of nervous tissue around the oesophagus in the head, and paired ganglia in each thoracic and abdominal segment, connected to each other in the midline, and anteriorly and posteriorly by connectives. The oesophageal ring includes concentrations of nervous tissue into supraoesophageal and suboesophageal ganglia. The supraoesophageal ganglion or brain receives sensory information from the eyes and antennae. It has an overriding excitatory or inhibiting effect on other centres of activity. The suboesophageal ganglion innervates the mouthparts and has a general excitatory effect on the thoracic ganglia. In higher insects there is a tendency to concentrate the paired segmental ganglia and in Cyclorrhapha they are fused into a single ganglionic mass in the thorax. There is considerable segmental autonomy, and decapitation does not lead to the immediate death of the insect. Providing the loss of haemolymph is prevented, a headless insect is able to walk and oviposit and, providing the critical stage has been passed before decapitation, it will moult.

Various endocrine glands are associated with the nervous system. Secretions of neurosecretory cells in the brain pass along axons to the corpus cardiacum and the corpora allata. The corpora allata secrete neotenin (the juvenile hormone) which controls the expression of immature characters during development, ensures the deposition of yolk in developing eggs, and influences reproductive behaviour. Under stimulation by neurosecretions the thoracic glands secrete ecdyson, the moulting hormone.

Sensory receptors play a major role in insect behaviour but their treatment is outside the scope of this book. Reference will be made to particular receptors when describing the biology of a taxonomic group.

Reproduction

In insects the sexes are separate, and the sex of an individual is determined by its complement of chromosomes. The genes for sex determination act directly on the cells in which they are located, and not via released chemicals (hormones) operating at a distance from the site of secretion. Consequently loss of sex chromosomes during development produces gynandromorphs with a mosaic of female and male characters, and not intersexes with intermediate characters.

The female reproductive organs (Fig. 5.9) consist of paired ovaries and oviducts, a common oviduct which opens into the vagina and spermathecae. Each ovary consists of a variable number of ovarioles, two in *Glossina*, and 2000 in certain termites. Each ovariole is a tube containing a string of oocytes, of which the one nearest the oviduct is the first to mature. When the mature ovum is shed, the empty sac contracts, but often remains as a relict body.

At oviposition the egg moves down the oviduct and passes the opening to the spermatheca. As it does so sperm are extruded on the micropyle, an area where the chorion (shell) of the egg is very thin and penetration is easy. The male reproductive system (Fig. 5.10) consists of paired testes, paired vesicula seminales in which the spermatozoa are stored, accessory glands and an ejaculatory duct through which the spermatozoa are deposited in the female. In many insects the spermatozoa are transferred, enclosed in a spermatophore, a proteinaceous membrane formed from secretions of the accessory glands. After deposition the spermatozoa are finally stored in the spermathecae where they may remain viable for years. Queen bees are inseminated during their sole nuptial flight. Fertilized eggs give rise to females, workers or queens, while unfertilized eggs develop into males (drones). A queen may fertilize eggs for many years before producing only drones and causing the death of the colony. Single mating is common in insects.

Most insects are oviparous, laying eggs which undergo embryonic development before hatching. In *Oestrus ovis* and *Sarcophaga* the eggs are retained until embryonic development is complete. The larva emerges as soon as the egg is deposited. This is ovoviviparity. Viviparity involves retention of the egg until hatching and subsequent growth of the larva within the body of the female. *Musca bezzii* and *M. planiceps* retain larvae for one or two ecdyses and deposit them in the 2nd and 3rd instars respectively. These larvae are not full grown and have an active, free-living, feeding period. In *Glossina* the larva completes its development in the female and has only a very brief free-living period, during which it

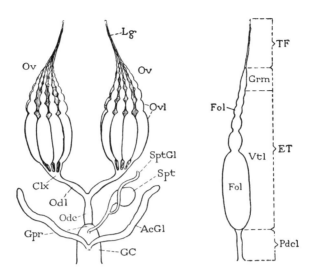

Fig. 5.9. Female reproductive system.
AcGl, accessory gland; *Clx*, calyx; *ET*, egg tube; *Fol*, follicle; *GC* vagina; *Gpr*, gonopore; *Grm*, germarium; *Lg*, ligament; *Odc*, common oviduct; *Odl*, lateral oviduct; *Ov*, ovary; *Ovl*, ovariole; *Pdcl*, pedicel; *Spt*, spermatheca; *SptGl*, spermathecal gland; *TF*, terminal filament; *Vtl*, vitellarium. Source: Reproduced with permission from Snodgrass, R.E. (1935) *Principles of Insect Morphology*. McGraw-Hill, New York.

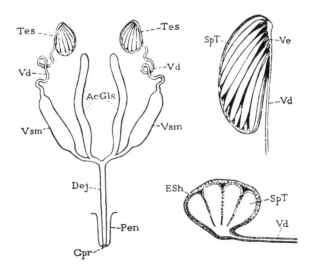

Fig. 5.10. Male reproductive system.
AcGls, accessory glands; *Dej*, ejaculatory duct; *Esh*, epithelial sheath; *Gpr*, gonopore; *Pen*, penis; *SpT*, spermatic tube; *Tes*, testis; *Vd*, vas deferens; *Ve*, vas efferens; *Vsm*, vesicula seminalis.
Source: Reproduced with permission from Snodgrass, R.E. (1935) *Principles of Insect Morphology*. McGraw-Hill, New York.

does not feed, but crawls away and buries itself. In the Pupipara, e.g. *Melophagus ovinus*, the larva is deposited fully grown. It is immobile and pupariates where it has been deposited.

Ageing Adult Diptera

In epidemiological studies of biologically transmitted pathogens it is important to know the age structure of the vector population in order to assess the rate of disease transmission. Only insects that have lived longer than the time taken for the pathogen to complete its cycle in the vector, will be potential transmitters of the pathogen. One method of ageing females depends on determining the number of completed ovarian cycles, which is given by the number of relict bodies in the ovarioles. This method has been used to determine the physiological age of individual insects and construct life tables of frequencies of different age classes in populations of *Anopheles* (Polovodova, 1949) and *Glossina* (Saunders, 1962).

More recently it has been shown that adult Cyclorrhapha can be aged by a fluorescence technique measuring the pteridine content of the head capsule. The rate of pteridine accumulation is dependent on the age of the fly and the temperature to which it has been exposed. This method has been applied successfully to determining the age of blood-sucking *Stomoxys calcitrans* (Lehane *et al.*, 1986), and four species of *Glossina* but was less reliable with *G. austeni* (Langley *et al.*, 1988; Lehane and Hargrove, 1988). This technique can be used to age both males and females, a distinct advantage when both sexes are capable of being vectors. It has also been applied successfully to the Old World and New World screw-worm flies, *Chrysomya bezziana* (Wall *et al.*, 1990) and *Cochliomyia hominivorax* (Thomas and Chen, 1989).

References

Beament, J.W.L. (1945) The cuticular lipoids of insects. *Journal of Experimental Biology* 21, 115–131.

Billingsley, P.F. (1990) The midgut ultrastructure of haematophagous insects. *Annual Review of Entomology* 35, 219–248.

Binnington, K. and Retnakaran, A. (1991) *Physiology of the Insect Epidermis*. CSIRO Publications, Melbourne.

Bursell, E. (1975) Substrates of oxidative metabolism in dipteran flight muscle. *Comparative Biochemistry and Physiology B* 52, 235–238.

Clements, A.N. (1992) *The Physiology of Mosquitoes*. Pergamon Press, Oxford.

Davis, J.G. and Slater, W.K. (1926) The aerobic and anaerobic metabolism of the common cockroach (*Periplaneta orientalis*). Part 1. *Biochemical Journal* 20, 1167–1172.

Dolmatova, A.V. (1942) The life cycle of *Phlebotomus papatasii* (Scopoli). *Meditsinskaya Parazitologiya* 11, 52–70.

Edney, E.B. (1947) Laboratory studies on the bionomics of the rat fleas, *Xenopsylla brasiliensis*, Baker, and *X. cheopis* Roths. II. Water relations during the cocoon period. *Bulletin of Entomological Research* 38, 263–280.

Eisemann, C.H. and Binnington, K.C. (1994) The peritrophic membrane: its formation, structure, chemical composition and permeability in relation to vaccination against ectoparasitic arthropods. *International Journal for Parasitology* 24, 15–26.

Gassner, G. (1989) Dipteran endocytobionts. In Schwemmler, W. and Gassner, G. (eds) *Insect Endocytobiosis: Morphology, Physiology, Genetics, Evolution.* CRC Press, Boca Raton, Florida, pp. 217–232.

Hackman, R.H. (1982) Structure and function in tick cuticle. *Annual Review of Entomology* 27, 75–95.

Kettle, D.S. (1948) The growth of *Anopheles sergenti* Theobald (Diptera: Culicidae), with special reference to the growth of the anal papillae in varying salinities. *Annals of Tropical Medicine and Parasitology* 42, 5–29.

Langley, P.A., Hall, M.J.R. and Felton, T. (1988) Determining the age of tsetse flies, *Glossina* spp. (Diptera: Glossinidae) an appraisal of the pteridine fluorescence technique. *Bulletin of Entomological Research* 78, 387–395.

Lehane, M.J. (1991) *Biology of Blood-sucking Insects.* Harper Collins Academic, London.

Lehane, M.J. and Hargrove, J. (1988) Field experiments on a new method for determining age in tsetse flies (Diptera: Glossinidae). *Ecological Entomology* 13, 319–322.

Lehane, M.J., Chadwick, J., Howe, M.A. and Mail, T.S. (1986) Improvements in the pteridine method for determining age in adult male and female *Stomoxys calcitrans* (Diptera: Muscidae). *Journal of Economic Entomology* 79, 1714–1719.

Lewis, D.J. (1953) *Simulium damnosum* and its relation to onchocerciasis in the Anglo-Egyptian Sudan. *Bulletin of Entomological Research* 43, 597–644.

Nikam, T.B. and Khole, V.V. (1989) *Insect Spiracular Systems.* Ellis Horwood, Chichester, England.

Peters W. (1992) *Peritrophic Membranes.* Springer-Verlag, Berlin.

Polovodova, V.P. (1949) Determination of the physiological age of female *Anopheles. Meditsinskaya Parazitologiya i Parazitarnye Bolezni* 18, 352–355.

Rice, M.J. (1970) Function of resilin in tsetse fly feeding mechanism. *Nature, London* 228, 1337–1338.

Saunders, D.S. (1962) Age determination for female tsetse flies and the age composition of samples of *Glossina pallidipes* Aust., *G. palpalis fuscipes* Newst and *G. brevipalpis* Newst. *Bulletin of Entomological Research* 53, 579–595.

Spence, K.D. (1991) Structure and physiology of the peritrophic membrane. In: Binnington, K. and Retnakaran, A. (eds) *Physiology of the Insect Epidermis.* CSIRO, Melbourne, pp. 77–93.

Thomas, D.B. and Chen, A.C. (1989) Age determination in the adult screwworm (Diptera: Calliphoridae) by pteridine levels. *Journal of Economic Entomology* 82, 1140–1144.

Wall, R., Lanley, P.A., Stevens, J. and Clarke, G.M. (1990) Age-determination in the old-world screw-worm fly *Chrysomya bezziana* by pteridine fluorescence. *Journal of Insect Physiology* 36, 213–218.

Wigglesworth, V.B. (1972) *The Principles of Insect Physiology.* Chapman & Hall, London.

6

Species Complexes, and Variation in *Aedes aegypti* and *Aedes albopictus*

In 1898 when Ronald Ross was investigating the transmission of malaria in India, knowledge of the systematics of mosquitoes was rudimentary. Ross recognized only grey and dapple winged mosquitoes and showed that *Plasmodium relictum*, a bird malaria, developed in the grey but not in the dapple winged mosquitoes. In the following year Grassi in Italy extended this observation to human malaria and showed that it developed in *Anopheles* mosquitoes. It was later shown that there were two tribes of mosquitoes of which the Anophelini were vectors of human malaria and the Culicini of bird malaria. Therefore control of malaria became equated with control of *Anopheles* mosquitoes. This led to intensive study of the Anophelinae and it was quickly appreciated that not all species of *Anopheles* were equally important. Some, such as *An minimus* and *An fluviatilis*, were dangerous vectors, and others, like *An subpictus*, were unimportant; the main difference being a simple feature of their behaviour: whether or not they bit humans, i.e. anthropophilic, or animals, i.e. zoophilic. This recognition gave rise to the concept of 'species sanitation', in which control was concentrated on the important species, after which it was often found that the minor species could be disregarded. Species sanitation gave excellent results in the tropics and subtropics but by 1920 there was some disquiet in Europe.

The European problem will be considered in some depth, because it was the first detailed study of a species complex in medical entomology and is the most complete, and not because it is necessarily the most important. Species complexes abound in medical entomology and some understanding of them is essential to the medical entomologist.

Anopheles maculipennis and Malaria in Europe (Hackett, 1937)

The malaria problem in Europe appeared to be relatively simple being very largely concerned with *Anopheles maculipennis*. This species is widely distributed in the

Palaearctic region from Japan to the Atlantic. The problem that European malario-logists encountered was that the distribution of malaria and the distribution of *An maculipennis* were not closely correlated. Malaria was largely coastal while *An maculipennis* was widely distributed both at the coast and in inland areas. This discrepancy undermined confidence in species sanitation. To resolve this problem the Rockefeller Foundation established a Malaria Institute in Rome and a field station in Albania across the Adriatic Sea. Even in coastal areas of Italy malaria was not uniformly present. In the coastal marshes of Salerno and Volturno, south and north of Naples respectively, and in the Pontine marshes south of Rome, *An maculipennis* was abundant. Thousands of them could be collected from animal shelters, the human population was obviously unhealthy, and malaria was rife. Yet further north around Pisa and Florence there were also thousands of *An maculipennis* present in stables, but the human population was vibrantly healthy, and there was no malaria. Hence the phrase 'Anophelism without malaria' was coined. What was the explanation of this quite striking difference?

Theories

In the absence of reliable data hypotheses and guesses abounded. There were the propagandists for healthy living, like Cirio, who argued that good food and drink would make people resistant to disease and the solution was to improve living standards. Cirio was prepared to test his hypothesis and he set up a community in the Pontine marshes which ended in disaster with the well-fed people becoming victims of malaria. Improved living standards will only be successful if they incidentally interfere with the habitat of the mosquito or reduce the *Anopheles*/human contact.

Grassi suggested that mosquitoes bred in salt water were larger and stronger. These super mosquitoes fed more avidly and lived longer. Greater longevity would permit longer survival after becoming infected and therefore would enhance malaria transmission. The view of Alessandrini was the direct opposite. He reasoned that mosquitoes reared in fresh water were larger and, being stronger, were resistant to malaria. He did not attempt the simple test of feeding mosquitoes reared from fresh water on persons suffering from malaria. Had he done so he would have found that fresh water *An maculipennis* were equally susceptible to infection by *Plasmodium*.

The French malariologist, Roubaud, came nearer to the truth, when in 1921 he proposed the existence of two races, differing in the number of teeth on their mandibles and maxillae. The multidentate race with more teeth fed on cattle, while the paucidentate race with fewer teeth was unable to pierce the skin of cattle and fed on humans. As an elaboration of this hypothesis the natural decline of malaria, which had been observed in some areas, was attributed to competition between the two races with the multidentate race replacing the paucidentate. This mechanical hypothesis gave rise to a great deal of work in which mouthparts were dissected and the numbers of teeth counted. This activity has extended to other blood-feeding Nematocera, but the rewards have been slight. There appears to be little correlation between the hosts of a species and the number of teeth on its mandibles and maxillae.

Malaria in Holland

The strain of malaria present in Holland was unusual in that the incubation period in people was many months compared with the usual 10 to 12 days. One leading malariologist, Swellengrebel, deliberately allowed himself to be bitten by an infected *An maculipennis* on 30 October but the primary attack of malaria did not occur until nine months later in July of the next year.

Normally in mosquitoes there is gonotrophic concordancy in which each blood meal is followed by the maturation of a batch of eggs. In the autumn mosquitoes that are preparing for hibernation, after a blood meal, develop fat body instead of eggs. They then seek a cool sheltered location in which to remain until the next spring. In Holland some *An maculipennis* hibernate in this fashion but others come indoors, stay in the warm and have dissociated blood-feeding and ovarian development. These individuals feed at intervals during the winter but develop no eggs until the spring. Malaria infections acquired in the winter do not produce clinical disease until the next summer. If there are both a carrier of malaria and feeding *Anopheles* present in a house, there is a high chance of the mosquitoes becoming infective and transmitting the disease to other members of the household during their winter feeding. This association of carrier and *Anopheles* gave rise to the so-called 'malaria houses', which occurred in Holland and in Norfolk in England. The existence of two behavioural patterns in adult *An maculipennis* recalls Roubaud's two races. Van Thiel made attempts to characterize them and by 1924 he was able to discriminate between a long-winged, hibernating form and a short-winged, non-hibernating form. The latter he named *atroparvus* (little, black), but size and colour are particularly variable characters being influenced by food and temperature. Wing length enabled populations to be separated but was less useful in categorizing individual mosquitoes.

Recognition of forms of *An maculipennis*

The precipitin test was developed in the 1920s enabling the source of a blood meal to be determined from the stomach contents of individual mosquitoes. It was applied to the malaria problem in Europe and by 1930 the precipitin test had confirmed that in malaria-free areas *An maculipennis* was zoophilic and in malarious areas it was anthropophilic. But there was still no means of differentiating the two forms.

In the meantime in Italy, Falleroni, a retired sanitary inspector, had collected female *Anopheles* as a hobby and allowed them to lay eggs. He found that the eggs of *An maculipennis* were patterned dark grey and silvery white (Falleroni, 1926). He figured five patterns but grouped them into two types – grey and dark eggs which he named *labranchiae* and *messeae*, respectively, after two former colleagues (Fig. 6.1). Falleroni's observations lay dormant for five years and were rediscovered in 1931. Could egg pattern be the vital character required to discriminate between the two forms of *An maculipennis*? To be a useful character, all the eggs of a female must conform to the same pattern and the patterns must be consistent throughout the range of *An maculipennis*. Surveys quickly revealed that malaria was associated with females that laid grey eggs and absent where females laid only dark eggs.

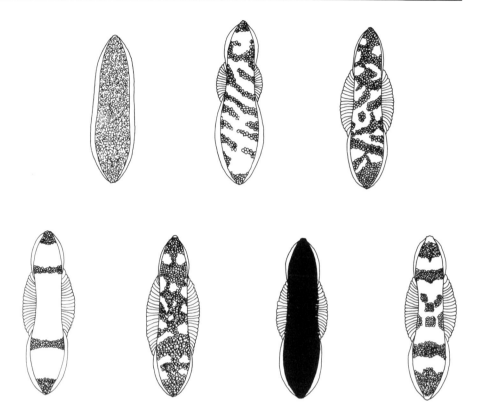

Fig. 6.1. Eggs of species of the Anopheles maculipennis complex. Top row from left to right: An sacharovi, An labranchiae, An atroparvus. Bottom row: An maculipennis, An messeae, An melanoon, An melanoon subalpinus. Redrawn from various sources.

The actual situation proved to be much more complicated than originally envisaged. There were not merely two different egg patterns but seven, all of which were consistent. Three types of grey eggs were named *labranchiae, atroparvus* and *sacharovi* and the four types of dark eggs were designated *typicus* (now *maculipennis*), *messeae, melanoon* and *subalpinus*. *Messeae* proved to be Van Thiel's long-winged hibernating species in Holland and his name *atroparvus* was retained for the egg pattern of the short-winged non-hibernating form. *Sacharovi* had long been recognized as a separate form of *An maculipennis* by its possession of a uniformly red-brown scutum compared with the longitudinal pale, broad, scutal band in *An maculipennis*. The ability to identify forms at the egg stage enabled accurately identified larvae and adults to be reared for critical examination. Such studies have revealed very minor differences between the forms which can only be appreciated by an expert in the group. For practical purposes only *sacharovi* can be readily distinguished in the adult and none in the larval stage.

Status of European forms of *An maculipennis*

There remains the problem as to the taxonomic status to be accorded these seven forms. Are they true species, subspecies or what? A species has been defined as a population of similar individuals, which breeds together and produces fertile off-spring. It is usually difficult to effect mating, let alone cross-mating, in Nematocera but in *An maculipennis* one form, *atroparvus*, proved very easy to colonize in the laboratory and it is now a standard laboratory animal. *Labranchiae* and *sacharovi* can be colonized with difficulty but the other forms are refractory. *Atroparvus* mates in small cages without initial swarming and its males were crossed with females of the other forms. The results of hybridization indicated a wide range of degrees of compatibility.

The most successful cross was of *atroparvus* and *labranchiae* in which all the females were normal but only some of the males. The *atroparvus/melanoon* cross gave females of which half were sterile as were all the males. Healthy but completely sterile adults resulted from the *atroparvus/maculipennis* cross. When *atroparvus* was crossed with *sacharovi* and *messeae* no adults were produced, death occurring in the larval and egg stages, respectively.

On the basis of these results six species and one subspecies were recognized within the *An maculipennis* complex (Knight and Stone, 1977). They were: *An atroparvus* Van Thiel; *An labranchiae* Falleroni; *An maculipennis* Meigen; *An melanoon melanoon* Hackett; *An melanoon subalpinus* Hackett and Lewis; *An messeae* Falleroni and *An sacharovi* Favre. To these Stegnii and Kabanova (1976) added *An beklemishevi* on cytological evidence (see below) and White (1978) considers that *An martinius* and *An sicaulti*, previously considered to be synonyms of *An sacharovi* and *An labranchiae* respectively, should be provisionally regarded as distinct species.

In passing it is worth while drawing attention to the difficulty of breeding many mosquitoes. To achieve success with *An messeae* and *An maculipennis* a large cage was built in Tirana, Albania, to accommodate a stable, donkey, pond and vegetation. Thousands of both species were released in the cage but only a few eggs of *An maculipennis* were laid and none of *An messeae*. *An messeae* is particularly refractory in mating in confinement. Although *An messeae* females mated with *An atroparvus* males and *An messeae* males mated with *An atroparvus* females they did not mate with each other. Some essential factor was lacking. These observations emphasize how complex the conditions for successful mating may be. If a species mates readily there is no problem; if it does not, a long, arduous study is likely to be required to give success.

Distribution of species of the *An maculipennis* complex in Europe

The European species of the *An maculipennis* complex have different geographical distributions with only four species occurring north of the Alps. *An sacharovi* occurs in the eastern Mediterranean, Iraq, Iran, southern Greece, southern Italy and Turkey. *An labranchiae* is found south of 45°N in Sicily, Sardinia, southern Italy and in North Africa, where it and *An sicaulti* are the only species of the complex present. The closely allied *An atroparvus* has very little overlap with *An labranchiae* being mainly found north of 45°, occurring in northern Europe, the UK, Holland, northern Italy and Spain. *An messeae* occurs in northern and central Europe, in the former USSR and in the UK, and *An maculipennis* in continental Europe, Spain and Iran. *An melanoon melanoon* has been recorded from Italy, Corsica, Sicily and Albania and *An m. subalpinus* from Albania, Turkey and Iran.

Although the breeding places may be characterized as brackish or fresh water, there is considerable selection of microhabitats within the larger setting. Thus the freshwater species differ in their oviposition sites, ovipositing in narrow drainage channels, at the lake edge and in vegetation remote from the edge. Although in Italy *An labranchiae* and *An sacharovi* are strictly coastal in distribution, when they are the only members of the complex present they extend into freshwater habitats as *An labranchiae* does in N Africa and *An sacharovi* in Israel. This expansion is considered to reflect lack of competition.

Status of vectors of malaria

An labranchiae and *An sacharovi* are important vectors of malaria throughout their ranges. *An atroparvus* is a vector of malaria but its importance depends on local conditions, as in Holland, or by being present in large numbers. *An messeae* is usually of no importance but can, when present in large numbers, be a vector of malaria. It was the most important vector of malaria in the former USSR. *An beklemishevi* is endemic to the cooler highlands and northern latitudes of Russia and the Baltic (White, 1978).

North American species of the *An maculipennis* complex

In North America *An freeborni*, *An occidentalis*, *An aztecus* and an unnamed species are found west of the continental divide, and *An earlei* in the north from the Atlantic to the Pacific Ocean. Of these only *An freeborni* and the unnamed species are considered to be potential vectors of malaria (Barr, 1988). In eastern and central North America the malaria vector, *An quadrimaculatus*, another member of the *An maculipennis* complex, is itself now recognized as being a complex of at least five species with no obvious morphological characters distinguishing the adults. The eggs of the five species do not have the distinctive colour patterns characteristic of the European members of the *An maculipennis* complex but they can be identified to species by scanning electron microscopy (Linley *et al.*, 1993).

Anopheles gambiae (Gillies and Coetzee, 1987)

Anopheles gambiae is one of the most efficient vectors of malaria and filariasis in the world, and is widely distributed throughout the Afrotropical region. Although in most of the region it breeds in small, sunlit, freshwater pools, which are often temporary and devoid of vegetation, larvae are also found at the coast in intertidal, salt water swamps. Morphological variation was noted in the banding of the palps, i.e. three or four pale bands, with larvae from salt water giving rise to a higher proportion of adults with four-banded palps, and eggs laid by adults from salt water were longer and broader than those laid by adults from fresh water. On the basis of these differences, salt water *An gambiae* were assigned to two new species, *An melas* in West Africa and *An merus* in East Africa. *An melas* plays a negligible role in the transmission of malaria and *An merus* is a vector of bancroftian filariasis.

There was still doubt about the homogeneity of freshwater *An gambiae*. In the absence of morphological differences two lines of investigation, one biological and the other cytological, were possible. The biological method involves the cross-mating of established laboratory strains, and the cytological method the study of giant (polytene) chromosomes in the population.

Laboratory colonies of *An gambiae* were established from individual, mated females, obtained from the wild. (The wild population may be mixed, and hence the need to found each colony from a single female.) In the laboratory, mated blood-fed females readily deposit fertile eggs, which can be reared to the adult. There is then the problem of mating. If this does not occur naturally, a form of artificial insemination can be achieved by lightly anaesthetizing a female and bringing a male to her in such a way that the terminalia make contact. This is usually sufficient to achieve coupling and subsequent insemination. Should there be any reluctance of males to mate or females to deposit eggs, inhibition can be removed by decapitation.

Having established colonies from single females from two or more localities their compatibility can be tested by crossing males of one colony with females from another and vice versa. Progeny of these matings have to be reared to adult and mated to determine the fertility of the F_1 generation. It will be apparent that this method requires meticulous attention to maintaining colony integrity. Any mixing would be disastrous. By this laborious method Davidson *et al.* (1967) and Davidson and Hunt (1973) have shown that freshwater *An gambiae* is a complex of four species, designated *An gambiae* s.s., *An arabiensis*, *An quadriannulatus* and *An bwambae* by White (1975). When *An gambiae* refers to the complex it will be followed by s.l. (*sensu lato*) and by s.s. (*sensu stricto*) when it refers to the species within the complex. Crosses between the four species gave rise to adults of which the females were reproductively normal, but the males were sterile. The validity of this separation has been confirmed by study of their polytene chromosomes by Coluzzi and Sabatini (1967, 1968, 1969) and Davidson and Hunt (1973) (Fig. 6.2). Coluzzi and Sabatini (1969) have also shown that the patterns on the polytene chromosomes of *An melas* and *An merus* are equally distinctive. The technique will be outlined below when considering the *Simulium damnosum* complex.

Anopheles gambiae s.s. and *An arabiensis* are vectors of malaria and bancroftian filariasis. They are widely distributed in the Afrotropical region with *An arabiensis* being absent from the more humid forested area of West Africa and *An gambiae* from the Sudan and Ethiopia. They may be present in the same locality, i.e. be sympatric. In one survey of 57 breeding sites *An gambiae* s.s. and *An arabiensis* were present together in 24 (42%) of them. Both species are endophilic, although this is more strongly developed in *An gambiae* s.s., and 100% anthropophilic in the absence of cattle. When cattle are present *An gambiae* s.s. is more anthropophilic (88%) than *An arabiensis* (39%), and more endophilic (White and Rosen, 1973). *An quadriannulatus* is zoophilic and not a vector of malaria or filariasis. It is tolerant of the relatively cool conditions on the highland plateaux

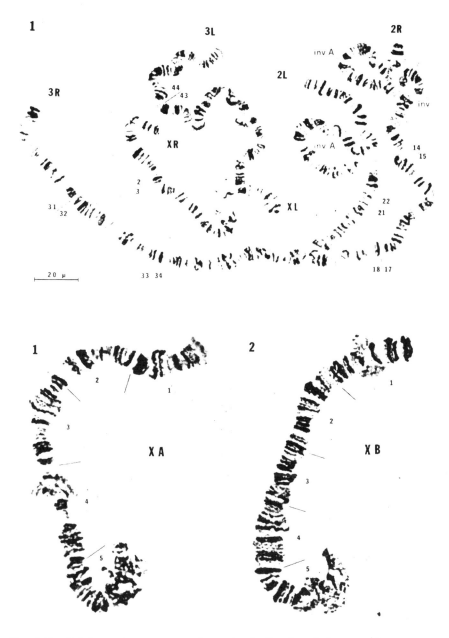

Fig. 6.2. Polytene chromosomes of species in the *Anopheles gambiae* complex. Top, salivary gland chromosomes of *An arabiensis*. Bottom, X chromosomes of female *An gambiae* (left) and *An arabiensis* (right). Source: reproduced by permission of Dr M. Coluzzi from *Parassitologia* 9, Plates 2 and 3.

of eastern Africa. *An bwambae* has a restricted distribution in the Bwamba Forest in Uganda. It breeds in mineralized water rather than fresh water. It is anthropophilic and, if huts are available, endophilic. It is a vector of malaria and probably of bancroftian filariasis (White, 1973, 1974).

Simulium damnosum

Simulium damnosum is widely distributed in the Afrotropical region, where in certain areas it is the vector of the filarial worm, *Onchocerca volvulus*, which may produce blindness in infected individuals. Like other simuliids, *S. damnosum* breeds in running water and the condition is known as river blindness. There is great variation throughout the range of *S. damnosum* in the incidence of onchocerciasis and blindness. The question therefore arose as to whether *S. damnosum* was a single species or a complex, and it has been subjected to a rigorous study of its polytene chromosomes.

In many Diptera polytene chromosomes occur in large cells of the salivary glands, Malpighian tubes and rectum. It is considered that the nuclei of these large cells are polyploid, possibly 1024 times: that is to say that the chromosomes have divided ten times (2^{10} = 1024) without cell division occurring (Sorsa, 1988). The homologous chromosomes align themselves with the same orientation, and when appropriately fixed and stained they show transverse banding, which is consistent from tissue to tissue of the same individual. The banding can be mapped and the maps made from different individuals compared. The pattern of banding is unique to each species.

Detailed study of the polytene chromosomes of *S. damnosum* has revealed a bewildering array of different patterns, referred to as sibling species or cytospecies of which more than 40 are now recognized. Among these Crosskey (1987) recognized three practical categories of taxa.

1. Formally named species based on chromosomal characters and to some degree identifiable morphologically or enzymatically. This group includes the well investigated sibling species of West Africa.
2. Formally named on morphological characters unsupported by chromosomal or enzymatic criteria, some of which are considered to be of dubious validity.
3. Vernacularly named taxa based on chromosomal characters including taxa from eastern Africa with restricted or scattered distributions and which are probably valid sibling species.

Forty or more taxa are included in the three groups – 10 in group 1, 8 in group 2, 22 in group 3 and 5 not yet described. Of these 13 are considered to be vectors of onchocerciasis in Africa and the Yemen and the remainder to be zoophilic and therefore not to be vectors (Crosskey, 1990).

Dunbar and Vajime (1981) have illustrated the relationship between 26 of these cytospecies in diagrammatic form in Fig. 6.3. Sixteen of these forms occur in eastern Africa, eight in West Africa and two species are present in both East and West Africa. The forms have been arranged in three species groups – Sanje, Kibwezi and Nile. The eight species of the Sanje group are found in East Africa, where they are not associated with onchocerciasis. The three species of the

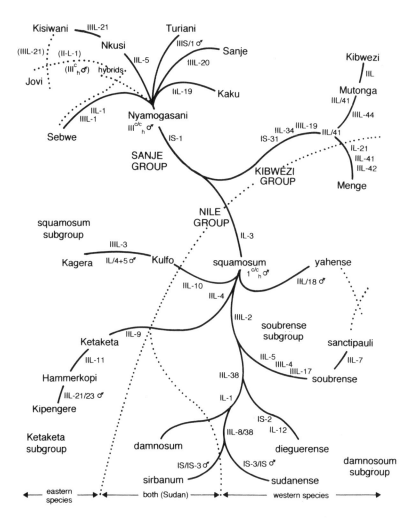

Fig. 6.3. Phylogenetic relationships of members of the *Simulium damnosum* complex. Twenty-four species are included. Formal names for eight West African cytospecies are shown in italics (bottom right) and provisional names for 16 mainly East African siblings are capitalized. Source: reproduced with permission from Durbar, R.W. and Vajime, C.G. (1981) in *Blackflies*, Academic Press, London.

Kibwezi group occur in East Africa and one of them has also been found in Cameroon, West Africa. Fourteen species are included in the Nile group, of which five occur in eastern Africa, eight in West Africa and one extends from West Africa into Uganda, East Africa. One species from Ethiopia has not been assigned to a group. The Nile group includes the vectors of onchocerciasis in West Africa, and Vajime and Dunbar (1975) have published formal specific names for seven of these species and redefined *S. damnosum* s.s.

Other Species Complexes and the Identification of Sibling Species

Species complexes are now widely recognized among vectors of human and animal diseases. In the Diptera they have been recorded in four families of Nematocera – Culicidae, Ceratopogonidae, Phlebotominae (Psychodidae) and Simuliidae; blood-sucking triatomine bugs (Hemiptera); trombiculid chigger mites (Acari); ixodid ticks (Service, 1988); and fleas of the genus *Citellophilus*, vectors of plague in northern China (Wang Shanqing *et al.*, 1994).

Many important vectors of malaria, in addition to the *An gambiae* and *An maculipennis* groups of species, have been shown to be species complexes. In China 13 species are recognized within the *An hyrcanus* complex of which *An anthropophagus* is the most important vector of malaria and brugian filariasis (Baolin *et al.*, 1993). Other complexes include in Papua New Guinea *An punctulatus* (six species; Foley *et al.*, 1993); in Thailand *An dirus* (five species; Baimai *et al.*, 1988; Green *et al.*, 1992b), *An maculatus* (eight species; Green *et al.*, 1992a), *An minimus* (two species; Green *et al.*, 1990); in India *An culicifacies* (four species; Subbarao *et al.*, 1988); in Indonesia *An barbirostris* (three putative species; Dharmawan, 1993).

In addition to cross-mating and the study of variation in the banding of polytene chromosomes, individuals within a species complex may be identified by gas chromatography and techniques developed in molecular biology. Differences in cuticular hydrocarbons have been used to identify members of the *Simulium ochraceum* complex (Millest, 1992), members of species complexes of *Anopheles* and *Simulium*, populations of *Phlebotomus* and *Glossina pallidipes* (Phillips *et al.*, 1988), and Kittayapong *et al.* (1990) have used cuticular lipids to separate members of the *An maculatus* complex. Four subspecies of *Citellophilus tesquorum* have been established on the basis of differences in the content and composition of their fatty acids, monosaccharides and cuticular hydrocarbons (Wang Shanqing *et al.*, 1994).

The polymerase chain reaction (PCR) and related techniques have been used to separate sibling species of *S. damnosum* (Brockhouse *et al.*, 1993), species of North American *Culicoides* (Raich *et al.*, 1993), members of the *Aedes scutellaris* group (Kambhampati *et al.*, 1992) and species of the *An gambiae* complex (Paskewitz and Collins, 1990; Taylor *et al.*, 1993; Wilkerson *et al.*, 1993). DNA probes have been developed to identify members of species complexes, e.g. *S. damnosum* (Post and Flook, 1992), *An farauti* (Booth *et al.*, 1991) and *An gambiae* (Hill *et al.*, 1991). Electrophoretic techniques have been used to separate members of the *An gambiae* complex (Coetzee *et al.*, 1993) and six species of the *An punctulatus* (= *farauti*) complex (Foley *et al.*, 1993), allowing in the latter case the creation of electrophoretic keys to identify the six species (Foley and Bryan, 1993).

The Implications of Species Complexes for the Medical Entomologist

The possibility that an abundant, widely distributed species may, on closer examination, prove to be a complex of species, has always to be borne in mind.

The existence of a complex is only likely to be recognized when a species has been subjected to detailed study, and therefore species complexes will appear to be commoner in economically important species.

The medical entomologist should appreciate the implications of species complexes. Firstly, as evident from the previous section, identification of the members of a complex is highly specialized requiring the services of an expert. This means that individuals are unlikely to be able to do their own identifications. Secondly, previous work done on the species may be inapplicable to the worker's current situation. It gives a new twist to the aphorism 'Every malaria problem is a local one', which was coined originally to emphasize that control must be based on intimate knowledge of the local situation. Now it implies that the local vector population may be different and results obtained elsewhere be inapplicable. Thirdly, differences in observations made in different areas may not be artefacts due to techniques or competence of the operator but reflect genuine differences in vector populations. Fourthly, ecological studies must be interpreted cautiously if the species in a complex are virtually indistinguishable. Ecological results may give a hybrid picture in which specific differences are masked.

Aedes aegypti and *Aedes albopictus* (see Fig. 6.4)

Aedes aegypti

The yellow fever mosquito, *Aedes aegypti* is widely but sporadically distributed throughout the tropical and subtropical areas of the world. Its distribution appears to be related to the 20°C isotherm which roughly correlates with latitudes 40°N and 40°S. It is a highly domesticated mosquito which can complete its entire life cycle within the confines of a single human dwelling. The female lays its eggs in small containers, such as flower vases, water storage jars and other containers holding water in houses. It will also lay eggs in small amounts of peridomestic water which collects in tyres, plastic containers and other debris associated with human settlement. When the embryo inside the egg of *Ae aegypti* has developed to a certain stage the egg becomes resistant to desiccation. It may then enter diapause in which it can remain for about a year. When the eggs are flooded they hatch and the larvae commence their development immediately. This ability of *Ae aegypti* to produce diapausing eggs enables the species to survive in areas with prolonged dry seasons while the rapid hatching of the eggs on flooding and the speedy development of the immature stages are adaptations to breeding in temporary collections of water.

Adult *Ae aegypti* emerging from breeding sites indoors can complete their cycle without going outside. Swarming is not an essential component of mating. Males orientate to females by responding to the female's wing beat and specific identification is achieved by a contact pheromone on the female. Although in domestic female *Ae aegypti* autogeny is low (Trpis, 1977), blood-feeding presents no problem because the female is strongly anthropophilic and feeds readily on the human inhabitants of the dwelling. The blood-fed female rests in the house while maturing her ovaries and then deposits her eggs in domestic water containers. The

Fig. 6.4. Female Aedes aegypti.

cycle is then complete. It is not surprising that a species with such modest requirements has readily adapted to laboratory colonization. The fact that *Ae aegypti* produces diapausing eggs made it easy to disseminate material widely throughout the world, even before the days of air transport, and colonies of *Ae aegypti* have been maintained for many years at centres in the northern hemisphere without the introduction of new genetic material. Such colonies form valuable material for studying biological processes but caution must be used in applying the results obtained on such material to 'natural populations'.

Taxonomic status of *Aedes aegypti*

The origin of *Ae aegypti* is uncertain but in the post-glacial period it was endemic to Africa from which it has been distributed around the world by humans. Not perhaps unexpectedly, considerable variation in form has been reported in this well-marked, abundant and widely distributed species, but there has been doubt as to whether *Ae aegypti* is polytypic or polymorphic, i.e. whether there is more than one type (form) of *Ae aegypti* or whether it is a single, very variable species. In a polymorphic species there would be free gene flow throughout the population whereas in a polytypic species there would be limited, if any, gene flow between the forms. On morphological criteria Jupp *et al.* (1991) conclude that in South Africa *Ae aegypti* is a single polymorphic species displaying plasticity in its man-biting behaviour.

Two forms, *Aedes aegypti aegypti* and *Ae aegypti formosus*, have been recognized (Knight and Stone, 1977). *Ae a. aegypti* is a brown or blackish form widely distributed throughout the range of the species but generally absent from inland areas of the Afrotropical region. *Ae a. formosus* is a black form confined to the Afrotropical region where it is the only form which occurs, except in coastal and a few limited inland areas. *Ae a. aegypti* is a domestic, anthropophilic form which feeds and breeds indoors while *Ae a. formosus* is a sylvan (feral), zoophilic form which feeds and breeds outdoors (Mattingly, 1957) but populations of *Ae aegypti* do not always fit neatly into this classification.

Electrophoretic analyses have been made on more than 100 collections of *Ae aegypti* from 85 geographic locations throughout the world with emphasis on the Afrotropical region (24 collections – 9 East Africa, 15 West Africa), Neotropical region (36 collections – 24 Caribbean, 12 South-Central America) and the USA (16 collections, 9 Texas, 7 south-east states) (Wallis *et al.*, 1983; Tabachnick, 1991). The genetic differentiation detected between populations was quite small, the largest being that between *Ae a. aegypti* and *Ae a. formosus* but even that was only one-fifth of the magnitude of the distance between subspecies of other insects, such as *Drosophila*. In East Africa there are distinct sympatric populations of *Ae a. aegypti* and *Ae a. formosus* but in West Africa the domestic *Ae a. aegypti* is not found and the genetically homogeneous *Ae a. formosus* exhibits domestic behaviour in urban settings. The Caribbean region maintains the most heterogeneous *Ae aegypti* populations.

Eight distinct genetic–geographic groupings of *Ae aegypti* were recognized. Populations from the Caribbean, South-Central America, Texas (eight collections) and Mexico were related to the domestic *Ae a. aegypti* from East Africa while *Ae aegypti* from Asia, West Africa, south-east USA including Beaumount in south-east Texas were more closely related to the sylvan form of East Africa (*Ae a. formosus*) (Tabachnick, 1991). These findings undermine acceptance of the simpler concept of two forms of *Ae aegypti* of which *Ae a. aegypti* was regarded as primarily a coastal subspecies widely but erratically distributed throughout the tropics and subtropics, where it was closely associated with humans, and *Ae a. formosus* as a dark, feral form which may have little connection with humans. In addition to electrophoretic differences there was variation in the competence of the different populations to be infected with the yellow fever virus (Powell, 1985).

Aedes albopictus (Rai, 1991)

Aedes albopictus has a wide distribution. It probably originated in south-east Asia from which it has spread westwards to Madagascar and the Seychelles, islands off the mainland of Africa, and eastwards to China, Japan, New Guinea and islands in the Pacific Ocean. In 1985 relatively large populations of *Ae albopictus* were found throughout Harris County in Texas. It is virtually certain that *Ae albopictus* entered North America as eggs/larvae in used tyres. Later, in one port of entry (Seattle) 1 in every 500 wet tyres was infested with *Ae albopictus* larvae. The recent dramatic increase in distribution of *Ae albopictus* around the world appears to be associated with the random movement of used tyres on a massive scale. From Harris County it has spread rapidly and by the summer of 1989 it was widely distributed in some 18 states of the USA and had spread into Mexico. It is estimated that the northern limit for overwintering *Ae albopictus* will be the 0°C isotherm and in summer its northward expansion be the −5°C isotherm, much further north than *Ae aegypti* can colonize.

The importance of this expansion of the range of *Ae albopictus* is its capacity as a vector of arboviral diseases. It is a primary vector of dengue and dengue haemorrhagic fever in rural areas of south-east Asia and may also be important in the epidemiology of other arboviruses, e.g. Japanese encephalitis. *Ae albopictus* is more susceptible than *Ae aegypti* to oral infection with all four serotypes of dengue but there is marked variation in susceptibility between geographic strains. General genetic variation in 57 populations of *Ae albopictus* collected from nine countries (USA 28, Brazil 4, Malaysia 9, Sri Lanka 1, Borneo 2, India 2, China 3, Japan 7 and Mauritius 1) has been studied by allozyme analysis. Each of the 57 populations was genetically distinct with populations from any given country being more closely related. Populations from Japan, China, Brazil and the USA were closely related, and formed a single loose cluster.

In the USA there has been a steady decline in populations of *Ae aegypti*, concomitant with a corresponding increase in *Ae albopictus* but this may not be a simple matter of competition because *Ae aegypti* populations were declining in the southern USA before the introduction of *Ae albopictus*. Laboratory experiments do not suggest that *Ae albopictus* was more competitive than *Ae aegypti*. Both species out-competed *Ae triseriatus* but between themselves intraspecific competition was more important than interspecific competition. Rai (1991) concludes that *Ae albopictus* is here to stay in the western hemisphere and has become part of the local fauna.

References

Baimai, V., Harbach, R.E. and Kijchalao, U. (1988) Cytogenetic evidence for a fifth species within the taxon *Anopheles dirus* in Thailand. *Journal of the American Mosquito Control Association* 4, 333–338.

Baolin, L., Tongyan, Z. and Wanmin, K. (1993) *Anopheles hyrcanus* group and their vector competence in China. *Acta Parasitologica et Medica Entomologica Sinica* 12, 55–63.

Barr, A.R. (1988) The *Anopheles maculipennis* complex (Diptera: Culicidae) in western

North America. In: Service, M.W. (ed.) *Biosystematics of Haematophagous Insects.* Clarendon Press, Oxford.

Booth, D.R., Mahon, R.J. and Sriprakash, K.S. (1991) DNA probes to identify members of the *Anopheles farauti* complex. *Medical and Veterinary Entomology* 5, 447–454.

Brockhouse, C.L., Vajime, C.G., Marin, R. and Tanguay, R.M. (1993) Molecular identification of onchocerciasis vector sibling species in black flies (Diptera: Simuliidae). *Biochemical and Biophysical Research Communications* 194, 628–634.

Coetzee, M., Hunt, R.H. and Braack, L.E.O. (1993) Enzyme variation at the aspartate aminotransferase locus in members of the *Anopheles gambiae* complex. *Journal of Medical Entomology* 30, 303–308.

Coluzzi, M. and Sabatini, A. (1967) Cytogenetic studies on species A and B of the *Anopheles gambiae* complex. *Parassitologia* 9, 73–88.

Coluzzi, M. and Sabatini, A. (1968). Cytogenetic studies on species C of the *Anopheles gambiae* complex. *Parassitologia* 10, 155–165.

Coluzzi, M. and Sabatini, A. (1969). Cytogenetic observations on the salt water species, *Anopheles merus* and *Anopheles melas*, of the gambiae complex. *Parassitologia* 11, 177–187.

Crosskey R.W. (1987) A taxa summary for the *Simulium damnosum* complex, with special reference to distribution outside the control areas of West Africa. *Annals of Tropical Medicine and Parasitology* 81, 181–192.

Crosskey R.W. (1990) *The Natural History of Blackflies.* John Wiley, Chichester, UK.

Davidson, G. and Hunt, R.H. (1973). The crossing and chromosome characteristics of a new, sixth species in the *Anopheles gambiae* complex. *Parassitologia* 15, 121–128.

Davidson, G., Paterson, H.E., Coluzzi, M., Mason, G.F. and Micks, D.W. (1967) The *Anopheles gambiae* complex. In: Wright, J.W. and Pal, R. (eds) *Genetics of Insect Vectors of Disease.* Elsevier, Amsterdam, pp. 211–250.

Dharmawan, R. (1993) Elektroforesis Lanjutan Beberapa enzim pada spesies kembar *Anopheles barbirostris. Majalah Komunikasi Dan Kedokteran* 2, 14–17 (in Indonesian).

Dunbar, R.W. and Vajime, C.G. (1981) Cytotaxonomy of the *Simulium damnosum* complex. In: Marshall Laird (ed.) *Blackflies.* Academic Press, London, pp. 31–43.

Falleroni, D. (1926) Fauna anofelica italiana e suo habitat (paludi, risaie, canali). Metodi di lotta contro la malaria. *Rivista di Malariologia* 5, 553–593.

Foley D.H. and Bryan, J.H. (1993) Electrophoretic keys to identify members of the *Anopheles punctulatus* complex of vector mosquitoes in Papua New Guinea. *Medical and Veterinary Entomology* 7, 49–53.

Foley, D.H., Paru, R., Dagoro, H. and Bryan, J.H. (1993) Alloenzyme analysis reveals six species within the *Anopheles punctulatus* complex of mosquitoes in Papua New Guinea. *Medical and Veterinary Entomology* 7, 37–48.

Gillies, M.T. and Coetzee, M. (1987) *A Supplement to the Anophelinae of Africa South of the Sahara (Afrotropical Region).* South African Institute for Medical Research, Johannesburg.

Green, C.A., Gass, R.F., Munstermann, L.E. and Baimai, V. (1990) Population – genetic evidence for two species in *Anopheles minimus* in Thailand. *Medical and Veterinary Entomology* 4, 25–34.

Green, C.A., Rattanarithikul, R and Charoensub, A. (1992a) Population genetic confirmation of species status of the malaria vectors *Anopheles willmori* and *An pseudowillmori* in Thailand and chromosome phylogeny of the Maculatus group of mosquitoes. *Medical and Veterinary Entomology* 6, 335–341.

Green, C.A., Munstermann, L.E., Tan, S.G., Panyim, S. and Baimai, V. (1992b) Population genetic evidence for species A, B, C and D of the *Anopheles dirus* complex in Thailand and enzyme electromorphs for their identification. *Medical and Veterinary Entomology* 6, 29–36.

Hackett, L.W. (1937) *Malaria in Europe*. Oxford University Press, London.

Hill, S.M., Urwin, R., Knapp, T.F. and Crampton, J.M. (1991) Synthetic DNA probes for the identification of sibling species in the *Anopheles gambiae* complex. *Medical and Veterinary Entomology* 5, 455–463.

Jupp, P.G., Kemp, A. and Frangos, C. (1991) The potential for dengue in South Africa: morphology and the taxonomic status of *Aedes aegypti* populations. *Mosquito Systematics* 23, 182–190.

Kambhampati, S., Black, W.C. and Rai, K. (1992) Random amplified polymorphic DNA of mosquito species and populations (Diptera: Culicidae): techniques, statistical analysis, and applications. *Journal of Medical Entomology* 29, 939–945.

Kittayapong, P., Clark, J.M., Edman, J.D., Potter, T.L., Lavine, B.K., Marion, J.R. and Brooks, M. (1990) Cuticular lipid differences between the malaria vector and non-vector forms of the *Anopheles maculatus* complex. *Medical and Veterinary Entomology* 4, 405–413.

Knight, K.L. and Stone, A. (1977) A catalog of the mosquitoes of the world. *The Thomas Say Foundation* 6, 1–611.

Linley, J.R., Kaiser, P.E. and Cockburn, A.F. (1993) A description and morphometric study of the eggs of species of the *Anopheles quadrimaculatus* complex (Diptera: Culicidae). *Mosquito Systematics* 25, 124–147.

Mattingly, P.F. (1957) Genetical aspects of the *Aedes aegypti* problem. I. Taxonomy and bionomics. *Annals of Tropical Medicine and Parasitology* 51, 392–408.

Millest, A.L. (1992) Identification of members of the *Simulium ochraceum* species complex in the three onchocerciasis foci in Mexico. *Medical and Veterinary Entomology* 6, 23–28.

Paskewitz, S.M. and Collins, F.H. (1990) Use of the polymerase chain reaction to identify mosquito species of the *Anopheles gambiae* complex. *Medical and Veterinary Entomology* 4, 367–373.

Phillips, A., Milligan, P.J.M., Broomfield, G. and Molyneux, D.H. (1988) Identification of medically important Diptera by analysis of cuticular hydrocarbons. In Service, M.W. (ed.) *Biosystematics of Haematophagous Insects*. Clarendon Press, Oxford.

Post, R.J. and Flook, P. (1992) DNA probes for the identification of members of the *Simulium damnosum* complex (Diptera: Simuliidae). *Medical and Veterinary Entomology* 6, 379–384.

Powell, J.R. (1985) Geographic genetic differentiation and arbovirus competency: *Aedes aegypti* and yellow fever. *Parassitologia* 27, 13–20.

Rai, K.S. (1991) *Aedes albopictus* in the Americas. *Annual Review of Entomology* 36, 459–484.

Raich, T.J., Archer, J.I, Robertson, M.A., Tabachnik, W.J. and Beaty, B.J. (1993) Polymerase chain reaction approaches to *Culicoides* (Diptera: Ceratopogonidae) identification. *Journal of Medical Entomology* 30, 228–232.

Service, M.W. (ed.) (1988) *Biosystematics of Haematophagous Insects*. Clarendon Press, Oxford.

Sorsa, V. (1988) *Polytene Chromosomes in Genetic Research*. Ellis Horwood Limited, Chichester, UK.

Stegnii, V.N. and Kabanova, V.M. (1976) Cytoecological study of natural populations of *Anopheles* in the territory of the USSR. Report I Isolation of a new species of *Anopheles* in the *maculipennis* complex by the cytodiagnostic method. *Meditsinskaya Parazitologiya i Parazitarnye Bolezni* 45, 192–198.

Subbarao, S.K., Vasantha, K. and Sharma, V.P. (1988) Cytotaxonomy of certain malaria vectors in India. In: Service, M.W. (ed.) *Biosystematics of Haematophagous Insects*. Clarendon Press, Oxford.

Tabachnik, W.J. (1991) The yellow fever mosquito. *American Entomologist* 37, 14–24.

Taylor, K.A., Paskewitz, S.M., Copeland, R.S., Koros, J., Beach, R.F., Githure, J.I. and Collins, F.H. (1993) Comparison of two ribosomal DNA-based methods for differentiating members of the *Anopheles gambiae* complex (Diptera: Culicidae). *Journal of Medical Entomology* 30, 457–461.

Trpis, M. (1977) Autogeny in diverse populations of *Aedes aegypti* from East Africa. *Tropenmedizin und Parasitologie* 28, 77–82.

Vajime, C.G. and Dunbar, R.W. (1975) Chromosomal identification of eight species of the subgenus *Edwardsellum* near and including *Simulium* (*Edwardsellum*) *damnosum* Theobald (Diptera: Simuliidae). *Tropenmedizin und Parasitologie* 26, 111–138.

Wallis, G.P., Tabachnik, W.J. and Powell, J.R. (1983) Macrogeographic variation in a human commensal: *Aedes aegypti*, the yellow fever mosquito. *Genetical Research* 41, 241–258.

Wang Shanqing, Wu Houyong, Chen Liyin, Zhou Fang, Liao Jie, Sheng Shishu and Jin Renci (1994) Comparisons between subspecies of *Citellophilus tesquorum* on their physiological and biochemical indexes. In: Wu Houyong and Liu Quan (eds) *Researches on Fleas*. China Science and Technology Press, Beijing, pp. 46–71.

White, G.B. (1973) Comparative studies on sibling species of the *Anopheles gambiae* Giles complex (Dipt.: Culicidae). III. The distribution, ecology, behaviour and vectorial importance of species D in Bwamba County, Uganda, with an analysis of biological, ecological, morphological and cytogenetic relationships of Uganda species D. *Bulletin of Entomological Research* 63, 65–97.

White, G.B. (1974) *Anopheles gambiae* complex and disease transmission in Africa. *Transactions of the Royal Society of Tropical Medicine and Hygiene* 68, 278–298.

White, G.B. (1975) Notes on a catalogue of Culicidae of the Ethiopian region. *Mosquito Systematics* 7, 303–344.

White, G.B. (1978) Systematic reappraisal of the *Anopheles maculipennis* complex. *Mosquito Systematics* 10, 13–44.

White, G.B. and Rosen, P. (1973) Comparative studies on sibling species of the *Anopheles gambiae* Giles complex (Dipt.: Culicidae). II. Ecology of species A and B in savanna around Kaduna, Nigeria, during the transition from wet to dry season. *Bulletin of Entomological Research* 62, 613–625.

Wilkerson, R.C., Parsons, T.J., Albright, D.G., Klein, T.A. and Braun, M.J. (1993) Random amplified polymorphic DNA (RAPD) markers readily distinguish cryptic mosquito species (Diptera: Culicidae: *Anopheles*). *Insect Molecular Biology* 1, 205–211.

INSECTS AND ACARINES OF MEDICAL AND VETERINARY IMPORTANCE

Culicidae (Mosquitoes) 7

There are two families of Nematocera, the Chaoboridae and the Dixidae, which are closely related to the Culicidae. They have been regarded as subfamilies of an enlarged Culicidae but are now treated as separate families. The blood-sucking habit is found only in the Culicidae, and it is adequate to recognize that there are these two other families which may be confused with mosquitoes.

- *Dixidae*. Adult dixids have short mouthparts; no scales on the wings and a wing venation, which is a distorted version of that of the Culicidae, with the second vein strongly arched (Fig. 7.1). Dixid larvae occur at the edges of streams and may be mistaken for *Anopheles* larvae, but in the dixid the three thoracic segments are not fused; there are pseudopods on abdominal segments 1 and 2, and ambulacral combs on segments 5, 6 and 7, with which it can climb out of water (Nowell, 1951; Fig. 7.2). Larvae of *An wellcomei* are unusual in that they can climb out of water (Gillett, 1971) but they lack pseudopods and ambulacral combs.
- *Chaoboridae* (*phantom midges*). Adult chaoborids have short mouthparts; scales on the wings forming a fringe on the posterior margin and a few on the veins; wing venation inseparable from that of the Culicidae with the second vein running parallel to the first and third veins (Fig. 7.1). Chaoborid larvae are aquatic predators.
- *Culicidae* (*mosquitoes*). Culicid adults have a long forwardly-directed proboscis, equal in length to the head and thorax combined (Fig. 7.3); wings with scales along the veins and forming a fringe; second vein not arched but running parallel to veins i and iii (Fig. 7.1). Culicid larvae are aquatic but in only a relatively small number of species are they predatory.

Culicidae

Throughout this chapter and elsewhere the generic and subgeneric abbreviations proposed by Reinert (1975) will be used. Knight and Stone (1977) and Knight

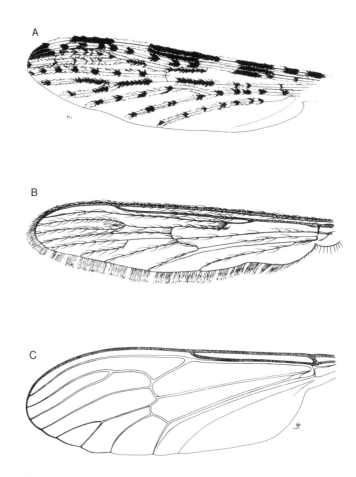

Fig. 7.1. Wings of: (A) an *Anopheles* mosquito; (B) a chaoborid and; (C) a dixid.

(1978) listed more than 3000 species in the Culicidae, and Ward (1984, 1992) and Gaffigan and Ward (1985) added another 203 (Table 7.1). This latest addition included 199 new species, 60 restored from subspecific level, less 56 placed in synonymy or classified as nomina dubia, giving an overall total of 3268 species. Service (1993) gives 'about 3450 species and *subspecies*'. Three subfamilies are recognized among the Culicidae: the Toxorhynchitinae, Anophelinae and Culicinae.

Toxorhynchitinae

This subfamily includes only one genus *Toxorhynchites* which occurs mainly in the tropics and subtropics of the Oriental, Neotropical and Afrotropical regions (Table 7.1). Steffan and Evenhuis (1981) reviewed the biology of *Toxorhynchites* and recognized 69 species, six less than in Table 7.1. *Toxorhynchites* are very large, metallic-coloured mosquitoes, which do not suck blood and in which the proboscis is markedly bent (Fig. 7.4). Morphologically they resemble culicine mosquitoes in

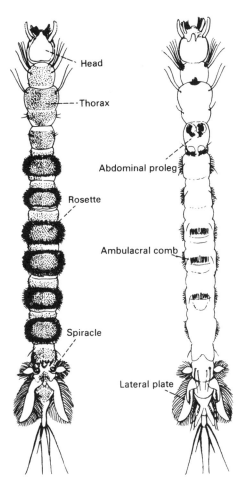

Fig. 7.2. Larva of *Dixa brevis*. Dorsal view (left) and ventral view (right). Source: from Nowell (1951).

the structure of the larva and the adult. They breed in small collections of water, e.g. tree holes, where their larvae are predacious on other mosquito larvae.

Anophelinae

Three genera of mosquitoes are included in the Anophelinae but only one, *Anopheles*, of which there are over 400 species, is widely distributed (Table 7.1). *Bironella* is confined to New Guinea and tropical Australia and *Chagasia* to the Neotropical region. Neither is of medical importance. *Anopheles* are medically important, being the sole vectors of malaria, and they play a substantial role in transmitting lymphatic filariasis due to *Wuchereria bancrofti*. The morphological characters distinguishing Anophelinae and Culicinae will be given later when their life cycles have been considered.

Table 7.1. Zoogeographical distribution of the Culicidae.

Subfamily	Tribe	Genus	No. of species	Number of species of each genus in each region					
				Australian	Nearctic	Neotropical	Oriental	Palaearctic	Afrotropical
Anophelinae			434						
		Anopheles	421	25	18	84	142	43	124
		Bironella	9	9					
		Chagasia	4			4			
Culicinae			2757						
	Aedeomyini		7						
		Aedeomyia	7	2		1	1		3
	Aedini		1183						
		Aedes	975	220	74	119	310	87	211
		Armigeres	49	6			43	1	
		Eretmapodites	45						45
		Haemagogus	28			28			
		Heizmannia	32				32		
		Opifex	1	1					
		Psorophora	47		14	41			
		Udaya	2				2		
		Zeugnomyia	4				4		
	Culicini		788						
		Culex	769	127	31	324	179	38	137
		Deinocerites	18		4	16			
		Galindomyia	1			1			
	Culisetini		36						
		Culiseta	36	14	8	1	4	15	2

Taxon	Total						
Ficalbini	51						
Ficalbia	7	1			3		4
Mimomyia	44	7			9		30
Hodgesini	11						
Hodgesia	11	4			4		4
Mansoniini	80						
Coquillettidia	57	15	1	13	8	2	22
Mansonia	23	6	2	13	5		2
Orthopodomyiini	29						
Orthopodomyia	29	1	3	7	10	1	10
Sabethini	377						
Johnbelkina	3			3			
Limatus	8			8			
Malaya	12	3			5		6
Maorigoeldia	1	1					
Phoniomyia	24			24			
Runchomyia	11			11			
Sabethes	29			29			
Shannoniana	3			3			
Topomyia	47	1			46		
Trichoprosopon	13			13			
Tripteroides	113	64			56	1	
Wyeomyia	113		4	111			
Uranotaeniini	195						
Uranotaenia	195	31	2	32	78	2	54
Toxorhynchitinae	75						
Toxorhynchites	75	5	1	17	35	3	17
TOTAL	3266	544	163	903	977	193	671

Note: Sum of regional totals (3453) exceeds number of species (3266) because some species occur in more than one region.
Source: Extracted from Knight and Stone (1977), Knight (1978), Ward (1984, 1992) and Gaffigan and Ward (1985).

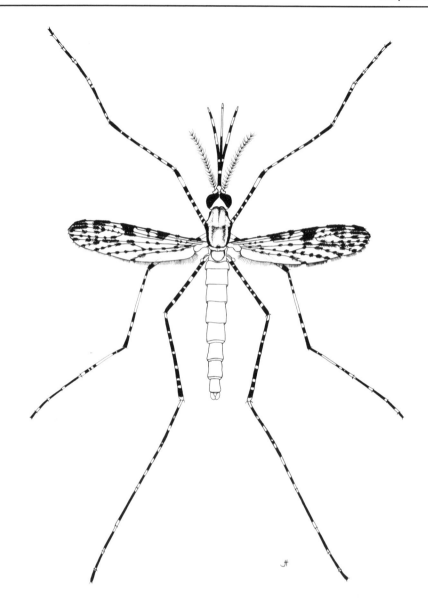

Fig. 7.3. Female *Anopheles annulipes*.

Culicinae

There are more than 2750 species of Culicinae of which the main genera are *Aedes* with nearly 1200 species, *Culex* with nearly 800 species and the medically unimportant *Uranotaenia* with nearly 200 species. Culicine genera are usually widespread, although *Eretmapodites* occurs only in the Afrotropical region, *Haemagogus* in the Neotropical region, *Psorophora* in the Nearctic and Neotropical regions and *Heizmannia*, with one exception, in the Oriental region.

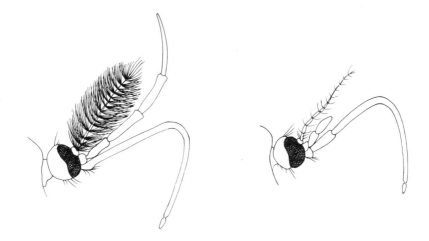

Fig. 7.4. Heads of *Toxorhynchites speciosus*: male (left) and female (right).

The Sabethini are predominantly Neotropical in distribution with seven of the twelve genera being restricted to that region (Table 7.1). *Wyeomyia* has four species in the adjoining Nearctic region; *Topomyia* is virtually confined to the Oriental region, and *Tripteroides* to the Oriental region and Australian region. The small genus *Malaya* has a wide distribution in the Old World tropics but is of no importance because its adults are not haematophagous. Female *Malaya* have an unusual method of feeding. The female stops foraging ants of the genus *Cremastogaster*, inserts its proboscis into the ant's mouth and sucks up honeydew which the ant has collected from plant-sucking bugs (Farquharson, 1918). Sabethines breed in small collections of water associated with plants, e.g. leaf axils of epiphytic bromeliads, pitcher plants, tree holes and bamboo. *Sabethes chloropterus* is an inefficient vector of yellow fever but it is long-lived and could be involved in the survival of the virus over the dry season (Galindo, 1958; Rodaniche *et al.*, 1959).

Mattingly (1969) classifies the Culicinae ecologically into four groups, which are more or less self-contained. The aedine genera *Aedes* and *Psorophora* have drought-resistant eggs, enabling them to breed in temporary ground pools and small containers. The quasi-sabethines exhibit both aedine and sabethine characters, breed in containers such as bamboo, tree holes and leaf axils, and include such genera as *Eretmapodites*, *Haemagogus* and *Armigeres*. A third group is associated with dense aquatic vegetation and includes *Ficalbia*, *Coquillettidia* and *Mansonia* The fourth group is an assemblage of miscellaneous genera of which the largest is *Culex* and one, *Deinocerites*, is associated with crab holes in the Nearctic and Neotropical regions (Horsfall, 1955; Belkin and Hogue, 1959).

Culicines are important vectors of human disease and are the major vectors of arboviruses and filariasis, e.g. *Ae aegypti* of yellow fever and dengue; *Cx tarsalis* of western equine encephalitis; *Cx quinquefasciatus* of *Wuchereria bancrofti*; and *Mansonia uniformis* of *Brugia malayi* (see Chapters 24 and 30).

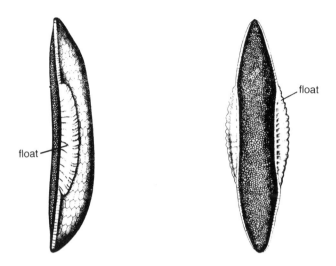

Fig. 7.5. Lateral (left) and dorsal (right) views of an anopheline egg.
Source: from Nuttall G.H.F. and Shipley, A.E. (1901) *Journal of Hygiene* 1.

Life Cycle

Egg

The female *Anopheles* mosquito lays a batch of 100–150 eggs usually at night on the surface of the water. They are boat-shaped objects about 1 mm long with a flattened upper surface, the deck, and a keel-shaped lower surface which is submerged (Fig. 7.5). A feature of anopheline eggs is the presence of paired lateral air-filled floats. These are missing in some species, e.g. *An sacharovi* lays eggs without floats in summer but in the spring its eggs have small floats (Mer, 1931). Scanning electron microscopy of the fine structure of the egg surface reveals differences between closely related species (Linley *et al.*, 1993). Hinton (1968) has suggested that the fine network of the outer layer of the chorion acts as a plastron facilitating respiration when the egg is submerged. The surface structure of the egg is such that eggs will attach end to end or side to side but never end to side. They become attached by surface forces to objects projecting from the water and this property prevents eggs from drifting into open stretches of water. The eggs of *An multicolor* are laid side by side in a row resembling cartridges in a bandolier (Horsfall, 1955). Anopheline eggs develop directly into larvae and only rarely undergo diapause. Ho Ch'i *et al.* (1962) have reported that *An lesteri* overwinters in the egg stage. Under optimal conditions anopheline eggs hatch in 1–2 days.

The eggs of *Aedes* are laid in a batch but are not attached to each other. They differ from those of *Anopheles* by the absence of floats. They are laid on the moist surface at the water's edge and not on the water itself. When the eggs are first laid they are permeable to water and increase their weight rapidly (Clements, 1992). At this stage they are susceptible to desiccation, and collapse and die if dried. Later, when the serosal cuticle is formed, the eggs can withstand desiccation

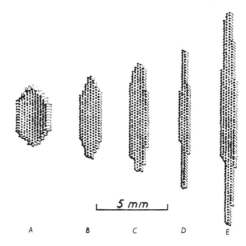

Fig. 7.6. Egg rafts of five species of *Coquillettidia*: (**A**) *Cq metallica*; (**B**) *Cq maculipennis*; (**C**) *Cq fraseri*; (**D**) *Cq fuscopennata*; (**E**) *Cq aurites*. Source: from Gillett (1961).

and remain viable in the dried state for many months, depending on the species (Harwood and Horsfall, 1959). The production of eggs resistant to desiccation makes *Aedes* species ideal colonizers of temporary collections of water, e.g. salt marshes subject to tidal inundation, tree holes, etc. When the eggs are flooded, most of them hatch immediately, but some will remain dormant and hatch at the second or third flooding. The early emergence of larvae is essential if the life cycle is to be completed before the habitat dries up. Similar dormant eggs are produced by other aedine genera including *Haemagogus* and *Psorophora*.

The egg raft, which is commonly regarded as typical of the Culicinae, is produced in only six genera: *Armigeres, Coquillettidia, Culex, Culiseta, Trichoprosopon* and *Uranotaenia*. Rafts commonly measure 3–4 mm long and 2–3 mm wide, but the precise form is a specific attribute. Gillett (1961) has shown that in six species of Afrotropical *Coquillettidia* the shape of the egg raft is a specific character (Fig. 7.6). The lower side of the raft on the water is convex and the other side concave. The eggs are orientated at right angles to the water surface, but arranged so that the larva is head down in the egg and emerges from the underside of the raft. The eggs cannot withstand desiccation and, if dried, they collapse and the embryos die.

The eggs of *Mansonia* are laid in clusters on the undersurface of floating leaves of aquatic plants such as *Salvinia*. The eggs are tapered at the free end, giving the cluster of eggs the appearance of a miniature pincushion (Wharton, 1962; Gillett, 1971). The eggs of *Sabethes chloropterus* are unusual in being rhomboidal in shape (Galindo, 1958).

Larva

The larva emerges from the egg by using a small egg tooth, placed posterodorsally on its head. Three regions can be differentiated in the body of the larva: a well

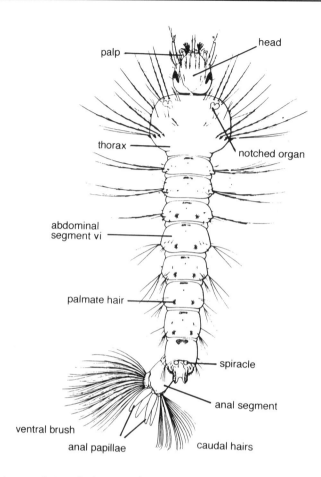

Fig. 7.7. Larva of *Anopheles maculipennis* viewed from above with anal
segment twisted round to display the ventral brush and caudal hairs.
Source: from Marshall (1938) *The British Mosquitoes*, British Museum (Natural
History), London.

developed sclerotized head; a broad thorax in which the three segments are fused;
and a segmented abdomen (Fig. 7.7). The larva is apodous. When anopheline
larvae come to the water surface they swim backwards by lateral movements of
the abdomen until the caudal setae on the anal segment of the abdomen are in
contact with a solid object, e.g. vegetation or stone. The larvae are positively
thigmotactic.

The anopheline larva lies parallel to the water surface supported by the paired
prothoracic notched organ, the posterior spiracular apparatus and paired palmate
hairs (Puri, 1931). Palmate hairs are present on most abdominal segments and occasionally on the thorax. They have short, stout bases from which radiate 10–20
leaflets and are unique to anopheline larvae (Fig. 7.8). When the palmate hair is
in contact with the water surface, the leaflets spread out and support the larva.

When the larva comes to the water surface its dorsal side is uppermost and

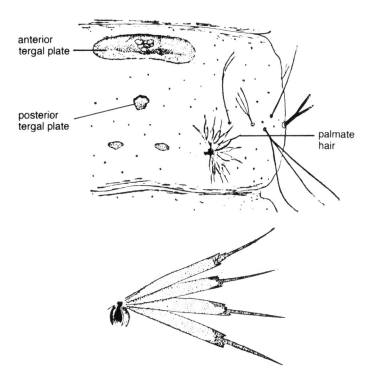

Fig. 7.8. Part of dorsal surface of abdominal segment V of *Anopheles culicifacies* (above) with a few leaflets from a palmate hair of segment IV (below). Source: from Puri (1931).

the mouthparts are directed downwards. Most *Anopheles* larvae feed by collecting/filtering at the air–water interface (Fig. 7.9A) (Merritt *et al.*, 1992a). The larva rotates its head through 180° and uses its lateral palatal brushes (LPBs), often referred to as mouth brushes, to create currents in the water comprising the surface film. These currents converge in the midline of the head and are directed medially by the antero-median palatal brush (APB) which beats at the same frequency but in opposite phase to the LPBs. As particles approach the head they accelerate and move between the LPBs, past the APB and into the feeding groove. The water comprising the flow moves laterally and then downwards between the labrum and the antennae forming a vertical plume which may extend several centimetres (Merritt *et al.*, 1992b).

Culicine larvae differ from those of *Anopheles* by the possession of a siphon on the penultimate segment of the abdomen (Fig. 7.10). The tracheae are continued into the siphon and the spiracles open at its tip. Within the subfamily there is great variety in the shape and size of siphons. In *Aedes* the siphon is typically short; in *Mansonia* it is short and highly specialized; and in *Culex* the siphon is long and slender. The larvae of *Mansonia* and *Coquillettidia* have short conical siphons with enlarged, heavily sclerotized valves (Fig. 7.13), which are used to

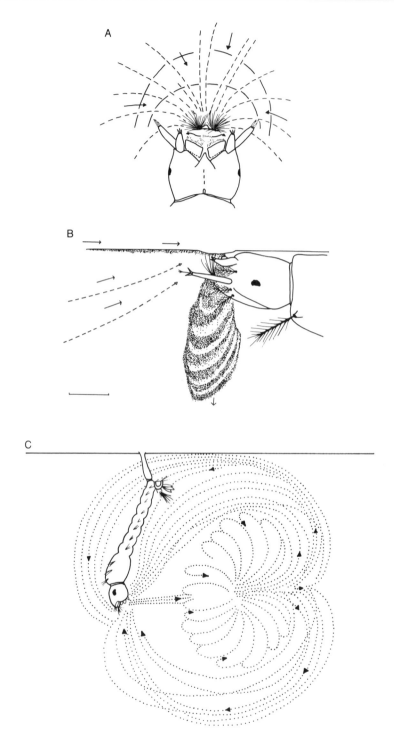

Fig. 7.9. Feeding currents generated by an *Anopheles* larva, viewed from: (**A**) above; and (**B**) the side. (**C**) Lateral view of feeding currents generated by a feeding culicine larva. Scale bar = 0.5 mm. Redrawn from Clements (1992).

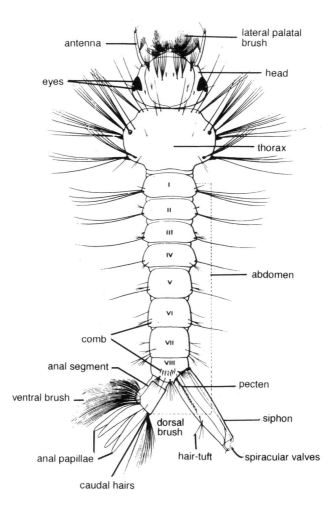

Fig. 7.10. Main features of a culicine larva. Source: Marshall (1935) *The British Mosquitoes*, British Museum (Natural History), London.

pierce aquatic plants and acquire air via the air canals of the plant. This behaviour renders the larva dependent upon emergent aquatic vegetation, but more or less immune to larvicides applied to the surface. Larvae of *Ma uniformis* and *Ma africana* show no selection for any particular plant species and in the laboratory will attach to brown paper (Laurence, 1960). These larvae can respire at the surface but are unable to moult, for which they need to be attached to an object when the exuviae is left *in situ* and the newly moulted larva reattaches. When other culicine larvae are submerged, the valves of the siphon close the opening and prevent the entry of water.

Possession of a siphon enables culicine larvae to hang head down from the surface film, and simultaneously respire and feed below the surface (Fig. 7.9 B). Such collecting/filtering feeding is used by most species of *Culex, Culiseta* and some *Aedes* to feed within the water column or, in the case of *Coquillettidia* and

Mansonia species, within the plant root zone. Species of *Psorophora*, *Haemogogus*, *Wyeomyia* and most *Aedes* feed more widely in the water column, on the surface of sediments and at the air–water interface while suspended from the surface. They are classified as collector-gatherers. *Aedes atropalpus* predominantly feeds by scraping microorganisms from mineral and organic surfaces, and a small number of species, including *Ae aegypti*, *Cx bitaeniorhynchus*, *Cs inornata* are classified as shredders feeding on detritus and dead invertebrates (Merritt *et al.*, 1992a). The larvae of *Psorophora* (*Psorophora*), *Aedes* (*Mucidus*) and *Culex* (*Lutzia*) are obligatory predators on larvae of their own or other species (Mattingly, 1969).

Eretmapodites larvae, inhabiting shallow collections of water in leaf axils, have short siphons but elongated abdomens, enabling them to breathe at the surface and feed upon the bottom at the same time. Larger larvae of *Eretmapodites* are commonly facultative predators upon smaller larvae within their restricted habitat (Haddow, 1946). The movement of culicine larvae is characteristic in some taxonomic groups and can assist field identification. Strickman (1989) defines the movement of *Culex* (*Culex*) as sustained irregular flexing, of *Culex* (*Melanoconion*) as sporadic irregular flexing and of *Haemogogus* as slow sinuous flexing.

When culicine larvae are feeding by collecting/filtering in the water column, the lateral palatal brushes beat towards the preoral cavity with their rows of filaments packed closely together and return more rapidly with the rows of filaments parted. The longer initial phase of the beat draws water towards the front of the preoral cavity from all directions while the return phase causes little water movement. There is a jet-like current directed perpendicular to the body axis of the larva at the level of the head (Dahl *et al.*, 1988; Widahl, 1992). Consequently, as the larva is freely suspended from the water surface, it moves slowly forward while feeding (Fig. 7.9C). As culicine larvae hang head down from the surface they do not possess palmate hairs or prothoracic notched organs and, feeding below the surface, are not vulnerable to toxins and pathogens applied to the surface. To be available to culicine larvae, toxins and pathogens have to be distributed throughout the water body or be present in the surface of the bottom sediment for ingestion by collector-gatherers.

In addition to filter-feeding culicine larvae also drink. For larvae living in a saline environment this intake is required to counteract the loss of water through the integument, but freshwater species, e.g. *Ae aegypti* and *Cx quinquefasciatus*, also drink and the rate is increased by phagostimulants and the presence of particulate food (Aly and Dadd, 1989). The difference in drinking rate and filtration rate explains the greater effectiveness of the particulate toxin produced by *Bacillus thuringiensis* var. *israelensis* which is 7000 times more effective than the same toxin in solution (Schnell *et al.*, 1984).

Like most small particle feeders size is an important factor in determining acceptability. The size of particle accepted increases as the larva grows, ranging from <2 µm in the first instar to up to 45 µm in fourth instar for *Cx pipiens* larvae feeding in a comparable manner, not at the surface but in the water body (Clements, 1992). Larger particles may be crushed between the prementum and the mandibles or rejected by a quick turn of the head. Although the larval feeding entails the indiscriminate ingestion of suitably sized material, it is affected by gustatory phagostimulants. The rate of beating of the LPBs does not increase but the proportion of time spent beating is increased (Merritt *et al.*, 1992a).

Such selective surface feeding offers the possibility of controlling *Anopheles* by the application of suitable toxins, e.g. *Bacillus sphaericus* or pathogens to the surface. In the past Paris green, a complex of copper metarsenite and copper acetate applied as a dust to the water surface, would be collected and ingested by feeding *Anopheles* larvae. The efficiency of Paris green as a larvicide against *Anopheles* is thoroughly attested by it being the only larvicide used by Soper and Wilson (1943) in their eradication of *An gambiae* from Brazil in the 1930s. The dosage applied selectively killed *Anopheles* larvae and left other organisms unharmed.

Culicid larvae breathe atmospheric air through a pair of spiracles on the eighth abdominal segment at the posterior end of the abdomen, i.e. the larvae are metapneustic. The opening of the spiracles is in the floor of the spiracular apparatus, which pierces the surface. To prevent water from entering the spiracles a film of oil is secreted by the perispiracular glands (Keilin *et al.*, 1935). (In passing it might be noted that larvae of the ephydrid fly, *Psilopa petrolei*, live in oil pools in southern California and Thorpe (1930) considers that they prevent oil entering their tracheae by the perispiracular glands secreting water.) The oil secretion repels water but allows oil to enter and this is the rationale for the use of oil as a mosquito larvicide. To be effective the oil has to have a low viscosity so that it readily enters the spiracles and penetrates the tracheae, and it has to be toxic. Respiration occurs both through the tracheal system and through the body surface. When the tracheal system is rendered inoperative, e.g. by a non-toxic oil, cutaneous respiration is adequate to permit larvae to complete development. A high proportion (44–83%) of fourth instar *Cx quinquefasciatus* larvae and a greater percentage of *Ae aegypti* larvae completed their development and emerged as adults in spite of the fact that their tracheal systems were filled with hexadecane, a straight chain hydrocarbon. Emergence of adults from such larvae was delayed for three to seven days but the adults were able to mate and produce viable eggs (Micks and Rougeau, 1976).

At intervals, the larva moults and, as described earlier (p. 75), the head capsule is rapidly inflated to the size characteristic of the next instar. Head size increases geometrically from instar to instar and often follows Dyar's law (Fig. 7.11). In *An sergenti* the head breadth increased by about 50% at each moult giving mean measurements in each of the four instars as 170, 260, 395 and 600 μm (Kettle, 1948). A linear increase of × 1.5 in head size at each moult implies an increase in volume of × 3.4. During the first three instars the head capsule increases in length by the addition of a collar but in the fourth instar the collar remains a narrow band. The only efficient criteria for identifying fourth instar larvae are head size and the extent of the collar. Other body dimensions, such as length, increase steadily throughout larval life and there is an overlap between instars. Under optimal conditions the fully grown fourth instar larva will pupate in 7–10 days but the duration of the larval stage is temperature dependent, and in the cool season in subtropical areas the larval stage may last several months, e.g. *An pharoensis* in Egypt. The effect of seasonal changes in temperature on *An merus* in summer is to produce adults with shorter wings and larvae with smaller head capsules, but the rate of increase of the head remains the same (1.6) (Le Sueur and Sharp, 1991).

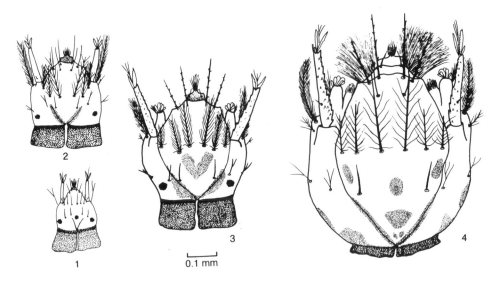

Fig. 7.11. Growth of head capsule of an *Anopheles maculatus* var. *willmori* larva. Note that the collar increases during instars 1–3 but not in instar 4. Redrawn from Puri (1931).

Pupa

The head and thorax of the pupa are combined into a single division, the cephalothorax, which is joined posteriorly to a segmented abdomen (Fig. 7.12). At rest the pupa floats at the water surface with its abdomen reflected under the cephalothorax. The pupa does not feed and is therefore unaffected by toxins which have to be ingested but it breathes through a pair of broad trumpets dorsally placed on the cephalothorax, i.e. it is propneustic, and is susceptible to oil treatment. The ninth segment of the abdomen carries a pair of broad, flat plates, the paddles. The pupa remains quiescent unless disturbed, when the abdomen is straightened out, the paddles spread widely, and the abdomen rapidly flexed. Depending upon the orientation of the cephalothorax this movement of the abdomen serves to drive the pupa forward or to dive below the surface where it may merely float to the surface again or stay down without any apparent further effort. Pupae can complete their development out of water; presumably either the pupa itself or the enclosed pharate adult has a complete waterproofing wax layer. On a dry surface pupae can jump about, which is possibly a defence mechanism against predators (Gillett, 1971).

The culicine pupa is very similar to that of the Anophelinae, apart from the differences in the shape of the respiratory horn which is tubular, compared with the distally expanded horn in the Anophelinae. In *Mansonia* and *Coquillettidia* the respiratory horns of the pupa are modified for penetrating plant tissue (Fig. 7.13) and, on pupation, the larval exuviae is not shed until the pupa has firmly attached itself to the plant. When this is achieved the pupa gives a quick flick and the larval

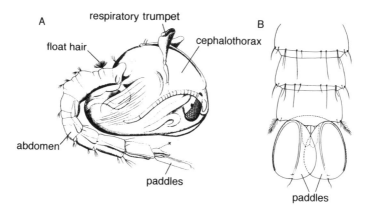

Fig. 7.12. (**A**) Pupa of *Anopheles maculipennis*; (**B**) dorsal view of terminal segments of abdomen of pupa of *Anopheles hilli*. Source: A from Nuttall G.H.F. and Shipley, A.E. (1901) *Journal of Hygiene* 1.

exuviae is both shed and detached from the plant. In these genera, at eclosion, the pupa frees itself from the plant, leaving behind the tips of the respiratory horns, and rises to the surface where it respires in the same manner as other culicine pupae (Gillett, 1971).

In general the development times for culicines are as short or shorter than those of *Anopheles*, but the life cycles of *Mansonia* and *Coquillettidia* are particularly protracted. At 23°C *Cq aurites* takes 40–50 days to develop from egg to adult while *Ae aegypti* completes development in 10 days at the same temperature, and at 26–30°C *Ma uniformis* and *Ma africana* develop in 25–40 days (Laurence, 1960; Gillett, 1971).

Control of the immature stages

In extensive breeding sites, control of the immature stages of mosquitoes has traditionally been effected by the application of insecticides in solution in oils, as emulsions, wettable powders or dusts. Oil-based insecticides will kill larvae and pupae of all surface-breathing culicids. Surface-applied dusts would selectively target *Anopheles* larvae. Temephos, an organophosphate, has proved to be a particularly efficient larvicide to use. It is often applied in a granular formulation which disintegrates in water, releasing insecticide slowly, thus prolonging its effect (Davidson, 1988).

In some situations non-toxic barriers can be effective control agents. Culicid pupae can be killed by the application of a monolayer of a water-insoluble surfactant, e.g. lecithin. The pupae are unable to pierce the monolayer to make contact with the air, and die from lack of oxygen. Other aquatic creatures, dependent for respiration on oxygen dissolved in the water, are unaffected (McMullen and Hill, 1971; McMullen *et al.*, 1977). Highly effective control of *Cx quinquefasciatus* breeding in wet pit latrines on the island of Zanzibar was achieved by the application of expanded polystyrene beads to produce a layer about 7 mm deep over the

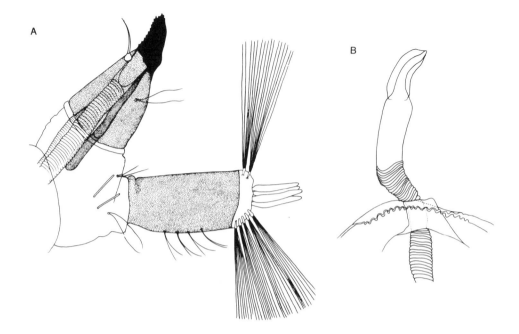

Fig. 7.13. *Mansonia uniformis:* (**A**) terminal segments of the abdomen of larva; (**B**) respiratory trumpet of pupa.

water surface. It was highly effective, producing a reduction in biting of 98% (Maxwell *et al.*, 1990).

Peridomestic water provides breeding sites for important disease vectors but as this water is used for household purposes there is consumer resistance to it being treated with insecticides. The introduction of invertebrate predators into peridomestic water containers is more acceptable. *Toxorhynchites amboinensis* has been introduced into French Polynesia and Samoa with the expectation that its predatory larvae will control *Ae aegypti* and *Ae polynesiensis* (Engber *et al.*, 1978; Rivière *et al.*, 1979). In French Polynesia this introduction was followed by that of *Mesocyclops aspericornis*, which has been highly successful in controlling both *Ae aegypti* and *Ae polynesiensis* in crab holes, tree holes, and artificial containers (Rivière *et al.*, 1987). *Mesocyclops* species are small and can only feed on first instar larvae (Brown *et al.*, 1993). In addition prey and predator do not necessarily share the same microhabitat. On a Torres Strait island *Ae aegypti* and *Cx quinquefasciatus* were found predominantly near the water surface while the copepods were found mainly at the bottom (Brown *et al.*, 1992). A major problem in controlling mosquito breeding in peridomestic water containers is a failure to understand fully the water management practices of the local population (Suarez, 1993).

The contributions which various fungal and protozoan parasites can make to the control of mosquitoes have been investigated, with the most promising candidate being the fungus, *Lagenidium giganteum*. The motile zoospores of this species encyst on the cuticle of mosquito larvae, usually in the head region, from which the fungus penetrates the cuticle and invades the haemocoele, causing death of the host in 48 h (Lacey and Undeen, 1986).

Two bacteria, *Bacillus sphaericus* and *B. thuringiensis* are highly toxic to some mosquito larvae. *B. sphaericus* is toxic to larvae of *Cx quinquefasciatus* and *Anopheles* but not to *Ae aegypti* or to non-target organisms. *B. thuringiensis* has a broader spectrum and is effective against *Ae aegypti* larvae. The bacteria are ingested by filter feeding, which concentrates the toxin. Bacterial toxins are most effective in clear, shallow, warm water and less effective in polluted, deeper waters. *B. thuringiensis* is more effective against culicines than anophelines, not because the latter are more resistant, but because their surface feeding brings them into less contact with the toxin (Lacey and Undeen, 1986).

Adult emergence

In preparation for emergence, gas appears between the pupal cuticle and the enclosed pharate adult. Then gas appears in the midgut, thus increasing the buoyancy of the pupa. The abdomen is raised into a horizontal position and a split appears mid-dorsally in the pupal cuticle. Pumping in of air raises the adult out of the pupal exuviae (Clements, 1992). Careful movements are required to ensure that the adult mosquito does not fall sideways and become trapped in the surface film. Adult *Opifex fuscus* are able to tumble around, on and in water, without becoming trapped (Marks, 1958). This danger is particularly acute when the adult is largely out of the pupal exuviae but the terminal segments of the appendages are still not free. The newly emerged adult expands its wings and legs by increasing haemolymph pressure. Eclosion takes about 15 min at the end of which the newly emerged adult is distended by air and fluid. Diuresis gets rid of excess fluid and air is dispersed from the midgut, enabling the adult to be able to fly more or less normally after about an hour (Clements, 1992). There follows a period of rest in which intense metabolic activity occurs involving the breakdown of larval muscles in the abdomen. There follows a second peak of diuresis which lasts for about 24 h in males and 36 h in females before mature flight and adult activities can be undertaken (Gillett, 1993).

Morphological Differences between Adult Anophelinae and Culicinae

Living adults can be readily recognized by the stance they adopt when resting on a flat surface. Adult *Anopheles* rest with the proboscis, head, thorax and abdomen in one straight line making an acute angle with the surface (Fig. 7.14). In some species, e.g. *An balabacensis*, the adult appears to stand almost on its head while other species, such as *An culicifacies*, as the specific name implies, have a more culicine-like stance. The culicine adult rests with its body angled and the abdomen directed back towards the surface on which it is resting (Fig. 7.14). The proboscis, head and anterior part of the thorax form one line and the posterior part of the thorax and the abdomen another. This difference enables selective hand collections to be made of living material, but does not apply to dead mosquitoes.

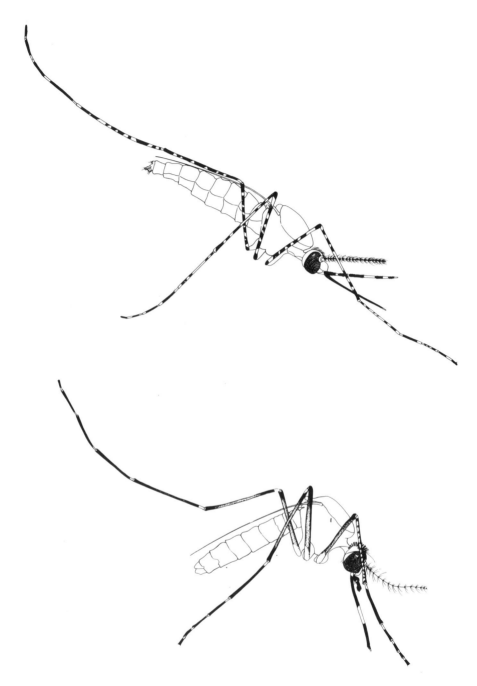

Fig. 7.14. Resting stances of living mosquitoes: above, *Anopheles*; and below, culicine.

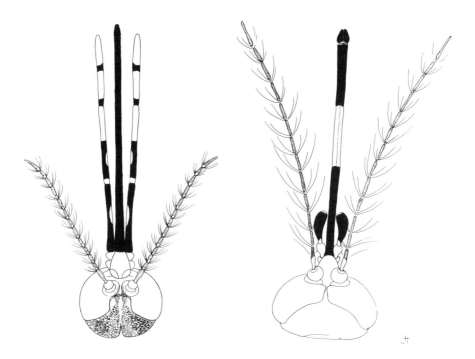

Fig. 7.15. Heads of female mosquitoes: left, *Anopheles annulipes*; and right, *Culex annulirostris*.

Dead adults

Dead adults can be classified after examination of the antennae and palps. Most male mosquitoes of all three subfamilies have plumose antennae, and females have pilose antennae with fewer, shorter hairs. This difference is a functional one, the antennae being sound receptors enabling the male to locate the female by the sound of her wing beat. In the female anopheline the palps are as long and straight as the proboscis, while the palps of the female culicine are considerably shorter, usually about one-quarter of the length of the proboscis (Fig. 7.15). The palps are often closely applied to the proboscis so that the culicine proboscis appears to have a thickened base and the anopheline proboscis to be trifid at the tip.

In the anopheline male the palps are as long as the proboscis and clubbed at the distal end where the last two segments are swollen (Fig. 7.16). In the typical culicine male the palps are as long as the proboscis and taper distally, but the tapering is sometimes obscured by the development of tufts of hair on the distal segments (Fig. 7.16). In a number of culicine genera, including *Sabethes*, *Uranotaenia* and *Wyeomyia*, the palps of the male are short and similar to those of the female (Mattingly, 1973). When the palps of the male are long, they are often upturned in Culicinae and laterally directed in the Anophelinae. In the Toxorhynchitinae the palps are of the culicine type, being short in the female and long and tapering in the male (Fig. 7.4).

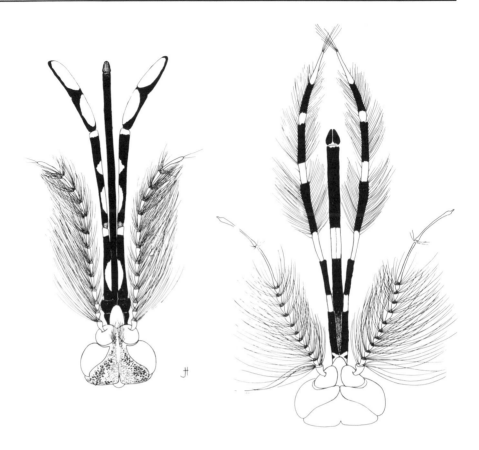

Fig. 7.16. Head of male mosquitoes: left, *Anopheles annulipes*; and right, *Culex annulirostris*.

There are also differences between the subfamilies in the distribution of scales on the body, the shape of the scutellum and the number of spermathecae. In the Anophelinae, the abdominal sterna and usually also the terga, are completely or largely devoid of scales; but in the Culicinae the abdomen is covered with a uniform layer of scales (Mattingly, 1973). The scutellum is evenly curved in the Anophelinae and there is a regular row of setae on the posterior border (Fig. 7.17). In the Culicinae the scutellum is trilobed and the setae are grouped on the lateral and median lobes. There are three spermathecae in most Culicinae, two in *Mansonia* and only a single one in the Anophelinae and *Uranotaenia* and *Aedeomyia* (Gillett, 1971).

General Culicid Bionomics

In this section a generalized account of the bionomics of mosquitoes will be given. Most of the features are common to all mosquitoes.

When the progeny of any one egg batch emerge as adults, the males emerge

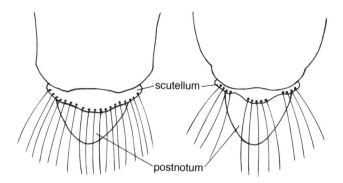

Fig. 7.17. Scutella of an *Anopheles* (left) and of a culicine (right).

first. This is not uncommon in insects, and in mosquitoes there is a special reason. The male terminalia have to rotate through 180° before the male is ready for mating. This process takes about 24 h so that by the time the females emerge the males are competent for mating. The males obtain additional energy for flight by taking nectar from plants, as do the females. In the Arctic tundra, both sexes of mosquitoes may be collected from flowers and be found with orchid pollinia on their heads, indicating that they have probed deeply into orchids (Twinn *et al.*, 1948).

Both sexes of the salt-marsh mosquito, *Ae taeniorhynchus*, show a marked circadian pattern of nectar feeding with peaks in the early morning and from late afternoon until nearly midnight. During the brief periods before sunrise and after sunset, when the males were swarming, no nectar feeding occurred (Haeger, 1955). Both sexes of *Ae aegypti* and *An arabiensis* responded to odours of composite flowers. The response of *Ae aegypti* was biphasic with dawn and dusk peaks (Jepson and Healy, 1988), and the nocturnal *An arabiensis* had a peak at 2000 h (Healy and Jepson, 1988).

Female mosquitoes are referred to as endophilic or exophilic depending on whether they rest indoors or outdoors, and as endophagic or exophagic depending on whether they feed indoors or outdoors. *An balabacensis* is both exophilic and exophagic; *An maculatus* is exophilic and endophagic, while *An minimus* is both endophilic and endophagic.

Mating

Mating is often preceded or accompanied by swarming, in which the males associate over a marker and fly in a particular manner. Markers are visually prominent objects such as the tops of tall trees or ground objects which contrast with the background (Downes, 1969). The males face into the wind and fly forward until they are vertically over the front edge of the marker, when they cease flying and are carried backwards. As they pass over the hind edge of the marker active flight is stimulated and the males fly forward again over the marker (Downes, 1969). Such oscillatory flight is characteristic of males in swarms. The long slender hairs or fibrillae on the plumose antennae of the male are erected from the shaft and increase the receptivity of the antenna to sound waves. In some species, e.g. *Ae aegypti*, the fibrillae remain erected permanently and in other species, e.g. *An gambiae*, the fibrillae undergo a circadian rhythm of being erect or recumbent (Charlwood and Jones, 1979). Sound waves impinging on the fibrillae cause the shaft of the antenna to vibrate and stimulate sensory cells in Johnston's organ in the pedicel, the swollen second segment of the antenna (Clements, 1963).

In *An melas* it is considered that the adults first orientate to an 'arena', an open flat area and then to markers within the arena. Male swarms form over the markers and virgin females orientating to markers will encounter swarming males. Males recognize a female from a short distance and fly directly towards her and attempt to couple and, if successful, the pair leave the swarm *in copulo* with insemination being completed elsewhere. Charlwood and Jones (1980) point out that a female spends less than 30 s in the swarm and at any one time the proportion of males to females would be of the order of 600 : 1 and hence copulating pairs are seen infrequently. During mating spermatozoa are deposited in the bursa copulatrix of the female, from which they move to the spermathecae. Wharton (1953) described the arenas used by anophelines in Malaya as 'relatively open, flat areas free from overhead obstructions and surrounded by trees or other objects', and the arena of *An melas* was similar.

In *An claviger* insemination occurs in flight, the couples separating when they reach the ground. This behaviour frustrated attempts to colonize this species until it was appreciated that the solution was to have a cage of sufficient height that the coupled pair did not reach the ground until insemination had occurred (Bates, 1949). Swarming occurs for a short period at dusk at 24 h intervals and Gillett (1993) has argued persuasively that such temporal discontinuity of swarming and mating increases and maintains the genetic diversity on which natural selection and survival depend.

Swarming requires sufficient illumination for the males to be able to see the marker, and a low enough wind speed so that they can orientate to it. Downes (1969) gives the range of wind speed for *Ae hexodontus* to swarm as 0.2–3.5 m s^{-1}. Swarming occurs commonly at sunset when the wind tends to die away, and the rapidly fading light has a stimulating effect on mosquito activity. Wharton (1962) observed swarming in five species of *Anopheles* in the same area. They varied in the height at which the swarm formed, with swarms of *An indiensis* occurring 0.3–0.6 m above the ground and swarms of *An maculatus* at 4.5–6.0 m. Most observations were made on *An philippinensis* which began to swarm about 5 min after sunset and continued for about 20 min. Copulating pairs were seen flying in the vicinity of swarms with the period of copulation lasting from 10 to 25 s. Similar observations were made by Russell and Rao (1942a) on *An culicifacies* with the difference that most (19/23) of the copulating females had already taken a blood meal whereas copulating female *An philippinensis* were newly emerged.

Some mosquitoes, such as *Ae aegypti*, mate without swarming – the male responding to the sound of the female's wing beat, which in 3–4-day-old females ranges from 450 to 600 c s^{-1} with an average of 500 c s^{-1}. Males show a strong response to sounds within the range of 400–600 c s^{-1} and particularly to sounds of 500–550 c s^{-1}, and are therefore adapted to respond to the flying female. Male

Ae aegypti can detect females up to a distance of 25 cm and respond to a sound of approximately 20 decibels (Wishart and Riordan, 1959). Male *Ae aegypti* have higher wing-beat frequencies than females, but as both sexes have lower frequencies early in adult life the wing beat of young males and old females are similar (Clements, 1963).

Although only female *Ae aegypti* are blood feeders, both sexes have similar bimodal, diurnal landing rates on humans with coinciding peaks before sunset between 17.00 and 18.00 h. In the morning the peak landing rate for females occurs just after sunrise, 06.00–07.00 h, while the male peak is about 2 h later (Corbet and Smith, 1974). This suggests that in nature mating occurs in the vicinity of the host. Discrimination between species and inseminated females is achieved by the presence of a contact pheromone on virgin female *Ae aegypti* (Nijhout and Craig, 1971). When mating and blood-feeding are separated both in time and space, a contact pheromone is not necessary and is not present in *An gambiae* (Charlwood and Jones, 1979). In *An stephensi* males become active at dusk and form mating swarms where they are joined by virgin females. Following mating females cease to be active at dusk and engage in nocturnal host-seeking and oviposition. Virgin females continue to be active at dusk unaffected by having taken a blood meal (Rowland, 1989). It was pointed out earlier (Chapter 6) that male *An atroparvus* would mate with females of other species in the *An maculipennis* complex and, in that case, either there is no contact pheromone or the pheromones are so similar that males fail to discriminate between virgin females in the complex.

In some species the male antennae are pilose, like the female's, and sexual recognition is based on other than auditory stimuli. Males of *Deinocerites cancer* search the surface of breeding sites for female pupae and mate with the female when she emerges. Female pupae are recognized through a contact pheromone which the male detects with its antennae (Haeger and Phinizee, 1959). Males of *Opifex fuscus* are capable of mating within an hour of emergence and will copulate with females, which are still within the pupal exuviae. *Op fuscus* males do not distinguish between male and female pupae but do not attempt to copulate with males (Kirk, 1923; Marks, 1958). In *Sabethes chloropterus* the male hovers over the resting female and tests her receptivity by tapping her hind legs with his mid-tarsi (Galindo, 1958).

During mating, secretions from the accessory glands of the male produce a mating plug in the female. In *Ae aegypti* it has been shown that the plug contains matrone, a substance which stimulates oviposition and inhibits mating. Matrone consists of two fractions, one of which stimulates oviposition while the other has no obvious effect on its own, but both fractions are required to inhibit further mating (Hinton, 1981).

Host Finding (See also Chapter 4 – Host Finding and Blood feeding, p. 70)

In some species, e.g. *Cx molestus*, the first egg batch may be matured autogenously, i.e. without a blood meal, but most mosquitoes require a blood meal for ovarian development. The source of the blood meal is a major factor in determining the potential of a species to be a vector of disease. Tempelis (1975) recognizes nine basic feeding patterns among mosquitoes, including species that feed: almost entirely on mammals, e.g. *An gambiae, Culiseta inornata*; almost entirely on birds, e.g. *Culiseta melanura*; readily on both birds and mammals, e.g. *Cx quinquefasciatus*; almost exclusively on amphibians, e.g. *Cx territans*; predominantly on reptiles, e.g. *Deinocerites dyari*; and a remarkable species, *Uranotaenia lateralis*, which feeds on the mudskipper, a fish, which lies out of the water on mud banks with its tail in the water. Within the mammal-feeding mosquitoes the feeding habits of *Anopheles* species can be further subdivided into those which are anthropophilic, feeding on humans, e.g. *An gambiae* s.s., and those which are zoophilic, such as *An maculatus* in Nepal (Garrett-Jones *et al.*, 1980).

There are several stages in the feeding behaviour of mosquitoes, which are designated activation, orientation, landing and probing. Most mosquitoes have a crepuscular or nocturnal circadian cycle of activity, which is stimulated by rapidly fading illumination. Orientation is both visual and chemical. The maximum distance at which orientation occurs is less than 20 m. Bidlingmayer and Hem (1980) found that most adult mosquitoes were attracted to large, unpainted plywood suction traps from 15 to 19 m. The anthropophilic *Ae vexans* was most responsive; the ornithophilic *Cs melanura* showed an average response for the ten species tested; but *Cx quinquefasciatus* was relatively unresponsive and its visual range was less than 7.5 m.

Gillies and Wilkes (1970) found that mosquitoes orientated to a single bait from similar distances. *An melas* and other *Anopheles* species responded from a distance of 13.5–18 m; *Cx tritaeniorhynchus* from 9–18 m, and members of the *Cx decens* group from only 5 m. For those species for which information was available orientation to carbon dioxide was from less than 15 m. Carbon dioxide acts as an attractant to mosquitoes which fly upwind in response. To be fully effective the carbon dioxide must be pulsed or accompanied by other host odours. In still air, carbon dioxide activates the mosquitoes but there is no orientation to the source (Gillies, 1980).

Kline *et al.* (1990) recognize three responses to octenol by host-seeking mosquitoes. Very few mosquito species, *Ae taeniorhynchus, Cq perturbans*, are attracted to octenol alone. In combination with CO_2 it greatly increases the catch of most species of *Aedes, Anopheles, Coquillettidia, Mansonia* and *Psorophora*, with the effect being a synergistic one. Species of *Culex* show little response to octenol in combination with CO_2. Most species of *Culex* are ornithophilic and octenol is a mammalian emanation. Butanone produced by cattle increases the catch of tsetse flies but has a depressing effect on the catch of mosquitoes.

Lactic acid, the product of muscular activity, in combination with CO_2 increases the catches of *Ae taeniorhynchus, Cx nigripalpus* and *An atropos*, but the increases were not statistically significant. The sensitivity of lactic acid receptors in female *Ae aegypti* varies throughout their life. They are non-responsive in newly emerged and blood-fed females and highly sensitive in females of any age that are exhibiting host-seeking behaviour. A similar difference in response is apparent in diapausing and post-diapausing female *Cx pipiens* (Bowen, 1991).

In its upwind movement a mosquito uses visual clues, but many mosquitoes are active on the darkest nights when visibility would be minimal and Gillett (1979) has proposed that mosquitoes orientate themselves upwind by adopting an undulating flight pattern. Gillies and Wilkes (1974), working in West Africa, found that a significant number of *Mansonia* (*Mansonioides*) spp. reached their host by flying downwind at a low level.

Host-seeking mosquitoes are more attracted to low intensity colours, such as blue, black and red than they are to high intensity colours such as white and yellow (Allan *et al.*, 1987). Convection currents given off by a warm body enable mosquitoes to orientate to a host very effectively over short distances (Clements, 1963); and Brown (1951) has shown that in the field *Aedes* mosquitoes were attracted to a warm body when the air temperature was less than 15°C.

Biting Cycle

Blood-feeding follows a circadian rhythm with most species of mosquitoes being nocturnal or crepuscular and a smaller number diurnal. Most species of *Anopheles* are nocturnal and many, including major disease vectors such as *An gambiae*, *An minimus* and *An farauti* have their peak biting rate in the early hours of the morning after midnight (Muirhead-Thomson, 1951). *Ae africanus* is crepuscular with its biting activity being largely concentrated into a 20-min period following sunset, but there is no similar period of activity at sunrise (Gillett, 1971). Species of *Aedes* (*Stegomyia*), including the disease vectors *Ae aegypti*, *Ae polynesiensis* and *Ae scutellaris*, are diurnal. *Ae aegypti* has two peaks of activity, one just after dawn 06.00–07.00 h and before sunset 17.00–18.00 h. Similar proportions of nulliparous and parous females were present at both times, indicating that the bimodality was a general property of the species and not due to females of different ages feeding at different times.

The biting cycle has practical implications in that mosquito nets are effective against nocturnal species, especially those which feed after midnight, whereas they give little protection against crepuscular species, which require house screening or the use of repellents and protective clothing.

Mosquitoes are not limited in their activity to ground level. In forests, potential hosts are to be found from ground level up to the canopy, the domain of birds and primates. In Zika Forest, a tropical rainforest in Uganda, the ground level is heavily shaded and covered with shrubs and small trees; the canopy lies between 12 and 18 m; the zone between at 6 m is void of vegetation, being above the shrubs and below the canopy; emergent trees rise above the canopy with the tallest reaching to 36 m. A tower was built to provide catching stations from ground level to 36 m at 6 m intervals. Collections were made on human bait at all levels over the full 24 h period. Haddow and Ssenkubuge (1965) gave some of the results obtained. *An implexus* was rarely taken above ground level where 98.7% (387/392) of this species were collected; *Cq aurites* was taken at all levels with a maximum at 36 m; and 66% (7550/11,410) were taken above the canopy at 24–36 m. *Ae africanus* was also collected at all levels but with a maximum (63%) in the canopy.

Although circadian rhythms are characteristic of species they can be modified by environmental conditions. *Mansonia uniformis* and *Ma africana* showed marked

peaks of activity immediately after sunset and before sunrise at the catching stations above the canopy (24–36 m). At 6–18 m the bimodality was less marked with
increased activity occurring throughout the night, and at ground level bimodality
was lost and a high level of activity occurred from sunset to sunrise. *Ae ingrami*
showed even greater changes in rhythm with height. It was not taken above
24 m and at that height it was crepuscular with greatest activity at sunset and none
between sunrise and sunset; at ground level activity was almost reversed, being
virtually confined to sunrise to sunset and with very little activity after sunset.
When the catches of *Ae ingrami* during specified periods of the diel are plotted
against height, maximum activity shifts from ground level at 09.00–15.00 h to
18 m between 19.00 and 06.00 h and back again to ground level by 09.00 h. It recalls
the vertical movement of plankton over the diel but in this case there is no evidence
of movement of *Ae ingrami* from one level to another. The same result could be
produced by insects at the various levels showing differing biting rhythms. Nevertheless, the important conclusion to be drawn is that circadian rhythms can be
modified by local conditions and that, particularly in forests, ground level observations give little indication as to what is occurring at higher levels.

Blood-feeding and Ovarian Development

Blood-feeding takes only a few minutes. When pool feeding the female waits until
sufficient blood has collected, before it is ingested rapidly and the stomach becomes
visibly distended. Some species, e.g. *Ae aegypti*, pass drops of clear fluid from the
anus while they are feeding, and others, e.g. *An stephensi*, pass apparently
unchanged blood from the anus. *Ae aegypti* ingests 4.2 mm^3 of blood, of which
1.5 mm^3 of clear fluid is passed within 15 min of feeding. The larger *Cx quinquefasciatus* ingests 10.2 mm^3 of blood when feeding on a chicken (Clements,
1963).

In the tropics the gonotrophic cycle of blood ingestion and ovarian development takes about 48 h. Following nocturnal activity four categories of resting
females can be recognized in the early morning: those which had fed the previous
night and are full of blood; those which had fed the night before and contained
some dark blood and developing eggs; fully gravid females, which will oviposit
during the succeeding night; and empty females either nulliparous or parous
females, which have oviposited but not yet fed again. The feeding and oviposition
cycles have a minimum duration of 2–3 days.

Most mosquitoes show gonotrophic concordancy in which there is strict alter-
nation of blood-feeding and oviposition, with each blood meal being followed by
the maturation and oviposition of a batch of eggs. Gillies (1955) has shown that
An gambiae and *An funestus* require two blood meals to mature the first batch of
eggs but only 2% and 4%, respectively, feed twice in subsequent ovarian cycles.
Ae aegypti also feeds twice during the maturation of its first batch of eggs (Corbet
and Smith, 1974) and a proportion of the population may take a second blood meal
during later ovarian cycles (Sheppard *et al.*, 1969; McClelland and Conway, 1971).
In the calculations of Conway *et al.* (1974) this proportion was variously estimated
as 6% and 49%.

In *Ae aegypti* most nullipars (58%) were inseminated before the first meal, some (17%) were mated between feeds and 25% were inseminated between the second feed and the first oviposition (Corbet and Smith, 1974). In *An gambiae* and *An funestus* 26% and 65%, respectively, were inseminated before the first blood meal and the rest between meals (Gillies, 1955). In *An freeborni* blood-feeding during development of the ovaries is inhibited by the hormone ecdysone which is produced by the ovaries during oogenesis (Beach, 1979).

In anautogenous *Ae aegypti* maturation of the ovaries is under hormonal control. Juvenile hormone (JH) is secreted soon after the female has emerged from the pupa, and under its influence the terminal oocyte in each ovariole develops to the previtellogenic resting stage. *Ae triseriatus* is an exception with JH being secreted in the pupal stage (Clements, 1992). Initially JH is produced by the corpora allata, endocrine glands in the front of the thorax and later by the ovary (Borovsky *et al.*, 1994a). The ovaries produce ecdysone which converts immediately to 20-hydroxyecdysone (20-HE) (Hagedorn, 1985). JH and 20-HE are both required to promote the sustained release of vitellogenin from the fat body. In females fed only on sugar the ovaries do not develop beyond the resting stage. Three hours after a blood meal, ecdysone is synthesized by the ovary under the stimulus of the egg development neurosecretory hormone from the medial neurosecretory cells of the brain (Steel and Davey, 1985). When the follicles approach maturity, the follicular epithelium of the ovary secretes a trypsin modulating oostatic factor, which terminates trypsin synthesis by the midgut after the blood has been digested and egg protein synthesis completed (Borovsky *et al.*, 1994b). Following the maturation of the terminal follicle the adjoining follicle is brought to the resting stage by the action of JH and 20-HE.

In each ovarian cycle only one egg is developed in each ovariole and therefore the number of eggs matured in an egg batch cannot exceed the number of ovarioles. The number of eggs matured will depend upon the species and on the source of the blood meal. *An melanoon* laid up to 500 eggs in its first gonotrophic cycle (Clements, 1963) and *Cx pipiens pipiens* developed an average of 121 eggs on 3.0 mg of human blood and 255 eggs from 3.1 mg of canary blood (Woke, 1937).

In the epidemiology of vector-borne diseases two features of a species' behaviour and ecology are particularly relevant: the host on which the females commonly feed and their survival between feeds. As there is usually one blood meal per ovarian cycle, survival between blood meals is the same as that between ovarian cycles. Various serological methods have been developed for identification of the source of a blood meal. The precipitin test has been the one most widely used in the past, but more sensitive ($\times$ 1000) methods, e.g. enzyme-linked immunosorbent assay (ELISA), are now being widely used (Washino and Tempelis, 1983). These tests enable the host species to be identified but it is possible to use more elaborate DNA analyses to identify the individual host (Gokool *et al.*, 1993).

It has been a common assumption that mortality of female mosquitoes was independent of age, i.e. mortality was exponential. This is true for *Ae aegypti* but in many important vectors of malaria, including *An arabiensis*, *An darlingi*, *An farauti* and *An vagus*, mortality increases with age and is better fitted by the Gompertz function. Although mortality of *An flavirostris* and *An gambiae* is better fitted by a Gompertz function, an exponential model is applicable to part of the age range (Clements and Paterson, 1981). Both exponential survival between ovarian cycles and length of the ovarian cycle can be calculated from a continuous time series of daily samples (Holmes and Birley, 1987). By using that technique it was found that, in Dubai, *Cx quinquefasciatus* had an ovarian cycle of 2 days with 31% survival between cycles.

Oviposition

Oviposition follows a circadian rhythm in *Ae aegypti* and probably does so in other mosquitoes. In the laboratory, peak oviposition of *Ae aegypti* occurs just before sunset and remains so even when the period of daylight is varied substantially. The factor which controls oviposition is the onset of darkness and not the length of daylight. Oviposition occurs 21–23 h after the onset of darkness the previous day. If the species is reared in continuous light, oviposition becomes arrhythmic and eggs are deposited when they are mature. If arrhythmic *Ae. aegypti* are then plunged into continuous darkness oviposition becomes rhythmic and the first egg batches are laid 22 h later and thereafter at approximately 24 h intervals for the next five days, i.e. the cycle is said to be 'free running'. Similarly, adults maintained in continuous darkness become rhythmic after exposure to light, which may be as brief as 5 s and become free running (Gillett *et al.*, 1961; Gillett, 1971).

In Trinidad *Ae aegypti* has a biphasic cycle of oviposition with peaks occurring after dawn between 0600 and 0800 h and before dusk at 1600 to 1800 h. If the oviposition site is closed during the night the early morning peak is missing. This suggests that gravid females aggregate near potential oviposition sites during the night (Chadee and Corbet, 1990). Under simulated tropical conditions newly blood-fed, inseminated female *An stephensi* are inactive for two nights and on the third night, when their eggs are mature, they become active at dusk and seek out oviposition sites during the night. Following oviposition, inseminated females resume nocturnal host-seeking (Rowland, 1989).

Gravid female *Cx tarsalis* are attracted to oviposit in water containing egg rafts of the same species or to water in which egg rafts have been laid previously. The attractant is an ether-soluble pheromone which was not attractive to *Ae aegypti* or *Cx pipiens pipiens* (Osgood, 1971). Female *Cx quinquefasciatus* respond in a similar manner to egg rafts of the same species and deposit their eggs nearby on the water surface. They are responding to the presence of a volatile pheromone secreted in the droplet found at the apex of each egg. Egg rafts that have hatched 24 and 48 h previously continue to attract ovipositing females (Bruno and Laurence, 1979). The major volatile component in the droplet is erytho-6-acetoxy-5-hexadecanolide (Laurence and Pickett, 1985). In the field in Kenya gravid female *Cx quinquefasciatus* were able to distinguish between sites with and without pheromone from a distance of up to 10 m (Otieno *et al.*, 1988).

Selection of Breeding Site

Mosquito larvae are found in habitats with a wide range of salinity, varying from the oligotrophic waters produced by melting snow to salinities greater than sea water in evaporating tidal pools. The cuticle of freshwater larvae is permeable to water and the stability of the osmotic pressure and ionic composition of the haemolymph is regulated by the production of a hyposmotic urine involving salt absorption in the rectum and the uptake of ions by the anal papillae (Bradley, 1987). Such hyper-regulation occurs in saline-tolerant larvae, such as *Culiseta inornata* and *Cx tarsalis* when reared in low salinities, i.e. equivalent of 40% sea water;

but in more saline media, e.g. 70% sea water, they maintain the osmotic pressure of the haemolymph at the level of the external medium, i.e. they osmoconform. The increase in osmotic pressure of the haemolymph is mainly due to the addition of amino acids and, to a lesser extent, sugars and other soluble organic compounds (Garrett and Bradley, 1984a, b, 1987). *Cs inornata* is limited to salinities equivalent to concentrations of sea water below 80%, which is well above the tolerance of freshwater species, e.g. *Ae aegypti*, of 50%, and well below that of saline species, e.g. *Ae taeniorhynchus*, which may occur in water with a salinity higher than that of sea water. In high salinities, larvae of salt water species maintain the haemolymph at a lower osmotic pressure by producing hyperosmotic urine by the addition of salts to the urine in the rectum. Although their cuticle is less permeable, there is still a considerable loss of water by osmosis. This is restored by drinking the external medium at substantial rates, e.g. 130% of body volume per day by *Ae dorsalis* and 240% per day by *Ae taeniorhynchus* (Bradley, 1987).

Gravid females seek a suitable site for oviposition. General descriptions of the breeding sites of any particular species can readily be given but exceptions will be encountered. The experienced entomologist will know which breeding sites to examine for a particular species. The criteria used by the entomologist, while effective, are almost certainly quite different from those used by the gravid female in search of an oviposition site, as Muirhead-Thomson (1951) elegantly showed for *An minimus* in Assam.

Larvae of *An minimus* are to be found in the grassy margins of lightly shaded, flowing streams. These sites are readily recognizable during the daytime on the criteria of shade, and air and water temperatures in the breeding site and outside, but these differences do not apply when gravid *An minimus* are searching for an oviposition site after midnight on moonless nights. There was no difference in illumination under the light shade of the breeding site and in the open, and differences in air and water temperatures, which had been so marked in the afternoon, were no longer present. *An minimus* only oviposited in these sites and only in these sites in the field could its larvae survive.

Larvae of *An minimus* have a relatively low thermal death point and would be killed if they were exposed in shallow, sunlit pools, and hence they occur in cooler running water. However, they are weak swimmers and can only maintain their position in running water if there is grass to provide shelter and reduce the current. In heavily shaded sections of hill streams there is inadequate illumination to maintain grass and the banks are bare and larvae of *An minimus* are absent. These requirements restrict the immatures of *An minimus* to lightly-shaded portions of flowing water.

The larval habitat is selected by the ovipositing female, and since conditions in the breeding site are likely to change after oviposition it would be anticipated that the female will select a narrower range of conditions than those which can be tolerated by the larvae. Thus larvae, which are normally found in strongly saline waters, are able in the laboratory to complete their development in tap water, e.g. *An multicolor*; and the larvae of *An sergenti* which occur in nature in fresh water, actually develop more quickly in the laboratory in water with a salinity equivalent to 0.5–0.75% NaCl.

It seems that there will be some species of mosquito which will breed in virtually every naturally occurring collection of water, with the exception of excessively hot springs and large bodies of water with clean edges. Mosquito larvae cannot withstand wave action, and this was used to eliminate *Anopheles* breeding from the River Jordan, by daily fluctuations in the water level by varying the discharge through the hydroelectric turbines. Daily fluctuations in water level eliminated the marginal vegetation and mosquito breeding. Modification of breeding sites is a powerful weapon in mosquito control but it requires knowledge about all species in the area. For example, malaria was reduced in parts of Malaya by exposing hill streams to sunlight and eliminating the vector, *An umbrosus*. In other areas this technique failed because the sunlit streams provided suitable breeding sites for *An maculatus*, another vector of malaria (Muirhead-Thomson, 1951). In the last situation the attempt at control had merely changed one vector for another with little effect on malaria transmission.

Larvae of *An culicifacies* occur in rice fields when the plants are small, but they disappear as the rice develops into tall plants. The water is still suitable for larval development because larvae introduced into the water among tall rice plants develop normally. Their absence, as the crop matures, is the result of the behaviour of the ovipositing female which drops its eggs while hovering 5–10 cm above the water surface. As plants grow taller the female is forced to hover at higher and higher levels and, when this reaches 40–50 cm, the female does not oviposit (Russell and Rao, 1942a, b).

Although it is possible to describe the typical breeding site characteristic of a species, there is a good deal of variation from the norm and species are always being found in unusual habitats. Breeding sites can be classified in a number of ways and Laird (1988) has reviewed previous classifications and proposed a major division into above-ground and subterranean waters, the former having nine subdivisions, and the latter two. Laird claims that this is a rational system phasing downwards from large water bodies of conventional limnology to smaller, more specialized ones. The sequence descends from flowing streams, ponded streams, lake edges, swamps and marshes, shallow permanent ponds, shallow temporary pools, intermittent ephemeral puddles, natural containers and artificial containers. Subterranean waters are classified as natural or artificial.

In considering different habitats, species will be cited as examples of mosquitoes which breed in such habitats, but each species has a limited geographical distribution. Running water habitats are important sources of *Anopheles* mosquitoes. The grassy edges of sunlit, flowing water breed species such as *An maculatus* and *An fluviatilis*. *An minimus* and *An umbrosus* are found in the grassy edges of shaded streams. Larvae of *An superpictus* hide among pebbles at the edges of shallow, bare, hill streams.

Permanent ground habitats, such as large swamps and vegetated lakes, are the breeding sites of species of *Mansonia*, *Coquillettidia*, *An funestus* and *An hyrcanus*. Rice fields are semi-permanent man-made swamps and are suitable for *An pharoensis* and *Cx tritaeniorhynchus*. *An funestus* occurs only where the water is clear and unpolluted but *Mansonia* and *Cx tritaeniorhynchus* thrive in moderately polluted water. *Cs melanura* breeds in oligotrophic waters of swamps and *Ae communis* in pools made by melting snow on the tundra.

In Indonesia *An sundaicus*, a vector of malaria, breeds in permanent saltwater fish ponds. The ponds carry a luxuriant growth of floating green algae, which form the food of the herbivorous fish and the breeding site for *An sundaicus*. The local economy was dependent on its fish culture and removal of the green algae for malaria control was unacceptable. An ingenious solution involved the digging of a deep trench at the edges of the ponds. When the main body of water was drained out, the fish survived in the water-filled trench but most of the green algal mat was stranded on the pond floor. Exposure to the sun killed the green algae. The ponds were then flooded to a depth adequate to support the growth of blue-green algae at the bottom of the pond. The fish fed on the blue-green algae and there was no floating algae to support *An sundaicus* (Walch and Schuurman, 1929).

Temporary ground pools of salt water are formed in coastal salt marshes and are a prolific source of anthropophilic *Aedes* mosquitoes in almost all parts of the world. In the Caribbean and the eastern coast of the USA, salt marshes breed *Ae taeniorhynchus* and *Ae sollicitans*, in the Palaearctic region they breed species such as *Ae caspius* and *Ae detritus* and in the Australian region *Ae vigilax*. Species that breed in temporary pools show marked fluctuations in the density of adults. There are periods of the year when the *Aedes* population in a coastal locality may consist only of dormant eggs. Following flooding of the breeding sites there is mass hatching of the eggs and ten days later there is a massive emergence of blood-seeking females, which invade adjoining residential areas.

Temporary freshwater ground pools are the main breeding sites of *An gambiae* and *Cx annulirostris*, although both breed in more permanent waters particularly during the dry season. Early exploitation of temporary pools enables immature *Cx annulirostris* to minimize predation. Survival from first instar to adult in temporary pools was 35%, compared with 4% and 2% respectively in the overflow from a permanent pond and in a semi-permanent pool. Mortality due to predation was 43% in the temporary pool and 69% in the other two sites (Mottram and Kettle, in preparation).

Container habitats are prolific sources of mosquitoes. All the Sabethini and Toxorhynchitinae breed in them and many Culicini but few Anophelinae. Adaptations to these specialized habitats involve morphological changes in larva and adult, and modifications of oviposition behaviour. Larval modifications include reduced head setae, enlarged maxillary spines and numerous large stellate hairs on the thorax and abdomen. *Armigeres dolichocephalus* has a long narrow thorax, which enables the adult female to enter small holes in bamboo for oviposition (Gillett, 1971). Ovipositing *Sabethes chloropterus* hover in front of vertical holes in bamboo and flick eggs through the hole, one at a time (Galindo, 1958).

Container habitats include tree holes, a source of *An plumbeus, Ae aegypti formosus* and *Ae africanus*; collections of water in bamboo internodes breed *Sa chloropterus*; leaf axils are a source of *Ae simpsoni*; pitchers of plants breed species of *Tripteroides* (*Rachisoura*); epiphytic bromeliads are a source of *Anopheles* (*Kerteszia*), including *An bellator*, fruits and husks breed *Ae albopictus* and *Eretmapodites* species; artificial containers, such as vases, water jars, tyres, etc., breed *Ae aegypti*; and snail shells, *Eretmapodites*.

Underground water habitats are mostly artefacts, such as storage tanks and wells. In the Mediterranean area *An claviger* breeds in wells and cisterns, and its counterpart in the Oriental region is *An stephensi*. Both species are anthropophilic and, living in close association with man, they are locally important vectors of malaria. Neither species is restricted to underground water and *An claviger* breeds in cool waters in other habitats and *An stephensi* occurs in irrigation ditches in Basra. *Cx p. pipiens* breeds in storage tanks of clean water and *Cx quinquefasciatus* in highly polluted water, such as that found in soakage pits, flooded latrines and the settling ponds of sewage works. Crab holes are natural underground water habitats and in the Neotropical region breed *Deinocerites cancer*.

Dispersal

In still air in the laboratory, adult *Ae aegypti* are randomly orientated and have an average ground speed of 17 cm s^{-1}. In moving air the substrate appears to move forwards and the mosquito responds by facing into the wind and increasing its flight speed, so that in a wind of 33 cm s^{-1} *Ae aegypti* still maintained a ground air speed of 16 cm s^{-1} indicating that its average air speed was 49 cm s^{-1}. *Ae aegypti* is able to maintain forward movement, but at a steadily reducing rate up to wind speeds of 150 cm s^{-1} ($= 5.45$ km h^{-1}) (Clements, 1963). Above this limit mosquitoes either settle or are carried downwind.

In the field species vary greatly in the extent to which they disperse from their breeding sites. This can be an important factor in mosquito control because it indicates the distance larval control measures need to be extended in order to protect a community. Assuming, unrealistically, that breeding sites are uniformly distributed, then doubling the distance over which larval control must be carried out, quadruples the area to be treated and the cost.

The flight range of a species can be considered in terms of probability. It is not a limit beyond which a species will not fly but an indication of the distance beyond which the species will be present only in insignificant numbers. Such an estimate is obviously subjective and it would be preferable to give the probability of an individual dispersing more than a fixed distance from the breeding site. This approach emphasizes the fact that the greater the population emerging from the site, the larger the number of mosquitoes that will reach a fixed distance, and the greater the maximum distance reached by some members of the population. Dispersal is influenced by the prevailing wind, by longevity of the species and by the presence of suitable hosts.

Mattingly (1969) comments that forest mosquitoes appear to have a more restricted flight range than those that breed in open situations. Where suitable domestic breeding sites are available *Ae aegypti* is unlikely to disperse in significant numbers more than 0.5 km. Sheppard *et al.* (1969) estimated that the average distance covered by *Ae aegypti* in 24 h was 37 m but Conway *et al.* (1974), working in a different continent, considered that this might be an overestimate and that their 1 ha site might have contained several, relatively distinct, small populations of *Ae aegypti*.

In the tropics larval control against *Anopheles* mosquitoes is usually effective if extended for 3 km. Many species occur abundantly 1 km from their breeding site but rarely reach 5 km, e.g. only 20% of a population of *An funestus* was considered to disperse further than 0.8 km from the breeding site with a practical limit of dispersal of 7 km (Horsfall, 1955); and few female *An gambiae* s.s. dispersed further than 3 km with the average being 1.0 to 1.6 km (White, 1974).

Anopheles pharoensis is a large mosquito which breeds abundantly in the Delta of Egypt and regularly disperses 6 km from its breeding site and, on occasions, disperses with the wind for distances of 100 km. In 1942 in the Western desert, remote from any water, there were two occasions when troops were attacked by large numbers of *An pharoensis* which, it is considered, had flown on a south-easterly wind, an unusual direction, from Wadi el Natrun and the Nile Delta and on another occasion of this species dispersing 280 km with the prevailing wind (Garrett-Jones, 1950, 1962).

Salt marsh mosquitoes seem to be particularly addicted to long-distance dispersal flights. The New World species *Ae taeniorhynchus* and *Ae sollicitans* emerge in vast numbers after their breeding sites have been inundated and disperse tens of kilometres. Hocking (1953) studied the intrinsic range of flight of four species of *Aedes* and found that three of them were capable of flying nearly 50 km in still air after one feed of nectar and the other species exceeded 20 km. Other aspects of behaviour will decide whether this potential is ever realized.

Female *Aedes taeniorhynchus* begin to migrate when they are 6–14 h old and their ovaries are immature (Haeger,1960). They may continue to disperse during the period of normal activity for four consecutive days (Provost, 1952) carrying large numbers of mosquitoes into urban areas to the dismay of the inhabitants. Female *Ae taeniorhynchus* have been recovered up to 40 km from their site of origin (Provost, 1957). At low wind velocities (0.25 m s^{-1}) they disperse upwind and at high velocities (1.5–2.0 m s^{-1}) downwind (Haeger, 1960). In eastern Australia *Ae vigilax* behaves in a comparable manner. In the Middle East *An sacharovi* has a prehibernation flight from Lake Huleh to Rosh Pinah, 14 km away (Kligler and Mer, 1930) and during this flight it feeds and lays down fat body for hibernation, requiring three or four blood meals to develop the fat body to its maximum (Mer, 1931).

In *Ae taeniorhynchus* mating occurs either at the breeding site or in the early stages of migration, because males rarely disperse more than 5 km. This illustrates a general feature of mosquito biology that males do not disperse as far as females. Indeed, if males are found, it is an indication that the breeding site is close at hand, often within 200 m.

Hibernation, Aestivation and Seasonal Cycles

In tropical regions breeding of most species continues throughout the year and seasonal fluctuations in numbers are related to the rainy and dry seasons. In the severe dry season, breeding either continues in relation to reduced bodies of permanent water or the population survives this hostile period as dormant eggs or aestivating adults. Aestivation is either a rare phenomenon or has been poorly studied. In the valley of the White Nile in Sudan, *An arabiensis* (*An gambiae auct.*) maintains itself through the dry season by low-level breeding, but in the more arid areas, 20 km from the Nile valley, there is evidence that the females aestivate (Omer and Cloudsley-Thompson, 1970; White, 1974). In the USA *Cs inornata* enters into winter diapause in the north and in the south it aestivates during the hot, dry season (Mitchell, 1988).

There is much more information available on the effect of low temperature on mosquito biology. In the subtropics and adjoining areas of the warmer temperate regions the main response to lower temperature in the cool season is to slow down the rate of biological processes. Ovarian development may take 10–14 days, compared with 2–3 days under optimum conditions and development from egg to adult takes 2–3 months. In the temperate high-veld region of South Africa *Cx pipiens* and *Cx theileri* overwinter by quiescence (Jupp, 1975), as does *Cx tarsalis* in southern California (Nelson, 1971) but in the cooler central California *Cx tarsalis* enters diapause (Bellamy and Reeves, 1963).

Hibernation is induced in a developing generation by exposure to decreasing hours of daylight, reinforced by lower temperatures. In species with an extensive N–S distribution, e.g. *An freeborni*, *Ae triseriatus* and *Wy smithii*, there is clinal variation in the length of the critical photoperiod to induce diapause. Diapause in the egg stage is found in all *Aedes*, *Psorophora* and *Haemogogus*, in which it may be a response to dryness as well as a photoperiod-induced diapause. Diapause in the egg stage occurs in *An walkeri*, *Ae albopictus* and *Cs morsitans*. Larval diapause in winter is a feature of *Culiseta* species, e.g. *Cs melaneura*, and is also found in *An barberi*, *Ae triseriatus* and *Wy smithii*. Larvae of *Wy smithii* are able to survive freezing but not for prolonged periods. Other species pass the winter as larvae in waters that are protected from freezing, e.g. *An plumbeus* and *Ae triseriatus* in tree holes, and *Cs melanura* in bogs. Diapausing larvae are as active as non-diapausing larvae but feed less (Mitchell, 1988).

The commonest way of surviving winter in species which do not produce dormant eggs is by overwintering as inseminated, nulliparous females. Two mechanisms have been recognized. In gonotrophic dissociation, e.g. *An atroparvus* and *An superpictus*, the female feeds at intervals during the winter but does not develop eggs. In gonotrophic concordancy the female feeds on plant juices, develops fat body and its ovaries remain undeveloped (Bowen, 1991). Washino (1977) considers that these definitions require clarification and regards blood-feeding during hibernation as being opportunistic. Dissociation is well recognized in the Californian population of *An freeborni* but not elsewhere in the western USA. Similarly, throughout much of its range in south-eastern USA, overwintering *An quadrimaculatus* take infrequent blood meals. The development of fat body and the survival rate over the winter was the same for female *An freeborni*, independent of whether they followed gonotrophic concordancy or gonotrophic dissociation (Washino, 1977). Diapausing *An freeborni* and *Cx tarsalis* develop fat body during the autumn and utilize the stored lipids for energy during winter. High humidity is necessary for most overwintering adult mosquitoes. Female *Cs alaskensis* have been found at the bases of dense stands of grass under snow, and in France several species have been recovered from dense vegetation in reed swamps. Some species, e.g. *An sacharovi*, *An freeborni* and *Cx tarsalis*, undertake a prehibernation flight from their breeding sites to their winter resting places or hibernacula (Mitchell, 1988). It is considered that diapause decreases in intensity with time and may end in mid-winter and the insect remain quiescent until favourable conditions return in the spring. Diapause is under hormonal control and that of *Ae triseriatus* larvae is broken by exposure to ecdysone and for female *An freeborni* by juvenile hormone (Washino, 1977; Mitchell, 1988).

In the humid tropics breeding may be continuous throughout the year with little variation in the size of the adult populations. In Bangkok Sheppard *et al.* (1969) found that differences in the monthly population of *Ae aegypti* were non-significant until movement was taken into account, when the differences were significant but only at the 5% level. It was not possible to attribute these fluctuations to either rainfall or temperature. Females survived better than males with a daily survival of 81–84% compared with 70–72% for the males. In the subtropics, with hot dry summers, mosquito populations are commonly bimodal with peaks in the spring and autumn and lower populations during the summer. Mer (1931) found that there was some decline in fecundity of *An sacharovi* in the late spring. In addition lower survival during the hot dry summer months would lead to fewer egg batches being laid and lower populations. In more temperate regions one or two generations may occur during the warmer months of each year.

References

Allan, S.A., Day, J.F. and Edman, J.D. (1987) Visual ecology of biting flies. *Annual Review of Entomology* 32, 297–316.

Aly, C. and Dadd, R.H. (1989) Drinking rate regulation in some fresh-water mosquito larvae. *Physiological Entomology* 14, 241–256.

Bates, M. (1949) *The Natural History of Mosquitoes.* Macmillan, New York.

Beach, R. (1979) Mosquitoes: biting behavior inhibited by ecdysone. *Science* 205, 829.

Belkin, J.N. and Hogue, C.L. (1959) A review of the crabhole mosquitoes of the genus *Deinocerites* (Diptera: Culicidae). *University of California Publications in Entomology* 14, 411–458.

Bellamy, R.E. and Reeves, W.C. (1963) The winter biology of *Culex tarsalis* (Diptera: Culicidae) in Kern County, California. *Annals of the Entomological Society of America* 56, 314–323.

Bidlingmayer, W.L. and Hem, D.G. (1980) The range of visual attraction and the effect of competitive visual attractants upon mosquito (Diptera: Culicidae) flight. *Bulletin of Entomological Research* 70, 321–342.

Borovsky, D., Carlson, D.A., Ujváry, I and Prestwich, G.D. (1994a) Biosynthesis of (10R)-juvenile hormone III from farnesoic acid by *Aedes aegypti* ovary. *Archives of Insect Biochemistry and Physiology* 27, 11–25.

Borovsky, D., Song, Q., Ma, M.C. and Carlson, D.A. (1994b) Biosynthesis, secretion and immunocytochemistry of trypsin modulating oostatic factor of *Aedes aegypti*. *Archives of Insect Biochemistry and Physiology* 27, 27–38.

Bowen, M.F. (1991) The sensory physiology of host-seeking behavior in mosquitoes. *Annual Review of Entomology* 36, 139–158.

Bradley, T.J. (1987) Physiology of osmoregulation in mosquitoes. *Annual Review of Entomology* 32, 439–462.

Brown, A.W.A. (1951) Studies of the responses of the female *Aedes* mosquito. Part IV. *Bulletin of Entomological Research* 42, 575–582.

Brown, M.D., Mottram, P., Fanning, I.D. and Kay, B.H. (1992) The peridomestic container-breeding mosquito fauna of Darnley Is. (Torres Strait) (Diptera: Culicidae), and the potential for its control by predacious *Mesocyclops* copepods. *Journal of the Australian Entomological Society* 31, 305–310.

Brown, M.D., Mottram, P., Santaguliana, G. and Kay, B.H. (1993) Natural control of peridomestic mosquitoes on Darnley Is., by an indigenous *Mesocyclops* (Crustacea, Copepoda) species. *Proceedings of the Sixth Symposium – Arbovirus Research in Australia, 7–11 December 1992, Brisbane*, pp. 105–110.

Bruno, D.W. and Laurence, B.R. (1979) The influence of the apical droplet of *Culex* egg rafts on oviposition of *Culex pipiens fatigans* (Diptera: Culicidae). *Journal of Medical Entomology* 16, 300–305.

Chadee, D.D and Corbet, P.S. (1990) A night-time role of the oviposition site of the mosquito, *Ae aegypti* (L.) (Diptera: Culicidae). *Annals of Tropical Medicine and Parasitology* 84, 429–433.

Charlwood, J.D. and Jones, M.D.R. (1979) Mating behaviour in the mosquito *Anopheles gambiae* s l. I. Close range and contact behaviour. *Physiological Entomology* 4, 111–120.

Charlwood, J.D. and Jones, M.D.R. (1980) Mating in the mosquito, *Anopheles gambiae* s.l. II. Swarming behaviour. *Physiological Entomology* 5, 315–320.

Clements, A.N. (1963) *The Physiology of Mosquitoes*. Pergamon Press, Oxford.

Clements, A.N. (1992) *The Biology of Mosquitoes*, vol. 1. Chapman & Hall, London.

Clements, A.N. and Paterson, G.D. (1981) The analysis of mortality and survival rates in wild populations of mosquitoes. *Journal of Applied Ecology* 18, 373–399.

Conway, G.R., Trpis, M. and McClelland, G.A.H. (1974) Population parameters of the mosquito *Aedes aegypti* (L.) estimated by mark–release–recapture in a suburban habitat in Tanzania. *Journal of Animal Ecology* 43, 289–304.

Corbet, P.S. and Smith, S.M. (1974) Diel periodicities of landing of nulliparous and parous *Aedes aegypti* (L.) at Dar es Salaam, Tanzania (Diptera: Culicidae). *Bulletin of Entomological Research* 64, 111–121.

Dahl, C., Widahl, L. and Nilsson, C. (1988) Functional analysis of the suspension feeding system in mosquitoes (Diptera: Culicidae). *Annals of the Entomological Society of America* 81, 105–127.

Davidson, G. (1988) Insecticides. Bulletin No. 1, Ross Institute of Tropical Hygiene, London.

Downes, J.A. (1969) The swarming and mating flight of Diptera. *Annual Review of Entomology* 14, 271–298.

Engber, B., Stone, P.F. and Pittai, J.S. (1978) The occurrence of *Toxorhynchites amboinensis* in Western Samoa. *Mosquito News* 38, 295–296.

Farquharson, C.O. (1918) *Harpagomyia* and other Diptera fed by *Cremastogaster* ants in southern Nigeria. *Proceedings of the Entomological Society of London* 29–39.

Gaffigan, T.V. and Ward, R.A. (1985) Index to the second supplement to 'A catalog of the mosquitoes of the world' with corrections and additions (Diptera: Culicidae). *Mosquito Systematics* 17, 52–63.

Galindo, P. (1958) Bionomics of *Sabethes chloropterus* Humboldt, a vector of sylvan yellow fever in middle America. *American Journal of Tropical Medicine and Hygiene* 7, 429–440.

Garrett, M. and Bradley, T.J. (1984a) The pattern of osmotic regulation in larvae of the mosquito *Culiseta inornata*. *Journal of Experimental Biology* 113, 133–141.

Garrett, M. and Bradley, T.J. (1984b) Ultrastructure of osmoregulatory organs in larvae of the brackish-water mosquito, *Culiseta inornata* (Williston). *Journal of Morphology* 182, 257–277.

Garrett, M. and Bradley, T.J. (1987) Extracellular accumulation of proline, serine and trehalose in the haemolymph of osmoconforming brackish-water mosquitoes. *Journal of Experimental Biology* 129, 211–218.

Garrett-Jones, C. (1950) A dispersion of mosquitoes by wind. *Nature, London* 165, 285.

Garrett-Jones, C. (1962) The possibility of active long-distance migrations by *Anopheles pharoensis* Theobald. *Bulletin of the World Health Organisation* 27, 299–302.

Garrett-Jones, C., Boreham P.F.L and Pant, C.P. (1980) Feeding habits of anophelines (Diptera: Culicidae) in 1971–78, with reference to the human blood index: a review. *Bulletin of Entomological Research* 70, 165–185.

Gillett, J.D. (1961) Laboratory observations on the life-history and ethology of *Mansonia* mosquitoes. *Bulletin of Entomological Research* 52, 23–30.

Gillett, J.D. (1971) *Mosquitoes*. Weidenfeld and Nicolson, London.

Gillett, J.D. (1979) Out for blood; flight orientation up-wind in the absence of visual clues *Mosquito News* 39, 222–229.

Gillett, J.D. (1993) The essential role of temporal discontinuity in swarming insects. *Biologist* 40, 53–57.

Gillett, J.D., Corbet, P.S. and Haddow, A.J. (1961) Observations on the oviposition cycle of *Aedes (Stegomyia) aegypti* (Linnaeus). VI. *Annals of Tropical Medicine and Parasitology* 55, 427–431.

Gillies, M.T. (1955) The pre-gravid phase of ovarian development in *Anopheles funestus*. *Annals of Tropical Medicine and Parasitology* 49, 320–325.

Gillies, M.T. (1980) The role of carbon dioxide in host-finding by mosquitoes (Diptera: Culicidae): a review. *Bulletin of Entomological Research* 70, 525–532.

Gillies, M.T. and Wilkes, T.J. (1970) The range of attraction of single baits for some West African mosquitoes. *Bulletin of Entomological Research* 60, 225–235.

Gillies, M.T. and Wilkes T.J. (1974) Evidence for downwind flight by host seeking mosquitoes. *Nature, London* 25, 388–389.

Gokool, S, Curtis, C.F. and Smith, D.F. (1993) Analysis of mosquito bloodmeals by DNA profiling. *Medical and Veterinary Entomology* 7, 208–215.

Haddow A.J. (1946) The mosquitoes of Bwamba County, Uganda. IV. Studies on the genus *Eretmapodites* Theobald. *Bulletin of Entomological Research* 37, 57–82.

Haddow, A.J. and Ssenkubuge, Y (1965) Entomological studies from a high steel tower in Zika Forest, Uganda. Part 1. The biting activity of mosquitoes and tabanids as shown by twenty-four-hour catches. *Transactions of the Royal Entomological Society of London* 117, 215–243.

Haeger, J.S. (1955) The non-blood feeding habits of *Aedes taeniorhynchus* (Diptera: Culicidae) on Sanibel Island. *Florida Mosquito News* 15, 21–26.

Haeger, J.S. (1960) Behaviour preceding migration in the salt-marsh mosquito, *Aedes taeniorhynchus* (Wiedemann). *Mosquito News* 20, 136–147.

Haeger, J.S. and Phinizee, J. (1959) The biology of the crab hole mosquito *Deinocerites cancer* Theobald. *Report of the Florida Antimosquito Association* 30, 34–37.

Hagedorn, H.H. (1985) The role of ecdysteroids in reproduction. In: Kerkut, G.A. and Gilbert, L.I. (eds) *Comparative Insect Physiology and Biochemistry and Pharmacology*. Pergamon Press, Oxford, pp. 205–262.

Harwood, R.F. and Horsfall, W.R. (1959) Development, structure, and function of coverings of eggs of floodwater mosquitoes. III. Functions of coverings. *Annals of the Entomological Society of America* 52, 113–116.

Healy, T.P. and Jepson, P.C. (1988) The location of floral nectar sources by mosquitoes: the long-range responses of *Anopheles arabiensis* Patton (Diptera: Culicidae) to *Achillea millefolium* flowers and isolated floral odour. *Bulletin of Entomological Research* 78, 651–657.

Hinton, H.E. (1968) Observations on the biology and taxonomy of the eggs of *Anopheles* mosquitoes. *Bulletin of Entomological Research* 57, 495–508.

Hinton, H.E. (1981) *Biology of Insect Eggs*, vol. 1. Pergamon Press, Oxford, pp. 43–47.

Ho Ch'i, Chou Tsu-chieh, Ch'en Teng'hung and Hsuch Ai-tseng (1962) The *Anopheles hyrcanus* group and its relation to malaria in east China. *Chinese Medical Journal* 81, 71–78.

Hocking, B. (1953) The intrinsic range and speed of flight in insects. *Transactions of the Royal Entomological Society of London* 104, 223–345.

Holmes, P.R. and Birley, M.H. (1987) An improved method for survival rate analysis from time series of haematophagous dipteran populations. *Journal of Animal Ecology* 56, 427–440.

Horsfall, W.R. (1955) *Mosquitoes: their Bionomics and Relation to Disease*. Constable, London.

Jepson, P.C. and Healy, T.P. (1988) The location of floral nectar sources by mosquitoes: an advanced bioassay for volatile plant odours and initial studies with *Aedes aegypti* (L.) (Diptera: Culicidae). *Bulletin of Entomological Research* 78, 641–650.

Jupp, P.G. (1975) Further studies on the overwintering stages of *Culex* mosquitoes (Diptera: Culicidae) in the highveld region of South Africa. *Journal of the Entomological Society of Southern Africa* 38, 89–97.

Keilin, D., Tate, P. and Vincent, M. (1935) The perispiracular glands of mosquito larvae. *Parasitology* 27, 257–262.

Kettle, D.S. (1948) The growth of *Anopheles sergenti* Theobald (Diptera: Culicidae) with special reference to the growth of the anal papillae in varying salinities. *Annals of Tropical Medicine and Parasitology* 42, 5–29.

Kirk, H.B. (1923) Notes on the mating-habits and early life-history of the culicid *Opifex fuscus* Hutton. *Transactions and Proceedings of the New Zealand Institute* 54, 400–406.

Kligler, J. and Mer, G. (1930) Studies on malaria: VI: long-range dispersion of *Anopheles* during the prehibernating period. *Rivista di Malariologia* 9, 363–374.

Kline, D.L., Takken, W., Wood, J.R. and Carlson, D.A. (1990) Field studies on the potential of butanone, carbon dioxide, honey extract, 1-octen-3-ol, L-lactic acid and phenols as attractants for mosquitoes. *Medical and Veterinary Entomology* 4, 383–391.

Knight, K.L. (1978) Supplement to a catalog of the mosquitoes of the world. *Thomas Say Foundation Supplement* to vol. 6, 1–107.

Knight, K.L. and Stone, A. (1977) A catalog of the mosquitoes of the world (Diptera: Culicidae). *Thomas Say Foundation* 6, 1–611.

Lacey, L.S. and Undeen, A.H. (1986) Microbial control of black flies and mosquitoes. *Annual Review of Entomology* 31, 265–296.

Laird, M. (1988) *The Natural History of Larval Mosquito Habitats*. Academic Press, London.

Laurence, B.R. (1960) The biology of two species of mosquito, *Mansonia africana* (Theobald) and *Mansonia uniformis* (Theobald) belonging to the subgenus *Mansonioides* (Diptera: Culicidae) *Bulletin of Entomological Research* 51, 491–517.

Laurence, B.R. and Pickett, J.A. (1985) An oviposition attractant pheromone in *Culex quinquefasciatus* Say (Diptera: Culicidae). *Bulletin of Entomological Research* 75, 283–290.

Le Sueur, D. and Sharp, B.L. (1991) Temperature-dependent variation in *Anopheles merus* larval head capsule width and adult wing length: implications for anopheline taxonomy. *Journal of Medical Entomology* 5, 55–62.

Linley, J.R., Kaiser, P.E. and Cockburn, A.F. (1993) A description and morphometric study of the eggs of species of the *Anopheles quadrimaculatus* complex (Diptera: Culicidae). *Mosquito Systematics* 25, 124–147.

Marks, E.N. (1958) Notes on *Opifex fuscus* Hutton (Diptera: Culicidae) and the scope for further research on it. *New Zealand Entomologist* 2, 20–25.

Mattingly, P.F. (1969) *The Biology of Mosquito-Borne Disease*. Allen & Unwin, London.

Mattingly, P.F. (1973) Culicidae (Mosquitoes). In: Smith K.G.V. (ed.) *Insects and Other*

Arthropods of Medical Importance. British Museum (Natural History), London, pp. 37–107.

Maxwell, C.A., Curtis, C.F., Haji, H., Kisumku, S., Thalib, A.T. and Yahya, S.A. (1990) Control of Bancroftian filariasis by integrating therapy with vector control using polystyrene beads in wet pit latrines. *Transactions of the Royal Society of Tropical Medicine and Hygiene* 84, 709–714.

McClelland, G.A.H. and Conway, G.R. (1971) Frequency of blood feeding in the mosquito *Aedes aegypti*. *Nature, London* 232, 485–486.

McMullen, A.I. and Hill, M.N. (1971) Anoxia in mosquito pupae under insoluble monolayers. *Nature, London* 234, 51–52.

McMullen, A.I., Reiter, P. and Phillips, M.C. (1977) Mode of action of insoluble monolayers on mosquito pupal respiration. *Nature, London* 267, 244–245.

Mer, G. (1931) Notes on the bionomics of *Anopheles elutus*, Edw. (Diptera: Culicidae). *Bulletin of Entomological Research* 22, 137–145.

Merritt, R.W., Dadd, R.H. and Walker, E.D. (1992a) Feeding behavior, natural food, and nutritional relationships of larval mosquitoes. *Annual Review of Entomology* 37, 349–376.

Merritt, R.W., Craig, D.A., Walker, E.D., Vanderploeg, H.A. and Wotton, R.S. (1992b) Interfacial feeding behavior and particle flow patterns of *Anopheles quadrimaculatus* (Diptera: Culicidae). *Journal of Insect Behavior* 5, 741–761.

Micks, D.W. and Rougeau, D. (1976) Entry and movement of petroleum derivatives in the tracheal system of mosquito larvae. *Mosquito News* 36, 449–454.

Mitchell, C.J. (1988) Occurrence, biology, and physiology of diapause in overwintering mosquitoes. In: Monath, T.P. (ed.) *The Arboviruses: Epidemiology and Ecology,* volume 1. CRC Press, Boca Raton, Florida USA, pp. 191–215.

Mottram, P. and Kettle, D.S. (in preparation) Development and survival of immature *Culex annulirostris* Skuse (Diptera: Culicidae) in three different breeding sites in south-east Queensland.

Muirhead-Thomson, R.C. (1951) *Mosquito Behaviour in Relation to Malaria Transmission and Control in the Tropics.* Edward Arnold. London.

Nelson, M.J. (1971) Mosquito studies (Diptera: Culicidae) XXVI. Winter biology of *Culex tarsalis* in Imperial Valley, California. *Contributions of the American Entomological Institute* 7(6), 1–56.

Nijhout, H.F. and Craig, G.B. (1971) Reproductive isolation in *Stegomyia* mosquitoes III. Evidence for a sexual pheromone. *Entomologia Experimentalis et Applicata* 14, 399–412.

Nowell, W.R. (1951) The dipterous family Dixidae in western North America (Insecta: Diptera). *Microentomology* 16, 187–270.

Omer, S.M. and Cloudsley-Thompson, J.L. (1970) Survival of female *Anopheles gambiae* Giles through a 9-month dry season in Sudan. *Bulletin of the World Health Organization* 42, 319–330.

Osgood, C.E. (1971) An oviposition pheromone associated with the egg rafts of *Culex tarsalis*. *Journal of Economic Entomology* 64, 1038–1041.

Otieno, W.A., Onyango, T.O., Pile, M.M., Laurence, B.R., Dawson, G.W., Wadhams, L.J. and Pickett, J.A. (1988) A field trial of the synthetic oviposition pheromone with *Cx quinquefasciatus* Say (Diptera: Culicidae) in Kenya. *Bulletin of Entomological Research* 78, 463–478.

Provost, M.W. (1952) The dispersal of *Aedes taeniorhynchus* I Preliminary studies. *Mosquito News* 12, 174–190.

Provost, M.W. (1957) The dispersal of *Aedes taeniorhynchus* II. The second experiment. *Mosquito News* 17, 233–247.

Puri, I.M. (1931) Larvae of anopheline mosquitoes, with full descriptions of those of the Indian species. *Indian Medical Research Memoirs* 21, 1–225.

Reinert, J.F. (1975) Mosquito generic and subgeneric abbreviations (Diptera: Culicidae). *Mosquito Systematics* 7, 105–110.

Rivière, F., Pichon, G., Duval, J., Thirel, R. and Toudic, A. (1979) Introduction de *Toxorhynchites* (*Toxorhynchites*) *amboinensis* (Dobschall, 1857) (Diptera: Culicidae) en Polynésie Française. *Cahiers ORSTOM Entomologie médicale et Parasitologie* 17, 225–234.

Rivière, F., Kay, B.H., Klein, J.M. and Séchan, Y. (1987) *Mesocyclops aspericornis* (Copepoda) and *Bacillus thuringiensis* var. *israelensis* for the biological control of *Aedes* and *Culex* vectors (Diptera: Culicidae) breeding in crab holes, tree holes, and artificial containers. *Journal of Medical Entomology* 24, 425–430.

Rodaniche, E. de, Galindo, P. and Johnson, C.M. (1959) Further studies on the experimental transmission of yellow fever by *Sabethes chloropterus*. *American Journal of Tropical Medicine and Hygiene* 8, 190–194.

Rowland, M. (1989) Changes in the circadian flight activity of the mosquito *Anopheles stephensi* associated with insemination, blood-feeding, oviposition and nocturnal light intensity. *Physiological Entomology* 14, 77–84.

Russell, P.F. and Rao, T.R. (1942a) On the swarming, mating and oviposition behaviour of *Anopheles culicifacies*. *American Journal of Tropical Medicine* 22, 417–427.

Russell, P.F. and Rao, T.R. (1942b) On relation of mechanical obstruction and shade to ovipositing of *Anopheles culicifacies*. *Journal of Experimental Zoology* 91, 303–329.

Schnell, D.J., Pfasnnenstiel, M.A. and Nickerson, K.W. (1984) Bioassay of solubilized *Bacillus thuringiensis* var. *israelensis* crystals by attachment to latex beads. *Science* 223, 1191–1193.

Service, M.W. (1993) Mosquitoes (Culicidae). In: Lane, R.P. and Crosskey, R.W. (eds) *Medical Insects and Arachnids*. Chapman & Hall, London, pp. 120–240.

Sheppard, P.M., Macdonald, W.W., Tonn, R.J. and Grab, B. (1969) The dynamics of an adult population of *Aedes aegypti* in relation to dengue haemorrhagic fever in Bangkok. *Journal of Animal Ecology* 38, 661–702.

Soper, F.L. and Wilson, D.B. (1943) *Anopheles gambiae in Brazil 1930 to 1940*. The Rockefeller Foundation, New York.

Steffan, W.A. and Evenhuis, N.L. (1981) Biology of *Toxorhynchites*. *Annual Review of Entomology* 26, 159–181.

Steel, C.G.H. and Davey, K.G. (1985) Integration in the insect endocrine system. In: Kerkut, G.A. and Gilbert, L.I. (eds) *Comparative Insect Physiology and Biochemistry and Pharmacology*. Pergamon Press, Oxford, pp. 1–35.

Strickman, D. (1989) Biosystematics of larval movement of Central American mosquitoes and its use for field identification. *Journal of the American Mosquito Control Association* 5, 208–218.

Suarez, M.F. (1993) *Mesocyclops aspericornis* as an alternative of *Aedes aegypti* control in Puerto Rico and Anguilla. *Proceedings of the Sixth Symposium – Arbovirus Research in Australia*, 7–11 December 1992, Brisbane, p. 115.

Tempelis, C.H. (1975) Host feeding patterns of mosquitoes, with a review of advances in analysis of blood meals by serology. *Journal of Medical Entomology* 11, 635–653.

Thorpe, W.H. (1930) The biology of the petroleum fly (*Psilopa petrolei* Coq). *Transactions of the Entomological Society of London* 78, 331–343.

Twinn, C.R. Hocking, B., McDuffie, W.C. and Cross, H.F. (1948) Preliminary account of the biting flies at Churchill, Manitoba. *Canadian Journal of Research D* 26, 334–357.

Walch, E.W. and Schuurman, C.J. (1929) Saltwater fishponds and malaria. *Mededelingen van den Dienst der volksgezondheid in Nederlandsch-Indie* 18, 341–366.

Ward, R.A. (1984) Second supplement to 'A catalog of the mosquitoes of the world' (Diptera: Culicidae). *Mosquito Systematics* 16, 227–270.

Ward, R.A. (1992) Third supplement to 'A catalog of the mosquitoes of the world' (Diptera: Culicidae). *Mosquito Systematics* 24, 177–230.

Washino, R.K. (1977) The physiological ecology of gonotrophic dissociation and related phenomena in mosquitoes. *Journal of Medical Entomology* 13, 381–388.

Washino, R.K. and Tempelis, C.H. (1983) Mosquito host bloodmeal identification: methodology and data analysis. *Annual Review of Entomology* 28, 179–201.

Wharton, R.H. (1953) The habits of adult mosquitoes in Malaya. IV. Swarming of anophelines in nature. *Annals of Tropical Medicine and Parasitology* 47, 285–290.

Wharton, R.H. (1962) The biology of *Mansonia* mosquitoes in relation to the transmission of filariasis in Malaya. *Institute for Medical Research, Federation of Malaya, Bulletin* 11, 1–114.

White, G.B. (1974) *Anopheles gambiae* complex and disease transmission in Africa. *Transactions of the Royal Society of Tropical Medicine and Hygiene* 68, 278–298.

Widahl, L.E. (1992) Flow patterns around suspension-feeding mosquito larvae (Diptera: Culicidae). *Annals of the Entomological Society of America* 85, 91–95.

Wishart, G. and Riordan, D.F. (1959) Flight responses to various sounds by adult males of *Aedes aegypti* (L.) (Diptera: Culicidae). *Canadian Entomologist* 91, 181–191.

Woke, P.A. (1937) Comparative effects of the blood of man and of canary on egg production of *Culex pipiens* Linn. *Journal of Parasitology* 23, 311–313.

Ceratopogonidae
(Biting Midges)

8

Ceratopogonids are popularly referred to as biting midges in Scotland and sandflies in the Caribbean and Australia. They are distinguished from the closely-related dancing midges or Chironomidae by the following characters. Ceratopogonids are generally smaller with blood-sucking species rarely having a wing length greater than 2 mm (many tropical species have wing lengths less than 1 mm); at rest they fold their wings scissor-like over the abdomen (Fig. 8.1); and their swarms are small and inconspicuous. Chironomids are larger; at rest they hold their wings rooflike over the abdomen; and often form large obvious swarms near water. Morphologically ceratopogonids have a forked media vein (M_1, M_2) (Fig. 8.2); piercing mouthparts; front pair of legs not lengthened; and postnotum gently rounded and without longitudinal groove. Chironomids have an unbranched media; reduced mouthparts; front pair of legs lengthened; and postnotum more prominent and usually with a median longitudinal groove (Edwards, 1926).

The Ceratopogonidae is a large family, which includes more than 60 genera and nearly 4000 species (Wirth *et al.*, 1974). The females are mostly predatory on other insects; ectoparasitic on insects or blood-sucking on vertebrates. Both males and females feed on nectar, and females of several species of *Forcipomyia* are pollinators of the cocoa tree (*Theobroma cocao*) (Winder, 1978). Most females require a protein meal for maturation of the ovaries and female *Atrichopogon* obtain theirs by feeding on pollen. In only four genera is the protein meal obtained by feeding on warm-blooded animals and therefore relevant to medical entomology. The largest of these is the genus *Culicoides*, of which nearly 1000 species have been described, and the smallest genus is *Austroconops* with only one species. About 80 species are included in the genus *Leptoconops*, of which several subgenera are recognized – *Holoconops, Styloconops* and *Leptoconops*. In the genus *Forcipomyia*, blood-sucking is restricted to the subgenus *Lasiohelea* which is represented by about 50 species.

The Ceratopogonidae is divided into four subfamilies: Leptoconopinae, Forcipomyiinae, Dasyheleinae and Ceratopogoninae. The Leptoconopinae includes only one genus, *Leptoconops*, which is distinguished by possession of milky-white

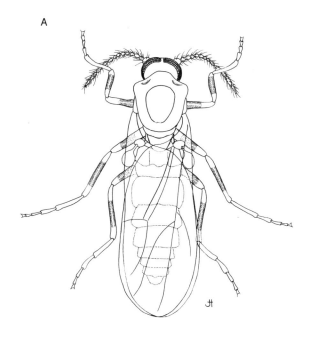

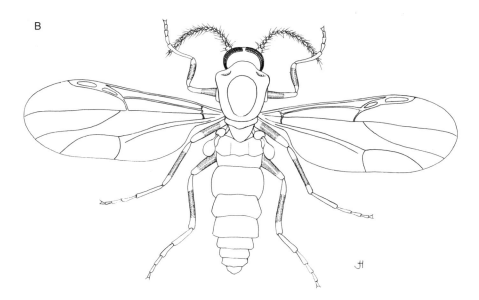

Fig. 8.1. Female *Culicoides* at rest with wings folded scissor-like (**A**) and with wings spread laterally (**B**) (x 55).

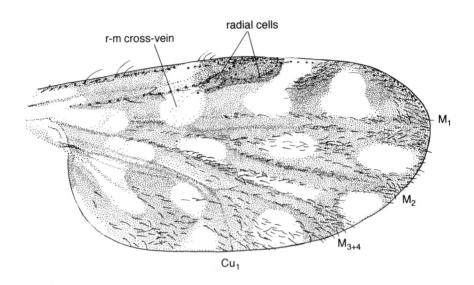

Fig. 8.2. Wing of female *Culicoides marmoratus* (×88).

wings, which contrast sharply with the black head and thorax; there are no macrotrichia (obvious hairs) on the wings; no r-m cross-vein (Fig. 8.3); and in the female the antenna has 12–14 segments. In other ceratopogonids the r-m cross-vein is present and the antennae are composed of 15 segments. The Forcipomyiinae contains the subgenus *Forcipomyia* (*Lasiohelea*) which has densely hairy wings, a well-developed empodium on the last tarsal segment, and a long second radial cell (Fig. 8.3). No member of the Dasyheleinae is of medical or veterinary importance. They have hairy wings, short second radial cell, vestigial empodium, and sculptured antennal segments. There is considerable variation within the Ceratopogoninae, which contains the important genus *Culicoides*, but they have vestigial empodia and their antennal segments are not sculptured. The antennae possess hair-like sensilla which are considered to be olfactory and mechano-receptors, and pit-like sensilla which are probably olfactory or thermo-receptors. The bushy appearance of the antennae of male *Culicoides* is due to abundant setaceous mechano-receptors (sensilla chaetica) which are almost certainly used to locate females by sound (Blackwell *et al.*, 1992a).

The wings of most species of *Culicoides* are patterned dark and light (Fig. 8.2). The pattern is due to pigmentation in the wing membrane and therefore cannot be rubbed off, as can the coloured scales, which form the wing pattern in the Culicidae. However, the pigment fades in specimens stored in alcohol and exposed to light. *Culicoides* have a petiolate media, i.e. vein M forks distally of the r-m cross-vein; the costa extends more than half way and less than two-thirds along the front margin of the wing; the radius forms two, small, more-or-less equal cells (the radial cells), one or both of which may be obliterated; distinct humeral pits are located anterolaterally on the scutum; and the claws of both sexes are small, equal and simple (Fig. 8.4).

The genus *Culicoides* is widely distributed in the world from the tropics to

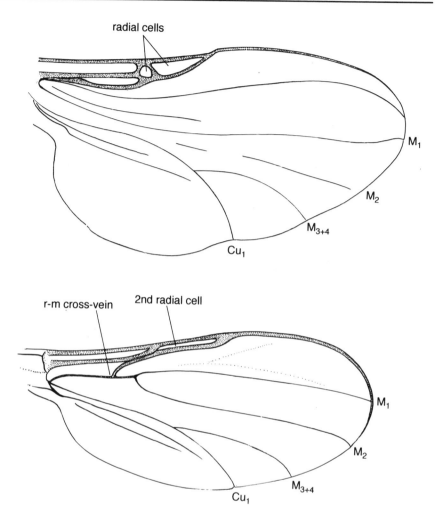

Fig. 8.3. Top, Wing of female *Leptoconops (Styloconops) australiensis* (x 88). Bottom, wing of *Forcipomyia (Lasiohelea) townsvillensis* (x 105).

the tundra, and from sea level to 4200 m in Tibet, where members of the 1921 Everest expedition complained of being bitten persistently (Howard-Bury, 1922). *Austroconops* has only been recorded from near Perth in Western Australia. *Lasiohelea* is associated with tropical and subtropical rainforests, although *F. (L.) sibirica* is a pest at Krasnoyarsk in Siberia (Gornostaeva, 1967). *Leptoconops* species are largely restricted to the warmer areas of the world with the subgenera *Holoconops* and *Styloconops* being associated with sandy coasts, e.g. *L. (H.) becquaerti* in the Caribbean (Linley and Davies, 1971). *L. (S.) spinosifrons* in the Indian Ocean (Laurence and Mathias, 1972; Duval *et al.*, 1974); while the subgenus *Leptoconops* is more prevalent in inland locations, e.g. *L. (L.) torrens* in Utah, USA (Smith and Lowe, 1948).

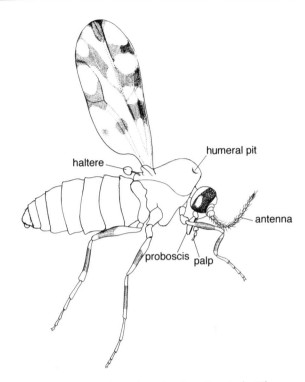

Fig. 8.4. Lateral view of female *Culicoides brevitarsis* (× 48).

Life Cycle

The life cycle of *Culicoides* will be considered first and then differences shown by *Leptoconops* and *Lasiohelea* will be indicated. Eggs of *Culicoides* are laid in batches, which vary from 30–40 in *C. brevitarsis* (Campbell and Kettle, 1975) and up to 450 with a mean of 250 in *C. circumscriptus* (Becker, 1961), The eggs are small, dark in colour, and slender, measuring 350–500 μm in length and 65–80 μm in breadth (Hill, 1947). They are covered with small projections (ansulae), which are particularly evident on the concave side and probably function as a plastron by retaining a film of air in contact with the egg, facilitating diffusion of oxygen for respiration when the egg is covered by water. In most species the eggs hatch in a few days at favourable temperatures, but those of the northern species, *C. grisescens*, do not hatch for 7–8 months in the laboratory and this species probably overwinters in the egg stage (Parker, 1950). Another Palaearctic species, *C. vexans*, breeds in temporary, open pools and has a single generation in the spring. Its eggs lie dormant over the summer and hatch in the autumn, when the breeding site is unlikely to dry up (Jobling, 1953). This behaviour is comparable to that of *Aedes* eggs (pp. 116–117).

The larva, which emerges from the egg, is a typical nematoceran larva, with a well-sclerotized head, 11 body segments, and no appendages (Fig. 8.5). The three thoracic segments are similar to the eight abdominal segments, being not fused

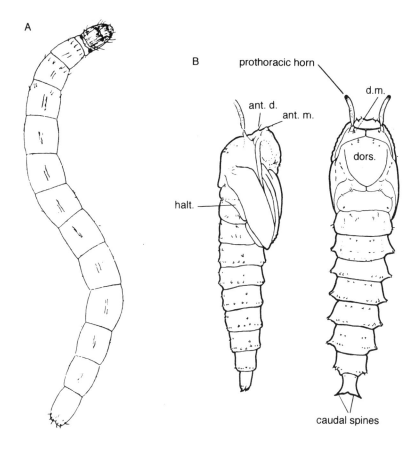

Fig. 8.5. Left, larva of *Culicoides impunctatus* (length 5.0 mm). Right, lateral and dorsal views of pupa of *Culicoides nubeculosus*. *ant.d., ant. m., d.m.,* and *dors.* indicate various setae on the pupa; *halt.,* haltere. Sources: larva from Hill (1947), and pupa from Lawson (1951).

together and no broader than the abdominal segments. In late fourth instar larvae the abdomen becomes full of an opaque white fat body, which does not extend into the thorax, where the imaginal wing and limb buds are developing. There are paired tracheae but the spiracles are closed, and respiration is cutaneous. Two pairs of narrow, bifid anal papillae can be extruded from the anus or retracted into the rectum (Fig. 8.6C). Presumably, as in the Culicidae, they are concerned with the absorption of salts from the surrounding medium.

The larva swims by rapid sinuous flexions of the body, reaching speeds of 16 mm s^{-1} in water at a temperature of 20–28°C. When a larva encounters a more viscous medium, e.g. mud or sand, the movement changes from rapid to slow sinuous flexion and the rate of movement through the medium is slower than in water (Linley, 1986). Unlike culicid larvae with their highly developed chaetotaxy, *Culicoides* larvae have only scanty, inconspicuous, unbranched setae. A few species, notably those that breed in tree holes, e.g. the Australian *C. angularis* (Kettle and

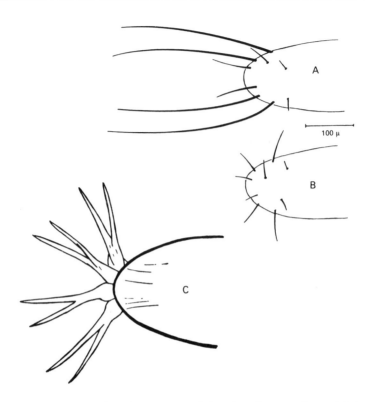

Fig. 8.6. Terminations of anal segments. (**A**) *Culicoides angularis*. (**B**) *C. marmoratus*. (**C**) Extruded bifid anal papillae of *C. nubeculosus*. Sources: (**A**) and (**B**) from Kettle and Elson (1976); (**C**) from Lawson (1951).

Elson, 1976) and the Neotropical *C. hoffmani* (Linley and Kettle, 1964) have four pairs of long perianal setae (Fig. 8.6A). Tree hole species probably spend more time free in water than do the larvae of other species and the long perianal setae may serve to amplify the body oscillations and increase the larva's speed of movement, enabling it to catch prey or to avoid predators.

The most prominent, internal structures in the head are the epipharynges, of which two main forms, heavy or light, occur (Figs 8.7, 8.8). In heavy epipharynges the lateral arms, body and combs are strongly sclerotized. Together with the hypopharynx, they act as a crushing structure, functioning like a pestle and mortar moving in one plane. Heavy epipharynges have, so far, only been reported in a small number of species belonging to the subgenus *Monoculicoides*, including the Palaearctic *C. nubeculosus* (Lawson, 1951) the closely-related Nearctic *C. variipennis* (Jamnback, 1965) and the Afrotropical *C. cornutus* (Khamala, 1975). Larvae of these species occur in muddy substrates and Megahed (1956) has described how larvae of *C. nubeculosus* browse upon the surface bacterial film, and on algal and fungal growth. *C. variipennis* has been mass colonized for many years with the larvae being fed on microorganisms (Jones *et al.*, 1969).

In light epipharynges the arms and medium body are only moderately sclerotized, and the combs, of which there are usually two to four, are finely and

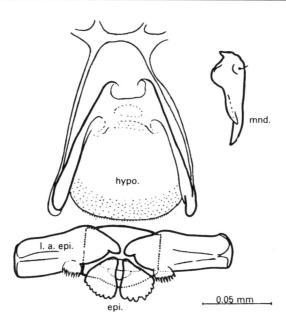

Fig. 8.7. Hypopharynx (*hypo.*); Mandible (*mnd.*); and 'heavy' epipharynges (*epi.*) of larva of *Culicoides nubeculosus*. Lateral arms of epipharynx (*l.a.epi.*). Source: from Lawson (1951).

delicately toothed (Fig. 8.8). In spite of the fact that the mandibles are not opposable, larvae with light pharynges are predators. Several species, e.g. *C. austropalpalis* (Kettle *et al.*, 1975), *C. furens* (Linley, 1966), have been reared from egg to adult on free-living nematodes. In the laboratory the activity of *C. furens* larvae increased in darkness and well-fed larvae were more active than starved or underfed larvae (Aussel and Linley, 1993). *C. furens* larvae are negatively phototactic, generalist feeders, feeding below the surface on nematodes during daylight hours and feeding at the surface on algae during the night (Aussel and Linley, 1994).

There are four larval instars, and the fourth ecdysis gives rise to the pupa, which is culicid in appearance (Fig. 8.5). The head and thorax are fused and bear a pair of moderately long, tubular prothoracic horns for respiration. These open to the atmosphere through a number of terminal and a smaller number of lateral openings, i.e. the pupa is propneustic with open spiracles. The segmented abdomen ends in a pair of caudal spines by which the pupa moves over the substrate. The pupa is a short-lived, non-feeding stage which gives rise to the winged adult. Before emergence the pupa, which is usually buried in the substrate with only the prothoracic horns reaching the surface, moves upwards to facilitate eclosion.

The development of *Culicoides* is generally much slower than in the Culicidae under comparable conditions. The egg stage and pupa are, with few exceptions, of short duration. In *C. variipennis* these stages each take 12% of the development time and the intervening larval stage 76%. Development is temperature dependent,

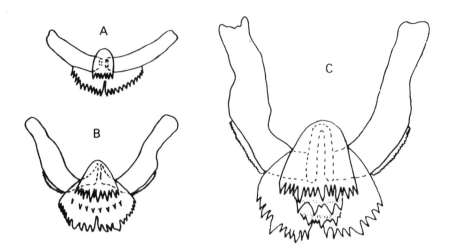

Fig. 8.8. 'Light' epipharynges of some Australian *Culicoides* larvae. (**A**) *C. brevitarsis*; (**B**) *C. subimmaculatus*; (**C**) *C. marmoratus*. Source: from Kettle and Elson (1976).

ranging from 7 weeks at 17°C to 2 weeks at 30°C in *C. variipennis* (Mullens and Rutz, 1983), and somewhat longer in *C. nubeculosus*, averaging 6 weeks at 25°C (Fahrner and Barthelmess, 1988). In the field the life cycle may be as short as 2 to 3 weeks in dung-breeding *C. brevitarsis* (Campbell and Kettle, 1976), 2 to 3 months in the subtropical, salt marsh *C. subimmaculatus* (Edwards, 1977), a year in the temperate region, bogland species, *C. impunctatus*, and nearly 2 years in some Arctic species (Downes, 1962). In temperate regions species are commonly univoltine with one generation a year, and even in the tropics there may be only three or four generations in a year. There is no record of hibernation or aestivation in adult *Culicoides*.

Other ceratopogonid larvae

Two forms of larvae occur in the Ceratopogonidae, the vermiform, apodous, prognathous larva described above for *Culicoides* and the crawling larva of the Forcipomyiinae. The latter has anterior and posterior pseudopods or prolegs (Fig. 8.9), comparable to those found in larvae of the Chironomidae; hypognathous head with its long axis at right angles to that of the body and the mouth directed ventrally; and prominent pectinate setae on the head and body (Saunders, 1964). Larvae of *Lasiohelea* occur in drier habitats than the vermiform larvae of other ceratopogonids, and have been found resting on liverworts and mosses (Lien, 1989). In the laboratory *F. (L.) taiwana* and *F. (L.) anabaenae* have been reared on pure cultures of the blue-green alga, *Anabaena* HA101. Larval development on *Anabaena* is unusually quick, being completed in 6 to 8 days at 25°C (Lien *et al.*, 1988). In the Dasyheleinae, the anterior pseudopod has been lost and the posterior

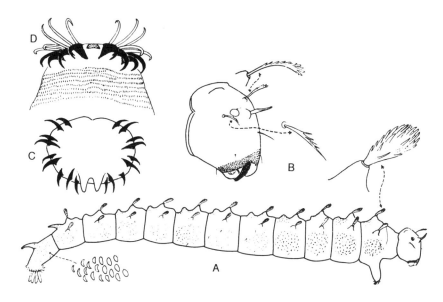

Fig. 8.9. Larva of *Forcipomyia (Lasiohelea) cornuta*. (**A**) lateral view; (**B**) head; (**C**) posterior pseudopod; (**D**) prothoracic pseudopod. Source: from Saunders (1964).

pseudopod has been reduced to a circlet of hooks, which can be extruded from and withdrawn into the rectum. Remnants of an anterior pseudopod are found in first instar larvae of some, but not all, *Culicoides*, e.g. *C. nubeculosus* (Lawson, 1951), *C. vexans* (Jobling, 1953), and vestiges of a posterior proleg have been found in first instar larvae of *C. cordiger*.

Larvae of *Leptoconops* appear to have more than the expected number of body segments due to the inclusion of intercalary segments, giving the appearance of 21 or 23 body segments and the larval head (Fig. 8.11) has only localized sclerotized thickenings and a system of internal, sclerotized rods which extend into the prothorax (Smith and Lowe, 1948). This may be an adaptation to life in a sandy habitat because sclerotization of the head capsule is much reduced in the sand-dwelling larvae of the Nearctic *C. melleus*, and the Australian *C. molestus* and *C. subimmaculatus* (Kettle and Elson, 1976). The pupae of *Lasiohelea* (Fig. 8.10) and *Leptoconops* (Fig. 8.11) are very similar to that of *Culicoides* (Fig. 8.5), but with a marked tendency for their prothoracic horns to be shorter.

Bionomics

In their bionomics blood-sucking midges have many features in common with mosquitoes. The males commonly emerge before the females but there is no permanent rotation of the terminalia, although during mating they are inverted through 180° (Downes, 1978). Male *C. melleus* are competent to mate within minutes of emergence, with male potency reaching a peak 4–8 h after emergence and mate

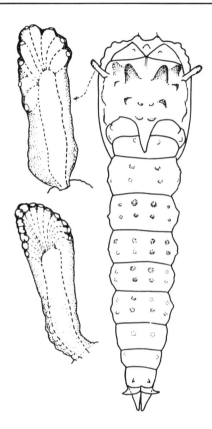

Fig. 8.10. Right, dorsal view of pupa of *Forcipomyia (Lasiohelea) cornuta*. Left, two variations on its prothoracic horn. Source: from Saunders (1964).

without swarming (Linley and Adams, 1972). In *C. nubeculosus*, mating can occur with or without swarming (Downes, 1955) but in most species, e.g. *C. brevitarsis* (Campbell and Kettle, 1979) and *C. impunctatus* (Blackwell *et al.*, 1992b), swarming occurs around sunset, forming columnar or ovoid groups of males over well-defined markers. Swarm size varied from 10 to 1000 individuals but was more commonly 50 in *C. brevitarsis* and 200 in *C. impunctatus*.

Autogeny has been shown to occur commonly among *Culicoides* species, including anthropophilic species such as *C. furens*, *C. subimmaculatus* (Linley *et al.*, 1970; Edwards, 1977) but all require a blood meal to mature second and subsequent egg batches and many species, e.g. *C. brevitarsis* (Campbell and Kettle, 1975) and *C. algecirensis* (= *C. puncticollis*, auct.) (Glukhova and Dubrovskaya, 1972), require a blood meal to mature the first egg batch. In *L. becquaerti* there exist two forms, a short-winged, autogenous form, and a long-winged, anautogenous form (Linley, 1968). The latter is responsible for most of the nuisance caused by this species to humans.

Species feed on a range of hosts: some, e.g. *C. furens*, *L. albiventris* (Aussel, 1993a) and *F.* (*L.*) *taiwana* (Lien, 1989) are anthropophilic; others, such as

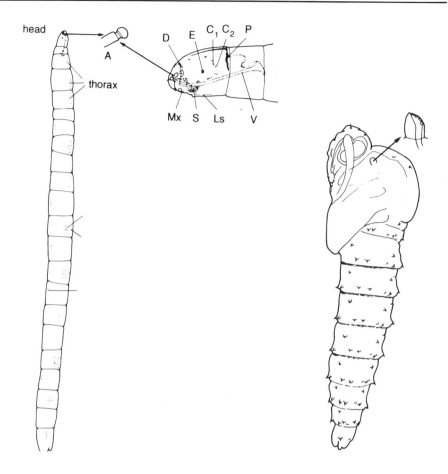

Fig. 8.11. Left, lateral view of larva of *Leptoconops spinosifrons*. Centre, lateral view of larval head. Right, lateral view of pupa of *L. spinosifrons*. A, antenna; C_1, C_2, epipharynges; D, S, sclerites; E, eye; Ls, labial sclerite; Mx, maxilla; P, pigment; V, ventral rod. Source: from Laurence and Mathias (1972).

C. brevitarsis, feed mainly on cattle; and yet others are ornithophilic, including the chicken-biting species *C. arakawae* and *C. odibilis* (Kitaoka and Morii, 1964). *L. specialis* feeds on lizards (Auezova *et al.*, 1990) and *F.* (*L.*) *phototropia* on Amphibia (Lien, 1989). *C. anophelis* is said to feed on mosquitoes, but this may be a case of phoresy (Das Gupta, 1964). *C. furens* is attracted to traps emitting CO_2 or the mammalian emanation, octenol. In combination the two compounds have a synergistsic effect greatly increasing catches (Kline *et al.*, 1990).

Blood-sucking midges are exophilic and exophagic, although in areas of high density biting midges will enter houses but in much smaller numbers than are present outside. *Culicoides* adults mainly rest among the herbage. *L. lucidus* has been found in cracks on the trunks of trees and in the upper layer of sand, and *L. mediterraneus* on the roots of *Limonium caspium*, a desert plant (Auezova, 1990).

Feeding preferences may be so strongly marked as to make subjective impressions of non-anthropophilic species very misleading. At Onderstepoort in South Africa human beings were quite untroubled by biting midges, but light traps revealed very large populations of *C. imicola* (= *C. pallidipennis* auct.), which were feeding on horses kept on the pasture at night. Such was the exophagy of this species that the risk of African horse sickness (see Chapter 24) could be greatly reduced, simply by stabling horses at night.

The biting cycle follows a circadian rhythm. Species of *Leptoconops* and *Lasiohelea* are diurnal, as are a small number of *Culicoides*, e.g. *C. nubeculosus* and *C. heliophilus*, but most *Culicoides* are crepuscular and/or nocturnal. Diurnal species commonly show two peaks of activity, in the morning and in the afternoon. The morning peak occurs 2 to 3 h after sunrise and is commonly the larger of the two peaks, e.g. *L. becquaerti* and *C. phlebotomus* (Nathan, 1981b). The afternoon peak occurs close to sunset, and in the case of *F. (L.) sibirica* may occur as late as 2100 h in the long summer days of high latitudes (Gornostaeva, 1967). The diurnal biting cycle of *F. (L.) taiwana* is unimodal with a peak between 1300 and 1500 h (Chen *et al.* 1981).

In Jamaica the pest species *C. furens* and *C. barbosai* showed peaks of activity at sunrise and sunset, presumably being initiated by rapidly changing light intensity. Between these peaks biting continued at a reduced rate throughout the night with a smaller peak around midnight (Kettle, 1969). Both species continued to bite during the morning until such activity was terminated by adverse meteorological conditions, especially increasing wind speed and temperature. Environmental conditions may modify basic activity patterns and Reubens (1963) caught *C. impunctatus*, mainly a crepuscular and nocturnal species, throughout the 24 h. Even at times when females were not actively seeking a host, the arrival of a host induced activity in nearby resting, hungry females.

Female *C. variipennis* become active at sunset reaching a peak just after sunset and have minor peaks in the middle of the night and at sunrise. This pattern is not followed by all females. Nulliparous females predominate in catches between sunrise and sunset, which is probably the time when mating occurs. The catch of gravid females reached a maximum at sunset, which may be the peak time for oviposition, and parous females dominated catches after the end of civil twilight. Parous females were more active at new moon and gravid females at full moon (Akey and Barnard, 1983).

When *C. arakawae* feeds on poultry the size of its blood meal has been estimated at 0.36 mg (Fujisaki *et al.*, 1987) about 12% of that taken by *Cx pipiens* when feeding on a bird. Female *C. v. variipennis* are anautogenous and develop the first batch of eggs to the resting stage in less than 2 days at 22°C. Following a blood meal, egg development is completed in a further 3 days (Mullens and Schmidtmann, 1982). Oogenesis is temperature-dependent and in *C. v. sonorensis* takes 2 to 10 days at 30°C and 13°C respectively (Mullens and Holbrook, 1991). Fecundity of female *C. variipennis* is positively correlated with size, which in turn is correlated with temperature. Females emerging in late summer (September) have a wing length of 1.48 mm, markedly shorter than the 2.02 mm wing length of females emerging in late winter (February) (Mullens, 1987).

Oviposition has not been observed in the field and only the eggs of *L.*

spinosifrons have been found in nature, where they were mostly found at a depth of 30–60 mm in the sand of the breeding site (Duval *et al.*, 1974). Female *Leptoconops* burrow into the substrate in the laboratory and presumably do so in the field to rest and oviposit. Gravid female *C. nubeculosus* oviposit more readily when in groups, than when held singly in tubes, and this may indicate the release of a pheromone at this time. Ismail and Kremer (1980) have described the secretion of a pheromone by virgin females, when they are unfed and again when they are gravid. This pheromone attracts males and stimulates mating, but its effect on female behaviour has not been studied (Kremer *et al.*, 1979). Linley and Carlson (1978) have found a contact mating pheromone in *C. melleus*.

Breeding sites

The breeding sites of blood-sucking ceratopogonids are commonly in wet soil in the ecotone between aquatic and terrestrial habitats, or in moist decaying vegetable material. *Culicoides* larvae burrow into the surface of the substrate and only rarely swim freely in the overlying water. Open, muddy sites, often contaminated with animal excreta, are the breeding grounds of *C. nubeculosus* and *C. variipennis*. The latter is an economically important species in the USA where it is a vector of bluetongue virus. It breeds in shallow water along the muddy shore lines of dairy waste water ponds of low salinity and accessible to cattle which pollute the water, leading to an increase in plankton. Factors that restrict breeding are deep water, steep banks, low pollution, and less plankton (Mullens, 1989). Late instar *C. variipennis* are to be found below, and eggs and early instars above the shore line (Mullens and Lip, 1987). When the water level is suddenly dropped larvae are stranded on the drying surface and all but pupae and late fourth instar larvae are killed. When the water line is suddenly raised the immature stages relocate at the new shore line (Mullens and Rodriguez, 1989).

Several coastal species of *Culicoides* breed in sand, e.g. *C. melleus*, *C. hollensis* and *C. molestus*. As the mud content in the substrate increases there is a range of habitats, culminating in mangrove swamps in the tropics and mud flats in the temperate regions. In the more sandy habitats of coastal eastern Australia *C. subimmaculatus* breeds in association with a surface-tunnelling soldier crab, *Mictyris livingstonei* (Marks and Reye, 1982). In the same area, larvae of *C. longior* occurred in peaty muds under mangroves where they were associated with larvae of *C. marmoratus*, and on their own in fluid muds subject to heavy water movement. *C. marmoratus* larvae were also found in algal mats on the soil surface (Hagan and Kettle, 1990). In the Caribbean *C. furens* and *C. barbosai* breed in association with mangroves but only *C. barbosai* is dependent on them and disappears when the mangroves are felled. *C. furens* continues to breed in high density in the absence of mangroves (Davies, 1969). The point is worth making that although there is an association between mangroves and many tropical anthropophilic *Culicoides*, the breeding sites of the midges form only a minor part of the mangrove forest. Felling mangroves to control *Culicoides* is economically wasteful and ecologically damaging. The use of selective measures against the restricted larval habitat is more efficient and less costly.

In Europe salt mud flats breed *C. halophilus* and *C. circumscriptus*, and in vegetated salt marshes, *C. maritimus* (Kettle and Lawson, 1952). There is an even greater range of freshwater breeding sites. Open, muddy sites, often contaminated with animal excreta, are the breeding grounds of the subgenus *Monoculicoides*, and temporary pools in pasture in Europe produce *C. vexans*. Vegetated swamps where the water table is above the soil surface, are the breeding grounds of *C. pulicaris* and *C. odibilis* in Europe (Kettle and Lawson, 1952) while in Japan the latter species occurs in the same type of habitat and also in rice fields, along with *C. arakawae*, which appears to be restricted to that habitat (Kitaoka and Morii, 1964). Marshland areas, where the water in winter is below the soil level, breed *C. pallidicornis* in Europe and *C. marksi* in Australia. Edges of lakes breed *C. austropalpalis* in Australia and *C. fascipennis*, *C. achrayi* and *C. duddingstoni* in Europe.

In Australia larvae of *C. bundyensis* and *C. bunroensis* occur in sandy creek beds. In Canada *C. denningi* breeds in the Saskatchewan River, and hibernates as a larva under ice in winter (Fredeen, 1969). In oligotrophic peaty areas, characterized by the mosses *Sphagnum* and *Polytrichum*, the *Culicoides* fauna is endemic and includes in Europe *C. impunctatus*, *C. truncorum* and *C. albicans*, and in North America *C. sphagnumensis*. The faunas of salt water, eutrophic fresh water, and oligotrophic bogland habitats are largely exclusive. This was demonstrated in north-west Scotland, where an outcrop of limestone supported a typical eutrophic flora and *Culicoides* fauna, sharply differentiated from the surrounding flora and fauna, in spite of it being isolated from other eutrophic areas, by a vast sea of acid bogland.

The immature stages of many species of *Culicoides* are found only in small, specialized habitats, usually of vegetable origin. These include tree holes, producing *C. fagineus*, *C. angularis*, *C. hoffmani* and *C. guttipennis*. A comparable habitat to tree holes is water collecting in dugout canoes, from which in West Africa, Carter *et al.* (1921) bred two species of *Culicoides*. Species of the subgenus *Avaritia*, such as *C. brevitarsis*, *C. imicola*, *C. dewulfi* and *C. chiopterus*, breed in dung, especially that of cattle. The New World *C. copiosus* group of species breed in rotting cacti, and *C. loughnani* has been introduced into Australia with the moth *Cactoblastis cactorum*, when it was introduced to control prickly pear (Dyce, 1969a).

In West Africa *Culicoides* have been reared from the rotting stems of bananas; in Trinidad Williams (1964) has recorded ten species, including the anthropophilic *C. paraensis*, breeding in decaying cocoa pods; and Buxton (1960) bred *C. scoticus* from large fungi. *C. heliconiae* breeds in the axils of the epiphytic bromeliad, *Heliconia* (Fox and Hoffman, 1944). Woodland leaf litter, which in other countries has proved an unrewarding habitat to examine for *Culicoides*, in the eastern USA breeds the anthropophilic *C. sanguisuga* (Jamnback, 1965).

From the foregoing, it is clear that species of *Culicoides* are able to exploit a wide range of moist habitats, but individual species utilize only a very limited range of breeding sites. This is of great practical importance because, where it is necessary to carry out larval control, measures can be restricted to the breeding sites of the target species, minimizing costs and environmental damage.

The same necessity to know the range of species in a locality before attempting control by habitat modification, applies to ceratopogonids as it did to culicids. In

Jamaica an enterprising hotelier cut down the mangroves in the adjoining swamp, burnt the trash, and filled the swamp with sand dredged from offshore. It had the desired effect of eliminating *C. furens*, the original pest species, but the sand-filled swamp created ideal conditions for the breeding of *L. becquaerti* which had previously been a rarity in the area. *L. becquaerti* was a far greater threat to the hotel's trade because its biting activity coincided with the periods of maximum tourist relaxation on the beach. The new situation was worse than the original. Effective control required the swamp to be filled to a depth greater than that to which subsoil water can ascend by capillarity (750–1100 mm) and/or covering the surface with marl and establishing and binding the surface with grass (Linley and Davies, 1971).

In French Polynesia larvae of *L. albiventris* occur in a narrow strip a few metres wide above the high tide level and near creeping vegetation (*Ipomoea pescaprae*). They are found down to a depth of 24 cm although most (78%) are found in the top 6 cm. The breeding sites are composed of fine well-sorted sand, particle size 180–190 μm, with low conductivity and humidity (Aussel, 1993b,c). Larvae of *Holoconops* and *Styloconops* occur in coastal areas in almost pure sand at the high spring tide level or even above it in sites inundated by exceptionally high tides (Smith and Lowe, 1948; Laurence and Mathias 1972; Duval *et al.*, 1974). Larvae of the subgenus *Leptoconops* occur inland in clay–silt soils where they may occur at considerable depths, e.g. *L. torrens* in the Sacramento Valley in California. When the soil cracks during the dry season the larvae of *L. torrens* pupate, and the adults emerge to feed and oviposit deep within the soil (Smith and Lowe, 1948). This has led to control by treating the soil so that cracking no longer occurs and adult emergence is therefore prevented.

Flight range

In general, most species disperse only short distances from their breeding sites, and control of all breeding sites within 500 m is enough to reduce substantially the nuisance caused by species such as *C. molestus* and *C. subimmaculatus*. There is a difference between dispersal by active flight and passive wind carriage. In woodland *C. impunctatus* disperses only a short distance from its breeding site, the density decreasing rapidly to a tenth of the initial value 70 m from the breeding site (Kettle, 1951). This observation appeared to offer the prospect of protecting midge-infested areas by limited larval control. Unfortunately in the open, *C. impunctatus* disperses downwind over 1000 m without any noticeable reduction in density. Indeed the population density was more a function of the availability of suitable hosts than distance from the breeding site (Kettle, 1960). Marked specimens of *C. mississippiensis* dispersed on average 2 km from the point of release. Wind did not appear to be involved in this dispersal (Lillie *et al.*, 1985). Wind plays an integral part in the dispersal of *C. brevitarsis*, a species which feeds on cattle and breeds in their dung. By day the midges rest in the ground herbage and at sunset become airborne with substantial numbers of all stages, males and females, being captured 4 and 6 m above the ground. This behaviour allows the species to 'keep in touch with the host' (Murray, 1987a). Using wind carriage has dispersed *C. brevitarsis* 130–200 km from the Hunter Valley to cause outbreaks of Akabane disease in drought-stricken areas of inland New South Wales (Murray, 1987b).

Culicoides are mainly troublesome under calm conditions and numbers decline rapidly with increasing wind speed until few are encountered at wind speed exceeding 2.5 m s^{-1}. Female *F. (L.) sibirica* cease to bite at wind speeds in excess of 1.3 m s^{-1}, (Gornostaeva, 1967) while those of *C. phlebotomus*, another day-biting species, remain active at wind speeds of 3.3 m s^{-1}, (Nathan, 1981b) and female *L. becquaerti* continue biting in wind speeds up to 5.0 m s^{-1} and have been collected in wind traps at greater wind speeds. Such small creatures are unlikely to be able to orientate to a host in other than winds of low speed. There is some evidence of a dispersal flight by *C. furens*, when large numbers were trapped in a truck trap, but only negligible numbers were biting (Bidlingmayer, 1961).

Specific behaviour in biting

Even when several species feed on the same host at the same time their behaviour is likely to be specific and different. In Jamaica three species, *C. furens*, *C. barbosai* and *L. becquaerti* were anthropophilic and all actively biting at the same time in the one locality (Kettle and Linley, 1967, 1969a, b). They differed in their relative densities on the limb exposed (leg or arm), although the 'bait' was seated on the ground and only offering an arm or a leg. In the morning, the arm : leg ratio for catches of *C. barbosai* was 3 : 2 and for *C. furens* 2 : 3 and for *L. becquaerti* 1 : 4. There was therefore a sixfold difference in the distribution of *C. barbosai* and *L. becquaerti*. Although *C. furens* and *L. becquaerti* were commoner on the leg, *L. becquaerti* fed on the sunlit upper surface while *C. furens* fed on the shady underside.

Four 'baits' were used, of whom two were Caucasian (K, L) and two were dark-skinned West Indians (C, S). S caught fewest of all these species but the relative catches made by C, K and L differed according to the species. They all caught equal numbers of *C. barbosai*; K and L caught significantly more *L. becquaerti* than C but the catches of *C. furens* made by each 'bait' were significantly different with C catching the most, K less and L the least. The three species were responding differently to the three baits under exactly the same conditions. This emphasizes the specific nature of the biting response and the inadequacy of any simple explanation implying attraction to pale or dark skins. Other evidence showed that an individual's 'attractiveness' was not constant, but changed with conditions.

Survival and frequency of feeding

Study of the age structure of populations of *Culicoides* has benefited greatly from Dyce (1969b) recognizing that a burgundy-red pigment is laid down in the surface layers of the abdomen during the first ovarian cycle. This has now been confirmed for many species of *Culicoides*. The ability of a species to be a vector of blood-dwelling pathogens will be a function of the frequency with which blood meals are taken and the daily survival rate. Holmes and Birley (1987) developed an improved method for determining the length of the ovarian cycle and survival between blood meals. They found that in the UK *C. obsoletus* had a 4-day ovarian cycle with 0.38 survival, and in Israel *C. imicola*, a 2-day cycle with 0.42 survival, i.e. daily survival of 0.65 which is considerably lower than that obtained for the same species in East Africa (0.80) by Walker (1977a). Braverman *et al.* (1985), also in Israel, found a longer ovarian cycle (4 day) and for much of the year a similar survival rate (0.3–0.4). In late August the survival rate of *C. imicola* rose to 0.75 and was followed by an increase in bluetongue infections in September. Using a similar analysis, Work *et al.* (1991) obtained a 3-day gonotrophic cycle for *C. variipennis* in California with 0.24 survival (= 0.62 daily survival). In Trinidad Nathan (1981a) found that the daily survival rate of *C. phlebotomus* was about 0.90 in the first 3 days of life declining to 0.69 at 6 days; but these observations were made on midges infected with *Mansonella ozzardi*, which may have affected their survival.

Changes in population

It has already been pointed out that, in temperate regions, many species have one generation a year with the adults usually emerging in the summer. Other species may have several generations. *C. obsoletus* has been reported as having two generations in the north of England (Hill, 1947) and three generations in southern England with a generation time of 7 weeks (Birley and Boorman, 1982). In the warm summers of north-eastern Colorado the generation time of *C. variipennis* may be as short as 2 weeks with seven generations being completed in a year (Barnard and Jones, 1980). In the tropics and subtropics generations are likely to overlap and adults be present all the year round. In subtropical Florida adult *C. mississippienis* was present throughout the year and *C. furens* and *C. barbosai* absent in winter (Lillie *et al.*, 1987). In subtropical eastern Australia *C. brevitarsis*, *C. marmoratus* and *C. victoriae* were taken in truck traps throughout the year, while *C. longior* was absent in winter (Kettle *et al.*, in preparation). In Kenya, Walker (1977b) found that the five species he was studying were present all the year round, and although the populations fluctuated, there was no clear correlation with rainfall.

Activity at all times of the diel is adversely affected by wind and positively by temperature. Truck trap catches of female *C. brevitarsis* were inversely and linearly related to wind speed and curvilinearly to temperature. Many nocturnal species are more active when the moon is shining, e.g. *C. variipennis* (Linhares and Anderson, 1990), *C. nipponensis* (Guan-Hong, 1983), *C. subimmaculatus* (Edwards *et al.*, 1987) and *C. mississippiensis* (Lillie *et al.*, 1987). It is commonplace for coastal species to show a lunar periodicity, determined by the tides. In Oceania *C. peleliouensis* has two peaks of emergence each month, corresponding to the neap tides (Dorsey, 1947). Similar tidal dependence has been shown for *C. subimmaculatus*, *C. austeni*, *L. spinosifrons*, *C. furens* and *C. barbosai*. Although *L. becquaerti* is coastal in distribution, the numbers of this species biting were more dependent upon rainfall than tidal movements.

The limits of the distribution of a species within a region are constantly changing. In Portugal *C. imicola* has been found 2.5 degrees further north than its previous limit (Capela *et al.*, 1993). In south-eastern Australia *C. brevitarsis* is established on the coastal plains from which, in wet years, it extends further south and inland where its continuing survival is dependent on the winter temperature (Murray and Nix, 1987).

Medical and Veterinary Importance

Species of *Culicoides* are economically important as vectors of arboviruses (see Chapter 24), blood-dwelling Protozoa (see Chapter 27) and filarial worms (see Chapter 32). The impact of biting midges on humans has been reviewed by Linley *et al.* (1983). To most people, biting midges are synonymous with acute discomfort and irritation on calm, humid summer days. In spite of their small size they often cause severe local reaction. Species of *Lasiohelea* and *Leptoconops* produce particularly persistent reactions, which may blister and weep serum from the site of the bite in sensitive people. The impact of midges is greatest on newcomers to an infested area and hence the greater sensitivity of tourists to local pest species. Control of biting midges has been essential in many areas for the development of an expanding tourist industry. In many parts of the world the midge problem is a coastal one and hence the popular, but misleading name of 'sandflies'. 'Sandflies' infest coastal areas of the Caribbean, eastern USA and South America, coastal areas of Australia, the Pacific islands and islands of the Indian Ocean. In all areas of the world where horses are subject to intense attack by *Culicoides* they suffer from *Culicoides* hypersensitivity (sweet itch). Various species are involved – in eastern Australia *C. brevitarsis* (Riek, 1954), in western Canada *C. obsoletus* (Anderson *et al.*, 1991) and in Florida *C. insignis* (Greiner *et al.*, 1990).

Populations of biting midges reach pest proportion in the northern temperate regions, extending to high latitudes. In the highlands of Scotland *C. impunctatus* can make life miserable for residents and visitors. In Siberia *F.* (*L.*) *sibirica* hampered construction of the Krasnoyarsk Power Station (Gornostaeva, 1967). Reference has already been made to biting midges at high altitudes in Tibet (p. 155), and Spencer Chapman in his book *The Jungle is Neutral* (BT Reprint Society, London 1950, p. 62), describing personal experiences of guerrilla warfare in Malaya, gave pride of place to biting midges as troublesome nocturnal pests.

References

Akey, D.H. and Barnard, D.R. (1983) Parity in airborne populations of the biting gnat *Culicoides variipennis* (Diptera; Ceratopogonidae) in northeastern Colorado. *Environmental Entomology* 12, 91–95.

Anderson, G.S., Belton, P. and Kleider, N. (1991) *Culicoides obsoletus* (Diptera: Ceratopogonidae) as a causal agent of *Culicoides* hypersensitivity (Sweet Itch) in British Columbia. *Journal of Medical Entomology* 28, 685–693.

Auezova, G. (1990) Day shelters of biting midges in Dzhungaria. *Abstracts 2nd International Congress on Dipterology*, Bratislava, 27 August–1 September 1990, p.11.

Auezova, G., Brushko, Z. and Kubykin, R. (1990) Feeding of biting midges (Leptoconopidae) on reptiles. *Abstracts 2nd International Congress on Dipterology*, Bratislava, 27 August–1 September 1990, p. 12.

Aussel, J.P. (1993a) Ecology of the biting midge *Leptoconops albiventris* in French Polynesia. I. Biting cycle and influence of climatic factors. *Medical and Veterinary Entomology* 7, 73–79.

Aussel, J.P. (1993b) Ecology of the biting midge *Leptoconops albiventris* in French Polynesia. II. Location of breeding sites and larval microdistribution *Medical and Veterinary Entomology* 7, 80–86.

Aussel, J.P. (1993c) Ecology of the biting midge *Leptoconops albiventris* in French Polynesia. III. Influence of abiotic factors on breeding sites. Towards ecological control? *Medical and Veterinary Entomology* 7, 87–93.

Aussel, J.P. and Linley, J.R. (1993) Activity of *Culicoides furens* larvae (Diptera: Ceratopogonidae) as related to light and nutritional state. *Journal of Medical Entomology* 30, 878–882.

Aussel, J.P. and Linley, J.R. (1994) Natural food and feeding behaviour of *Culicoides furens* larvae (Diptera: Ceratopogonidae). *Journal of Medical Entomology* 31, 99–104.

Barnard, D.R. and Jones, R.H. (1980) *Culicoides variipennis* seasonal abundance, overwintering, and voltinism in northeastern Colorado. *Environmental Entomology* 9, 709–712.

Becker, P. (1961) Observations on the life cycle and immature stages of *Culicoides circumscriptus* Kieff. (Diptera: Ceratopogonidae). *Proceedings of the Royal Society of Edinburgh B* 67, 363–386.

Bidlingmayer, W.L. (1961) Field activity studies on adult *Culicoides furens*. *Annals of the Entomological Society of America* 54, 149–156.

Birley, M.H. and Boorman, J.P.T. (1982) Estimating the survival and biting rates of haematophagous insects, with particular reference to the *Culicoides obsoletus* group (Diptera: Ceratopogonidae) in southern England. *Journal of Animal Ecology* 51, 135–148.

Blackwell, A., Mordue, A.J. and Mordue, W. (1992a) Morphology of the antennae of two species of biting midge: *Culicoides impunctatus* (Goetghebuer) and *Culicoides nubeculosus* (Meigen) (Diptera: Ceratopogonidae). *Journal of Morphology* 213, 85–103.

Blackwell, A., Mordue, A.J., Young, M.R. and Mordue, W. (1992b) The swarming behaviour of the Scottish biting midge, *Culicoides impunctatus* (Diptera: Ceratopogonidae). *Ecological Entomology* 17, 319–325.

Braverman, Y., Linley, J.R., Marcus, R. and Frish, K. (1985) Seasonal survival and expectation of infective life of *Culicoides* spp. (Diptera: Ceratopogonidae) in Israel, with implications for bluetongue virus transmission and a comparison of the parous rate in *C. imicola* from Israel and Zimbabwe. *Journal of Medical Entomology* 22, 476–484.

Buxton, P.A. (1960) British Diptera associated with fungi. III. Flies of all families reared from about 150 species of fungi. *Entomologist's Monthly Magazine* 96, 61–94.

Campbell, M.M. and Kettle, D.S (1975) Oogenesis in *Culicoides brevitarsis* Kieffer (Diptera: Ceratopogonidae) and the development of a plastron-like layer on the egg. *Australian Journal of Zoology* 23, 203–218.

Campbell, M.M. and Kettle, D.S. (1976) Numbers of adult *Culicoides brevitarsis* Kieffer (Diptera: Ceratopogonidae) emerging from bovine dung exposed under different conditions in the field. *Australian Journal of Zoology* 24, 75–85.

Campbell, M.M. and Kettle, D.S. (1979) Swarming of *Culicoides brevitarsis* Kieffer (Diptera: Ceratopogonidae) with reference to markers, swarm size, proximity of cattle, and weather. *Australian Journal of Zoology* 27, 17–30.

Capela, R., Sousa, C., Pena, I. and Caeiro, V. (1993) Preliminary note on the distribution and ecology of *Culicoides imicola* in Portugal. *Medical and Veterinary Entomology* 7, 23–26.

Carter, H.F., lngram, A. and Macfie, J.W.S. (1921) Observations on the ceratopogonine midges of the Gold Coast with descriptions of new species. Part 1. *Annals of Tropical Medicine and Parasitology* 14, 187–210.

Chen, C.S., Lien, J.C., Lin, Y.N. and Hsu, S.J. (1981) The diurnal biting pattern of a blood-sucking midge *Forcipomyia* (*Lasiohelea*) *taiwana* (Shiraki) (Diptera: Ceratopogonidae). *Chinese Journal of Microbiology and Immunology* 14, 54–56.

Das Gupta, S.K. (1964) *Culicoides* (*Trithecoides*) *anophelis* Edwards (Insecta: Diptera: Ceratopogonidae) as an ectoparasite of insect vectors. *Proceedings of the Zoological Society* 17, 1–20.

Davies, J.B. (1969) Effect of felling mangroves on emergence of *Culicoides* spp. in Jamaica. *Mosquito News* 29, 566–571.

Dorsey, C.K. (1947) Population and control studies of the Palau gnat on Peleliu, Western Caroline Islands. *Journal of Economic Entomology* 40, 805–814.

Downes, J.A. (1955) Observations on the swarming flight and mating of *Culicoides* (Diptera: Ceratopogonidae). *Transactions of the Royal Entomological Society of London* 106, 213–236.

Downes, J.A. (1962) What is an arctic insect? *Canadian Entomologist* 94, 143–162.

Downes, J.A. (1978) Feeding and mating in the insectivorous Ceratopogoninae (Diptera). *Memoirs of the Entomological Society of Canada* 104, 1–62.

Duval, J., Rajaonarivelo, E. and Rabenirainy, L. (1974) Écologie de *Styloconops spinosifrons* (Carter, 1921) (Diptera: Ceratopogonidae) sur les plages de la côte Est de Madagascar. *Cahiers ORSTOM Entomologie Médicale et Parasitologie* 12, 245–258.

Dyce, A.L. (1969a) Biting midges (Diptera: Ceratopogonidae) reared from rotting cactus in Australia. *Mosquito News* 29, 644–649.

Dyce, A.L. (1969b) The recognition of nulliparous and parous *Culicoides* (Diptera: Ceratopogonidae) without dissection. *Journal of the Australian Entomological Society* 8, 11–15.

Edwards F.W. (1926) On the British biting midges (Diptera: Ceratopogonidae). *Transactions of the Entomological Society of London* 74, 389–426.

Edwards, P.B. (1977) Biology and bionomics of the biting midge *Culicoides subimmaculatus* Lee and Reye (Diptera: Ceratopogonidae) and other coastal *Culicoides* in southeast Queensland. PhD thesis, University of Queensland, Brisbane, Australia

Edwards, P.B., Kettle, D.S. and Barnes, A. (1987) Factors affecting the numbers of *Culicoides* (Diptera: Ceratopogonidae) in traps in coastal south-east Queensland, with particular reference to collections of *C. subimmaculatus* in light traps. *Australian Journal of Zoology* 35, 469–486.

Fahrner, J. and Barthelmess, C. (1988) Rearing of *Culicoides nubeculosus* (Diptera: Ceratopogonidae) by natural or artificial feeding in the laboratory. *Veterinary Parasitology* 28, 307–313.

Fox, I. and Hoffman, W.A. (1944) New neotropical biting sandflies of the genus *Culicoides* (Diptera: Ceratopogonidae). *Puerto Rico Journal of Public Health and Tropical Medicine* 20, 108–111.

Fredeen, F.J.H. (1969) *Culicoides* (*Selfia*) *denningi* a unique river-breeding species. *Canadian Entomologist* 101, 539–544.

Fujisaki, K., Kamio, T., Kitaoka, S. and Morii, T. (1987) Quantitation of the blood meal ingested by *Culicoides arakawae* (Diptera: Ceratopogonidae). *Journal of Medical Entomology* 24, 702–703.

Glukhova, V.M. and Dubrovskaya, V.V. (1972) On autogenic maturation of eggs of blood-sucking midges. *Parazitologiya* 6, 309–319.

Gornostaeva, R.M. (1967) The diurnal activity of attacks of *Lasiohelea sibirica* Bujan midges in the area of Krasnoyarsk hydropower station construction. *Meditsinskaya Parazitologiya i Parazitarnye Bolezni* 36, 11–17.

Greiner, E.C., Fadok, V.A. and Rabin, E.B. (1990) Equine *Culicoides* hypersensitivity in Florida: biting midges aspirated from horses. *Medical and Veterinary Entomology* 4, 375–381.

Guan-Hong, S. (1983) Effect of meteorological conditions on the nocturnal activity of *Culicoides nipponensis* Tokunaga. *Acta Entomologica Sinica* 26, 49–58.

Hagan, C. E. and Kettle, D.S. (1990) Habitats of *Culicoides* spp. in an intertidal zone of southeast Queensland, Australia. *Medical and Veterinary Entomology* 4, 105–115.

Hill, M.A (1947) The life-cycle and habits of *Culicoides impunctatus* Goetghebuer and *Culicoides obsoletus* Meigen, together with some observations on the life-cycle of *Culicoides odibilis* Austen, *Culicoides pallidicornis* Kieffer, *Culicoides cubitalis* Edwards and *Culicoides chiopterus* Meigen. *Annals of Tropical Medicine and Parasitology* 41, 55–115.

Holmes, P.R. and Birley, M.H. (1987) An improved method for survival rate analysis from time series of haematophagous dipteran populations. *Journal of Animal Ecology* 56, 427–440.

Howard-Bury, C.K. (1922) *Mount Everest, the Reconnaissance, 1921.* Edward Arnold, London.

Ismail, M.T. and Kremer, M. (1980) L'effet du repas sanguin sur la production de phéromone par les femelles de *C. nubeculosus* (Diptera). *Annales de Parasitologie* 55, 455–466.

Jamnback, H. (1965) The *Culicoides* of New York State (Diptera: Ceratopogonidae). *New York State Museum and Science Service Bulletin* 399, 1–154.

Jobling, B. (1953) On the blood-sucking midge *Culicoides vexans* Staeger, including the description of its eggs and the first-stage larva. *Parasitology* 43, 148–159.

Jones, R.H., Potter, H.W. and Baker, S.K. (1969) An improved larval medium for colonised *Culicoides variipennis. Journal of Economic Entomology* 62, 1483–1486.

Kettle, D.S. (1951) The spatial distribution of *Culicoides impunctatus* Goet. under woodland and moorland conditions and its flight range through woodland. *Bulletin of Entomological Research* 42, 239–291.

Kettle, D.S. (1960) The flight of *Culicoides impunctatus* Goetghebuer (Diptera: Ceratopogonidae) over moorland and its bearing on midge control. *Bulletin of Entomological Research* 51, 461–489.

Kettle, D.S. (1969) The biting habits of *Culicoides furens* (Poey) and *C. barbosai* Wirth and Blanton. 1. The 24-h cycle, with a note on differences between collectors. *Bulletin of Entomological Research* 59, 21–31.

Kettle, D.S. and Elson, M.M. (1976) The immature stages of some Australian *Culicoides* Latreille (Diptera: Ceratopogonidae). *Journal of the Australian Entomological Society* 15, 303–332.

Kettle, D.S. and Lawson, J.W.H. (1952) The early stages of British biting midges: *Culicoides* Latreille (Diptera: Ceratopogonidae) and allied genera. *Bulletin of Entomological Research* 43, 421–467.

Kettle, D.S. and Linley, J.R. (1967) The biting habits of *Leptoconops bequaerti.* I. Methods; standardisation of technique; preference for individuals, limbs and positions. *Journal of Applied Ecology* 4, 379–395.

Kettle, D.S. and Linley, J.R. (1969a) The biting habits of some Jamaican *Culicoides.* I. *C. barbosai* Wirth and Blanton, *Bulletin of Entomological Research* 58, 729–753.

Kettle, D.S. and Linley, J.R. (1969b) The biting habits of some Jamaican *Culicoides.* II. *C. furens* (Poey). *Bulletin of Entomological Research* 59, 1–20.

Kettle, D.S. Wild, C.H. and Elson, M.M. (1975) A new technique for rearing individual *Culicoides* larvae (Diptera: Ceratopogonidae). *Journal of Medical Entomology* 12, 263–264.

Kettle, D.S., Edwards, E.B. and Barnes, A. (in preparation) Factors affecting the numbers of *Culicoides* in truck traps in coastal south-east Queensland.

Khamala, C.P.M. (1975) Investigations of seasonal and environmental influences on biting and immature populations of *Culicoides cornutus* in Kenya, East Africa. *East African Journal of Medical Research* 2, 283–292.

Kitaoka, S. and Morii, T. (1964) Chicken-biting ceratopogonid midges in Japan with special reference to *Culicoides odibilis* Austen. *National Institute of Animal Health Quarterly* 4, 167–175.

Kline, D.L., Takken, W., Wood, J.R. and Carlson, D.A. (1990) Field studies of the potential of butanone, carbon dioxide, honey extracts, 1-octen-3-ol, L-lactic acid and phenols as attractants for mosquitoes. *Medical and Veterinary Entomology* 4, 383–391.

Kremer, M., Ismail, M.T. and Rebholtz, C. (1979) Detection of a pheromone released by females of *Culicoides nubeculosus* (Diptera: Ceratopogonidae) attracting the males and stimulating copulation. *Mosquito News* 39, 627–631.

Laurence, B.R. and Mathias, P.L. (1972) The biology of *Leptoconops* (*Styloconops*) *spinosifrons* (Carter) (Diptera: Ceratopogonidae) in the Seychelles Islands, with descriptions of the immature stages. *Journal of Medical Entomology* 9, 51–59.

Lawson, J.W.H. (1951) The anatomy and morphology of the early stages of *Culicoides nubeculosus* Meigen (Diptera: Ceratopogonidae = Heleidae). *Transactions of the Royal Entomological Society of London* 102, 511–570.

Lien, J.C. (1989) Taxonomic and ecological studies on the biting midges of the subgenus *Lasiohelea*, genus *Forcipomyia* from Taiwan. *Journal of Taiwan Museum* 42, 37–77.

Lien, J.C., Huang, T.C., Lin, Y.N. and Lu, L.C. (1988) Rearing of the larvae of *Forcipomyia* species with BG-11 agar-plate culture of the blue-green alga, *Anabaena* HA101. *Chinese Journal of Parasitology* 1, 183–184.

Lillie, T.H., Kline, D.L. and Hall, D.W. (1985) The dispersal of *Culicoides mississippiensis* (Diptera: Ceratopogonidae) in a salt marsh near Yankeetown, Florida. *Journal of the American Mosquito Control Association* 1, 463–467.

Lillie, T.H., Kline, D.L. and Hall, D.W. (1987) Diel and seasonal activity of *Culicoides* spp. (Diptera: Ceratopogonidae) near Yankeetown, Florida monitored with a vehicle-mounted insect trap. *Journal of Medical Entomology* 24, 503–511.

Linhares, A.X. and Anderson, J.R. (1990) The influence of temperature and moonlight on the flight activity of *Culicoides variipennis* (Coquillett) (Diptera: Ceratopogonidae) in northern California. *Pan-Pacific Entomologist* 66, 199–207.

Linley, J.R. (1966) Field and laboratory observations on the behavior of the immature stages of *Culicoides furens* Poey (Diptera: Ceratopogonidae). *Journal of Medical Entomology* 2, 385–391.

Linley, J.R. (1968) Autogeny and polymorphism for wing length in *Leptoconops becquaerti* (Kieff.) (Diptera: Ceratopogonidae). *Journal of Medical Entomology* 5, 53–66.

Linley, J.R. (1986) Swimming behavior of the larva of *Culicoides variipennis* (Diptera: Ceratopogonidae) and its relationship to temperature and viscosity. *Journal of Medical Entomology* 23, 473–483.

Linley, J.R. and Adams, G.M. (1972) A study of the mating behaviour of *Culicoides melleus* (Coquillett) (Diptera: Ceratopogonidae). *Transactions of the Royal Entomological Society of London* 124, 81–121.

Linley, J.R. and Carlson, D.A. (1978) A contact mating pheromone in the biting midge, *Culicoides melleus*. *Journal of Insect Physiology* 24, 423–427.

Linley, J.R. and Davies, J.B. (1971) Sandflies and tourism in Florida and the Bahamas and Caribbean area. *Journal of Economic Entomology* 64, 264–278.

Linley, J.R. and Kettle, D.S. (1964) A description of the larvae and pupae of *Culicoides furens* Poey and *Culicoides hoffmani* Fox (Diptera: Ceratopogonidae). *Annals and Magazine of Natural History Series* 13 (7), 129–149.

Linley, J.R., Evans, H.T. and Evans, F.D.S. (1970) A quantitative study of autogeny in a naturally occurring population of *Culicoides furens* (Poey) (Diptera: Ceratopogonidae). *Journal of Animal Ecology* 39, 169–183.

Linley, J.R., Hoch, A.L. and Pinheiro, F.P. (1983) Biting midges (Diptera: Ceratopogonidae) and human health. *Journal of Medical Entomology* 20, 347–364.

Marks, E.N. and Reye, E.J. (1982) *An Atlas of Common Queensland Mosquitoes with a Guide to Common Queensland Biting Midges.* Queensland Institute of Medical Research, Brisbane, Australia.

Megahed, M.M. (1956) A culture method for *Culicoides nubeculosus* (Meigen) (Diptera: Ceratopogonidae) in the laboratory, with notes on the biology. *Bulletin of Entomological Research* 47, 107–114.

Mullens, B.A. (1987) Seasonal size variability in *Culicoides variipennis* (Diptera: Ceratopogonidae) in southern California. *Journal of the American Mosquito Control Association* 3, 512–513.

Mullens, B.A. (1989) A quantitative survey of *Culicoides variipennis* (Diptera: Ceratopogonidae) in dairy wastewater ponds in southern California. *Journal of Medical Entomology* 26, 559–565.

Mullens, B.A. and Holbrook, F.R. (1991) Temperature effects on the gonotrophic cycle of *Culicoides variipennis* (Diptera: Ceratopogonidae). *Journal of the American Mosquito Control Association* 7, 588–592.

Mullens, B.A. and Lip, K.S. (1987) Larval population dynamics of *Culicoides variipennis* (Diptera: Ceratopogonidae) in southern California. *Journal of Medical Entomology* 24, 566–574.

Mullens, B.A. and Rodriguez, J.L. (1989) Response of *Culicoides variipennis* (Diptera: Ceratopogonidae) to water level fluctuations in experimental dairy wastewater ponds. *Journal of Medical Entomology* 26, 566–572.

Mullens, B.A. and Rutz, D.A. (1983) Development of immature *Culicoides variipennis* (Diptera: Ceratopogonidae) at constant laboratory temperatures. *Annals of the Entomological Society of America* 76, 747–751.

Mullens, B.A. and Schmidtmann, E.T. (1982) The gonotrophic cycle of *Culicoides variipennis* (Diptera: Ceratopogonidae) and its implications in age-grading field populations in New York State, USA. *Journal of Medical Entomology* 19, 340–349.

Murray, M.D. (1987a) Local dispersal of the biting midge *Culicoides brevitarsis* Kieffer (Diptera: Ceratopogonidae) in south-eastern Australia. *Australian Journal of Zoology* 35, 559–573.

Murray, M.D. (1987b) Akabane epizootics in New South Wales: evidence for long-distance dispersal of the biting midge *Culicoides brevitarsis*. *Australian Veterinary Journal* 64, 305–308.

Murray, M.D. and Nix, H.A. (1987) Southern limits of distribution and abundance of the biting midge *Culicoides brevitarsis* Kieffer (Diptera: Ceratopogonidae) in south-eastern Australia: an application of the GROWEST model. *Australian Journal of Zoology* 35, 575–585.

Nathan, M.B. (1981a) Transmission of the human filarial parasite *Mansonella ozzardi* by *Culicoides phlebotomus* (Williston) (Diptera: Ceratopogonidae) in coastal north Trinidad. *Bulletin of Entomological Research* 71, 97–105.

Nathan, M.B. (1981b) A study of the diurnal biting and flight activity of *Culicoides phlebotomus* (Williston) (Diptera: Ceratopogonidae) using three trapping methods. *Bulletin of Entomological Research* 71, 121–128.

Parker, A.H. (1950) Studies on the eggs of certain biting midges (*Culicoides* Latreille) occurring in Scotland. *Proceedings of the Royal Entomological Society of London A* 25, 43–52.

Reuben, R. (1963) A comparison of trap catches of *Culicoides impunctatus* Goetghebuer (Diptera: Ceratopogonidae) with meteorological data. *Proceedings of the Royal Entomological Society of London A* 38, 181–193.

Riek, E.F. (1954) Studies on allergic dermatitis (Queensland itch) of the horse; the aetiology of the disease. *Australian Journal of Agricultural Research* 5, 109–129.

Saunders, L.G. (1964). New species of *Forcipomyia* in the *Lasiohelea* complex described in all stages (Diptera: Ceratopogonidae). *Canadian Journal of Zoology* 42, 463–482.

Smith, L.M. and Lowe, H. (1948) The black gnats of California. *Hilgardia* 18, 157–183.

Walker, A.R. (1977a) Adult lifespan and reproductive status of *Culicoides* species (Diptera: Ceratopogonidae) in Kenya, with reference to virus transmission. *Bulletin of Entomological Research* 67, 205–215.

Walker, A.R. (1977b) Seasonal fluctuations of *Culicoides* species (Diptera: Ceratopogonidae) in Kenya. *Bulletin of Entomological Research* 67, 217–233.

Williams, R.W. (1964) Observations on habitats of *Culicoides* larvae in Trinidad, W.I. (Diptera: Ceratopogonidae). *Annals of the Entomological Society of America* 57, 462–466.

Winder, J.A. (1978) Cocoa flower Diptera: their identity, pollinating activity and breeding sites. *PANS* 24, 5–18.

Wirth, W.W., Ratanaworabhan, N.C. and Blanton, F.S. (1974) Synopsis of the genera of Ceratopogonidae (Diptera). *Annales de Parasitologie Humaine et Comparée* 49, 595–613.

Work, T.M., Mullens, B.A. and Jessup, D.A. (1991) Estimation of survival and gonotrophic cycle length of *Culicoides variipennis* (Diptera: Ceratopogonidae) in California. *Journal of the American Mosquito Control Association* 7, 242–249.

Psychodidae–Phlebotominae (Sandflies)

9

The phlebotomine sandflies are a well-defined group of species, which are sometimes accorded family rank as the Phlebotomidae, but more frequently are regarded as a subfamily, Phlebotominae, of the Psychodidae (Lane, 1993). Psychodids are small nematoceran flies with long antennae, pendulous palps, and hairy bodies and wings (Fig. 9.1). Their venation is characterized by more or less parallel, longitudinal veins, the radial sector 3- or 4-branched, and the media 4-branched (Fig. 9.2).

Phlebotomine sandflies are brownish, long-legged flies with narrow bodies; narrow, lanceolate wings, less than 3 mm long, which are held erect above the body; the radial sector is 4-branched; and the palps 5-segmented (Fig. 9.1). The fork of R_{2+3} and R_4 occurs about the middle of the wing (Fig. 9.2). The mouthparts are moderately long with functional mandibles in the haematophagous females. Mandibles are absent in the males, which are not blood-sucking. Phlebotomines are primarily inhabitants of the warmer areas of the world although they extend as far as 50°N in central Asia.

The majority of the other psychodids are in the Psychodinae, and are dark, squat flies with broad, oval wings held roof-like over the body at rest. The fork separating veins R_{2+3} and R_4 occurs towards the base of the wing (Fig. 9.2). The mouthparts are not piercing and they do not feed on blood. The antennae bear cupuliform whorls of hairs. Psychodine larvae feed on organic material, and some are important in facilitating filtration in the clinker beds of sewage works by keeping them free from accumulating organic matter. The larvae have a posterior siphon, and secondary annulations on the body, which bear tergal plates.

About 700 species of phlebotomines have been described and included in six genera. About half of the species are contained in the genus *Lutzomyia*, one-third in *Sergentomyia* and the majority of the remainder in the genus *Phlebotomus* with a small number of species in *Brumptomyia* and *Warileya*, and *Chinius* (one cave-dwelling species) (Lane, 1993). Phlebotomines are a geologically old group, being identified from the Lower Cretaceous, about 120 million years ago. They therefore originated before the mammals and must have fed originally on reptiles. This

Fig. 9.1. Lateral view of a female *Phlebotomus*.

ancient origin has led to the evolution of different genera in the Old and New Worlds. *Phlebotomus, Sergentomyia* and *Chinius* are confined to the Old World and *Lutzomyia, Brumptomyia* and *Warileya* to the New World, mainly the Neotropical region (Lewis, 1974).

Life Cycle

Phlebotomine eggs measure 300 to 400 μm long by 90 to 150 μm wide; have one side flat and the other convex and the poles rounded (Fig. 9.3). They are white, when laid, but in a few hours darken to various shades of brown to black, according to the species (Abonnenc, 1972). In the laboratory *P. longipes* lays an average of 52 eggs in a batch with a range of 11 to 95 (Gemetchu, 1976); *L. vexator occidentis* matures 70 eggs (Chaniotis, 1967) and *L. longipalpis* an average of 80 eggs (max. 146) (Killick-Kendrick *et al.*, 1977). The eggs of 13 species, examined by EM scanning, have sculptured chorions which probably act as plastrons when the eggs are covered by water (Ward and Ready, 1975). The pattern of the sculpturing on the chorion is a specific character (Fausto *et al.*, 1992). Both eggs and larvae of

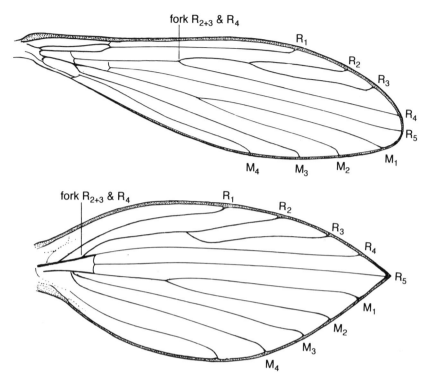

Fig. 9.2. Wing venation of a phlebotomine sandfly (above) and of a psychodine (below). Both (x 63)

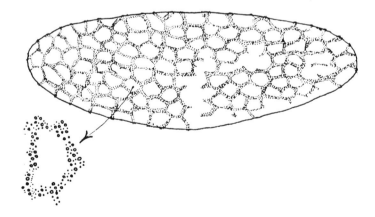

Fig. 9.3. Phlebotomine egg showing ornamentation on the chorion. Source: from Abonnenc (1972).

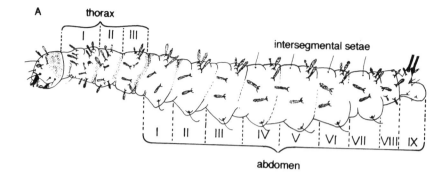

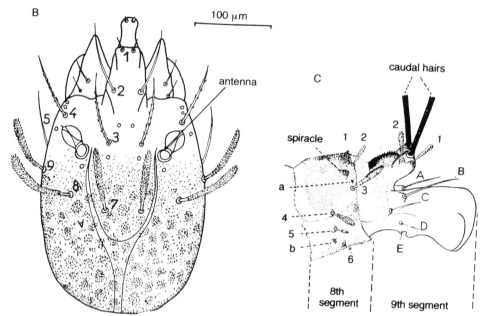

Fig. 9.4. (**A**) lateral view of a phlebotomine larva; (**B**) dorsal view of head (numbers 1–9 refer to setae); (**C**) lateral view of terminal abdominal segments. Source: from Abonnenc (1972).

L. v. occidentis need contact with water and are unable to survive even in a saturated atmosphere (Chaniotis, 1967).

The larva which emerges from the egg passes through 4 instars before pupating. The mature larva is greyish white with a dark head and no secondary annulations on the body (Fig. 9.4). The antennae are small and leaf-like. The thorax is not differentiated from the abdomen although the abdominal segments bear ventral pseudopods, i.e. unjointed evaginations from the body, which are used for progression. The body segments bear characteristic pinnate hairs, the function of which is unknown. A diagnostic feature of phlebotomine larvae is the possession of two or four long caudal setae. First instar larvae have only one pair of caudal

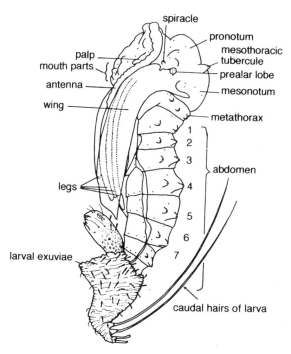

Fig. 9.5. Pupa of a phlebotomine. Source: Abonnenc (1972).

setae, but 2nd to 4th instars have two pairs. The larva is amphipneustic with spiracles opening on the prothorax and the 8th abdominal segment (Fig. 9.4) (Abonnenc, 1972). The head bears chewing mouthparts, which the larva uses to feed on decaying organic matter, leaf mould, insect bodies and, when living in animal burrows, faeces of the host animal. In the laboratory a most useful food consists of equal weights of rabbit faeces, potting compost, dried *Daphnia* and sand (Ward, 1990).

Pupa

The pupa stands upright being secured to the substrate by the larval exuviae, which is retained at the end of the abdomen (Lane, 1993). Therefore all the larval setal characters are available for identification of the pupa. The pupa is exarate with legs and wings free from the body (Fig. 9.5). It has short prothoracic respiratory horns. (The long horns of the culicid and ceratopogonid pupae are adaptations to an aquatic environment.) Although phlebotomines have terrestrial larvae and pupae they are very sensitive to desiccation. Pupae of *L. v. occidentis* are independent of free water, but require a relative humidity of 75 to 100% for survival (Chaniotis, 1967).

In the laboratory *P. longipes* reached the adult stage from the egg in 6 to 7 weeks at 25°C (Gemetchu, 1976), and Modi and Tesh (1983) mass reared

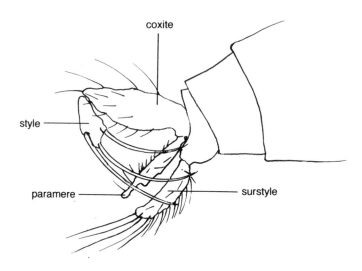

Fig. 9.6. Lateral view of terminal segments of the abdomen of a male *Phlebotomus*.

P. papatasi and *L. longipalpis* from egg to adult in 5 to 6 weeks. In the absence of diapause *P. papatasi* and *P. caucasicus* took 6 to 7 weeks to develop at 25–26°C, but 7 to 9 months when the larvae entered diapause (Dergacheva, 1972).

Adult

Mating

The 16 segmented antennae are pilose in both sexes, showing no sexual dimorphism. Males are, however, easily recognized by the possession of large terminalia of which the claspers (coxite and style) and surstyle are particularly prominent (Fig. 9.6). Absence of plumosity on the male antenna suggests that male phlebotomines do not form aerial swarms in which females are located by sound. Hertig (1949) has reported mating occurring in flight, and Ashford (1974) observed mating dances of *P. orientalis* on white surfaces at dusk. Male *P. orientalis* land and run around in all directions, stopping periodically to shake their wings before running on. Females land and behave similarly until pairing with a male. Presumably sexual recognition is achieved by pheromones emitted when the wings are shaken. Males of some species form two-dimensional swarms spaced out over the surface of the hosts on which the females feed. Male *L. longipalpis* swarm on the backs of humans (Quinnell *et al.*, 1992) and cattle (Dye *et al.*, 1991), and *P. argentipes* on the undersides of standing cattle (Lane *et al.*, 1990). Female *L. longipalpis* respond to a host quicker and in greater numbers in the presence of the male pheromone (Morton and Ward, 1989; Nigam and Ward, 1991), and *P. papatasi* by a pheromone produced by the palps and mouthparts of feeding females (Schlein *et al.*, 1984). The tropical rainforest *L. vespertilionis* and *L. ylephiletor* use tree buttresses as swarming sites (Memmott, 1992).

Male *L. v. occidentis* emerge earlier than the females and rotation of the male terminalia is completed 12 h after emergence. Mating often occurs shortly after the female has fed or while still feeding. When *L. v. occidentis* feeds on cold-blooded hosts, feeding is protracted, taking an hour (Chaniotis, 1967). Spermatozoa are sometimes injected directly into the spermathecal ducts (Hertig, 1949) and are stored in paired spermathecae of unusual appearance. The ducts are moderately long and Abonnenc (1972) distinguished three main groups among Afrotropical phlebotomines: spermathecae with annulated walls; with tomentose or folded walls; and with smooth walls.

Feeding

The palps are 5-segmented and, as in the Ceratopogonidae, the 3rd segment bears sensilla. The maxillae are hooked at the tip in mammal-feeding *Phlebotomus* and *Lutzomyia*, and ridge-tipped in reptile-feeding *Sergentomyia* (Lewis, 1978). In the female the cibarium often bears teeth, which are valuable in identification (Fig. 9.7).

As in other blood-sucking Nematocera only the females are haematophagous, but both sexes feed on plant juices. *P. ariasi* feeds on the honeydew produced by the oak-feeding aphis *Lachanus roboris* (Killick-Kendrick and Killick-Kendrick, 1987) and both sexes of *P. papatasi* pierce leaves and stems to imbibe sap from plants at night (Schlein and Warburg, 1986; Yuval and Schlein, 1986). When *L. longipalpis* feeds on free sugar solutions the ingested fluid passes to the crop, but when sugar solutions are imbibed by piercing a membrane they pass directly to the midgut. Blood passes directly to the midgut, and engorgement is not stimulated by the presence of ATP as in other blood-feeding insects (Ready, 1978; see also Chapter 4). The midgut consists of a narrow, anterior cardiac portion, and a broad, sac-like posterior portion. A peritrophic membrane is secreted around the blood meal, and its structure may be important in the development of the pathogenic protozoans of the genus *Leishmania*. In *P. longipes* the peritrophic membrane develops within 24 h of feeding, reaches a maximum in 48 h, before breaking up on the third day, and finally is excreted with the residue of the blood meal after six to seven days (Gemetchu, 1974).

Major Genera (Lewis, 1973; Lane, 1993)

The Old World genera *Phlebotomus* and *Sergentomyia* are separated from the New World genera by having the 5th palpal segment the longest, no post-spiracular setae, and no posterior bulge to the cibarium. In the New World genera palpal segment 3 is usually the longest, post-spiracular setae are present, and there is a posterior bulge to the cibarium. Old World phlebotomines are savanna and desert species, associated with areas of low rainfall, whereas New World phlebotomines are mainly inhabitants of forests and occur in areas of higher rainfall.

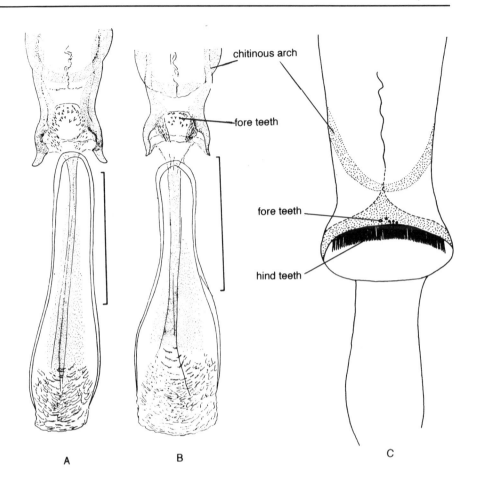

Fig. 9.7. Cibarium and pharynx of: (**A**) female *Phlebotomus papatasi*; (**B**) male *P. papatasi*; (**C**) *Sergentomyia queenslandi*. Sources: A and B from Quote (1964). *Journal of Medical Entomology* 1, 241.

Phlebotomus (13 subgenera)

This mammal-biting genus reaches its maximum development in the warmer temperate and subtropical regions with hot summers and cold winters. It is characterized by the absence of cibarial teeth in the female (Fig. 9.7A) and by the possession of erect hairs on the hind borders of abdominal tergites 2 to 6 (Fig. 9.1).

Sergentomyia (9 subgenera)

This is the dominant genus in the Old World tropics of Africa, India and Australia, where its species feed on reptiles and Amphibia. The genus is characterized by the possession of a posterior transverse row of cibarial teeth in the female (Fig. 9.7C) and recumbent setae on abdominal tergites 2 to 6.

Lutzomyia (14 subgenera, 11 species groups)

This genus is mainly Neotropical in distribution but a few species occur in the south of the Nearctic region. They feed on both mammals and reptiles. Species of *Lutzomyia* are characterized by having a transverse row of hind teeth and one or more rows of fore teeth on the cibarium of the female.

Bionomics of Phlebotominae

Lewis (1973) remarks that 'Adults [sandflies] are often hard to find and larvae usually impossible'. Consequently there are fewer observations on the bionomics of Phlebotominae than on the Culicidae or Ceratopogonidae. They are likely to show a comparable range of behaviour and ecology, aspects which are receiving much attention. Feeding on mammals is largely confined to female *Phlebotomus* and *Lutzomyia*, and species in these genera are the vectors of disease to humans and domestic animals.

In arid savanna and desert areas phlebotomines seek a favourable micro-climate, and may be found in the burrows of rodents or in termitaria, where females feed on the mammalian occupants or on hosts in the close vicinity, and lay their eggs. This habit, coupled with the short flight range, characteristic of the subfamily, leads to local concentrations of phlebotomines and the diseases they transmit, and is a good example of the concept of focality or nidality of disease, enunciated by Pavlovsky (1964).

Most species are exophilic but a few, and these are the most important, have become endophilic in human dwellings and anthropophilic, e.g. *P. papatasi* in the Mediterranean and Middle East to longitude 80°. E; *P. sergenti* in Iran; *P. argentipes* in India; and *L. longipalpis* in north-east Brazil (Lewis, 1974). Two distinct populations of *P. ariasi* have been recognized in southern France by gas chromatography of their cuticular hydrocarbons, a sylvatic population in oak woodlands and a domestic population in houses only 900 m away (Kamhawi *et al.*, 1987). Female *P. guggisbergi* and *L. longipalpis* feed on a wide range of hosts with both species showing a preference for larger animals (Quinnell *et al.*, 1992; Johnson *et al.*, 1993). In rural Ethiopia *P. longipes* is exophilic and a cattle feeder, but in Addis Ababa it has become endophilic and attacks people near cattle sheds (Lewis, 1974). *P. pedifer*, a sibling species of *P. longipes*, is a cave dwelling species, which on Mt Elgon in Kenya fed largely (87%) on cattle that entered the cave (Mutinga, 1975). Most of the remaining feeds (9%) were taken from hyraces. *P. orientalis* in Ethiopia is exophilic, but markedly anthropophilic (Ashford, 1974); and *L. wellcomei* in Brazil is largely anthropophilic (65%), but also feeds regularly on rodents (25%) (Ward *et al.*, 1973). Another forest dwelling phlebotomine *L. flaviscutellata* is zoophilic, feeding on ground dwelling rodents of the genera *Oryzomys* and *Proechimys* (Shaw *et al.*, 1972). *P. caucasicus* lives in the burrows of the giant gerbil *Rhombomys opimus*, on which it feeds.

On humans, phlebotomines feed on exposed areas of skin, and these are the sites for the development of ulcers of cutaneous leishmaniasis. Feeding is mainly nocturnal and crepuscular. In Belize, biting of ten species of *Lutzomyia* was at its highest between 1800 and 2400 h and a few species continued biting until 06.00 h (Williams, 1970). The nocturnal feeding cycle of *P. argentipes* was concentrated around midnight (22.00 to 03.00 h) in Bengal but in Sri Lanka it was mainly after midnight (02.00 to 06.00 h) (Lane *et al.*, 1990). In Ethiopia the feeding of *P. orientalis* on humans began after sunset, and rapidly reached a peak from which it declined and ceased when the temperature reached 16°C, which in the highlands was rarely more than 4 h after sunset (Ashford, 1974). Some species, e.g. *P. papatasi*, will feed during the day under shaded conditions indoors, and others will feed if disturbed from their resting places. Adult *L. longipalpis* spend the major part of their lives in animal sheds. Males spend about 2 days in a shed before leaving and are the first to colonize new sheds, to which females are attracted by a sex pheromone emitted by the males. Initially more females are attracted to the shed than males but after about 2 weeks a stable population is reached with males outnumbering females (Dye *et al.*, 1991).

Phlebotomines rest by day in dark, cool, humid niches where the microclimate is favourable for survival, e.g. *L. betrani* rests in caves (Williams, 1976). *P. longipes* occurs in a variety of cavities including caves, tree holes and burrows (Ashford, 1974). Three closely-related species of the subgenus *Paraphlebotomus* are associated with rodent burrows in the former USSR. *P. mongolensis* is found in oases where the conditions are relatively cool and humid, *P. caucasicus* in loess desert and *P. andrejevi* in very hot sandy deserts. The three species differ in their spiracular indices – the index is largest in *P. mongolensis* and smallest in *P. andrejevi*. Reduction in spiracular index would be an adaptation to an increasingly arid environment (Dergacheva, 1974).

Phlebotomines have a characteristic hopping flight and, if disturbed, quickly settle again a short distance away. In the open they are very sensitive to wind speed and feed only under near calm conditions. Perhaps to avoid higher wind speeds many species tend to fly close to the ground in the open. Even under forest conditions 80% of the blood meals of *L. olmeca olmeca* are taken near the ground (Williams, 1970); while *L. trapidoi* rests in the lower layers by day, and ascends into the canopy at night to feed on arboreal vertebrates (Chaniotis *et al.*, 1974). *L. trinidadensis* takes 60% of its feeds at 13 m, and less than 5% of them at ground level (Williams, 1970).

Most species are unable to develop eggs without a blood meal, e.g. *P. longipes* (Foster *et al.*, 1970); *L. californica* (Chaniotis, 1967), *L. longipalpis* (Killick-Kendrick *et al.*, 1977) and *L. trapidoi* (Chaniotis, 1975); but autogeny occurs in *P. papatasi* (El Kammah, 1972) and *L. gomezi* (Killick-Kendrick, 1978).

Females lay their eggs in soil, in burrows, in leaf litter on the forest floor, and around the bases of forest trees. The eggs of *L. longipalpis* have a soluble pheromone on their surface which attracts gravid females to oviposit and remains effective for at least 6 days, but to be effective more than 80 eggs have to be present (Elnaiem and Ward, 1991). In the laboratory extracts of commercial rabbit food, rabbit faeces and oviposition pheromone were equally attractive to ovipositing females. The effects were additive and significantly more eggs were laid in the presence of an extract combining oviposition pheromone and rabbit food (Dougherty *et al.*, 1993). *P. papatasi* breeds in a range of sites, with greatest numbers being reared from the burrows of *Rhombomys opimus* and cattle sheds (more than 1000 individuals per site), and fewer from burrows of *Meriones erythrourus* and unoccupied store rooms (100 to 250 per site) (Artem'ev *et al.*, 1972). In Egypt the breeding sites of *P. langeroni* and *P. papatasi* were characterized by being rich in organic matter, having a high moisture content, a high percentage of silt and a pH of about 7.5 (El Sawaf *et al.*, 1991). Rutledge and Ellenwood (1975a, b, c) carried out a detailed analysis of the breeding sites of phlebotomines on the forest floor in Panama. *L. trapidoi* was the dominant species and favoured hillside and streams in the vicinity of large lianas (*Orouparia* and *Sabicea*) while *L. pessoana* was associated with hilltops and larger trees, e.g. *Anacardium*.

The slow rate of development, already referred to, will limit the number of generations which may be produced each year. Palaearctic species are often bivoltine, e.g. early summer eggs of *P. papatasi* and *P. caucasicus* develop without diapause, but late summer eggs give rise to diapausing larvae thus producing two generations per annum (Dergacheva, 1972). Ward and Killick-Kendrick (1974) have reported diapausing eggs in a species of *Lutzomyia* provisionally referred to as *L.* sp. 260. In cooler climates species overwinter in the larval stage, which may be a true diapause or a temperature-induced cessation of growth (Dergacheva, 1972; Killick-Kendrick, 1978). Perhaps as an adaptation to survival in a hostile environment, emergence of a generation may be extended by delaying hatching of eggs and eclosion of adults from pupae (Abonnenc, 1972; Lewis, 1974). Abundance of *Lutzomyia* species seems to be dependent on rainfall and temperature. Numbers of *L. flaviscutellata* declined during the rains (January to May) and then increased to reach a peak in December to January (Shaw and Lainson, 1972).

Many species of *Lutzomyia* and *Phlebotomus* are now established in laboratory colonies but oviposition is still the stage in the life cycle with the highest mortality. Field estimates of the life expectancy of *P. ariasi* is 1.5 ovarian cycles (Ward, 1990). Detinova (1968) has recorded finding individual *P. papatasi*, which had completed six ovarian cycles. Killick-Kendrick (1978) notes that in nature infected *P. ariasi* can survive at least 29 days.

The flight range of phlebotomines is short, usually a matter of 100 to 200 m, but they may disperse 1 km or more. Mark–release–recapture experiments showed that unengorged *P. ariasi* dispersed more than 1 km from the point of release and 3 days after release one female was captured 2.2 km away. Engorged females remained more local and males dispersed no more than 600 m (Killick-Kendrick *et al.*, 1984). In a coastal desert area of North Africa *P. papatasi* dispersed a maximum of 1.5 km. In two experiments the average distance travelled by unfed females was 820 m compared to fed females (620 m) and males (600 m) (Doha *et al.*, 1991). In a semi-arid valley in Colombia *L. longipalpis* dispersed a maximum of 960 m but only a small percentage dispersed more than 470 m with males on average dispersing 95 m (Morrison *et al.*, 1993). When 20,000 marked *L. trapidoi* were released at ground and canopy (30 m) levels in forest, 90% of those recaptured were taken within 57 m of the point of release, and only four individuals were collected at the limit of observations (200 m) (Chaniotis *et al.*, 1974).

Medical and Veterinary Importance

Phlebotomines are rarely present in sufficient density to reach pest proportions and their importance is as vectors of various pathogens, the most important of which are species of *Leishmania* causing human cutaneous, visceral and mucocutaneous leishmaniases (see Chapter 29). These diseases are widely but patchily distributed throughout the warmer areas of the world. They are mostly zoonoses in which, in the Neotropical region, people become involved by entering the focus of the disease and hence infections are confined to forest workers. Some species of *Leishmania* also infect dogs, and domestic foci of disease may develop in association with endophilic vectors.

Other diseases of which phlebotomines are vectors include bartonellosis, a disease of humans living in certain high altitude valleys in the Andes of South America (see Chapter 25); and sandfly or papatasi fever, a virus disease spread by *P. papatasi* throughout much of its range (see Chapter 24). This virus is of particular interest because there is evidence of transovarian transmission from one generation of *P. papatasi* to the next. Phlebotomines are also implicated in the transmission of vesicular stomatitis, a virus disease of cattle and horses (see Chapter 24).

Phlebotomines are highly susceptible to DDT, and in areas where the vector is endophilic, house spraying with DDT for malaria control dramatically reduced phlebotomine populations and the diseases they spread. Reinfestation is slow and may take up to 3 years after the cessation of spraying. Resistance to DDT has been identified in *P. papatasi* in North Bihar (Davidson, 1988).

References

Abonnenc, E. (1972) Les phlébotomes de la région Ethiopienne (Diptera: Psychodidae). *Mémoires ORSTOM* 55, 1–289.

Artem'ev, M.M., Flerova, O.A. and Belyaev, A.E. (1972) Quantitative evaluation of the productivity of breeding places of sandflies in the wild and in villages. *Meditsinskaya Parazitologiya i Parazitarnye Bolezni* 41, 31–35.

Ashford, R.W. (1974) Sandflies (Diptera: Phlebotomidae) from Ethiopia: taxonomic and biological notes. *Journal of Medical Entomology* 11, 605–616.

Chaniotis, B.N. (1967) The biology of Californian *Phlebotomus* (Diptera: Psychodidae) under laboratory conditions. *Journal of Medical Entomology* 4, 221–233.

Chaniotis, B.N. (1975) A new method for rearing *Lutzomyia trapidoi* (Diptera: Psychodidae) with observations on its development and behaviour in the laboratory. *Journal of Medical Entomology* 12, 183–188.

Chaniotis, B.N., Correa, M.A., Tesh, R.B. and Johnson, K.M. (1974) Horizontal and vertical movements of phlebotomine sandflies in a Panamanian rain forest. *Journal of Medical Entomology* 11, 369–375.

Davidson, G. (1988) *Insecticides*. Publication No. 1, Ross Institute of Tropical Hygiene, London.

Dergacheva, T.I. (1972) The duration of development of the preimaginal stages of some species of sandflies (Diptera: Phlebotomidae) as observed in the laboratory. *Meditsinskaya Parazitologiya i Parazitarnye Bolezni* 41, 536–542.

Dergacheva, T.I. (1974) The ecological relations of certain species of the subgenus

Paraphlebotomus as observed in the Karshinskaya Steppe. *Zoologicheskii Zhurnal* 53, 1661–1668.

Detinova, T.S. (1968) Age structure of insect populations of medical importance. *Annual Review of Entomology* 13, 427–450.

Doha, S., Shehata, M.G., El Said, S. and El Sawaf, B. (1991) Dispersal of *Phlebotomus papatasi* (Scopoli) and *P. langeroni* Nitzulescu in El Hammam, Matrouh Governorate, Egypt. *Annales de Parasitologie Humaine et Comparée* 66, 69–76.

Dougherty, M.J., Hamilton, J.G.C. and Ward, R.D. (1993) Semiochemical mediation of oviposition by the phlebotomine sandfly *Lutzomyia longipalpis*. *Medical and Veterinary Entomology* 7, 219–224.

Dye, C., Davies, C.R. and Lainson, R. (1991) Communication among phlebotomine sand-flies: a field study of domesticated *Lutzomyia longipalpis* populations in Amazonian Brazil. *Animal Behaviour* 42, 183–192.

El Kammah, K.M. (1972) Frequency of autogeny in wild-caught Egyptian *Phlebotomus papatasi* (Scopoli) (Diptera: Psychodidae). *Journal of Medical Entomology* 9, 294.

Elnaiem, D.A. and Ward, R.D. (1991) Response of the sandfly *Lutzomyia longipalpis* to an oviposition pheromone associated with conspecific eggs. *Medical and Veterinary Entomology* 5, 87–91.

El Sawaf, B.M., Helmy, N., Kamal, H.A., Osman, A. and Shehata, M. (1991) Soil analysis of breeding sites of *Phlebotomus langeroni* Nitzulescu and *Phlebotomus papatasi* (Scopoli) in El Agamy, Egypt. *Annales de Parasitologie Humaine et Comparée* 66, 134–136.

Fausto, A.M., Maroli, M. and Mazzini, M. (1992) Ootaxonomy and eggshell ultrastructure of *Phlebotomus* sandflies. *Medical and Veterinary Entomology* 6, 201–208.

Foster, W.A., Tesfa-Yohannes, T.M. and Tecle, T. (1970) Studies on leishmaniasis in Ethiopia. II. Laboratory culture and biology of *Phlebotomus longipes* (Diptera: Psychodidae). *Annals of Tropical Medicine and Parasitology* 64, 403–409.

Gemetchu, T. (1974) The morphology and fine structure of the midgut and peritrophic membrane of the adult female, *Phlebotomus longipes* Parrot and Martin (Diptera: Psychodidae). *Annals of Tropical Medicine and Parasitology* 68, 111–124.

Gemetchu, T. (1976) The biology of a laboratory colony of *Phlebotomus longipes* Parrot and Martin (Diptera: Phlebotomidae). *Journal of Medical Entomology* 12, 661–671.

Hertig, M. (1949) The genital filament of *Phlebotomus* during copulation. *Proceedings of the Entomological Society of Washington* 51, 286–288.

Johnson, R.N., Ngumbi, P.M., Mwanyumba, J.P. and Roberts, C.R. (1993) Host feeding preference of *Phlebotomus guggisbergi*, a vector of *Leishmania tropica* in Kenya. *Medical and Veterinary Entomology* 7, 216–218.

Kamhawi, S., Molyneux, D.H., Killick-Kendrick, R., Milligan, P.J.M., Phillips, A., Wilkes, T.J. and Killick-Kendrick, M. (1987) Two populations of *Phlebotomus ariasi* in the Cévennes focus of leishmaniasis in the south of France revealed by analysis of cuticular hydrocarbons. *Medical and Veterinary Entomology* 1, 97–102.

Killick-Kendrick, R. (1978) Recent advances and outstanding problems in the biology of phlebotomine sandflies. *Acta Tropica* 35, 297–313.

Killick-Kendrick, R. and Killick-Kendrick, M. (1987) Honeydew of aphids as a source of sugar for *Phlebotomus ariasi*. *Medical and Veterinary Entomology* 1, 297–302.

Killick-Kendrick, R., Leaney, A.J. and Ready, P.D. (1977) The establishment, maintenance and productivity of a laboratory colony of *Lutzomyia longipalpis* (Diptera: Psychodidae). *Journal of Medical Entomology* 13, 429–440.

Killick-Kendrick, R., Rioux, J.A., Bailly, M., Guy, M.W., Wilkes, T.J., Guy, F.M., Davidson, I., Knechtli, R., Ward, R.D., Guilvard, E., Perieres, J. and Dubois, H. (1984) Ecology of leishmaniasis in the south of France 20. Dispersal of *Phlebotomus ariasi* Tonnoir, 1921 as a factor in the spread of visceral leishmaniasis in the

Cévennes. *Annales de Parasitologie Humaine et Comparée* 59, 555–572.

Lane, R.P. (1993) Sandflies (Phlebotominae). In: Lane, R.P. and Crosskey, R.W. (eds) *Medical Insects and Arachnids*. Chapman & Hall, London, pp. 78–109.

Lane, R.P., Pile, M.M. and Amerasinghe, F.P. (1990) Anthropophagy and aggregation behaviour of the sandfly *Phlebotomus argentipes* in Sri Lanka. *Medical and Veterinary Entomology* 4, 79–88.

Lewis, D.J. (1973) Phlebotomidae and Psychodidae. In: Smith K.G.V.(ed.) *Insects and Other Arthropods of Medical Importance*. British Museum (Natural History), London, pp. 155–179.

Lewis, D.J. (1974) The biology of the Phlebotomidae in relation to leishmaniasis. *Annual Review of Entomology* 19, 363–384.

Lewis, D.J. (1978) Phlebotomine sandfly research. In: Willmott, S. (ed.) *Medical Entomology Centenary Symposium Proceedings*. Royal Society of Tropical Medicine and Hygiene, London, pp. 94–99.

Memmott, J. (1992) Patterns of sandfly distribution in tropical forest: a causal hypothesis. *Medical and Veterinary Entomology* 6, 188–194.

Modi, G.B. and Tesh, R.B. (1983) A simple technique for mass rearing *Lutzomyia longipalpis* and *Phlebotomus papatasi* (Diptera: Psychodidae) in the laboratory. *Journal of Medical Entomology* 20, 568–569.

Morrison, A.C., Ferro, C., Morales, A., Tesh, R.B. and Wilson, M.L. (1993) Dispersal of the sandfly *Lutzomyia longipalpis* (Diptera: Psychodidae) at an endemic focus of visceral leishmaniasis in Colombia. *Journal of Medical Entomology* 30, 427–435.

Morton, I.E. and Ward R.D. (1989) Laboratory response of female *Lutzomyia longipalpis* sandflies to a host and male pheromone source over distance. *Medical and Veterinary Entomology* 3, 219–223.

Mutinga, M.J. (1975) The animal reservoir of cutaneous leishmaniasis on Mount Elgon, Kenya. *East African Medical Journal* 52, 142–151.

Nigam, Y. and Ward, R.D. (1991) The effect of male sandfly pheromone and host factors as attractants for female *Lutzomyia longipalpis* (Diptera: Psychodidae). *Physiological Entomology* 16, 305–312.

Pavlovsky, E.N. (1964) *Natural Nidality of Transmissive Diseases with Special Reference to the Landscape Epidemiology of Zooanthroponoses*. Nauka, Leningrad.

Quinnell, R.J., Dye, C. and Shaw, J.J. (1992) Host preferences of the phlebotomine sandfly *Lutzomyia longipalpis* in Amazonian Brazil. *Medical and Veterinary Entomology* 6, 195–200.

Ready, P.D. (1978) The feeding habits of laboratory-bred *Lutzomyia longipalpis* (Diptera: Psychodidae). *Journal of Medical Entomology* 14, 545–552.

Rutledge, L.C. and Ellenwood, D.A. (1975a) Production of phlebotomine sandflies on the open forest floor in Panama: the species complement. *Environmental Entomology* 4, 71–77.

Rutledge, L.C. and Ellenwood, D.A. (1975b) Production of phlebotomine sandflies on the open forest floor in Panama: hydrologic and physiographic relations. *Environmental Entomology* 4, 78–82.

Rutledge, L.C. and Ellenwood, D.A. (1975c) Production of phlebotomine sandflies on the open forest floor in Panama: phytologic and edaphic relations. *Environmental Entomology* 4, 83–89.

Schlein, Y. and Warburg, A. (1986) Phytophagy and the feeding cycle of *Phlebotomus papatasi* (Diptera: Psychodidae) under experimental conditions. *Journal of Medical Entomology* 23, 11–15.

Schlein, Y., Yuval, B. and Warburg, A. (1984). Aggregation pheromone released from the palps of feeding female *Phlebotomus papatasi* (Psychodidae). *Journal of Insect Physiology* 30, 153–156.

Shaw, J.J. and Lainson, R. (1972) Leishmaniasis in Brazil: VI. Observations on the seasonal variations of *Lutzomyia flaviscutellata* in different types of forest and its relationship to enzootic rodent leishmaniasis (*Leishmania mexicana amazonensis*). *Transactions of the Royal Society of Tropical Medicine and Hygiene* 66, 709–717.

Shaw, J.J., Lainson, R. and Ward, R.D. (1972) Leishmaniasis in Brazil: VII. Further observations on the feeding habitats of *Lutzomyia flaviscutellata* (Mangabeira) with particular reference to the biting habits at different heights. *Transactions of the Royal Society of Tropical Medicine and Hygiene* 66, 718–723.

Ward, R.D. (1990) Some aspects of the biology of phlebotomine sandfly vectors. *Advances in Disease Vector Research* 6, 93–126.

Ward, R.D. and Killick-Kendrick, R. (1974) Field and laboratory observations on *Psychodopygus lainsoni* Fraiha and Ward and other sandflies (Diptera: Phlebotomidae) from the Transamazonica highway, Para State, Brazil. *Bulletin of Entomological Research* 64, 213–221.

Ward, R.D. and Ready, P.A. (1975) Chorionic sculpturing in some sandfly eggs (Diptera: Psychodidae). *Journal of Entomology A* 50, 127–134.

Ward, R.D., Lainson, R. and Fraiha, H. (1973) Leishmaniasis in Brazil: VIII. Observations on the phlebotomine fauna of an area highly endemic for cutaneous leishmaniasis in the Serra Dos Carajas, Pará State. *Transactions of the Royal Society of Tropical Medicine and Hygiene* 67, 174–183.

Williams, P. (1970) Phlebotomine sandflies and leishmaniasis in British Honduras (Belize). *Transactions of the Royal Society of Tropical Medicine and Hygiene* 64, 317–364.

Williams, P. (1976) The phlebotomine sandflies (Diptera: Psychodidae) of caves in Belize, Central America. *Bulletin of Entomological Research* 65, 601–614.

Yuval, B. and Schlein, Y. (1986) Leishmaniasis in the Jordan Valley. III. Nocturnal activity of *Phlebotomus papatasi* (Diptera: Psychodidae) in relation to nutrition and ovarian development. *Journal of Medical Entomology* 23, 411–415.

Simuliidae (Blackflies) 10

Our knowledge of the Simuliidae has been summarized in two recent publications – *The Natural History of Blackflies* (Crosskey, 1990) and *Blackflies, Ecology, Population Management and Annotated World List* (Kim and Merritt, 1987).

Simuliids are often known as blackflies. They are small, dark, stout-bodied, hump-backed Nematocera (Fig. 10.1). They are larger than blood-sucking ceratopogonids and have wing lengths of 1.5 to 6.0 mm. Simuliids are largely diurnal and vision plays an important role in their behaviour. In the female the individual elements (ommatidia) of which the eyes are composed are small (10–15 µm) and the eyes are well separated above the antennae, i.e. the female is dichoptic (Fig. 10.2). In the male the eyes are larger and are broadly contiguous above the antennae, i.e. the male is holoptic, and the lower ommatidia are similar to those of the female but the upper ones are greatly enlarged, measuring 25–40 µm. The antennae are the same in both sexes and consist of small, globular segments, compacted together to give a beaded appearance (Fig. 10.2). The commonest number of antennal segments is 11; occasionally there are 10 (*Austrosimulium*) and rarely there are 9 in some North American prosimuliines. The 5-segmented, pendulous palps are considerably longer than the short proboscis and carry on the third segment Lutz's organ, a sensory pit. In males and in a few species in which the females do not bite, the mandibles and maxillae are not toothed (Crosskey, 1990).

The wings are short and broad with a large anal lobe (Fig. 10.3). The venation is characteristic with well-developed radial veins along the anterior margin of the wing and weaker median and cubital veins posteriorly. In spite of its weak appearance the wing is highly efficient and in still air simuliids are capable of flying in excess of 100 km (Hocking, 1953). The radial sector may be unbranched or have two branches. Between the median (M_2) and the cubital (Cu_1) veins there is a forked submedian fold. The male terminalia are compact and relatively inconspicuous, particularly when compared to the prominent terminalia of male phlebotomines. The female has a single subspherical spermatheca (Crosskey, 1990).

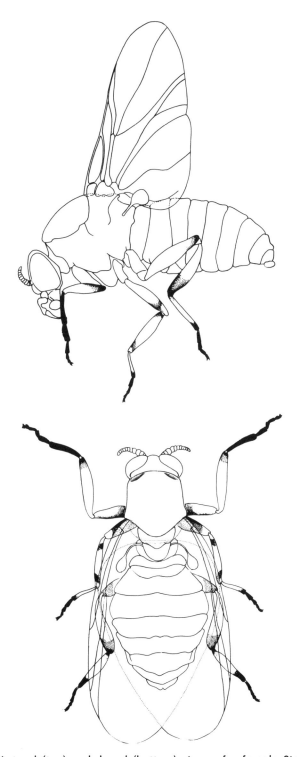

Fig. 10.1. Lateral (top) and dorsal (bottom) views of a female *Simulium*.

Between 1986 and 1989 the number of valid simuliid species increased from 1461 to 1554 arranged in 24 genera of which four, *Simulium, Prosimulium, Cnephia* and *Austrosimulium*, are of economic importance. The largest genus is *Simulium* with nearly 1200 species arranged in 43 subgenera. *Simulium* occurs in all zoogeographical regions (Table 10.1) with the greatest number (410) being found in the Palaearctic region. *Prosimulium* with 110 species assigned to six subgenera is largely confined to the Holarctic region, as is the genus *Cnephia*. *Austrosimulium* is restricted to Australia and New Zealand. The other large genera are *Gigantodax* with 65 species in the Neotropical region and *Metacnephia* with 51 species in the Holarctic (Crosskey, 1990). The Simuliidae is rich in species complexes and that of *S. damnosum* has been considered briefly in Chapter 6.

In *Simulium* and *Austrosimulium* the radial sector on the wing is unbranched (Fig. 10.3); the costa bears spiniform setae and hairs and the hind leg has a rounded lobe (calcipala) at the inner apex of the first tarsal segment and a dorsal groove (pedisulcus) near the base of the second tarsal segment. In *Prosimulium* the radial sector is branched (sometimes only slightly), the costa bears only hairs and there is neither calcipala nor pedisulcus on the hind leg. Species identification is difficult and uses, among other characters, the structure of the male and female terminalia, the pupal respiratory organ, and the larval head. Peterson and Dang (1981) list the morphological characters of the Simuliidae.

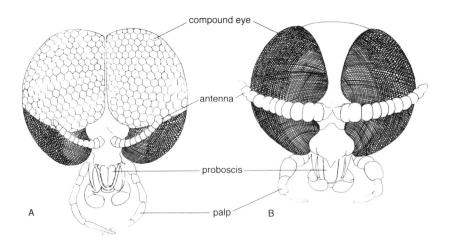

Fig. 10.2. Front view of heads of male (**A**) and female (**B**) *Simulium.*

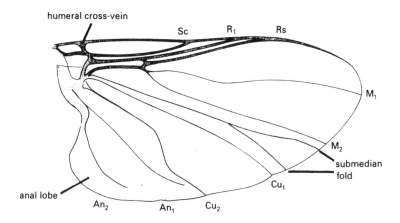

Fig. 10.3. Wing of female *Simulium* (x 50). *Rs*, radial sector.

Table 10.1. Geographical distribution of genera of Simuliidae.

Genera	Af.	Au.	Or.	Nt.	Na.	Pa.	World fauna
Simulium	189	86	169	269	85	410	1187
Prosimulium	10	0	0	0	44	59	110
Gigantodax	0	0	0	65	0	0	65
Metacnephia	0	0	0	0	7	44	51
Austrosimulium	0	25	0	0	0	0	25
Cnephia	0	0	0	0	5	5	10
Other genera	0	9	0	21	22	53	106
Total	199	120	169	355	163	571	1554

Source: Adapted from Crosskey (1990).
Figures indicate number of species in world and in each zoogeographical region (Af. = Afrotropical; Au. = Australasian: Or. = Oriental; Nt. = Neotropical. Na. = Nearctic; Pa. = Palaearctic). Some species occur in more than one region resulting in the total for the six regions (1577) exceeding the total for the world (1554).

Life Cycle

In their early stages simuliids are limited to fluvial ecosystems, breeding in running, often swiftly running, water. The eggs are commonly laid in batches of 200–300, in a range of 30 to 800, on objects in or near running water or directly into water (Crosskey, 1990). It is not uncommon for communal egg masses to be formed by several females ovipositing in close proximity. In the presence of freshly laid or one-day-old eggs the period between landing and oviposition is significantly reduced in gravid female *S. reptans*. A pheromone is likely to be involved (Coupland, 1991). Eggs are either dropped directly into the water and sink to the bottom or are laid on emergent objects close to the waterline, where they are either directly wetted by water or are in the splash zone. Females of several species crawl up to 15 cm below the water surface to oviposit on submerged substrates (Golini and Davies, 1987).

Egg

Eggs are 100 to 400 μm long and ovoid – triangular in shape. Their surface is comparatively smooth, lacking the patterned chorion found in the eggs of *Culicoides* and culicids. Williams (1974) considers that the gelatinous substance in which the eggs are embedded is formed by adherent outer membranes of the individual eggs, i.e. their exochorions. The apparent egg 'shell' is the inner egg membrane or endochorion and the chorionic plastron is poorly developed. A well-developed plastron would not be necessary in eggs laid near the surface of running water where the oxygen tension would be high. Simuliid eggs are sensitive to desiccation; even those of *A. pestilens*, which survive for many months in wet river deposits, desiccate rapidly when exposed to relative humidities of 96% or less (Colbo and Moorhouse, 1974). Eggs laid near the surface hatch when the embryo has completed development, a matter of days under favourable conditions. Other species produce dormant eggs in which the adverse conditions of summer and/or winter are passed (Colbo and Wotton, 1981).

Larva

The egg hatches to produce a larva, much of whose behaviour revolves around the secretion of silk by the long salivary glands, which are longer than the larva. The larva spins a web of silk on the substrate, which is continued into a silken thread on which the larva drifts downstream with the current in search of a suitable object on which to settle. When this has been found, the larva spins a patch of silk to which it anchors itself by its posterior circlet of hooks (Fig. 10.4). Larvae also produce copious amounts of protein glue which is used to attach them to the substrate. In four species of *Prosimulium* production of the glue is dependent on the larva being in running water (Brockhouse and Tanguay, 1992). Larvae remain near the surface of the water and are usually found at depths of less than 300 mm. The larva can change its location by drifting downstream on a silken thread, or by looping over the surface using the posterior circlet and the hooks on the anterior proleg to retain a hold on secreted silk. Some species disperse further from the oviposition site than others. Larvae of *S. ornatipes* are more sessile than those of *A. bancrofti*, which move from the quieter waters of the oviposition site to rapids (Colbo and Moorhouse, 1979). In very large rivers, with fast-flowing water, larvae have been found at depths of several metres (Colbo and Wotton, 1981).

The larva has a distinct, sclerotized head with paired, simple eyes (stemmata), and an elongated hour-glass shaped body, in which the thorax and posterior part of the abdomen are broader than the anterior segments of the abdomen (Fig. 10.4). The head bears a pair of cephalic (labral) fans, homologous structures to the lateral palatal brushes of the Culicidae. They do not create a current but filter water passing over the larva. Larvae are anchored posteriorly and extended in the direction of the current with the head leading. The body is twisted through 90–180° so that

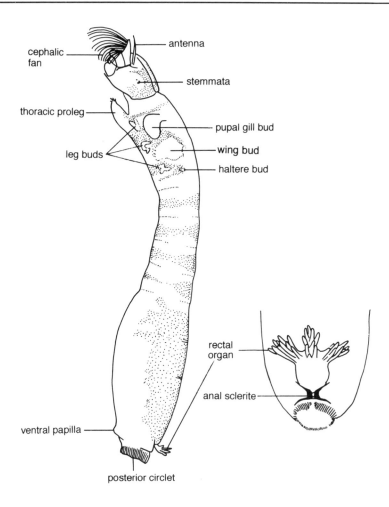

cephalic fan

antenna

stemmata

thoracic proleg

pupal gill bud

leg buds

wing bud

haltere bud

rectal organ

anal sclerite

ventral papilla

posterior circlet

Fig. 10.4. Lateral view of a *Simulium* larva and dorsal view of posterior end. Redrawn from Crosskey (1990).

the fans and mouthparts face towards the surface of the water (Fig. 10.5). The water current is divided by the proleg and directed towards the fans. A sticky secretion produced by the cibarial glands enables the fans to capture fine particles, which are transferred to the cibarium by the mandibular brushes (Colbo and Wotton, 1981). Larvae of *S. piperi* defend their territory and are aggressive to their upstream neighbours, who would be competing for the incoming food. Territorial defence declines dramatically when food is abundant (Hart, 1987).

Chance (1970) has found that although simuliid larvae ingest particles up to 350 μm, the most commonly ingested particles are 10–100 μm. In practice this means that simuliid larvae can be reared in the laboratory on bacteria, which, in certain streams, may form an important element in the feeding of blackfly larvae (Fredeen, 1964). Wotton (1976) has shown that simuliid larvae are capable of ingesting articles of colloidal size (0.091 μm). Algae pass apparently unchanged

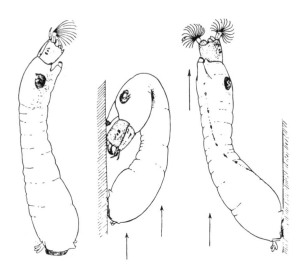

Fig. 10.5. Left, larva of *Simulium*. Centre, attitude of larva in response to sudden increase in current velocity or disturbance – head close to substrate in position to attach silken thread. Right, larva in feeding posture in a current (arrows indicate direction); Source: from Grenier P. (1948) *Physiologia Comparata et Oecologia* 1, 232.

through the blackfly gut but diatoms may form as much as 50% of the gut contents and are digested (Kurtak, 1979). Ladle (1972) found that larvae of *S. ornatum* and *S. equinum* ingested particles with a maximum diameter of 25–30 µm, with diatoms forming the main food early in the year and small particles of detritus predominating later in the year. Filter-feeding larvae may also browse and species of *Twinnia* and *Gymnopais* do not filter feed but graze on the substrate (Currie and Craig, 1987).

Simuliid larvae are particularly abundant where the water current accelerates, as at rapids, and where presumably larvae will strain a greater volume of water per unit time. Heavy larval concentrations are to be found at the outflows of large lakes, where the water will be rich in phytoplankton for larval food. This was true of the Ripon Falls in Uganda, where the Nile flowed out from Lake Victoria. Breeding was eliminated when a dam was built and the falls were submerged.

The larva has a single anterior proleg, surmounted by a circlet of hooks and the abdomen ends in a posterior circlet (Fig. 10.4). The anus opens dorsally of the posterior circlet, and from it may be extruded the rectal organ, which probably, by analogy with the anal papillae of culicid larvae, is concerned with chloride extraction from the water. There is often an X-shaped anal sclerite between the anus and the posterior circlet. Movement of water over the body surface provides the larva with adequate dissolved oxygen for respiration. In deoxygenated water larvae detach and drift downstream. Larvae pass through six to nine instars and the number is not constant even within a species; *P. mixtum* may have six or seven larval instars (Colbo and Wotton, 1981). Simuliid larvae reach a length of 4–12 mm, and being reasonably large and aggregated, are easily seen on submerged objects. Phoretic larvae of the subgenera *Lewisellum* and *Phoretomyia* of the genus *Simulium*, have microsculptured cuticles, which differentiate them from all except two species of free-living larvae (Williams, 1978).

The mature larva is actually a pharate pupa within the larval skin, and may move to a different site before pupating. Pupae of *A. bancrofti* occur on the downstream side of submerged substrates (Colbo and Moorhouse, 1979). In most species the pharate pupa spins a cocoon, often slipper-shaped with the closed end directed upstream and the open end downstream (Fig. 10.6). This alignment prevents the cocoon being torn off the substrate by the current. Construction of the cocoon takes about an hour and then the larval skin is shed.

Pupa

The head and thorax of the pupa are combined into a single cephalothorax, and there is a segmented abdomen (Fig. 10.6). The latter bears spines and hooks which engage with the threads of the cocoon and retain the pupa in place. The cephalothorax bears a pair of elongate, branched pupal gills, which trail downstream of the cocoon. They are homologous with the respiratory horns of the Culicidae and Ceratopogonidae, but they do not have open spiracles. The tubular branches of the gill bear vertical struts which support a very thin, outer, minutely perforated, trilaminate epicuticle and an inner fine meshwork. The enclosed air-filled space around the struts functions as a plastron, and Hinton (1976) considers that the water–air interface will be about 50% of the total plastron area. The shapes of the cocoon and gills are important characters in the identification of species.

The pupa, which does not feed, becomes progressively darker as the adult develops within, but the mature pupa takes on a silvery appearance as a film of air is secreted between the pharate adult and the pupal cuticle. When the pupal exuviae splits, the adult floats up to the surface in a bubble of air and immediately takes flight. Alternatively the newly emerged adult crawls up some emergent object to reach the air. The length of the life cycle varies with the species and environmental conditions. In temperate regions species may have one generation a year, while continuous breeding occurs in tropical species. The larval stage of *S. damnosum* can be completed in as little as 8 days. For that reason the Onchocerciasis Control Programme in West Africa treats breeding sites weekly (Walsh *et al.*, 1981). The life cycle from egg to adult can be completed in less than 2 weeks.

Adult

Adult emergence occurs predominantly in the daytime, depending on light and temperature. In *S. damnosum* 60–90% of the day's emergence has occurred by midday and there is no emergence at night. At 24–28°C peak emergence of *S. damnosum* occurs at 0600–0900 h and when the water temperature is lower (20–24°C) the peak is reached later in the morning between 0900 and 1200 h (Wenk, 1981).

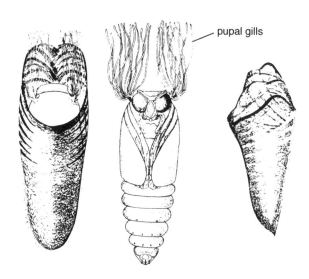

Fig. 10.6. Pupa and cocoon of *Simulium simile*. From left to right – dorsal view of cocoon containing a pupa; ventral view of pupa, removed from cocoon; lateral view of cocoon. The cocoon is arranged with its closed end directed into the current. Source: Reproduced by permission of the Minister, Supply and Services, Canada. From Cameron, A.E. (1922). *Agriculture Canada Entomological Bulletin No 20. The Morphology and Biology of a Canadian Cattle-infesting Blackfly, Simulium simile Mall. Diptera: Simuliidae.*

Mating

Mating occurs in close association with the breeding site, and in a few species occurs on the ground (Downes, 1962) but in the large majority of species it occurs on the wing when males form small swarms in association with visual markers. In *A. pestilens* male swarms orientate to *Callistemon viminalis* along the banks of the semi-permanent stream in which they breed (Moorhouse and Colbo, 1973). Male simuliids recognize the female up to a distance of 50 cm and pursue the female and attempt to couple. There appears to be no contact pheromone because males will attempt to mate with other males and with individuals of other species. In some species male swarms and mating occur in close proximity to the feeding sites of the females, which, for *S. ornatum* and *S. erythrocephalum*, are respectively the navel and ears of cattle. During mating a 2-chambered spermatophore is transferred to the female (Wenk, 1987).

Feeding

Ornithophilic and mammalophilic species are distinguished morphologically by the shape of the claws. In mammal feeders the claws are simple but in bird feeders they are toothed. This must be an adaptation to holding and penetrating feathers but Crosskey (1990) points out that this host preference is not absolute and that many species feed indiscriminately on both avian and mammalian hosts. Simuliids are exophilic, exophagic and largely diurnal, although Davies and Williams (1962) collected ten species in light traps in Scotland, mainly *S. ornatipes* and *S. tuberosum*. Some species will enter the natural openings of the body, nose, ear and eye, behaviour which is particularly worrying to livestock. As in other blood-sucking Nematocera both males and females feed on nectar, which is stored in the crop, and only the females are haematophagous with blood passing directly to the midgut.

Female simuliids can be classified reproductively as showing obligate auto-geny, i.e. females maturing all eggs without a blood meal; primiparous autogeny, i.e. females maturing the first egg batch without a blood meal but needing blood for each subsequent ovarian cycle; and obligate anautogeny when females need blood to mature every egg batch (Crosskey, 1990). When feeding, the female simuliid anchors its proboscis to the host by small hooks on the labrum and hypopharynx. The maxillae are protruded alternately, penetrating downwards and anchoring the proboscis more firmly. The mandibles cut into the skin with rapid scissor-like movements, penetrating to a depth of about 400 μm. Blood is ingested, using the cibarial and pharyngeal pumps. Blood-feeding takes about 4–5 min (Wenk, 1981).

Bionomics

Oviposition

Golini and Davies (1987) list the oviposition habits of 110 species. *S. damnosum* females oviposit communally in the short period between tropical sunset and darkness. Dense swarms of females lay their eggs on vegetation trailing in the water achieving densities of 2000–3000 eggs per square centimetre (Thomson, 1956). *A. pestilens* forms swarms of ovipositing females which scatter their eggs over the surface of the water, where they become incorporated in the sandy river bed and can survive two and a half years, providing they are kept permanently damp (Moorhouse and Colbo, 1973; Colbo and Moorhouse, 1974). Dalmat (1955) found that in Guatemala *S. ochraceum* dropped its eggs directly into the water; *S. callidum* laid its eggs one at a time on the inclined surfaces of rocks; and *S. metallicum*, in fast-flowing water, laid its eggs on leaves without landing, and in slower flowing water actually landed. Eggs of *S. argyreatum* can withstand dryness during autumn and winter when the temperatures are low, and they and the eggs of *S. pictipes* resist frost and ice to survive the winter and hatch in the spring (Kurtak, 1974; Rühm, 1975).

Breeding sites

Simuliids breed in running water, ranging from torrential mountain streams to slow-moving lowland rivers, and a few species are adapted to streams in which there is little perceptible current (Burger, 1987). Ross and Merritt (1987) list the major environmental factors which have been reported as affecting the larval distribution of 12 genera. In Newfoundland Lewis and Bennett (1975) found that the most significant factors affecting the distribution of simuliid larvae were current velocity, substrate type and water depth. Grunewald (1976) investigated the distribution of larvae of the *S. damnosum* complex and found that their breeding sites could be classified on the pH and conductivity of the water.

A small number of species (28) have evolved a phoretic association with decapod Crustacea (crabs, prawns) or Ephemeroptera (mayflies) in Africa and the Himalayan region (Crosskey, 1990). Larvae and pupae of *S. nyasalandicum* and *S. woodi* occur on the sides, the chelipeds, and the basal segments of the walking legs of the crab *Potamonautes pseudoperlatus*. They also occur on other species of crab (Raybould, 1968; Raybould and Yagunga, 1969). Eggs are not laid on the crab and the young larva must find its own phoretic partner. The most important of these phoretic simuliid species is *S. neavei*, a vector of onchocerciasis (see Chapter 31). In Africa mayfly phoretics are found mainly in heavily shaded, forest streams and crab phoretics are found in small, forest streams and larger and more open rivers (Crosskey, 1990).

When eggs are deposited in dense masses it is essential that the first instar larvae disperse. Larvae drift downstream attached to a silken thread or they can break the thread and drift with the current (Colbo and Wotton, 1981). Larvae drift throughout the 24 h and early instars of *A. bancrofti* show a diurnal tendency with a greater proportion of older instars drifting at night (Colbo and Moorhouse, 1979).

Biting habits

The females of most simuliids require a blood meal to develop eggs. *A. pestilens* feeds on many mammals including humans and is probably an opportunistic feeder, which may be related to the fact that it remains close to the breeding site. The closely related *A. bancrofti* is more selective and did not bite any of the six species offered – four mammals, including a man, and two birds (Hunter and Moorhouse, 1976a). In Scotland Davies *et al.* (1962) found that *S. tuberosum* fed on a wide range of hosts including humans, other mammals and birds; *S. latipes* fed largely on birds; and *S. reptans* and *S. monticola* took over 90% of their feeds on bovines.

The gonotrophic cycle may be remarkably short, being completed in 24 h in *A. pestilens* (Hunter and Moorhouse, 1976a), and in 2 days in *S. metallicum* (Ramirez-Pérez *et al.*, 1976). In *S. damnosum* the first cycle from blood meal to oviposition takes 3 or 4 days and thereafter eggs are laid at intervals of 4 to 5 days, the additional time being required for nectar feeding before the blood meal (Wenk, 1981).

Simuliids are essentially diurnal species and in open sunny situations *S. damnosum* tends to have a bimodal pattern of activity with peaks around 0900 h in the morning and 1700 h in the afternoon, but in shaded areas biting is more evenly distributed throughout the day (Kaneko *et al.*, 1973). The circadian rhythm of biting activity varies with the age of the flies, with parous females feeding earlier in the day than nullipars (Lewis, 1956). Although mainly diurnal, ten species of simuliids were taken in light traps in Scotland (Davies and Williams, 1962), three different species were trapped in Norway (Raastad and Mehl, 1972), and Service (1979) collected large numbers (3600 per night) of ovipositing *S. squamosum*, a member of the *S. damnosum* complex, in light traps in Ghana. Activity of simuliids is influenced by barometric pressure (Wellington, 1974) and in Ghana the numbers of simuliids biting increase with rising humidity and lowering barometric pressure (Crisp, 1956).

Host finding

Sutcliffe (1986) recognized three phases in host location by female blackflies. The first involved nectar feeding, mating and dispersal to bring the insect into the host's habitat. The second phase involved host searching, non-orientated flight driven by endogenous activity rhythms and hunger, and the third phase host location proper, orientated by external stimuli. Bradbury and Bennett (1974a) recognized three stages within the third phase: long range attraction initiated by host odour leading to an upwind response by the fly; nearer the host, orientation to carbon dioxide emitted by the host, and within 1.8 m orientation was visual. Simuliids can discriminate between colours of the same reflectance with *S. venustum* being attracted to blue, *P. mixtum* to black and *S. vittatum* to black, red and blue but not to yellow (Bradbury and Bennett, 1974b).

The visual component in host finding is supported by the fact that *S. erythrocephalum*, which feeds on the ears of cattle, will attack protruding parts of a dummy; and *S. ornatum*, which feeds near the navel, attacks the flat underparts of a model (Wenk, 1981). Higher numbers of *A. bancrofti*, another cattle feeder, were taken on a horizontally elongated trap than on a similar trap arranged vertically or on a square trap of the same surface area. The horizontal trap being the one most closely resembling cattle (Ballard and Barnes, 1988). Most (93%) *S. damnosum* land on the ankles of humans, but then ascend the leg and feed on the calves. Females of the anthropophilic *S. ochraceum* complex feed mainly on the back and shoulders, and very little on the legs below knee level (Crosskey, 1990). After landing, other stimuli, such as odour, sweat and other chemicals, are likely to be involved in probing (Wenk, 1981). The North American cattle-feeding *S. arcticum* is attracted by the emission of carbon dioxide and less so by cattle urine on its own, but the combination of CO_2 and urine is highly effective, increasing catches by $\times 3$ to $\times 20$ (Sutcliffe *et al.*, 1995). The ornithophilic species, *S. euryadminiculum*, is strongly attracted to extracts of the uropygial gland of the common loon (*Gavia immer*) (Bennett *et al.*, 1972).

Dispersal

Adult females of many species of simuliids disperse far from their breeding sites. Hocking (1953) has shown by laboratory experiments that *S. venustum* is capable of flying 116 km in still air, following a sugar meal. In the field, Baldwin *et al.* (1975) found that this species dispersed on average 9–13 km, but a few individuals covered 35 km in 2 days. Females of *S. arcticum* dispersed for distances of at least 150 km from the Athabasca River in western Canada in sufficient numbers to be a pest (Charnetski and Haufe, 1981). Wenk (1981) distinguishes different categories of dispersal in *S. damnosum*. There is linear dispersal along river courses in the gallery forest of the West African savanna; radial dispersal in the savanna during the rainy season, and in the forest region throughout the year; and differential dispersal, when nullipars disperse further than parous females, which tend to remain near the breeding sites.

Only parous females can transmit *Onchocerca volvulus*, and if they stay close to the breeding site then transmission of onchocerciasis will be concentrated in the vicinity of rivers and streams (Duke, 1975). There is also wind-borne dispersal of blackflies, and Garms *et al.* (1979) found that flies which invaded the Onchocerciasis Control Programme in West Africa were parous flies, many of which were infected with *Onchocerca* larvae. This indicates that, under certain conditions, parous females can disperse as widely, if not more widely, than nullipars.

Response to varying conditions

Simuliids have to survive periods when the temperature is too low to sustain normal activities, and when rivers cease to flow in the dry season, which may be of indeterminate length. Larvae of *P. mixtum* and *P. fuscum* grow actively during the winter at temperatures near freezing point (Lewis and Bennett, 1974) but larvae of *P. mysticum* are dormant below 4°C reducing their respiratory rates substantially and replacing trehalose in the haemolymph by high polyhydric alcohols. When the temperature rises above 4°C the reverse process is rapidly completed and growth resumes (Mansingh and Steele, 1973).

Simulium latipes was taken in light traps in Scotland in every month of the year in 1957, and in other years was present from March or April to November (Davies and Williams, 1962). Perhaps this species has overwintering adults, which can be active or dormant, depending upon the temperature. In the severe climate of high latitudes simuliids show various adaptations to survival, involving reducing the time spent in the adult stage, which is the most vulnerable to low temperatures and high winds. Eight out of the nine species restricted to the Canadian tundra, e.g. *S. baffinense*, are autogenous and have reduced mouthparts; their females do not need to seek a blood meal and are unable to feed. The risk to the species' survival if the males had to swarm is avoided by mating occurring on the ground where adults cluster near the breeding site. As an adaptation to that behaviour the eyes of the male are sometimes dichoptic as in the female, e.g. *Gymnopais dichopticus*. As a further adaptation, a species may be both autogenous and parthenogenetic with the adult female becoming gravid in the pupa, e.g. *P. ursinum*, and there is virtually no free adult life (Downes, 1962).

Aridity poses other problems. *A. pestilens* survives a dry season of uncertain length as viable eggs deep in the moist, sandy beds of transient rivers (Colbo and Moorhouse, 1979). Eggs of *C. pecuarum* laid in April remained dormant until November before developing and hatching in December. This carried the species over the summer months when many breeding places ceased to flow (Bradley, 1935a). Adult *S. damnosum* are regarded as having a maximum life span of 3 to 4 weeks (Le Berre, 1966) but this does not explain their appearance before the rivers begin to flow again after the dry season. Have these adults arrived from unidentified breeding sites which have persisted during the dry season? Or have these adults been aestivating (Crisp, 1956)?

Many species of *Simulium* have several generations a year. When such species overwinter as growing larvae, as in *S. monticola*, there are likely to be large size differences between adults produced in the different seasons. Winter larvae produce larger adults than summer larvae. Changes in the total biovolume of the adult were found to be inversely related to mean water temperature (Neveu, 1973). In addition size was also influenced by the quantity of food available and the photoperiod.

Medical and Veterinary Importance

Simuliids are important as pests in their own right and also as vectors of pathogens. Vectors need not be present in high enough density to be regarded as pests.

Vectors of disease

The most serious human disease associated with simuliids is onchocerciasis or river blindness, due to infection with the filarial worm *Onchocerca volvulus*. This disease exists in the Afrotropical and Neotropical regions and is dealt with in Chapter 31. Eight species of *Simulium* are involved in the transmission of human onchocerciasis in Latin America and 13 species plus four unnamed forms in Africa (Crosskey, 1990). Along the Transamazon highway in Brazil a haemorrhagic syndrome, associated with some deaths, occurred among immigrants to forested areas, and has been attributed to the intense biting of simuliids (Pinheiro *et al.*, 1974).

Simuliids are vectors of *Onchocerca* spp. to livestock (see Chapter 32) and transmit blood-dwelling Protozoa of the genus *Leucocytozoon* among birds including domestic poultry (see Chapter 27). They play only a minor role in the transmission of arboviruses, e.g. Venezuelan equine encephalitis in Colombia in 1967 (Cupp, 1987) and vesicular stomatitis in Colorado in 1982 (Francy *et al.*, 1988) but are capable of transmitting myxomatosis mechanically among rabbits (Joubert and Monnet, 1975).

Pest species

Simuliids have a well-deserved reputation as pests, particularly of livestock, in many areas of the world. Crosskey (1990) lists 29 species of Simuliidae as pests of humans or domestic animals of which 24 are species of *Simulium*, 3 of *Austrosimulium* and 1 each of *Prosimulium* and *Cnephia*. Along the lower reaches of the Mississippi River *C. pecuarum* caused the death of large numbers of livestock prior to 1897. This loss was reduced as the flooding of the river was contained by the building of levees, but even as late as 1931 the deaths of more than 1000 mules was attributed to *C. pecuarum* (Bradley, 1935b). Flood control has now eliminated *C. pecuarum* as a pest.

In Yugoslavia and Rumania the production of vast numbers of the Golubatz fly, *S. columbaschense* in the 1930s, caused thousands of deaths of livestock over a very wide area. This species bred in the Danube at the Iron Gate at Golubatz, from which it dispersed 100–250 km. Outbreaks of pest proportions required dry, warm weather and low water levels in the Danube in spring, creating extensive breeding sites (Baranov, 1935, 1937). Environmental changes have eliminated this pest (Crosskey, 1990). Even when the numbers of simuliids biting do not cause deaths, livestock are less thrifty, with lower weight gains in beef cattle and decreased milk production in dairy cattle. Following the floods of 1974, *A. pestilens* reduced milk yields by up to 15% in parts of Queensland (Hunter and Moorhouse, 1976b).

Control

Control of vectors of human onchocerciasis will be considered in Chapter 31. Since the early 1980s *B. t. israelensis* has been increasingly used to control pest blackflies, against which it has proved to be highly effective with little effect on non-target species (Molloy, 1990). Control of *S. arcticum* has been prosecuted in Saskatchewan for many years. It was originally achieved by the application of DDT to breeding sites. In the late 1960s DDT was replaced by the ecologically more acceptable methoxychlor. Changes in river volume resulted in *S. arcticum* being replaced by *S. luggeri*, which not only attacked livestock but also, unlike *S. arcticum*, severely harassed people (Fredeen, 1987).

References

Baldwin, W., West, A.S. and Gomery, J. (1975) Dispersal patterns of blackflies (Diptera: Simuliidae) tagged with ^{32}P. *Canadian Entomologist* 107, 113–118.

Ballard, J.W.O. and Barnes, A. (1988) Factors influencing silhouette trap captures of the blackfly, *Austrosimulium bancrofti* in Queensland, Australia. *Medical and Veterinary Entomology* 2, 371–378.

Baranov, N. (1935) New information on the Golubatz fly, *S. columbaczense*. *Review of Applied Entomology B* 23, 275–276.

Baranov, N. (1937) Contribution to the knowledge of the Golubatz fly. V. Study of the epidemiology of the fly in 1936. *Review of Applied Entomology B* 25, 249–250.

Bennett, G.F., Fallis, A.M. and Campbell, A.G. (1972) The response of *Simulium (Eusimulium) euryadminiculum* Davies (Diptera: Simuliidae) to some olfactory and visual stimuli. *Canadian Journal of Zoology* 50, 793–800.

Bradbury, W.C. and Bennett, G.F. (1974a) Behavior of adult Simuliidae (Diptera). II. Vision and olfaction in near-orientation and landing. *Canadian Journal of Zoology* 52, 1355–1364.

Bradbury, W.C. and Bennett, G.F. (1974b) Behaviour of adult Simuliidae (Diptera). I. Response to color and shape. *Canadian Journal of Zoology* 52, 251–259.

Bradley, G.H. (1935a) The hatching of eggs of the southern buffalo gnat. *Science* 82, 277–278.

Bradley, G.H. (1935b) Notes on the southern buffalo gnat *Eusimulium pecuarum* (Riley) (Diptera: Simuliidae). *Proceedings of the Entomological Society of Washington* 37, 60–64.

Brockhouse, C.L. and Tanguay, R.M. (1992) Characterization of the aquatic glue proteins of black flies (Diptera; Simuliidae). *Proceedings of the XIX International Congress of Entomology. Abstracts* 517.

Burger, J.F. (1987) Specialized habitat selection by black flies. In: Kim, K.C. and Merritt, R.W. (eds) *Black Flies. Ecology, Population Management and Annotated World List.* The Pennsylvania State University, University Park, Pennsylvania, pp. 129–145.

Chance, M.M. (1970) The functional morphology of the mouthparts of blackfly larvae (Diptera: Simuliidae). *Quaestiones Entomologicae* 6, 245–284.

Charnetski, W.A. and Haufe, W.O. (1981) Control of *Simulium arcticum* Malloch in northern Alberta, Canada. In: Marshall Laird (ed.) *Blackflies: The Future for Biological Methods in Integrated Control.* Academic Press, London, pp. 117–132.

Colbo, M.H. and Moorhouse, D.E. (1974). The survival of eggs of *Austrosimulium pestilens* Mack. and Mack. (Diptera: Simuliidae). *Bulletin of Entomological Research* 64, 629–632.

Colbo, M.H. and Moorhouse, D.E. (1979) The ecology of pre-imaginal Simuliidae (Diptera) in south-east Queensland, Australia. *Hydrobiologia* 63, 63–79.

Colbo, M.H. and Wotton, R.S. (1981) Preimaginal blackfly bionomics. In: Marshall Laird (ed.) *Blackflies The Future for Biological Methods in Integrated Control.* Academic Press, London, pp. 209–226.

Coupland, J.B. (1991) Oviposition response of *Simulium reptans* (Diptera: Simuliidae) to the presence of conspecific eggs. *Ecological Entomology* 16, 11–18.

Crisp, G. (1956) Simulium *and Onchocerciasis in the Northern Territories of the Gold Coast.* H.K. Lewis, London.

Crosskey, R.W. (1990) *The Natural History of Blackflies.* John Wiley, Chichester, UK.

Currie, D.C. and Craig, D.A. (1987) Feeding strategies of larval black flies. In: Kim, K.C. and Merritt, R.W. (eds) *Black Flies. Ecology, Population Management and Annotated World List.* The Pennsylvania State University, University Park, Pennsylvania, pp. 155–170.

Cupp, E.W. (1987) The epizootiology of livestock and poultry diseases associated with black flies. In: Kim, K.C. and Merritt, R.W. (eds) *Black Flies. Ecology, Population Management and Annotated World List.* The Pennsylvania State University, University Park, Pennsylvania, pp. 387–395.

Dalmat, H.T. (1955) The blackflies (Diptera: Simuliidae) of Guatemala and their role as vectors of onchocerciasis. *Smithsonian Miscellaneous Collections* 125, 1–425.

Davies, L., Downe, A.E.R., Weitz, B. and Williams, C.B. (1962) Studies on black flies (Diptera: Simuliidae) taken in a light trap in Scotland. II. Blood-meal identification by precipitin tests. *Transactions of the Royal Entomological Society of London* 114, 21–27.

Davies, L. and Williams, C.B. (1962) Studies on black flies (Diptera: Simuliidae) taken in a light trap in Scotland. I. Seasonal distribution, sex ratio and internal condition of catches. *Transactions of the Royal Entomological Society of London* 114, 1–20.

Downes, J.A. (1962) What is an arctic insect? *Canadian Entomologist* 94, 143–162.

Duke, B.O.L. (1975) The differential dispersal of nulliparous and parous *Simulium damnosum. Tropenmedizin und Parasitologie* 26, 88–97.

Francy, D.B., Moore, C.G., Smith, G.C., Jakob, W.L., Taylor, S.A. and Calisher, C.H. (1988) Epizootic vesicular stomatitis in Colorado, 1982: Isolation of virus from insects collected along the northern Colorado Rocky Mountain Front Range. *Journal of Medical Entomology* 25, 343–347.

Fredeen, F.J.H. (1964) Bacteria as food for blackfly larvae (Diptera: Simuliidae) in laboratory cultures and in natural streams. *Canadian Journal of Zoology* 42, 527–548.

Fredeen, F.J.H. (1987) Black flies: approaches to population management in a large temperate-zone river system. In: Kim, K.C. and Merritt, R.W. (eds) *Black Flies. Ecology, Population Management and Annotated World List.* The Pennsylvania State University, University Park, Pennsylvania, pp. 295–304.

Garms, R., Walsh, J.F. and Davies, J.B. (1979) Studies on the reinvasion of the Onchocerciasis Control Programme in the Volta River basin by *Simulium damnosum* s.l. with emphasis on the south-western areas. *Tropenmedizin und Parasitologie* 30, 345–362.

Golini, V.I. and Davies, D.M. (1987) Oviposition of black flies. In: Kim, K.C. and Merritt, R.W. (eds) *Black Flies. Ecology, Population Management and Annotated World List.* The Pennsylvania State University, University Park, Pennsylvania, pp. 261–275.

Grunewald, J. (1976) The hydrochemical and physical conditions of the development of the immature stages of some species of the *Simulium (Edwardsellum) damnosum* complex (Diptera). *Tropenmedizin und Parasitologie* 27, 438–454.

Hart, D.D. (1987) Processes and patterns of larval competition in larval black flies. In: Kim, K.C. and Merritt, R.W. (eds) *Black Flies. Ecology, Population Management and Annotated World List.* The Pennsylvania State University, University Park, Pennsylvania, pp. 109–128.

Hinton, H.E. (1976) The fine structure of the pupal plastron of simuliid flies. *Journal of Insect Physiology* 22, 1061–1070.

Hocking, B. (1953) The intrinsic range and speed of flight of insects. *Transactions of the Royal Entomological Society of London* 104, 223–345.

Hunter, D.M and Moorhouse, D.E. (1976a) Comparative bionomics of adult *Austrosimulium pestilens* Mackerras and Mackerras and *A. bancrofti* (Taylor) (Diptera: Simuliidae). *Bulletin of Entomological Research* 66, 453–467.

Hunter, D.M. and Moorhouse, D.E. (1976b) The effects of *Austrosimulium pestilens* on the milk production of dairy cattle. *Australian Veterinary Journal* 52, 97–99.

Joubert, L. and Monnet, P. (1975) Vérification expfimentale du rôle des simulies (*Tetisimulium bezzii* Corti, 1914 et *Odagmia* group *ornatum*) dans la transmission du virus myxomateux en Haute-Provence. *Revue de Médecine Vétérinaire* 126, 617–634.

Kaneko, K., Saito, K. and Wonde, T. (1973) Observations on the diurnal rhythm of the biting activity of *Simulium damnosum* in Omo-Gibe and Gojjeb Rivers, south-west Ethiopia. *Japanese Journal of Sanitary Zoology* 24, 175–180.

Kim, K.C. and Merritt, R.W. (eds) (1987) *Black Flies. Ecology, Population Management and Annotated World List.* The Pennsylvania State University, University Park, Pennsylvania.

Kurtak, D. (1974) Overwintering of *Simulium pictipes* Hagen (Diptera: Simuliidae) as eggs. *Journal of Medical Entomology* 11, 383–384.

Kurtak, D.C. (1979) Food of black fly larvae (Diptera: Simuliidae): seasonal changes in gut contents and suspended material at several sites in a single watershed. *Quaestiones Entomologicae* 15, 357–374.

Ladle, M. (1972) Larval Simuliidae as detritus feeders in chalk streams. *Memorie dell'Instituto Italiano di Idrobiologia* 29 (Supplement), 429–439.

Le Berre, R. (1966) Contribution a l'étude biologique et écologique de *Simulium damnosum* Theobald, 1903 (Diptera: Simuliidae). *Memoires ORSTOM* 17, 1–204.

Lewis, D.J. (1956). Biting times of parous and nulliparous *Simulium damnosum*. *Nature, London* 178, 98–99.

Lewis, D.J. and Bennett, G.F. (1974) The blackflies (Diptera: Simuliidae) of insular Newfoundland. II. Seasonal succession and abundance in a complex of small streams on the Avalon Peninsula. *Canadian Journal of Zoology* 52, 1107–1113.

Lewis, D.J. and Bennett, G.F. (1975) The blackflies (Diptera: Simuliidae) of insular Newfoundland. III. Factors affecting the distribution and migration of larval simuliids in small streams on the Avalon Peninsula. *Canadian Journal of Zoology* 53, 114–123.

Mansingh, A. and Steele, R.W. (1973) Studies on insect dormancy. 1. Physiology of hibernation in the larvae of the blackfly *Prosimulium mysticum* Peterson. *Canadian Journal of Zoology* 51, 611–618.

Molloy, D.P. (1990) Progress in the biological control of black flies with *Bacillus thuringiensis israelensis*, with emphasis on temperate climates. In: Barjac, H. de and Sutherland, D.J. (eds) *Bacterial Control of Mosquitoes and Blackflies. Biochemistry, Genetics and Applications of* Bacillus thuringiensis israelensis *and* Bacillus sphaericus. Rutgers University Press, New Brunswick, pp. 161–186.

Moorhouse, D.E. and Colbo, M.H. (1973) On the swarming of *Austrosimulium pestilens* Mackerras and Mackerras (Diptera: Simuliidae). *Journal of the Australian Entomological Society* 12, 127–130.

Neveu, A. (1973) Variations biométriques saisonnières chez les adultes de quelques especes de Simuliidae (Diptera: Nematocera). *Archives de Zoologie Expérimentale et Generale* 114, 261–270.

Peterson, B.V. and Dang, P.T. (1981) Morphological means of separating siblings of the *Simulium damnosum* complex (Diptera: Simuliidae). In: Marshall Laird (ed.) *Blackflies: The Future for Biological Methods in Integrated Control*. Academic Press, London, pp. 45–56.

Pinheiro, F.P., Bensabath, G., Costa, D., Maroja, O.M., Lins, Z.C. and Andrade, A.H.P. (1974) Haemorrhagic syndrome of Altamira. *Lancet* 1, 639–642.

Raastad, J.E. and Mehl, R. (1972) Night activity of black flies (Diptera: Simuliidae) in Norway. *Norsk Entomologisk Tidsskrift* 19, 172–173.

Ramirez-Pérez, J., Rassi, E., Convit, J. and Ramirez, A. (1976) Importancia epedemiológica de los grupos de edad en las poblaciones de *Simulium metallicum* (Diptera: Simuliidae) en Venezuela. *Boletin de la Oficina Sanitaria Panamericana* 80, 105–122.

Raybould, J.N. (1968) Studies on the immature stages of the *Simulium neavei* Roubaud complex and their associated crabs in the Eastern Usambara Mountains in Tanzania. I. Investigations in rivers and large streams. *Annals of Tropical Medicine and Parasitology* 63, 269–287.

Raybould, J.N. and Yagunga, A.S.K. (1969) Studies on the immature stages of the *Simulium neavei* Roubaud complex and their associated crabs in the Eastern Usambara Mountains in Tanzania. II. Investigations in small heavily shaded streams. *Annals of Tropical Medicine and Parasitology* 63, 289–300.

Ross, D.H. and Merritt, R.W. (1987) Factors affecting larval black fly distributions and population dynamics. In: Kim, K.C. and Merritt, R.W. (eds) *Black Flies. Ecology, Population Management and Annotated World List*. The Pennsylvania State University, University Park, Pennsylvania, pp. 90–108.

Rühm, W. (1975) Freilandbeobachtungen zum Funktionskreis der Eiablage verschiedener Simuliidenarten unter besonderer Beruchsichtigung von *Simulium argyreatum* Meig. (Dipt.: Simuliidae). *Zeitschrift für Angewandte Entomologie* 78, 321–334.

Service, M.W. (1979) Light trap collections of ovipositing *Simulium squamosum* in Ghana. *Annals of Tropical Medicine and Parasitology* 73, 487–490.

Sutcliffe, J.F. (1986) Black fly host location: a review. *Canadian Journal of Zoology* 64, 1041–1053.

Sutcliffe, J.F., Shemanchuk, J.A. and McKeown, D.B. (1995) Preliminary survey of odours that attract the black fly, *Simulium arcticum* (IIS-10.11) (Diptera: Simuliidae) to its cattle hosts in the Athabasca region of Alberta. *Insect Science and its Application* (in press).

Thomson, R.C.M. (1956) Communal oviposition in *Simulium damnosum* Theobald (Diptera: Simuliidae). *Nature, London* 178, 1297–1299.

Walsh, J.F., Davies, J.B. and Cliff, B. (1981) World Health Organization Onchocerciasis

Control Programme in the Volta River basin. In: Marshall Laird (ed.) *Blackflies: The Future for Biological Methods in Integrated Control*. Academic Press, London, pp. 85–103.

Wellington, W.G. (1974) Black-fly activity during cumulus-induced pressure fluctuations. *Environmental Entomology* 3, 351–353.

Wenk, P. (1981) Bionomics of adult blackflies. In: Marshall Laird (ed.) *Blackflies: The Future for Biological Methods in Integrated Control*. Academic Press, London, pp. 259–279.

Wenk, P. (1987) Swarming and mating behavior of black flies. In: Kim, K.C. and Merritt, R.W. (eds) *Black Flies. Ecology, Population Management and Annotated World List*. The Pennsylvania State University, University Park, Pennsylvania pp. 215–227.

Williams, T.R. (1974) Egg membranes of Simuliidae. *Transactions of the Royal Society of Tropical Medicine and Hygiene* 68, 15–16.

Williams, T.R. (1978) Cuticular microsculpture in larval blackflies (Diptera: Simuliidae). *Journal of Anatomy* 127, 214–215.

Wotton, R.S. (1976) Evidence that blackfly larvae can feed on particles of colloidal size. *Nature, London* 261, 697.

Tabanidae (Horseflies, Deer Flies, Clegs)

11

Tabanids are large, stout-bodied Brachycera (Fig. 11.1), ranging in wing length from 6 to 30 mm. The large, bean-shaped head is much broader than long and the eyes are particularly well developed, as befits predominantly diurnal creatures. As in the Simuliidae, the sexes can be differentiated on eye size with the males being holoptic and the females dichoptic. The frons, separating the eyes in the female, is wider in *Haematopota* (cleg) than in *Tabanus* (horsefly). The eyes are often brilliantly coloured, being spotted in *Chrysops*, with zigzag bands in *Haematopota*, and unicolorous or horizontally banded in *Tabanus* (Oldroyd, 1973). The eye colours disappear shortly after death.

The palps are 2-jointed with the second segment being particularly prominent. The antennae are porrect, i.e. stiffly projecting forwards, and consist of scape, pedicel and flagellum. The 3rd segment of the antenna, i.e. 1st of the flagellum, is greatly enlarged. The number of flagellar segments varies from four to eight. The mouthparts are adapted for both blood-sucking and lapping (see Chapter 4). Both sexes feed on nectar and in most species the female also feeds on blood but in some species the females have reduced mandibles and are solely nectar feeders. This occurs particularly in the Pangoniinae.

The wing venation is well developed with all wing veins being obvious (Fig. 11.2). The venation is characteristic, but not diagnostic, similar venations occurring in related brachyceran families. Both the radius and the media are 4-branched with R_4 and R_5 straddling the apex of the wing. The discal cell is located more or less in the centre of the wing. The anal cell may be open or closed, i.e. the first anal vein may join Cu, before the wing margin. The squamae are large and obvious. The stout legs end in three pads because the empodium is pad-like and similar to the pulvilli (Fig. 2.10).

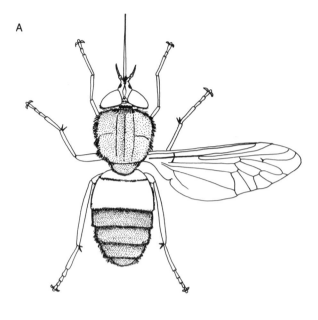

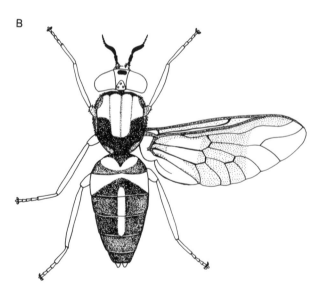

Fig. 11.1. (and opposite) Female Tabanidae: (**A**) *Pangonia*; (**B**) *Chrysops*; (**C**) *Haematopota*; and (**D**) *Tabanus parvicallosus*. Drawn from pinned specimens supplemented by colour plates in Chvala *et al.* (1972).

C

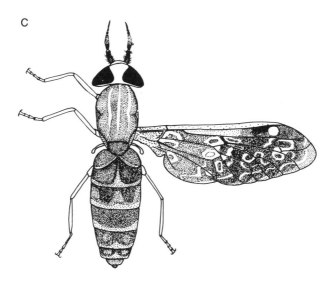

D

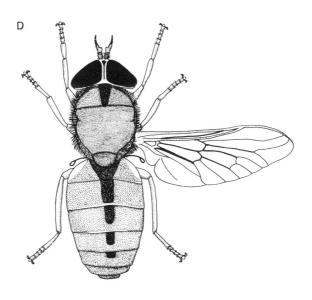

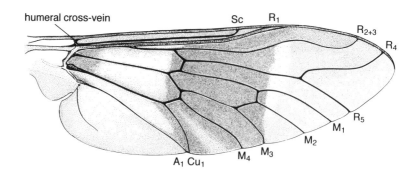

Fig. 11.2. Wing of *Chrysops australis* (x 11).

Classification

More than 4000 species of tabanids have been described. The classification of
Mackerras (1954) is generally accepted (Oldroyd, 1952–57; Chvala *et al.*, 1972;
Chainey, 1993) and is followed here. Four subfamilies are recognized on the basis
of the male and female terminalia. One of these, the Scepsidinae, includes only
eight species of non-blood-sucking tabanids of which seven are found in Africa and
one in South America. The Pangoniinae are of little economic importance but the
Chrysopsinae, main genus *Chrysops*, and the Tabaninae, main genera *Tabanus* and
Haematopota, are of considerable economic significance. These three genera and
the Pangoniinae may be distinguished by the following characters.

Pangoniinae

Head with functional ocelli; proboscis longer than head and may exceed length
of insect (Fig. 11.1A); antennae with seven or eight segments in the flagellum; a
pair of large spurs apically on the hind tibiae (all tabanids have apical spurs on
the mid tibiae); wings clear or dusky. Largely tropical or subtropical in
distribution.

Chrysops

Head with functional ocelli; proboscis not longer than head (Fig. 11.1B); antennal
flagellum with five segments; apical spurs on hind tibiae small, and may be hidden
in hair; wings usually with costal region dark and a single, broad, transverse, dark
band (Fig. 11.2). Mainly Holarctic and Oriental.

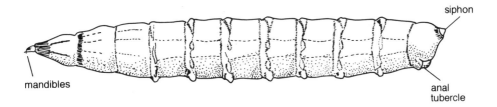

Fig. 11.3. Ninth larval instar of *Haematopota pluvialis* (x 23). Redrawn from Cameron (1934).

Tabanus

Head with, at most, vestigial ocelli; proboscis shorter than head; antennal flagellum with five segments; no apical spurs on hind tibiae; wings usually clear but may be dark or banded. Worldwide.

Haematopota

As for *Tabanus* but only four segments in the antennal flagellum and wings mottled (Fig. 11.1C). Palaearctic, Afrotropical, Oriental. Absent from Australia and rare in the Americas.

Life Cycle

Eggs are laid in masses of special design on leaves or rocks or debris overhanging water. An egg mass may contain 200 to 1000 eggs. Eggs of *Chrysops* are often laid in a single layer, and those of *Tabanus* and *Haematopota* stratified into three to four and two to three layers, respectively (Chvala *et al.*, 1972). The creamy white eggs darken with age and are dark grey to black at hatching, which occurs in four or more days depending on temperature (Hafez *et al.*, 1970c).

Larva

The larva emerges from the egg using a hatching spine and moults soon after emergence. The 2nd-stage larva does not feed and is positively phototactic, moving over the surface of the substrate in close association with water. After 3 to 6 days it moults into the 3rd instar which is negatively photactic and it burrows into the substrate, where it will spend many months. The number of moults is variable even in the same species and 7 to 11 instars may occur during larval development (Chvala *et al.*, 1972).

The mature larva is a greyish white, soft bodied, cylindrical grub (Fig 11.3). In common with the larvae of other Brachycera the head is reduced and retractile into the thorax. It bears a pair of simple eyes and a pair of piercing mandibles. The larvae of *Tabanus* and *Haematopota* are carnivorous and cannibalistic whereas those of *Chrysops* feed on plant remains (Oldroyd, 1973). Consequently, *Chrysops* larvae are often found in considerable density compared to carnivorous larvae. The larval thorax and abdomen merge imperceptibly. Ventral abdominal pads and smaller dorsal pads facilitate movement of the larva.

The larva is metapneustic with the spiracles opening at the end of a siphon, located dorsally on the 8th abdominal segment. The siphon is variable in shape, being short and conical in *H. pluvialis* (Cameron, 1934) and moderately long and blunt in *T. septentrionalis* (Cameron, 1926). Graber's organ is an internal structure, located dorsally of the gut and anterior to the siphon. Its function is unknown, and it appears as a longitudinal row of small, paired, dark bodies. The number of pairs is approximately related to the number of moults which the larva has undergone, but this relationship is not consistent (Cameron, 1934; Chvala *et al.*, 1972).

Pupa

At pupation the larva moves to the edge of aquatic habitats or to the surface of edaphic habitats. *T. biguttatus*, which breeds in temporary ponds, has a pattern of behaviour at pupation, which is designed to avoid being exposed to predators and parasites when the pond dries out and the mud cracks. The larva comes near to the surface, and then descends on a spiral course to a depth of 8–10 cm, isolating a central core of mud. It then moves upwards on the outside of the core, and near the surface burrows into it, hollowing out the interior to form a pupal cell. The entrance to the cell is blocked to deter predators. When the mud dries the soil cracks at the line of weakness made by the spiral but the mud cell remains intact. At emergence the pupa rasps its way through the cap of the cell and the adult emerges (Lamborn, 1930; Oldroyd, 1952–57).

The tabanid pupa is obtect (Fig. 11.4) and with limited movement. There is no compound cephalothorax but head, thorax and abdomen are distinct. Respiration occurs through kidney-shaped, thoracic spiracles and seven pairs of abdominal spiracles on short, lateral projections (Fig. 11.4). The abdominal segments are each fringed with a row of stout bristles and the terminal segment bears a spiny aster (Fig. 11.4). The function of these spines is to give the pupa purchase, to enable it to move up and down in the substrate or pupal cell, permitting it to move away from adverse conditions at the surface, and to move up for emergence at the appropriate time. The head is often rugose and is used to effect escape from the soil. The pupal stage lasts about 1 to 3 weeks.

Adult

The newly emerged female mates before seeking a blood meal. Males form swarms, usually early in the morning, and virgin females entering the swarms are seized by the males with copulation beginning in the air and insemination being completed on the ground, taking about 5 min. In forests, male swarms occur above the canopy. Recognition of the female by the male is visual (Allan *et al.*, 1987). The mated female seeks a blood meal which is required in most species for the development of the ovaries, although some species, e.g. *T. nigrovittatus* (Bosler and Hansens, 1974), are autogenous for the first batch of eggs, requiring a blood meal for second and subsequent batches, and some genera, e.g. *Pangonia* (Chvala

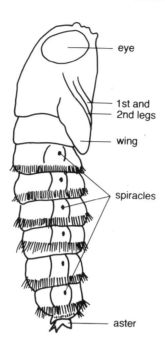

Fig. 11.4. Lateral view of pupa of a species of *Hybomitra*. Redrawn from Chvala *et al.* (1972).

et al., 1972) and *Scaptia* (Mackerras, 1955) have many flower-feeding, non-blood-sucking species.

Most species are diurnal, but a few, e.g. *T. paradoxus* are nocturnal. Activity is higher on warm, sunny days with low wind speeds. European species have a temperature threshold of 13 to 15°C and reach maximum activity at 25°C, providing that the wind speed is less than 4 m s^{-1} (Chvala *et al.*, 1972). Tabanids have evolved in close association with the ungulates (Oldroyd, 1973) and feed mainly on mammals and only rarely on birds. In feeding, the female will imbibe 20 to 200 mg of blood (Chvala *et al.*, 1972) but the loss of blood from the host is frequently greater as the mouthparts of the tabanid make a relatively large puncture, which continues to ooze blood after the mouthparts have been withdrawn. An average-sized female *T. fuscicostatus* (wing length = 8.80 mm) matures 174 eggs on a bovine blood meal of 55 mg (Leprince and Foil, 1993).

Bionomics

Oviposition

Gravid female tabanids are selective of the plants on which they oviposit. Thus *C. discalis* oviposits only on *Scirpus americanus* and not on other plants (Gjullin and Mote, 1945), while various species of *Chrysops* selected *Pontederia cordata* and three other plants for oviposition out of 13 common plants in the oviposition site at the lake edge (Foster *et al.*, 1973). In the field *T. taeniola* laid an average of 259 eggs (Hafez *et al.*, 1970c). The eggs of *C. discalis* hatched in 5 to 6 days and the young larvae dispersed widely from the site of hatching, being carried by water movement as they floated at the surface. Mature larvae of *C. discalis* have been found 10 m from the shore in shallow, alkaline lakes under 60 cm of water and at a depth of 5 to 10 cm in the underlying mud (Gjullin and Mote, 1945).

Breeding sites and immature stages

Three broad divisions of breeding sites have been recognized depending on the proportion of water which they contain. *Chrysops* larvae are generally hydrobionts, occurring in the wettest situations; *Tabanus* larvae are hemihydrobionts, occurring in soil near water; while *Haematopota* larvae are edaphic, being found in soil (Chvala *et al.*, 1972; Soboleva and Bodrova, 1973). Lane (1976) found *Chrysops* larvae to be associated with large amounts of decaying organic matter, on which presumably they fed: *Hybomitra* in mosses; larvae of *Sylvius* in sand and silt above the margins of flowing water; while larvae of *Tabanus* were generally more widely distributed, with those of *T. punctifer* occurring in every semi-aquatic habitat except tree holes. Within a habitat larvae may be generally distributed as are those of *C. fuliginosus* and *T. nigrovittatus* in *Spartina* salt marshes (Dukes *et al.*, 1974).

Larvae of *T. taeniola* pass through seven instars usually but the range is six to nine instars before pupation (Hafez *et al.*, 1970c). Although the larvae of *C. fuliginosus* are widespread through the salt marsh, adult emergences occur within 9 m of the ecotone where the influence of fresh water is apparent (Hansens and Robinson, 1973). Pupae of *T. taeniola* were able to reach the surface from a depth of 2 m in dry sand but from only 10 cm when the sand was wet, the more natural situation (Hafez *et al.*, 1970a). Speed of development is temperature-dependent. At a constant high temperature, i.e. 32 and 35°C, *T. taeniola* completes its life cycle from egg to adult in 10 to 11 weeks but at 22°C it takes 42 weeks. This 'hibernation' is a function of temperature and season. At 32 and 35°C there is no 'hibernation' while at 22°C, 'hibernation' always occurs with the prepupal stage being prolonged. At 27°C 'hibernation' occurs in larvae, which emerge from eggs laid after August, but is absent in larvae hatching from eggs laid in May to August (Hafez *et al.*, 1970c). Pavlova (1974) calculated the threshold temperatures and total day-degrees required for the development of the pupal stage of ten species of tabanids. Thresholds varied from 5.7 to 10.1°C and day-degrees from 92 to 192. For *T. autumnalis*, *H. pluvialis* and *C. relictus* the respective figures were: 8.9, 7.2, 8.7°C and 176, 170 and 104 day-degrees.

Adult mating and feeding

Males of *C. fuliginosus* become active earlier in the day than females. Mating and blood-feeding are temperature-dependent with mating occurring at 19 to 20°C, and females becoming host-seeking at 24 to 25°C (Catts and Olkowski, 1972). The effect of this is for mating to occur early in the morning and blood-feeding later in the day.

Sugars play a large part in the survival of tabanids, although they are inadequate for egg production in haematophagous species. With constant access to water and either blood or sugar solution, *C. silacea* survived considerably longer on sugar but laid no eggs (Crewe and Beesley, 1963). *T. taeniola* fed on sugar within a few hours of emergence but did not take its first blood meal for several days (Hafez *et al.*, 1970a); 59 to 97% of *T. iyoensis* coming to take a blood meal had previously fed on sugar (Watanabe and Kamimura, 1975). Both sexes of *T. sackeni* were taken dipping at pools of water; a pattern of behaviour noted in other species of tabanid. The female *T. sackeni* were developing eggs and had little in their crops (Taylor and Smith, 1989).

Visual cues were the most important to host-seeking tabanids when the host was moving, while a stationary host was located either visually or olfactorily, the two stimuli reinforcing each other (Vale and Phelps, 1974). The presence of dry ice doubled the catch of tabanids, mainly *T. molestus*, compared to octenol. In combination the two compounds had a synergistic effect with the catch being three times that of dry ice on its own (French and Kline, 1989). Species feed preferentially on different parts of the host. *C. fuliginosus* attacks the head of humans and *C. atlanticus* the upper parts of a walking person, but the lower limbs when the person is seated (Anderson, 1973). In Zimbabwe most tabanid species (9/16) alighted preferentially on the lower legs of cattle, *T. unilineatus* landed widely on the underside of a standing ox and *H. albihirta* on the upper parts of the body (Phelps and Holloway, 1990).

Adult activity and dispersal

The activity of tabanids is greatly influenced by meteorological conditions, especially light intensity and temperature (Polyakov and Polyakov, 1973). The flight activity of three species of salt marsh tabanids responded differently to light intensity. *C. fuliginosus* was most active under very bright conditions (100,000 lux); *T. nigrovittatus* under bright, warm conditions (40,000 lux; 25°C); while *C. atlanticus* was most active under hot, overcast conditions (5000 lux; 30°C) (Dale and Axtell, 1975). Burnett and Hays (1974) found barometric pressure to be the most important meteorological factor influencing the flight of *T. pallidescens* and *T. fulvulus*, which formed 60% of their total catch of tabanids. Both species were influenced by evaporation, but the catches of *T. fulvulus* were more closely related to changes in rate of evaporation, and those of *T. pallidescens* to the actual rate of evaporation. Three species of *Tabanus* showed marked bimodal, diurnal activity in sunny weather, but activity became irregular under cloudy conditions unless the air temperature was above 25°C, when it remained bimodal (Sasakawa *et al.*, 1969). Some species, e.g. *C. fuliginosus*, and *T. taeniola*, normally have only one period of activity during the day (Anderson, 1973; Kangwagye, 1973).

In the open, tabanids, such as *T. nigrovittatus*, remain close to the ground (Schultze *et al.*, 1975) but in the forest many species occur in the canopy, e.g. *C. langi*, while *C. silacea* feeds in the canopy and only comes to the forest floor in clearings or when attracted by wood smoke. Wood smoke increases the biting of *C. silacea* at ground level by more than tenfold, and that of *C. dimidiata* by nearly fivefold (Duke, 1959). The richness of the canopy fauna was appreciated when a nest of *Bembix bequaerti dira* was examined. This wasp builds its nest at ground level but hunts in the canopy. The nest contained 26 species of tabanids, of which four were species new to science, two had previously been rare, and there were also the unknown males of ten more species (Oldroyd, 1952–57). Clearly our limited view of the tabanid fauna from the forest floor is but a pale reflection of the teeming life of the canopy.

Tabanid flies differ in the habitats they haunt. Inaoka (1975) recognized three groups: (i) species that were confined to forest; (ii) those that occurred on open land; and (iii) those that were widespread, i.e. eurytopic. With the extension of pasture and rice production in recent years in Hokkaido, there has been a change in the tabanid fauna from one dominated by forest/eurytopic species to an openland/eurytopic fauna, with a corresponding decrease in *H. tristis*, and an increase in *T. nipponicus*.

Although tabanid flies have an intrinsic flight range of over 50 km (Hocking, 1953) they are not noted for their wide dispersal. The salt marsh tabanids *T. nigrovittatus*, *C. fuliginosus* and *C. atlanticus* dispersed less than 200 m from the marshes from which they emerged (Hansens and Robinson, 1973; Schultze *et al.*, 1975), *C. silacea* and *T. iyoensis* dispersed 1 to 3 km (Beesley and Crewe, 1963; Inoue *et al.*, 1973) and *C. discalis* can be numerous 7 km from its source (Gjullin and Mote, 1945) but these flights represent only a small fraction of the potential flight range of tabanids.

Survival

In the field, adults live a maximum of 3 to 4 weeks, and produce five to six batches of eggs (Beesley and Crewe, 1963; Hafez *et al.*, 1970b). The survival rate of *T. iyoensis* has been calculated to be 0.73 per day which gives a survival of 11% after one week, and 0.13% after 3 weeks. Making certain reasonable assumptions, Inoue *et al.* (1973) have calculated the emergence of *T. iyoensis* in a linear habitat to be 14,000 per metre of river.

With the larval stage being very long, tabanids pass adverse conditions in that stage. Little is known about the ways in which tabanids survive drought, except for the pupal mud cells constructed by *T. biguttatus* and *T. conspicuus* (Parsons, 1971) but presumably larvae survive deep in soil, where the atmosphere is moist. The cold season can be passed in the larval stage, but a problem arises when the soil freezes. *T. autumnalis* overwinters as a larva at a depth of 2 to 20 cm, and 50 to 100 cm above the waterline of small lakes, where it is warmed by the winter sun. Larvae of this species can survive $-4°C$ and, as the temperature 10 cm deep in the soil remains above that threshold, the larvae survive. After cold acclimatization, 60% of *T. autumnalis* larvae can survive $-6°C$. *C. caecutiens* larvae remain in the soil below water, where they are protected by a layer of ice (Boshko and Shevtsova, 1975).

Medical and Veterinary Importance

The role of tabanids as vectors of disease has been reviewed by Krinsky (1976) and Foil (1989). Tabanids are biological vectors of three species of filaroid worms:

Elaeophora schneideri, the arterial worm of sheep; *Loa loa*, the cause of Calabar or fugitive swellings in humans in West Africa; and *Dirofilaria roemeri*, a parasite of macropodid marsupials (see Chapter 32). Tabanids are also biological vectors of the blood-dwelling sporozoan, *Haematoproteus metchnikovi* of turtles, and there is evidence that tabanids are biological vectors of *Trypanosoma theileri*, a benign parasite of cattle (Böse and Heister, 1993) (see Chapters 27 and 29).

The mouthparts of tabanids are particularly well suited to the mechanical transmission of blood-dwelling pathogens from host to host (Chapter 4). This ability is enhanced when the species is a determined feeder such as *T. taeniola* which passes readily from host to host (Wiesenhutter, 1975). Leclercq (1952) has summarized the extensive experiments of Nieschultz made between 1925 and 1930, on the mechanical transmission of *Trypanosoma evansi* which causes surra in camels and horses (see Chapter 29). Species of *Tabanus* were more efficient vectors than those of *Chrysops* and *Haematopota*, and all tabanids were better vectors than mosquitoes or biting muscids, such as *Stomoxys*. The probability of transmission occurring declined rapidly with time between successive feeds on an infected and a susceptible host. It was as high as 0.5, when the two feeds were separated by less than 15 min, but declined to 0.04, 0.003, 0.001 and 0.0003 when the intervals were 1, 3, 6 and 24 h, respectively.

Trypanosoma evansi and *Tr. vivax viennei* are only transmitted mechanically, and tabanids are the most important of the mechanical vectors; but it is more difficult to evaluate the importance of mechanical transmission where there are alternative routes of infection. Thus the role of tabanids in the transmission of pathogenic trypanosomes in tsetse fly (*Glossina*) areas is unknown. With regard to other diseases, there is evidence that tabanids are important mechanical vectors of *Anaplasma marginale* to cattle; of *Francisella tularensis*, the causative organism of tularaemia, to humans and domestic stock; and of *Bacillus anthracis*, which causes anthrax in man and animals. Tabanids are also the vectors of three viral diseases, bovine leukaemia, equine infectious anaemia and hog cholera, and they may play a significant role in the transmission of the rinderpest virus.

Independent of their role in spreading disease, when tabanids are abundant, they worry stock and make them less thrifty, but few attempts have been made to control them. When *T. fuscicostatus* fed on cattle fitted with lambda-cyhalothrin impregnated ear tags, 96% were killed after 15 s exposure. Only 16% of the much larger *T. americanus* were killed after 15 s exposure, but both species suffered 100% mortality after one minute's exposure (Leprince *et al.*, 1992).

References

Allan, A.S., Day, J.F. and Edman, J.K.D. (1987) Visual ecology of biting flies. *Annual Review of Entomology* 32, 297–316.

Anderson, J.F. (1973) Biting behavior of salt marsh deerflies. *Annals of the Entomological Society of America* 66, 21–23.

Beesley, W.N. and Crewe, W. (1963) The bionomics of *Chrysops silacea* Austen 1907. II. The biting rhythm and dispersal in rainforest. *Annals of Tropical Medicine and Parasitology* 57, 191–203.

Böse, R. and Heister, N.C. (1993) Development of *Trypanosoma* (*M.*) *theileri* in tabanids. *Journal of Eukaryotic Microbiology* 40, 788–792.

Boshko, G.V. and Shevtsova, N.P. (1975) Hibernation of tabanid larvae in the Ukrainian SSR. *Vestnik Zoologii* 5, 71–74.

Bosler, E.M. and Hansens, E.J. (1974) Natural feeding behavior of adult saltmarsh greenheads, and its relation to oogenesis. *Annals of the Entomological Society of America* 67, 321–324.

Burnett, A.M. and Hays, K.L. (1974) Some influences of meteorological factors on flight activity of female horse flies (Diptera: Tabanidae). *Environmental Entomology* 3, 515–521.

Cameron, A.E. (1926) Bionomics of the Tabanidae (Diptera) of the Canadian prairie. *Bulletin of Entomological Research* 17, 1–42.

Cameron, A.E. (1934) The life history and structure of *Haematopota pluvialis* Linne (Tabanidae). *Transactions of the Royal Society of Edinburgh* 58, 211–250.

Catts, E.P. and Olkowski, W. (1972) Biology of Tabanidae (Diptera): mating and feeding behavior of *Chrysops fuliginosus*. *Environmental Entomology* 1, 448–453.

Chainey, J.E. (1993) Horse-flies, deer-flies and clegs (Tabanidae). In: Lane, R.P. and Crosskey, R.W. (eds) *Medical Insects and Arachnids*. Chapman & Hall, London, pp. 310–332.

Chvala, M., Lyneborg, L. and Moucha, J. (1972). *The Horse Flies of Europe* (*Diptera, Tabanidae*). Entomological Society of Copenhagen, Copenhagen.

Crewe, W. and Beesley, W.N. (1963). The bionomics of *Chrysops silacea* Austen 1907. I. The longevity and food requirements of the adult fly. *Annals of Tropical Medicine and Parasitology* 57, 1–6.

Dale, W.E. and Axtell, R.C. (1975) Flight of the salt marsh Tabanidae (Diptera), *Tabanus nigrovittatus*, *Chrysops atlanticus* and *C. fuliginosus*. Correlation with temperature, light, moisture and wind velocity. *Journal of Medical Entomology* 12, 551–557.

Duke, B.O.L (1959) Studies on the biting habits of *Chrysops* VI A comparison of the biting habits, monthly biting densities and infection rates of *C. silacea* and *C. dimidiata* (Bombe form) in the rain forest at Kumba, Southern Cameroons, U.U.K.A. *Annals of Tropical Medicine and Parasitology* 53, 203–214.

Dukes, J.C., Edwards, T.D. and Axtell, R.C. (1974) Distribution of larval Tabanidae (Diptera) in a *Spartina alterniflora* salt marsh. *Journal of Medical Entomology* 11, 79–83.

Foil, L.D. (1989) Tabanids as vectors of disease agents. *Parasitology Today* 5, 88–96.

Foster, C.H., Renaud, G.D. and Hays, K.L. (1973) Some effects of the environment on oviposition by *Chrysops* (Diptera: Tabanidae). *Environmental Entomology* 2, 1048–1050.

French, F.E. and Kline, D.L. (1989) 1-Octen-3-ol, an effective attractant for Tabanidae (Diptera). *Journal of Medical Entomology* 26, 459–461.

Gjullin, C.M. and Mote, D.C. (1945) Notes on the biology and control of *Chrysops discalis* Williston (Diptera: Tabanidae). *Proceedings of the Entomological Society of Washington* 47, 236–244.

Hafez, M., El-Ziady, S. and Hefnawy, T. (1970a) Biological studies on *Tabanus taeniola* P. de B. (Diptera: Tabanidae) adults. *Bulletin de la Société Entomologique d'Egypte* 54, 327–344.

Hafez, M., El-Ziady, S. and Hefnawy, T. (1970b) Studies on the feeding habits of *Tabanus taeniola* P. de B. (Diptera: Tabanidae). *Bulletin de la Société Entomologique d'Egypte* 54, 365–376.

Hafez, M., El-Ziady, S. and Hefnawy, T. (1970c) Biological studies of the immature stages of *Tabanus taeniola* P. de B. (Diptera: Tabanidae) in Egypt. *Bulletin de la Société Entomologique d'Egypte* 54, 465–493.

Hansens, E.J. and Robinson, J.W. (1973) Emergence and movement of the saltmarsh deerflies *Chrysops fuliginosus* and *Chrysops atlanticus*. *Annals of the Entomological Society of America* 66, 1215–1218.

Hocking, B. (1953) The intrinsic range and speed of flight of insects. *Transactions of the Royal Entomological Society of London* 104, 223–345.

Inaoka, T. (1975) Habitat preference of tabanid flies in Hokkaido based upon the collection of female adults. *Journal of the Faculty of Science Hokkaido University Series VI Zoology* 20, 77–92.

Inoue, T., Kamimura, K. and Watanabe, M. (1973) A quantitative analysis of dispersal in a horse fly *Tabanus iyoensis* Shiraki and its application to estimate the population size. *Research on Population Ecology* 14, 209–233.

Kangwagye, T.N. (1973) Diurnal and nocturnal biting activity of flies (Diptera) in western Uganda. *Bulletin of Entomological Research* 63, 17–29.

Krinsky, W.L. (1976) Animal disease agents transmitted by horse flies and deer flies (Diptera: Tabanidae). *Journal of Medical Entomology* 13, 225–275.

Lamborn, W.A. (1930) The remarkable adaptation by which a dipterous pupa (Tabanidae) is preserved from the dangers of fissures in drying mud. *Proceedings of the Royal Society of London B* 106, 83–87.

Lane, R.S. (1976) Density and diversity of immature Tabanidae (Diptera) in relation to habitat type in Mendocino County, California. *Journal of Medical Entomology* 12, 683–691.

Leclercq, M. (1952) Introduction a l'étude des tabanides et revision des espèces de Belgique. *Memoires de l'Institut Royale des Sciences Naturelles de Belgique* 123, 1–80.

Leprince, D.J. and Foil, L.D. (1993) Relationship among body size, blood meal size, egg volume, and egg production of *Tabanus fuscicostatus* (Diptera: Tabanidae). *Journal of Medical Entomology* 30, 865–871.

Leprince, D.J., Hribar, L.J. and Foil, L.D. (1992) Evaluation of the toxicity of lambda-cyhalothrin against horse flies (Diptera: Tabanidae) via bioassays and exposure to treated hosts. *Bulletin of Entomological Research* 82, 493–497.

Mackerras, I.M. (1954) The classification and distribution of Tabanidae (Diptera). I. General Review. *Australian Journal of Zoology* 2, 431–454.

Mackerras, I.M. (1955) The classification and distribution of Tabanidae (Diptera). II. History; morphology; classification: subfamily Pangoniinae. *Australian Journal of Zoology* 3, 439–511.

Oldroyd, H. (1952–57) *Horseflies of the Ethiopian Region: I* Haematopota *and* Hippocentrum. *II* Tabanus *and related genera. III* Chrysopinae, Scepsidinae, Pangoniinae. British Museum (Natural History), London.

Oldroyd, H. (1973) Tabanidae (horseflies, clegs, deerflies, etc.). In: Smith K.G.V. (ed.) *Insects and Arthropods of Medical Importance*. British Museum (Natural History), London, pp. 195–208.

Parsons, B.T. (1971) Construction of mud cylinders by species of the genus *Tabanus* (Dipt: Tabanidae). *Entomologist's Monthly Magazine* 107, 89–90.

Pavlova, R.P. (1974) Influence of the temperature of the surrounding environment on the length of the pupal phase of tabanids. *Parazitologiya* 8, 243–248.

Phelps, R.J. and Holloway, M.T.P. (1990) Alighting sites of female Tabanidae (Diptera) at Rekomitjie, Zimbabwe. *Medical and Veterinary Entomology* 4, 349–356.

Polyakov, V.A. and Polyakov, M.A. (1973) Statistical analysis of the influence of environmental factors on the flight of tabanids in the conditions of Chukotka. *Sel'skokhozyaistvennaya Biologiya* 8, 690–694.

Sasakawa, M., Yoshida, A., Yamouchi, T. and Kuriyama, M. (1969) Studies on the bionomics and control of the horseflies attacking the grazing Japanese cattle. III. Diurnal biting activity in cloudy weather and control of the flies by trapping. *Review of Applied Entomology B* 61, 2360 (Abstract).

Schultze, T.L., Hansens, E.J. and Trout, J.R. (1975) Some environmental factors affecting the daily and seasonal movements of the salt marsh greenhead *Tabanus nigrovittatus*. *Environmental Entomology* 4, 965–971.

Soboleva, R.G. and Bodrova, Yu.D. (1973) The breeding places of gadflies (Diptera: Tabanidae) in the southern part of the Maritime Territory. *Entomological Researches in the Far East, Issue 2 Diptera of the Far East* 57–77.

Taylor, P.D. and Smith, S.M. (1989) Activities and physiological states of male and female *Tabanus sackeni*. *Medical and Veterinary Entomology* 3, 203–212.

Vale. G.A. and Phelps, R.J. (1974) Notes on the host-finding behaviour of Tabanidae (Diptera). *Arnoldia, Rhodesia* 6, 1–6.

Watanabe, M. and Kamimura K. (1975) Nectar sucking behaviour of *Tabanus iyoensis* (Diptera: Tabanidae). *Japanese Journal of Sanitary Zoology* 26, 41–47.

Wiesenhutter, E. (1975) Research into the relative importance of Tabanidae (Diptera) in mechanical disease transmission. II. Investigation of the behaviour and feeding habits of Tabanidae in relation to cattle. *Journal of Natural History* 9, 385–392.

Glossinidae (Tsetse Flies) 12

Tsetse flies (Glossinidae) are readily recognized. They are medium to large brown flies with a long, forwardly directed proboscis, sheathed by equally long palps (Fig. 12.1), measuring 6 to 14 mm long, excluding the proboscis. The antenna has the typical cyclorrhaphan structure with an elongated 3rd segment but is distinctive in that the rays of the arista are feathered and are only present on the dorsal side (Fig. 12.2). When viewed laterally the 3rd segment bears a fringe of hairs, the length of which is a specific character. The eyes are dichoptic in both sexes. The wings are folded scissor-like at rest and extend a short distance beyond the end of the abdomen. The wing venation is characterized by the discal cell being 'hatchet' shaped (Fig. 12.3). The abdomen of the male bears posteriorly on its ventral surface a button-like structure, the folded male terminalia. Both sexes are haematophagous and therefore both are potential vectors of trypanosomiasis to man and domestic animals.

Classification and Distribution

Only one genus, *Glossina*, is included in the *Glossinidae*. At the present day *Glossina* is virtually confined to the Afrotropical region although two species have been found recently in south-west Saudi Arabia (Elsen *et al.*, 1990), but in the past it occurred in the Nearctic since four species of fossil *Glossina* have been found in beds of Oligocene age in Colorado. The genus *Glossina* contains 31 living taxa, 23 species and 8 subspecies (Jordan, 1993). The species are assigned to three subgenera (*Glossina, Nemorhina, Austenina*), which are also referred to respectively as the *morsitans, palpalis* and *fusca* groups, named after the commonest species in each group. The status of the subspecies is unresolved but there is evidence of genetic incompatibility causing sterility when subspecies of *G. morsitans* are crossmated (Curtis, 1972). Subspecies of *G. morsitans* and *G. palpalis* will be prefixed by *G. m.* and *G. p.*, e.g. *G. m. centralis* and *G. p. gambiensis*.

When discussing tsetse distribution Jordan (1993) recognizes four subdivisions

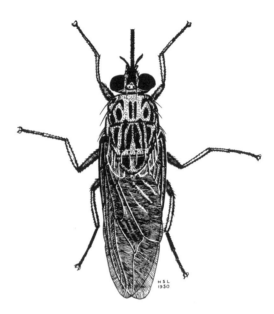

Fig. 12.1. Female *Glossina fuscipes* viewed from above resting on a vertical surface. Source: from Buxton (1955).

of the Afrotropical region: Western Africa, which is approximately the area north of the Gulf of Guinea; Central Africa, which extends from western Africa eastwards to the Sudan and south to Angola and includes the densely forested belt west of the western Rift Valley; eastern Africa, which extends from the Sudan to Tanzania and borders the Indian Ocean; and southern Africa which includes Africa south of Zaire and Tanzania, and is roughly the area south of 10°S.

The subgenus *Glossina* contains five species of which two, *G. morsitans* and *G. pallidipes*, are of major economic importance, and *G. swynnertoni* and *G. austeni* are of local significance. They are small to medium sized tsetse, 6 to 11 mm long, with distinct bands on the abdomen (except in *G. austeni*); the distal segments of the hind tarsus are dark dorsally; and the male claspers are swollen distally. They are species which are commonly found in savanna woodland and evergreen thickets, except for *G. austeni* which is restricted to coastal forest and relict forest. The three subspecies of *G. morsitans* (*morsitans, centralis and submorsitans*) range widely in tropical Africa; *G. longipalpis* is found in western and central Africa; *G. pallidipes* occurs in Zaire and eastern and southern Africa; *G. swynnertoni* in eastern Africa and *G. austeni* in eastern and southern Africa.

The subgenus *Nemorhina* also contains five species of which three, *G. palpalis*, *G. fuscipes* and *G. tachinoides*, occur in riverine and lakeside habitats and are particularly important as vectors of human trypanosomiasis. They are small to medium sized tsetse, 6 to 11 mm long, with the dorsum of the abdomen dark brown (except in *G. tachinoides*, which has a banded abdomen of the *morsitans* type), all hind tarsi dark dorsally, and male claspers not swollen distally but joined by a

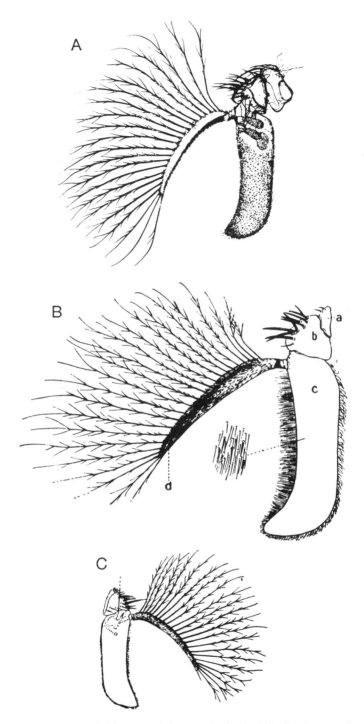

Fig. 12.2. Antennae of *Glossina*: (**A**) *G. palpalis* (x 50); (**B**) *G. nigrofusca* (x 41); (**C**) *G. morsitans* (x 40). *a*, *b*, *c* first, second and third antennal segments; *d*, arista. Source: from Buxton (1955).

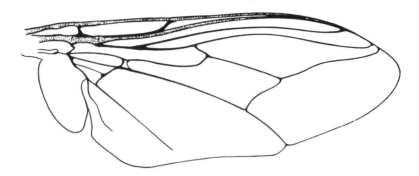

Fig. 12.3. Wing of *Glossina* (x 18).

membrane. This subgenus occurs mainly in western and central Africa, with *G. tachinoides* extending into eastern Africa and *G. pallicera* occurring in Angola. *G. fuscipes* is mainly found in central and eastern Africa, but also occurs in Chad, Angola and Zambia.

The 13 species of the subgenus *Austenina* are, except for *G. brevipalpis* and *G. longipennis*, forest dwellers found mainly in western and central Africa with lesser representation in eastern Africa and Angola. *G. longipennis* is found only in eastern Africa and *G. longipalpis* in eastern and southern Africa. These species have little contact with humans or livestock, and are of little economic importance, except when cattle are moved into forested areas, as for example in western Uganda (Bikingi-Wataaka, 1975). *Austenina* are large tsetse, 11 to 14 mm in length, with strong bristles on the anepimeron (pteropleuron) as well as on the katepisternum (sternopleuron) and in which male claspers are neither swollen distally nor joined by a membrane. In the other subgenera, *Glossina* and *Nemorhina*, there are no strong bristles on the anepimeron (Jordan, 1993).

Life Cycle

Tsetse flies are viviparous, the female producing fully grown larvae. Each ovary is composed of two polytrophic ovarioles. Ovaries and ovarioles produce ova alternately, beginning with the right ovary. Since the relict body left in the ovariole after ovulation can be easily recognized, it is possible to age (physiologically) tsetse flies accurately over the first four cycles by observing the condition of the ovaries and the contents of the uterus (Saunders, 1960). Using these criteria 14 different stages can be recognized, covering the equivalent of 40 days of adult life at 25°C. The technique has been elaborated to enable female flies to be classified up to 80 days (Challier, 1965). When the egg passes down the common oviduct to the uterus (Fig. 12.4) its micropyle comes opposite the opening of the spermathecal duct. The release of sperm is regulated by a sphincter on the duct, which opens at ovulation but is closed during the rest of the breeding cycle (Roberts, 1973a).

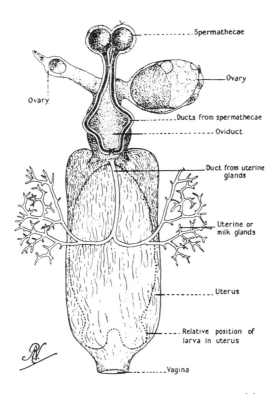

Fig. 12.4. Reproductive organs of female *Glossina*. Dotted line in the uterus indicates the position of larva with its anterior end lying near the duct from the uterine or milk gland. Source: from Buxton (1955).

Larva

The larva uses its labral tooth to emerge from the egg, and is nourished by secretions of the uterine gland, which are discharged on to the oviductal shelf (Fig. 12.4). The larva is held in position by the choriothete, which secretes a sticky material. Although the chorion and exuviae may adhere to the sticky material the choriothete plays no role in hatching or ecdysis (Roberts, 1973b). As the larva grows in size the role of the choriothete declines and the 2nd instar larva is held in place by uterine ridges, and the 3rd instar by the walls of the uterus, which are stretched by the size of the larva. A fully-grown 3rd instar larva at deposition weighs more than the containing female. At 25°C the egg stage lasts 4 days, and the three instars, 1, 1.5 and 2.5 days, respectively, in a 9-day developmental cycle. (All times quoted in this chapter will refer to 25°C unless stated otherwise.)

The larva grows steadily throughout its time in the uterus feeding on secreted 'milk'. To meet the growing larval needs the uterine gland goes through a cycle during which its cells increase in volume 100-fold, i.e. × 4.6 linearly. Maximum size of the cells is reached about two-thirds of the way through pregnancy, just before the larva reaches the 3rd instar (Ma *et al.*, 1975). The secretion is mainly composed of acidic lipids at first but changes later to protein. Unassimilated food material in the gut of the fully-grown larva comprises two-thirds of the larva's live weight. At deposition the larval gut contains 80% of the fat, 70% of the water and 50% of the protein in the whole larva (Langley, 1977).

Symbionts

Two different Gram-negative bacteroids are found in the gut and ovaries of *Glossina*. The larger bacteroid in the midgut is extracellular and assigned to the gamma subdivision of the Proteobacteria while the smaller ovarian bacteroid is intracellular and assigned to the alpha subdivision, possibly the genus *Wolbachia* (Rickettsiaceae) (Gassner, 1989; O'Neill *et al.*, 1993).

Tsetse flies whose symbionts have been reduced by antibiotics have reduced fecundity and fertility but their longevity is not affected (Hill *et al.*, 1973).

Respiration

First and 2nd instar larvae respire through posterior spiracles, which become elaborated into polypneustic lobes in the 3rd instar (Figs 12.4 and 12.5a). In each lobe there are three air chambers which open via numerous supernumerary stigmata. The 2nd instar tracheal system remains within the 3rd instar system, except posteriorly, where it opens to the exterior between the polypneustic lobes. The three outer chambers are connected individually to the inner air chambers by felt chambers. The felt chambers act as pistons, and when they are drawn forwards, air enters the supernumerary stigmata, and, at the same time, air is driven forwards between the 2nd and 3rd instar tracheae, backwards through the 2nd instar tracheae, and out through the 2nd instar spiracles. Reversal of air flow is prevented by the stigmata being closed by valves (Bursell, 1955).

Puparium

The polypneustic lobes darken and harden 24 to 48 h before the larva is deposited (Fig. 12.5). In *G. m. centralis* parturition appears to have a circadian basis with larvae being deposited in the late afternoon (Zdarek *et al.*, 1992). The larva is able to move forward and burrow into the soil by peristaltic contractions. The length of free larval life is influenced by time of deposition, presence or absence of light, and nature of substrate. Pupariation is accelerated by darkness and mechanical stimulation of a particulate substrate. The larva is negatively phototactic, positively thigmotactic, and seeks a high humidity but moves away from free water (Zdarek and Denlinger, 1991). It burrows rapidly into the substrate, and pupariation occurs within 1 to 5 h of deposition (Langley, 1977). During barrelling and contraction an acidic fluid is discharged from the anus, which rapidly coats the puparium. Its function is uncertain, but it may include a pheromone attracting females to larviposit in the same site (Nash *et al.*, 1976) or be a defensive secretion against predators and pathogens. The puparium, measuring 3 to 8 mm, darkens rapidly becoming light brown in 10 min and dark brown in several hours. During this process the endocuticle of the larval cuticle is sclerotized to form the puparium. Pupal apolysis occurs 2 to 3 days later, when the pupal cuticle is formed. At first

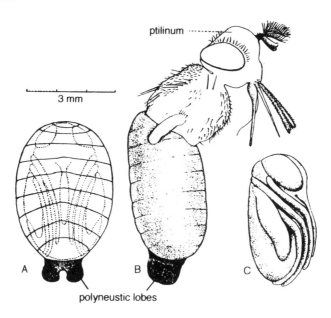

Fig. 12.5. Pupa and puparium of *Glossina morsitans*: (**A**) puparium, showing the pupa by dotted lines; (**B**) adult fly emerging from puparium; (**C**) pupa removed from puparium. Source: from Buxton (1955).

the pupa is cryptocephalic, but it becomes phanerocephalic 5 days later when the appendages are everted. Two days later the pupal/adult apolysis occurs, but another 3 weeks will elapse before adult development is complete, and the adult emerges (Langley, 1977). The polypneustic lobes cease to function after pupal formation. The pupal spiracles are located in the same position as the adult prothoracic spiracles, and they continue to function until adult emergence (Bursell, 1958).

Adult

The adult emerges using its ptilinum to force off the cap from the puparium (Fig. 12.5b), and then to burrow upwards to the surface. The wings are expanded, but for the first 10 to 14 days of its life the adult is soft to handle, and is said to be the 'teneral'. During this phase the endocuticle is secreted, the exocuticle hardened, and the flight muscles developed. In *G. pallidipes* the residual dry weight of the thorax increases linearly with age to the end of the teneral stage (Langley *et al.*, 1990).

Mating (Wall and Langley, 1993)

Non-inseminated females are rare in the field and most females are mated at the time of their first blood meal. In the laboratory maximum insemination rates are achieved with 3-day-old *G. m. morsitans* and *G. p. palpalis* and somewhat later (7–9 days) in *G. pallidipes*. Males will copulate early in life but are more successful in achieving insemination after a blood meal. Older males (5–10 days old) are more successful in achieving insemination than younger males with the exception of the males of *G. austeni* which are capable of inseminating females within 24 h of emergence. During copulation the male transfers successively a diffuse secretion, material to form the spermatophore coat, and finally the spermatozoa, some of which may be introduced directly into the spermathecal duct, but most stream into the spermathecae after cessation of copulation (Pollock, 1974).

Male *G. m. morsitans* responded visually to a target fly at a maximum distance of 50 cm, the cue for response being the rate at which the approaching object affected more than one ommatidium (Brady, 1991). In the field aggregations of males, known as following swarms, are associated with large, slowly moving, conspicuous objects. In this situation mating pairs of *G. m. morsitans* and *G. longipennis* may be seen but mating pairs of *G. pallidipes* are rare. It is uncertain whether coupling is initiated in the air or whether the male follows the female until it lands. Movement is not essential for a male to recognize a female and males will respond to suitable decoys and dead females. Final recognition is achieved by a non-volatile, species-specific, contact pheromone, which appears in the cuticular waxes of a female 2 days before eclosion and remains for the rest of her life (Huyton *et al.*, 1980a). Males detect the pheromone through chemoreceptors present on the tarsi and tibiae of all legs.

The sexual activity of male *G. m. morsitans* and *G. pallidipes* increases with time after a meal and falls after a blood meal. Mated females are refractory to further mating and although multiple mating has been observed in the laboratory it is considered to be unlikely in the field. Males responded similarly to both dead and alive conspecific females, and showed varying responses to females of other species, except for female *G. austeni* to which only male *G. austeni* of the seven species tested, responded (Huyton *et al.*, 1980b). Male *G. m. morsitans* will 'copulate' with suitably sized decoys dosed with the appropriate pheromone, the males being attracted to the vicinity of the decoys by a synthetic ox odour of carbon dioxide, acetone and octenol (Hall, 1988). This opens up the possibility in the field of using males to transfer a chemosterilant from a decoy to wild females during mating.

Reproduction

With free access to a host for feeding a female will have a fully developed egg ready for ovulation 7 to 9 days after emergence. Ovulation appears to be independent of any material transferred by the male but controlled by hormones released as a result of prolonged copulation. In *G. morsitans* ovulation is dependent on copulation exceeding 60 min and can be achieved by several interrupted matings during which no spermatozoa are transferred. It can also be induced in a virgin female by the injection of haemolymph from a mated female (Saunders and Dodd, 1972; Chaudhury and Dhadialla, 1976). Males may mate repeatedly, but females require only one insemination to be fully fertile. If a female is mated twice, the products of the first mating fertilize 75% of the progeny (Langley, 1977).

Subsequent ovulations occur within an hour of parturition, and therefore occur at 9- to 10-day intervals. Since the first ovulation occurs when the female is 9 to 10 days old, the first larviposition will be made at about 20 days, followed by subsequent larvipositions at 10-day intervals. Therefore a female must live for 30 days to produce two progeny as replacements for herself and her mate. Equal numbers of each sex occur at emergence. This implies that tsetse flies must be long lived if the population is not to die out, and females may survive for several months in the field. Female *G. palpalis* have been recaptured 6 months after having been caught in the wild, marked and released. A reasonable average length of life is 6 weeks for males and 14 weeks for females (Nash, 1969). Jackson (1941) calculated that the length of life of male *G. morsitans* varied from 2 weeks in the dry season to 6 weeks in the wet season.

The low reproduction rate of tsetse flies implies that populations of *Glossina* will be relatively steady, changing only slowly with time. The sustained maximum recorded rate of increase of tsetse in the field would permit a 20-fold increase in population size in 10 months and a 60-fold increase in less than 14 months. Bimonthly catches of *G. swynnertoni* in Tanzania showed a 20-fold range over 10 years and the annual mean catch the same variation over 20 years. Monthly catches of *G. palpalis* in northern Nigeria over 24 years showed a 60-fold range (Rogers and Randolph, 1985). Tsetse fly populations are classified as *K*-selected and considered to live in a habitat which is either constant or predictably seasonal in time while *r*-selected insects such as mosquitoes and houseflies, whose populations show rapid fluctuations, are considered to live in a habitat which is either unpredictable or ephemeral (Begon *et al.*, 1986).

Feeding and digestion

When *Glossina* feeds, blood passes at first into the midgut and, when the midgut is distended, blood passes into the more capacious crop. Blood is transferred from the crop to the midgut, as required, by the coordinated action of valves, and the posterior section of the foregut, which functions as an 'oesophageal pump' (Rice, 1970). In the midgut the blood is enclosed in the peritrophic membrane, secreted at the proventriculus. In young tsetse flies the size of the meal may be limited by the development of the peritrophic membrane.

There are no air sacs to compensate for the volume of blood ingested and meal size is probably regulated by stretch receptors in the ventral wall of the abdomen, which becomes greatly distended (Langley, 1977). Freshly fed *G. morsitans* and *G. pallidipes*, collected in the field, were estimated to have ingested 37 to 76 mg of blood. *G. pallidipes* ingested more than *G. morsitans* and in both species females took larger meals than males (Taylor, 1976). The actual size of the meal in the female will depend upon the size of the ovaries and uterus. The meal is rapidly reduced in volume by excretion of water, and within 30 min all the blood stored in the crop will have been transferred to the midgut, and the blood meal reduced by half. This concentration occurs in the anterior portion of the midgut, an area where no digestive enzymes are secreted. Enzymes are produced in the middle section of the midgut and include a powerful protease. Absorption of the products of digestion occurs in the posterior section of the midgut. Digestion is 60% complete in 24 h and 90% by 48 h (Langley, 1977).

Energy production

Bursell *et al.* (1974) distinguish three stages in the hunger cycle of *Glossina*. A lipogenic phase, lasting about 24 h, follows immediately after a blood meal and is accompanied by a rapid rise in oxygen consumption. The energy, which becomes available, is used for growth, for excretion of excess nitrogen, and for lipid formation, which reaches a peak 12 h after the feed. The first lipolytic phase begins on the second day after feeding, and is marked by an increase in proline, mobilization of the lipid reserves, and a reduced respiratory rate. The second lipolytic phase begins on the 3rd to 4th day, when digestion is complete. The rate of lipid mobilization is reduced and the concentration of proline declines. The insect now suffers nitrogen depletion.

In flight proline can only supply the short-term energy needs of the tsetse fly. It is converted into alanine in the flight muscles and the reverse process (alanine to proline) occurs in the abdominal fat body utilizing available lipids. This process limits continuous flight to about 2 min in *Glossina*. After 2 to 3 min continuous flight the wing beat falls from 220 to 180 c s^{-1} and is incapable of lifting a load. Teneral flies have a low wing beat because their flight muscles are underdeveloped, and consequently they cannot lift heavy loads and they ingest smaller meals. The greater part of meals taken early in life is used to develop the thoracic musculature, and the build-up of fat reserves is undertaken later (Bursell *et al.*, 1974).

The times given above for the various stages in the hunger cycle are appropriate for a temperature of 25°C which may be considered optimal since, at that temperature, fat reserves are used most economically, and mortality is minimal in *G. morsitans* (Phelps and Burrows, 1969). At 25°C the cycle of larviposition is 9 to 10 days, at 18°C it is extended to 25 days, and at 30°C it is reduced to 7 days. Similar temperature-related changes occur in the duration of the puparium before emergence of the adult. At 16, 25 and 32°C the durations are 100, 30 and 20 days respectively. The relationship is linear from 18 to 27°C, but above 27°C development is slower than expected, and thermal stress is becoming evident. One hour's exposure to 40.6°C and 6 h to 39.7°C produce death of the puparium (Langley, 1977).

Behaviour of *Glossina*

The need to increase world food production and pressure for economic growth in the countries of Africa have focused attention on the tsetse fly as a constraint on agricultural development. A great deal of work has been, and is being, done in the laboratory and field on *Glossina*, providing knowledge which is essential to the development of rational, economic control measures.

Activity cycles

Tsetse flies are active by day and Brady (1975b) has shown that in *G. m. morsitans* six out of seven unrelated behavioural responses were bimodal, with morning and evening peaks separated by a decline of activity at noon. This pattern is comparable to the activity pattern of their hosts (Ford, 1971). Two of the responses of G. m. morsitans are true circadian rhythms, which persist under constant conditions (Brady, 1975b). Such rhythms are modified by environmental conditions, and the bimodal response is temperature dependent, being bimodal at 22°C, but reduced to a single morning peak at 19°C (Brady, 1974). The feeding response is reduced by high illumination (25,000 lux), while the phototactic response to light varies with temperature. At 26°C G. m. morsitans is positively phototactic, but at 34 to 38°C it seeks the dark, and the threshold for this response is lowered by 2.2°C for a tenfold increase in illumination (Huyton and Brady, 1975).

Field observations indicate specific differences in activity patterns, with G. pallidipes showing increased feeding activity throughout the day and reaching a peak in the evening. This pattern applies to catches made off humans in a fly round, and off an ox, but in a stationary, unbaited trap the maximum was after midday. G. f. fuscipes showed a midday peak with low activity both in the morning and in the late afternoon. On the contrary, G. brevipalpis had peaks before dawn and after sunset (Harley, 1965). Activity patterns are also modified by environmental conditions in the field. G. longipalpis is inactive below 22°C (Thomson, 1968). while G. pallidipes feeds at temperatures between 18 and 32°C (Ford, 1971). Eclosion of G. morsitans occurs in the same temperature range (18–32°C), which produces a bimodal pattern in summer, when midday temperatures exceed 32°C, and a unimodal pattern in winter, when the temperature remains below 32°C (Phelps and Jackson, 1971). In the hot season G. m. morsitans is active early in the day, becoming inactive when the temperature reaches 32°C; in the cool season activity begins later in the day, and in the wet season activity is more evenly distributed throughout the day (Jordan, 1974).

Host-seeking behaviour (Colvin and Gibson, 1992)

Host finding by tsetse flies involves visual and chemosensory stimuli. In the laboratory G. m. morsitans shows a bimodal rhythm of response to a visual stimulus with peak activity in the morning and afternoon being five times that at midday (Brady, 1972). The response increases steadily with time since the last feed until flies become moribund after five days' starvation. Visual responsiveness correlated most closely with the fly's total body weight ($r = -0.89$) (Brady, 1975a). As might be expected, gravid females were only half as responsive as males to visual stimuli.

Some aspects of host finding and blood feeding have been mentioned in Chapter 4. It is important to appreciate that species differ in their response to hosts, as may different populations of the same species. It has long been known that tsetse flies can locate hidden hosts from distances of 50 m, and G. pallidipes and G. m. morsitans have been attracted to a stationary ox from 90 m, the host's presence being revealed by its odour being carried downwind (Vale, 1977). In typical tsetse woodland wind meanders at low speeds (< 1 m s^{-1}), and wind-carried odours can reach a particular point from any direction with only about one-third of the wind 'packets' indicating the true direction of the source ($\cdot 10°$) (Brady et al., 1989). In the presence of ox odour in a wind tunnel G. pallidipes is active in bursts of 22 s followed by 86 s of inactivity. The active period involves several flights averaging 3.3 s. Fewer flights are made when ox odour is replaced by carbon dioxide (Warnes, 1989).

Tsetse flies respond to a number of bovine emanations – carbon dioxide, acetone, octenol and phenols. Three of the four phenols that occur in ox urine caused upwind flight of *G. pallidipes* and *G. m. morsitans* in the laboratory, and in the field increased the catch of traps baited with acetone and octenol (Bursell *et al.*, 1988). At field concentrations acetone and octenol do not activate resting tsetse but increase their responsiveness to visual stimuli. Carbon dioxide will activate resting tsetse and with phenols, increase upwind flight towards the source. In odour plumes in the field *G. pallidipes* and *G. m. morsitans* fly in a general upwind direction at a height of less than 50 cm above the ground in shallow curves, 1–2 m in diameter, across the odour plume at speeds in excess of 4 m s^{-1}. Within 2–3 m of leaving a plume they make an inflight turn and an upwind turn when they re-enter the odour plume. Approaching the odour source more female *G. pallidipes* and *G. m. morsitans* landed on a vertical net than on a horizontal one, while more males of both species landed on the horizontal net (Torr, 1988c). Tsetse flies use visual cues to maintain their direction while in flight, and when responding to a visual target they fly downwind of the target, presumably to detect its odour.

The possibility of olfactory and visual stimuli activating resting tsetse flies is dependent on the stage they have reached in various endogenous cycles, e.g. hunger, circadian rhythm and, for females, the stage of pregnancy. Only a fraction of the resting tsetse fly population will be capable of responding to presented stimuli. Resting tsetse are not attracted to a stationary host but, in the field, about one-third will respond to ox odour or a moving visual stimulus. The natural odour of an ox caught twice as many female *G. pallidipes* as a synthetic ox odour consisting of carbon dioxide, acetone and octenol (Torr, 1988a,b). Some of this difference might be due to the absence of phenols in the synthetic odour. In addition to movement host recognition involves shape and colour. The more conspicuous an object is the greater its attraction and blue, black or white objects are more attractive than yellow or green. *G. pallidipes* and *G. m. morsitans* flying upwind in an odour plume were diverted by different coloured and shaped visual targets. Black and blue targets were equally attractive and yellow the least. Circular targets were more attractive than square ones, which were more attractive than oblong targets (Torr, 1989). Male *G. tachinoides* landed in greater numbers on phthalogen blue targets and females on UV reflecting white targets. A simple blue and UV reflecting white target was 70% as effective as a much larger, flanked blue target. The response of *G. p. palpalis* was similar but it landed less readily than *G. tachinoides*. Tsetse flies do not land immediately they reach an object but encircle it and even then they do not necessarily land. It is estimated that two-thirds of *G. palpalis* attracted to a black screen left without landing on it (Green, 1990). The addition of host odours to a visually attractive object greatly increases the landing rate.

Sampling

Tsetse flies are diurnally active blood-sucking flies which rest and feed out of doors. They are present in low density with total populations estimated to range from 4 to 18 tsetse ha^{-1} (Nash, 1969), but see below. From the marked difference in lengths of life of the two sexes it is generally agreed that, although equal numbers of each sex emerge from puparia, field populations contain 70 to 80% females and only 20 to 30% males; yet on human or animal baits more males than females are captured, and the percentage of males may be as high as 98% (Thomson, 1968). Female *G. pallidipes* feed every 3 days and, when pregnant, feed earlier in the day, and are more active on day five of the pregnancy cycle when the larva is in the first instar (Randolph *et al.*, 1991).

Traditionally the density of tsetse flies in an area was estimated by catches off a stationary animal, usually an ox, or during a fly round when an animal was led along a fixed circuit, stopping at intervals to collect the tsetse flies attracted to the mobile host. A disadvantage of this method is that humans are not attractive hosts to *G. morsitans* and *G. pallidipes*, and indeed there is evidence that they are actually repellent (Bursell, 1973; Vale, 1974). Electric traps are superior in that they can be operated in the absence of humans. Present-day traps incorporate colour, electric grids and bait in the form of a selection of carbon dioxide, acetone, octenol and phenols. They need to both attract and capture tsetse flies. Plain flat traps are made more effective by being flanked by an electric grid, which catches tsetse flies circling around the target. The catch size depends on the composition of the tsetse fly population. Teneral flies and females within 48 h of parturition are less active and males more active with activity increasing with time since the previous blood meal (Rogers and Randolph, 1985).

Biconical traps baited with acetone and cow urine caught more representative samples of *G. pallidipes* than unbaited traps. The catch included more females in the later stages of pregnancy and males showing a full range of fat/haematin content; a measure of time since their last blood meal (Randolph *et al.*, 1989). Electric traps and unaccompanied traps baited with acetone, octenol and 4-methylphenol caught the biggest daily catches of *G. m. morsitans* and *G. pallidipes* than were caught on a stationary ox, an odour-baited trap in the presence of humans or a mobile ox fly-round. At all ages *G. pallidipes* was more active than *G. morsitans* and was more attracted to the baited trap. Evening catches of both species were several times larger than the morning catches and, although in the natural population *G. pallidipes* was considered to be only three times as numerous as *G. morsitans*, the ratio between the two species attracted to an unaccompanied baited trap was 15:1, and as the odour increased so did the proportion of *G. pallidipes* (Hargrove, 1991). Male *G. pallidipes* of all ages were caught by a mobile electrified net and a stationary trap baited with acetone, octenol and phenol. Nulliparous female *G. pallidipes* were not attracted to the stationary trap (Langley *et al.*, 1990). Baited traps can be so well designed that it is anticipated that they will replace aerial and ground application of insecticides to control *G. m. morsitans* and *G. pallidipes* (Vale, 1993).

The powerful attraction of the odour of cattle to female *G. m. morsitans* and both sexes of *G. pallidipes* has been dramatically shown by the demonstration that the number of tsetse caught was linearly related to the weight of livestock in the pit trap (using the logarithms of both variables). This relationship applied up to the highest level tested (60,000 kg). Seven thousand tsetse flies were caught in 3 h attracted by the odour of 11,500 kg of livestock and large catches continued to be made for 60 days (Colvin and Gibson, 1992). Such a large catch implies either that the density or the mobility of these species was greater than expected.

Host selection

Teneral flies are low in energy reserves, particularly if their puparia have been exposed to high temperatures, and the mortality of teneral flies is high in the first week of adult life (Hargrove, 1981). They need a blood meal in the minimum time, and young *G. morsitans* will feed on humans, hunger overcoming repellency. Non-teneral flies feeding on people have fewer energy reserves than those feeding on cattle which, in turn, have fewer energy reserves than those in swarms following a host. The length of the feeding cycle is usually 2 to 3 days but can be extended in the absence of hosts to 5 or possibly more days.

Choice of host is only partly a matter of availability. *G. swynnertoni* and *G. morsitans* fed on warthog, giraffe and buffalo, in an area where zebra, wildebeeste and impala formed 80% of the available hosts (Jordan, 1974). Only one of the identified blood meals had been taken from the more abundant mammals. This may be due in part to the reaction of the host to the presence of the flies, because *G. morsitans* and *G. pallidipes* are said to avoid mammals, such as baboons, impala and humans, which resist attack, and to feed on relatively quiescent animals, such as pigs and cattle (Bursell, 1973).

Species of *Nemorhina* are opportunist feeders and frequently feed on reptiles, including crocodiles, and *G. f. fuscipes* feeds readily on monitor lizards (*Varanus* spp.) (Jordan, 1974). Change of habit, plus the preference of *Glossina* for warthog, operates against the success of tsetse control by game eradication. The larger game are relatively easy to exclude or shoot, but the tsetse population may not be markedly affected (Wilson, 1975).

Evolution of feeding preferences

Ford (1971) has outlined a possible evolution of tsetse flies based on their feeding preferences. The genus probably arose in the Mesozoic when it would have fed on large reptiles. This ancestral habit is reflected in the *Nemorhina*, which inhabit riverine and lacustrine environments and feed largely on reptiles. With the decline of reptiles in the Tertiary and the emergence of mammals it is postulated that tsetse flies transferred their feeding to early mammals, which would have included the Suidae or pigs, and the ancestors of large mammals, such as hippopotami, which predominated in the Eocene and Oligocene. Pigs were, and still largely are, inhabitants of forest, to which the subgenus *Austenina* is largely restricted. The ancestors of the subgenus *Glossina* moved into the savanna with the ancestors of the savanna-dwelling warthog. Ruminants, such as antelope, giraffe and buffalo, only became dominant in the Pliocene, and the shift of tsetse to these hosts is

Table 12.1. Summary of blood meal analyses, excluding human, domestic animals and blood meals not fully identified.

Habitat	Host	%
Lacustrine and riverine feeders	Reptiles	56
subgenus *Nemorhina*	Bushbuck	22
	Remainder	22
Forest thicket or forest edge feeders		
A. G. (G.) *austeni*	Bushpig and forest hog	74
G. (A.) *tabaniformis*	Remainder	26
B. G. (G.) *pallidipes*	Bushbuck	61
G. (A.) *fusca*	Bushpig	12
	Remainder	27
Savanna feeders		
G. (G.) *morsitans*	Wart hog	57
G. (G.) *swynnertoni*	Buffalo	21
	Giraffe, kudu and remainder	22
Specialized East African *Austenina*		
G. (A.) *brevipalpis*	Elephant, rhino, hippo, buffalo	71
G. (A.) *longipennis*	Bushpig	18
	Remainder	11

Source: Adapted from Ford (1971).

relatively recent. Evidence for this is to be found in their response to infection with trypanosomes (see Chapter 29).

Data on the blood meals of *Glossina* on wild hosts, i.e. excluding people and domestic animals, are summarized in Table 12.1. Species of *Nemorhina* fed largely on reptiles, which are common in their riverine and lakeside habitats. Forest dwelling species can be separated into those that fed mainly on pigs, e.g. *G. (G.) austeni*, *G. (A.) tabaniformis*, and those that fed mainly on bushbuck with some feeding on bushpig, e.g. *G. (G.) pallidipes*, *G. (A.) fusca*. The savanna species of subgenus *Glossina* feed largely on warthog, buffalo and large bovids. Lastly there are the specialized feeders among the *Austenina*, which have adapted to feeding on large mammals, which frequent thickets and secondary forest.

Ecology of Tsetse Flies

Climate and population

Calculated mortality rates of *G. m. submorsitans* in the Yankari Game Reserve in Nigeria varied seasonally and were closely correlated with saturation deficit and temperature. Using these two meteorological variables it was possible to draw contours enclosing areas of equal mortality. The bioclimatic optimum being the region enclosed by the lowest mortality contour. Similar data from Zambia for the subspecies *G. m. morsitans* fell within the predicted bioclimatic limits (Rogers and Randolph, 1985). In the Ivory Coast over an 18 month period the monthly mortality of *G. p. palpalis* was significantly correlated with temperature and saturation deficit (Rogers *et al.*, 1984).

There was seasonal elimination of young male *G. m. morsitans* in Zimbabwe due to the depletion of their puparial fat reserves in periods of high temperature. Mortality was zero between January and April when temperatures were close to those allowing emerging flies to have maximum fat reserves, and in November mortality was 75% (Phelps and Clarke, 1974). In Kenya seasonal changes in the size of nulliparous females and young males was correlated with the relative humidity 2 months earlier. There was a size-dependent mortality at emergence leading to field-caught flies being significantly larger than those emerging from field-collected puparia (Dransfield *et al.*, 1989).

Density dependence of *Glossina* populations (Rogers and Randolph, 1985)

Predation of puparia by ants, especially species of *Pheidole*, and of adults by birds is strongly density dependent, success favouring continued searching. Under stable climatic conditions the number of *G. p. palpalis* and *G. m. submorsitans* were dependent on the catch in the previous month, the number falling if the catch was high and rising if it were low. The population increase of *G. morsitans* in Botswana was inversely related to the non-teneral, i.e. older female catch, in all but the most extreme seasons. In Burkina Faso the impact of released sterile male *G. p. gambiensis* on the sterility of wild females was inversely correlated with the number of sterile males released.

Habitats

There is a trend towards survival in habitats of increasing aridity. *Austenina* species inhabit humid forests, and even *G. (A.) brevipalpis*, which extends into dry regions of eastern Africa, requires a relative humidity in excess of 70% for its puparia to produce an adult, and, as a result, it is restricted to residual or secondary forest. The puparia of the savanna species, *G. (G.) swynnertoni* and *G. (G.) m. centralis*, can complete development in humidities as low as 10%. Other species of the subgenus *Glossina* and those of *Nemorhina* have intermediate requirements. Rather contrary to this view is the fact that *G. longipennis*, the species best adapted to aridity, is a member of the forest-dwelling subgenus *Austenina*, and yet its puparia are 'completely viable at 0% RH' (Bursell, 1958). Of course, this series can be reversed and be read as a movement from arid to more humid habitats. Puparia are found in holes in the ground, under logs, under fallen leaves and bark, and in tufts of grass. In the hot season the microclimate under logs becomes too hot for the survival of puparia of *G. m. centralis* and they are deposited in holes in the ground, which are more humid (Jordan, 1974).

Tsetse are considered to concentrate in favourable habitats in the dry season. This is marked in *G. m. submorsitans* during the hot, dry season in northern Nigeria but is less evident in the less extreme climate of the south (Jordan, 1974). In Zimbabwe in the hot, dry season the sexually appetitive males concentrate in riverine areas, but other adults are more evenly distributed (Bursell, 1966). Removal of selected vegetation, which formed the dry season resting sites for *G. palpalis* in northern Ghana, virtually eliminated this species, but success was dependent in part on the severity of the dry season (Morris, 1946). At one time considerable attention was paid to the vegetation in which tsetse were found, but it is more likely that the observed tsetse–vegetation associations reflect habitat selection by the host rather than by the tsetse (Ford, 1971). *G. tachinoides* has adapted to a peridomestic mode of life in southern Nigeria, where its puparia are deposited in areas of human activity and in lantana thickets. *G. f. fuscipes* has also established a population in lantana, away from its typical lacustrine habitat (Jordan, 1974).

Resting places

Bursell and Taylor (1980) calculate that female *G. m. morsitans* spend on average only 5 min per day in flight, and the more active males about 15 min in the hot season and more than twice that in the cold season. Consequently most of their time is spent resting. The daytime and nocturnal resting places are different. *G. m. morsitans* rests on branches, boles and rot holes of trees by day, and at night on twigs and leaves (Jordan, 1974). This may be an adaptation to avoid nocturnal predators. By day tsetse could see an approaching predator, and by night detect the vibration caused by a predator on an easily disturbed leaf or twig. *G. pallidipes* occupies similar resting places but nearer the ground, i.e. mostly below 3 m, whereas *G. morsitans* extends up to 6 m and on occasions to 12 m. Both species show a change in resting site during the day in the hot season, when they move from branches to rot holes and boles (Thomson, 1968; Jordan, 1974). Hargrove and Coates (1990) have produced biochemical evidence to indicate that in the field *G. m. morsitans* lives at temperatures two to six degrees below the average recorded in the Stevenson screen.

Another response to harsh conditions is for tsetse to rest nearer the ground. In the Cameroons the behaviour of *G. tachinoides* is temperature dependent. Above 30°C it rests nearer the ground, and when the temperature exceeds 35°C it moves to a cooler site. In intense heat, it seeks the shelter of dense stands of *Mimosa nigra* (Gruvel, 1975). This response has an effect on control measures. The severity of the hot dry season in West Africa increases with distance from the coast. In the severer climate it is sufficient to treat vegetation with insecticides to a height of 1.5 m to control *G. m. submorsitans*, whereas nearer the coast, treatment must be applied up to 3.6 m (Jordan, 1974).

Dispersal

Bursell and Taylor (1980) calculated that, flying at 11 km h^{-1} for 18 min, a male *G. m. morsitans* would fly a distance of 3.3 km. However, such a flight is not made in one direction, but is composed of a series of short, random flights, each occupying about 5 s. For male *G. m. morsitans* the average length of a step was determined to be 15.9 m and the number of steps per day as 208. The effect of such a large number of short, random movements is to disperse male *G. f. fuscipes* 338 m day^{-1}, and the average of 13 observations on four species, mostly *G. morsitans*, was 252 m day^{-1} (Rogers, 1977). These are average figures and Hargrove (1981) considers that in the first week of life male *G. morsitans* disperse only a short distance and that this rises steadily as the flies age, to reach a peak in the fourth to sixth weeks, after which dispersion declines rapidly in older flies.

Such modest dispersal is consistent with the long-term extension of *G. morsitans* in Nigeria, where its spread along the cattle routes has been calculated at 5 km year^{-1} or about 100 m week^{-1}. This is about half the rate of dispersal (180 m week^{-1}) observed for individual flies (Riordan, 1976). Nevertheless, when continued over many years tsetse have considerable potential as invaders of suitable uninfested areas. In a trapping-out experiment against *G. palpalis* in a village in the Ivory Coast, for a period of 11 weeks the daily catch of flies hardly changed, but the characteristics of the captured females changed. It was evident that the removal of the original population by trapping was rapidly compensated for by the immigration of females from the bush (Rogers *et al.*, 1984).

Control of Tsetse Flies

Tsetse flies are exophilic and exophagic and are not amenable to control by the treatment of houses and animal shelters with residual insecticides, but such chemicals have been used to control tsetse flies by the selective treatment of their resting places. In West Africa this treatment has been used to treat the known resting sites of *G. palpalis* and *G. tachinoides* in their restricted riverine habitats. Such a restricted spraying technique is more difficult to apply to the control of widespread savanna species, such as *G. morsitans*. However, large areas (6000 km^2) have been treated successfully in Botswana and Zimbabwe by the application of low dosage (e.g. endosulphan 10 g ha^{-1}) non-residual aerosols using fixed wing aircraft. Five or six applications not more than 20 days apart produced excellent control but re-invasion occurred from the surrounding untreated area (Allsopp, 1984; Turner and Brightwell, 1986). The failure of aerial spraying to eradicate *G. pallidipes* from the Lambwe Valley in western Kenya was attributed to the habitats in Botswana and Zimbabwe being marginal for tsetse survival compared to the more favourable habitat of the Lambwe Valley (Turner and Brightwell, 1986).

Wall and Langley (1991) have reviewed the control of tsetse flies with emphasis on the design and use of traps and targets. Control of *G. p. palpalis* in a 1500 km^2 area in central Nigeria was achieved by biconical traps reducing the tsetse population by 90%, followed by the release of sufficient sterilized males to outnumber wild males by 10 : 1, and the use of blue cotton insecticide-impregnated screens to combat re-invasion of the area (Takken *et al.*, 1986). A similar trial against *G. m. morsitans* in Tanzania used two aerial applications of endosulphan (20 g ha^{-1}) 28 days apart followed by sterile male release. Over the 15 months of the experiment 81% control was achieved in the 195 km^2 treated area (Wiliamson *et al.*, 1983). In Zimbabwe *G. m. morsitans* and *G. pallidipes* were eradicated from an island in Lake Kariba by the use of odour baited traps, in some of which the flies were chemosterilized and then released. When both populations had been reduced very substantially (90%), odour baited, insecticide-impregnated, black targets eliminated the remaining *G. pallidipes* in 11 weeks and *G. m. morsitans* in 9 months (Vale *et al.*, 1986). In the Zambezi Valley similarly baited black targets at a density of 3–5 traps km^{-2} in six months reduced populations of *G. m. morsitans* and *G. pallidipes* to less than 0.01% at the centre of the 600 km^2 block (Vale *et al.*, 1988). Similar success was achieved in Kenya against *G. pallidipes* (98–99% reduction) over 100 km^2 using traps made and maintained by the local community (Dransfield *et al.*, 1990).

Medical and Veterinary Importance

Species of *Glossina* are present in about 40% of tropical Africa, an area of 10 million km^2, larger than the United States of America, but they are probably never present in sufficient density to pose a biting fly problem. Their economic importance resides in their role as biological vectors of pathogenic trypanosomes, which cause severe disease in humans and domestic animals.

Two forms of *Trypanosoma brucei*, *T. b. gambiense* and *T. b. rhodesiense*, cause sleeping sickness in people in western, central and eastern Africa and are transmitted, respectively, by tsetse flies of the subgenera *Nemorhina* and *Glossina*. Devastating epidemics of sleeping sickness have occurred in the past. Animal trypanosomiases, sometimes referred to as nagana, are caused by several species of *Trypanosoma* (see Chapter 29) and are associated with tsetse flies of the subgenus *Glossina*. Wherever there are tsetse flies it is impossible to keep cattle without regular chemotherapy. Trypanosomiasis has inhibited the development of a cattle-raising industry in much of tropical Africa, and even today, cattle ranching is impractical in much of Africa's savanna lands.

References

Allsopp, R. (1984) Control of tsetse flies (Diptera: Glossinidae) using insecticides: a review and future prospects. *Bulletin of Entomological Research* 74, 1–23.

Begon, M., Harper, J.L. and Townsend, C.R. (1986) *Ecology, Individuals, Populations and Communities*. Blackwell Scientific Publications, Oxford.

Bikingi-Wataaka, S.C.U. (1975) The incidence of trypanosomes in *Glossina fusca congolensis* in Bunyoro District, western Uganda. *East African Journal of Medical Research* 2, 13–16.

Brady, J. (1972) The visual responsiveness of the tsetse fly *Glossina morsitans* Westw. (Glossinidae) to moving objects: the effect of hunger, sex, host odour and stimulus characteristics. *Bulletin of Entomological Research* 62, 257–279.

Brady, J. (1974) The pattern of spontaneous activity in the tsetse fly *Glossina morsitans* Westw. (Diptera: Glossinidae) at low temperatures. *Bulletin of Entomological Research* 63, 441–444.

Brady, J. (1975a) 'Hunger' in the tsetse fly: the nutritional correlates of behaviour. *Journal of Insect Physiology* 21, 807–829.

Brady, J. (1975b) Circadian changes in central excitability – the origin of behavioural rhythms in tsetse flies and other animals? *Journal of Entomology A* 50, 79–95.

Brady, J. (1991) Flying mate detection and chasing by tsetse flies (*Glossina*). *Physiological Entomology* 16, 153–161.

Brady, J., Gibson, G.A. and Packer, M.J. (1989) Odour movement, wind direction, and the problem of host-finding by tsetse flies. *Physiological Entomology* 14, 369–380.

Bursell, E. (1955) The polypneustic lobes of the tsetse larva (*Glossina*: Diptera). *Proceedings of the Royal Society of London B* 144, 275–286.

Bursell, E. (1958) The water balance of tsetse pupae. *Philosophical Transactions of the Royal Society of London B* 241: 179–210.

Bursell, E. (1966) The nutritional state of tsetse flies from different vegetation types in Rhodesia. *Bulletin of Entomological Research* 57, 171–180.

Bursell, E. (1973) Entomological aspects of the epidemiology of sleeping sickness. *Central African Journal of Medicine* 19, 201–204.

Bursell, E. and Taylor, P. (1980) An energy budget for *Glossina* (Diptera: Glossinidae). *Bulletin of Entomological Research* 70, 187–196.

Bursell, E., Billing, K.C., Hargrove, J.W., McCabe, C.T. and Slack, E. (1974) Metabolism of the bloodmeal in tsetse flies (a review). *Acta Tropica* 31, 297–320.

Bursell, E., Gough, A.J.E., Beevor, P.S., Cork, A., Hall, D.R. and Vale, G.A. (1988) Identification of components of cattle urine attractive to tsetse flies, *Glossina* spp. (Diptera: Glossinidae). *Bulletin of Entomological Research* 78, 281–291.

Buxton, P.A. (1955) *The Natural History of Tsetse Flies*. H.K. Lewis, London.

Challier, A. (1965) Amélioration de la methode de détermination de l'age physiologique des glossines. Etudes faites sur *Glossina palpalis gambiensis* Vanderplank 1949. *Bulletin de la Société de Pathologie Exotique* 58, 250–259.

Chaudhury, M.F.B. and Dhadialla, T.S. (1976) Evidence of hormonal control of ovulation in tsetse flies. *Nature, London* 260, 243–244.

Colvin, J. and Gibson, G. (1992) Host-seeking behavior and management of tsetse. *Annual Review of Entomology* 37, 21–40.

Curtis, C.F. (1972) Sterility from crosses between subspecies of the tsetse fly *Glossina morsitans*. *Acta Tropica* 29, 250–268.

Dransfield, R.D., Brightwell, R., Kilu, J., Chaudhury, M.F. and Adabie, D.A. (1989) Size and mortality rates of *Glossina pallidipes* in the semi-arid zone of southwestern Kenya. *Medical and Veterinary Entomology* 3, 83–95.

Dransfield, R.D., Brightwell, R., Kyorku, C. and Williams, B. (1990) Control of tsetse fly (Diptera: Glossinidae) populations using traps at Nguruman, south-west Kenya. *Bulletin of Entomological Research* 80, 265–276.

Elsen, P., Amoudi, M.A. and Leclercq, M. (1990) First record of *Glossina fuscipes fuscipes* Newstead, 1910 and *G. morsitans submorsitans* Newstead, 1910 in southwestern Saudi Arabia. *Annales de la Société Belge Médecine Tropicale* 70, 281–287.

Ford, J. (1971) *The Role of the Trypanosomiases in African Ecology; a Study of the Tsetse Fly Problem*. Clarendon Press, Oxford.

Gassner, G. (1989) Dipteran endocytobionts. In: Schwemmler, W. and Gassner, G. (eds) *Insect Endocytobiosis: Morphology, Physiology, Genetics, Evolution*. CRC Press, Boca Raton, Florida, USA, pp. 217–232.

Green, C.H. (1990) The effect of colour on the numbers, age and nutritional status of *Glossina tachinoides* (Diptera: Glossinidae) attracted to targets. *Physiological Entomology* 15, 317–329.

Gruvel, J. (1975) Lieux de repos de *Glossina tachinoides* W. *Revue d 'Elevage et de Médecine Vétérinaire des Pays Tropicaux* 28, 153–172.

Hall, M.J.R. (1988) Characterization of the sexual response of male tsetse flies, *Glossina morsitans morsitans*, to pheromone-baited decoy 'females' in the field. *Physiological Entomology* 13, 49–58.

Hargrove, J.W. (1981) Tsetse dispersal reconsidered. *Journal of Animal Ecology* 50, 351–373.

Hargrove, J.W. (1991) Ovarian ages of tsetse flies (Diptera: Glossinidae) caught from mobile and stationary baits in the presence and absence of humans. *Bulletin of Entomological Research* 81, 43–50.

Hargrove, J.W. and Coates, T.W. (1990) Metabolic rates of tsetse flies in the field as measured by the excretion of injected caesium. *Physiological Entomology* 15, 157–166.

Harley, J.M.B. (1965) Activity cycles of *Glossina pallidipes* Aust. *G. palpalis fuscipes* Newst. and *G. brevipalpis* Newst. *Bulletin of Entomological Research* 56, 141–160.

Hill, P., Saunders, D.S. and Campbell, J.A. (1973) The production of 'symbiont-free' *Glossina morsitans* and an associated loss of female fertility. *Transactions of the Royal Society of Tropical Medicine and Hygiene* 67, 727–728.

Huyton, P.M. and Brady, J. (1975) Some effects of light and heat on the feeding and resting behaviour of tsetse flies, *Glossina morsitans* Westwood. *Journal of Entomology A* 50, 23–30.

Huyton, P.M., Langley, P.A., Carlson, D.A. and Coates, T.W. (1980a) The role of sex pheromones in initiation of copulatory behaviour by male tsetse flies, *Glossina morsitans morsitans. Physiological Entomology* 5, 243–252.

Huyton, P.M., Langley, P.A., Carlson, D.A. and Schwarz, M. (1980b) Specificity of contact sex pheromones in tsetse flies, *Glossina* spp. *Physiological Entomology* 5, 253–264.

Jackson, C.H.N. (1941) The analysis of a tsetse-fly population. *Annals of Eugenics* 10, 332–369.

Jordan, A.M. (1974) Recent developments in the ecology and methods of control of tsetse flies (*Glossina* spp.) (Dipt.: Glossinidae) – a review. *Bulletin of Entomological Research* 63, 361–399.

Jordan, A.M. (1993) Tsetse-flies (Glossinidae). In: Lane, R.P. and Crosskey, R.W. (eds) *Medical Insects and Arachnids.* Chapman & Hall, London, pp. 333–388.

Langley, P.A. (1977) Physiology of tsetse flies (*Glossina* spp.) (Diptera: Glossinidae): a review. *Bulletin of Entomological Research* 67, 523–574.

Langley, P.A., Hargrove, J.W. and Wall, R.L. (1990) Maturation of the tsetse fly *Glossina pallidipes* (Diptera: Glossinidae) in relation to trap-orientated behaviour. *Physiological Entomology* 15, 179–186.

Ma, W.C., Denlinger, D.L., Jarlfors, U. and Smith, D.S. (1975) Structural modifications in the tsetse fly milk gland during a pregnancy cycle. *Tissue and Cell* 7, 319–330.

Morris, K.R.S. (1946) The control of trypanosomiasis by entomological means. *Bulletin of Entomological Research* 37, 201–250.

Nash, T.A.M. (1969) *Africa's Bane: the Tsetse Fly.* Collins, London.

Nash, T.A.M., Trewern, M.A. and Moloo, S.K. (1976) Observations on the free larval stage of *Glossina morsitans morsitans* Westw. (Diptera: Glossinidae): the possibility of a larval pheromone. *Bulletin of Entomological Research* 66, 17–24.

O'Neill, S.L., Gooding, R.H. and Aksoy, S. (1993) Phylogenetically distant symbiotic microorganisms reside in *Glossina* midgut and ovary tissues. *Medical and Veterinary Entomology* 7, 337–383.

Phelps, R.J. and Burrows, P.M. (1969) Puparial duration in *Glossina morsitans orientalis* under conditions of constant temperature. *Entomologia Experimentalis et Applicata* 12, 33–43.

Phelps, R.J. and Clarke, G.P.Y. (1974) Seasonal elimination of some size classes in males of *Glossina morsitans morsitans* Westw. (Diptera: Glossinidae). *Bulletin of Entomological Research* 64, 313–324.

Phelps, R.J. and Jackson, P.J. (1971) Factors influencing the moment of larviposition and eclosion in *Glossina morsitans orientalis* Vanderplank (Diptera: Muscidae). *Journal of the Entomological Society of Southern Africa* 34, 145–157.

Pollock, J.N. (1974) Anatomical relations during sperm transfer in *Glossina austeni* Newstead (Glossinidae: Diptera). *Transactions of the Royal Entomological Society of London* 125, 489–501.

Randolph, S.E., Dransfield, R.D. and Rogers, D.J. (1989) Effect of host odours on trap composition of *Glossina pallidipes* in Kenya. *Medical and Veterinary Entomology* 3, 297–306.

Randolph, S.E., Rogers, D.J. and Kilu, J. (1991) The feeding behaviour, activity and trapability of wild *Glossina pallidipes* in relation to their pregnancy cycle. *Medical and Veterinary Entomology* 5, 335–350.

Rice, M.J. (1970) A Study of the Innervation, Structure and Function of the Anterior Alimentary Canal of the Adult Tsetse Fly (*Glossina austeni*) and other Diptera. PhD thesis, University of Birmingham.

Riordan, K. (1976). Rate of linear advance by *Glossina morsitans submorsitans* Newst. (Diptera: Glossinidae) on a trade cattle route in south-western Nigeria. *Bulletin of Entomological Research* 66, 365–372.

Roberts, M.J. (1973a) The control of fertilisation in tsetse flies. *Annals of Tropical Medicine and Parasitology* 67, 117–123.

Roberts, M.J. (1973b) Observations on the function of the choriothete and on egg hatching in *Glossina spp.* (Dipt: Glossinidae). *Bulletin of Entomological Research* 62, 371–374.

Rogers, D. (1977) Study of a natural population of *Glossina fuscipes fuscipes* Newstead and a model of fly movement. *Journal of Animal Ecology* 46, 309–330.

Rogers, D.J. and Randolph, S.E. (1985) Population ecology of tsetse. *Annual Review of Entomology* 30, 197–216.

Rogers, D.J., Randolph, S.E. and Kuzoe, F.A.S. (1984) Local variation in the population dynamics of *Glossina palpalis palpalis* (Robineau-Desvoidy) (Diptera: Glossinidae). I. Natural population regulation. *Bulletin of Entomological Research* 74, 403–423.

Saunders, D.S. (1960) The ovulation cycle in *Glossina morsitans* Westwood (Diptera: Muscidae) and a possible method of age determination for female tsetse flies by the examination of their ovaries. *Transactions of the Royal Entomological Society of London* 112, 221–238.

Saunders, D.S. and Dodd, C.W.H. (1972) Mating, insemination and ovulation in the tsetse fly *Glossina morsitans*. *Journal of Insect Physiology* 18, 187–198.

Takken, W., Oladunmade, M.A., Dengwat, L., Feldmann, H.U., Onah, J.A., Tenabe, S.O. and Hamann, H.J. (1986) The eradication of *Glossina palpalis palpalis* (Robineau-Desvoidy) (Diptera: Glossinidae) using traps, insecticide-impregnated targets and the sterile insect technique in central Nigeria. *Bulletin of Entomological Research* 76, 275–286.

Taylor, P. (1976) Blood-meal size of *Glossina morsitans* Westw. and *G. pallidipes* Austen (Diptera: Glossinidae) under field conditions. *Transactions of the Rhodesia Scientific Association* 57, 29–34.

Thomson, R.C.M. (1968) *Ecology of Insect Vector Populations*. Academic Press, London.

Torr, S.J. (1988a) The activation of resting tsetse flies (*Glossina*) in response to visual and olfactory stimuli in the field. *Physiological Entomology* 13, 315–325.

Torr, S.J. (1988b) The flight and landing of tsetse (*Glossina*) in response to components of host odour in the field. *Physiological Entomology* 13, 453–465.

Torr, S.J. (1988c) Behaviour of tsetse flies (*Glossina*) in host odour plumes in the field. *Physiological Entomology* 13, 467–478.

Torr, S.J. (1989) The host-orientated behaviour of tsetse flies (*Glossina*): the interaction of visual and olfactory stimuli. *Physiological Entomology* 14, 325–340.

Turner, D.A. and Brightwell, R. (1986) An evaluation of a sequential aerial spraying operation against *Glossina pallidipes* Austen (Diptera: Glossinidae) in the Lambwe Valley of Kenya: aspects of post-spray recovery and evidence of natural population regulation. *Bulletin of Entomological Research* 76, 331–349.

Vale, G.A. (1974) The responses of tsetse flies (Diptera: Glossinidae) to mobile and stationary baits. *Bulletin of Entomological Research* 64, 545–588.

Vale, G.A. (1977) The flight of tsetse flies (Diptera: Glossinidae) to and from a stationary ox. *Bulletin of Entomological Research* 67, 297–303.

Vale, G.A. (1993) Development of baits for tsetse flies (Diptera: Glossinidae) in Zimbabwe. *Journal of Medical Entomology* 30, 831–842.

Vale, G.A., Hargrove, J.W., Cockbill, G.F. and Phelps, R.J. (1986) Field trials of baits to control populations of *Glossina morsitans morsitans* Westwood and *G. pallidipes* Austen (Diptera: Glossinidae). *Bulletin of Entomological Research* 76, 179–193.

Vale, G.A., Lovemore, D.F., Flint, S. and Cockbill, G.F. (1988) Odour-baited targets to control tsetse flies *Glossina* spp. (Diptera: Glossinidae), in Zimbabwe. *Bulletin of Entomological Research* 78, 31–49.

Wall, R. and Langley, P. (1991) From behaviour to control: the development of trap and target techniques for tsetse fly population management. *Agricultural Zoology Reviews* 4, 137–158.

Wall, R. and Langley, P.A. (1993) The mating behaviour of tsetse flies (*Glossina*): a review. *Physiological Entomology* 18, 211–218.

Warnes, M.L. (1989) Responses of the tsetse fly, *Glossina pallidipes*, to ox odour, carbon dioxide and a visual stimulus in the laboratory. *Entomologia Experimentalis et Applicata* 50, 245–253.

Williamson, D.L., Dame, D.A., Gates, D.B., Cobb, P.E., Bakuli, B. and Warner, P.V. (1983) Integration of insect sterility and insecticides for control of *Glossina morsitans morsitans* Westwood (Diptera: Glossinidae) in Tanzania. V. The impact of sequential releases of sterilised tsetse flies. *Bulletin of Entomological Research* 73, 391–404.

Wilson, V.J. (1975) Game and tsetse fly in eastern Zambia. *Occasional Papers of the National Museums and Monuments of Rhodesia B* 5, 339–404.

Zdarek, J. and Denlinger, D.L. (1991) Wandering behaviour and pupariation in tsetse larvae. *Physiological Entomology* 16, 523–529.

Zdarek, J., Denlinger, D.L. and Otieno, L.H. (1992) Does the tsetse parturition rhythm have a circadian basis? *Physiological Entomology* 17, 305–307.

Muscidae and Fanniidae (Houseflies, Stableflies)

13

The Muscidae is a large family of nearly 4000 species and includes the houseflies and the blood-sucking stableflies. The Fanniidae is a small family of some 265 species of which 220 are in the genus *Fannia*. In the Muscidae the squamae are conspicuous with the lower much larger than the upper. In the Fanniiidae the squamae are small, subcircular and of similar size. The medically important members of both families are dark coloured medium-sized flies, whose immature stages occur in fermenting organic material of vegetable origin and are often associated with the dung of herbivorous mammals.

Classification (Crosskey and Lane, 1993)

Muscidae

Skidmore (1985) recognizes ten subfamilies in the Muscidae of which two, the Stomoxyinae and Muscinae, are medically important. The Stomoxyinae are haematophagous muscids readily recognized by the possession of an elongate, sclerotized proboscis (Fig. 13.1B, C), and wing vein iv being gently curved towards vein iii (Fig. 13.2). It includes the stablefly, *Stomoxys calcitrans* (Fig. 13.1B), the horn fly, *Haematobia irritans* and the buffalo fly, *H. exigua* (Fig. 13.1C). In the Muscinae the mouthparts are of the lapping type and can be folded into the subcranial cavity. Vein iv is strongly bent towards vein iii (Fig. 13.2), and the lower squama is broad with its posterior margin almost straight and at right angles to the long axis of the body. This subfamily includes the cosmopolitan housefly, *Musca domestica*, and the Old World *Musca sorbens* complex. Skidmore (1985) has described the early stage of 440 species, including all the medically important ones, and summarized their biology and geographical distribution.

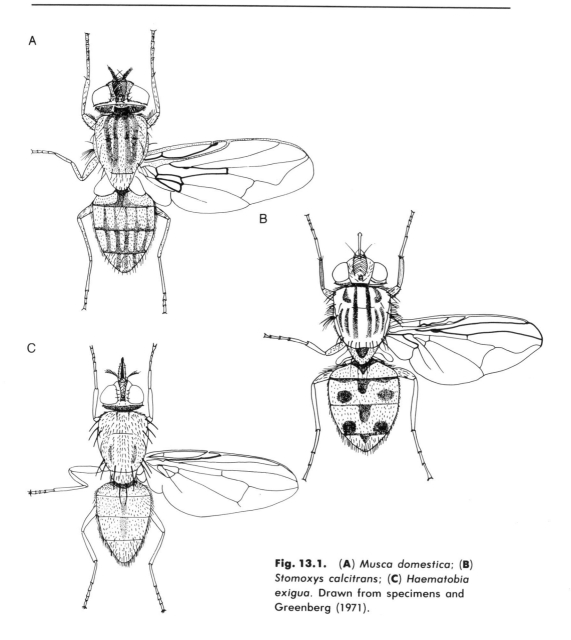

Fig. 13.1. (**A**) *Musca domestica*; (**B**) *Stomoxys calcitrans*; (**C**) *Haematobia exigua*. Drawn from specimens and Greenberg (1971).

Fanniidae

Fannia is the only genus of importance in this family. They are small to medium-sized flies with bare arista, and a characteristic venation, in which vein iv is straight, vein vi is short and vein vii curved so that if extended it would intersect with an extended vein vi (Fig. 13.3). The males are holoptic and the females dichoptic. A few species are of importance: *F. canicularis*, the lesser or little housefly, is a worldwide endophilic synanthrope; *F. scalaris*, the latrine fly, is a

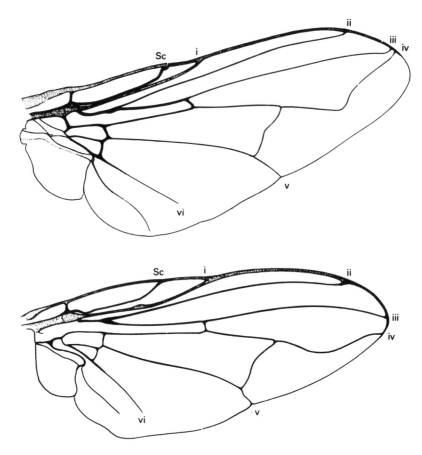

Fig. 13.2. Wings of *Musca domestica* (above) (x 20) and *Stomoxys calcitrans* (below) (x 17).

worldwide exophilic synanthrope. In North America *F. femoralis* is an occasional pest of humans.

Flies and other animals which live closely with man are said to be synanthropic. Greenberg (1971) recognizes various forms of synanthropy. Eusynanthropes can complete their entire development within the residences of humans and their domestic animals. Many of these species have become cosmopolitan, spreading throughout the world with people. *M. domestica* and *F. canicularis* are endophilic, eusynanthropes which are trophically and microclimatically related to humans. Exophilic eusynanthropes, such as *Lucilia sericata* (Calliphoridae) and *F. scalaris*, are less closely related to humans both trophically and microclimatically. Symbovines are linked through the excreta of domestic herbivores and fall into two groups depending on whether their association is with animals confined in stables and feedlots, e.g. *S. calcitrans*, or with free ranging stock on pasture, e.g. *H. exigua*. Hemisynanthropes, e.g. *Hydrotaea irritans*, live independently of humans but interact with them when their habitat is entered.

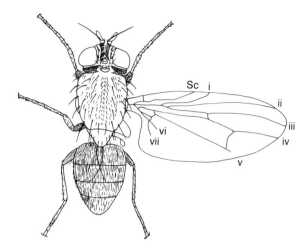

Fig. 13.3. Adult *Fannia canicularis*.

Annotated bibliographies

Annotated bibliographies on *M. domestica* have been prepared by West and Peters (1973) and Keiding (1986), on *Stomoxys* by Morgan *et al.* (1983) and on *Haematobia* by Morgan and Thomas (1974, 1977).

Musca domestica (**Housefly**) (West, 1951; Pont, 1973; Ferrar, 1979)

Three subspecies are recognized in this widely distributed species. *M. domestica domestica* occurs worldwide but is least abundant in Africa, where the two other subspecies occur – the endophilic *M. d. curviforceps* and the exophilic *M. d. calleva* (Crosskey and Lane, 1993). All three may be recognized by the presence of four dark, longitudinal stripes or vittae on the scutum (Fig. 13.1A). The arista is plumose with branches above and below. The arrangement of bristles on the thorax includes, on the scutum, two or three dorsocentrals anterior to the suture and four posterior, and a single acrostichal bristle posterior to the suture; three sterno-pleurals on the katepisternum, one anterior and two posterior.

Life cycle

The pearly-white eggs are long and narrow, measuring about 1.20 × 0.25 mm. Under optimal conditions (37°C) they hatch in about 8 h to give rise to legless, saprophagous larvae (maggots). There are three larval instars, of which the first two last about 24 h, and the third for 3 or more days.

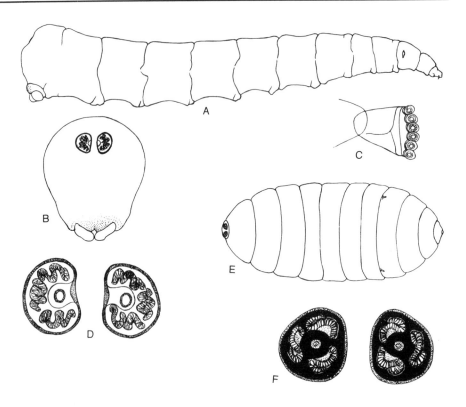

Fig. 13.4. (**A**) Third stage larva of *Musca domestica*; (**B**) posterior view of larva; (**C**) anterior spiracle; (**D**) posterior spiracles; (**E**) puparium showing pupal respiratory horns; (**F**) posterior spiracles of *Stomoxys calcitrans*.

Larva

The third stage larva (Fig. 13.4) measures 6–12 × 1–2 mm, and has 12 visible segments (one head segment, three thoracic and eight abdominal). On the ventral side of the first segment there are two oral lobes transversed by parallel tubes, which converge on the mouth. The tubes function in a comparable manner to the pseudotrachae of the labellum of the adult fly. The head is retracted into the thorax and its dark cephalopharyngeal skeleton can be seen through the translucent body of the larva. This skeleton supports a pair of retractable mouth-hooks, which can be extended through the mouth and used for progression and tearing at the substrate. In *Musca* the mouth-hooks are closely apposed, with the right one being markedly larger than the left.

Respiration is amphipneustic with fan-shaped anterior spiracles on the second segment, undeveloped in first instar larvae, and dark, flat, plate-like spiracles on the posterior surface of the body. In the 3rd instar the anterior spiracles have five to seven openings and the posterior spiracles are D-shaped with three sinuous slits and a button in the middle of the straight side of the D (Fig. 13.4D). The button is the scar left at the moult from 2nd to 3rd instar. The posterior spiracle has one simple, reniform opening in the 1st instar, and two nearly straight slits in the 2nd instar (Skidmore, 1985).

Pupa

The fully grown larva ceases to feed, and empties its gut to become a prepupa, which moves into drier conditions and buries itself into the substrate, where it pupates within the last larval skin, which forms the puparium (Fig. 13.3E). Immediately after moulting the puparium is creamy white, but it steadily darkens through shades of reddish brown until the mature puparium is almost black. The external larval structures are retained on the outside of the puparium, which measures 4–6 × 2–2.5 mm. The larval spiracles are non-functional and the pupa respires through pupal horns which pierce the puparium between the fifth and sixth segments (Hewitt, 1914).

When the adult fly has completed its development, it emerges from the puparium by using its ptilinum (Fig. 12.5) to force off a hemispherical cap from the anterior end of the puparium and make its way up to the surface of the soil when the ptilinum is withdrawn into the head, and the frontal suture closed. The body of the fly is expanded by taking air into the gut, and the wings extended by pumping haemolymph through the veins. This ability of the newly emerged fly to move to the surface can render the burying of infested material ineffective as a fly control measure. Flies are able to emerge from material buried under 1.2 m of clay, loam or sand because most of the prepupae move to within 30 cm of the surface before pupating, and from this distance adults can easily reach the surface.

Adult

Adult *M. domestica* are diurnal, and activity is favoured by high temperatures and low humidities, but, as the name housefly implies, they are more active in shade than in sunlight. Females emerge before the males, and mating takes place soon after emergence. Males will mate on the day of emergence, and the mating response is highest in 3-day-old females (Sacca, 1964). Two pheromones are involved in mating, one produced by the female attracts males and the other, produced by males, induces aggregation and receptivity in virgin females (Crosskey and Lane, 1993). Maturation of the eggs depends upon the female having access to a diet of protein, and a batch of eggs may be laid as early as 54 h after emergence of the female (Patton and Evans, 1929).

The ovipositing female deposits her eggs in clumps in cracks and crevices of a suitable medium, and sometimes the whole batch may be deposited in a single clump. By inserting the eggs into a moist medium the female protects them from desiccation. The ancestral breeding site was probably horse dung, but houseflies now breed in the dung of a wide range of herbivores, in fowl manure, in fermenting kitchen waste, and in rubbish tips (Pont, 1973). Indeed separate populations of *M. domestica* may develop in association with stabled animals, and breed in urine and dung-contaminated stable refuse, but *M. domestica* does not breed in cow dung, which is the main source of other species of *Musca*.

Temperature and development

The duration of the larval instars is a function of temperature and the quality of the larval medium. When dung quality is not limiting, the development time of the larval stages is 145 day-degrees above a threshold of 12°C and up to an optimum of 36°C, above which development is adversely affected. This relationship implies that at 22°C, i.e. 10°C above the threshold, the larval duration would be 14.5 days. Activity of the prepupa is optimal at 29°C. Susceptibility to high temperatures is lowest in the egg and highest in the pupa, and all stages are killed by exposure to 50°C or higher; a fact used in fly control. Part of the rationale for the tight packing of refuse dumps containing organic material is to attain temperatures lethal to the immature stages of the housefly through the fermentation and decay of the organic material.

Population growth

On the average a female will mature 120 eggs in a batch (range 100–150) and deposit four to six batches of eggs during a lifetime of 2 to 4 weeks in summer (Hewitt, 1914). Patton and Evans (1929) recorded a particularly fecund female which laid 2387 eggs in 21 batches over a period of 31 days after emergence. This is exceptional but *M. domestica* is an *r*-selected species with a high potential rate of increase. Using conservative rates of development and fecundity Howard, cited by Hewitt (1914) calculated that, over the northern hemisphere summer, a single female ovipositing on 15 April would have produced 5.6×10^{12} progeny by 10 September, if they had all lived. Obviously they do not, but the species' potential for exploiting favourable conditions is outstanding. With a cycle from egg to egg of 3 weeks, ten to twelve generations can be produced a year in the warmer temperate regions of the world. In colder regions breeding will be restricted to the warmer months, and the winter passed as slowly growing larvae and pupae, some of which will survive to emerge, when warmer conditions return (Sacca, 1964).

Houseflies and Disease

Houseflies have adapted to domestic living, feeding and breeding on human food, organic wastes and faeces. The movement of houseflies between faeces and food makes them ideal transmitters of human disease. A vast literature has developed on this subject and was summarized by West (1951) for work published before 1950. Greenberg (1971, 1973) has given a full list of the organisms which have been recovered from houseflies, and a detailed consideration of their relationship to human and animal diseases.

Pathogens involved

Houseflies have been found to harbour about 100 different pathogens and charged with transmitting 65 of these (Greenberg, 1965). The pathogens recovered from flies range from viruses to helminths, and include the viruses of poliomyelitis and infectious hepatitis, the bacteria associated with cholera (*Vibrio*), enteric infections caused by species of *Salmonella* and *Shigella*, pathogenic *Escherichia coli*, haemolytic streptococci, *Staphylococcus aureus*, agents of trachoma, bacterial conjunctivitis, anthrax, diphtheria, tuberculosis, leprosy and yaws. In addition flies can carry the cysts of Protozoa, including those of *Entamoeba histolytica* which causes amoebic dysentery, and the eggs of the threadworm *Trichuris trichiura*, the hookworm *Ancylostoma duodenale* and of other nematodes and cestodes.

The housefly is the biological vector and intermediate host of certain cestodes of poultry, and of nematodes, which cause habronemiasis in horses (see Chapter 32). Fly larvae have only a tenuous association with myiasis, the invasion of living tissue by dipterous larvae. When they occur in an advanced stage of myiasis in sheep they do not feed on living tissue, but on the exudate and matted wool (Pont, 1973). In this connection it is of interest that fly larvae were only found in carcasses when the vegetable contents of the gut were exposed (Hepburn, 1943).

The recovery of pathogens from houseflies does not necessarily involve them in the transmission of disease. There are other routes of infection and, depending upon circumstances, houseflies may play a major or minor or no role at all in disease transmission. A comparison between two groups of towns in Texas, one of which was sprayed with DDT for housefly control and the other left untreated, produced a reduction in acute diarrhoeal infection of children, with infections due to *Shigella* declining, but those due to *Salmonella* were much less affected (Greenberg, 1965).

Methods of transmission

There are three ways in which houseflies can disseminate pathogens. The surface of the body of the fly, particularly its legs and proboscis, can be contaminated; pathogens can be regurgitated on to food via the vomit drop; or pass through the gut of the fly and be deposited in its faeces. Infective material picked up on the body hairs and tarsi of houseflies may survive only a short period. They will be subject to the cleaning behaviour of the fly, in which it seeks to rid itself of foreign material, and organisms exposed on the surface will be subject to desiccation, particularly in flight, and to UV sterilization in sunlight (Greenberg, 1973). Greater survival would be expected for organisms trapped between the lobes of the labellum. There is a minimum number of pathogens needed to infect a human being, but a lesser number of organisms deposited into a medium, e.g. milk, in which they can multiply, can reach an infective density. The human infective dose is 10^5 virus particles and 10^6 for bacteria. Only the latter could multiply in human food; viruses require a host cell (Greenberg, 1965).

When a housefly feeds, the filtering function of the pseudotracheae will exclude protozoan cysts and eggs of helminths, but a fly can ingest larger particles directly through the prestomum at the distal end of the food canal. The vomit drop is formed from the contents of the crop, which represents the most recently ingested food. The vomit drop is therefore an important method of disseminating pathogens. Small pathogens pass out freely with the vomit drop, but larger cysts and eggs are held back by the pseudotracheae (Greenberg, 1973). The contents of the crop are passed to the midgut, and viable cysts and eggs may appear in the fly's faeces.

Pathogens and the fly's gut fauna

Pathogens that pass through the gut of an adult housefly have an opportunity to multiply before being deposited in the faeces. Their ability to develop an infection in the housefly depends upon the number ingested. When inputs were below 10^3 *Salmonella*, no organisms appeared in the fly's faeces. A higher intake is required in the presence of the fly's normal gut flora. Interspecific antagonism may lead to the rapid elimination of *Salmonella*, and rapid multiplication of *Salmonella* occurs in houseflies freed from their normal gut flora. In comparison with the green blowfly, *Lucilia sericata*, the housefly was a superior host for *Salmonella* (Greenberg, 1973).

The fermenting material in which housefly larvae live is teeming with bacteria. For growth, fly larvae require microorganisms or one or more of their products. On emergence the adult fly is virtually free from microorganisms. Several factors account for this change. Firstly, in the prepupa the fly larva ceases to feed, and reduces its microbe population to less than 1%, completely eliminating a substantial population of *Salmonella*. There are two main reasons for this elimination. There is competition from the fly's normal commensal bacterium, *Proteus mirabilis*, and high mortality in the midgut. Most of the fly's intestinal tract is alkaline but the midgut is strongly acid with a pH of 3.0–3.5. Passage through the midgut reduces the microbial population to less than 2%. The speed with which this reduction takes place indicates the existence of a factor other than pH in producing mortality. In the alkaline rectum the *Proteus* population may recover but that of *Salmonella* rarely does. The microbial population is further reduced at pupation when the linings of the fore- and hindgut and most of their contents are shed (Greenberg, 1965).

Musca sorbens Complex

The *Musca sorbens* complex of species is widely distributed in the tropics and sub-tropics of the Afrotropical, Oriental, Australian regions and in the southern Palaearctic. *Musca sorbens* s.s. occurs throughout the range with the exception of Australia where the only species present is *M. vetustissima*. A third member of the complex *M. biseta* occurs in Africa (Crosskey and Lane, 1993). Flies of the *M. sorbens* complex have two broad, dark longitudinal vittae on the scutum (Fig. 13.1B). The wings are clear with white squamae, and the first abdominal segment black. The proboscis is of the normal lapping type with small prestomal teeth. The female is dichoptic and the male nearly holoptic. The primary breeding

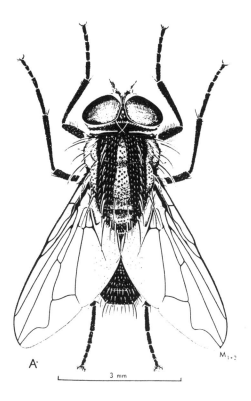

Fig. 13.5. *Musca vetustissima.* Source: from *Insects of Australia,* Melbourne University Press.

site of members of the *M. sorbens* complex is cow dung, but in Tadzhikhistan it breeds mainly in human faeces (Zimin, 1948, quoted in Skidmore, 1985).

Musca vetustissima (Australian bushfly)

The Australian bushfly (Fig. 13.5) has been studied in depth by Hughes and his colleagues (Hughes *et al.*, 1972). The eggs of *M. vetustissima* are larger than those of *M. domestica*, measuring 1.7 × 0.3 mm (Ferrar, 1979). They have the typical muscine shape with a hatching strip on the concave surface, which functions as a plastron facilitating respiration when the eggs are in water. Eggs of *M. vetustissima* will hatch in water but not if totally immersed in faeces, when the plastron is unable to function. The eggs are very susceptible to desiccation and some die when exposed for as short a time as 1 h to 90% RH. Development is rapid being completed in 7 h at 32°C and 17 h at 21°C. Larval growth is even more rapid than in *M. domestica*, the three instars being completed in 8, 10 and 49 h, respectively, at 32°C. For rapid development larvae need access to moist dung for feeding, and air to which they expose their posterior spiracles for respiration. The 3rd instar larva is very similar to that of *M. domestica*, being of comparable size, with the right mouth-hook longer than the left, the anterior spiracle with about six openings, and three sinuous slits in the posterior spiracle.

Mature larvae are sensitive to exposure to excess water, and no pupation occurs in waterlogged soil. Even temporary waterlogging reduces pupal survival, particularly if exposed early in their lives before the respiratory horns have pierced the puparium. Prepupae leave the dung between midnight and dawn, and bury themselves into the substrate, where they pupate at depths of 20–30 mm. Moving from the dung during the hours of darkness must reduce predation. The duration of the pupal stage is about 80% of that of the combined larval stages, taking 3 days at 39°C and 18 days at 18°C. Eclosion occurs around dawn, which is appropriate for a diurnally active adult. Adults emerging from dung pads with high densities of immatures are smaller than those emerging from pads of low density. Mortality is greater when the surviving flies have head widths of less than 2 mm (Ridsdill-Smith, 1991).

Females develop faster than males and emerge first. At 27°C mating occurs on the third day after emergence and it is believed that a second mating occurs after the second ovarian cycle. Copulation is a long process lasting about 80 min during which spermatozoa are transferred together with an oviposition stimulant and monocoitic substances, which temporarily restrict further mating. The female requires a protein meal to develop the ovaries. In the field females complete two to three oviposition cycles during their lifetime (Ridsdill-Smith, 1991). Wild flies develop 4–48 eggs in a cycle, the number of eggs maturing being reduced in each subsequent cycle. Ovarian development depends upon the nutritional state of the female, and where that is inadequate fewer eggs are matured. The female is attracted to fresh cow dung to which it flies upwind and deposits its eggs in crevices in the dung. Oviposition occurs during daylight and an ovipositing female attracts other gravid females, probably through the emission of an egg-laying pheromone. Mortality of the immatures is high, averaging 94% in the field (Ridsdill-Smith, 1991).

Adult survival is inversely related to temperature, being 11 days at 29°C and 7 or more weeks at 12°C, the threshold temperature for development. Survival is dependent on the adult having access to free water, and an easily metabolized energy source such as sugar. They are active in temperatures up to 35°C and wind speeds of 8 km h^{-1}. They can be displaced hundreds of kilometres in a day on hot, strong winds, and are also regularly dispersed via human transport.

They are mainly nuisance flies, feeding at the eyes, mouths and wounds of people and domestic animals, and also on dung. They settle out of the wind, and can be present on the backs of men and women in large numbers. Their wide dispersal from the pastures in which they originate allows them to be a pest in the suburbs of large cities.

Phenology

Several features of the bushfly biology determine its seasonal distribution. There is no diapausing stage in the life cycle, and the temperature threshold for activity, and immature and adult development is 12°C. At low temperatures all stages may survive for a long time but temperatures below freezing are lethal. The effect of this is that the bushfly dies out in the southern parts of Australia during the winter. Rapid development of the immature stages, and high fecundity in the females, is dependent on the quality of bovine dung available. In addition temporary waterlogging reduces survival of both prepupae and pupae. The net result of these different responses is that the bushfly population increases dramatically after the wet season when the cattle are feeding on rapidly growing grasses and producing rich dung.

The bushfly populations are highest in the north in the autumn following the summer rains and decline in the winter and spring as the pasture deteriorates. They move south on warm winds and are able to exploit the spring flush of vegetation following the winter rains in the south of Australia. There, bushfly populations reach a peak in late spring and early summer, and then wane as the summer dry season advances. Like many other Australian animals the bushfly is nomadic, moving around to exploit favourable conditions.

Control

In view of its wide dispersal local control measures will be ineffective against the bushfly. The ready availability of bovine dung to bushflies in Australia results partly from the absence, over the greater part of the country, of any dung beetles able to cope with bovine dung. To redress this situation a range of dung beetles has been introduced into Australia. *Onthophagus gazella* has proved to be highly effective over a limited area of the country, successfully competing with bushflies, and spreading widely from centres of release. The effect of *O. gazella* on *M. vetustissima* in the laboratory has been studied by Bornemissza (1970). The small beetle *O. binotis* causes increased mortalities of the eggs and young larvae, while a larger beetle, *O. ferox*, causes increased mortality of larger larvae (Ridsdill-Smith *et al.*, 1987).

Stomoxyinae (Zumpt, 1973)

Species of *Stomoxys* can be recognized by their palps being less than half the length of the proboscis (Fig. 4.6), and the katepisternum (sternopleuron) having one posterior bristle. *S. calcitrans* is found worldwide, *S. niger* is restricted to the Afrotropical region, and *S. sitiens* is found in both the Afrotropical and the Oriental regions. In *Haematobia* the palps, which are grooved internally, are as long as the proboscis, and the katepisternum has both an anterior and a posterior bristle. *H. irritans* is found in the New World, Hawaii and Japan, *H. minuta* in the Afrotropical and Oriental regions and *H. exigua* in the Oriental and Australian regions (Crosskey, 1993). In both *Stomoxys* and *Haematobia*, but not in all Stomoxyinae, the arista carries hairs only on the dorsal side. Stomoxyines show strong phototactic responses to UV and blue radiation (Allan *et al.*, 1987) which explains the greater attractiveness of traps using UV reflective fibreglass panels (Alsynite®) and phthalogen blue dyed cloth (Holloway and Phelps, 1991).

Stomoxys calcitrans (stablefly, biting housefly, dog-fly)

Stomoxys calcitrans has four dark, longitudinal vittae on the thorax, the lateral ones being interrupted at the suture (Fig. 13.1B) and the abdomen has a grey and dark brown pollinosity, forming a variable but characteristic pattern of dark spots on a lighter background. Its life cycle is that of a typical muscid fly. The elongate, white egg hatches into a saprophagous maggot, which undergoes three moults. In the third stage larva the right mouth-hook is larger than the left, the anterior spiracles have about six openings, and the posterior spiracles have three S-shaped slits surrounding a central button (Fig. 13.4F) (Ferrar, 1979). At 26.7°C the egg stage lasts 23 h, and the three instars 23 h, 27 h and approximately 7 days respectively (Parr, 1962a). At 30–31°C the pupal stage lasts 5 days. The female is anautogenous, requiring several blood meals to complete ovarian development, and Parr (1962b) reports that the average blood meal (25.8 mg) is three times the average body weight (8.6 mg).

Stomoxys calcitrans has a bimodal diurnal pattern of feeding, locating hosts by responding to carbon dioxide and octenol (Holloway and Phelps, 1991). Individual *S. calcitrans* may feed more than once a day, biting their host low down. They attack the ankles of humans, and the belly, lower body and limbs of domestic stock, particularly cattle and horses. In summer the adults survive for 3 to 4 weeks, and considerably longer in the cooler times of the year (Roberts, 1952). The duration of damaging populations of *S. calcitrans* at feedlots on the great plains of western Kansas varies from 6 to 18 weeks depending on rainfall (Greene, 1989).

Adults assemble on sunlit light objects from which males dart out after flying insects. Males mount females in the air or on the ground with copulation occurring on a perch (Buschman and Patterson, 1981). Insolation can raise the body temperature by more than 10°C to 22–28°C but when the body temperature reaches 31–34°C they seek shade (Allan *et al.*, 1987). Under ordinary conditions *S. calcitrans* can disperse 5 km but much further (maximum recorded 225 km) under exceptional weather conditions. They are regularly carried by northerly winds to the beaches of west Florida, travelling above the ground at a height of about 90 cm. Cold fronts carry large numbers of *S. calcitrans* long distances at heights of 30 to 60 m (Hogsette *et al.*, 1989).

The number of eggs matured by an individual female *S. calcitrans* ranges from 200 to 600 (Roberts, 1952). The female scatters eggs throughout a suitable medium such as straw contaminated with the urine and dung of cattle and horses. Larvae may be found in animal bedding, lawn cuttings and rotting vegetables. In Uganda the breeding sites of *S. calcitrans* were characterized by the presence of rotted cattle manure, rotten straw, grass or leaves, and shade (Parr, 1962b). Pupation occurs in the drier parts of stable refuse (Ferrar, 1979).

Economic importance

Stomoxys calcitrans worries stock particularly around stables and feedlots, but less in pastures. It may reduce milk yield by 25%, or as much as 40–60% (Greenberg, 1973). In the summer in Nebraska densities of *S. calcitrans*, above five per front leg, reduce weight gain and feeding efficiency in stabled cattle (Campbell and Berry, 1989). It can affect the behaviour of wild animals; in the Ngorongoro Crater in Tanzania lions climbed trees to avoid being bitten (Fosbrooke, 1963).

Pathogens

A large number of pathogens have been recorded from *S. calcitrans* (Greenberg, 1971). Stableflies act as both biological and mechanical vectors of disease. They are the intermediate host of nematode worms, including *Setaria cervi* a parasite of cattle, and of several species of *Habronema*, parasites of horses (Greenberg, 1973). They are persistent biters, and often engage in interrupted feeding. This, together with the fact that they feed more than once per day, fits them to be mechanical vectors of blood-dwelling pathogens. They contribute to the spread of *Trypanosoma evansi*, which causes the disease surra in a wide range of hosts (Zumpt, 1973).

Stomoxys niger

In Mauritius *S. niger* breeds in the decaying trash of canefields, with the population being high from October to December and low from May to July (Ramsamy, 1978, 1981). In Zanzibar the highest concentrations of *S. niger* are found in heavily forested areas on poorly drained soils. They are particularly active just after sunrise and before sunset. They fly close to the ground with maximum numbers being caught in traps at a height of 30 cm and there is evidence that they can locate hosts 2–5 km away (Patterson, 1989).

Haematobia irritans, H. exigua (Fig. 13.1C)

Haematobia irritans, the horn fly, and *H. exigua*, the buffalo fly, are small, brownish, obligate parasites of cattle, variable in colour being darker in northern and paler in southern populations (Zumpt, 1973). They have a typical muscid life cycle. The eggs are bright yellowish brown in colour and hatch in 18–24 h, being followed by three larval instars, and a pupa within a puparium. The larva is a small, slender maggot with the right mouth-hook larger than the left, the anterior spiracle with about five lobes and the posterior spiracle with three very sinuous slits. The duration of the larval stages is 3 to 5 days and a similar time is spent in the pupa. Pupariation occurs at the dung pat or in nearby soil (Roberts, 1952). The mortality of the immatures in dung in the field is high, averaging over 90% (Thomas and Morgan, 1972). In the northern temperate regions *H. irritans* over-winters as a diapausing pupa.

    Both sexes are haematophagous and feed mainly on cattle and buffaloes, and occasionally on other animals, including humans, closely associated with bovines. They rest on cattle, being present in the greatest numbers on the withers, shoulders and flanks, but also on the neck, ribs and back (Roberts, 1952). They do not walk over the surface of the bovid but fly to change position. There is a single mating which occurs on the host. They take frequent small blood meals. The female oviposits only into fresh dung, leaving the host when dung is dropped, to deposit small batches of eggs (12–20) on to the underside of the dung pat, before returning after a few minutes.

Emergence of adult *H. exigua* occurs in the afternoon and early evening (12.00 to 20.00 h). Unfed flies, with or without access to water, live less than 24 h. Most flies arrived at an isolated host between 16.00 and 08.00 h and fed immediately. They were mostly newly emerged flies, the females being nulliparous but a variable proportion were parous, indicating that they had left a previous host. Released marked flies quickly found hosts, the majority travelling downwind but about 20% of the flies, predominantly females, found hosts 400 and 800 m upwind and others located water buffalo 200 m across wind (Macqueen and Doube, 1988). These observations help to account for *H. exigua* invading Magnetic Island, 7 km off the Queensland coast (Ferrar, 1969). *H. exigua* was introduced into Australia from Timor in 1838 on imported water buffalo, since when it has established itself in the tropical northern area and spread southwards. Its distribution is determined by an annual rainfall of 500 mm, and a temperature of 22°C. In addition the immature stages require dung with a moisture content of 68% (Pont, 1973).

Economic importance

It is considered that 100–300 buffalo flies per beast can be tolerated without adverse effect but densities of 500–1000 and up to 5000 are found, which are considered to lower weight gains in beef cattle, and milk yield in dairy herds. However the adverse effects of *H. irritans* may be compensated for by the animal's immune system and ample forage (Hogsette *et al.*, 1991). *H. irritans* is the intermediate host of *Stephanofilaria stilesi* a parasite of cattle in North America (Zumpt, 1973). The sedentary nature of these species would operate against their being important mechanical vectors of pathogens.

Other Muscids

Hydrotaea irritans

This is known as the sheep headfly and is notorious for its predilection for mammalian perspiration. It worries domestic stock and people by feeding on secretions from the mouth, nose, ears, eyes and wounds. Adults are active under calm, humid, sultry conditions particularly before and after rain, when their attacks can be intolerable. In Denmark the highest densities of *Hy. irritans* and *Hae. irritans* occur in permanent, low-lying, fairly sheltered grassland compared to temporary, dry, wind-exposed pastures (Jensen *et al.*, 1993). Adults also feed on honeydew, flowers, carrion and faeces from which they presumably acquire the necessary protein for ovarian development. The threshold temperature for activity is about 12°C, similar to that for *M. vetustissima*, and activity ceases at wind speeds above 3.6 km h^{-1}. Activity is bimodal with peaks in the morning and evening. Although there are records of this species breeding in cow dung, the normal larval biotope of this univoltine species is in pasture soil, under long grass or on woodland edges (Skidmore, 1985). There is circumstantial evidence for *Hy. irritans* playing a central role in transmitting summer mastitis pathogens to the teats of healthy cattle but experimental transmission of summer mastitis from sick to healthy cattle was unsuccessful (Madsen *et al.*, 1991).

Musca crassirostris (= *Philaematomyia insignis*)

The Indian cattle-fly occurs from Africa through the Middle East to south-east Asia. It has large prestomal teeth which can rasp away skin and enable the fly to feed on blood. It deposits small batches (40–50) of large (2 mm long) eggs into freshly dropped cow dung (Crosskey, 1993).

Musca autumnalis

Another feeder on mammalian secretions is known as the face fly because it attacks the head of cattle. It has well-developed prestomal teeth which can be used to re-open healing wounds but cannot penetrate the unbroken skin. It breeds in cow dung and is found in pastures close to woodland (Skidmore, 1985). It is a Palaearctic and Afrotropical species which was introduced, probably in 1951, into North America (Sabrosky, 1961).

Fannia

The larvae of *Fannia* (Fig. 13.6) are flattened, tapering anteriorly and bearing prominent lateral processes on most segments. In the mature larva the anterior spiracles are prominent and the posterior ones elevated with three lobes. The larvae of *F. canicularis* have long lateral and dorsal processes with short basal spines; and those of *F. scalaris* (Fig. 13.6) have short dorsal processes and pinnate lateral processes (James, 1947). They breed in decaying animal and vegetable matter, especially faeces, e.g. *F. scalaris* in semi-fluid faeces of humans and pigs (Greenberg, 1971). The adults are more abundant in the cooler months, declining in the summer, and where necessary the species overwinter as pupae buried 50–80 mm in the soil (Greenberg, 1973). *F. canicularis*, which is attracted indoors, does not readily settle on human food, and is therefore less annoying than *M. domestica*.

Filth Flies in Poultry Houses (Axtell and Arends, 1990)

Filth flies are commonest in commercial egg-production cage-layered houses; less in breeder houses and least in broiler houses. *M. domestica* is the most abundant and pestiferous fly species in poultry houses, followed by species of *Fannia*. In southern California *F. canicularis* and *F. femoralis* breed in chicken manure, causing problems for poultry producers. *F. canicularis* is the more important pest and is tolerant of lower temperatures than *F. femoralis* although both are intolerant of temperatures above 30°C and hence they are cool season pests (Meyer and Mullens, 1988). *Ophyra* spp. are sometimes abundant and occasionally *Muscina*

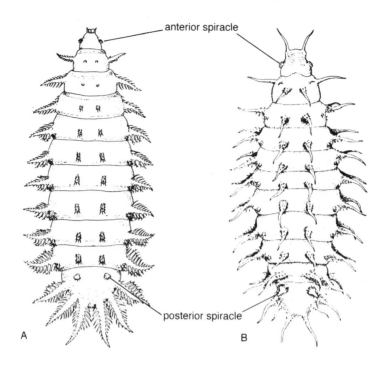

Fig. 13.6. (**A**) Third stage larvae of *Fannia scalaris*; and (**B**) *Fannia canicularis*. Source from Hewitt, C.G. (1912) *Parasitology* 5, 164.

stabulans. *Ophyra aenescens* has been introduced from the New World into Europe where it is becoming a pest. A beneficial side effect of infestations of species of *Ophyra* is that their larvae are predators on *M. domestica* larvae.

As the manure accumulates and ages, predatory mites exert some biological control. Species of *Poecilochirus* invade the manure early, followed by *Macrocheles muscaedomesticae* and later *Fuscuropoda vegetans*. Other predators that may invade, include histerid beetles of the genus *Carcinops*, and hymenopterous pupal parasites (Pteromalidae). Routine chemical larviciding has an adverse effect on biological control but use of the insect growth regulator, cyromazine, which is relatively non-toxic to predacious mites, is an acceptable treatment. Successful control has been achieved in southern California by feeding cyromazine to caged poultry at 5 p.p.m. in the poultry feed. This reduced adult fly emergence by more than 90% (Meyer *et al.*, 1987).

Outline of Muscid Control (Mock and Greene, 1989)

The most effective method of controlling flies that breed in domestic wastes is efficient garbage and sewage disposal systems. Similar methods can be used to dispose of waste materials coming from stabled animals but they are not applicable to the control of species which breed in field-deposited dung. Residual insecticides applied to known resting sites around houses and stables are initially effective but muscids readily develop resistance to persistent insecticides. Space sprays can be used to produce temporary relief in confined spaces and on special occasions ultra low volume sprays have been used to provide an effective but short-lived reduction in the muscid population.

The application of chemical larvicides is inefficient and expensive and is reserved for extreme situations. Juvenile hormone analogues have been applied to feedlot debris and eel grass drifts on Florida beaches, producing a reduction in pest density. The biological control of muscoid flies has been comprehensively reviewed by Patterson and Rutz (1986). Sprays and wipe-on products containing permethrin and/or pyrethrin, applied to cattle have given variable results. Eartags impregnated with insecticide have proved particularly effective in protecting large numbers of cattle in feedlots. A more recent technique involves the addition of insect growth regulators to the fodder of the dung-producing animals or introducing ivermectin, subcutaneously or orally. One drawback to this method is that it controls not only the pest breeding in the dung but also the beneficial insects responsible for the dung's decomposition (Sommer *et al.*, 1992).

References

Allan, S.A., Day, J.F. and Edman, J.D. (1987) Visual ecology of biting flies. *Annual Review of Entomology* 32, 297–316.

Axtell, R.C. and Arends, J.J. (1990) Ecology and management of arthropod pests of poultry. *Annual Review of Entomology* 35, 101–126.

Bornemissza, G.F. (1970) Insectary studies on the control of dung breeding flies by the activity of the dung beetle, *Onthophagus gazella* F. (Coleoptera: Scarabaeinae). *Journal of the Australian Entomological Society* 9, 31–41.

Buschman, L.L. and Patterson, R.S. (1981) Assembly, mating and thermoregulating behavior of stable flies under field conditions. *Environmental Entomology* 10, 16–21.

Campbell, J.B. and Berry, I.L. (1989) Economic threshold for stable flies on confined livestock. *Miscellaneous Publications of the Entomological Society of America* 74, 18–22.

Crosskey, R.W. (1993) Stable-flies and horn-flies (bloodsucking Muscidae). In: Lane, R.P. and Crosskey, R.W. (eds) *Medical Insects and Arachnids.* Chapman & Hall, London, pp. 389–402.

Crosskey, R.W. and Lane, R.P. (1993) House-flies, blow-flies and their allies (calypterate Diptera). In: Lane, R.P. and Crosskey, R.W. (eds) *Medical Insects and Arachnids.* Chapman & Hall, London, pp. 403–428.

Ferrar, P. (1969) Colonisation of an island by the buffalo fly, *Haematobia exigua. Australian Veterinary Journal* 45, 290–292.

Ferrar, P. (1979) The immature stages of dung-breeding muscoid flies in Australia, with notes on the species, and keys to larvae and puparia. *Australian Journal of Zoology Supplementary Series* 73, 1–106.

Fosbrooke, H.A. (1963) The stomoxys plague in Ngorongoro, 1962. *East African Wildlife Journal* 1, 124–126.

Greenberg, B. (1965) Flies and disease. *Scientific American* 213(1), 92–99.

Greenberg, B. (1971) *Flies and Disease. I. Ecology, Classification and Biotic Associations.* Princeton University Press, Princeton, New Jersey.

Greenberg, B. (1973) *Flies and Disease. 11. Biology and Disease Transmission.* Princeton University Press, Princeton, New Jersey.

Greene, G.L. (1989) Seasonal population trends of adult stable flies. *Miscellaneous Publications of the Entomological Society of America* 74, 12–17.

Hepburn, G.A. (1943) Sheep blowfly research. V. Carcases as sources of blowflies. *Ondersterpoort Journal of Veterinary Science and Animal Industry* 18, 59–72.

Hewitt, C.G. (1914) *The Housefly.* Cambridge University Press, Cambridge.

Hogsette, J.A., Ruff, J.P. and Jones, C.J. (1989) Dispersal behaviour of stable flies (Diptera: Muscidae). *Miscellaneous Publications of the Entomological Society of America* 74, 23–32.

Hogsette, J.A., Prichard, D.L. and Ruff, J.P. (1991) Economic effects of horn fly (Diptera: Muscidae) populations on beef cattle exposed to three pesticide treatment regimes. *Journal of Economic Entomology* 84, 1270–1274.

Holloway, M.T.P. and Phelps, R.J. (1991) The responses of *Stomoxys* spp. (Diptera: Muscidae) to traps and artificial host odours in the field. *Bulletin of Entomological Research* 81, 51–55.

Hughes, R.D., Greenham, P.M., Tyndale-Biscoe, M. and Walker, J.M. (1972) A synopsis of observations on the biology of the Australian bushfly (*Musca vetustissima* Walker). *Journal of the Australian Entomological Society* 11, 311–331.

James, M.T. (1947) The flies that cause myiasis in man. *Miscellaneous Publications, United States Department of Agriculture* 631, 1–175.

Jensen, K.M.V., Jespersen, J.B. and Nielsen, B.O. (1993) Variation in density of cattle-visiting muscid flies between Danish inland pastures. *Medical and Veterinary Entomology* 7, 17–22.

Keiding, J. (1986) The house-fly – biology and control. WHO/VBC/86.937, World Health Organization, Geneva.

Macqueen, A. and Doube, B.M. (1988) Emergence, host-finding and longevity of adult *Haematobia irritans exigua* de Meijere (Diptera: Muscidae). *Journal of the Australian Entomological Society* 27, 167–174.

Madsen, M., Sorensen, G.H. and Nielsen, S.A. (1991) Studies on the possible role of cattle nuisance flies, especially *Hydrotaea irritans*, in the transmission of summer mastitis in Denmark. *Medical and Veterinary Entomology* 5, 421–429.

Meyer, J.A. and Mullens, B.A. (1988) Development of immature *Fannia* spp. (Diptera: Muscidae) at constant laboratory temperatures. *Journal of Medical Entomology* 25, 165–171.

Meyer, J.A., McKeen, W.D. and Mullens B.A. (1987) Factors affecting control of *Fannia* spp. (Diptera: Muscidae) with cyromazine feed-through on caged-layer facilities in southern California. *Journal of Economic Entomology* 80, 817–821.

Mock, D.E. and Greene, G.L. (1989) Current approaches in chemical control of stable flies. *Miscellaneous Publications of the Entomological Society of America* 74, 46–53.

Morgan, C.E. and Thomas, G.D. (1974) Annotated bibliography of the horn fly, *Haematobia irritans* (L.), including references to the buffalo fly, *H. exigua* (de Meijere), and other species belonging to the genus *Haematobia*. *United States Department of Agriculture Miscellaneous Publication* 1278, 1–134.

Morgan, C.E. and Thomas, G.D. (1977) Supplement 1: Annotated bibliography of the horn fly, *Haematobia irritans irritans* (L.), including references on the buffalo fly, *H. irritans exigua* (de Meijere), and other species belonging to the genus *Haematobia*. *United States Department of Agriculture Miscellaneous Publication* 1278 (Supplement 1), 1–38.

Morgan, C.E., Thomas, G.D. and Hall, R.D. (1983) Annotated bibliography of the stable fly, *Stomoxys calcitrans* (L.), including references on other species belonging to the genus *Stomoxys*. *University of Missouri-Columbia Agricultural Experiment Station Research Bulletin* 1049, 1–190.

Parr, H.C.M. (1962a) Studies on *Stomoxys calcitrans* (L.) in Uganda (Diptera). I. The morphological development of the cephalopharyngeal sclerites of *S. calcitrans. Journal of the Entomological Society of Southern Africa* 25, 73–81.

Parr, H.C.M (1962b) Studies on *Stomoxys calcitrans* (L.) in Uganda, East Africa. II. Notes on life-history and behaviour. *Bulletin of Entomological Research* 53, 437–443.

Patterson, R.S. (1989) Biology and ecology of *Stomoxys nigra* and *S. calcitrans* on Zanzibar, Tanzania. *Miscellaneous Publications of the Entomological Society of America* 74, 2–11.

Patterson, R.S. and Rutz, D.A. (eds) (1986) Biological control of muscoid flies. *Miscellaneous Publications of the Entomological Society of America* 61, 1–174.

Patton, W.S. and Evans, A.M. (1929) *Insects, Ticks, Mites and Venomous Animals of Medical and Veterinary Importance. I. Medical.* Grubb, Croydon.

Pont, A.C. (1973) Studies on Australian Muscidae (Diptera). IV. A revision of the subfamilies Muscinae and Stomoxyinae. *Australian Journal of Zoology Supplementary Series* 21, 129–296.

Ramsamy M. (1978) Some aspects of stable fly (*Stomoxys nigra* Macquart) control by the sterile insect release method PhD thesis, University of London.

Ramsamy M. (1981) Development of a sampling plan for estimating the absolute population of *Stomoxys nigra* Macquart (Diptera: Muscidae) in Mauritius. *Insect Science and its Application* 1, 133–137.

Ridsdill-Smith, J. (1991) Competition in dung-breeding insects. In: Bailey, W.J. and Ridsdill-Smith, J. (eds) *Reproductive Behaviour of Insects, Individuals and Populations.* Chapman & Hall, London, pp. 264–292.

Ridsdill-Smith, T.J., Hayles, L. and Palmer, M.J. (1987) Mortality of eggs and larvae of the bush fly, *Musca vetustissima* Walker (Diptera: Muscidae), caused by scarabaeine dung beetles (Coleoptera: Scarabaeidae) in favourable cattle-dung. *Bulletin of Entomological Research* 77, 731–736.

Roberts, F.H.S. (1952) *Insects Affecting Livestock.* Angus and Robertson, Sydney.

Sabrosky, C.W. (1961) Our first decade with the face fly, *Musca autumnalis. Journal of Economic Entomology* 54, 761–763.

Sacca, G. (1964) Comparative bionomics in the genus *Musca. Annual Review of Entomology* 9, 341–358.

Skidmore, P. (1985) *The Biology of the Muscidae of the World.* W. Junk, Dordrecht.

Sommer, C., Steffansen, B., Nielsen, B.O., Gronvold, J., Jensen, K.M.V., Brockner-Jespersen, J., Springborg, J. and Nansen, P. (1992) Ivermectin excreted in cattle dung after subcutaneous injection or pour-on treatment: concentrations and impact on dung fauna. *Bulletin of Entomological Research* 82, 257–264.

Thomas, G.D. and Morgan, C.E. (1972) Field-mortality studies of the immature stages of the horn fly in Missouri. *Environmental Entomology* 1, 455–459.

West, L.S. (1951) *The Housefly.* Comstock Publishing Company, New York.

West, L.S. and Peters, O.B. (1973) *An Annotated Bibliography of* Musca domestica *Linnaeus.* Dawsons of Pall Mall, London.

Zumpt, F. (1973) *The Stomoxyine Biting Flies of the World (Diptera: Muscidae).* Gustav Fischer Verlag, Stuttgart.

Calliphoridae, Sarcophagidae (Blowflies) and Myiasis

The subject of myiasis will be dealt with in this and the succeeding chapter. This division is somewhat arbitrary. Myiasis is the invasion of living tissue of animals by larvae of Diptera. The Calliphoridae and Sarcophagidae are large families in which the adults have functional mouthparts. A few species are obligatory agents of myiasis; rather more are facultative agents but the majority of the species breed in carrion. The next chapter will deal with a number of highly specialized families, each containing relatively few species. In these families the mouthparts of the adult are non-functional and the larvae are obligatory endoparasites of mammals often parasitizing specific hosts.

Classification (Colless and McAlpine, 1991; Crosskey and Lane, 1993)

Adult Calliphoridae and Sarcophagidae are medium to large flies with a row of bristles on the meron and one or more bristles on the anepimeron. The related Tachinidae have those same two characters but they are stout-bodied, strongly bristled flies with a prominent subscutellum, whose larvae are endoparasites of insects. The Calliphoridae are metallic (Fig. 14.1) or testaceous flies and the Sarcophagidae are grey-black, non-metallic flies with prominent stripes on the scutum (Fig. 14.9).

The Calliphoridae is a large family of over 1000 species, divided into several subfamilies of which two, the Chrysomyinae and Calliphorinae, are of particular medical and veterinary importance. The Chrysomyinae, containing the genera *Cochliomyia* and *Chrysomya*, are found in the warmer parts of the world and may be distinguished by the stem vein (stem of veins i, ii and iii) being ciliated, i.e. bearing a row of short, fine hairs posteriorly, the bristles on the scutum poorly developed (Fig. 14.3), and there is no external posthumeral bristle. *Cochliomyia* are green to violet-green blowflies with tiny palps, three prominent black, longitudinal vittae on the scutum, and the upper surface of the lower squama mostly bare. *Chrysomya* are green to bluish-black blowflies with normal palps, no bold, black

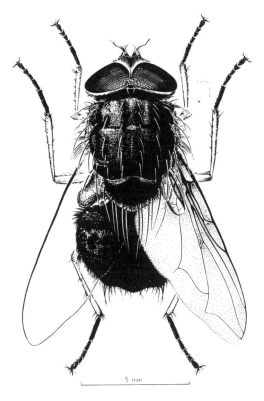

Fig. 14.1. Male *Calliphora stygia*. Source: from Colless and McAlpine (1991).

stripes on the scutum and the upper surface of the lower squama hairy. *Cochliomyia* was restricted to the New World and *Chrysomya* to the Old World until recent transcontinental introductions.

In the Calliphorinae the radial stem vein is bare (Fig. 14.2); the bristles on the scutum are well developed (Fig. 14.1); and there are either two notopleural bristles and two anterior plus one posterior sternopleural bristles, or three notopleurals and one plus one sternopleurals. The external posthumeral is located lateral to the level of the presutural. The Calliphorinae contains the worldwide genera *Lucilia* (= *Phaenicia*) and *Calliphora*, known respectively as greenbottles and bluebottles. *Lucilia* have glossy green or coppery green thorax and abdomen, a bare lower squama, and measure 6–9 mm. *Calliphora* are larger flies (10–14 mm) with black thorax, a steely blue to blue-black slightly metallic abdomen, and the lower squama having a hairy upper surface.

The testaceous calliphorids, *Cordylobia* and *Auchmeromyia*, are found in tropical Africa. They are reddish yellow or reddish brown flies with reddish yellow legs. The Phormiinae, containing the monotypic genera *Protophormia* (*Pr. terraenovae*) and *Phormia* (*Ph. regina*) are dark blue to black holarctic calliphorids with a ciliated stem vein and a bare lower squama.

The Sarcophagidae is a large family with over 2000 species distributed

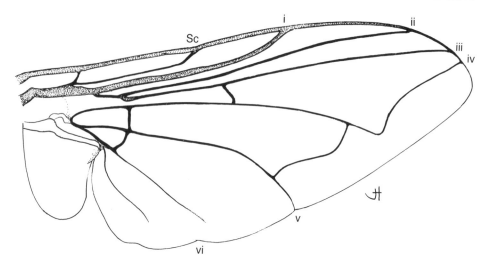

Fig. 14.2. Wing of *Lucilia cuprina* (x 11).

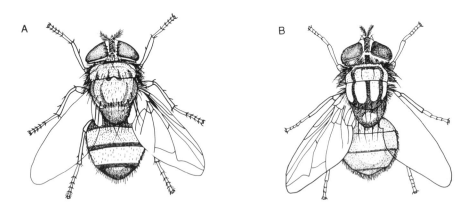

Fig. 14.3. (**A**) Female *Chrysomya bezziana*; and (**B**) *Cochliomyia hominivorax*. Source: (**A**) redrawn from James (1947). (**B**) redrawn from Spradbery (1991).

throughout the world. It includes the genus *Wohlfahrtia* in which the arista is bare, the notopleuron and katepisternum bear two setae each and the abdomen is grey with a pattern of black spots which are unaffected by the angle at which they are viewed (Fig. 14.9B). In *Sarcophaga* the arista is plumose, notopleuron and katepisternum each bear three or four setae and the abdomen has a tesselated pattern of silver grey and black markings, which varies with the angle of the incident light (Fig. 14.9A). *Sarcophaga* is a very large genus which some taxonomists would divide into hundreds of genera.

Myiasis (Zumot, 1966; Leclercq, 1990)

The various forms of myiasis may be classified from an entomological or a clinical point of view. Entomologically flies may be classified into three groups: obligatory; facultative; and accidental. An example of an obligatory agent of myiasis is *Chrysomya bezziana*, the larvae of which are found in wounds; *Lucilia cuprina* is a facultative agent, which causes myiasis in sheep and breeds in carrion. Eggs or maggots of *Musca domestica* or *Sarcophaga* consumed with food and which survive in the alimentary tract, can be regarded as accidental agents of myiasis. The group of facultative myiasis agents may be further refined into primary flies which initiate myiasis, secondary flies which are unable to initiate myiasis but which readily participate once an animal has been infested, and tertiary flies which become involved in myiasis at a late stage when the host animal is almost dead. *L. cuprina* is a primary fly and *Chrysomya rufifacies* a secondary fly. Many carrion-breeding blowflies may act as tertiary flies.

The ability of calliphorid maggots to feed on decaying organic matter of animal origin was turned to good use by Baer (1931) who used maggots in the treatment of chronic osteomyelitis before the advent of antibiotics. This treatment is still used in cases where antibiotics are ineffective and surgery is impracticable (Sherman and Pechter, 1988).

Clinically myiasis can be classified according to the tissue and part of the body affected. Dermal and subdermal myiasis includes wound or traumatic myiasis and furuncular myiasis in which a boil-like condition is produced, e.g. *Cordylobia anthropophaga*. Nasopharyngeal myiasis, including aural and ocular myiases, involves invasion of the head cavities of the outer ear, nose, mouth and accessory sinuses. Intestinal and urogenital myiases involve invasion of the alimentary tract or the urogenital system. The last category, sanguinivorous, is atypical and includes blood-sucking larvae of Diptera. West (1951) gives a long table classifying the different species involved in myiasis on both clinical and entomological grounds and includes their geographical distribution.

Chrysomyinae

The Chrysomyinae includes two important species, *Chrysomya bezziana* and *Cochliomyia hominivorax*, which are obligatory agents of myiasis. Their larvae are armed with broad, encircling bands of spines, which give them an undulating outline (Fig. 14.4) and the common name of screwworms. The posterior spiracles consist of three straight slits surrounded by an incomplete peritreme with an indistinct button in the unsclerotized zone.

Cochliomyia hominivorax (= *Callitroga americana*)

An annotated bibliography of *C. hominivorax* has been produced by Snow *et al.* (1981). As the specific name implies this species will attack humans and untreated infestations can be fatal. Its distribution extended from the USA to southern Brazil. The closely related *C. macellaria* has a rather wider distribution from

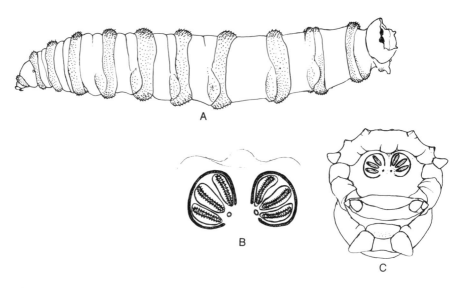

Fig. 14.4. Third stage larva of *Chrysomya bezziana*: (**A**) lateral view of larva; (**B**) posterior spiracles; (**C**) posterior view of larva.

southern Canada to Chile and Argentina (Eads, 1979). The eggs of *C. hominivorax* are laid in batches on the dry surfaces at the edge of 2- to 10-day-old wounds. They hatch in 11–21 h and the larvae bunch together to feed with their posterior spiracles exposed. They are fully developed in 4 to 8 days, leave the host in the morning (0900–1400) and pupariate in the surface layers of the soil (Eads, 1979). Prepupae bury themselves rather deeper in cold weather and if they are exposed to temperatures below 9.5°C for 3 months they die. They also require a soil moisture of less than 16% (Norris, 1965).

The adults emerge around dawn (0400–0700). Females mate only once and after a preoviposition period of 5 to 10 days deposit batches of about 300 eggs. A particularly fecund female may produce nearly 3000 eggs (Hall, 1948). Their activity is reduced by hot, dry conditions, strong winds and by rain, but increases after rain (Norris, 1965). They feed on wounds, dung and fresh meat, presumably to obtain protein for ovarian development (Hall, 1948). *C. hominivorax* has a symbiotic relationship with *Providencia rettgeri*, a microorganism which is abundant in larvae and pupae, and present in folds of the ovipositor. *P. rettgeri* is introduced into the wound, where it produces substances that attract more gravid flies to oviposit. Within 2 days of oviposition the wound is supporting an apurulent, monoculture of *P. rettgeri* (Gassner, 1989).

Cochliomyia hominivorax is active all the year round in areas where the temperature is above 16°C, and during the summer disperses widely from its overwintering areas moving 56 km week^{-1} (Barrett, 1937) and a maximum of nearly 300 km (Hightower *et al.*, 1965). Populations of *C. hominivorax* in Texas are depressed in the summer, increase in the autumn and are low in the winter. These fluctuations follow the incidence of myiasis (Norris, 1965). In the USA populations of *C. hominivorax* overwintered in the south, especially in Texas. Control of

C. hominivorax is complicated by the fact that it causes myiasis not only in cattle and humans but also in wild animals including opossums, cottontail and jack rabbits and the white-tailed deer *Odocoileus virginianus texanus* (Lindquist, 1937). Both *C. hominivorax* and *Ch. bezziana* may be trapped using swormlure, a mixture of 11 organic compounds of which dimethyl disulphide is the most important. The attracted flies may be trapped on adhesive plates or electrified black targets (Spradbery, 1991; Green *et al.*, 1993).

Cochliomyia hominivorax was the first insect to be eradicated from an area by the use of the sterile insect technique, which requires the release of sterilized flies in numbers adequate to swamp the wild population (Knipling, 1955). *C. hominivorax* was a good test insect to use because its populations are much lower than those of other pest insects, e.g. a few hundred per 250 ha, cf. thousands or hundreds of thousands (Lindquist, 1955). The initial technical problem was to develop a technique for sterilizing flies without markedly reducing their ability to compete with normal wild males. This was achieved by irradiating 5-day-old pupae at a dosage of 5000 R. (Bushland and Hopkins, 1953).

The first trial on Sanibel Island in Florida showed that the technique had possibilities and a major experiment was conducted on Curaçao in the Netherlands Antilles. A sterilizing dose of 7500 R was used because this prevented oviposition in sterilized females. Sterilized adult flies were distributed by aircraft once a week. A release of 100 sterilized males per 250 ha per week for 6 weeks produced 15% sterility among egg batches laid on wounded goats. From 9 August to 3 October 1954 the entire island was treated at an average rate of 435 males per 250 ha per week and in 8 weeks all egg masses deposited were sterile; and from early October only two sterile egg batches were laid in the next 3 months. The experiment was an outstanding success. The screw-worm had been eradicated from Curaçao (Baumhover *et al.*, 1955).

This success was followed by the eradication of *C. hominivorax* from Florida in 17 months ending in November 1959 (Meadows, 1985). The challenge to deal with the overwintering areas along the Mexico–USA border was much more demanding involving a very large area and a long border over which flies were free to migrate. The programme began in 1962 and achieved encouraging, but limited success. Thereafter fluctuating results were obtained until the protective barrier extended 80 to 100 km into Mexico. Due to special circumstances 1972 was a 'disaster' year, but from then on the situation improved with California, Arizona and New Mexico becoming screw-worm-free in 1979 and Texas in 1982 (Bushland, 1985). This mammoth programme involved the production of 100 million sterile male flies per week and their release along a barrier 400 km wide and more than 3000 km in length (Snow and Whitten, 1979).

The Governments of Mexico and the United States of America entrusted the Mexico–American Screwworm Eradication Commission with the task of eradicating *C. hominivorax* from Mexico southwards to the Isthmus of Tehuantepec with the intention of establishing a sterile fly area 400 km wide, extending approximately from 92° to 96° west. Full-scale operations were begun in 1977 and successful eradication was achieved by 1984. During this period the area treated increased from 773,000 to almost 2 million km² and involved the liberation of more than 150 billion flies. Only in Quintana Roo can appreciable numbers of

C. hominivorax now be found in Mexico and that State is outside the control area (Pineda-Vargas, 1985).

Cochyliomyia hominivorax became established in the Old World in Libya during 1988 leading to FAO and the Libyan Government undertaking a massive eradication programme. It used the sterile insect technique, which had proved so effective in North America. An area of 25,000 km^2 around Tripoli on the North African Mediterranean coast was put under quarantine. Livestock within the area were inspected and, if necessary, treated and movement of infested livestock out of the area prohibited. The first release of sterile flies from the Mexico–American Commission was made in December 1990 at a rate of 3.5 million flies per week, increasing to 40 million per week by May 1991. Release of sterile flies ceased in October 1991, 6 months after the last detected case of screw-worm myiasis in Libya. Another successful chapter in the control of the new world screw-worm had been concluded (Lindquist *et al.*, 1992).

Control of *C. hominivorax* has had unexpected consequences in some areas. The screw-worm attacked both deer and cattle and was an important factor in limiting the deer population. Control of the screw-worm fly has led to a substantial increase in the deer population. Deer and cattle are parasitized by the same ixodid tick, *Amblyomma maculatum*. An increase in the deer population has led to an increase in *A. maculatum* and a higher incidence of ticks on cattle.

Chrysomya bezziana (Spradbery, 1991)

Chrysomya bezziana is widely distributed in the Afrotropical and Oriental regions extending as far south as Papua New Guinea, but not to Australia. In view of the introduction of *C. hominivorax* into Libya it is of concern that *Ch. bezziana* has the potential to colonize a large part of the warmer areas of the world (Sutherst *et al.*, 1989). It attacks a wide range of hosts, but there are few records from wild animals. A more realistic picture is probably given by infestations among animals in a zoo in Malaysia, where in a period of 15 years there were 91 attacks on 21 species of mammals, resulting in 12 deaths (Spradbery and Vanniasingham, 1980). The economic importance of *Ch. bezziana* stems from it causing myiasis in cattle. Cases of human myiasis are common in the Oriental region but rare in Africa (Zumpt, 1965).

Females are attracted to wounds several days old, for oviposition (Norris, 1965) and eggs are laid on the upper, dry side of wounds in a shingle-like mass, giving the inner eggs protection from solar radiation. Oviposition occurs in the late afternoon, 2 to 3 h before dusk when 100–250 eggs are deposited in a single batch, and they hatch 12–16 h later (Spradbery *et al.*, 1976; Spradbery, 1979). The importance of this timing and location of oviposition enables egg development to be completed in the hours of darkness. Eggs exposed to solar radiation for more than 2 h suffer a high mortality and all are dead after 6 h. There is also a low hatch when eggs are kept moist, hence the value of ovipositing on the dry upper edge of the wound (Spradbery, 1979).

Larvae feed initially on blood and serum and later lacerate tissue with their mouth-hooks. They bunch together and tunnel deeply (15 cm) into the host's tissue causing considerable destruction. Several females may oviposit at the same site, probably attracted by pheromones emitted by the first ovipositing female. As a result 3000 larvae may occur in a wound (Spradbery *et al.*, 1976). In 6 to 8 days the larvae are fully developed, leave the host as prepupae, and pupariate in the ground, the pupal period lasting 8 to 10 days (Spradbery *et al.*, 1976).

The female is autogenous and can develop the first batch of eggs without a protein meal but ingestion of protein increases the number of eggs matured and the rate of ovarian maturation, which is maximal at 33°C (Spradbery *et al.*, 1991). Peak sexual activity occurs on the third day and the female is gravid in 6 to 8 days (Sands, 1979). The female can take in sufficient protein to mature a second batch of eggs in 13 s, and this can be done while ovipositing the first batch (Spradbery and Schweizer, 1979). In the field mean life expectancy was estimated to be 9 days, not long enough to complete two ovarian cycles (Spradbery and Vogt, 1993). The population of *Ch. bezziana* is very low, of the order of 1 to 200 in every 25 ha (Sands, 1979). They range widely and labelled females have deposited egg masses 100 km from their point of release (Spradbery, 1991).

In domestic stock the areas most susceptible to attack are the navels of newborn animals, surgical wounds produced during castration, docking and dehorning, and tick bites; the condition is complicated by secondary infections. Preventive measures include delaying surgery to the cooler season of the year when *Ch. bezziana* is less active, dressing wounds, and twice weekly inspection of livestock (Blood and Radostits, 1989).

Other chrysomyines

Chrysomya rufifacies in the Australian and Oriental regions and *Ch. albiceps* in the Palaearctic and Afrotropical regions are facultative agents of myiasis. Their larvae, known as hairy maggots, have a row of fleshy tubercles and feed on the host and are predatory on primary blowfly larvae. *Cochliomyia macellaria* is a minor secondary fly in myiasis and more important as a fly of the market place where it oviposits on meat. *Ch. marginalis* and *Ch. megacephala* are common bazaar flies (Greenberg, 1971). Four species of *Chrysomyia*, *Ch. albiceps*, *Ch. rufifacies*, *Ch. putoria* and *Ch. megacephala*, have become established in the Neotropical region, where the dispersal rates of the first three species have been estimated at 1.8 to 3.2 km day^{-1}. Their introduction has led to the suppression of the indigenous *C. macellaria* (Baumgartner and Greenberg, 1984). *Ch. putoria* and *Ch. megacephala* breed in wet faecal material and are commonly found breeding in latrines (Crosskey and Lane, 1993). In south-east Asia *Ch. megacephala* is a major cause of loss in the salted, dried fish industry (Esser, 1991).

Metallic Calliphorinae

The majority of metallic Calliphorinae breed in carrion but some are facultative myiasis flies of which the most important are primary flies of the genus *Lucilia* and *Calliphora*. Calliphorine larvae are smooth-bodied maggots in which the

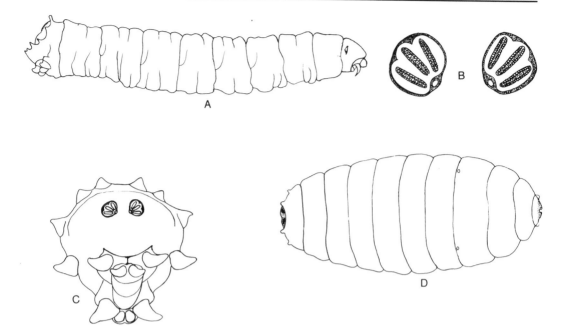

Fig. 14.5. (**A**) Third stage larva of *Calliphora* (lateral view); (**B**) posterior spiracles of larva; (**C**) posterior view of larva; (**D**) puparium of *Calliphora*.

posterior spiracles have three slits surrounded by a continuous peritreme ring, which includes the button (Fig. 14.5). Larvae of *Calliphora* have an additional accessory sclerite between the mouth-hooks. This is lacking in *Lucilia*.

Sheep myiasis

Species of *Lucilia* and *Calliphora* are the cause of cutaneous myiasis in sheep in which they may be associated with species of *Chrysomya*. This condition is associated with sheep with heavily wrinkled skins, e.g. merino, and is particularly important in Australia, South Africa and Britain, and of lesser importance in New Zealand and the USA. In a bad outbreak 30% of a flock may die and in 1985 sheep myiasis was estimated to have cost Australia A\$200 million (Blood and Radostits, 1989).

Oviposition on the host is referred to as 'blow' and the establishment of larvae as 'strike'. In Australia and South Africa the main species is *L. cuprina* which is involved in 90% of all strikes and was the only species present in 55–71% of infestations (Hepburn, 1943; Watts *et al.*, 1976). The closely related *L. sericata*, the sheep blowfly of Great Britain, plays only a minor role in sheep myiasis in the southern hemisphere, and is less important than species of *Calliphora* such as *C. stygia*, which is responsible for 95% of strikes in New Zealand compared to 5% for *L. sericata* (Waterhouse and Paramonov, 1950; Norris, 1965; Zumpt, 1965).

Lucilia cuprina (Foster *et al.*, 1975)

Gravid female *L. cuprina* lay their eggs in carrion or on sheep. They are attracted to sheep which have areas of soiled fleece or are suffering from bacterial decomposition of the fleece in fleece rot. The eggs are deposited in a cluster of 100–300, depending upon the size of the female. For development they require a temperature above 30 °C, which is found on the skin and in the fleece of sheep, and a humidity of more than 90%. Under these conditions the eggs hatch in 8–12 h. *L. cuprina* is diurnal in activity and eggs laid in the mid-afternoon complete their development during the night when humidities are high. Eggs placed at skin level failed to hatch unless the area was kept saturated. Fertility of the eggs is of the order of 70–86%.

The process by which *L. cuprina* orientates to sheep and suitable oviposition sites involves olfactory, visual, and gustatory stimuli. *L. cuprina* is attracted to fly-struck sheep from 20 m upwind but from only 10 m by sound, dry sheep, in response to volatile kairomones (Eisemann, 1988). These are mainly produced by the bacterium *Pseudomonas aeruginosa* which is largely responsible for the dermatitis which occurs in fleece-rot lesions (Burrell, 1985). In exploring a sheep for oviposition, the female probes potential sites with her proboscis, feeds, and then shuffles backwards, probing with her ovipositor, thrusting it deep within the fleece. There are taste receptors on the ovipositor, and this probing process is repeated before each egg is deposited (Rice, 1982).

Newly hatched larvae are very susceptible to desiccation and move down to the skin to feed on the protein-rich exudate produced by a skin irritated by fouling and fleece rot. Second and third stage larvae attack the skin causing and extending lesions. The mature third stage larvae drop from the host at night and burrow into the soil to a depth of 1–2 cm where they pupariate, providing the temperature is above 10 °C and the humidity low. There is no larval diapause as occurs in *L. sericata*, which is adapted to a cooler climate. The pupal stage lasts 6 days at 30 °C and 25 days at 15 °C with a survival of 75–95%. At the height of the season (October–February), survival of 1st instar larvae to adults is 20–25% and the larval and pupal stages occupy 2 to 3 weeks out of a generation time of 3 to 4 weeks. There are about eight generations a year.

In adult life both sexes of *L. cuprina* need sources of carbohydrate, e.g. nectar, honeydew, and protein, which can be obtained from carrion, from scalded sheep, wounds and to a lesser extent, dung. A protein meal is necessary for ovarian development to proceed beyond the resting stage, after which the female becomes sexually receptive and mating occurs (Barton-Browne *et al.*, 1987). *L. cuprina* occurs in low density in the field and sources of protein may serve to bring the sexes together. In the field females that have completed more than three cycles are rare.

Bionomics

Lucilia cuprina is an early colonizer of fresh sheep carcasses but even so fewer adults are produced than from living sheep. Under optimal conditions the average emergence of *L. cuprina* from a sheep carcass was 304 out of a total emergence of more than 36,000 calliphorids, and in less favourable conditions only 4 out of 10,000. This failure is attributed to competition from other blowflies, especially *C. augur* in spring and *Chrysomya rufifacies* in summer, the adverse effect of high temperature generated in carcasses, and predation. Prepupae are subject to predation by *Ch. rufifacies* within the carcass and by beetles after leaving it. In contrast about 90% of the flies bred from struck sheep were *L. cuprina*, averaging 1220 adults per strike (Waterhouse, 1947). In eastern coastal Australia *L. cuprina* also breeds in refuse.

Population densities of *L. cuprina* are low but comparable to that of its host. Densities of females which have completed one ovarian cycle, and could therefore have attacked sheep, range from 0.2 at the start of the season to $22\,\mathrm{ha}^{-1}$, which compares with a sheep density of $10\,\mathrm{ha}^{-1}$. Dispersal of the adults is affected by the habitat and in favourable areas adult flies may disperse only a few kilometres. This is supported by the existence of localized insecticide-resistant populations which spread slowly.

Other species

In Australia the species of blowflies attacking sheep have different seasonal and geographical distributions. *L. cuprina* is commoner north of 30°S and *C. stygia*, *C. augur* and *C. dubia* (= *nociva*) are commoner south of 25°S (Norris, 1959). Adult *C. stygia* may be present in warm periods in winter when the bulk of its population is present as prepupae and pupae. It has been calculated that 10% of strikes are initiated by *C. stygia* but it also occurs in strikes commenced by other species. *C. augur* is ovoviviparous depositing about 50 eggs which hatch immediately. This species has more varied breeding sites, causing myiasis in birds and other mammals. It and *C. dubia* are associated with wound myiases (Norris, 1959).

Chrysomya rufifacies has a higher temperature threshold than *L. cuprina* and is most abundant in the hottest months of the year. It is a secondary blowfly whose larvae are predatory on primary blowfly larvae and it oviposits in carcasses a few days old when other blowflies have established themselves. *Ch. rufifacies* larvae do not leave the carcass as prepupae but pupariate in or on it (Norris, 1959). Large infestations of *Ch. rufifacies* may cause the death of a sheep in a short time (Spradbery, 1991).

The effect on sheep

The common site of blowfly attack on sheep is the breech, tail and crutch area, where the wool has been fouled and remains moist from faeces and urine. Less common sites are the prepuce of rams and wethers, i.e. pizzle strike; poll strike on the dorsum of the head where there is excessive skin folding; body strike in wet seasons particularly in poor, dense pasture; and wound strike. Young sheep are more susceptible (Blood and Radostits, 1989).

In untreated infestations the maggots burrow into the subcutaneous tissue and, when primary strikes are invaded by secondary flies, particularly *Ch. rufifacies*, the affected area is extended and the maggots may burrow deeply into the tissue (Blood and Radostits, 1989). The optimum pH for development of

L. cuprina larvae is 8–9 (Guerrini *et al.*, 1988). This alkalinity produces ammonia toxicity in sheep which is fatal when the ammonia concentration exceeds 200 μmol l^{-1} in venous blood (Guerrini, 1988). Such deaths can occur early in an infestation.

Control

The predisposition to strike can be offset by keeping the fleece short in susceptible areas. Removal of the fleece around the breech is only effective for about 6 weeks. More permanent protection is given by Mules' operation in which woolled skin is removed in a strip from both sides of the breech with the intention of increasing the width of the woolless area. This procedure reduces breech strike by 80–90% (Blood and Radostits, 1989). Suppression of helminth infections can reduce breech strike from 50% to 5% by reducing fouling of the breech (Morley *et al.*, 1976).

In 1931 Mules' operation was developed to minimize breech strike but it was replaced by insecticidal protection using chlorinated hydrocarbons, organophosphorus compounds and carbamates in the period 1949–1965. *L. cuprina* readily developed resistance to these insecticides and consequently Mules' operation has been reintroduced (Morley and Johnstone, 1984) and attention is now being given to immunological techniques directed against *P. aeruginosa* and *L. cuprina*. Sheep vaccinated against *P. aeruginosa* are protected against severe fleece-rot lesions and consequently against blowfly strike (Burrell, 1985). The best targets for antibodies against *L. cuprina* larvae would be components of the anterior midgut and the peritrophic membrane where the activity of an ingested antibody remains high (Eisemann *et al.*, 1993).

Phormiinae

Phormia regina, the black blowfly, and *Pr. terraenovae* breed mainly in carrion but can be facultative agents of myiasis. *Ph. regina*, a common sheep blowfly in the south-west USA, has been associated with wound myiasis in humans. *Pr. terraenovae* is a more northern species capable of causing fatal wound myiasis in cattle sheep and reindeer (James, 1947).

Testaceous Calliphorids

Two species of testaceous calliphorids, *Auchmeromyia senegalensis*, the Congo floor maggot, and *Cordylobia anthropophaga*, the Tumbu or mango fly, cause myiasis in humans. The adults may occur in huts or houses and be distinguished by the second visible abdominal segment being obviously longer than the third segment in *A. senegalensis*, and the two segments being of similar length in *C. anthropophaga*. The eyes of male *A. senegalensis* are widely separated and those of *C. anthropophaga* almost contiguous (James, 1947).

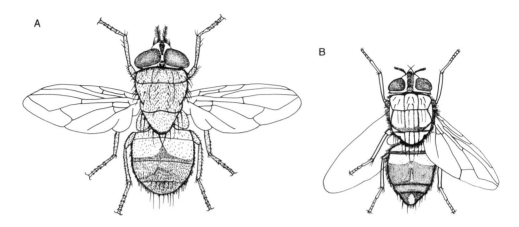

Fig. 14.6. (**A**) Female *Cordylobia anthropophaga*; (**B**) female *Auchmeromyia senegalensis*. Source: *C. anthropophaga* redrawn from James (1947) and *A. senegalensis* from Crosskey and Lane (1993).

Auchmeromyia senegalensis (= *A. luteola*) (see Fig. 14.6)

Publications on this species used the name *A. luteola* until 1980 when it was changed to *A. senegalensis*. Adult *A. senegalensis* have lapping mouthparts and feed on faeces, particularly human faeces, and fermenting fruit. The larvae are haematophagous. A female lays batches of about 50 eggs, and in her lifetime may lay a total of 300 eggs. They are laid in dry, dusty soil or sand in the earth floor of huts and hatch in 36–60 h at 26–28°C and 50–60% RH (Zumpt, 1965). There are three larval stages (Fig. 14.7) which feed at night on the inhabitants of the hut, sleeping on the ground. The larvae are unable to climb vertical surfaces and protection against being bitten is provided by a bed raising the occupant 10 cm above ground level. Larvae take at least two feeds during each larval instar and 6–20 feeds in the course of development. The pupal stage lasts 11 days at 28.5°C, and in the laboratory the generation time is 10 weeks giving five generations per annum (Garrett-Jones, 1951).

Each blood meal takes about 20 min. When a larva feeds it attaches itself almost at right angles to the skin, and makes an incision using the mouth-hooks and the minute toothed maxillary plates in front of them (Zumpt, 1965). Blood is said to be taken in by contractions of the crop in which it is stored before being passed to the foregut and on to the convoluted midgut with digestion being completed in the rectum (Dutton *et al.*, 1904 cited by Boreham and Geigy, 1976). There are two unusual features in this description made over 80 years ago. Firstly, the crop is usually a storage organ and not a sucking apparatus and the rectum is usually lined with cuticle and the site of the absorption of water and salts but not of digestion. Garrett-Jones (1951) considers that larvae will feed daily, but in the field digestion is slow and at least 5 days would elapse between blood meals.

Larvae of *A. senegalensis* are associated with humans, warthogs, *Phacochoerus aethiopicus*, and aardvarks, *Orycteropus afer* (Zumpt, 1965). In the Serengeti

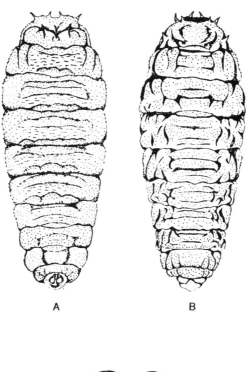

Fig. 14.7. Third stage larva of *Auchmeromyia senegalensis*. (**A**) dorsal view; (**B**) ventral view; (**C**) posterior view. Source: from Zumpt (1965).

National Park larvae and adults of *A. senegalensis* were common in culverts in the complete absence of people; 98% of the feeds were on suids, and those meals which could be identified to species were from warthogs. A small number of larvae had also fed on hyenas, probably the spotted hyena *Crocuta crocuta*, and are the first records of *A. senegalensis* feeding on a hairy mammal (Boreham and Geigy, 1976).

A high incidence of *Trypanosoma brucei* is found in the spotted hyena and the potential of larvae of *A. senegalensis* acting as vectors was investigated. There was no evidence of cyclical development of the trypanosomes which remained viable for only 21 h in larvae but there was some evidence that *Auchmeromyia* larvae might act as mechanical vectors (Boreham and Geigy, 1976; Geigy and Kauffmann, 1977). There are four other species of *Auchmeromyia* of which one, *A. bequaerti*, also occurred in the Serengeti culverts. Larval populations die out if the ground is damp or wet but areas are rapidly repopulated when they dry out.

Cordylobia anthropophaga (Blacklock and Thompson, 1923) (see Fig. 14.6)

Adult *C. anthropophaga* are diurnal with peaks of activity in the early morning (0700–0900) and late afternoon (1600–1800). They seek shade, being more able to resist cold and damp than exposure to sunlight. They feed on fermenting fruit and, given the opportunity, liver. Females mate soon after emergence and may mate several times on the first day. Two batches of eggs are laid. The first contains about 300 eggs and the second 100–200. Oviposition takes about 30 min and the female seeks out dry sand soiled by urine and excreta for oviposition and avoids sand that is too moist. The female lives about 2 weeks.

The egg is rather smaller than those of other muscoid flies, measuring 0.8 mm. The eggs hatch in 24–48 h and the larva remains buried in the sand until responding to vibrations, heat and carbon dioxide, which could signify the arrival of a host. The larva then raises its front end in the air and searches around for a host. On a suitable host the larva will penetrate the unbroken skin and bury itself in less than a minute. It will penetrate the skin of many different mammals and also fowl but it does not develop in the last-named. The larva develops in a boil-like swelling in the skin. The swelling has an opening through which the larva breathes using its posterior spiracles in which there are three slightly sinuous slits in a weakly sclerotized peritreme (Fig. 14.8). The three larval stages are completed in 8 days, and then the prepupa leaves the host and pupariates in the ground. The puparium is tolerant of dry conditions but killed by continuous exposure to 37°C.

The main hosts of *C. anthropophaga* are black and brown rats and among domestic animals, dogs, particularly puppies. Larvae penetrate the feet, genitals, tail and axillary regions of their host. On humans lesions are found mainly on areas of the body covered by clothing and are commoner in children. Breeding of *C. anthropophaga* continues throughout the year and the seasonal nature of human infections, which are more numerous in the wet season, is attributed to rat burrows being flooded and rats being more closely associated with human habitations. Females will oviposit on soiled or inefficiently washed clothing or bedding and hence larvae are distributed over the areas of the body covered by clothing. The boil-like swelling causes considerable discomfort as the larva increases in size and there is a copious exudate of serum, blood and larval faeces. The danger of infestation from clothing can be avoided by drying clothes in full sunlight out of contact with the ground, and ironing the clothing to kill any eggs or larvae before storing in covered receptacles.

Persons infected with *C. anthropophaga* have been recorded in many parts of the world – USA, Saudi Arabia, Europe (Leclercq, 1989), and Australia

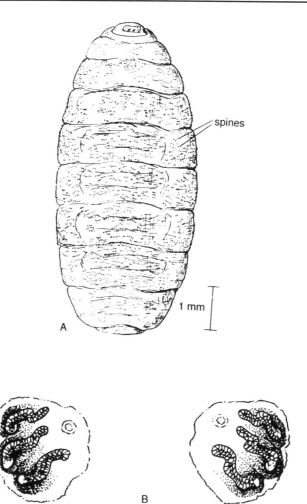

spines

1 mm

A

B

Fig. 14.8. Third stage larva of *Cordylobia anthropophaga*.
(**A**) ventral view; (**B**) posterior spiracles of larva. Source: larva from Bertram,
D.S. (1938) *Annals of Tropical Medicine* 32, 433. Spiracles drawn from various
sources.

(Moorhouse, 1982, personal communication). Laurence and Herman (1973)
reported an infection in a woman who had never been to Africa and who appeared
to have acquired the infection in Spain. It is possible that a focus of *C.*
anthropophaga could be established outside tropical Africa. The related *C. rodhaini*
is found in the moister parts of tropical Africa and infests humans less frequently
(Hall and Smith, 1993). Two human infections have been recorded in France in
travellers from the Cameroons (Leclercq, 1989).

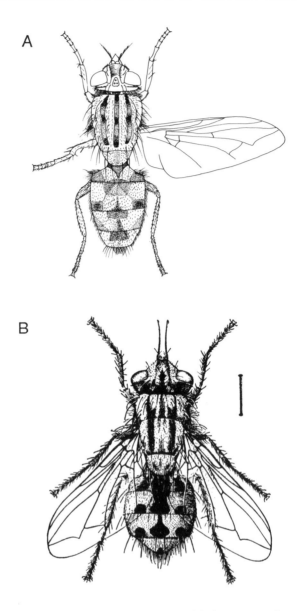

Fig. 14.9. (**A**) Female *Sarcophaga* sp. (**B**) *Wohlfahrtia magnifica*. Source: B from James (1947).

Sarcophagidae (James, 1947; Zumpt, 1965) (see Fig. 14.9)

The Sarcophagidae are viviparous or ovoviviparous, producing either 1st instar larvae or eggs which hatch immediately on deposition. This gives members of this family considerable advantage in competing for carrion with calliphorids which are oviparous. The mature sarcophagid larva has its posterior spiracles recessed

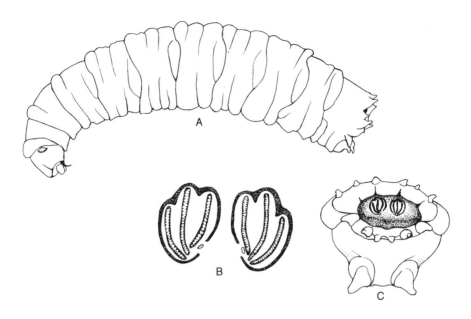

Fig. 14.10. Third stage larva of *Sarcophaga*: (**A**) lateral view; (**B**) posterior spiracles; (**C**) posterior view.

in a posterior depression and hidden from view (Fig. 14.10). The spiracles have three slits which are orientated more or less dorsoventrally and surrounded by an incomplete peritreme. Two species, *Wohlfahrtia magnifica* and *Wohlfahrtia vigil*, are obligatory agents of myiasis.

Wohlfahrtia magnifica

Wolfahrtia magnifica is adapted to the desert areas of North Africa, Asiatic Russia and Asia Minor where it causes wound myiasis in humans and domestic stock including horses, donkeys, cattle, buffalo, sheep, goats, pigs and even geese. According to Zumpt (1965) there are no records of this species from wild animals. Adults occur in fields and orchards where they feed on flowers. They are active during the bright, hot period of the day from 1000 to 1600 h. Females larviposit in wounds and body openings, ear, nose and eyes. A female may contain 120–170 larvae. Once deposited the larvae grow rapidly, burrow into the tissues and cause extensive damage which may prove fatal.

Infection of the ears can lead to deafness, loss of an eye and cause severe damage to the tissue of the nasal region. The larvae are full grown in 6 to 7 days when they leave the host, pupariate in the ground, and, if the soil temperature is above 12–14°C, emerge as adults. In Kazakhstan *W. magnifica* overwinters as diapausing pupae, a stage which lasts 6 to 10 months (Akhmetov, 1990). The larvae are extremely hardy, pupating and producing normal adults after exposure for 1 h to 95% alcohol. They survive for considerable time in concentrated hydrochloric acid and corrosive sublimate.

Other species of *Wohlfahrtia*

Wohlfahrtia vigil is an obligatory myiasis fly, occurring in North America where it is a parasite of mink. Larvae are deposited in groups and can only penetrate thin skin, and for that reason human cases are restricted to young babies in whom they cause a furuncular myiasis. Prevention is easy by fly screening prams left out of doors, containing sleeping infants. *W. meigeni* (= *W. opaca*) may be a subspecies of *W. vigil* (Eads, 1979). It causes human furuncular myiasis in North America where its larvae penetrate unbroken skin causing boil-like swellings on exposed parts of the body. There are usually 12–14 lesions and rarely as many as 40 (Haufe and Nelson, 1957). *W. nuba* infests wounds of livestock in North Africa and the Middle East, feeding on dead or diseased tissues and has been used in maggot therapy (Zumpt, 1965).

Other sarcophagids

The genus *Sarcophaga* is widely distributed throughout the world and species are very difficult to identify. They develop in carrion, excrement and any kind of decomposing organic matter. A number of species have been associated with myiasis, usually as tertiary flies or as accidental agents, but some species of *Sarcophaga* are facultative agents of myiasis.

Medicolegal Forensic Entomology (Smith, 1986; Marchenko, 1991; Catts and Goff, 1992)

Forensic entomology is now a recognized discipline concerned with the application of the study of insects and other arthropods to legal issues. It has three categories – urban, stored-products and medicolegal. It is the last which is related to medical entomology. Its most common application is determining time since death, movement of the corpse, manner of death and other aspects of the investigation. Insects occurring in carrion can be assigned to one of four categories. There are those that feed on the carrion, of which blowflies are the first to appear, arriving within a few hours of death. They are followed by predators and parasites of the necrophagous species; omnivorous species which feed on both the corpse and its inhabitants; and adventive species which use the corpse as an extension of their environment. As a result of seepage changes also occur in the fauna under the corpse.

Various stages are recognized in the decomposition of the corpse and are important in fixing the time of death. The stages are:

1. Autolyic and microbial decomposition during which blowflies oviposit.
2. Active decomposition by insects.
3. Advanced decomposition mainly by coleopteran larvae.
4. Microbial decomposition takes over when the coleopteran larvae leave and ends in mummification.

5. The skeleton separates into individual bones and the time of death can no longer be determined.

Stages one to three pass rapidly while stages four and five are prolonged (Marchenko, 1991).

The entomologist has to deduce from the stages and species of arthropods present on a corpse the time at which the insect attack began. This will be some hours after death. The material must be identified to species and stage of development. The time taken to reach a particular stage is temperature-dependent but allowance must be made for the fact that the temperature in the cadaver will be higher than air temperature due to the production of metabolic heat. Account must be taken of the location of the cadaver, e.g whether it was exposed to the sun or sheltered from the wind. Time for development of a species must be known in degree-hours above a threshold temperature. Then it should be possible to calculate when the first insect eggs were laid. With a widely distributed species account must be taken of the possibility of the local population differing from the norm. To confirm findings it may be necessary to simulate conditions using a freshly killed pig as the cadaver.

Leclercq and Verstraeten (1988) recorded 68 species of arthropods being associated with 49 cadavers. This number included 30 species of Diptera; 25 Coleoptera; 8 Acarina and 5 others. For each species it is necessary to know its seasonal cycle, its geographical distribution and behaviour. The presence of larvae of *L. sericata* on a corpse in Finland indicated that the corpse had been infested in a city in southern Finland and had been exposed to sunshine, while on another corpse in shade *Calliphora erythrocephala* but not *L. sericata* had oviposited (Nuorteva *et al.*, 1967). Insect attack will be delayed when a corpse has been covered, hidden in an enclosed space, burned or covered by soil. A covering of 2.5 cm of soil is enough to exclude blowflies from a corpse but *Muscina*, a muscid fly, lays its eggs on the surface and the larvae are able to penetrate soil to a depth of 10 cm in search of carrion. Decomposition of a corpse in water will be slow because of loss of heat to the surrounding medium. Indications that a corpse has been moved from the place of death may be given by the fauna of the corpse differing from that of the locality in which it has been found and the fauna below the corpse being less advanced than would be expected from the stage of decomposition.

Catts and Goff (1992) and Marchenko (1991) have produced overall views of medicolegal forensic entomology. Smith (1986) does the same but also includes case studies and provides illustrations, keys and descriptions to enable the reader to make identifications of material. Lord (1990) describes an unusual case in which a recently deceased human and two mummified bodies were found in different rooms of a dwelling. One was completely skeletonized and associated with hundreds of calliphorid puparia while the other showed minimal skeletalization. It was deduced that the first had died in the autumn and the other in early winter. These times were confirmed subsequently. Insects can assist in determining the cause of death by the accumulation in the insect of heavy metals and toxic compounds, or their degradation products. Lethal doses of cocaine accelerated the growth of sarcophagid maggots but had no detectable effect on the adult flies.

References

Akhmetov, A. (1990) Peculiarities of the diapuse of *Wohlfahrtia magnifica* in Kazakhstan. *Abstracts of the 2nd International Congress of Dipterology*, Bratislava, 27 August–1 September 1990, p. 3.

Baer, W.S. (1931) The treatment of chronic osteomyelitis with the maggot (larva of the blowfly). *Journal of Bone and Joint Surgery* 13, 438–475.

Barrett, W.L. (1937) Natural dispersion of *Cochliomyia americana*. *Journal of Economic Entomology* 30, 873–876.

Barton-Browne, L., van Gerwen, A.C.M. and Smith, P.H. (1987) Relationship between mated status of females and their stage of ovarian development in field populations of the Australian sheep blowfly, *Lucilia cuprina* (Wiedemann) (Diptera: Calliphoridae). *Bulletin of Entomological Research* 77, 609–615.

Baumgartner, D.L. and Greenberg, B. (1984) The genus *Chrysomya* (Diptera: Calliphoridae) in the New World. *Journal of Medical Entomology* 21, 105–113.

Baumhover, A.H., Graham, A.J., Bitter, B.A., Hopkins, D.E., New, W.D., Dudley, F.H. and Bushland, R.C. (1955) Screw-worm control through the release of sterilised flies. *Journal of Economic Entomology* 48, 462–466.

Blacklock, B. and Thompson, M.G. (1923). A study of the tumbu-fly, *Cordylobia anthropophaga* Grünberg in Sierra Leone. *Annals of Tropical Medicine and Parasitology* 17, 443–502.

Blood, D.C. and Radostits, O.M. (1989) *Veterinary Medicine – A Textbook of the Diseases of Cattle, Sheep, Pigs and Horses*. Baillière-Tindall, London.

Boreham, P.F.L. and Geigy, R. (1976) Studies on the genus *Auchmeromyia* Brauer and Bergenstamm (Diptera: Calliphoridae). *Acta Tropica* 33, 74–87.

Burrell, D.H. (1985) Immunisation of sheep against experimental *Pseudomonas aeruginosa* dermatitis and fleece-rot associated body strike. *Australian Veterinary Journal* 62, 55–57.

Bushland, R.C. (1985) Eradication program in the southwestern United States. *Miscellaneous Publications of the Entomological Society of America* 62, 12–15.

Bushland, R.C. and Hopkins, D.E. (1953). Sterilisation of screw-worm flies with X-rays and gamma-rays. *Journal of Economic Entomology* 46, 648–656.

Catts, E.P. and Goff, M.L. (1992) Forensic entomology in criminal investigations. *Annual Review of Entomology* 37, 253–272.

Colless, D.H. and McAlpine, D.K. (1991) Diptera (Flies). In: *The Insects of Australia*, Vol. 2. Melbourne University Press, pp. 717–786.

Crosskey, R.W. and Lane, R.P. (1993) House-flies, blow-flies and their allies (calyptrate Diptera). In: Lane, R.P. and Crosskey, R.W. (eds) *Medical Insects and Arachnids*. Chapman & Hall, London, pp. 403–428.

Eads, R.B. (1979) Notes on muscoid Diptera of public health interest. *Mosquito News* 39, 674–675.

Eisemann, C.H. (1988) Upwind flight by gravid Australian sheep blowflies, *Lucilia cuprina* (Wiedemann) (Diptera: Calliphoridae), in response to stimuli from sheep. *Bulletin of Entomological Research* 78, 273–279.

Eisemann, C.H., Pearson, R.D., Donaldson, R.A., Cadogan, L.C. and Vuocolo, T. (1993) Uptake and fate of specific antibody in feeding larvae of the sheep blowfly, *Lucilia cuprina*. *Medical and Veterinary Entomology* 7, 177–185.

Esser, J.R. (1991) Biology of *Chrysomya megacephala* (Diptera: Calliphoridae) and reduction of losses caused to the salted-dried fish industry in south-east Asia. *Bulletin of Entomological Research* 81, 33–41.

Foster, G.G., Kitching, R.L. Vogt, W.G. and Whitten, M.J. (1975) Sheep blowfly and its control in the pastoral ecosystem of Australia. *Proceedings of the Ecological Society of Australia* 9, 213–229.

Garrett-Jones, C. (1951) The Congo floor maggot, *Auchmeromyia luteola* (F.), in a laboratory culture. *Bulletin of Entomological Research* 41, 679–708.

Gassner, G. (1989) Dipteran endocytobionts. In: Schwemmler, W. and Gassner, G. (eds) *Insect Endocytobiosis: Morphology, Physiology, Genetics, Evolution.* CRC Press, Boca Raton, Florida, USA, pp. 218–232.

Geigy, R. and Kauffmann, M. (1977) Experimental mechanical transmission of *Trypanosoma brucei* by *Auchmeromyia* larvae. *Protozoology* 3, 103–107.

Green, C.H., Hall, M.J.R., Fergiani, M., Chirico, J. and Husni, M. (1993) Attracting adult New World screwworm, *Cochliomyia hominivorax*, to odour baited targets in the field. *Medical and Veterinary Entomology* 7, 59–65.

Greenberg, B. (1971) *Flies and Disease. I. Ecology, Classification and Biotic Associations.* Princeton University Press, Princeton, New Jersey.

Guerrini, V.H. (1988) Ammonia toxicity and alkalosis in sheep infested by *Lucilia cuprina* larvae. *International Journal for Parasitology* 18, 79–81.

Guerrini, V.H., Murphy, G.M. and Broadmeadow, M. (1988) The role of pH in the infestation of sheep by *Lucilia cuprina* larvae. *International Journal for Parasitology* 18, 407–409.

Hall, D.G. (1948) *The Blowflies of North America.* The Thomas Say Foundation, USA.

Hall, M.J.R. and Smith, K.G.V. (1993) Diptera causing myiasis in man. In: Lane, R.P. and Crosskey, R.W. (eds) *Medical Insects and Arachnids.* Chapman & Hall, London, pp. 429–469.

Haufe, W.O. and Nelson, W.A. (1957) Human furuncular myiasis caused by the flesh fly *Wohlfahrtia opaca* (Coq.) (Sarcophagidae: Diptera). *Canadian Entomologist* 89, 325–327.

Hepburn, G.A. (1943) Sheep blowfly research 1. A survey of maggot collections from live sheep and a note on the trapping of blowflies. *Onderstepoort Journal of Veterinary Science and Animal Industry* 18, 13–17.

Hightower, R.G., Adams, A.L. and Alley, D.A. (1965) Dispersal of released irradiated laboratory-reared screw-worm flies. *Journal of Economic Entomology* 58, 373–374.

James, M.T. (1947) The flies that cause myiasis in man. *United States Department of Agriculture Miscellaneous Publication* 631, 1–175.

Knipling, E.F. (1955) Possibilities of insect control or eradication through the use of sexually sterile males. *Journal of Economic Entomology* 48, 459–462.

Laurence, B.R. and Herman, F.G. (1973) Tumbu fly *(Cordylobia)* infection outside Africa. *Transactions of the Royal Society of Tropical Medicine and Hygiene* 67, 888.

Leclercq, M. (1989) Importation de myiases cutanées tropicales humains. *Revue médicale de Liège* 44, 28–32.

Leclercq, M. (1990) Les myiases. *Annales de la Société Entomologique de France* (NS) 26, 335–350.

Leclercq, M. and Verstraeten, C. (1988) Entomologie et medicine legale, datation de la mort: insectes et autres arthropodes trouves sur les cadavres humaines. *Bulletin et Annales de la Société Royal Entomologique de Belgique,* 124, 311–317.

Lindquist, A.W. (1937) Myiasis in wild animals in southwestern Texas. *Journal of Economic Entomology* 30, 735–740.

Lindquist, A.W. (1955) The use of gamma radiation for control or eradication of the screwworm. *Journal of Economic Entomology* 48, 467–469.

Lindquist, D.A., Abusowa, M. and Hall, M.J.R. (1992) The New World screwworm fly in Libya: a review of its introduction and eradication. *Medical and Veterinary Entomology* 6, 2–8.

Lord, W.D. (1990) Case histories of the use of insects in investigations. In: Catts, E.P. and

Haskell, N.H. (eds) *Entomology and Death: a Procedural Guide.* Joyce's Print Shop, Clemson, South Carolina, pp. 9–37.

Marchenko, M.I. (1991) Forensic entomology. In: Weismann, L., Országh, I. and Pont, A.C. (eds) *Proceedings of the Second International Congress of Dipterology*, Bratislava, Czechoslovakia 27 August to 1 September, 1990. SPB Academic Publishing, The Hague, pp. 183–199.

Meadows. M.E. (1985) Eradication program in the southeastern United States. *Miscellaneous Publications of the Entomological Society of America* 62, 8–11.

Morley, F.H.W. and Johnstone, I.L. (1984) Development and use of Mules operation. *The Journal of the Australian Institute of Agricultural Sciences*, 50, 86–97.

Morley, F.H.W., Donald, A.D., Donnelly, I.R., Axelsen, A. and Waller, P.J. (1976) Blowfly strike in the breech region of sheep in relation to helminth infection. *Australian Veterinary Journal* 52, 325–329.

Norris, K.R. (1959) The ecology of sheep blowflies in Australia. In: Keast, A., Crocker, R.L. and Christian, C.S. (eds) *Biogeography and Ecology in Australia.* Junk, The Hague, pp. 514–544.

Norris, K.R. (1965) The bionomics of blow flies. *Annual Review of Entomology* 10, 47–68.

Nuorteva, P., Isokoski, M. and Laiho, K. (1967) Studies on the possibilities of using blowflies (Dipt.) as medicolegal indicators in Finland. *Annales Entomologici Fennici* 33, 217–225.

Pineda-Vargas, N. (1985) Screwworm eradication in Mexico: activities of the Mexico–American Screwworm Eradication Commission 1977–1984. *Miscellaneous Publications of the Entomological Society of America* 62, 22–27.

Sands, D.P.A. (1979) Address to the Entomological Society of Queensland, 10 September.

Sherman, R.A. and Pechter, E.A. (1988) Maggot therapy: a review of the therapeutic applications of fly larvae in human medicine, especially for treating osteomyelitis. *Medical and Veterinary Entomology* 2, 225–230.

Smith, K.G.V. (1986) *A Manual of Forensic Entomology.* Trustees of the British Museum (Natural History), London.

Snow J.W. and Whitten, C.J. (1979) Status of the screwworm (Diptera: Calliphoridae) control program in the southwestern United States during 1977. *Journal of Medical Entomology* 15, 518–520.

Snow J.W., Sienbenaler A.J. and Newell F.G. (1981) *Annotated Bibliography of the Screwworm* Cochliomyia hominivorax *(Coquerel).* United States Department of Agriculture, Science and Education Administration, Agricultural Reviews and Manuals ARM-S-14.

Spradbery, J.P. (1979) Daily oviposition activity and its adaptive significance in the screwworm fly, *Chrysomya bezziana* (Diptera: Calliphoridae). *Journal of the Australian Entomological Society* 18, 63–66.

Spradbery, J.P. (1991) *A Manual for the Diagnosis of Screwworm Fly.* Commonwealth of Australia, Canberra.

Spradbery, J.P. and Schweizer, G. (1979) Ingestion of food by the adult screw-worm fly, *Chrysomya bezziana* (Diptera: Calliphoridae). *Entomologia Experimentalis et Applicata* 25, 75–85.

Spradbery, J.P. and Vanniasingham, J.A. (1980) Incidence of the screw-worm fly, *Chrysomya bezziana*, at the Zoo Negara, Malaysia. *Malaysian Medical Journal* 7, 28–32.

Spradbery, J.P. and Vogt, W.G. (1993) Mean life expectancy of Old World screwworm fly, *Chrysomya bezziana*, inferred from the reproductive structure of native females caught on sworm lure-baited stick traps. *Medical and Veterinary Entomology* 7, 147–154.

Spradbery, J.P., Sands, D.P.A. and Bakker, P. (1976) Evaluation of insecticide smears for the control of screw-worm fly, *Chrysomya bezziana*, in Papua New Guinea. *Australian Veterinary Journal* 52, 280–284.

Spradbery, J.P., Vogt, W.G., Sands, D.P.A. and Drewett, N. (1991) Ovarian development rates in the Old World screw-worm fly, *Chrysomya bezziana*. *Entomologia Experimentalis et Applicata* 58, 261–265.

Sutherst, R.W., Spradbery, J.P. and Maywald, G.F. (1989) The potential geographical distribution of the Old World screw-worm fly, *Chrysomya bezziana*. *Medical and Veterinary Entomology* 3, 273–280.

Waterhouse, D.F. (1947) The relative importance of live sheep and of carrion as breeding grounds for the Australian sheep blowfly *Lucilia cuprina*. *Council for Scientific and Industrial Research (Australia) Bulletin* 217, 1–31.

Waterhouse, D.F. and Paramonov, S.J. (1950) The status of two species of *Lucilia* (Diptera: Calliphoridae) attacking sheep in Australia. *Australian Journal of Scientific Research B* 3, 310–336.

Watts, J.E., Muller, M.J., Dyce, A.L. and Norris, K.R. (1976) The species of flies reared from struck sheep in south-eastern Australia. *Australian Veterinary Journal* 52, 488–489.

West, L.S. (1951) *The Housefly*. Comstock Publishing Company, New York.

Zumpt, F. (1965) *Myiasis in Man and Animals in the Old World*. Butterworths, London.

15

Oestridae (Gad Flies, Warble Flies and Stomach Bots)

The Oestridae are moderately large, bee-like flies with vestigial mouthparts and small eyes, giving a large interocular space (Fig. 15.1). Their stout, thick larvae are obligatory endoparasites of mammals, and are referred to as grubs. Four distinct groups are readily discernible but there is no agreement as to whether they merit family or subfamily status. Following Hall and Smith (1993) they will be treated here as subfamilies – Oestrinae, Hypodermatinae, Gasterophilinae, Cuterebrinae – of an enlarged Oestridae.

The Oestrinae develop in the nasopharyngeal cavities of Perissodactyla and Artiodactyla, although one species (*Tracheomyia macropi*) parasitizes the red kangaroo, and another (*Pharyngobolus africanus*) the African elephant (Zumpt, 1965). Three species are of importance, *Oestrus ovis*, *Rhinoestrus purpureus* and *Cephalopina titillator*, which respectively parasitize sheep and goats, equines and camels. Adult oestrines have a distinct postscutellum, large squamae and the apical cell is closed by vein iv joining vein iii before the wing margin (Fig. 15.2). The larvae have well-developed mouth-hooks and the posterior spiracles are large plates with numerous small openings (Fig. 15.3).

The Hypodermatinae are dermal parasites of Artiodactyla, Lagomorpha and Rodentia. Two species, *Hypoderma bovis* and *H. lineatum*, are important parasites of cattle. They share the following characters with the Oestrinae: distinct postscutellum, large squamae and the posterior spiracles of the larvae with numerous small openings (Fig. 15.6), but differ in having the apical cell of the wing open, i.e. vein iv meets the wing margin independently of vein iii (Fig. 15.5), and the larva having only rudimentary mouth-hooks (Fig. 15.6).

The Gasterophilinae are mainly parasites of the alimentary tract of Equidae, but three species have been described from Asiatic and African rhinoceroses, and three species from African and Indian elephants (Zumpt, 1965). Species of *Gasterophilus* (sometimes misspelt without the 'e' as *Gastrophilus*) are important parasites of horses and donkeys. In the adult the postscutellum is undeveloped, the squamae small and the apical cell wide open because vein iv does not bend towards vein iii (Fig. 15.11). The larva has well-developed

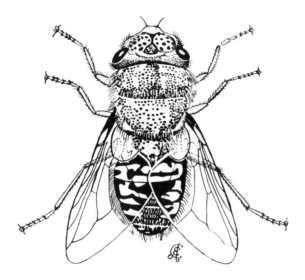

Fig. 15.1. Female *Oestrus ovis*. Source: from Cameron A.E. (1942) *Transactions of the Highland and Agricultural Society.*

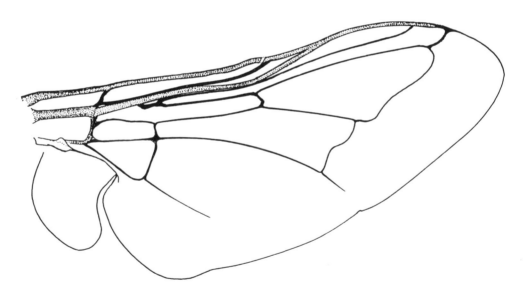

Fig. 15.2. Wing of *Oestrus ovis* (x 18).

mouth-hooks and the posterior spiracles open by three bent slits in a shallow concavity.

The Cuterebrinae are dermal parasites of rodents and rabbits with one species, *Dermatobia hominis*, parasitizing wild and domestic animals including cattle and humans. In cuterebrines the postscutellum is undeveloped, the squamae large and the apical cell narrowed by vein iv turning towards vein iii (Fig. 15.12), but not

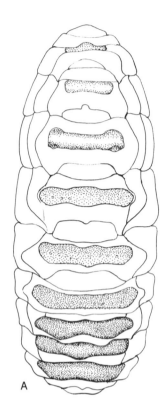

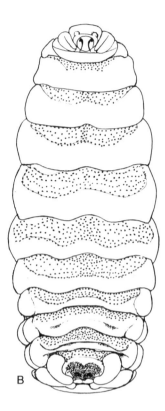

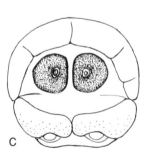

Fig. 15.3. Third instar larva of *Oestrus ovis*: (**A**) dorsal view; (**B**) ventral view; (**C**) posterior view.

joining it. In the larva the mouth-hooks are well developed (Fig. 15.14) and the posterior spiracles, which are deeply sunk, have three straight slits.

For further information the reader is referred to James (1947) and Zumpt (1965), for entomological aspects and to Blood and Radostits (1989), and Seddon and Albiston (1976) for their veterinary importance. The account which follows has drawn extensively on these four works.

Oestrinae

Oestrus ovis (sheep nostril fly)

Five species are included in the genus *Oestrus* of which four are parasites of antelopes and one, *Oestrus ovis*, parasitizes sheep and goats, but not antelopes (Howard, 1980). A comprehensive list of the literature on *O. ovis* from 1686 to 1973 is given by Papavero (1977). *O. ovis* is considered to be a Palaearctic species which has achieved a worldwide distribution by being taken with their hosts throughout the world by humans. Adult *O. ovis* have black pits dorsally between the eyes on the frons and black tubercles among the yellow hairs on the yellow–brown scutum and scutellum (Fig. 15.1). The abdomen is black with an irregular pattern of lighter marking which varies with the angle of illumination.

Life cycle

The female is ovoviviparous and matures about 500 eggs, which are deposited as newly hatched larvae in small batches of less than 50 at a time. Sheep react to the presence of *O. ovis* by pushing their noses into the soil or into the fleece of others in the flock. They may run about erratically. When an opportunity presents itself *O. ovis* deposits larvae into the nasal cavity. They may develop there for a month before moving into the frontal, and sometimes the maxillary sinuses, where the larvae complete their development. The mature larva moves forward, is sneezed out by the host, burrows into the soil and pupariates.

The 1st instar larva has gently curved mouth-hooks and 22–25 terminal spines arranged in two groups. These characters enable this larva to be separated from that of the 1st instar *R. purpureus* which can also causes temporary myiasis, especially ocular myiasis, in humans. The mature 3rd instar larva is about 25 mm long, white or yellowish in colour with darker transverse dorsal bands and transverse rows of spines on the ventral surface of each segment (Fig. 15.3). The posterior spiracles are exposed, flat, D-shaped plates with the button enclosed by numerous small openings.

Bionomics

Two related nasopharyngeal parasites (*Cephenemyia* spp.) of cervids deposited larvae on the lips of a model baited with carbon dioxide, octenol and deer trail scent but not on an unbaited model (Anderson, 1989). Similar kairomones may attract *O. ovis* to sheep.

Development of the 1st instar larva may be delayed, and individuals from the same larviposition may spend from 1 to 9 months in the 1st instar. This plays a role in the overwintering cycle. The pupal stage lasts 1–2 months depending upon temperature. Bukshtynov (1978) recorded a steady increase in rate of development from 12.5 to 35°C with adult emergence occurring after 14 days at the highest temperature, but Rogers and Knapp (1973) found that temperatures above 32°C were fatal. Breev *et al.* (1980) found that there was increased pupal mortality at a constant temperature of 34°C, but that a fluctuating daily temperature of

21–38°C did not affect mortality. The same authors calculated a threshold of 12°C for pupal development, and its duration to be 243 day-degrees for males and 279 for females.

In areas with warm winters year-round breeding of *O. ovis* would be possible, but in many sheep areas there is a definite cool season and two generations per annum are considered to occur. In Pretoria there is no larviposition by *O. ovis* in the three winter months July–September (Horak, 1977). The common pattern in the northern hemisphere is that adults emerge in late spring in June, mate and larviposit, with the larvae developing rapidly and mature 3rd instars leaving the host in July and August. These pupariate to produce an autumn generation larvipositing in September and October. These larvae may remain in the 1st instar for a long period before developing to become mature larvae in March, when they leave the host, pupariate and remain dormant until emergence in June. Mortality among the immature stages has been calculated at 90–94% in the first generation and 99% in the second generation (Rogers and Knapp, 1973).

Veterinary and medical importance

High infestations of *O. ovis* in sheep are commonly recorded, e.g. 73% in Pretoria (Horak, 1977), more than 90% in Kentucky (Rogers and Knapp, 1973), with an average infestation, respectively, of 15 and 22 larvae per sheep. Infestation with *O. ovis* is regarded as relatively benign (Rich, 1965). Annoyance by adult *O. ovis* causes sheep to lose valuable grazing time, and the presence of larvae irritates the mucosa, resulting in a mucopurulent discharge and difficult snoring respiration. Control can be achieved in late summer and/or winter with various drugs including ivermectin (Blood and Radostits, 1989). Merino rams treated against infestation showed reduced nasal discharge and increased weight gain (Horak and Snijders, 1974).

Although the main host of *O. ovis* is sheep and goats, infestation occasionally occurs in dogs (Rich, 1965) and humans but larvae do not complete their development in these hosts. In humans *O. ovis* is associated with ocular myiasis and 80 such cases were treated over 2 years at a clinic in Benghazi (Dar *et al.*, 1980).

Rhinoestrus and *Cephalopina*

Eleven species of *Rhinoestrus* have been described of which four are restricted to equines and seven are host specific for a wide range of wild animals including giraffe, warthog, bushpig and antelope. Of the four species in equines, one is restricted to zebras. One, *R. usbekistanicus*, parasitizes horses and donkeys in the Palaearctic region, and occurs in zebras in Africa. *R. latifrons* parasitizes domestic horses in Central Asia and adjoining regions. The most important species is *R. purpureus* which parasitizes horses, donkeys and their cross-breeds. *Cephalopina* is a monotypic genus with one species, *C. titillator*.

Rhinoestrus purpureus (see Fig. 15.4)

This Palaearctic species has been introduced into the Afrotropical and Oriental regions where it occurs sporadically. Rastagaev (1978) found that in Mongolia and

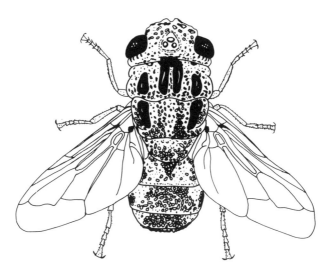

Fig. 15.4. Female *Rhinoestrus purpureus*. Redrawn from Zumpt (1965).

Buryat ASSR, virtually all (98–99%) horses were infested with *R. purpureus* with infestations ranging from 34 to 731 larvae in a single host. There is one generation a year. The female matures 700 to 800 eggs which are deposited as larvae in batches of 8–40 at a time into the nostrils or eyes of a horse. This species causes ocular myiasis in humans; and the 1st instar can be separated from that of *O. ovis*, which also causes this condition, by the possession of more strongly curved mouth-hooks and only 8–12 terminal hooklets in a single row. In the third-stage larva the spiracles are crescent-shaped and do not completely surround the button. As with *O. ovis* 1st instar larvae may remain in the nasal cavity for periods of several months before moving to the sinuses to complete their development.

Cephalopina titillator

In the Sudan all camels are normally infested, containing up to 250 larvae per beast. Female *C. titillator* hover above the host before rapidly larvipositing in the nostrils. Mature larvae leave the nostril towards sunset and pupariate in the soil. Heavy infestations cause nasopharyngeal lesions, providing a suitable environment for pathogens leading to pyogenic infections and pneumonia (Zumpt, 1965; Musa *et al.*,1989)

Hypodermatinae – *Hypoderma* (Warble Flies)

Six species are included in the genus *Hypoderma*. Two species, *H. bovis* and *H. lineatum*, are parasites of cattle, and the other four, including *H. diana*, are parasites of deer. The cattle species are bumble-bee-like flies (Fig. 15.5) with reddish yellow pile at the end of the abdomen. In *H. bovis* the hairs anterior to the suture are variously described as whitish-yellow or reddish-yellow contrasting with those posterior to the suture which are black. In *H. lineatum* the hairs on the scutum

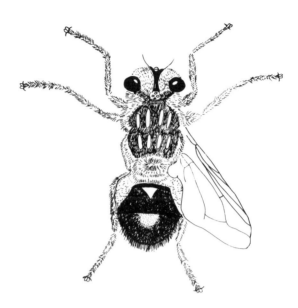

Fig. 15.5. Female *Hypoderma lineatum* Redrawn from Cameron A.E. (1942) Insect pests of 1942. *Transactions of the Highland and Agricultural Society.*

are white and yellow with a predominance of white anteriorly and yellow posteriorly. *H. bovis* and *H. lineatum* are widespread in the northern hemisphere between the latitudes of 25° and 60°N. In North America *H. lineatum* ranges from northern Mexico to northern Canada and *H. bovis* north of a line from northern California through Kansas to the Carolinas. In recent years *H. bovis* has virtually disappeared west of the Mississippi River (Scholl, 1993). They have not established themselves in the southern hemisphere, although frequently introduced in infested cattle, but there has been a recent record of genuine endemic cases in Chile (Beesley, 1974).

Biology and life cycle of *H. bovis* and *H. lineatum* (Beesley, 1961, 1962, 1966, 1974, 1977; Scholl, 1993)

The females emerge with a fully developed complement of 300–650 eggs having matured two oocytes within the same ovariole. The female attaches its eggs firmly to the host's hair. The base of the egg is connected to an attachment organ by a flexible petiole. The attachment organ has a central groove filled with adhesive and a pair of adhesive-coated lateral flanges, which nearly meet around the hair. The adhesive solidifies and firmly attaches the egg to the hair, and the flexible petiole enables the egg to adjust its alignment and reduce stress (Cogley *et al.*, 1981).

Hypoderma bovis lays its eggs while in flight or when walking over the host, causing cattle to gad. It attaches its eggs singly on the rump and upper parts of the hind leg. Ovipositing *H. lineatum* do not disturb the host, and several eggs are deposited in line along the shaft of a single hair. It oviposits on the legs and lower parts of the body on resting and standing cattle. The eggs of both species hatch in about 4 days, and the larvae crawl down the hair and penetrate the skin using proteolytic enzymes secreted from the blind midgut. The exact route taken by migrating larvae in the host is not known with certainty. Several months elapse before the larva reaches its final site on the back of the host.

Migrating larvae of *H. lineatum* are found in the wall of the oesophagus, and those of *H. bovis* in the epidural fat of the spinal canal, with most larvae being found in the region of the lumbar and posterior thoracic vertebrae. It has been suggested that larvae reach the spinal canal by migrating along nerve trunks or through muscles. Although it is considered that most larvae of *H. bovis* migrate via the spinal canal, that may not be the only route as the larval densities there are lower than on the back (Beesley, 1962).

Early in the year the larvae move to their final site on the back in an area 25 cm either side of the midline from shoulder to tail where the cysts or warbles, from which the flies get their popular name, are formed (Wright, 1979). The larva is now about 15 mm long but still in the 1st instar, and moults into the 2nd soon after reaching the back. The larva cuts a hole in the skin through which it respires and develops through the 3rd instar to the prepupa.

The mature larva is about 30 mm long, with a convex ventral surface and a flat dorsal surface (Fig. 15.6). Most of the body segments carry on their ventral surfaces an anterior row of larger, backwardly-directed spines and a posterior band of smaller, forwardly-directed spines (Zumpt, 1965). The spines are less well developed on the dorsal surface. In *H. lineatum* the spiracles are flat, crescent-shaped and with a considerable gap between the arms of the crescent surrounding the button (Fig. 15.6D). In *H. bovis* the posterior spiracles are funnel-shaped with a much smaller gap between the arms (Fig. 15.6C). In addition the openings on the posterior spiracles of *H. bovis* are more numerous and more densely packed than in *H. lineatum*. After several weeks, varying from 4 to 11 (Gregson, 1958; Beesley, 1974), the yellowish-brown prepupa forces its way through the breathing hole in the skin and drops to the ground, where it moves around actively seeking cover; but Gregson (1958) observed only prepupae of *H. bovis* burrowing beneath loose soil.

Dropping of prepupae from cattle occurred early in the morning in southern England (Beesley, 1974) but nearer noon under colder Canadian conditions, where prepupae may be covered by snow and exposed to subzero temperatures. Early in life, puparia can survive cooling to $-15°C$, and following loss of water, late puparia can survive $-28°C$ (Gregson, 1958). At 20°C pupal development of *H. lineatum* is independent of humidity over the range 0 to 98% and that of *H. bovis* optimal at 76% RH (Gregson, 1958). The duration of the pupal stage varies from 3 to 10 weeks depending upon external conditions. Adult *H. lineatum* appear about 4 weeks earlier than *H. bovis* (Beesley, 1974).

There is one generation a year, and Beesley (1966) has summarized the life cycles of both species for southern England (Fig. 15.7). With some minor adjustment on timing the main features will apply to other locations. Adult *H. lineatum* are on the wing from late March to the end of May. First-stage larvae are migrating from April until September, and can be found in the oesophagus from September to March with the peak from November to January. Warbles appear in the back

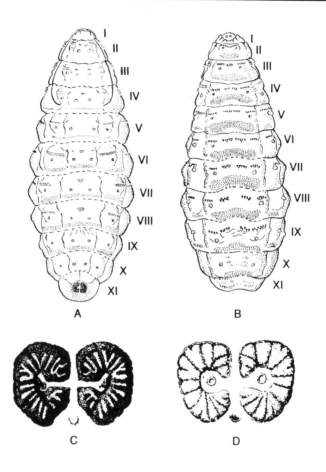

Fig. 15.6. Third instar larva of *Hypoderma lineatum*: (**A**) dorsal view; (**B**) ventral view; (**C**) and (**D**) enlarged posterior spiracles of *Hypoderma bovis* (**C**) and *H. lineatum* (**D**). Source: **A** and **B** from Cameron A.E. (1937) *Transactions of the Highland and Agricultural Society*. **C** and **D** from James (1947).

from January to April, with a maximum in February and March, and prepupae leave the host and pupariate from March to early May.

The timing of *H. bovis* is slightly later with adults being present from June to mid-September; migrating larvae from June to November, arriving in the spinal canal from November to May (main period December to March); warbles from March to July and puparia from May to August. Knowledge of this cycle is important in the timing of control measures.

Adults are active on sunny days when the temperature is above 18°C (Zumpt, 1965). Being unable to feed they live only 3 to 5 days. In the presence of suitable hosts they probably do not fly very far, but marked flies have been recovered 300 m from the point of release after 95 min; and laboratory flight mill studies indicate that *H. lineatum* is capable of flying up to 16 km. Adults emerge early in the day (0730–0830) and mate within 1 h of emergence (Gregson, 1958).

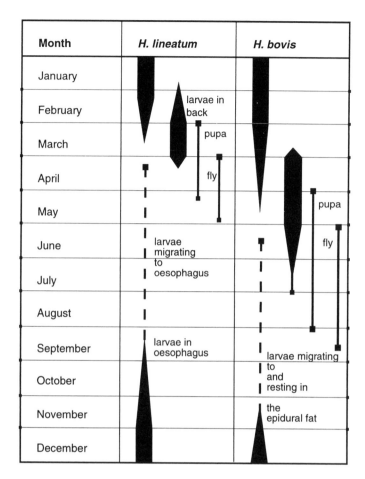

Fig. 15.7. Annual life cycles of *Hypoderma lineatum* and *H. bovis* in southern England. Source: Beesley (1966).

Veterinary and medical importance

Some idea of the intensity of infestation is given by the data of Rich (1965), who examined over 6000 beasts in a 9-year period (1956–1964) in Canada, of which 67% were infested, with the number of warbles ranging from 0 to 229 in individual beasts. The overall average number of warbles was 14.3 per beast examined and 20.8 per infected beast. The larvae were not identified to species but, on other evidence, the ratio of *H. lineatum* to *H. bovis* was estimated as 3:1. The mortality of larvae in warbles was estimated as 51% and 59% in two consecutive years, and the emergence of adult *H. lineatum* from dropped larvae varied from 8 to 80% depending upon location (Gregson, 1958).

Infestations with immature stages have been determined by Beesley (1961, 1962, 1966) by examining bovine gullets and spines. In the three seasons (1957–1960) nearly 7000 gullets were examined of which 7% were infested with

6.6 larvae per infested gullet (September–February); and in the next four seasons (1960–1964) during the main period (November–January) the infestation rate was lower, 3.8% in more than 5000 beasts, but the number of larvae (7.0) per infested gullet remained virtually unchanged. In the same area, living *H. bovis* larvae were found in 23.5% of bovine spines during November 1960 to April 1961 with 2.3 larvae per infested spine. In 1963 and 1964 the number of larvae per infested spine was of the same order, 2.2 and 3.1, respectively, but for comparable months the percentage of cattle infested was lower in 1964.

Losses due to *Hypoderma* arise from a number of causes. There is the disturbance called 'gadding', caused by actively ovipositing flies, especially *H. bovis*. It reduces weight gain and produces losses of 10 to 15% in milk production (Beesley, 1977). The passage of larvae results in the formation of jelly-like tracks in muscle which have to be removed at the abattoir. Destruction of larvae can result in the collapse of an infested beast. This may be due to anaphylaxis or a reaction to toxins liberated from dead larvae. To avoid these reactions systemic insecticides should not be used between mid-November and mid-March when larvae are in the spinal canal or oesophagus (Wright, 1979). In addition the breathing holes made by larvae weaken the hide for use as leather. In 1982 the cost of warble fly damage in Great Britain was estimated as £35 million (Blood and Radostits, 1989).

On rare occasions humans who are closely associated with cattle become infested with *Hypoderma* and suffer from a creeping myiasis due to wandering 1st instar larvae, abscesses, and on occasions the eye may be invaded and destroyed (Zumpt, 1965). *Hypoderma* larvae do not complete their development in humans. Sometimes the infestation of abnormal hosts by *Hypoderma* larvae can be fatal. Invasion of the brain of a horse by *H. bovis* in the USA led to its death (Hadlow *et al.*, 1977).

An attempt was made to introduce reindeer into Scotland. The introduced animals were quarantined to prevent the introduction of the reindeer warble fly, *Hypoderma* (= *Oedemagena*) *tarandi*, but when a small number of reindeer were released into a 100 ha paddock, six became infested with the deer warble fly, *Hypoderma diana*, and two died (Kettle and Utsi, 1955). Infestation of unusual hosts, if successful, can be more dangerous than in the normal host.

Control

Organophosphorus insecticides have proved very successful in killing migrating larvae before their arrival in the back. The avermectins have proved to be even better systemic materials being highly effective against second and third stage larvae in warbles and producing 100% mortality in migrating warble larvae at low dosage levels (Scholl, 1993). Using avermectins, Britain appeared to be close to total eradication of hypodermosis in cattle in 1992 (Tarry, 1992), but the relaxation of import restrictions with other EC countries, allowed the importation in the first 6 months of 1993 of 7000 cattle, of which almost 20% were infested. It is believed that the speedy action taken may have countered that threat (Tarry, 1994, personal communication). Ivermectin, administered with food, eliminated three species of oestrids from red deer (*Cervus elephas hippelaphus*) and roe deer (*Capreolus capreolus capreolus*) (Kutzer, 1994). A North American sterile insect release programme was

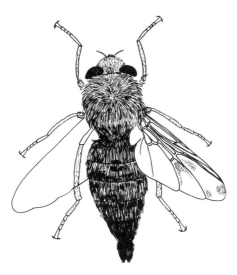

Fig. 15.8. Female *Gasterophilus intestinalis*.

frustrated by the lack of an efficient technique for large-scale *in vitro* rearing of *Hypoderma* spp.

Beasts appear to acquire some immunity with age and vaccines are being prepared using selected protein antigens of 1st instar larvae. Vaccines would complement chemical control. Evaluation of control methods requires an estimation of the *Hypoderma* population before and after treatment and Lysyk *et al.* (1991) have developed a model to determine the number of grubs in the backs of calves from the proportion of uninfested calves. It is important that a control programme should be continued to the stage of eradication because populations are able to recover. An isolated herd, in which the number of grubs per head had been reduced from 30 to 0.2 by the application of control measures, rose to 10.2 in two generations when treatment was relaxed (Beesley, 1974).

Gasterophilinae – *Gasterophilus* (Stomach Bots of Equines)

Nine species of *Gasterophilus* have been described from equines of which six parasitize domestic horses and donkeys, two are restricted to zebras and the host of one species is unknown. Of the six horse parasites *G. nigricornis* has the most limited distribution, occurring in the southern Asiatic part of the Palaearctic region, and *G. intestinalis* is the most important and most widely distributed horse bot. Four species, *G. nasalis*, *G. haemorrhoidalis*, *G. pecorum* and *G. inermis*, parasitize both horses and zebras, and two others are found only in zebras (Zumpt, 1965).

Gasterophilus intestinalis (Fig. 15.8) was originally a Palaearctic species which has been introduced to many parts of the world with the horse. It is now the most important horse bot in the USA (Drudge *et al.*, 1975) and in Australia (Seddon and Albiston, 1976). The next commonest is *G. nasalis*, followed by

G. haemorrhoidalis, and both are widely distributed. In Mongolia and Buryat ASSR, east of Lake Baikal, all six species are present and virtually all horses are parasitized. The relative proportions of the species, based on larval numbers, were *G. intestinalis* (40%), *G. haemorrhoidalis* (20%), *G. nasalis* and *G. pecorum* (10–15%) and *G. inermis* and *G. nigricornis* (5–8%) (Rastagaev, 1979).

Biology and life cycle

In the northern temperate regions there is one generation per annum, but in the tropics and subtropics there may be continuous breeding. In a typical life cycle the eggs are laid on the host; the 1st instar larvae are found in the tissues of the oral cavity; the 2nd and 3rd instars are attached to the intestinal tract for many months, before the prepupae are voided in the faeces and pupariate. Some of the specific variations on this pattern will be considered.

Egg

Fecundity is roughly correlated with the size of the adult. *G. haemorrhoidalis* matures about 160 eggs, *G. nasalis* and *G. inermis* 300 to 500 eggs, *G. intestinalis* 400 to 700, and the largest species, *G. pecorum*, 1300 to 2400 eggs. The eggs are laid in specific locations. *G. pecorum* lays its glossy black eggs in batches of 10 to 15 on vegetation, mainly grasses; the dark eggs of *G. haemorrhoidalis* are laid on the hairs around the lips; and the yellowish eggs of *G. nasalis*, *G. inermis*, and *G. intestinalis* laid, respectively, on the intermandibular space below the jaws, on the cheeks and on the front legs (Zumpt,1965; Blood and Radostits, 1989). The eggs of *G. haemorrhoidalis* have a long corrugated stalk-like pedicel, and the flanges which attach the eggs of *G. intestinalis* and *G. nasalis* to the hairs extend for half the length of the egg or the full length, respectively (Fig. 15.9). It is possible, therefore, to identify gasterophiline eggs to species.

The method of attachment of the egg of *G. intestinalis* is similar to that described for *Hypoderma* above (Gregson, 1958). There is an attachment groove filled with an adhesive material produced by or from the follicle cells, and lateral extensions which completely surround the hair. The egg is waterproofed by layers of wax on the outer surface of the vitelline membrane surrounding the ovum and on the inner membrane of the endochorion of the eggshell. The egg respires through a free air space between the inner membrane and an outer tanned protein layer of the endochorion (Tatchell, 1961).

Eggs of *G. pecorum* and *G. intestinalis* do not hatch until ingested by a horse or stimulated by warmth, moisture and friction. At 25–30°C embryonic development of *G. intestinalis* is completed in 2 to 4 days, but no hatching occurs for another 3 to 6 days, giving a minimum period for larval emergence of 5 to 10 days. The viability of embryonated eggs was inversely related to humidity and temperature from 10 to 30°C. At 10°C eggs are viable for 8 weeks at 100% RH and for 12 weeks at 25% RH. Eggs laid late in the autumn will retain the ability to hatch for several months until the advent of subzero temperatures. Viability of field-collected eggs remained virtually unchanged until the end of December after

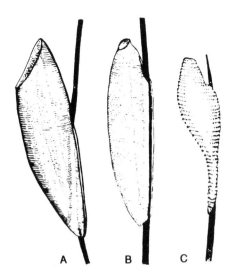

Fig. 15.9. Eggs of *Gasterophilus* species: (**A**) *G. intestinalis*; (**B**) *G. nasalis*; (**C**) *G. haemorrhoidalis*. Source: Cameron A.E. (1942) *Transactions of the Highland and Agricultural Society.*

which it rapidly diminished at a time when minimum daily temperatures were below −10°C (Sukhapesna *et al.*, 1975).

Larva

Eggs of *G. inermis* hatch spontaneously and the larvae burrow into the cheek of the horse causing a condition known as 'summer dermatitis' (Sukhapesna *et al.*, 1975). Eggs of *G. nasalis* also hatch spontaneously and the larvae migrate towards the mouth and enter the oral cavity between the lips (Wells, 1931). First instar *G. intestinalis* burrow into the mucous membrane of the tongue (Sukhapesna *et al.*, 1975), but they, and larvae of *G. nasalis*, have been found in the alveolar space between the teeth and below the gum line, where they cause necrosis and pus formation (Schroeder, 1940; Tolliver *et al.*, 1974). In this location the larvae spend up to a month and develop to the 2nd instar. Eggs of *G. haemorrhoidalis* hatch under the stimulus of moisture and burrow into the epidermis of the lip, migrating into the mouth through the subepithelial layer (Sukhapesna *et al.*, 1975).

The 2nd and 3rd instars are free in the intestinal tract, attached to the wall of the gut by well-developed mouth-hooks. Species have different distributions. The larvae of *G. pecorum* occur on the soft palate and at the root of the tongue, but older 3rd instars pass to the stomach; those of *G. intestinalis* are in the cardiac region of the stomach; and *G. nasalis* in the pyloric region of the stomach. Larvae of *G. haemorrhoidalis* occur in the fundus of the stomach, the duodenum and 3rd instars re-attach in the rectum where they may occur with *G. inermis* (Zumot, 1965; Blood and Radostits, 1989).

In an area where all six horse-infesting species were present, Rastagaev (1978)

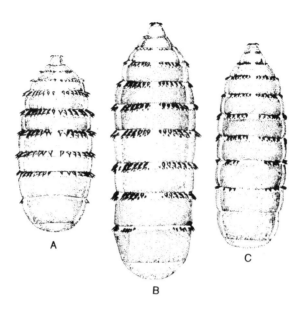

Fig. 15.10. Larvae of *Gasterophilus* species: (**A**) *G. nasalis* (16 mm); (**B**) *G. intestinalis* (18 mm); (**C**) *G. haemorrhoidalis* (18 mm). Source: Cameron A.E. (1942) *Transactions of the Highland and Agricultural Society.*

found 7% of larvae in the oral cavity, 56% in the stomach, 25% in the intestine and 12% in the rectum. The distribution within the stomach was 83–92% in the cardiac region, 7–15% in the fundus and 1 to 2% in the pylorus. Most larvae have two rows of stout spines anteriorly on most segments but in *G. nasalis* there is only one row (Fig. 15.10). In the other common horse bots, the spines are sharply pointed in *G. haemorrhoidalis*, and blunt in *G. intestinalis* (Zumpt, 1965). In all species the two posterior spiracles are united along their inner margins.

Larvae feed on tissue exudates and do not draw blood. Larvae of *G. haemorrhoidalis* contain haemoglobin but it is produced by the larva itself. Larvae pupariate in the soil and the duration of the pupal stage varies with temperature. For *G. intestinalis* it is 8 weeks at 21°C and 18–20 days at 27–32°C and the most favourable conditions tested were 29°C and 80–92% RH with no pupation occurring at 5°C, and pupariation but no eclosion at 38°C (Knapp *et al.*, 1979). There was 77% survival from prepupa to adult among 269 prepupae of several species (Schroeder, 1940).

Adult

Adult gasterophilines are bee-like flies with hairy head and thorax but few bristles (Fig. 15.8). In the female more abdominal segments are exposed than is usual in the Cyclorrhapha and the abdomen is characteristically recurved ventrally. The wing of *G. pecorum* is very dark, that of *G. intestinalis* has a broad transverse median band and dark areas at the end of vein iv and the wing apex (Fig. 15.11). The wings of *G. nasalis* and *G. haemorrhoidalis* are hyaline with the two cross-veins

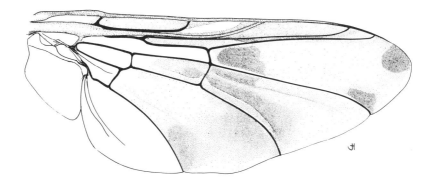

Fig. 15.11. Wing of *Gasterophilus intestinalis* (x 18).

near the middle of the wing almost meeting on vein iv in *G. nasalis*, but widely separated in *G. haemorrhoidalis*.

The adults are diurnal with peak activity occurring in the early afternoon in warm, sunny weather; and with no activity on cloudy days, in strong winds or in heavy rain (Rastagaev, 1979). Adults are short-lived and, given favourable conditions, may live only 1 day (Catts, 1979). They do not feed and the mouthparts are greatly reduced. The maxillary palps are less reduced and carry sensilla which Faucheux (1977) has not found in other Cyclorrhapha, and he considers them to function as olfactory receptors, being used to seek out the gasterophilines' equine hosts.

Adult gasterophilines buzz, a habit shared by other oestrids. This is associated with endothermic heat production which may raise the temperature of the thorax 12°C above ambient, and is the prelude to flight. Heat loss is controlled by insulation provided by the thoracic hair, and by restricting haemolymph circulation to the abdomen. The preferred ambient temperature for flight is 20–24°C when the thoracic temperature is 31–32°C. The energy required to produce this heat would exhaust the fly's energy reserves in a few hours, supporting a very short adult life (Humphreys and Reynolds, 1980).

Mating occurs in the vicinity of horses when solitary hovering males will establish and defend a territory, of either a single horse or a small group of horses. Mating also occurs on hilltops where the males hover and aggressively pursue passing objects, presumably in search of females. Males can hover in winds of 15–20 km h^{-1} and are active at temperatures of 19–34°C. Individuals can hover for half an hour without landing; males make contact with a female on the wing, couple, and sink to the ground where copulation is completed in 3 to 4 min (Catts, 1979).

Veterinary importance

Horses are infested with gasterophiline larvae all the year round and a typical pattern has been described by Drudge *et al.* (1975) for Kentucky. Virtually all horses are infested with *G. intestinalis* and 81% with *G. nasalis*. The average number of

G. intestinalis varies from a low of 50 in September to a high of 229 in March, and the corresponding figures for *G. nasalis* are 14 in September and 82 in February. Second instar *G. intestinalis* from the previous season continue to reach the stomach until April, and are not voided in large numbers as prepupae until August, when 2nd instars of the present season which reached the stomach in July will have moulted to 3rd instars.

A similar overlap in generations occurs with *G. nasalis* when prepupae of the previous year's egg laying are voided from March to August, and 2nd instars of the new generation reach the stomach in July and become 3rd instars in 5 to 7 weeks, before the previous generation's infestation has been exhausted. Hatching of the eggs of *G. nasalis* is not delayed and consequently horses become infected only between May and November. Eggs of *G. intestinalis* can remain viable for long periods and although oviposition only extends from early May till late October the acquisition of infection is almost a year-long process with the exception of April (Drudge *et al.*, 1975). This is somewhat at variance with other work in Kentucky in which no viable eggs of *G. intestinalis* were recovered in the field after the end of January (Sukhapesna *et al.*, 1975).

In the presence of ovipositing gasterophilines, horses become more difficult to control and may injure themselves. Infestations with second and third stage larvae are usually in the hundreds with maximal infestations reaching nearly 1500 (Drudge *et al.*, 1975; Rastagaev, 1979). In Kentucky only one horse out of 476 was infested with more than 500 *G. nasalis* while 13 had more than 500 *G. intestinalis* (Drudge *et al.*, 1975). In Ireland lower maximal infestations were found in examinations of 2500 horses. The maximum recorded for *G. nasalis* larvae was 120 and for *G. intestinalis* was 513 (Hatch, *et al.*, 1976).

Considerable swelling occurs around the point of attachment of the *Gasterophilus* larva, leaving a ring-like swelling when the larva is removed. Heavy infestations result in chronic gastritis, loss of condition and in rare cases, perforation and death (Blood and Radostits, 1989). In a survey in eastern Australia Waddell (1972) found 64% of horses infested with *G. intestinalis*, 19% had ulcers in the oesophageal region of the stomach, and 92% of the ulcerated stomachs were associated with infestations of *G. intestinalis*. Ulcers are most common in early summer when the deeply-embedded third stage larvae are almost fully grown. It is postulated that infestation with *G. intestinalis* may lead to subserosal abscess formation and death by peritonitis.

Infestation of stomach bots in horses can be controlled by the oral administration of organophosphorus insecticides or ivermectin. Both agents also act against the intestinal helminth parasites and care must be taken that the dose administered to control stomach bots does not induce resistance in helminths that may be present. Infestation with gasterophilines can be assessed by detecting anti-*Gasterophilus* circulating antibodies, for which the two most sensitive techniques were thin layer immunoassay and diffusion in gel ELISA (Escartin-Peña and Bautista-Garfias, 1993).

Very occasionally larvae of *Gasterophilus* penetrate the human skin and cause a creeping myiasis in which the larva tunnels in the epidermis causing considerable irritation as the larva advances up to 20 mm a day. The infestation may end spontaneously or the larva may be excised (James, 1947). Experimentally it has been

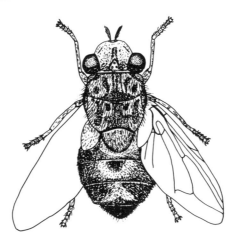

Fig. 15.12. Female *Dermatobia hominis*. Redrawn from James (1947).

shown that the 1st instar larva of *G. nasalis* cannot penetrate the intact skin; that *G. intestinalis* can penetrate damaged human skin; and that 1st instar larvae of the other four species of horse bots, including *G. haemorrhoidalis* and *G. pecorum*, can penetrate intact human skin (Rastagaev, 1978).

Dermatobia hominis (Catts, 1982)

Dermatobia hominis is the only species in the genus. It is a bluebottle-like fly with yellow to orange head and legs, and a feathered arista (Fig. 15.12). It is endemic to the Neotropical region where it occurs from 18°S to 25°N, being associated with moist, cool, tropical highlands between 160 and 2000 m above sea level, especially the coffee growing areas between 600 and 1000 m above sea level. It causes cutaneous myiasis in a wide range of mammalian hosts, and is particularly important as a parasite of cattle. It has also been reported from chickens, turkeys and toucans (Mateus, 1975; Catts, 1982).

Dermatobia hominis has a unique method of ensuring that its progeny reach a range of hosts. The female uses other insects as carriers of its eggs. These are carefully glued on to the carrier in such a way as not to affect its flight efficiency adversely (Fig. 15.13). Nearly 50 species of carriers have been reported of which about half are mosquitoes, and a third are muscoids (Anthomyiidae). In Costa Rica five carrier species were involved of which the most important was *Sarcopromusca arcuata*, which carried an average load of 28 eggs. Fewer eggs (6–10) were laid on mosquitoes, and Mateus (1975) states that there is no evidence that eggs are laid on plants or directly on hosts in the field.

A female *D. hominis* will produce 800 to 1000 eggs. Development of the egg requires 4 to 9 days at a temperature of 20–30°C, but hatching is delayed until the stimulus of a sudden increase in temperature, which would occur when a carrier insect visits a warm-blooded host. The larva transfers to the host, and either enters through the feeding puncture made by the carrier, or penetrates the

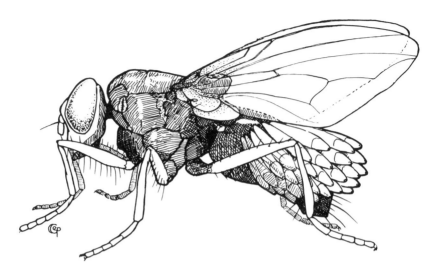

Fig. 15.13. *Sarcopromusca arcuata* with the eggs of *Dermatobia hominis* glued to its abdomen (length of egg 1 mm). Source: Catts (1982). Reproduced by permission of *Annual Reviews Inc.*

unbroken skin, which it can do in 5–10 min. Each larva penetrates individually and a boil-like swelling develops around it. The larva feeds on tissue exudate and grows slowly, requiring 4 to 18 weeks to complete development. The swelling has an opening through which the larva respires and secondary infections occur. Often the discharge from the opening has a foetid odour, which may attract other myiasis-causing flies (James, 1947).

The first-stage larva is subcylindrical with small spines on segments three and four, and stouter spines on segments five to seven, arranged in two dorsal rows and one ventral row. The second-stage larva is pyriform with stout spines on the globular anterior portion and no spines on the narrower posterior part (Fig. 15.14A). The third-stage larva is elongate ovate, with prominent flower-like anterior spiracles, reduced spines, and prominent mouth-hooks (Fig. 15.14B, C) (James, 1947). The posterior spiracles have three slits, no button and are sunk in a pit (Kremer *et al.*, 1978).

The prepupae emerge from the host in the early morning, before 0800 h, and burrow into the soil where they pupariate. The pupal stage is long, taking 4 to 11 weeks. Females mate 24 h after emergence. The adults do not feed, and are comparatively inactive, living 1 to 9 days (Mateus, 1975). They frequent the forest edge but will pursue large hosts 1.5 km into open cleared land (Catts, 1982).

Veterinary and medical importance

Dermatobia hominis causes myiasis in a wide range of wild and domestic hosts, of which the most important economically is cattle. The loss caused by *D. hominis* was estimated at US$260 million in the early 1970s as a result of reduced calf growth and weight gains, lowered milk yields, and hide damage. Infestations in

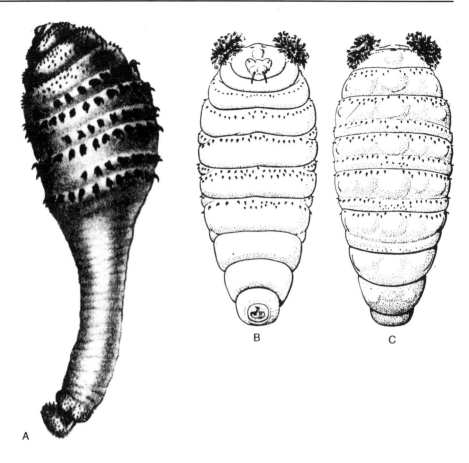

Fig. 15.14. Larvae of *Dermatobia hominis*: (**A**) second instar larva; (**B**) and (**C**) ventral and dorsal views of a mature third instar larva. Source: James (1947).

susceptible animals can exceed 1000 warbles per beast and result in death of the host. Zebu cattle are very resistant, but Holstein and Brown Swiss breeds are highly susceptible. Infestation with *D. hominis* was so great in Panama at one time that the practice of purchasing cattle for fattening on readily available lush pasture had to be abandoned in favour of purchasing fat cattle for immediate slaughter (Dunn, 1934; Mateus, 1975). Sheep may become heavily infested and develop severe abscesses (Dunn, 1934); dogs are frequently attacked; cats and rabbits less frequently attacked; and Equidae are troubled little (James, 1947).

Humans can also be hosts to *D. hominis* and suffer from painful, discharging, cutaneous swellings on the body. Larvae are able to penetrate clothing and boils may be found on all parts of the body (Dunn, 1934). When the larva is removed, in the absence of secondary infection, the condition clears spontaneously in about a week (Rosen and Neuberger, 1977). Rarely, a larva infesting the scalp may penetrate the skull into the brain with fatal results. This rare condition has been reported in a 5-month-old baby (Rossi and Zucoloto, 1973) and an 18-month-old child (Dunn, 1934). This is facilitated by the ossification of the skull being incomplete at that age. The long period of development of larvae in the host favours the introduction of *D. hominis* into other parts of the world. Travellers from South America have introduced *D. hominis* into Canada (Rosen and Neuberger, 1977) and Australia (Moorehouse, 1982, private communication). If suitable conditions exist it is conceivable that *D. hominis* could establish itself in other parts of the tropics.

References

Anderson, J.B. (1989) Use of deer models to study larviposition by wild nasopharyngeal bot flies (Diptera: Oestridae). *Journal of Medical Entomology* 26, 234–236.

Beesley, W.N. (1961) Observations on the development of *Hypoderma lineatum* De Villiers (Diptera: Oestridae) in the bovine host. *Annals of Tropical Medicine and Parasitology* 55, 18–24.

Beesley, W.N. (1962) Observations on the development of *Hypoderma bovis* de Geer (Diptera: Oestridae) in the bovine host. *Research in Veterinary Science* 3, 203–208.

Beesley, W.N. (1966) Further observations on the development of *Hypoderma lineatum* De Villiers and *Hypoderma bovis* Degeer (Diptera: Oestridae) in the bovine host. *British Veterinary Journal* 122, 91–98.

Beesley, W.N. (1974) Economics and progress of warble fly eradication in Britain. *Veterinary Medical Review* 4, 334–347.

Beesley, W.N. (1977) Practical relationships between the biology and control of cattle grubs. *Veterinary Parasitology* 3, 251–257.

Blood, D.C. and Radostits, O.M. (1989) *Veterinary Medicine – A Textbook of the Diseases of Cattle, Sheep, Pigs and Horses*, Baillière-Tindall, London.

Breev, K.A., Zagretdinov, R.G. and Minar, J. (1980) Influence of constant and variable temperatures on pupal development of the sheep bot fly (*Oestrus ovis* L.). *Folia Parazitologica* 27, 359–365.

Bukshtynov, V.l. (1978) Determining the time of development of *Oestrus ovis*. *Veterinariya Moscow* 9, 60–62.

Catts, E.P. (1979) Hilltop aggregation and mating behaviour by *Gasterophilus intestinalis* (Diptera: Gasterophilidae). *Journal of Medical Entomology* 16, 461–464.

Catts E.P. (1982) Biology of New World bot flies: Cuterebridae. *Annual Review of Entomology* 27, 313–338.

Cogley, T.P., Anderson, J.R. and Weintraub, J. (1981) Ultrastructure and function of the attachment organ of warble fly eggs (Diptera: Oestridae: Hypodermatinae). *International Journal of Insect Morphology and Embryology* 10, 7–18.

Dar, M.S., Amer, M.B., Dar, F.K. and Papazotos, V. (1980) Ophthalmomyiasis caused by the sheep nasal bot, *Oestrus ovis* (Oestridae) larvae, in the Benghazi area of eastern Libya. *Transactions of the Royal Society of Tropical Medicine and Hygiene* 74, 303–306.

Drudge, J.H., Lyons, E.T., Wyant, Z.N. and Tolliver, S.C. (1975) Occurrence of second and third instars of *Gasterophilus intestinalis* and *Gasterophilus nasalis* in stomachs of horses in Kentucky. *American Journal of Veterinary Research* 36, 1585–1588.

Dunn, L.H. (1934) Prevalence and importance of the tropical warble fly, *Dermatobia hominis* Linn., in Panama. *Journal of Parasitology* 20, 219–226.

Escartin-Peña, M. and Bautista-Garfias, C.R. (1993) Comparison of five tests for the serologic diagnosis of myiasis by *Gasterophilus* spp. larvae (Diptera: Gasterophilidae) in horses and donkeys: a preliminary study. *Medical and Veterinary Entomology* 7, 233–237.

Faucheux, M.J. (1977) Les pièces buccales vestigiales de l'imago femelle de gastrophile (*Gasterophilus intestinalis* De Geer). *Bulletin de la Société des Sciences Naturelles de l' Ouest de la France* 75, 7–10.

Gregson, J.D. (1958) Recent cattle grub life-history studies at Kamloops, British Columbia, and Lethbridge, Alberta. *Proceedings of the 10th International Congress of Entomology* 3, 725–734.

Hadlow, W.J., Ward, J.K. and Krinsky, W.L. (1977) Intracranial myiasis by *Hypoderma bovis* (Linnaeus) in a horse, *Cornell Veterinarian* 67, 272–281.

Hall, M.J.R. and Smith, K.G.V. (1993) Diptera causing myiasis in man. In: Lane, R.P. and Crosskey, R.W. (eds) *Medical Insects and Arachnids*, Chapman & Hall, London, pp. 429–469.

Hatch, C., McCaughey, W.J. and O'Brien, J.J. (1976) The prevalence of *Gasterophilus intestinalis* and *G. nasalis* in horses in Ireland. *Veterinary Record* 98, 274–276.

Horak, I.G. (1977) Parasites of domestic and wild animals in South Africa. 1. *Oestrus ovis* in sheep. *Onderstepoort Journal of Veterinary Research* 44, 55–63.

Horak, I.G. and Snijders, A.J. (1974) The effect of *Oestrus ovis* infestation on merino lambs. *Veterinary Record* 94, 12–16.

Howard, G.W. (1980) Second stage larvae of nasal botflies (Oestridae) from African antelopes. *Systematic Entomology* 5, 167–177.

Humphreys, W.F. and Reynolds, S.E. (1980) Sound production and endothermy in the horse botfly, *Gasterophilus intestinalis*. *Physiological Entomology* 5, 235–242.

James, M.T. (1947) The flies that cause myiasis in man. *United States Department of Agriculture Miscellaneous Publication* 631, 1–175.

Kettle, D.S. and Utsi, M.N.P. (1955) *Hypoderma diana* (Diptera: Oestridae) *Lipoptena cervi* (Diptera: Hippoboscidae) as parasites of reindeer (*Rangifer tarandus*) in Scotland with notes on the second stage larva of *Hypoderma diana*. *Parasitology* 45, 116–120.

Knapp, F.W., Sukhapesna, V., Lyons, E.T. and Drudge, J.H. (1979) Development of third-instar *Gasterophilus intestinalis* artificially removed from the stomachs of horses. *Annals of the Entomological Society of America* 72, 331–333.

Kremer, M., Rebholtz, C. and Rieb, J.P. (1978) Iconographie des plaques stigmatiques de *Dermatobia hominis* Linné Jr (= *D. cyaniventris* Macquart 1843). *Annales de Parasitologie Humaine et Comparée* 53, 439–440.

Kutzer, E. (1994) Die Bekämpfung der Öestrinose und Hyperdermose bei Cerviden (Rothirsh, Reh) mittels Ivermectin (Ivomec®). *Mitteilungen der Deutschen Gesellschaft für allegemeine und angwandte Entomologie* 9 (in press).

Lysyk, T.J., Colwell, D.D. and Baron, R.W. (1991) A model for estimating abundance of cattle grub (Diptera: Oestridae) from the proportion of uninfested cattle as determined by serology. *Medical and Veterinary Entomology* 5, 253–258.

Mateus, G. (1975) Ecology and control of *Dermatobia hominis* in Colombia. In: Thompson, K.G. (ed.) *Workshop on the Ecology and Control of External Parasites of Economic Importance of Bovines in Latin America*, Centro Internacional de Agricultura Tropical, Cali, Colombia, pp. 117–123.

Musa, M.F., Harrison, M., Ibrahim, A.M. and Taha, T.O. (1989) Observations on Sudanese camel nasal myiasis caused by the larvae of *Cephalopina titillator*. *Revue d'Elevage et de Médecine Vétérinaire des Pays Tropicaux*, 42, 27–31.

Papavero, N.P. (1977) *The World Oestridae (Diptera), Mammals and Continental Drift*. W. Junk, The Hague.

Rastagaev, Yu.M. (1978) Subcutaneous myiasis in man caused by larvae of the horse bot-fly. *Meditsinskaya Parazitologiya i Parazitarnye Bolezni* 47(6), 72–73.

Rastagaev, Yu.M. (1979) The distribution and species composition of botflies of horses in the Buryat ASSR and the Mongolian People's Republic (Oestridae: Gasterophilidae). *Parazitologiya* 13, 547–548.

Rich, G.B. (1965) Post-treatment reactions in cattle during extensive field tests of systemic organophosphate insecticides. *Canadian Journal of Comparative Medicine and Veterinary Science* 29, 30–37.

Rogers, C.E. and Knapp, F.W. (1973) Bionomics of the sheep bot fly *Oestrus ovis*. *Environmental Entomology* 2, 11–23.

Rosen, I.J. and Neuberger, D. (1977) Myiasis *Dermatobia hominis*, Linn. Report of a case and review of the literature. *Cutis* 19, 63–66.

Rossi, M.A. and Zucoloto, S. (1973) Fatal cerebral myiasis by the tropical warble fly, *Dermatobia hominis*. *American Journal of Tropical Medicine and Hygiene* 22, 267–269.

Scholl, P.J. (1993) Biology and control of cattle grubs. *Annual Review of Entomology* 39, 53–70.

Schroeder, H.O. (1940) Habits of the larvae of *Gasterophilus nasalis* (L.) in the mouth of the horse. *Journal of Economic Entomology* 33, 382–384.

Seddon, H.R. and Albiston, H.E. (1976) *Diseases of Domestic Animals in Australia Part 2. Arthropod Infestations (Flies, Lice and Fleas)*. Department of Health, Commonwealth of Australia.

Sukhapesna, V., Knapp, F.W., Lyons, E.T. and Drudge, J.H. (1975) Effect of temperature on embryonic development and egg hatchability of the horse bot *Gasterophilus intestinalis* (Diptera: Gasterophilidae). *Journal of Medical Entomology* 12, 391–392.

Tarry, D.W. (1992) Eradication of hypodermosis in northern Europe and the problems arising from international livestock movements. *Abstracts XIXth Congress of Entomology*, Beijing, 28 June–4 July 1992, p. 496.

Tatchell, R.I. (1961) Studies on the egg of the horse bot-fly. *Gasterophilus intestinalis* (De Geer). *Parasitology* 51, 385–394.

Tolliver, S.C., Lyons, E.T. and Drudge, J.H. (1974) Observations on the specific location of *Gasterophilus* spp. larvae in the mouth of the horse. *Journal of Parasitology* 60, 891–892.

Waddell, A.H. (1972) The pathogenicity of *Gasterophilus intestinalis* larvae in the stomach of the horse. *Australian Veterinary Journal* 48, 332–335.

Wells, R.W. (1931) The method of ingress of newly hatched larvae of the throat bot of horses, *Gasterophilus nasalis* L. *Journal of Economic Entomology* 24(1), 311.

Wright, A.I. (1979) Warble fly eradication. *Veterinary Annual* 19, 54–60.

Zumpt, F. (1965) *Myiasis in Man and Animals in the Old World*. Butterworths, London.

Hippoboscidae (Keds, Louse Flies)

16

The Hippoboscidae are blood-sucking ectoparasites of birds and mammals, probably related to the blood-sucking muscids. They are leathery, dorsoventrally flattened flies with porrect mouthparts and robust legs ending in large recurved claws
(Fig. 16.1). Wings, when present, have the anterior veins strongly developed. The
antennae are apparently immovable and placed far forwards in a deep antennal
pit. The second antennal segment forms the greater part of the antenna and houses
the greatly reduced third antennal segment in a ventral recess from which the arista
protrudes (Bequaert, 1953). The Hippoboscidae are similar to *Glossina* in the structure of their mouthparts, method of reproduction and the possession of a
bacteriome on the midgut. In the Hippoboscidae the bulb of the haustellum is
withdrawn into the head and the narrow terminal portion concealed in grooves
on the inner side of the palps. The female is viviparous and retains the larva within
a modified common oviduct until it is fully grown and deposits it as an immobile
prepupa which pupariates *in situ*. Intracellular bacteroids in the bacteriome produce B-group vitamins and play a role in reproduction. When the bacteroids are
depleted there is a loss of ability to produce viable prepupae (Gassner, 1989). The
structure, physiology and natural history of this family have been reviewed by
Bequaert (1953).

Classification (Maa, 1963, 1966, 1969)

About 200 species are now recognized in this family and arranged in three
subfamilies – Ornithomyinae, Melophaginae (=Lipopteninae) and Hippoboscinae. The Ornithomyinae is the largest with over 150 species, mostly parasites
of birds, but it also includes five species parasitic on wallabies and one species
on lemurs in Madagascar. The Melophaginae contains about 30 species parasitic
on bovids and cervids including the economically important sheep ked, *Melophagus
ovinus*. The Hippoboscinae contains eight species of which six parasitize equines

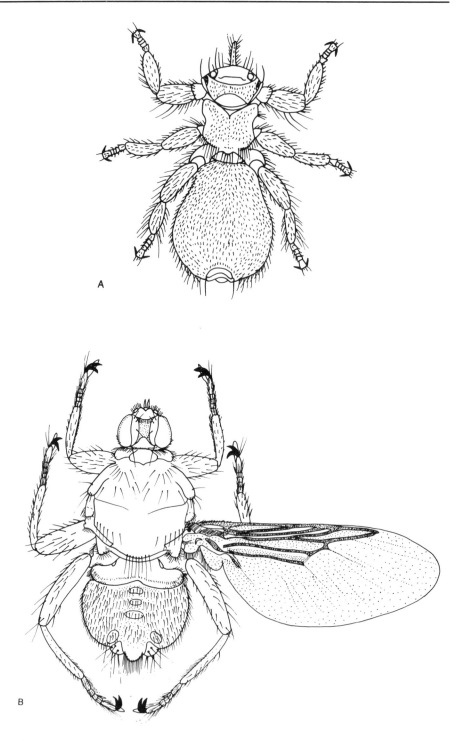

Fig. 16.1. (**A**) *Melophagus ovinus*, the sheep ked; (**B**) *Hippobosca equina*.
Source: (**B**) based on Fig. 6 in Ma (1963).

and ruminants, mainly bovids; one species, *Hippobosca longipennis*, parasitizes carnivores and another species is found only on ostriches.

In view of the fact that about 80% of hippoboscid species occur on birds it is a pleasant surprise to find that domesticated birds, with the exception of the pigeon, are free from these blood-sucking parasites. There are no records of a hippoboscid breeding on poultry, turkeys, ducks, geese, guinea fowl or canaries. Domestic pigeons are regularly infested with *Pseudolynchia canariensis*. *P. canariensis* is widely distributed geographically in the tropics and subtropics of the Old World where it is recorded from 33 genera of birds, breeding mainly on pigeons and raptors, but perhaps also on cuckoos. *P. canariensis* has been introduced into the New World where it is confined to domestic pigeons.

The three main genera which occur on mammals are readily distinguished. Adult *Melophagus* are wingless (Fig. 16. 1A), the wings being reduced to tiny, veinless, opaque knobs and there are no halteres. In *Lipoptena* the newly emerged fly has fully developed and functional wings, which break off close to the base after the final host is reached. They are said to be caducous. Adult *Hippobosca* are permanently winged (Fig. 16.1B). They are distinguished from other winged genera by the pronotum being large and clearly visible, forming an easily observable neck-like segment between the scutum and the head. In all mammal-infesting hippoboscids the paired claws are simple while in the majority of bird-infesting species the claws have two separate teeth. Maa (1963) points out that care must be take not to confuse the basal lobe, which is present on the claws of all hippoboscids, with a tooth. The basal lobe is distinguished by being unevenly pigmented and sclerotized, and not pointed.

Melophagus ovinus (Sheep Ked)

Melophagus ovinus is a Palaearctic species, which has spread with sheep widely throughout the world and established itself in temperate countries and in the cooler highlands of the tropics but is absent from the hot, humid tropics.

Life cycle

The life cycle of the ked has been studied in several parts of the world with substantially the same conclusions. The account which follows is based very largely on Evans (1950). The newly emerged female *M. ovinus* mates within 24 h of emergence, but the ovaries have to be matured before an egg is available for fertilization. This process takes 6 to 7 days and further development within the female takes an additional 7 days so that the first fully developed larva is deposited when the female is 13–14 days old. Thereafter additional larvae are deposited every 7 to 8 days so that in a lifetime of 4 to 5 months a female will produce about 15 larvae, a comparatively slow rate of increase for an insect. The deposited larva pupates within 6 h and the duration of the pupal stage is 20–26 days. The cycle from newly emerged adult female to the emergence of an adult of the next generation is 5 weeks. *M. ovinus* is a permanent ectoparasite on a homoiothermic host and therefore living under very constant conditions, which accounts for the narrow range in the durations of the different stages.

Pupae develop over a relatively narrow range of temperature (25–34°C) with optimal development at 30°C. The puparia are glued to the fleece and carried away from the skin as the fleece grows. The temperature at the skin surface will be 37°C and near the surface of the fleece it will be nearer to air temperature, say 15°C. It is advantageous to the species to deposit puparia in areas of the fleece where a suitable range of temperature (25–34°C) will be found during the 3 week development period of the pupa. This is found most easily in the neck region where the wool staple lies parallel to the skin, and temperature varies slowly with increasing wool length (=time). In hoggs (yearling sheep) over 50% of puparia were found in the neck region while nearly 60% of the adults were found on the region of the forelegs and flanks. On lambs, puparia were concentrated on the hind legs, neck and belly although substantial numbers of adults were found on the flanks and forelegs.

Bionomics

Populations of *M. ovinus* show seasonal changes and, at the same time of the year, different levels of infestation on sheep of different ages. Populations of *M. ovinus* are at their highest in the winter and lowest in summer. At the start of the year in the northern hemisphere there is an increase in keds on both hoggs and 2- and 3-year-old ewes but with substantially higher infestation (× 5–6) on the hoggs. Later when the sheep are penned for lambing there is a rapid rise in infestation on ewes due to transfer from hoggs as a result of their close association. After lambing (late April) the ked population on hoggs and ewes decreases sharply coincidental with a rapid rise in infestation on lambs (Fig. 16.2).

This continues until shearing in late June, when 80–90% of the puparia and keds may be removed with the fleece. Similar reductions (77% – Pfadt *et al.*, 1975; 71–98% – Pfadt, 1974) have been reported as a result of the shearing in Wyoming, where in summer the ked population on lambs decreased by 35–69%, but densities of keds still ranged from 36 to 66 adults per lamb (Pfadt, 1976).

Transfer of keds from one sheep to another occurs when they move to the surface of the fleece in response to temperature. When the air temperature was 15°C only a small number of keds (average four) were on the surface of the fleece, but when the air temperature increased to 23°C the number of keds on the surface soared (average 98). Keds are vulnerable when they are on the surface of the fleece. They may be dislodged and fall to the ground where they will survive only 2 to 5 days, and are unlikely to find another host unless sheep are densely crowded. Keds on the surface of the fleece are subject to predation by birds such as magpies and starlings (Evans, 1950). They are also ingested by sheep biting their fleece, and this is probably the route by which sheep become infected with the benign trypanosome, *Trypanosoma melophagium*.

Several factors contribute to the fluctuations in natural populations of *M. ovinus*. Nelson (1958) and Nelson and Sten (1968) state that only newly emerged keds go on lambs, and that the older keds stay on the ewes and suffer considerable mortality from infection with *T. melophagium* (Nelson, 1956) but the latter

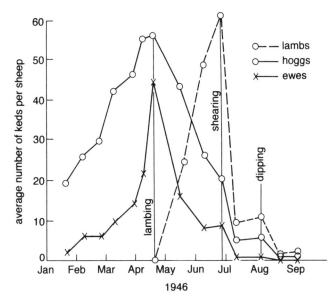

Fig. 16.2. Fluctuations in numbers of *Melophagus ovinus* on sheep in Wales between January and September 1946. Source: Evans (1950).

conclusion is disputed by Hoare (1972). Undoubtedly older animals have fewer keds, and in the Wyoming study ked numbers increased on lambs from March to May, and then a natural decrease occurred in unshorn lambs until September, after which numbers increased to reach a maximum in February, before declining until May (Pfadt, 1976).

Yearlings, which have lambs in March, have similar densities of keds in the following summer and autumn as older ewes (Pfadt, 1976). Two factors contribute to this: temperature and the development of resistance in individual sheep. Nelson and Bainborough (1963) and Nelson and Hironaka (1966) have shown that there are at least two factors in the development of resistance. There is a long-lasting cutaneous arteriolar vasoconstriction which reduces the ability of keds to obtain blood so that they die from starvation, and a need for adequate amounts of vitamin A in the diet. The fact that *M. ovinus* is more abundant in winter than summer and has not established itself in tropical areas, suggests that temperature may play an important part in the decline of ked populations in the summer.

Economic importance

Infestation with *M. ovinus* does not produce any very marked change in the sheep. Presence of keds leads to wool biting and staining of the wool by ked faeces. Both responses lead to downgrading of the fleece. Very heavy infestations may cause severe anaemia and produce skin blemishes which are costly to the leather industry (Blood and Radostits, 1989). Experiments with ked-free and ked-infested lambs showed that on a diet of alfalfa ked-free lambs gained 3.6 kg more in 4 months and produced 13% more wool. The latter was attributed to a reduction in cutaneous blood flow in lambs developing resistance. When the two groups of lambs were fed a high energy concentrate, differences in weight gain were highly significant after 1 month, but by 4 months, although the ked-free lambs had gained more than the ked-infested group, the difference was not statistically significant (Nelson and Sten, 1968).

Other Hippoboscids of Veterinary Importance

Geographical distribution

Hippobosca longipennis occurs in the western Oriental, southern Palaearctic and Afrotropical regions, excluding West Africa. It parasitizes carnivores, including domestic dogs. *H. equina* is primarily an ectoparasite of horses and cattle in the Palaearctic and western Oriental regions, but has been introduced and become more widely established in south-east Asia, and in some island groups in the Pacific. *H. variegata* parasitizes equines and cattle in the Afrotropical and Oriental regions. No wild hosts of this species are known. *H. rufipes* occurs in the Afrotropical region where it parasitizes wild bovids and domestic cattle, and less frequently, domestic and wild equines. *H. camelina* occurs where camels are present in the northern part of eastern Africa, the Mediterranean region and the southern part of the eastern Palaearctic. *Lipoptena capreoli* parasitizes domestic goats in the eastern Mediterranean region and eastwards through the desert countries to north-west India (Maa, 1963, 1969).

Studies have been made on the biology of *H. longipennis*, and the biology and ecology of *H. equina* by Hafez and his colleagues in Egypt (Hafez and Hilali, 1978; Hafez *et al.*, 1977, 1979). There are many similarities between the two species of *Hippobosca* and they make an interesting comparison with studies on *M. ovinus*. Both species of *Hippobosca* are more abundant in summer and are at low numbers during the winter. Larvae are deposited off the host: those of *H. equina*, in crevices in mud walls of stables; and in keeping with this, *H. equina* is more abundant on stabled animals than free ranging animals. The newly deposited larva is creamy in colour with its flattened posterior end bearing dark spiracular plates. The larva pupariates in 4 to 6 h. The puparium rapidly darkens to a dark red-black colour. It is broadly oval with posterolateral spiracular lobes.

The adults are winged and fly directly to a host. The newly emerged adult does not feed for the first 24 h but thereafter feeds frequently, several times a day in the case of *H. longipennis*. Adult *H. equina* aggregate on the host in areas where the skin is thinner and comparatively hairless. On horses two-thirds of the louse flies were under the tail and around the genitalia, and on cows under the tail and on the udder. On buffalo, 84% of *H. equina* were on the genitalia and inner thighs. The average density of *H. equina* on horses was six to ten times that on cows and buffaloes. Significantly more females than males were bred from puparia of both species of *Hippobosca*. Newly emerged adults take several days (4–11) to become sexually mature, and in *H. equina* males mature more quickly than females, but the reverse is true for *H. longipennis*.

Bionomics

The longevity of both species was about 6 weeks in summer (August) and 8 to 9 weeks in winter, with females living slightly longer than males. The interval between successive larvipositions was shorter in summer than in winter, but combined with the shorter longevity of *H. longipennis* in summer, the average number of larvae produced per female was the same all the year round; but in *H. equina* females were more fecund in winter. Both species showed a marked seasonal change in the duration of the puparium, which increased rapidly during October and November from about 3 weeks to more than 4 months. Thereafter there was a steady decline in its duration from December to June. This change was not closely correlated with temperature.

From the end of November until mid-April the mean weekly temperature ranged from 13 to 18°C without any particular pattern, while the duration of the puparium declined more or less steadily from 130 to 33 days in *H. longipennis*, and a similar change occurred in *H. equina*. The responses of puparia of both species to constant temperature were similar. No adults emerged from puparia kept at temperatures above 32°C. The response over the range of 20–32°C was markedly different between 20–27°C and 25–32°C. A change of 5°C from 30–32°C to 25–27°C increased the duration of the pupa by 36% in *H. longipennis* and 45% in *H. equina*. A similar change from 25–27°C to 20–22°C increased the pupal duration by a factor of three in *H. equina* and by four in *H. longipennis*.

The number of *H. equina* on horses increased rapidly from February to reach a peak in July. This coincided with the highest percentage (57%) of females with third-stage larvae in the uterus. One would have expected the population to go even higher but it did not. The population declined steadily from July to December. In spite of the high percentage of females maturing larvae and the short duration of the puparium at this time of the year, recruitment to the population of *H. equina* on domestic animals was insufficient to maintain the population, which steadily declined. The mean monthly temperature was 29.6°C, the highest of the year, in July and it is possible that there were prolonged periods with temperatures above 32°C which led to a low emergence of adults from puparia.

Control of Hippoboscids

Numbers of *M. ovinus* are reduced substantially by shearing and the remaining keds controlled by the external application of organophosphorus insecticides, synthetic pyrethroids or the oral application of ivermectin. It is important that the insecticide used persists long enough to kill adults emerging from retained puparia (Blood and Radostits, 1989). In India 12 months' control of *H. maculata* on equines was obtained by the application of two litres of 0.005% deltamethrin over the entire body surface (Parashar *et al.*, 1991). Ivermectin controlled both oestrids and *L. cervi* on red and roe deer (Kutzer, 1988).

References

Bequaert, J.C. (1953) The Hippoboscidae or louse flies (Diptera) of mammals and birds. *Entomologica Americana* 32 (33), 1–442.

Blood, D.C. and Radostits, O.M. (1989) *Veterinary Medicine – A Textbook of the Diseases of Cattle, Sheep, Pigs and Horses.* Baillière-Tindall, London.

Evans, G.O. (1950) Studies on the bionomics of sheep ked, *Melophagus ovinus* L., in west Wales. *Bulletin of Entomological Research* 40, 459–478.

Gassner, G. (1989) Dipteran endocytobionts. In: Schwemmler, W. and Gassner, G. (eds) *Insect Endocytobiosis: Morphology, Physiology, Genetics, Evolution.* CRC Press, Boca Raton, Florida, USA, pp. 217–232.

Hafez, M. and Hilali, M. (1978) Biology of *Hippobosca longipennis* (Fabricius, 1805) in Egypt (Diptera: Hippoboscidae). *Veterinary Parasitology* 4, 275–288.

Hafez, M., Hilali, M. and Fouda, M. (1977) Biological studies on *Hippobosca equina* (L.) (Diptera: Hippoboscidae) infesting domestic animals in Egypt. *Zeitschrift für Angewandte Entomologie* 83, 426–441.

Hafez, M., Hilali, M. and Fouda, M. (1979) Ecological studies on *Hippobosca equina* (Linnaeus, 1758) (Diptera: Hippoboscidae) infesting domestic animals in Egypt. *Zeitschrift für Angewandte Entomologie* 87, 327–335.

Hoare, C.A. (1972) *The Trypanosomes of Mammals.* Blackwell, Oxford.

Kutzer, E. (1988) Ektoparasitenbekämpfung mit ivermectin (Ivomec®) bei schalenwild (Rothirsch, Reh, Wildschwein). *Mitteilungen der Deutschen Gesellschaft für allegemeine und angwandte Entomologie* 6, 217–222.

Maa, T.C. (1963) Genera and species of Hippoboscidae (Diptera): types, synonymy, habitats and natural groupings. *Pacific Insects Monograph* 6, 1–186.

Maa, T.C. (1966) Studies in Hippoboscidae (Diptera). *Pacific Insects Monograph* 10, 1–148.

Maa, T.C. (1969) Studies in Hippoboscidae (Diptera). Part 2 *Pacific Insects Monograph* 20, 1–312.

Nelson, W.A. (1956) Mortality in the sheep ked, *Melophagus ovinus* (L.) caused by *Trypanosoma melophagium* Flu. *Nature, London* 178, 750.

Nelson, W.A. (1958) Transfer of sheep keds, *Melophagus ovinus* (L.), from ewes to their lambs. *Nature, London* 181, 56.

Nelson, W.A. and Bainborough, A.R. (1963) Development in sheep of resistance to the ked *Melophagus ovinus* (L.) III. Histopathology of sheep skin as a clue to the nature of resistance. *Experimental Parasitology* 13, 118–127.

Nelson, W.A. and Hironaka, R. (1966) Effect of protein and vitamin A intake of sheep on numbers of the sheep ked, *Melophagus ovinus* (L.). *Experimental Parasitology* 18, 274–280.

Nelson, W.A. and Sten, S.B. (1968) Weight gains and wool growth in sheep infested with the sheep ked *Melophagus ovinus.* *Experimental Parasitology* 22, 223–226.

Parashar, B.D., Gupta, G.P. and Rao, K.M. (1991) Control of the haematophagous fly *Hippobosca maculata*, a serious pest of equines, by deltamethrin. *Medical and Veterinary Entomology* 5, 363–367.

Pfadt, R.E. (1976) Sheep ked populations on a small farm. *Journal of Economic Entomology* 69, 313–316.

Pfadt, R.E., Lloyd, J.E. and Spackman, E.W. (1975) Power dusting with organophosphorus insecticides to control the sheep ked. *Journal of Economic Entomology* 68, 468–470.

Siphonaptera (Fleas) **17**

The Siphonaptera or fleas are wingless, ectoparasites of mammals and birds. In common with other ectoparasites, fleas are flattened to minimize damage from the host's reaction to their presence. Fleas are laterally flattened compared with lice which are dorsoventrally flattened. While lice are host specific, fleas generally parasitize a range of hosts and it is this ability to transfer from one host species to another that makes them of medical importance by transmitting disease from animals, mostly rats, to humans (see Chapter 26). Adult fleas are readily recognized by their habit of jumping when disturbed, their mahogany brown colour, and size (1–6 mm long) (Fig. 17.1). Females are larger than males of the same species.

Traub (1985) considers the hosts of 1790 species of fleas of which 1667 (93%) occur on mammals and only 123 (7%) on birds. [Lewis (1993) gives 2500 species and subspecies.] Fleas that infest mammals can be classified on their range of hosts. About 25% occur on only a single mammalian species or genus. Others have an increasingly wide range of hosts. The medically important fleas are among those that infest more than one mammalian order, e.g. *Tunga penetrans* and *Xenopsylla cheopis*, and those with a very broad host range, e.g. *Echidnophaga gallinacea*, *Ctenocephalides felis* and *Pulex irritans*.

All fleas have basically the same structure and their identification is a matter for the specialist. Some idea as to the complexity of the taxonomy of the Siphonaptera is given by the history of the catalogue of the Rothschild collection of fleas in the British Museum. It is considered to include 85% of the known species (Holland, 1964). It was estimated that it would require five volumes and take eight years (Hopkins and Rothschild, 1953). More than 30 years later, seven volumes, totalling nearly 3200 pages, have been produced, leaving the species of the largest family, the Ceratophyllidae, considered to contain more than 20% of all flea species (Lewis, 1993), still to be described. This would require two more volumes. A companion volume by Traub *et al.* (1983) provides keys to the genera of Ceratophyllidae and notes on species but not descriptions.

Fleas are classified into 15 or 17 families of which the medically most important are the Ceratophyllidae, Ctenophthalmidae, Leptopsyllidae, Pulicidae and

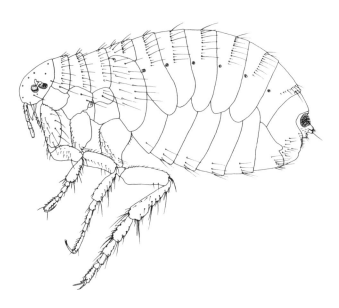

Fig. 17.1. Lateral view of female *Pulex irritans*.

Tungidae. They originated as parasites of mammals, a small number of species, mainly of the genus *Ceratophyllus* have become secondarily adapted to birds (mostly Passeriformes and sea birds). There are about 60 species and subspecies of *Ceratophyllus* in the Holarctic region parasitizing many families of birds but especially the Hirundinidae, swallows and martins (Holland, 1964).

Fleas occur on a wide range of terrestrial mammals, and their life cycle is such that they are particularly associated with mammals that spend part of their life in nests, dens, holes or caves. Fleas are therefore common on rodents, carnivores, bats and rabbits and virtually absent from free-ranging ungulates and primates. It is considered that fleas evolved from a mecopteran-like ancestor in the late Mesozoic and have evolved with the mammals. A study of the coevolution of fleas and their mammalian hosts throws light on the evolution of both groups (Traub, 1985).

The so-called human flea, *Pulex irritans*, is a normal parasite of pigs (Holland, 1964).The genus *Pulex*, which infests diurnal mammals especially porcines, originated in the New World, to which most species in the genus are restricted, with the exception of *P. simulans*, which also occurs in Hawaii, and *P. irritans* which has become cosmopolitan. *P. irritans* is considered to be a late arrival in western Europe (< 14,000 years ago) and to have come from North America via the Beringian land bridge into Asia and on to Europe (Hopla, 1980; Buckland and Sadler, 1989).

The main fleas of medical importance are the tropical rat flea, *X. cheopis*, the main vector of plague and murine typhus to humans; the sand-flea *T. penetrans*, the female of which develops as an endoparasite under the skin of people, particularly on the feet and ankles; and *P. irritans* which breeds in human habitations. Fleas of veterinary importance include the sticktight flea of poultry, *E. gallinacea*;

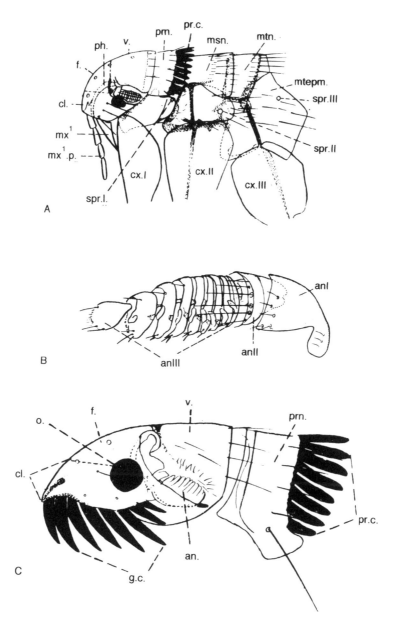

Fig. 17.2. (**A**) Head and thorax of *Nosopsyllus fasciatus* female. (**B**) Antenna of male *N. fasciatus.* (**C**) Head and pronotum of *Ctenocephalides felis.* an, antenna; *anl, anll, anlll,* first, second and third segments of antenna; *cl,* clypeus; *cx.I, cx.II, cx.III,* coxae of legs; *f.,* frons; *g.c.* genal comb; *msn.,* mesonotum; *mtepm.,* metepimeron; *mtn.,* metanotum; *mx¹* maxilla (stipes); *mx¹.p.,* maxillary palp; *o.,* ocellus; *ph.,* pharynx; *pr.c.,* pronotal comb; *prn.,* pronotum; *spr.I, spr.II, spr.III,* thoracic and first two abdominal spiracles; *v.,* vertex. Source: from Patton, W.S. and Evans, A.M. (1929) *Insects, Ticks, Mites and Venomous Animals. Part I: Medical.* H.R. Grubb, Croydon.

Ceratophyllus gallinacea, a pest of poultry and many other species of birds; and the cat flea, *Ct. felis*.

Structure of Adult Flea

Fleas show many adaptations to an ectoparasitic mode of life. Each antenna is recessed in a deep antennal fossa (Fig. 17.2); the neck is foreshortened so that the head is sessile on the prothorax; and the body is covered with backwardly-directed setae (Fig. 17.1) and, in many cases, combs. These features, together with the lateral flattening, enable fleas to move forwards easily through the pelage or feathers of their hosts. The setae and combs would impede the flea being dragged backwards by activities of the host.

Head

The antennae are 3-segmented with the third segment being elaborately developed (Fig. 17.2). In many species the antennae of the males have adhesive disks on the inner surface which are used to hold the female during mating. The maxillary palps are well developed with four obvious segments. There are no compound eyes but lateral ocelli are present on either side of the head. Their development is variable, being particularly large in species of *Xenopsylla*, and reduced in many species or absent, as in the house-mouse flea *Leptopsylla segnis*.

Thorax

The thorax bears three pairs of legs, of which the third pair is particularly well developed for jumping and consequently the metathorax supporting these legs is well developed. In *Xenopsylla* and some other genera the mesopleuron above the coxa of the second pair of legs is divided by the pleural rod into an anterior mesepisternum and a posterior mesepimeron (Fig. 17.3). The pleural rod is absent in *Pulex*, enabling these two combless genera to be distinguished.

Abdomen

The shape of the abdomen may be used to distinguish between the sexes. In female fleas both the ventral and dorsal surfaces are convex and in the male the dorsal surface is more or less flat and the ventral surface greatly curved. In addition male fleas may be distinguished by possession of complex copulatory apparatus posteroventrally in the abdomen. In both sexes there is a sensilium (= pygidium) posteriorly on the dorsal surface (Figs 17.4, 17.5). The antesensilial seta is immediately anterior to the sensilium.

The abdomen has ten segments of which eight are easily recognizable externally and each of these bears a pair of spiracles (Fig. 17.1). In addition there are two pairs of spiracles on the thorax. The ninth abdominal segment is much

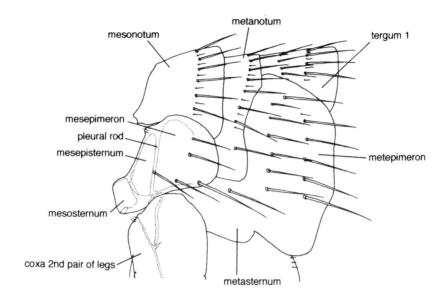

mesonotum
metanotum
tergum 1
mesepimeron
pleural rod
mesepisternum
metepimeron
mesosternum
coxa 2nd pair of legs
metasternum

Fig. 17.3. Lateral view of mesothorax, metathorax and tergum 1 of *Xenopsylla cheopis*.

modified in the male with tergum IX forming paired manubria and articulating claspers, and sternum IX forming an L-shaped clasping organ (Fig. 17.4), the apical arm of which is a useful character in the separation of species of *Xenopsylla*.

There is a single spermatheca in the female (Fig. 17.5). The spermathecal duct opens into the head of the spermatheca which is separated by a small constriction from the tail. The relative sizes of the head and tail are useful characters in separating species of *Xenopsylla*. They are of similar size in *X. cheopis*; the head is considerably larger than the tail in *X. brasiliensis*; and the reverse is true in *X. astia*.

Combs or ctenidia

Many fleas possess combs. The Pulicidae are relatively combless and there are no combs in *Xenopsylla*, *Pulex*, *Echidnophaga* or *Tunga* (Figs 17.1, 17.7) but there are both genal and pronotal combs with backwardly-directed teeth in *Ctenocephalides* (Fig. 17.2). Combs are particularly well developed in fleas that parasitize bats where there may be metathoracic and abdominal combs present. In the ceratophyllid genera, *Ceratophyllus* and *Nosopsyllus*, only the pronotal comb is developed (Fig. 17.2). The number of spines in the pronotal comb varies with the type of host. In bird fleas, e.g. *Ceratophyllus* spp., the spines are narrower and more numerous, exceeding 24, while in parasites of mammals, e.g. *Nosopsyllus* spp., the spines are broader and less numerous, less than 24 (Holland, 1964).

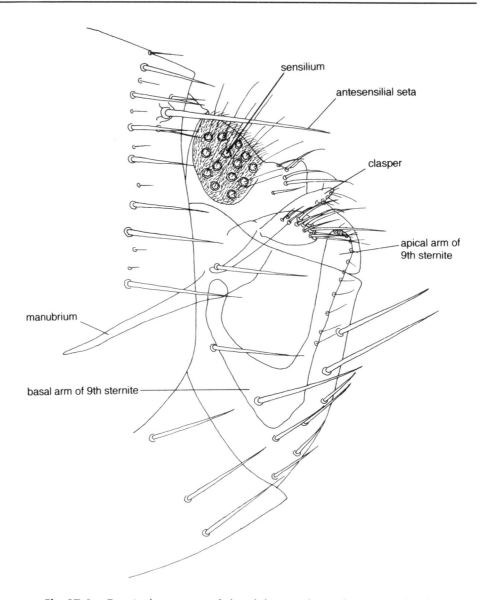

Fig. 17.4. Terminal segments of the abdomen of a male *Xenopsylla cheopis*.

Life Cycle (see Fig. 17.6)

Egg

The female flea produces relatively large (0.3–0.5 mm) whitish, oval eggs, which are sticky in *X. cheopis* and dry in *T. penetrans* and *E. gallinacea* (Suter, 1964). Eggs are deposited in the nest, or on the host from where they fall to the ground. Female *Ct. felis* have six ovarioles in each ovary of which half contain mature

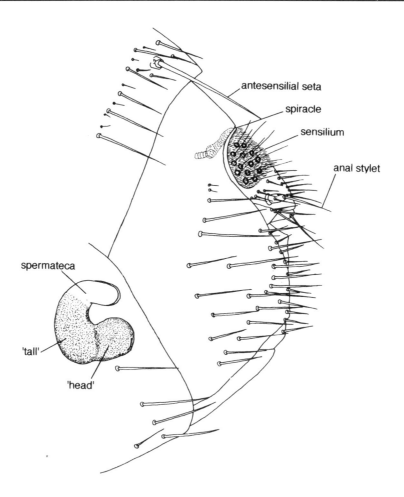

Fig. 17.5. Terminal segments of the abdomen of a female *Xenopsylla cheopis.*

oocytes. Egg production begins 2 days after the first blood meal and reaches a maximum in 6 to 7 days with a female producing on average 13.5 eggs per day and 158 in a lifetime (Osbrink and Rust, 1984). These results were obtained from single females confined with five males in microcells on a cat. Unconfined fleas on a female cat averaged 24 eggs day^{-1} (Hinkle *et al.*, 1991). The majority of the eggs of *Ct. felis* are laid during the last 8 h of the scotophase (Dryden and Rust, 1994).

Eggs hatch in a few days provided that the humidity is above 70%. At 80 % RH eggs of *X. brasiliensis* hatch in 6 days at 24°C and 4 days at 35°C (Edney, 1945) and those of *E. gallinacea* in three to four days at 26°C and 85 %RH. Eggs that lose water when exposed to humidities less than 70% RH are unable to recover when placed in a higher humidity (Suter, 1964). The threshold temperature for egg development is a specific character, being 12°C for the tropical rat flea, *X. cheopis*, and 5°C for the temperate region European rat flea, *Nosopsyllus fasciatus* (Bacot, 1914).

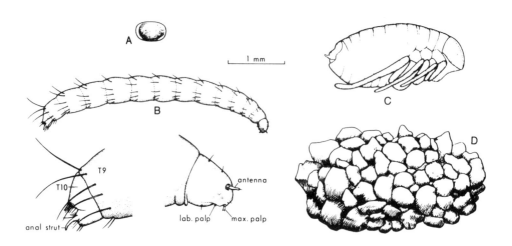

Fig. 17.6. Life cycle of *Ctenocephalides felis*: (**A**) egg; (**B**) larva with caudal and head ends enlarged; (**C**) pupa; (**D**) sand-encrusted cocoon. Source: from Dunnet and Mardon (1991).

Larva (see Fig. 17.6B)

The larva uses a hatching spine to emerge from the egg. It has a distinct head and 13 body segments with no distinction between thoracic and abdominal segments, and no appendages. Although the larva has no eyes it is negatively phototactic, burrowing into the material of the nest or substrate. The whitish, vermiform larva measures 4–10 mm when fully grown and the body segments bear a circlet of backwardly-directed bristles which, together with the anal struts on the last segment (Fig. 17.6B), enable the larva to be vigorously active. Pilgrim (1991, 1992) has found a range of external morphological features which can be used to separate flea larvae into families, species and subspecies. There are three larval instars in most species but only two in *T. penetrans* (Suter, 1964).

Larvae feed on organic debris in their environment and this is supplemented in many species by adults passing undigested blood through the anus when feeding and by the faeces of the adult. The faeces of adult *Ct. felis* contain 7 to 11% protein (Hinkle *et al.*, 1991). In *N. fasciatus* the interaction of larva and adult flea has become much closer. Larvae actively pursue and seize adult fleas with their mandibles in the region of the sensilium. The adult responds by defecating after which the larva releases its hold on the adult and sucks the excreted faecal blood. The pharynx of the larva is muscular and larvae of *N. fasciatus* imbibe blood, water and rat urine. These larvae are semipredatory and will attack damaged adults and kill them (Molyneux, 1967). Adult *Spilopsyllus cuniculi* normally defecate every 20 min, but the frequency is greatly increased shortly before oviposition, presumably to provide a more favourable environment for larval development in the rabbit burrow (Rothschild, 1965). There are specific differences in the nutritional requirements of flea larvae of the same genus. Larvae of *X. astia* require a more nutritive diet than those of *X. cheopis* and *X. brasiliensis*. Larvae of *X.*

cheopis cannot develop solely on a diet of blood, and need to supplement it with food containing vitamins of the B group (Sharif, 1948).

Larvae of *X. cheopis* lack a closing mechanism on their spiracles (Mellanby, 1934), and consequently they require high humidities for development. Exposure to 0% RH and 22°C for 24 h is lethal but at 90% RH the lethal temperature is increased to 36°C (Mellanby, 1932). Larvae of *X. cheopis* are hygropositive and move to zones of high humidity (Yinon *et al.*, 1967). The length of the larval stage is dependent on temperature and humidity, and at 24°C its duration in *X. brasiliensis* increases from 12 to 25 days as the relative humidity decreases from 93 to 70% (Edney, 1945). At 25°C and 85% RH, Suter (1964) found similar rates of development in four species of Pulicidae with the larval stages being completed in 1 to 2 weeks. Bacot (1914) found considerable variation in the speed of development of four species of fleas even when larvae from the same batch of eggs were reared under identical conditions.

Pupa

The mature third stage larva empties its gut, enters the prepupal phase, and constructs a thin, loosely woven cocoon, which is typically ovoid, measuring about 3 mm long by 1 mm in width and height. Humphries (1967b) followed the development of three species of *Ceratophyllus* and *Ct. felis* within the cocoon. He found that for several days the larva remained motionless in a doubled up position. It then pupated and remained quiescent until it emerged as an adult. The main threat to adult emergence was desiccation, reducing body volume and preventing the shedding of the pupal exuviae. If these shrunken pharate adults were liberated from their pupal exuviae they were capable of normal locomotion, surviving for several days (one even took a blood meal).

Maintenance of a high water content is a key factor in the successful emergence of the adult flea. Although flea larvae readily lose water in a dry atmosphere, prepupae of *X. brasiliensis* actually gain water when maintained in an atmosphere of 50–90% RH but desiccate and die at all temperatures when the humidity falls below 45%. The average gain in weight before pupation is 14%. The cocoon itself offers no protection against desiccation. Pupae of *X. brasiliensis* lose water but at a very slow rate (Edney, 1947). If the same is true for pupae of *X. cheopis* then the rate of loss must be exceptionally low because adults emerge from pupae kept at 0% RH and 30°C (Mellanby, 1933).

Temperature is the other factor affecting development. Successful pupation and emergence of *X. cheopis* occurred at 18–35°C and 60% or higher RH, with the prepupa lasting 4 days at 35°C and 8 days at 18°C (Mellanby, 1933). At lower temperatures pupation was variable with none occurring at 10 and 14°C, and reduced pupation at 15°C and 80% RH with many prepupal deaths (Margalit and Shulov, 1972). As with the threshold temperature for egg development, the optimum temperature for pupal development is about 10°C higher for *X. cheopis* than *N. fasciatus* (Bacot, 1914).

Adult

Adults of *C. gallinae* emerge from the cocoon by using the frontal tubercle on the head to weaken the fibres of the cocoon (Humphries, 1967b). The tubercle may be lost later in adult life. Females of *X. cheopis*, *X. brasiliensis* and *E. gallinacea* emerge 3 to 4 days before the males (Edney, 1945; Suter, 1964). In view of the importance of water conservation to survival of the adult flea, it is of interest to learn that *C. gallinae* is able to take up water from air with a humidity in excess of 82% RH, but this ability is only present during the first day of adult life (Humphries, 1967d). Edney (1945, 1947) has shown that unfed, newly emerged adult *X. brasiliensis* are shorter lived if they have been reared at high temperatures or low humidities. The duration of the period spent in the cocoon between prepupa and the emergence of the adult varies from 1 week to 6 to 12 months. This is a major factor in the survival of flea populations during the absence of a host or adverse climatic conditions. Adult fleas may be long lived, e.g unfed *S. cuniculi* can survive for 9 months at $-1°C$ (Rothschild, 1965). *Ct. felis* must feed every 12 h in order to survive and reproduce (Hinkle *et al.*, 1991). In microcells the average life of adult *Ct. felis* was 11.2 days for females and 7.2 days for males. Few (10%) female *Ct. felis* confined on a cat lived longer than 6 weeks (Osbrink and Rust, 1984).

Unusual life cycle of *Uropsylla tasmanica*

The life cycle of the flea *Uropsylla tasmanica*, a parasite of dasyurids (marsupials), is most unusual. The eggs are cemented on to the hairs of the host and the newly emerged larvae penetrate the host's skin with their large mandibles and are endoparasitic on their host, living in burrows extending into the dermis. They possess well-developed mandibular glands which are not found in free-living larvae (Williams, 1986). When mature, they drop to the ground and spin a cocoon in the normal way (Dunnet and Mardon, 1991).

Adult Behaviour and Bionomics

Host finding

Adult fleas feed only on blood, and the newly emerged flea must find a host. This may be achieved by dispersing actively, e.g. *C. gallinae* and *C. styx*, in search of a host, or by waiting in the nest or burrow for the return of the host, e.g. *N. fasciatus* and the rabbit flea *S. cuniculi*. Both groups of fleas need to be able to detect a host when it is near, to orientate to it and achieve contact. Various attributes of the hosts provide stimuli to which the flea may respond. These include vibration, warmth, exhalation of carbon dioxide, characteristic odour and, in lighted situations, casting a shadow. According to Askew (1971) fleas are able to detect air currents by receptors on the sensilium. Female *Ct. felis* respond to visual and thermal stimuli. Heat with air currents created by a warm moving body attracted fleas in the absence of visual stimuli. Females less than 4 days old were less responsive than 5- to 6-day-old females (Osbrink and Rust, 1985b). Suter (1964) found that three species of fleas, including *X. cheopis*, were positively phototactic when unfed but became negatively phototactic when fed. This does not occur in the female sticktight flea, *E. gallinacea*, which may remain attached to heads of poultry for several weeks. *Ct. felis* is most responsive to green light (510–550 nm) (Dryden and Rust, 1994).

In an olfactometer adult *X. cheopis* responded positively to the odour from a white rat at a distance of 30 cm, and distinguished between its odour and that of three other murids to which *X. cheopis* do not respond (Shulov and Naor, 1964). Some indication of the ability of fleas to find their hosts is given by an experiment with *S. cuniculi* in which 270 marked adults were released in an enclosed area of 1800 m². Three rabbits released into the area had picked up 45% of the marked fleas within a few days (Rothschild, 1965). The same species was released on wild rabbits in various locations in New South Wales, Australia, for the spread of myxomatosis virus among rabbit populations. The flea has spread slowly, the fastest rate observed being 8–10 km a year (Rothschild, 1975).

The host-finding behaviour of *C. gallinae* and *C. styx* has been studied in detail. *C. gallinae* parasitizes more than 75 species of birds, while *C. styx* is restricted to the sand martin. *C. gallinae* overwinters as an adult in the cocoon and emerges in the spring in response to a sharp rise in temperature and/or tactile stimuli. The flea emerges in an old nest which will not be reoccupied. At first the newly emerged adult is negatively phototactic but after 3 or 4 days becomes positively phototactic, which, associated with negative geotaxis, ensures that the adults crawl up trees and bushes. In search of a host the flea stops periodically and faces towards the brightest source of light, and jumps when the light is suddenly obscured. Readiness to jump rises to a peak 4 days after emergence, coinciding with the positive response to light, but falls off later due in part to water loss (Humphries, 1968). In new housing estates in the west of Scotland, where trees and shrubs were scarce and small, dispersing *C. gallinae* attacked people out of doors (Hosie, 1980).

Adult *C. styx* emerging from old sand martin burrows, congregate in the entrance to the burrows in the spring at the time of arrival of the migrant host. Sometimes sand martins hover in front of old burrows, and *C. styx* responds to the vibration and jumps on to the hovering bird. Burrows are not reoccupied and the fleas disperse both laterally and vertically up to 34 m from the old burrow. They appear to respond to the horizontal floor of the burrow, but as they do not collect on the cliff top they must distinguish between the two horizontal surfaces, but the way in which that is done is not known (Bates, 1962).

Feeding

Adult fleas are capillary feeders. Lavoipierre and Hamachi (1961) studied the feeding of three species of fleas including *X. cheopis* and *P. irritans*. They found that the maxillae are used to penetrate the host's skin, and the tip of the labrum epipharynx enters a capillary from which the flea imbibes blood. Saliva is passed into the host by the salivary pump and appears as clear drops of fluid outside the capillary. The saliva of *X. cheopis* contains an anticoagulant, and a material of low molecular weight which provokes allergenic activity in the host (Rothschild, 1975).

In *Ct. canis* there are three pumps, cibarial, precerebral and postcerebral, to convey blood to the midgut (Munshi, 1960). *S. cuniculi* is probably a pool feeder and at times *X. cheopis* may feed in the same manner. *Ct. felis* feeds to repletion in 10 min imbibing 7 µl of blood, and doubling its weight. Without further feeding, the increased weight was lost in 12 h but the protein content was double that found in the unfed flea (Hinkle *et al.*, 1991). Feeding is more frequent at higher temperatures as a result of accelerated physiological activity and increased rate of water loss. The latter can be replaced by drinking water and Humphries (1966) has observed seven species of fleas, belonging to five genera, drinking water and increasing their weight by 13%. The volume of liquid imbibed varies with its composition. When *X. cheopis* was offered blood, plasma or distilled water the relative respective weights imbibed were $4:2:1$ with the increased uptake of whole blood being attributed to the presence of ATP on the red cells (Galun, 1966).

Role of proventriculus in feeding and digestion

Blood passes directly to the midgut, which has the dual functions of initial storage and digestion. The proventriculus is particularly well developed, and is easily recognized in whole mounts of fleas as a mass of needle-like spines at the junction of the thorax and abdomen. Morphologically these are not true spines but modified setae (Rothschild, 1975). Three regions are readily discernible in the proventriculus of *X. cheopis*: a ridged, spineless anterior zone; a middle spined region; and a posterior spineless region which protrudes into the midgut. The spines are arranged in a regular series. In females there are 15 rows of 30 spines each, and in males 12 rows of 22 spines each. The most anterior spines are short, being only 10 µm long. The spines steadily increase to a maximum of 70 µm in the central and posterior regions with the last two rows being somewhat shorter (Munshi, 1960).

The function of the spines appears to be to keep an open passage from foregut to midgut facilitating rapid feeding. In addition the spines play a role in the fragmentation of red blood cells. Three to five peristaltic waves are followed by one antiperistaltic wave which thrusts the blood forwards against the spines. The spines do not penetrate into the midgut but only into the posterior spineless zone of the proventriculus. Other actions which contribute to breaking up the erythrocytes are the to and fro shifting movements of the proventriculus itself, and the contractile action of the posterior region of the proventriculus. The anterior half of the spined region of the proventriculus serves as an effective barrier to regurgitation.

Mating

Some fleas, especially bird fleas, mate on emergence before taking a blood meal, but most fleas, especially females, require a considerable period of feeding before mating (Suter, 1964; Rothschild, 1975). In *C. gallinae* the initial contact is fortuitous, and recognition is achieved by a contact pheromone detected by receptors on the male palps. The pheromone is species-specific and is present in both sexes. The male erects his antennae and moves under the female grasping her with the adhesive organs on their inner surfaces. Correct alignment of the pair, and successful coupling, is probably assisted by receptors on the sensilium and the antesensilial seta. Movement of the female is inhibited by pressure of the ninth sternum of the male on hairs of the female's sensilium. The aedeagus or intromittent organ of the male is used to dilate the female's genital chamber, and one penis rod is inserted into the spermathecal duct and may reach the spermatheca. Penetration of the rod is slow, and copulation may last for up to 9 h with the average time in this species being 3 h (Humphries, 1967a, c).

In *S. cuniculi* the two penis rods operate together with the thicker rod, which penetrates to the female genital chamber, acting as a guide to the thinner rod which enters the spermathecal duct with sperm wound around its terminal portion (Rothschild, 1965). In species where the female is sessile or semi-sessile, e.g. *E. gallinacea* and *S. cuniculi*, the antennae of the male lack adhesive disks; copulation occurs while the female is feeding (Geigy and Suter, 1960; Rothschild and Hinton, 1968). After the male *E. gallinacea* is coupled with the female, it may lose all contact with the host's skin and its legs be free in the air. When, after mating, the male is returned to the skin surface it may take a blood meal. In *T. penetrans* mating occurs when the female is endoparasitic within the skin of its host. The male genitalia are modified to achieve mating with an almost inaccessible female. During copulation the male takes a blood meal from the host. Geigy and Suter (1960) give times of mating for *X. cheopis* (10 min), *E. gallinacea* (15 min) and *T. penetrans* (20 min).

Control of mating in *S. cuniculi* is a complex process dependent on the physiological state of the host. Maturation of the female and maximum maturation of the male only occur when the fleas feed on a pregnant doe rabbit or its newborn young, 1 to 10 days old (Rothschild, 1975). The reproductive cycles of both flea and host are therefore closely coordinated, a possibly unique relationship.

Reproduction

In Portugal the numbers of *S. cuniculi* on wild-caught rabbits reach a peak in February, coinciding with the hormonal cycle of the rabbits which peaks between January and March. These rabbits are also parasitized by *X. cunicularis* which attains its maximum numbers in September and October. This indicates either that *X. cunicularis* reproduces independently of the hormonal cycle of its host, or that its dependency on the host cycle is completely different. The first explanation is the more likely which would bring *X. cunicularis* in line with *X. cheopis* and *X. astia* (Abreu, 1980).

In the tropics and subtropics it is likely that fleas breed continuously throughout the year, although it may be moderated by very hot or very dry conditions. In southern California *Ct. felis* is more abundant from May to November than in the cooler part of the year, December to April (Osbrink and Rust, 1985a). In China *Citellophilus tesquorum* overwinters on its ground squirrel host (*Citellus dauricus*) feeding intermittently when the host temporarily wakes up and its body

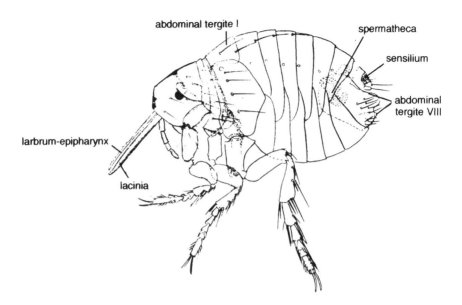

Fig. 17.7. Female *Echidnophaga gallinacea*. Source: from Patton, W.S. and Evans, A.M. (1929). *Insects, Ticks, Mites and Venomous Animals. Part I: Medical*. H.R. Grubb, Croydon.

temperature rises. This enables the flea to mature and deposit eggs while its host is hibernating (Li Chengyi *et al.*, 1994).

In *E. gallinacea* egg production rose for 8 days after mating to a peak of more than 20 eggs per day. Egg production then declined steadily for the next 3 weeks to less than five eggs per day. After a second mating, egg production rose for nearly 3 weeks when the experiment was terminated (Suter, 1964). Oviposition ceased when all the sperm had been used.

Relationship with hosts

Fleas that infest diurnal hosts have well-developed eyes, and those that remain in the host's nest have reduced eyes, thorax and legs in keeping with their more sedentary life style. The geographical distribution of a flea species is limited by the distribution of its hosts, the need for suitable larval habitats and its evolutionary history (Holland, 1964).

In sticktight fleas the mouthparts are relatively much longer than they are in more mobile fleas. In *E. gallinacea* (Fig. 17.7) the mouthparts are one-third the length of the body, whereas in *P. irritans* and *X. cheopis* their length is only 10–20% of the body length. In addition the laciniae in *Echidnophaga* are strongly toothed and serve to anchor the flea in place (Fig. 17.7). These adaptations are highly successful and when a chicken was artificially infested with *E. gallinacea*, 50% of the females were still attached at the end of 2 weeks, and some remained attached for periods up to 6 weeks (Suter, 1964).

The relationship between a flea, its host and environmental conditions is complex. Under three different sets of conditions – 16°C (60% RH), 24°C (50% RH) and 28°C (70% RH) – populations of *L. segnis* varied according to the rodent host. On the house mouse (*Mus musculus*) and *Apodemus sylvaticus* the populations were markedly higher at 16°C and 24°C than at 28°C; on *Apodemus agrarius* they declined at 16°C but thrived at 24°C and 28°C; while on *Mastomys natalensis* the population declined at 16°C and 24°C, and at 28°C produced the highest growth rate recorded for any of the nine hosts used in these experiments (Krampitz, 1980).

Fleas have evolved together with their hosts and show certain adaptations, but there is no unanimity on their interpretation. Marshall (1980) considers that the main structures for attachment are the claws and mouthparts; the function of the combs being not to prevent dislodgment but to protect highly mobile joints and their associated membranes. Traub (1980, 1985) considers that the combs and bristles on fleas not only protect and aid progress through the host's pelage but maintain hold on the host. Fleas with well-developed genal combs have flying or gliding hosts, or hosts that are both nocturnal and climbing, or have vast home ranges. Hosts such as hedgehogs and porcupines have fleas with bristles which are widely spaced and/or exceptionally thickened and may also have highly modified combs.

Jumping

One of the most distinctive features in the behaviour of fleas is their ability to jump. The rat flea (? *X. cheopis*) averages 18 cm, presumably horizontally, and a maximum of 31 cm in a single jump (Rothschild, 1965). The hen flea, *C. gallinae*, jumps up to 24 cm horizontally and 11 cm vertically (Humphries, 1968). There is considerable variation in the ability of different flea species to jump. The nestdwelling, semi-sessile rabbit flea, *S. cuniculi*, jumps a mere 3.5 cm vertically while the human flea, *P. irritans*, was observed to jump a vertical height of 13 cm, and is considered to be able to reach 20 cm.

A flea cannot jump 'to order' when prodded, because direct muscular action is not able to deliver the required energy in the time, and over the distance involved. The flea has to be pre-set before being able to jump. Jumping is carried out by the third pair of legs with the other two pairs of legs acting chiefly as supports, at least in *X. cheopis*. The femur of the third pair of legs is rotated to a vertical position and connected to the substrate by the trochanter and tibia. On jumping, the femur rotates downwards transmitting its thrust via the tibia to the substrate, and the flea jumps. The tarsi and claws play no part in jumping. Energy for jumping is stored in the pleural arches which are laterally placed pads of resilin. In *S. cuniculi* 2.1 ergs can be stored in each pad giving a total of 4.2 ergs, compared with 2.25 ergs required for the flea to jump 3.5 cm vertically (Bennet-Clark and Lucey, 1967; Rothschild *et al.*, 1972).

The flea is pre-set by muscular contractions which engage certain cuticular catches, and compress the pleural arch. When the catches are engaged the muscle which compressed the resilin can be relaxed. The jump is initiated by the relaxation of certain muscles which enable the femur to descend, and simultaneously the muscles holding the catches in place relax, releasing the energy stored in the

resilin, and arched pleural and coxal walls (Rothschild *et al.*, 1972). During jumping the flea may turn over and the legs be held out. In *N. fasciatus* the second
pair of legs may extend above the dorsal surface of the body (Rothschild, 1965).
This will increase the probability of the flea holding on to a host should it
encounter one during its jump.

Biology and Control of *Ctenocephalides felis*

In the laboratory the life cycle of *Ct. felis* takes 14 days at 32°C and 140 days
at 13°C, providing the humidity is above 50%. Larvae are most susceptible to
humidity and require a minimum of 50% RH for 50% survival. The eggs are more
tolerant, requiring 33% RH, and while prepupae have the same humidity
requirements as larvae, adults will emerge from 80% of pupae at 2% RH (Silverman *et al.*, 1981). At 30°C and 78% RH maximum pupation occurred 7 days after
eggs, less than 24 h old, were added to larval medium (Hinkle *et al.*, 1992). Under
slightly different conditions (27°C; 75% RH) development from egg to adult was
17 to 21 days with 79% emergence (Moser *et al.*, 1991).

The cat flea has become a major domestic pest in North America and Australia
with pest control operators in Florida providing routine domestic treatments at
intervals of 1 to 3 months (Hinkle, 1991). Dogs are more allergic to flea saliva
than cats but both can be severely affected by infestations of *Ct. felis* (Whiteley,
1987a, b). Control measures involve treating the infested animal, its contacts, and
its environment. The host exerts its own pressure on the parasite and cats remove
50% of their flea population in a week (Wade and Georgi, 1988). The pelage of
an infested animal can be treated by applying an insecticidal dust, spray, shampoo
or dip. In severe cases 20% fenthion may be applied to the dorsum of a dog's back
after warning those associated with the animal of the toxicity of the application
(Whiteley, 1987a, b). Systemically-active insecticides, e.g. cythioate, and insect
growth regulators, such as lufenuron, can effectively stop flea reproduction for
6 weeks. Larvae are extremely sensitive to desiccation and most outdoor environments do not provide suitable conditions for their survival, which requires a
relative humidity of more than 50%, a soil moisture content of less than 20% and
temperature within the range of 4 to 35°C (Dryden and Rust, 1994).

Indoors pressurized sprays containing an insect growth regulator (IGR), e.g.
fenoxycarb or methoprene, and a organophosphorus or pyrethroid insecticide gave
excellent control for 2 months (Osbrink *et al.*, 1986). The effect of methoprene
is to kill larvae in the third instar while another IGR, diflubenzuron which inhibits
chitin synthesis, kills in all instars but must have been acquired by early second
instar (Moser *et al.*, 1992). Pre-emerged adults in cocoons provide a source of
reinfestation. Some directed sprays and aerosols achieved more than 70% control
of pre-emerged adults with organophosphorus insecticides being more effective
than carbamates, which in turn were superior to pyrethroids. In many cases the
effect of spraying was to accelerate emergence. This was not due to disturbance
because the control, a water spray, did not have this effect (Rust and Reierson,
1989).

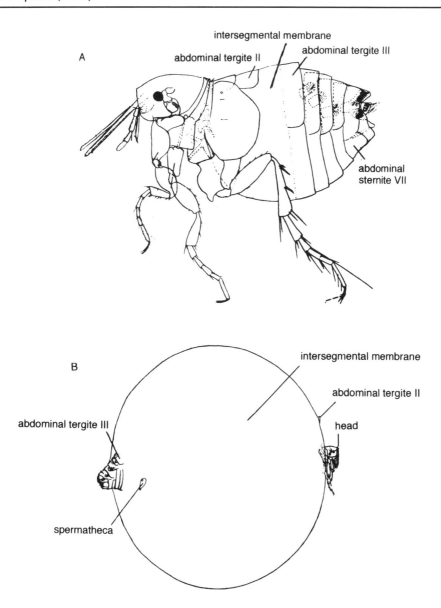

Fig. 17.8. Female *Tunga penetrans*: (**A**) recently fertilized female with the intersegmental membrane between segments II and III beginning to be stretched; (**B**) gravid female, fully expanded. Source: from Patton, W.S. and Evans, A.M. (1929). *Insects, Ticks, Mites and Venomous Animals. Part I: Medical.* H.R. Grubb, Croydon.

Medical and Veterinary Importance

The most important human diseases associated with fleas are plague and murine typhus, which are dealt with in Chapters 25 and 26. A number of fleas, e.g. *P. irritans*, can achieve pest status through the irritation caused by their bites. When domestic cats and dogs are present in a house *Ct. canis* and *Ct. felis* can reach pest proportions. These fleas can also act as intermediate hosts of certain cestodes, including the dog and cat tapeworm *Dipylidium caninum*. Severe infestations of calves, lambs and kids by *Ct. felis* were found among young ruminants maintained in barns on straw bedding, which provided ideal breeding sites for fleas introduced by farm dogs and cats. Lambs and kids were more severely affected than young calves but deaths occurred in all three unusual hosts (Yeruham *et al.*, 1989).

Ceratophyllus gallinae and *E. gallinacea* can be important pests of poultry. Both species have a wide range of natural hosts. *C. gallinae* evolved in the northern hemisphere, but has been taken round the world by humans with poultry. *E. gallinacea* is widely distributed in the warmer countries of the world. These fleas attach themselves to the heads of poultry and may occur in clusters of 100 or more on the comb, wattles, back of the head and round the eyes and beak. This sticktight flea is a serious pest as large numbers cause a progressive anaemia and emaciation leading to lowered egg production in laying hens, and death in young birds (Seddon and Albiston, 1967).

Tunga penetrans (Fig. 17.8), the sand flea, jigger or chigoe, is an important parasite of humans in the Neotropical and Afrotropical regions where people often walk about barefooted. *T. penetrans* was originally a parasite of pigs in South America and was introduced into West Africa in the middle of the nineteenth century, probably with the slave trade. It is now widespread in tropical Africa and has reached Madagascar. The male *T. penetrans* is very small and free-living. The female burrows into the skin of the feet and ankles of human beings. It begins as one of the smallest of the fleas, but then undergoes considerable hypertrophy of the abdomen, particularly the second and third abdominal segments, becoming the shape and size of a pea (Fig. 17.8). Such a radical change during a stage in the life cycle, called neosomy, is associated with the development of giant polyploid cells (Audy *et al.*, 1972; Rothschild, 1988). The female feeds head down and consequently the spiracles on abdominal segments 5–8 are very large, while the other abdominal spiracles are not developed. The female matures about 200 eggs which are passed out to the exterior. They hatch to produce a larva which follows the normal cycle of development. The presence of a number of adult *T. penetrans* in the foot can be crippling, and the damage to the skin can facilitate the entry of other pathogens leading to secondary infection and ulceration.

References

Abreu, M.H. (1980) Quelques aspects particuliers du cycle annuel d'infestation du lapin de garenne par deux espèces de puces. In: Traub, R. and Starcke, H. (eds) *Fleas*. A.A. Balkema, Rotterdam, pp. 391–396.

Askew, R.R. (1971) *Parasitic Insects*. Heinemann, London.

Audy, J.R., Radovsky, F.J. and Vercammen-Grandjean, P.H. (1972) Neosomy: radical intrastadial metamorphosis associated with arthropod symbioses. *Journal of Medical Entomology* 9, 487–494.

Bacot, A. (1914) A study of the bionomics of the common rat fleas and other species associated with human habitations, with special reference to the influence of temperature and humidity at various periods of the life history of the insect. *Journal of Hygiene* 13 *Plague Supplement* III, 447–652.

Bates, J.K. (1962) Field studies on the behaviour of bird fleas. I. Behaviour of the adults of three species of bird fleas in the field. *Parasitology* 52, 113–132.

Bennet-Clark, H.C. and Lucey, E.C.A. (1967) The jump of the flea: a study of the energetics and a model of the mechanism. *Journal of Experimental Biology* 47, 59–76.

Buckland, P.C. and Sadler, J.P. (1989) A biogeography of the human flea, *Pulex irritans* L. (Siphonaptera: Pulicidae). *Journal of Biogeography* 16, 115–120.

Dryden, M.W. and Rust, M.K. (1994) The cat flea – biology, ecology and control. *Veterinary Parasitology* 52, 1–19.

Dunnet, G.M. and Mardon, D.K. (1991) Siphonaptera (Fleas). In: *The Insects of Australia: A Textbook for Students and Research Workers*. Melbourne University Press, Melbourne, pp. 705–716.

Edney, E.B. (1945) Laboratory studies on the bionomics of the rat fleas, *Xenopsylla brasiliensis*, Baker and *X. cheopis*, Roths I. Certain effects of light, temperature and humidity on the rate of development and on adult longevity. *Bulletin of Entomological Research* 35, 399–416.

Edney, E.B. (1947) Laboratory studies on the bionomics of the rat fleas, *Xenopsylla brasiliensis*, Baker and *X. cheopis*, Roths. II. Water relations during the cocoon period. *Bulletin of Entomological Research* 38, 263–280.

Galun, R. (1966) Feeding stimulants of the rat flea *Xenopsylla cheopis* Roth. *Life Sciences* 5, 1335–1342.

Geigy, R. and Suter, P. (1960) Zur Copulation de Flöhe. *Revue Suisse de Zoologie* 67, 206–210.

Hinkle, N.C. (1991) Flea control: attitudes and experiences of PCOs. *Pest Management* June, 15.

Hinkle, N.C., Koehler, P.G. and Kern, W.H. (1991) Haematophagous strategies of the cat flea (Siphonaptera: Pulicidae). *Florida Entomologist* 74, 377–385.

Hinkle, N.C., Koehler, P.G. and Patterson, R.S. (1992) Flea rearing *in vivo* and *in vitro* for basic and applied research. In: Houck, T. and Knecht, T. (eds) *Advances in Insect Rearing for Research and Pest Management*. Westview Press, San Francisco, pp. 119–131.

Holland, G.P. (1964) Evolution, classification, and host relationships of Siphonaptera. *Annual Review of Entomology* 9, 123–146.

Hopkins, G.H.E. and Rothschild, M. (1953) *An Illustrated Catalogue of the Rothschild Collection of Fleas (Siphonaptera) in the British Museum (Natural History). I. Tungidae and Pulicidae*. British Museum, Natural History, London.

Hopla, C.E. (1980) A study of the host associations and zoogeography of *Pulex*. In: Traub, R. and Starcke, H. (eds) *Fleas*. A.A. Balkema, Rotterdam, pp. 185–207.

Hosie, G. (1980) Observations on the occurrence of *Ceratophyllus gallinae* around new housing estates in the west of Scotland. In: Traub, R. and Starcke, H. (eds) *Fleas*. A.A. Balkema, Rotterdam, pp. 415–420.

Humphries, D.A. (1966) Drinking of water by fleas. *Entomologist's Monthly Magazine* 102, 200–201.

Humphries, D.A. (1967a) The mating behaviour of the hen flea *Ceratophyllus gallinae* (Schrank) (Siphonaptera: Insecta). *Animal Behaviour* 15, 82–90.

Humphries, D.A. (1967b) The behaviour of fleas (Siphonaptera) within the cocoon. *Proceedings of the Royal Entomological Society of London A* 42, 62–70.

Humphries, D.A. (1967c) The action of the male genitalia during the copulation of the hen flea, *Ceratophyllus gallinae* (Schrank). *Proceedings of the Royal Entomological Society of London A* 42, 101–106.

Humphries, D.A. (1967d) Uptake of atmospheric water by the hen flea *Ceratophyllus gallinae* (Schrank). *Nature, London* 214, 426.

Humphries, D.A. (1968) The host-finding behaviour of the hen flea *Ceratophyllus gallinae* (Schrank) (Siphonaptera). *Parasitology* 58, 403–414.

Krampitz, H.E. (1980) Host preference, sessility and mating behaviour of *Leptopsylla segnis* reared in captivity. In: Traub, R. and Starcke, H. (eds) *Fleas*. A.A. Balkema, Rotterdam, pp. 371–378.

Lavoipierre, M.M.J. and Hamachi, M. (1961) An apparatus for observations on the feeding mechanism of the flea. *Nature, London* 192, 998–999.

Lewis, R.E. (1993) Fleas (Siphonaptera). In: Lane, R.P. and Crosskey, R.W. (eds) *Medical Insects and Arachnids*, Chapman & Hall, London, pp. 529–575.

Li Chengyi, Wu Houyong, Fei Rongzhong and Wang Zhigang (1994) Studies on overwintering of the flea *Citellophilus tesquorum sungaris*. In: Wu Houyong and Liu Quan (eds) *Researches on Fleas*. China Science and Technology Press, Beijing, pp. 12–25.

Margalit, J. and Shulov, A.S. (1972) Effect of temperature on the development of prepupa and pupa of the rat flea, *Xenopsylla cheopis* Rothschild. *Journal of Medical Entomology* 9, 117–125.

Marshall, A.G. (1980) The function of combs in ectoparasitic insects. In: Traub, R. and Starcke, H. (eds) *Fleas*. A.A. Balkema, Rotterdam, pp. 79–87.

Mellanby, K. (1932) The influence of atmospheric humidity on the thermal death point of a number of insects. *Journal of Experimental Biology* 9, 222–231.

Mellanby, K. (1933) The influence of temperature and humidity on the pupation of *Xenopsylla cheopis*. *Bulletin of Entomological Research* 24, 197–202.

Mellanby, K. (1934) The site of loss of water from insects. *Proceedings of the Royal Society of London B* 116, 139–149.

Molyneux, D.H. (1967) Feeding behaviour of the larval rat flea *Nosopsyllus fasciatus* Bosc. *Nature, London* 215, 779.

Moser, B.A., Koehler, P.G. and Patterson, R.S. (1991) Effect of larval diet on cat flea (Siphonaptera: Pulicidae) developmental times and adult emergence. *Journal of Economic Entomology* 84, 1257–1261.

Moser, B.A., Koehler, P.G. and Patterson, R.S. (1992) Effect of methoprene and diflubenzuron on larval development of the cat flea (Siphonaptera: Pulicidae). *Journal of Economic Entomology*, 85, 112–116.

Munshi, D.M. (1960) Micro-anatomy of the proventriculus of the common rat flea *Xenopsylla cheopis* (Rothschild). *Journal of Parasitology* 46, 362–372.

Osbrink, W.L.A. and Rust, M.K. (1984) Fecundity and longevity of the adult cat flea, *Ctenocephalides felis felis* (Siphonaptera: Pulicidae). *Journal of Medical Entomology* 21, 727–731.

Osbrink, W.L.A. and Rust, M.K. (1985a) Seasonal abundance of adult cat fleas, *Ctenocephalides felis* (Siphonaptera: Pulicidae), on domestic cats in southern California. *Bulletin of the Society for Vector Ecology* 10, 30–35.

Osbrink, W.L.A. and Rust, M.K. (1985b) Cat flea (Siphonaptera: Pulicidae): factors influencing host-finding behavior in the laboratory. *Annals of the Entomological Society of America* 78, 29–34.

Osbrink, W.L.A., Rust, M.K. and Reierson, D.A. (1986) Distribution and control of cat fleas in homes in southern California (Siphonaptera: Pulicidae). *Journal of Economic Entomology*, 79, 135–140.

Pilgrim, R.L.C. (1991) External morphology of flea larvae (Siphonaptera) and its significance in taxonomy. *Florida Entomologist* 74, 386–395.

Pilgrim, R.L.C. (1992) Taxonomy of flea larvae. *Abstracts of the 19th International Congress of Entomology*, Beijing, June 1992, pp. 487–488.

Rothschild, M. (1965) Fleas. *Scientific American* 213(6), 44–53.

Rothschild, M. (1975) Recent advances in our knowledge of the order Siphonaptera. *Annual Review of Entomology* 20, 241–259.

Rothschild, M. (1988) Giant polyploid cells in *Tunga monositus* (Siphonaptera: Tungidae). In: Service, M.W. (ed.) *Biosystematics of Haematophagous Insects*. Clarendon Press, Oxford, pp. 313–323.

Rothschild, M. and Hinton, H.E. (1968) Holding organs on the antennae of male fleas. *Proceedings of the Royal Entomological Society of London A* 43, 105–107.

Rothschild, M., Schlein, Y., Parker K. and Sternberg, S. (1972) Jump of the oriental rat flea *Xenopsylla cheopis* (Roths.). Nature, London, 239, 45–48.

Rust, M.K. and Reierson, D.A. (1989) Activity of insecticides against the preemerged adult cat flea in the cocoon (Siphonaptera: Pulicidae). *Journal of Medical Entomology* 26, 301–305.

Seddon, H.R. and Albiston, H.E. (1967) *Disease of Domestic Animals in Australia. Part 2. Arthropod Infestations (Flies, Lice and Fleas)*. Department of Health, Commonwealth of Australia.

Sharif, M. (1948) Nutritional requirements of flea larvae, and their bearing on the specific distribution and host preferences of the three Indian species of *Xenopsylla* (Siphonaptera). *Parasitology* 38, 253–263.

Shulov, A. and Naor, D. (1964) Experiments on the olfactory responses and host specificity of the oriental rat flea (*Xenopsylla cheopis*), (Siphonaptera: Pulicidae). *Parasitology* 54, 225–231.

Silverman, J., Rust, M.K. and Reierson, D.A. (1981) Influence of temperature and humidity on survival and development of the cat flea, *Ctenocephalides felis* (Siphonaptera: Pulicidae). *Journal of Medical Entomology* 18, 78–83.

Suter, P.R. (1964) Biologie von *Echidnophaga gallinacea* (Westw.) und Vergleich mit andern Verhaltenstypen bei Flöhen. *Acta Tropica* 21, 193–238.

Traub, R. (1980) Some adaptive modifications in fleas. In: Traub, R. and Starcke, H. (eds) *Fleas*. A.A. Balkema, Rotterdam, pp. 33–67.

Traub, R. (1985) Coevolution of fleas and mammals. In: Kim, K.C. (ed.) *Coevolution of Parasitic Arthropods and Mammals*. John Wiley, New York, pp. 295–437.

Traub, R., Rothschild, M. and Haddow, J.F. (1983) *The Rothschild Collection of Fleas. The Ceratophyllidae: Key to the Genera and Host Relationships. With notes on their Evolution, Zoogeography and Medical Importance*. Published privately, distributed by Academic Press.

Wade, S. and Georgi, J.R. (1988) Survival and reproduction of artificially fed cat fleas *Ctenocephalides felis* Bouché (Siphonaptera: Pulicidae). *Journal of Medical Entomology* 25, 186–190.

Whiteley, H.E. (1987a) Flea-control tips from the experts. *Veterinary Medicine* 82, 913–916.

Whiteley, H.E. (1987b) Five flea-control programs for cats. *Veterinary Medicine* 82, 1022–1026.

Williams, B. (1986) Mandibular glands in the endoparasitic larva of *Uropsylla tasmanica* Rothschild (Siphonaptera: Pygiopsyllidae). *International Journal of Insect Morphology and Embryology* 15, 263–268.

Yeruham, I., Rosen, S. and Hadani, A. (1989) Mortality in calves, lambs and kids caused by severe infestations with the cat flea, *Ctenocephalides felis felis* (Bouché, 1835) in Israel. *Veterinary Parasitology*, 30, 351–356.

Yinon, U., Shulov, A. and Margalit, J. (1967) The hygroreaction of the larvae of the oriental rat flea *Xenopsylla cheopis* Rothsch. (Siphonaptera: Pulicidae). *Parasitology* 57, 315–319.

Blood-sucking Hemiptera (Bugs) 18

Two families of Hemiptera, the Cimicidae and Polyctenidae, and the subfamily Triatominae of the Reduviidae (assassin bugs), are blood-sucking. The Polyctenidae are small, ectoparasites of bats and are of no medical importance. The Cimicidae are blood-sucking, temporary ectoparasites of birds and mammals. They have oval, flattened bodies, and at first sight appear to be wingless, but they are micropterous with the forewings reduced to hemelytral pads and the hind wings absent. The Triatominae are blood-sucking on mammals and birds and are fully winged in the adult stage, as are other reduviids. Members of the other 22 subfamilies of Reduviidae are predacious (Schofield and Dolling, 1993).

Cimicidae (Ryckman *et al.*, 1981; Schofield and Dolling, 1993)

The Cimicidae is particularly well represented in the northern hemisphere. There are no native cimicids in Australia and no bird-feeding cimicids in tropical Africa or Central America. The 91 species within the Cimicidae are arranged in 23 genera, most of which are parasites of bats or birds, and the genus *Cimex* which parasitizes both mammals and birds. Twelve of the genera are found only in the New World, nine only in the Old World and two, *Cimex* and *Oeciacus*, in both.

The 21 species of *Cimex* are mainly parasites of bats, with one species parasitic on birds; and the bedbugs, *C. lectularius* and *C. hemipterus*, which feed on humans. *C. lectularius* is a cosmopolitan species of temperate and subtropical regions. It is a parasite not only of humans but of bats, chickens and other domestic animals. *C. hemipterus* is tropicopolitan occurring in the warmer areas of the world. It is a parasite of humans and chickens, but only rarely of bats. *Leptocimex boueti* is a parasite of bats and humans in West Africa.

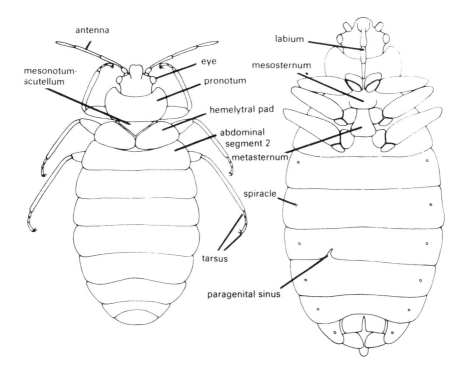

Fig. 18.1. Female *Cimex lectularius*. Dorsal view (left). Ventral view (right).

Cimex lectularius and *Cimex hemipterus* (Usinger, 1966)

Cimex lectularius and *C. hemipterus* are mostly allopatric in their geographical distributions but where they are sympatric, as in KwaZulu, South Africa, cross-mating occurs. Male *C. hemipterus* and female *C. lectularius* mate as readily with each other as they do with their conspecific partners. The few eggs produced from such matings are infertile. The reverse mating (male *C. lectularius* and female *C. hemipterus*) occurs readily in the laboratory resulting in a high proportion of abnormal eggs (Walpole, 1988; Coetzee *et al.*, 1994).

External structure (see Fig. 18.1)

The two major bedbug species will be considered together and specific differences indicated. The adult bedbug measures 5–7 mm with females being slightly longer than males, and *C. hemipterus* being about 25% longer than *C. lectularius*. They are red-brown in colour.

Head

The head bears long 4-segmented antennae, of which the last three segments are long and slender, and a pair of widely separated compound eyes, laterally placed

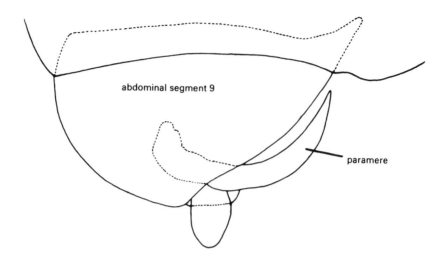

Fig. 18.2. Ventral view of the terminalia of a male *Cimex lectularius*.

at the sides of the head. There are no ocelli; the labium has three obvious segments
and is reflected backwards under the head reaching as far as the coxae of the first
pair of legs.

Thorax

The prothorax is recessed anteriorly and its sides surround the posterior part of
the head. In *C. lectularius* the breadth of the pronotum is more than two-and-a-half
times the length of the prothorax in the midline, less in *C. hemipterus*. The
mesonotum–scutellum is triangular in shape with the base adjoining the pronotum
and the apex backwardly directed. Laterally there are the hemelytral pads. The
tarsus has three segments, the last segment bearing a pair of simple claws. The
metasternum is a more or less square, flat plate between the coxae, with rounded
posterior corners. The mesosternum is rectangular being wider than it is long.

Abdomen

The abdomen is 11-segmented with segments 2 to 9 being easily recognizable dorsally. When the bedbug engorges, the abdomen greatly increases in volume by
exposing the intersegmental membranes and expanding the so-called 'hunger
folds', membranous areas in the midventral line of the second to fifth abdominal
segments. There are seven pairs of spiracles located ventrally on abdominal
segments 2 to 8. Ventrally on the right side of the female there is a notch or
paragenital sinus on the posterior margin of the fifth segment. It opens into the
ectospermalege. There is no corresponding structure on the left side of the female.
The male is similarly asymmetrical, and only the left paramere is developed at the
posterior end of the abdomen (Fig. 18.2). It is directed towards the left side. There
is no right paramere.

Internal structure

The midgut is divided into three ventriculi. The first is a large and bulbous zone in which the blood is stored and concentrated. This section is separated by a sphincter from the second and third ventriculi which are concerned with digestion and absorption which is so complete that only a little haematin enters the hindgut. The ganglia of the thorax and abdomen are concentrated in a single ganglionic mass in the metathorax.

Mycetome

Laterally in the abdomen between the fourth and the fifth segments there are the mycetomes, which contain symbiotic microorganisms. Two different organisms occur in *C. lectularius*, one rod-shaped and the other 'pleomorphic'. These organisms are transmitted transovarially from generation to generation. Mycetomes and symbiotic microorganisms are common in insects which only ever feed on blood. The microorganisms provide certain essential compounds effecting fertility. When *C. lectularius* is rendered nearly symbiote-free by heat treatment (37°C) only a few eggs are produced and they are infertile (Chang, 1974).

Life cycle

Egg

The female oviposits on rough rather than smooth surfaces, inserting eggs into cracks and crevices. They are laid individually and held in place by a transparent cement. The eggs are white in colour, little more than 1 mm in length and less than 0.5 mm in breadth. The eggs are fertilized while still in the ovary, and the embryos have already undergone some development when the eggs are laid. The minimum time for development is 4 to 5 days at 30–35°C. No eggs hatch at 37°C or at temperatures below 13°C. Below 13°C eggs remain viable for shorter periods as the temperature approaches 0°C, and none survives for 3 months. Therefore in temperate climates, eggs laid in the autumn are likely to have died before the temperature rises above the threshold in spring, and consequently the species will not survive in the egg stage over the winter (Johnson, 1941).

Nymph

The egg has a distinct cap at one end and this is forced off when the nymph emerges. In appearance the nymph is very similar to the adult and feeds on blood in the same way. Nymphs will feed within 24 h of emergence or of moulting to the next instar. There are five nymphal instars in the life cycle with the body weight increasing by 30–40% at each instar. When feeding, the bug grasps the skin with its forelegs as a prelude to piercing it. Saliva injected during feeding contains an anticoagulant and the blood does not clot until it reaches the second ventriculus. Feeding takes 5 to 10 min, during which time nymphs will take in two to five times their own body weight of blood. The duration of the instars is very similar for the first four, but the fifth is somewhat longer. At 30°C development from egg to adult takes 3 weeks (Usinger, 1966).

Mating

The method of insemination in *Cimex* is most unusual. Males recognize another member of the same species at distances less than 15 mm, but do not distinguish between the sexes. The male climbs on to the back of the female with his head on the left side of the pronotum and the abdomen tucked under the right side of the female. There are three stages in insemination of *C. lectularius*. In the spermalege stage, the male penetrates the ectospermalege and injects a mass of sperm into the adjacent mesospermalege. The spermatozoa become mobile in about 30 min and after 3 to 4 h move into the haemocoele to begin the second phase (Usinger, 1966).

The spermatozoa concentrate at the base of the genital apparatus near the junction of the paired and median oviducts. They penetrate into the seminal receptacle which has no direct communication with the lumen of the oviduct. The spermatozoa move up in the wall of the oviduct to the pedicel of the ovariole, and concentrate in the syncitial tissue at the distal end of the ovariole. The egg is fertilized in the ovariole before the chorion is formed. This is the intragenital phase (Usinger, 1966). This traumatic method of insemination may be a method of transferring nutrient materials from the male to the female which could be of value in survival of the species under adverse conditions.

Behaviour

Bedbugs are nocturnal creatures which reach peak activity before dawn. They are negatively phototactic, which combined with positive thigmotaxis ensures that they hide away in cracks and crevices during the day. In their search for hosts bedbugs respond to warmth and carbon dioxide, but not to odours. A bedbug will respond to a body which is two or more degrees above ambient temperature, but there is disagreement as to the distance from which it can respond with values of 3–4 cm and 150 cm being claimed (Usinger, 1966). Temperature receptors are probably located on the basal segments of the antennae (Levinson *et al.*, 1974).

Adult bedbugs have a scent gland which opens ventrally on the metathorax on to an evaporative area. The main components of the secretion are two aldehydes (octenal and hexenal) which are present in the ratio of 7:3 (Levinson and Bar Ilan, 1971). They function as alarm pheromones and cause dispersal of aggregated bugs. The receptors for this pheromone are found on the terminal segment of the antenna (Levinson *et al.*, 1974). When alarmed bedbugs can move at a rate of 2 cm s^{-1} (Usinger, 1966).

Bedbugs also produce an aggregation pheromone which brings them together, when thigmotaxis will ensure that they stay grouped. The origin of this pheromone is unknown but the receptors are on the antennae. This is not a sex pheromone because the scent of males or females attracts both sexes equally. It is a pheromone to which adults are particularly sensitive, and to which 5th instars respond only slightly although they are present in the aggregations (Levinson and Bar Ilan,

1971). Nymphs have scent glands which open in the mid-dorsal line on segments 3, 4 and 5 of the abdomen but their function is not known.

Fecundity, rate of development and survival

Fecundity

The frequency of oviposition and the number of eggs laid are dependent on the ease with which blood meals are available. At 23°C and 75% RH, newly moulted female *C. lectularius* took on average one-and-a-half times their own body weight of blood in their first meal and then laid nine eggs 6 to 12 days later. When they were fed twice a week egg laying was continuous (Johnson, 1941).

At 23°C and 90% RH females kept with males and fed once a week, produced on average nearly three eggs in the first week, between seven and eight eggs in the second and third weeks, and more than eight eggs in the fourth week. Thereafter egg production continued to average six to seven eggs per week for the next 13 weeks, after which it declined rapidly. Rather surprisingly, *C. lectularius* kept under the same conditions, but at 10% RH, were more fecund; and averaged more than ten eggs in the fourth week, and for the next 13 weeks the average egg production was 8.6 eggs per female, cf. 6.5 for females kept at 90% RH (Johnson, 1941).

Rate of development

Omori (cited from Usinger, 1966) gives comparative data for the development of *C. lectularius* and *C. hemipterus* at various temperatures, and the survival of once-fed bedbugs of all stages at different temperatures. In all cases the performance of *C. lectularius* was superior to that of *C. hemipterus*. *C. lectularius* developed twice as fast as *C. hemipterus* at 18°C and at 33°C. This is strange when it is realized that *C. hemipterus* has a tropical and subtropical distribution. At 27°C and 30°C the two species developed at their fastest, and there was no difference between them.

Survival

In longevity experiments the 1st instar of both species survived for the shortest time. At all temperatures (10, 18, 27 and 37°C) adult *C. lectularius* survived much longer than adult *C. hemipterus*. At 37°C all stages of *C. lectularius* survived considerably longer than *C. hemipterus* and at 10°C, *C. lectularius* survived for almost twice as long (Usinger, 1966).

Medical importance

The nocturnal biting of bedbugs can be debilitating to humans whose sleep is disturbed every night. The presence of bedbugs in a dwelling can be recognized from specks of faeces left by the bugs and by their odour, a complex mixture of octenal, hexenal and minor amounts of other components (Levinson *et al.*, 1974).

Although their own powers of dispersal are very limited they have been carried throughout the world by humans in their possessions. The status of bedbugs as vectors of disease was summarized by Usinger (1966) as 'Cimicidae have been suspected in the transmission of many diseases or disease organisms of man and bats, but in most cases conclusive evidence is lacking.'

Hepatitis B surface antigen has been recovered from unengorged *C. hemipterus* (Wills *et al.*, 1977) and *C. lectularius* (Jupp *et al.*, 1978). The antigen has been shown to persist in both species for up to 6 weeks after a blood meal, and to be excreted in the faeces of *C. hemipterus* from which it 'could infect a susceptible person by contamination of skin lesions or mucosal surfaces, or by inhalation of dust' (Ogston and London, 1980). Lyons *et al.* (1986) found that HIV could survive for 1 h in *C. lectularius* offering the possibility that mechanical transmission could occur, but Lindsay (1992) showed that bedbugs were not a major route for the transmission of hepatitis B virus among children in The Gambia and were even less likely to transmit other blood-borne viruses, such as HIV, which is less infectious the HBV.

Reduviidae – Triatominae (Lent and Wygodzinsky, 1979)

Adult triatomines are large insects with broad abdomens, commonly measuring 20–28 mm in length and 8–10 mm wide. They have long, thin, 4-segmented antennae (Fig. 18.3). The compound eyes are placed laterally, and the ocelli are located dorsally behind the eyes (Fig. 18.4). In front of the eyes the head is narrowed and forwardly produced (Fig. 18.3). The rostrum (labium) is 3-segmented and straight, not arched (Fig. 18.4), and extends back to the prosternal stridulatory groove. The tarsi are 3-segmented. The forewings are hemelytra with a sclerotized basal area (corium and clavus) and a distal membranous portion (Fig. 18.5). The hind wings are entirely membranous, and in repose, are folded beneath the hemelytra.

In the unfed state the abdomen is almost flat and the hinged dorsal and ventral connexival plates are close together and almost parallel. On feeding, the abdomen becomes greatly distended and the connexival plates rotate on the hinge and become widely separated. In some triatomines, e.g. *Rhodnius*, abdominal expansion is further increased by the unfolding of an additional longitudinally pleated membrane.

The 118 species of the Triatominae are classified into five tribes and 14 genera (Schofield and Dolling, 1993). The Triatominae are largely confined to the western hemisphere with one tropicopolitan species (*T. rubrofasciata*), a single genus (*Linshcosteus*) in India, and a group of seven species of *Triatoma* in south-east Asia. In the New World the triatomines are found from approximately 42°S to 42°N with their greatest diversity occurring in South and Central America and only five species extending from Mexico into the Nearctic region (Schofield, 1988).

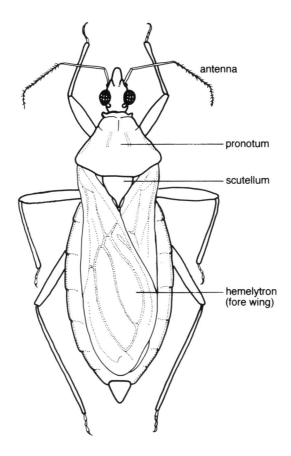

Fig. 18.3. Female *Panstrongylus megistus*. Redrawn from Schofield and Dolling (1993).

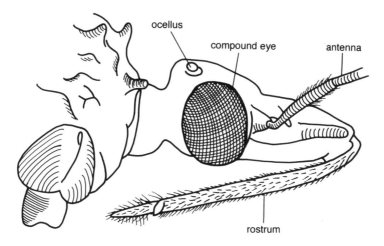

Fig. 18.4. Lateral view of head of *Panstrongylus megistus*. Redrawn from Schofield and Dolling (1993).

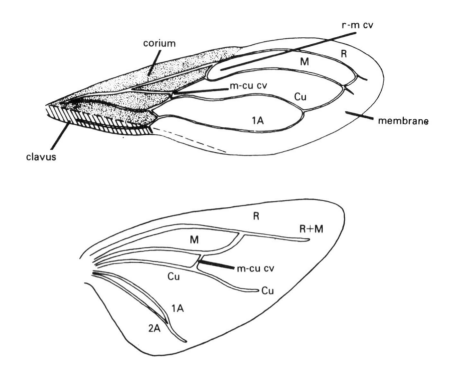

Fig. 18.5. Wings of a triatomine bug. Above, hemelytron. Below, hind wing.

Biology and Bionomics

Life cycle

Egg

Oviposition follows a circadian rhythm in *T. infestans* and *P. megistus* with eggs being deposited early in the scotophase (Constantinou, 1984). The eggs are oval, about 2.5 mm in length with an obvious operculum at one end. Eggs of *R. prolixus* become a bright lobster red within a few hours of being deposited (Uribe, 1926) and those of other species are pearly white when laid and become red as the embryo develops (Schofield and Dolling, 1993). They are laid in cracks and crevices in houses, and in wild populations, eggs of *R. prolixus* may be glued on to palm fronds (Zeledón and Rabinovich, 1981). A female will produce 200–300 eggs in her lifetime, laid in a number of batches. In *T. infestans* peak fecundity is reached 11 weeks after the first oviposition (Rabinovich, 1972).

Eggs of *R. prolixus* hatch over the range 16–34°C with the highest fertility (82%) occurring at 21–32°C. Fertility is also related to humidity. At 25–27°C all eggs hatched at humidities of 50% RH or higher, but at 32°C a humidity of 80% was required. Eggs are resistant to desiccation and 50% hatched at 20% humidity

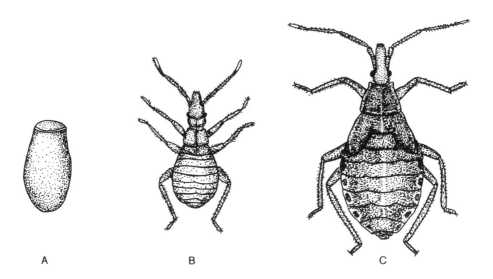

Fig. 18.6. Early stages of *Triatoma rubrofasciata*. Egg (left). First stage nymph (centre). Third stage nymph (right). Redrawn from Patton, W.S. and Evans, A.M. (1929). *Insects, Ticks, Mites and Venomous Animals. Part I: Medical.* H.R. Grubb, Croydon.

and 25°C (Clark, 1935). The egg stage is relatively long, taking 12 days at 32°C and nearly a month at 22°C in *R. prolixus* (Gómez-Núñez, 1964) and about 3 weeks in *T. infestans* at 26–27°C with 86% fertility (Rabinovich, 1972).

Nymph

The nymph emerges by pushing off the operculum from the egg. It is a miniature version of the adult except for the absence of wings (Fig. 18.6). There are five nymphal stages all of which feed on blood. If a nymph engorges fully, only one blood meal is required before moulting to the next stage. The first-stage nymph of *R. prolixus* imbibes 12 times its own body weight of blood, and the other four nymphal stages six times their body weight, while adults take in only one-and-a-half times (Buxton, 1930). During development the nymph increases by 1.6 times at each moult, growing from 2.7 mm in the 1st instar to 17 mm in the 5th instar (Uribe, 1926). In mass rearing at 28°C and 60–70% RH and fed every 2 weeks on hens, the life cycle of *R. prolixus* from egg to adult took 80 days with 42% survival. The 'mortality' included deaths and individuals that did not feed which was about 18% of each instar (Gómez-Núñez, 1964). At 26–27°C and 60% RH the life cycle of *T. infestans* takes about 23 weeks with the 5th instar being the longest (7 weeks) and the 4th instar the shortest (11 days). Mortality is high in the 1st (19%) and 5th (16%) instars and lowest in the 3rd instar (1.6%) (Rabinovich, 1972).

Population growth

Significantly, more female *T. infestans* reach the adult stage than males (5 : 4), but the males are longer-lived averaging 26 weeks compared with 16 weeks for females. Assuming a 1 : 1 ratio of males to females at the egg stage then about half the female eggs give rise to adults. *R. prolixus* and *T. infestans* had the shortest life cycles among nine species of Triatominae, and *T. dimidiata* and *P. megistus* took about twice as long to complete development under the same conditions (Zeledón and Rabinovich, 1981). Rabinovich (1972) calculated a generation time of 31 weeks for *T. infestans* with a net reproductive rate (R_0) of 25 and an intrinsic rate of increase (*r*) of 0.101. That is, the population increased 25 times from one generation to the next or, put another way, a stable-aged population would increase by 10% per week.

Relationship to humans

Triatomines can be arranged in five grades of increasing association with humans. Grade 1 includes sylvatic species with specialized hosts and habitats, e.g. *Psammolestes*. Grade 2 are fairly specialized but adults are occasionally found around houses, e.g. *T. protracta*. Grade 3 are more usually found in a sylvatic ecotope but occasionally colonize houses, e.g. *R. ecuadoriensis*. Grade 4 are commonly domestic in chicken houses and animal shelters but still retain a sylvatic ecotope, e.g. *T. sordida*. This grade includes several locally important vectors of Chagas' disease. Grade 5 are highly adapted to the domestic environment and includes the major vectors of Chagas' disease (Schofield, 1988).

Feeding

Most triatomines are nocturnal, and in houses feed on sleeping inhabitants. For that reason they are most abundant in bedrooms, which may contain 50% of the total population of *R. prolixus* in a house (Rabinovich *et al.*, 1979). They are attracted to a host by warmth, carbon dioxide and odour, detected in *T. infestans* by sensory receptors on the antennae (Mayer, 1968; Wigglesworth, 1972). Orientation on the host and probing involves additional receptors on the tarsi and labium (Wiesinger, 1956; Zeledón and Rabinovich, 1981).

In feeding the mandibles pierce the skin and then cease to move. They are barbed and hold the insect in position. The flexible maxillae penetrate deeply and move about freely until they contact a capillary. The right maxilla is longer than the left and penetrates into the capillary, the left maxilla may enter the capillary or remain closely attached on the outside because the junction between the two maxillae is the effective functional mouth (Lavoipierre *et al.*, 1959).

During feeding saliva is injected containing an anticoagulant and blood is pumped into the midgut, the first section of which acts as a storage organ. Two anticoagulants have been recovered from *R. prolixus*, one from the salivary glands, the other from the gut (Hellman and Hawkins, 1965). Although the precise location of the gut anticoagulant was not stated it is probably the storage section where blood remains unclotted. This section of the midgut contains in its lumen symbionts which are important in the successful development of a triatomine (Wigglesworth, 1936). In their absence development stops in the 4th or 5th instar. A coagulant in this section would prevent the ingested blood from trapping symbionts in the clot. It would also ensure that the blood remained fluid for easier passage into the digestive section of the midgut. In development the nymph acquires its symbionts by probing the contaminated surface of the egg from which it has emerged or from excreta of other members of the species (Brecher and Wigglesworth, 1944).

The bites of domesticated triatomine bugs are relatively painless, and causing pain is regarded as a primitive character indicating a relatively recent association of bug and humans (Schofield and Dolling, 1993). On repeated exposure some individuals may have a delayed reaction 24–48 h after being bitten, and in one case a more severe response. It is possible that longer periods of exposure would produce an immediate response (Lavoipierre *et al.*, 1959).

Triatomines feed on a wide range of hosts, and domestic species feed on humans, dogs, cats, chickens, and rodents infesting the house. They will feed on more than one host, particularly if their feeding has been interrupted, and will often feed again before the previous meal has been completely digested. More than one-third of *T. dimidiata* had fed on more than one host and the gut contents of one adult reacted positively to antisera of six hosts – human, dog, cat, mouse, cow and opossum (Zeledón *et al.*, 1973). In Argentina the biting rate of *T. infestans* is controlled by temperature during the cool season and by population density in the hot season (Catalá, 1991).

The time to take a full blood meal varies with the instar and size of the insect (Zeledón and Rabinovich, 1981). Among five species the times taken to engorge varied from 3 to 30 min, with *T. protracta* being the slowest feeder (Wood, 1951). *R. prolixus* and *T. infestans* fed faster and with fewer interruptions than did *T. dimidiata* (Zeledón *et al.*, 1977). Interruptions involved the bug withdrawing its mouthparts in the course of a single meal. Male and female *T. dimidiata* ingested over 40% and 50%, respectively, of their body weight at each feed and in the course of adult life a female may consume 4–10 g of blood and males about half that (Zeledón *et al.*, 1977). Fifth instar nymphs of *P. megistus* took twice as much blood as the same stage of *R. prolixus* (300 cf. 600 mg) (Miles *et al.*, 1975). Nymphs of *T. infestans* consumed about 1000 mg of blood in developing to an adult (Zeledón and Rabinovich, 1981).

When a heavily infested rural house in Venezuela was demolished, nearly 8000 *R. prolixus* of all stages were recovered. The information was analysed in great detail and it was calculated that in that house the feeding rate was 58 *R. prolixus* per person per day, and in 13 other houses it was nine, ranging from 0.2 to 33; and the loss of blood per person per month in the 13 houses ranged from 0.7 to 40 ml; and in the heavily infested house exceeded 100 ml (Rabinovich *et al.*, 1979).

Defecation

Infection with *Trypanosoma cruzi* depends upon infective forms in the faeces of the bug being deposited on the host and gaining access via a wound or moist mucosa. Therefore the earlier a bug defecates during or after feeding, the more likely the species is to be an efficient vector. *R. prolixus* and *T. infestans* defecate within 10 min of finishing meal when the insect is still likely to be on the host and 8% of *R. prolixus* defecate during the meal. Only two-thirds of *T. dimidiata* defecate within 10 min of finishing a meal but, possibly because of the longer feeding time, 13% defecate during the meal (Zeledón *et al.*, 1977).

Zeledón *et al.* (1977) proposed a defecation index which, averaged for all instars and sexes, rates *R. prolixus* at 2.3, *T. infestans* at 1.1 and *T. dimidiata* 0.6. The index of *T. protracta*, which often delays defecation, was 0.2 and for *T. rubida uhleri*, which defecates frequently and early, 9.0 (based on data from Wood, 1951). *T. rubida* frequents the nests of woodrats (*Neotoma* spp.) which are sylvatic reservoirs of *T. cruzi* in the USA (Lent and Wygodzinsky, 1979).

Resting places

In South America many species occur in the crowns of palms but the populations are low (5–70 per tree) because blood meals depend on visiting birds and mammals, and natural enemies are present. Other biotopes occupied by triatomines include bromeliads, under bark of trees, in hollow trees and fence posts and in ground burrows (Zeledón and Rabinovich, 1981). *R. prolixus* occurs in both palms and houses, and in the latter occurs equally in the walls and roof, particularly in palm thatch roofs (Rabinovich *et al.*, 1979). *T. infestans* and *P. megistus* infest cracks in unplastered mud and cane walls (Wiesinger, 1956), and are less common in mud-brick (Zeledón and Rabinovich, 1981). *T. dimidiata* occurs lower down than the other two species, rarely being more than a metre above the floor; and in houses raised on supports it occurs on the ground or on the foundations that support the floor. Here they feed on animals that shelter under the house, and are able to enter the house at night through cracks in the floor and feed on its sleeping occupants (Zeledón *et al.*, 1973). The bugs congregate near a food source, hence the greater concentration in the walls of bedrooms. This concentration is assisted in *T. infestans* and *R. prolixus* by a pheromone in the faeces which attracts the unfed nymphs of both species. Fed nymphs are not attracted by the faecal material but their locomotory activity is arrested in its presence which leads to them congregating in the same area (Schofield and Patterson, 1977).

Dispersal

Wild *R. prolixus* disperses mainly as an ectoparasite of birds and this is considered to be the means by which it has spread to Mexico. *T. infestans*, the most domesticated triatomine, is considered to have originated as a parasite of wild guinea-pigs in the Cochabamba Valley in the Bolivian Andes. When, in pre-Columbian times, guinea-pigs were domesticated, *T. infestans* went with them. It is now the most widespread domestic species occurring from the temperate areas of southern Argentina to the dry tropics of north-eastern Brazil (Schofield, 1988).

Houses are invaded by bugs being introduced in palm fronds, in firewood and in household articles. Nearly one-third of *R. prolixus* released into a house found shelter in household articles (Gómez-Núñez, 1969). In 40 days, labelled *R. prolixus* dispersed less than 4 m in houses and less than 15 m outside. Adult *R. prolixus* appear to fly very little and there was movement between palm trees and houses only when the two habitats were close together. Other workers have found some nymphs and adults of *R. prolixus* dispersing 100–500 m and *P. megistus* moved 400 m from a natural to an artificial biotope (Zeledón and Rabinovich, 1981).

Triatomines are attracted to light which would favour their establishment in houses. There can be two-way movement between natural and artificial biotopes. The range of blood meals identified in *T. dimidiata* in natural and artificial biotopes indicated free movement between the two, with 22% of the bugs in natural biotopes and some of the bugs found in houses containing opossum blood (Zeledón *et al.*, 1973).

Survival and longevity

In the laboratory, triatomines are long-lived and able to withstand long periods of starvation. The greatest resistance is shown by 4th and 5th instar nymphs which in *T. dimidiata* survived 6 months unfed, whereas adults survived only 4–5 months (Zeledón *et al.*, 1970). Similar powers of survival were shown by nymphs of *T. infestans* and *R. prolixus* (Zeledón and Rabinovich, 1981). In the field, survival of *R. prolixus* was considerably lower, being measured in weeks rather than months; predation by domestic fowls, rodents and other predators being the main cause of early mortality. It was calculated that *R. prolixus* had a one in four chance of successfully moving from a palm tree to an adjoining house (Gómez-Núñez, 1969). Perhaps as a defence against predation, nymphs of *T. dimidiata* camouflage themselves by covering their bodies with debris (Zeledón *et al.*, 1973).

Control of Blood-sucking Bugs

Bugs shelter in cracks and crevices of human and animal habitations and it is desirable that they should have a minimum of recesses in which bugs can shelter. That is the counsel of perfection with older houses and makeshift animal shelters offering ideal conditions for colonization by bedbugs and triatomines. The application of residual insecticides to the bugs' daytime resting places in bedrooms and animal shelters with the organochlorine insecticide BHC (benzene hexachloride) and synthetic pyrethroids has proved highly effective, the latter giving much greater residual activity on mud walls than other insecticides. When treatment ceases there is the possibility that the original vector may recover or be replaced by another vector, e.g. *T. infestans* being replaced by *P. megistus* and *T. sordida* (Dias, 1987).

Medical Importance

Triatomines are of considerable medical importance as vectors of Chagas' disease caused by *Trypanosoma cruzi* (see Chapter 29). Natural infections with *T. cruzi* have been found in 65 species belonging to eight genera and including 41 out of 68 species of *Triatoma*, 8/12 species of *Rhodnius* and 10/13 species of *Panstrongylus*

(Schofield and Dolling, 1993). The main vectors of *T. cruzi* are *T. infestans*, *T. dimidiata*, *T. brasiliensis*, *P. megistus* and *R. prolixus*.

References

Brecher, G. and Wigglesworth, V.B. (1944) The transmission of *Actinomyces rhodnii* Erikson in *Rhodnius prolixus* Stål (Hemiptera) and its influence on the growth of the host. *Parasitology* 35, 220–224.

Buxton, P.A. (1930) The biology of a blood-sucking bug, *Rhodnius prolixus. Transactions of the Entomological Society of London* 78, 227–236.

Catalá, S. (1991) The biting rate of *Triatoma infestans* in Argentina. *Medical and Veterinary Entomology* 8, 325–333.

Chang, K.P. (1974) Effects of elevated temperature on the mycetome and symbiotes of the bed bug *Cimex lectularius* (Heteroptera). *Journal of Invertebrate Pathology* 23, 333–340.

Clark, N. (1935) The effect of temperature and humidity upon the eggs of the bug *Rhodnius prolixus* (Heteroptera: Reduviidae). *Journal of Animal Ecology* 4, 82–87.

Coetzee, M., Hunt, R.H. and Walpole, D.E. (1994) Interpretation of mating between two bedbug taxa in a zone of sympatry in KwaZulu, South Africa. In: Lambert, D.M. and Spencer, H.G. (eds) *Speciation and the Recognition Concept: Theory and Application.* The Johns Hopkins University Press, Baltimore, Maryland, pp. 175–190.

Constantinou, C. (1984) Circadian rhythm of oviposition in the blood sucking bugs, *Triatoma phyllosoma, T. infestans* and *Panstrongylus megistus* (Hemiptera: Reduviidae). *Journal of Interdisciplinary Cycle Research* 15, 203–212.

Dias, J.C.P. (1987) Control of Chagas' disease in Brazil. *Parasitology Today* 3, 336–341.

Gómez-Núñez J.C. (1964) Mass rearing of *Rhodnius prolixus. Bulletin of the World Health Organization* 31, 565–567.

Gómez-Núñez J.C. (1969) Resting places, dispersal and survival of Co_{60}-tagged adult *Rhodnius prolixus. Journal of Medical Entomology* 6, 83–86.

Hellmann, K. and Hawkins, R.l. (1965) Prolixin-S and prolixin-G: two anticoagulants from *Rhodnius prolixus* Stål, *Nature, London* 207, 265–267.

Johnson, C.G. (1941) The ecology of the bed-bug, *Cimex lectularius* L., in Britain. *Journal of Hygiene* 41, 345–461.

Jupp, P.G., Prozesky, O.W., McElligott, S.E. and van Wyk, L.A.S. (1978) Infection of the common bedbug (*Cimex lectularius* L.) with hepatitis B virus in South Africa. *South African Medical Journal* 53, 598–600.

Lavoipierre, M.M.J., Dickerson, G. and Gordon, R.M. (1959) Studies on the methods of feeding of blood-sucking arthropods. I. The manner in which triatomine bugs obtain their blood-meal, as observed in the tissues of the living rodent with some remarks on the effects of the bite on human volunteers. *Annals of Tropical Medicine and Parasitology* 53, 235–250.

Lent, H. and Wygodzinsky, P. (1979) Revision of the Triatominae (Hemiptera: Reduviidae), and their significance as vectors of Chagas' disease. *Bulletin of the American Museum of Natural History* 163, 125–520.

Levinson, H.Z. and Bar Ilan, A.R. (1971) Assembling and alerting scents produced by the bedbug *Cimex lectularius* L. *Experientia* 27, 102–103.

Levinson, A.R., Muller, B. and Steinbrecht, R.A. (1974) Structure of sensilla, olfactory perception, and behaviour of the bedbug, *Cimex lectularius*, in response to its alarm pheromone. *Journal of Insect Physiology* 20, 1231–1248.

Lindsay, S.W. (1992) The possibility of transmission of the human immuno-deficiency virus or hepatitis B virus by bedbugs or mosquitoes. *Abstracts XIIIth International Congress for Tropical Medicine and Malaria Jomtien, Pattaya, Thailand,* 29 November–4 December 1992, vol. 1, p. 61.

Lyons, S.F., Jupp, P.G. and Schoub, B.D. (1986) Survival of HIV in the common bedbug. *The Lancet* 2, 45.

Mayer, M.S. (1968) Response of single olfactory cell of *Triatoma infestans* to human breath. *Nature, London* 220, 924–925.

Miles, M.A., Patterson, J.W., Marsden, P.D. and Minter, D.M. (1975) A comparison of *Rhodnius prolixus, Triatoma infestans* and *Panstrongylus megistus* in the xenodiagnosis of a chronic *Trypanosoma (Schizotrypanum) cruzi* infection in a rhesus monkey (*Macaca mullatta*). *Transactions of the Society of Tropical Medicine and Hygiene* 69, 377–382.

Ogston, C.W. and London, W.T. (1980) Excretion of hepatitis B surface antigen by the bedbug *Cimex hemipterus* Fabr. *Transactions of the Royal Society of Tropical Medicine and Hygiene* 74, 823–825.

Rabinovich, J.E. (1972) Vital statistics of Triatominae (Hemiptera: Reduviidae) under laboratory conditions. *Journal of Medical Entomology* 9, 351–370.

Rabinovich, J.E., Leal, J.A. and Feliciangeli de Pinero, D. (1979). Domiciliary biting frequency and blood ingestion of the Chagas's disease vector *Rhodnius prolixus* Stahl (Hemiptera: Reduviidae), in Venezuela. *Transactions of the Royal Society of Tropical Medicine and Hygiene* 73, 272–283.

Ryckman, R.E., Bentley, D.G. and Archbold, E.F. (1981) The Cimicidae of the Americas and Oceanic Islands, a checklist and bibliography. *Bulletin of the Society for Vector Ecology* 6, 93–142.

Schofield, C.J. (1988) Biosystematics of the Triatominae. In: Service, M.W. (ed.) *Biosystematics of Haematophagous Insects.* Clarendon Press, Oxford, pp. 284–312.

Schofield, C.J. and Dolling, W.R. (1993) Bedbugs and kissing-bugs (blood-sucking Hemiptera). In: Lane, R.P. and Crosskey, R.W. (eds) *Medical Insects and Arachnids.* Chapman & Hall, London, pp. 483–516.

Schofield, C.J. and Patterson, J.W. (1977) Assembly pheromone of *Triatoma infestans* and *Rhodnius prolixus* nymphs (Hemiptera: Reduviidae). *Journal of Medical Entomology* 13, 727–734.

Uribe, C. (1926) On the biology and life history of *Rhodnius prolixus* Stahl. *Journal of Parasitology* 13, 129–136.

Usinger, R.L. (1966) *Monograph of Cimicidae (Hemiptera–Heteroptera).* The Thomas Say Foundation 7, 1–585.

Walpole. D.E. (1988) Cross-mating studies between two species of bedbugs (Hemiptera: Cimicidae) with a description of a marker of interspecific mating. *South Africa Journal of Science,* 84, 215–216.

Wiesinger, D. (1956) Die Bedeutung der Umweltfaktoren für den Saugakt von *Triatoma infestans. Acta Tropica* 13, 98–141.

Wigglesworth, V.B. (1936) Symbiotic bacteria in a blood-sucking insect, *Rhodnius prolixus* Stål. (Hemiptera: Triatomidae). *Parasitology* 28, 284–289.

Wigglesworth, V.B. (1972) *The Principles of Insect Physiology.* Chapman & Hall, London.

Wills, W., London, W.T., Werner, B.G., Pourtaghva, M., Larouze, B., Millman, I. Ogston, W., Diallo, S. and Blumberg, B.S. (1977) Hepatitis-B virus in bedbugs (*Cimex hemipterus*) from Senegal. *Lancet* 2, 217–219.

Wood, S.F. (1951) Importance of feeding and defecation times of insect vectors in transmission of Chagas' disease. *Journal of Economic Entomology* 44, 52–54.

Zeledón, R. and Rabinovich, J.E. (1981) Chagas' disease: an ecological appraisal with special emphasis on its insect vectors. *Annual Review of Entomology* 26, 101–133.

Zeledón, R., Guardia, V.M., Zúñiga, A. and Swartzwelder, J.C. (1970) Biology and ethology of *Triatoma dimidiata* (Latreille, 1811). I. Life cycle, amount of blood ingested, resistance to starvation and size of adults. *Journal of Medical Entomology* 7, 313–319.

Zeledón, R., Solano, G., Zúñiga, A and Swartzwelder, C. (1973) Biology and ethology of *Triatoma dimidiata* (Latreille, 1811). III. Habitat and blood sources. *Journal of Medical Entomology* 10, 363–370.

Zeledón, R., Alvarado, R. and Jirón, L.F. (1977) Observations on the feeding and defecation patterns of three triatomine species (Hemiptera: Reduviidae). *Acta Tropica* 34, 65–77.

Phthiraptera (Lice)   **19**

The Phthiraptera or lice are wingless, dorsoventrally flattened, permanent ectoparasites of birds and mammals with a high degree of host specificity to a single species or higher taxon. Four groups of Phthiraptera are readily recognizable – Anoplura, Rhynchophthirina, Amblycera and Ischnocera. Lyal (1985) using cladistic analysis classifies all four groups as suborders of the Phthiraptera with the Anoplura being most closely related to the Rhynchophthirina and most distant from the Amblycera.

The Anoplura are blood-sucking ectoparasites of mammals. The Rhynchophthirina includes just two species, *Haematomyzus elephantis*, on the Indian and African elephant and the closely related *H. hopkinsi* on wart hogs. They have a long forwardly directed proboscis, which bears small cutting mandibles at its tip (Ferris, 1931; Jeu Minghwa *et al.*, 1990). The term Mallophaga will be used to include the Amblycera and Ischnocera, which have chewing mouthparts and feed on skin debris on birds and mammals.

Over 3000 species of lice have been described of which the greater number are mallophagan parasites of birds. The evolution of lice and their mammalian hosts has been dealt with by Kim (1985a) for the Anoplura and Emerson and Price (1985) for the Mallophaga and the geographical distribution of the avian lice and their hosts by Clay (1974). The species of lice parasitizing domestic mammals are listed in Table 19.1.

Adaptations to an ectoparasitic mode of life include a prognathous head, i.e. the head being horizontal and the mouthparts directed forwards, a reduction in the numbers of antennal segments, thoracic and abdominal spiracles, tarsal segments and nymphal stages to three (Kim 1985a,b). The characters distinguishing the Anoplura from the Mallophaga have been given in Chapter 2.

Anoplura

It has been postulated that there are more than 1000 species of Anoplura of which about half have been described and arranged in 15 families and 47 genera.

Table 19.1. Lice found on domestic animals.

Host	Anoplura	Mallophaga
Cattle	Haematopinus eurysternus Haematopinus quadripertusus Haematopinus tuberculatus Linognathus vituli Solenopotes capillatus	Damalinia bovis
Horse	Haematopinus asini	Damalinia equi
Pig	Haematopinus suis	None
Sheep	Linognathus ovillus Linognathus pedalis Linognathus africanus Linognathus stenopsis	Damalinia ovis
Goat	Linognathus africanus Linognathus stenopsis	Damalinia caprae Damalinia crassipes Damalinia limbata
Dog	Linognathus setosus	Heterodoxus spiniger Trichodectes canis
Cat	None	Felicola subrostrata

Two-thirds of them, mainly in the Hoplopleuridae and Polyplacidae, are parasites of rodents. They are highly host specific with 63% being found on a single host species and only 13% occurring on four or more host species. The medically important genera, *Pediculus and Pthirus*, each have two species. Both species of *Pediculus* are parasites of humans with one also occurring on New World monkeys, gibbons and the great apes. The two distinct species of *Pthirus* occur on separate hosts, humans and the gorilla. The 16 species of the related genus *Pedicinus* occur on Old World monkeys. Three genera are of veterinary importance. *Haematopinus* (22 species) and *Linognathus* (51 species) parasitize Artiodactyla and Perissodactyla with another four species of *Linognathus* parasitizing the Canidae. The majority of species of *Linognathus* are parasites of the Bovidae with a high rate (72%) of host specificity. *Solenopotes* (10 species) parasitizes Cervidae and Bovidae (Kim 1985a, 1988).

External morphology

Anoplura are small insects, ranging from less than 0.5 mm to 8 mm in the adult and 2 mm would be an average length. The antennae are usually 5-segmented; the eyes are reduced and usually absent; and there are no ocelli (Fig. 19.1). The mouth opening is terminal. The highly specialized mouthparts are not visible externally and have been described in Chapter 4. There are no palps. The three thoracic segments are fused. There is only a single tarsal segment and a single claw. When the claw is retracted it makes contact with a thumb-like process on the tibia, the enclosed space having the diameter of the hairs of the host, and enables the louse to maintain itself on an active host (Fig. 19.4C). There is one pair of spiracles,

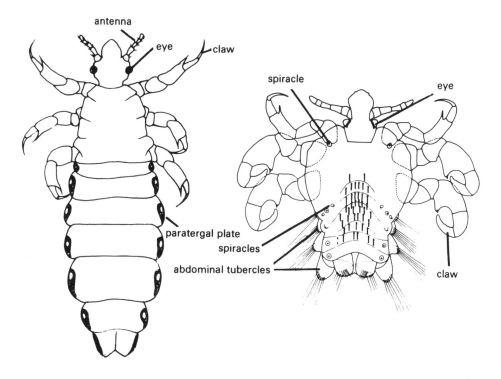

Fig. 19.1. Two common human lice. *Pediculus capitis* (left) and *Pthirus pubis* (right) from dorsal view. Setae omitted from *P. capitis* and only selected setae shown in *P. pubis*.

the mesothoracic, on the thorax, and six pairs on segments 3–8 of the abdomen. The abdomen has nine segments.

The sexes can be easily distinguished. In the male the end of the abdomen is rounded and ventrally the sclerotized genitalia are prominent in the midline posteriorly. In the female the end of the abdomen is bilobed and ventrally the paired lateral gonopods and the sternal plate of the eighth abdominal segment are sclerotized to varying degrees (Fig. 19.2) (Buxton, 1950).

Internal structure

The following account is largely based on *Pediculus* on which most work has been done. The thoracic and abdominal ganglia are fused into a single ganglionic mass in the thorax. The oesophagus opens into a huge midgut dominated by a capacious ventriculus. A short narrow posterior section of the midgut connects the ventriculus to the hindgut. With the posterior section being so short the ventriculus functions as both a storage and a digestive organ. Buxton (1950) comments that 'with regard to digestion nothing is known', and that still appears to be true.

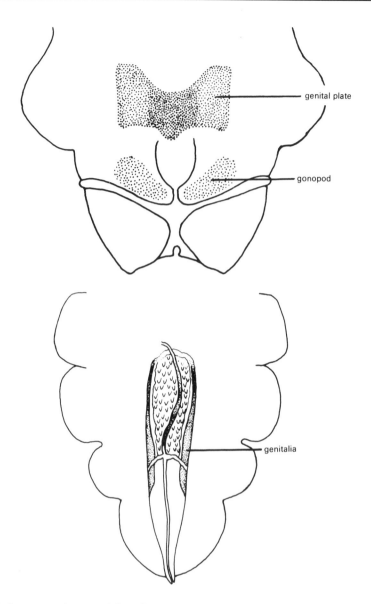

Fig. 19.2. Ventral view of female terminalia (above), and male genitalia (below) of *Pediculus humanus.*

Mycetome

On the ventral surface of the ventriculus there lies the mycetome, containing symbionts. In development the mycetome arises as a pouch off the midgut and symbionts, which are in the gut of the embryo, enter the mycetome. In nymphs and males they remain there throughout the life of the individual, but in females they migrate to the ovary and there is transovarian transmission of symbionts from one generation to another. In the absence of symbionts nymphs live for only a

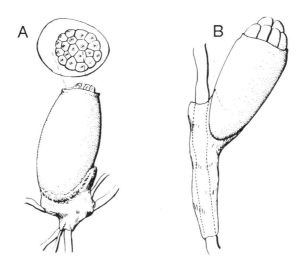

Fig. 19.3. Left, egg of *Pediculus humanus* with operculum viewed from above. Right, egg of *Pthirus pubis*. Source: from Ferris (1951).

few days and females are sterile (Buxton, 1950). The loss of symbionts can be counteracted by a single dose of the vitamin B complex (Puchta, 1955).

Biology and behaviour

Anoplura firmly attach their eggs individually to the hairs of their host, or in the case of *Pediculus humanus* (Fig. 19.3A), the body louse, to the clothing of its host. The eggs hatch to produce a nymph which is clearly recognizable as a small edition of the adult, living and feeding in the same way. There are three nymphal stages before the adult. All stages feed on blood and maintain contact with their host by a number of relatively simple responses.

Both *Pediculus* and *Haematopinus* respond to warmth and smell, and *Haematopinus* will distinguish between a finger and a glass rod at the same temperature. Many receptors are located on the antennae, but heat receptors are more generally distributed over the body. Humidity receptors are present on the antennae, and the louse avoids high humidities, but once adapted to a given humidity it turns away from higher or lower humidities. The responses of lice to stimuli are kineses and not taxes, that is lice are not attracted directly to the source of the stimulus but show increased turning when they move away from the source (Wigglesworth, 1941, 1972).

In addition lice are positively thigmotactic, moving less on rough surfaces, and both negatively phototactic and positively skototactic, moving towards dark objects. The preferred temperature is 29–30°C and movement into areas of higher or lower temperature results in more frequent turning of the louse to bring it back to the preferred temperature. Being ectoparasitic on warm-blooded animals, lice live at a relatively high ambient temperature, and *Pediculus* does not oviposit at temperatures below 25°C (Wigglesworth, 1941, 1972).

Pediculus humanus, P. capitis (see Fig. 19.1)

Pediculus humanus is the body or clothing louse of humans and *P. capitis*, the head louse. Adult *Pediculus* measure 2–3 mm in the male and 2.4–3.6 mm in the female. The simple, lateral eyes are well developed, for a louse. The legs are essentially the same size and shape; the margins of the abdomen are more or less strongly lobed, the lobes on segments 3–8 being covered by sclerotized paratergal plates; and there is a sclerotized sternal plate on the thorax.

Egg

The egg measures 0.8×0.3 mm and is glued to the hair in the case of *P. capitis* and to the inner clothing by *P. humanus* (Fig. 19.3A). Eggs will hatch over the temperature range 24–37°C with the highest hatching rate of 70–90% occurring at 29–32°C. Outside that range the hatching rate declines reaching 10% at 24 and 37°C. The percentage of eggs hatching, but not the duration of the egg stage, is affected by humidity with the highest rate occurring at 75% RH. The egg stage lasts 7 to 10 days at 29–32°C, and the maximum time that eggs can survive unhatched is 3 to 4 weeks, which may be important when considering the survival of lice in infested clothing (Buxton, 1950).

Nymph

At hatching the nymph swallows the amniotic fluid and air, forces the operculum (Fig. 19.3A) off the egg and tears the vitelline membrane, using an elaborate hatching device on a ridged area of embryonic cuticle on the front of the head. When kept on the skin all day the three nymphal stages are passed in 8 to 9 days but when removed at night the duration of the nymphal stages extends to 16–19 days. Adults mate frequently throughout life, beginning soon after the final moult. Human lice survive off the host for only a few days, the duration varying inversely with the temperature, but at low temperatures lice are inactive which reduces their chance of finding another host (Buxton, 1950).

Longevity and fecundity

Three sets of data on the longevity and fecundity of *P. humanus* will be considered. When maintained as monogamous pairs on the skin permanently, the longevity of females was 34 ± 13 days and for males 31 ±12 days. Under these conditions the preoviposition period was 24–36 h, and on average a female laid 270 to 300 eggs at a rate of nine to ten eggs per day. When lice were removed from the body at night and subjected to fluctuating ambient temperatures, the longevity was very similar with that of the females being 29 ± 13 days, and of the males 31 ± 12 days. The preoviposition period was 2 days and fecundity was greatly reduced, ranging from 80 to 212 eggs per female at a rate of three to five eggs per female per day (Buxton, 1940, 1950).

Evans and Smith (1952) reared *P. humanus* in groups of 100 at 30°C and 30–55% RH. The lice were fed twice a day on volunteers. The mean length of life was 17.6 days for both sexes with standard deviations of 9.2 (females) and 8.0 (males). The average fecundity was 82 eggs per female laid at a rate of 5 eggs per female per day, and was comparable to the findings for lice removed from the body at night (Buxton, 1950).

Population dynamics

Evans and Smith (1952) found that the individual instars lasted 3 to 5 days, and the total nymphal life averaged 12.8 days with a coefficient of variation of only 5%. Using their observed hatching rate (87.7%) and survival of nymphs to adults (86%), they calculated the net reproductive rate (R_0) as 31, and the intrinsic rate of natural increase (r) at 0.111 per day. In the absence of other limiting factors the population would increase 31 times in one generation, and would double itself in 6 days. It might be noticed that the value for r for *P. humanus* (0.111) is similar to that obtained by Rabinovich (1972) for *Triatoma infestans* for which he obtained a value of 0.101, a major difference being that the r value for *P. humanus* was per day, and for *T. infestans* per week. This is comparable to the longevity of the two forms, being weeks for lice and months for triatomines. Both Buxton (1950) and Evans and Smith (1952) agree that in a stable-aged population, over two-thirds of the lice will be present as eggs, about a quarter as nymphs and only 6–7% as adults.

Medical importance

Infestations of *Pediculus* on humans cause considerable irritation and scratching; the irritation disturbs the infested person's rest, and the scratching leads to skin lesions and secondary infections. The medical importance of *Pediculus* does not reside in its direct effects on the human host, but in its role as the vector of epidemic typhus (*Rickettsia prowazekii*) and relapsing fever (*Borrelia recurrentis*). The transmission of these pathogens is dealt with in Chapters 25 and 26.

Pthirus pubis (see Fig. 19.1)

Pthirus pubis, the crab louse, has simple eyes and is shorter (1.5–2 mm) and broader than the more slender *Pediculus*. Its body is less than twice as long as wide. The thorax is very wide and passes imperceptibly into the short abdomen. Compared to the first pair of legs the second and third pairs are strongly developed; there is no thoracic sternal plate; the abdomen bears four pairs of lateral sclerotized tubercles; and, as a result of compression, the first three pairs of abdominal spiracles are in an almost straight, transverse row.

    Pthirus pubis occurs on hair in the pubic and perianal regions of the body, and occasionally in the axillae, eyebrows and beard (Buxton, 1950). The incidence of *P. pubis* has been increasing among the human population in recent years, and pediculosis pubis, as the condition is known, is the most contagious sexually transmitted disease according to Felman and Nikitas (1980). Transmission can also occur between individuals sleeping in the same bed, one of whom is infested. A common feature of infestation with *P. pubis* is the presence of blue or slate-grey macules on the skin which are considered to be either altered patient blood pigments or substances excreted from the louse's salivary glands. Fortunately *P. pubis* is not involved in the transmission of any pathogenic organisms. It has a similar life cycle to that of *P. humanus* but it attaches its eggs (Fig. 19.3B) to hairs in the area of infestation. *P. pubis* is considered to be less mobile than *P. humanus*, moving only 100 mm per day compared with 175 mm per hour for the latter. It survives for even shorter periods off the host, dying in less than 48 h at 15°C (Felman and Nikitas, 1980).

Control of lice on humans

Washing the hair reduces the number of nymphs and adult *P. capitis* in an infestation and combing with a finely toothed comb will remove the eggs and further reduce the infestation. Washing has the added advantage of removing the louse's faeces which are a source of irritation. In heavy infestations the application of an insecticidal lotion incorporating malathion, carbaryl or synthetic pyrethroids is preferable to an insecticidal shampoo. When a malathion lotion is left to dry on the hair, it kills the eggs of *P. capitis* in 12 h. Insecticidal lotions are effective against *P. pubis* on the body and infestations on the eye lashes can be dealt with by the application of petroleum jelly twice a day for 10 days. Reinfestation can still occur from contact with untreated people. *P. humanus* can be combated by boiling infested clothing to destroy eggs and lice or by heating clothing to a temperature in excess of 60°C for 15 min, as in a tumble drier. These methods are inapplicable to refugee situations where body lice can multiply and spread unchecked. In such situations DDT dust applied to clothing has been highly effective but resistance to DDT has arisen and other insecticides must be substituted (Ibarra, 1993).

Haematopinus (see Fig. 19.4)

Species of *Haematopinus* are large lice measuring about 4 mm with prominent ocular points but without eyes. The thoracic sternal plate is well developed; the legs are all of similar size; the paratergal plates are strongly sclerotized on abdominal segments 2 or 3 to 8; and there is a sclerotized plate at the base of the tarsal segment, which is referred to as the pretarsal sclerite or discotibial process (Fig. 19.4C).

Twenty-two species of *Haematopinus* have been described and they are all parasites of ungulates. Those occurring on domestic animals include *H. suis* on pigs; *H. asini* on equines; and three species on cattle: *H. eurysternus*, the short-nosed sucking louse; *H. quadripertusus*, the tail louse; and *H. tuberculatus*, the buffalo louse (Meleny and Kim, 1974).

On cattle

Louse populations build up during the winter when the animal's coat is longer and thicker. In Arizona cattle not given supplementary feeding lose weight during winter, and this was shown to be significantly increased when cattle had heavy

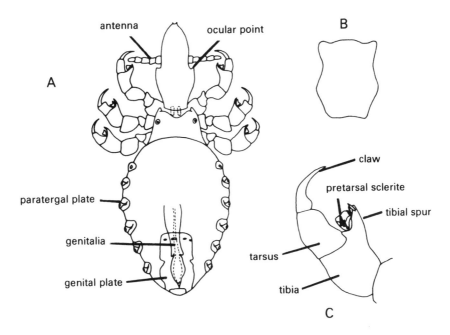

Fig. 19.4. Male *Haematopinus asini:* (**A**) in dorsal view; (**B**) sternal plate; (**C**) terminal segments of leg.

infestations of *H. eurysternus*. Heavy infestations were also associated with marked anaemia. The effect was more pronounced on heifers than bulls, with heifers losing significantly more weight when moderately infested (Collins and Dewhirst, 1965).

The female *H. eurysternus* measures about 3.5 mm in length and occurs on cattle worldwide. The main areas of infestation are the head and neck, spreading in heavy infestations to other parts of the body. The life cycle from egg to egg averages 4 weeks with females living up to 16 days and laying 35–50 eggs. *H. quadripertusus* occurs on cattle in tropical areas. It is a rather larger louse measuring 4.5 mm and occurring mainly in the tail-switch where the eggs are laid almost exclusively. The nymphs migrate to the soft skin around the anus, vulva and eyes. *H. tuberculatus* is the largest of the three measuring 5.5 mm and was originally described from the Indian buffalo. In Australia it has been found infesting camels and cattle but is not considered of any great importance (Roberts, 1952).

On pigs and horses

Haematopinus suis is the largest anopluran found on domestic animals and occurs in folds of the neck and jowl, and around the ears of pigs. It causes severe irritation resulting in a depressed growth rate (Seddon and Albiston, 1967). *H. asini* is about 3.5 mm in length and favours the roots of the mane, the forelock, round the butt of the tail and above the hooves (Roberts, 1952). Louse populations in domestic animals are reduced at the beginning of summer by the shedding of the coat. This has least effect on populations of *H. asini* because the coarse hairs of the mane and tail to which they attach their eggs are not shed (Murray, 1957c).

Linognathidae (see Figs 2.14 and 19.5)

Members of this family are distinguished by the absence of eyes and ocular points; by the second and third pairs of legs, which end in large stout claws, being considerably larger than the first pair; by the thoracic sternal plate being absent or weakly developed in *Linognathus*, but distinct in *Solenopotes*, and with no paratergal plates on the abdomen. These two genera have species which parasitize domestic animals. Most species of *Linognathus* are found on Artiodactyla, and a few on carnivores. More than 50 species of *Linognathus* have been described, and six occur on domestic animals. *L. setosus* parasitizes dogs, particularly long-haired breeds on which it infests the neck and shoulders (Emerson *et al.*, 1973).

On sheep

Two species – *L. pedalis*, the foot louse, and *L. ovillus* (Fig. 19.5A), the face louse – are most commonly found on sheep. Both frequent the hairy parts of the body, *L. ovillus* occurring on the face and lower jaw from which it spreads to the body (Murray, 1955), and *L. pedalis* occurring on the lower hairy parts of the body, namely shanks, belly and scrotum (Roberts, 1952). *L. pedalis* is able to survive for several days off the host, 50% being alive after 7 days at 12°C and 75% RH. In comparison, all *L. ovillus* were dead in 4 days under the same conditions (Murray, 1963a).

The eggs of both species hatch over a relatively narrow range of temperature around 35°C, and few hatch at 38°C or higher temperatures (Murray, 1960a, 1963b). During the summer months in Australia the temperature near the skin of the sheep may rise to over 45°C, and within the fleece the temperature may exceed 50°C (Murray, 1968). Such conditions would have an adverse effect on louse populations.

The fact that *L. pedalis* can survive for several days off the host raises the possibility of infection being acquired from contaminated pasture (Seddon and Albiston, 1967). On sheep *L. pedalis* is more sedentary and congregates in clusters, behaviour which does not occur in *L. ovillus* (Murray, 1963a). *L. stenopsis*, the goat louse, also occurs on sheep. It is larger than the other lice on sheep and can cause scabby, bleeding areas on the host (Seddon and Albiston, 1967). *L. africanus* parasitizes domestic sheep and goats (Radostits *et al.*, 1994).

On cattle (Fig. 2.14)

Linognathus vituli is the long-nosed sucking louse of cattle. In a trial in New Zealand, cattle with moderate infestations of *L. vituli* gained slightly more weight than the control group but the trial was conducted in winter on good pasture (Kettle, 1974), which means that the results are not directly comparable with those obtained with *H. eurysternus* in Arizona (Collins and Dewhirst, 1965). *L. vituli* is more common on calves and young stock, and more important among dairy cattle (Seddon and Albiston, 1967).

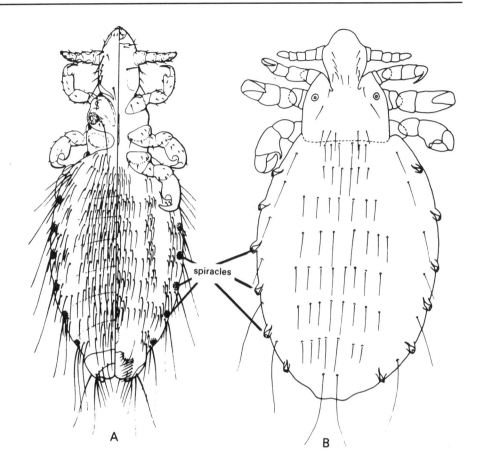

Fig. 19.5. (**A**) Female *Linognathus ovillus* – dorsal view presented in left half and the ventral view in the right half of the illustration. (**B**) Dorsal view of female *Solenopotes capillatus*. Source: *L. ovillus* from Ferris (1951).

Solenopotes capillatus is the smallest of the sucking lice on cattle and occurs in conspicuous clusters on the neck, head, shoulders, dewlap, back, anus and tail (Roberts, 1952). In *Solenopotes* the abdominal spiracles are borne on lightly sclerotized tubercles which project slightly from the body (Fig. 19.5B) (Ferris, 1951).

Mallophaga (Amblycera and Ischnocera)

The Mallophaga or chewing lice, of which about 2600 species have been described, are mainly (85%) ectoparasites of birds but they also occur on 9 of the 18 orders of living mammals (Emerson and Price, 1985; Kim, 1985c). They feed on fragments of feathers, hair and other epidermal products. In the Mallophaga the eyes are reduced or absent; there are no ocelli; the antennae are 3–5-segmented; the prothorax is free; and the mesothorax and metathorax may be fused. The mouthparts have been described in Chapter 4.

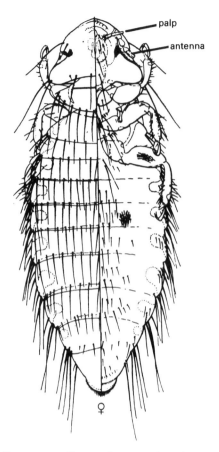

Fig. 19.6. Female *Menopon gallinae*, showing dorsal view on left and ventral view on right. Source: from Ferris, G.F. (1924) *Parasitology* 16, 58.

In the Amblycera the antennae are recessed in antennal grooves from which the last segment may protrude (Figs 19.7). The mandibles lie parallel to the ventral surface of the head and cut in a horizontal plane. The mesothorax and metathorax are usually separate; the antennae are 4-segmented; and the maxillary palps 2–4-segmented.

In the Ischnocera the antennae are 3- or 5-segmented, and are quite obvious because there is no antennal groove (Fig. 19.8). The mandibles are inserted more or less at right angles to the head and operate in a vertical plane. There are no maxillary palps, and the mesothorax and metathorax are fused to form the pterothorax.

Emerson (1956) lists 11 species of Mallophaga from the domestic chicken [4 Menoponidae (Amblycera) and 7 Philopteridae (Ischnocera)], of which 7 (2 Menoponidae and 5 Philopteridae) will be briefly considered briefly here.

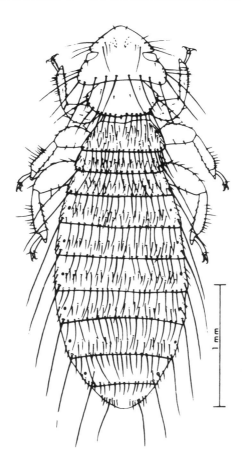

Fig. 19.7. Female *Menacanthus stramineus*. Source: Emerson *et al.* (1973).

Amblycera

The 836 species of Amblycera are arranged in seven families of which three (729 species) occur on birds and four (107 species) on marsupials and mammals in South America and Australia (Emerson and Price, 1985; Kim, 1985c). The Boopidae are parasites of marsupials, and are distinguished from other domestic mammal infesting lice by possessing two claws compared with one in the Anoplura and Trichodectidae (Figs 2.15 and 19.9). One species of Boopidae, *Heterodoxus spiniger*, occurs on domestic dogs in many parts of Australia, Africa, Asia and the Americas (Calaby and Murray, 1991). Guinea-pigs are frequently infested with two species of Gyropidae, *Gyropus ovalis* with an oval abdomen, broad in the middle, and *Gliricola porcelli*, a slender louse with the sides of the abdomen somewhat parallel (Emerson *et al.*, 1973).

Several species of Menoponidae occur on domestic birds of which the most important are *Menopon gallinae*, the shaft louse (Fig. 19.6), and *Menacanthus stramineus*, the chicken body louse (Fig. 19.7). *M. gallinae* is about 2 mm long and

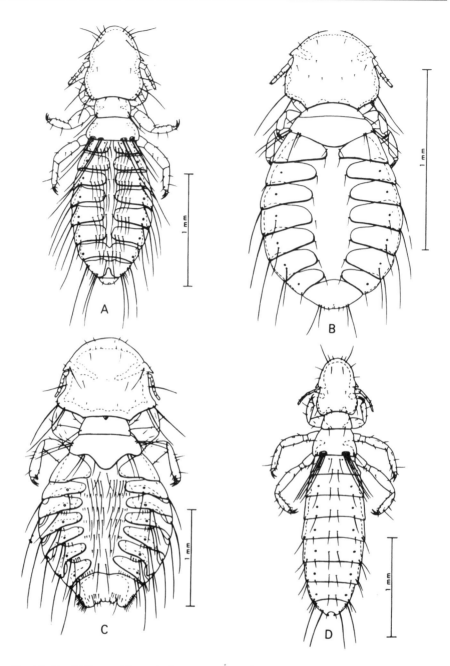

Fig. 19.8. Philopteridae which occur on poultry: (**A**) female *Cuclotogaster heterographus*; (**B**) female *Goniocotes gallinae*; (**C**) *Goniodes dissimilis*; (**D**) *Lipeurus caponis*. Source: from Emerson *et al.* (1973).

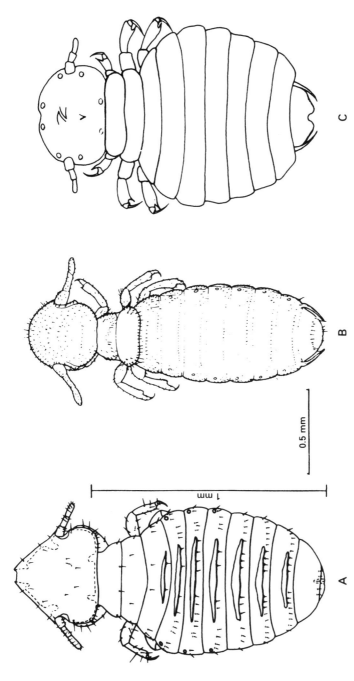

Fig. 19.9. (**A**) Female *Felicola subrostrata*. (**B**) *Damalinia ovis*. (**C**) *Trichodectes canis*. Source: (**A**) and (**C**) (redrawn) from Emerson *et al.* (1973). (**B**) from Calaby and Murray (1991).

lays its eggs singly at the base of a feather. It occurs on the thigh and breast feathers, and is claimed to be harmful to young fowls (Roberts, 1952; Emerson *et al.*, 1973). *M. stramineus* (Fig. 19.6) is the commonest and most destructive louse found on chickens and has a worldwide distribution. It is up to 3.5 mm long and deposits its eggs in masses at the base of feathers especially around the vent. It occurs on the breast, thighs and around the vent, causing a marked reddening of the skin, and 'sometimes gnaws through the skin or punctures the soft quills near the base and consumes the blood that oozes out' (Emerson *et al.*, 1973). Other species of Menoponidae occur on ducks, geese and pigeons but heavy infestations are rarely seen on these birds and little harm results.

Ischnocera

Three families of Ischnocera are recognized of which two are of veterinary importance: the Philopteridae (1460 species) on birds and the Trichodectidae (about 300 species) on mammals (Kim, 1985c). The Philopteridae have 5-segmented antennae and paired claws on the tarsi, and the Trichodectidae have 3-segmented antennae and single claws on the tarsi. The antennae of the trichodectid *Damalinia ovis* show sexual dimorphism but in both sexes the antennae contain olfactory and chemosensory pegs and a possible thermohygroreceptor (Clarke, 1990).

Philopteridae

On poultry

Five species of Philopteridae occur on poultry and have virtually a worldwide distribution. The chicken head louse, *Cuclotogaster heterographus* (Fig. 19.7A), occurs on the skin and feathers of the head and neck, where the lice feed on tissue debris and occasionally ingest blood. Severe infestations in young chickens are sometimes fatal (Emerson *et al.*, 1973). The fluff louse, *Goniocotes gallinae* (Fig. 19.7B), is a small louse which occurs on the down feathers anywhere on the body and generally causes little irritation. *Goniodes dissimilis* (Fig. 19.7C) and *G. gigas* are about 3 mm long and brown in colour. They are among the largest lice that are found on chickens, and occur anywhere on the body. In small numbers, they have little effect on their host. *G. gigas* is more prevalent in the tropics, and *G. dissimilis* in temperate regions (Emerson *et al.*, 1973). The fifth species, the wing louse, *Lipeurus caponis* (Fig. 19.7D), is not very active, and occurs on the underside of the wing and tail feathers (Roberts, 1952; Emerson *et al.*, 1973). Other philopterids occur on turkeys, ducks, pigeons and appear to do little harm.

Trichodectidae

On dogs and cats

Felicola subrostrata (Fig. 19.8A) is the only louse that occurs on cats. The head is triangular with the point directed forwards and notched at the apex. Ventrally there is a median longitudinal groove on the head which fits around the hair of the host. *F. subrostrata* is of minor importance, being found in large numbers only on elderly or sick cats especially if they are long-haired (Roberts, 1952). *Trichodectes canis* (Fig. 19.8C) is found on the dog and wild canids throughout the world. The head is broader than long being rectangular with rounded corners. It is found on the head, neck and tail attached to the base of hairs. Infestations are commoner on very young, very old or sick dogs. This louse can act as an intermediate host of the tapeworm *Dipylidium caninum* of which further details are given in Chapter 32.

On sheep

Damalinia ovis (Fig. 19.8B) is a small, pale species which occurs on sheep worldwide. For oviposition it requires both a suitable temperature, and fibres of an appropriate diameter to which eggs can be attached. The temperature at the skin surface of sheep is 37.5°C, and this is the temperature at which maximum oviposition of *D. ovis* occurs. The distribution of the eggs of *D. ovis* on sheep is governed by skin temperature. Low temperatures in certain areas of the body, e.g. legs and tail, inhibit egg laying. When the thickness of the fleece was 30–100 mm, most eggs (75%) were laid within 6 mm of the skin surface, and even when it was 100 mm deep few eggs (5%) were laid more than 12 mm from the skin surface (Murray, 1957a,b).

Eggs develop and hatch over the range 33–39°C, and are virtually independent of humidity over the range 7 to 75% RH. Very few eggs hatch at 92% RH, and this was attributed to humidity inhibiting hatching, because fully developed embryos were present in eggs which aborted (Murray, 1960b). In fleeces where the temperature ranged from 38°C at the skin surface to 15°C near the tip of the fleece, 69% of the mobile population (nymphs and adults) were within 6 mm of the skin surface and only 15% were more than 12 mm from the skin. The fleece depth varied from 25 to 75 mm (Murray and Gordon, 1969). When the tip of the fleece was shaded and warmed, adults and third-stage nymphs came to the surface. It is under these conditions that *D. ovis* spreads among a closely herded flock (Murray, 1968).

Populations of *D. ovis* are limited by a number of factors including shearing, when 30–50% of the population may be lost. Heavy rain can cause high mortality due to soaking the fleece, immersing all stages of the louse and maintaining a high humidity during the drying out period. In Australia in the summer, temperatures in a fleece exposed to the sun can reach 45°C at the skin surface in 5 to 10 min with temperatures near the fleece tip being 65–70°C. Such temperatures would quickly be lethal to all stages of the louse, and help to explain why louse populations are low in summer (Murray, 1963c, 1968).

On cattle

Damalinia bovis is a small, reddish-brown louse on cattle, particularly dairy cattle. This louse is commonest at the front end and on the back of cattle, spreading more widely in heavy infestations. Its effect on the host is minimal. In one experiment in New Zealand cattle developed far larger numbers of *D. bovis* than usual, and although their weight gain was less than in louse-free cattle, the difference was not significant (Kettle, 1974).

When two groups of Hereford cattle were fed on low and high planes of nutrition as part of an experiment with ticks, it was observed that heavy infestations of *D. bovis* occurred on the low nutrition group. This was attributed to reduced self-grooming and delayed shedding of the winter coat (Utech *et al.*, 1969). The difference between the levels of nutrition of the two groups is shown by the fact that over a period of 10 months the low nutrition group actually lost weight while the high nutrition group almost doubled in weight.

On horses and goats

Damalinia equi, which occurs on horses, cannot attach its eggs to the coarse hairs of the face, mane and tail, and consequently it suffers a loss of population when the coat is shed (Murray, 1957c). *D. caprae* parasitizes the common goat and several other species occur on the Angora goat.

Effect of Lice on Domestic Animals

The effect of lice on their host is a function of their density. A small number of lice on an individual presents no particular problem other than the prospect of a future population explosion. The life cycles of all lice are very similar with the duration of the egg stage being 1 to 2 weeks, the nymphal stages occupying 1 to 3 weeks, and the total time from egg to egg being 3 to 5 weeks (Table 19.2). Adults probably live for up to a month, although Arends (1991) considers that the normal life span of lice on poultry is several months.

The intrinsic rate of natural increase of *P. humanus* has been calculated at 0.111 (Evans and Smith, 1952) and for *D. ovis* 0.065 (Murray and Gordon, 1969). Both rates are per day and indicate a rate of increase of 11.7% and 6.7% per day, respectively. During winter when lice populations thrive, the numbers of *D. ovis* on a sheep are likely to increase from 4000 to more than 400,000 by the spring (Murray and Gordon, 1969). On sheep the main effect of lice is to lower the value of the wool crop, and this is sufficiently important for the Department of Agriculture in Western Australia to mount an intensive campaign against sheep lice (Wilkinson, 1978).

Heavy infestations of lice are associated with young animals or old animals in poor health and/or animals maintained in unhygienic conditions. Nevertheless, the irritation caused by modest populations of lice leads to animals scratching and rubbing, causing damage to fleece and hides, and heavily infested calves develop hairballs as a result of licking areas of irritation. In sheep *L. pedalis* can cause lameness, and *H. suis* spreads swine-pox among pigs. There is conflicting evidence on the effect of lice on production, i.e. milk, beef, and this is likely to be dependent upon other factors affecting the health of the host (Radostits *et al.*, 1994).

Treatment of louse infestations is by the application of the selected insecticide by dipping, jetting or the use of a pour-on preparation. A range of organo-

Table 19.2. Duration (in days) of various stages in the life cycle of various Phthiraptera.

Species	Egg	Nymph	Preoviposition	Egg–egg	Adult		Reference
					M	F	
Pediculus humanus	7–9	8–9	1–2	20	29	31	Buxton (1950)
Pediculus humanus	5–12	13	–	22–23	18	18	Evans and Smith (1952)
Haematopinus asini	12–14	11–12	–	–	–	–	Roberts (1952)
Haematopinus eurysternus	9–19	9–16	2–7	20–41	10	16	Roberts (1952)
Haematopinus quadripertusus	11	–	–	–	–	–	Roberts (1952)
Haematopinus suis	12–14	10	–	28–33	–	–	Roberts (1952)
Haematopinus tuberculatus	9–13	9–11	3	–	–	–	Roberts (1952)
Linognathus pedalis	17	21	5	43	–	–	Roberts (1952)
Linognathus setosus	5–12	–	–	–	–	–	Emmerson et al. (1973)
Linognathus vituli	8–13	–	–	21–30	–	–	Roberts (1952)
Menacanthus stramineus	7	17–30	–	–	–	–	Roberts (1952)
Menacanthus stramineus	4.5	9	–	14	–	12	Emmerson et al. (1973)
Cuclotogaster heterographus	–	–	–	14–21	months	months	Emmerson et al. (1973)
Goniodes gigas	7	–	–	28	19	24	Emmerson et al. (1973)
Lipeurus caponis	–	–	–	21–35	–	–	Emmerson et al. (1973)
Damalinia bovis	8	18	3	29	–	–	Roberts (1952)
Damalinia equi	8–10	–	–	–	–	–	Roberts (1952)
Damalinia ovis	9–10	21	3	34	–	–	Roberts (1952)
Damalinia ovis	10	21	3–4	34	–	–	Murray and Gordon (1969)
Felicola subrostrata	10–20	14–21	–	21–42	14–21	14–21	Emmerson et al. (1973)
Trichodectes canis	7–14	14	–	21–28	–	30	Emmerson et al. (1973)

phosphate insecticides, carbamates and synthetic pyrethroids are available but lice develop resistance and it is wise to change the insecticide regularly. Pour-on applications are easy to use but expensive and will not kill all lice on the host. On sheep they must be applied immediately after shearing. Following treatment for *L. pedalis*, sheep should be moved to paddocks which have been free of sheep for a month. Subcutaneous ivermectin will remove Anoplura but is not completely effective against Mallophaga (Radostits *et al.*, 1994), although, in the laboratory, the external application of avermectins has proved effective against *D. ovis* (Rugg and Thompson, 1993).

Host Specificity of Lice

Being permanent ectoparasites lice are only accidentally divorced from their host, and then their low powers of survival and high temperature threshold for activity severely limit the probability of their finding another host. Transference from one individual host to another occurs when animals are closely herded or penned, and in the close contact of mother and young with lice transferring from mother to young within a few hours of birth, e.g. lambs can become infested with *L. pedalis* within 48 h of birth (Roberts, 1952) and piglets have become infested with *H. suis* within 10 h of birth (Hiepe and Ribbeck, 1975).

Where lice occur on more than one host species the hosts are usually closely related. *L. stenopsis* and *L. africanus* occur on both sheep and goats; *T. canis* occurs on dogs and wild canids; and *H. tuberculatus* has been found not only on the Indian buffalo and cattle, but in Australia on the more distantly related camel (Roberts, 1952).

Where chickens and ducks are in close contact there may be transference of chicken lice to ducks, e.g. *M. gallinae*, *M. stramineus* and *C. heterographus* (Roberts, 1952). One unusual way in which lice may be spread from host to host is by carriage on another insect. Three *D. bovis* were found in a phoretic association with the hornfly, *Haematobia irritans*, but the number involved was very small, and this would appear to be a very minor route for dispersing *D. bovis* (Bay, 1977).

References

Arends, J.J. (1991) External parasites and poultry pests. In: Calnek, B.W. (ed.) *Diseases of Poultry*. Iowa State University Press, Ames, pp. 702–730.

Bay, D.E. (1977) Cattle biting louse, *Bovicola bovis* (Mallophaga: Trichodectidae), phoretic on the horn fly, *Haematobia irritans* (Diptera: Muscidae). *Journal of Medical Entomology* 13, 628.

Buxton, P.A. (1940) The biology of the body louse (*Pediculus humanus corporis:* Anoplura) under experimental conditions. *Parasitology* 32, 303–312.

Buxton, P.A. (1950) *The Louse*. Edward Arnold, London.

Calaby, J.H. and Murray, M.D. (1991) Phthiraptera (Lice). In: *The Insects of Australia, a Textbook for Students and Research Workers*. Melbourne University Press, Melbourne, pp. 421–428.

Clarke, A.R. (1990) External morphology of the antennae of *Damalinia ovis* (Phthiraptera: Trichodectidae). *Journal of Morphology* 203, 203-209.

Clay, T. (1974) Geographical distribution of the avian lice (Phthiraptera): a review. *Journal of the Bombay Natural History Society* 71, 536-547.

Collins, R.C. and Dewhirst, L.W. (1965) Some effects of the sucking louse, *Haematopinus eurysternus*, on cattle on unsupplemented range. *Journal of the American Veterinary Medical Association* 146, 129-132.

Emerson, K.C. (1956). Mallophaga (chewing lice) occurring on the domestic chicken. *Journal of the Kansas Entomological Society* 29, 63-79.

Emerson, K.C. and Price, R.D. (1985) Evolution of Mallophaga on mammals. In: Kim, K.C. (ed.) *Coevolution of Parasitic Arthropods and Mammals.* John Wiley, New York, pp. 233-255.

Emerson, K.C., Kim, K.C. and Price, R.D. (1973) Lice. In: Flynn R.J. (ed.) *Parasites of Laboratory Animals.* Iowa State University Press, Ames, pp. 376-397.

Evans, F.C. and Smith, F.E. (1952) The intrinsic rate of natural increase for the human louse. *Pediculus humanus* L. *American Naturalist* 86, 299-310.

Felman, Y.M. and Nikitas, J.A. (1980) Pediculosis pubis. *Cutis* 25, 482, 487-489, 559.

Ferris, G.F. (1931) The louse of elephants. *Haematomyzus elephantis* Piaget (Mallophaga: Haematomyzidae). *Parasitology* 23, 112-127.

Ferris, G.F. (1951) *The Sucking Lice.* Memoirs of the Pacific Coast Entomological Society, San Francisco.

Hiepe, T. von and Ribbeck, R. (1975). Die Schweinelaus (*Haematopinus suis*). *Angewandte Parasitologie 16, Merkblatt* No. 21, 1-13.

Ibarra, J. (1993) Lice (Anoplura). In: Lane, R.P. and Crosskey, R.W. (eds) *Medical Insects and Arachnids. Chapman & Hall,* London, pp. 517-528.

Jeu Minghwa, Fan Peifang and Jiang Fuming (1990) Morphological study of the adult stage of elephant louse *Haematomyzus elephantis* with light and scanning electron microscopy (Insecta: Rhyncophthiraptera). *Journal of the Shanghai Agricultural College* 8, 9-19.

Kettle, P.R. (1974) The influence of cattle lice (*Damalinia bovis* and *Linognathus vituli*) on weight gain in beef animals. *New Zealand Veterinary Journal* 22, 10-11.

Kim, K.C. (1985a) Evolution and host associations of Anoplura. In: Kim, K.C. (ed.) *Coevolution of Parasitic Arthropods and Mammals.* John Wiley, New York, pp. 197-231.

Kim, K.C. (1985b) Evolutionary aspects of the disjunct distribution of lice on Carnivora. In: Kim, K.C. (ed.) *Coevolution of Parasitic Arthropods and Mammals.* John Wiley, New York, pp. 257-294.

Kim, K.C. (1985c) Evolutionary relationships of parasitic arthropods and mammals. In: Kim, K.C. (ed.) *Coevolution of Parasitic Arthropods and Mammals.* John Wiley, New York, pp. 3-82.

Kim, K.C. (1988) Evolutionary parallelism in Anoplura and eutherian mammals. In: Service M.W. (ed.) *Biosystematics of Haematophagous Insects.* Clarendon Press, Oxford, pp. 91-114.

Lyal, C.H.C. (1985) Phylogeny and classification of the Psocodea, with particular reference to the lice (Psocodea: Phthiraptera). *Systematic Entomology* 10, 145-165.

Meleney, W.P. and Kim, K.C. (1974) A comparative study of cattle-infesting *Haematopinus*, with redescription of *H. quadripertusus* Fahrenholz, 1916 (Anoplura: Haematopinidae). *Journal of Parasitology* 60, 507-522.

Murray, M.D. (1955) Infestation of sheep with the face louse (*Linognathus ovillus*). *Australian Veterinary Journal* 31, 22-26.

Murray, M.D. (1957a) The distribution of the eggs of mammalian lice on their hosts. II.

Analysis of the oviposition behaviour of *Damalinia ovis* L. *Australian Journal of Zoology* 5, 19–29.

Murray, M.D. (1957b) The distribution of the eggs of mammalian lice on their hosts. III. The distribution of the eggs of *Damalinia ovis* (L.) on the sheep. *Australian Journal of Zoology* 5, 173–182.

Murray, M.D. (1957c) The distribution of the eggs of mammalian lice on their hosts. IV. The distribution of the eggs of *Damalinia equi* (Denny) and *Haematopinus asini* (L.) on the horse. *Australian Journal of Zoology* 5, 183–187.

Murray, M.D. (1960a) The ecology of lice on sheep. I. The influence of skin temperature on populations of *Linognathus pedalis* (Osborne). *Australian Journal of Zoology* 8, 349–356.

Murray, M.D. (1960b) The ecology of lice on sheep. II. The influence of temperature and humidity on the development and hatching of the eggs of *Damalinia ovis* (L.). *Australian Journal of Zoology* 8, 357–362.

Murray, M.D. (1963a) The ecology of lice on sheep. III. Differences between the biology of *Linognathus pedalis* (Osborne) and *L. ovillus* (Neumann). *Australian Journal of Zoology* 11, 153–156.

Murray, M.D. (1963b) The ecology of lice on sheep. IV. The establishment and maintenance of populations of *Linognathus ovillus* (Neumann). *Australian Journal of Zoology* 11, 157–172.

Murray, M.D. (1963c) The ecology of lice on sheep. V. Influence of heavy rain on populations of *Damalinia ovis* (L.). *Australian Journal of Zoology* 11, 173–182.

Murray, M.D. (1968) Ecology of lice on sheep. VI. The influence of shearing and solar radiation on populations and transmission of *Damalinia ovis*. *Australian Journal of Zoology* 16, 725–738.

Murray, M.D. and Gordon, G. (1969) Ecology of lice on sheep. VII. Population dynamics of *Damalinia ovis* (Schrank). *Australian Journal of Zoology* 17, 179–186.

Puchta, O. (1955) Experimentelle Untersuchungen über die Bedeutung der Symbiose der Kleiderlais *Pediculus vestimenti* Burm. *Zeitschrift für Parasitenkunde* 17, 1–40.

Rabinovich, J.E. (1972) Vital statistics of Triatominae (Hemiptera: Reduviidae) under laboratory conditions. *Journal of Medical Entomology* 9, 351–370.

Radostits, O.M., Blood, D.C. and Gay, G.C. (1994) *Veterinary Medicine – a Textbook of the Diseases of Cattle, Sheep, Pigs and Horses*. Baillière-Tindall, London.

Roberts, F.H.S. (1952) *Insects Affecting Livestock*. Angus and Robertson, Sydney.

Rugg, D. and Thompson, D.R. (1993) A laboratory assay for assessing the susceptibility of *Damalinia ovis* (Schrank) (Phthiraptera: Trichodectidae) to avermectins. *Journal of the Australian Entomological Society* 32, 1–3.

Seddon, H.R. and Albiston, H.E. (1967) *Diseases of Domestic Animals in Australia. Part 2. Arthropod Infestations (Flies, Lice and Fleas)*. Department of Health, Commonwealth of Australia.

Utech, K.B.W., Wharton, R.H. and Wooderson, L.A. (1969) Biting cattle-louse infestations related to cattle nutrition. *Australian Veterinary Journal* 45, 414–416.

Wigglesworth, V.B. (1941) The sensory physiology of the human louse *Pediculus humanus corporis* De Geer (Anoplura). *Parasitology* 33, 67–109.

Wigglesworth, V.B. (1972) *The Principles of Insect Physiology*. Chapman & Hall, London.

Wilkinson, F.C. (1978) New policy hits hard at sheep lice. *Journal of Agriculture, Western Australia* 19, 90.

Acari – Astigmata and Oribatida (Mange Mites, Beetle Mites)

The Acari are a group of arthropods which, apart from the parasitic species, has been largely neglected but this is being rectified with the publication of several comprehensive texts (Griffiths and Bowman, 1984; Woolley, 1988; Schuster and Murphy, 1991; Evans, 1992; Houck, 1994). Acarines are widely distributed throughout the world, being mainly terrestrial; but with a group of aquatic families in the Prostigmata. The freshwater aquatic mites (Hydrachnellae) form an ecological rather than a morphological group; the Halacaroidea includes over 300 species of marine mites (Woolley, 1988). About 30,000 species of Acari, belonging to more than 2000 genera, have been described; however, this is a small proportion of the half a million species which, it is believed, exist today (Krantz, 1978).

Classification

There is general agreement as to the main groups within the Acari but not on their names or classificatory level. The classification of Evans (1992) will be followed here and other terms in common use will be indicated in parentheses when a term is first mentioned. The Acari are a subclass of the Arachnida with two superorders, Anactinotrichida (=Parasitiformes) and Actinotrichida (=Acariformes). The tactile and chemosensory hairs of the Actinotrichida contain a layer of optically active material, actinochitin, which exhibits birefringence in polarized light. Actinochitin is lacking in the Anactinotrichida.

Krantz (1978) characterizes the Actinotrichida as being without visible stigmata posterior to the coxae of the second pair of legs, and with the coxae (epimeres) often fused to the ventral body wall (Fig. 20.3). The Anactinotrichida are defined as possessing one to four pairs of lateral stigmata posterior to the coxae of the second pair of legs, and with freely movable coxae (Fig. 23.1). The Actinotrichida contain three orders, the Astigmata (=Acaridida), the Prostigmata (=Actinedida) and the Oribatida (=Cryptostigmata); the Anactinotrichida has four orders, two of which have few species and are medically unimportant and two, the Mesostigmata (= Gamasida) and Ixodida (= Metastigmata), are of medical and veterinary importance.

The Astigmata form a well-defined natural group of slow-moving, weakly-sclerotized mites. It includes the economically important families Sarcoptidae and Psoroptidae, which cause mange and scab in domestic animals, a number of families containing parasites and commensals of mammals and birds, and families associated with stored foods.

The Prostigmata are the most heterogeneous of the acarine orders with adults ranging in size from 100 μm to 16 mm. Three of the 31 superfamilies recognized by Woolley (1988) are of medical or veterinary importance. The Trombidioidea includes the Trombiculidae, members of which are parasitic in the larval stage on vertebrates and some are vectors of scrub typhus, a rickettsial disease of humans. The Cheyletoidea are a heterogeneous assemblage of eight families which are, with few exceptions, parasites of arthropods and vertebrates including humans. The Pyemotoidea includes a few species of Pyemotidae which cause dermatitis in people and domestic animals.

The Oribatida are soil- or humus-dwelling mites which are of minor importance as intermediate hosts of certain tapeworms of domestic animals. The Mesostigmata are a large and successful group, most of which are predatory, but some are external or internal parasites of mammals, birds, reptiles and invertebrates. They range in size from 0.2 to more than 2.0 mm. Most of the mesostigmatic mites of medical and veterinary importance are in the Dermanyssoidea.

The Astigmata and Oribatida will be dealt with in this chapter, the Prostigmata and Mesostigmata in the next and the Ixodida or ticks in Chapters 22 and 23.

General Structure

Acarines are arachnids with a body typically composed of an anterior gnathosoma or capitulum and a posterior idiosoma (Fig. 20.1), which are separated by a circumcapitular suture. The gnathosoma resembles the head of a generalized arthropod only in that the mouthparts are appended to it. The brain lies in the idiosoma. The idiosoma can be subdivided into the area of the legs, the podosoma and the area behind the fourth pair of legs, the opisthosoma. In the Actinotrichida the four pairs of legs are usually arranged in anterior and posterior pairs, and the areas associated with them referred to respectively as the propodosoma and metapodosoma. Another term in use is prosoma which includes the gnathosoma and podosoma, but excludes the opisthosoma. The gnathosoma and propodosoma are known collectively as the proterosoma while the metapodosoma and opisthosoma are referred to as the hysterosoma.

The gnathosoma bears laterally the palps and more medially the chelicerae (Fig. 20.1). Typically, the palps are simple sensory appendages, which aid the acarine in locating its food. However, in some predatory Mesostigmata the palps are used to manipulate the prey. The palps are 1- or 2-segmented in most Astigmata and 5-segmented in the Mesostigmata. The chelicerae are 2-segmented in the Ixodida and 3-segmented in the other orders. The chelicerae are normally

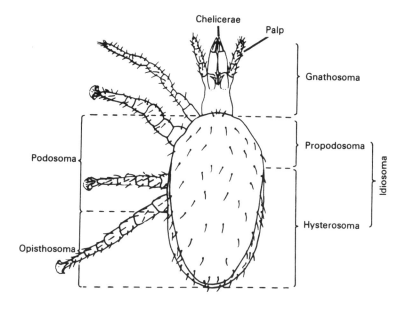

Fig. 20.1. Divisions of the body of a mite. Source: from Savory, T.H. (1977) *Arachnida*, Academic Press, London.

chelate with the third segment being movable (Fig. 20.3), but they are modified according to the feeding habits of the mite.

The legs of a typical mite are 6-segmented, seven if the pretarsus is included as a segment (Fig. 20.6). The other segments are the coxa, trochanter, femur, genu, tibia and tarsus. The tarsus carries the pretarsus or ambulacral stalk, bearing the ambulacrum composed of paired claws and/or a median empodium, and a terminal membranous pulvillus may also be present (Krantz, 1978). The empodium is variable and can be hair-like, pad-like, sucker-like or claw-like. The first pair of legs often differ from the other three pairs by being modified for special usage. Frequently they are longer and more slender, being used as sensory structures. They may also be modified in predatory species for capturing prey, and in some Rhinonyssidae (Mesostigmata) they function as surrogate chelicerae.

In small mites respiration may be entirely cutaneous but in larger ones gaseous exchange is facilitated by a complex tracheal system which opens to the exterior at the stigmata. In the Mesostigmata there is one pair of stigmata laterally located; in the Trombiculidae there are no stigmata in the larvae but one pair in the nymphs and adults, located between the bases of the chelicerae. Stigmata are absent in the Astigmata.

Transference of spermatozoa from the male to the female may be direct, as in the Astigmata; or by a complicated process known as podospermy in the Mesostigmata; while in the Trombiculidae, the male deposits a spermatophore which is subsequently picked up by the female.

Life Cycle

Female mites produce relatively large eggs and only a few, sometimes one or two, are laid at each oviposition. The size of the egg is conditioned by the fact that it must contain enough material to give rise to a larva which is capable of being self-sustaining. There is a minimum size for such an independent organism and it is this requirement which determines the minimum size of a mite egg.

When the egg hatches a hexapod larva emerges. Later it moults to become an octopod nymph. There may be one to three nymphal stages, usually two, before the production of the adult mite. The three nymphal stages are referred to as protonymph, deuteronymph and tritonymph. One or more of these developmental stages may be inactive and non-feeding. In the Astigmata the deuteronymph may be completely unlike the preceding and succeeding stages both in morphology and behaviour. Such heteromorphic nymphs (hypopus, plural hypopodes) are highly resistant to environmental stresses and some are adapted for dispersal by phoresy, i.e. attaching to passing animals (Houck and OConnor, 1991). A true tritonymph occurs in the Oribatida and certain Prostigmata and Astigmata. It is common current practice to restrict the term deuteronymph in the Astigmata to the heteromorphic nymph and refer to the two nymphal stages as protonymph and tritonymph (Arlian and Vyszenski-Moher, 1988; Burgess, 1994).

References to Mites and the Diseases They Cause

Excellent drawings, many of which have been reproduced in later works, are to be found in Hirst (1922). A wide selection of parasitic mites and the diseases they cause is given in Baker *et al.* (1956). The diseases of large farm animals are dealt with in Radostits *et al.* (1994) and those of small domestic animals in Muller *et al.* (1989). The chapter on 'Mites' by Yunker (1973) in Flynn's *Parasites of Laboratory Animals* has wider coverage than might be expected from the title of the book.

Astigmata

The Astigmata are small, thin-skinned mites lacking obvious shields. The coxae are sunk into the body and referred to as epimeres. The empodium is claw-like, and the membranous pulvillus is stalked (Fig. 20.2) or sessile. True paired claws are absent. Fertilized eggs are extruded through an anteroventral slit, the oviporus or genital opening. In both sexes the genital opening may be reinforced anteriorly by a pregenital plate or epigynium (Fig. 20.3). Most of the astigmatic mites of medical and veterinary importance are in the division Psoroptidia, and includes the Sarcoptidae and Psoroptidae in the Psoroptoidea; the Knemidokoptidae, Cytoditidae and Laminosoptidae in the Analgoidea; and the Pyroglyphidae in the Pyroglyphoidea. The division Acaridia contains the stored food mites which are important as sources of allergens.

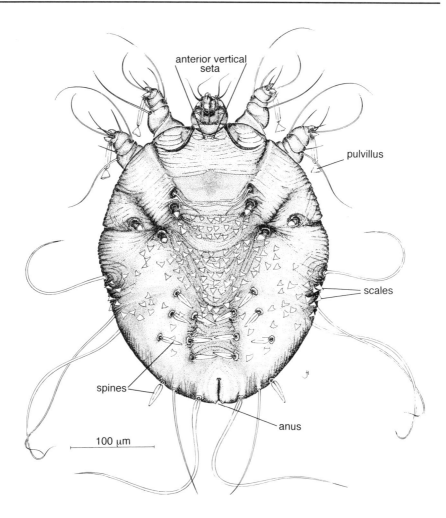

Fig. 20.2. Dorsal view of female *Sarcoptes scabiei*.

Sarcoptidae

Sarcoptid mites are parasitic throughout their life, burrowing into the skin of mammals. They are globose mites with the ventral surface somewhat flattened, the cuticle finely striated (Fig. 20.2) and the chelicerae adapted for cutting and paring. *Sarcoptes scabiei* is an important human and animal parasite and *Notoedres cati* is of minor veterinary importance.

Sarcoptes scabiei (see Figs 20.2 and 20.3)

Sarcoptes scabiei, the itch mite, causes scabies in humans and mange in a wide range of domestic and wild mammals throughout the world, and has been found on 17

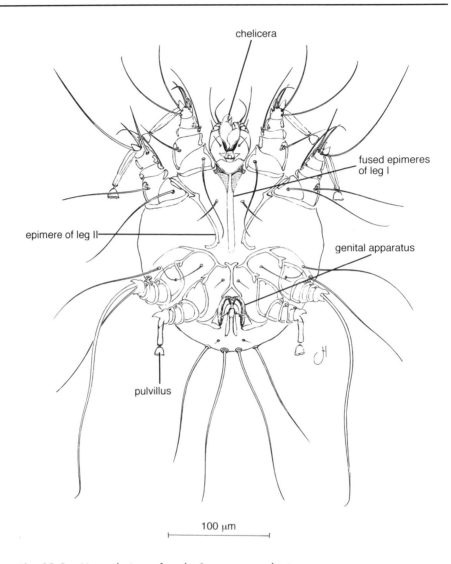

chelicera

fused epimeres
of leg I

epimere of leg II

genital apparatus

pulvillus

100 μm

Fig. 20.3. Ventral view of male *Sarcoptes scabiei.*

families belonging to 7 orders of mammals (Arlian, 1989). Hosts include chimpanzees and gibbons in the Primates; horses and tapirs in the Perissodactyla; domestic and wild bovids – cattle, sheep, goats, kudu and hartebeest – Old World Camels, South American llamas, pigs in the Artiodactyla; lions, foxes, wolves, dogs and ferrets among the Carnivora; rabbits and guinea-pigs in the Rodentia; and koalas and wombats in the Marsupialia.

In literature from eastern Europe *S. scabiei* is sometimes referred to as *Acarus siro*. In the West *Acarus siro* is the name given to a free-living mite associated with grain and grain products. The reason for the confusion is that Linnaeus thought that the itch mite and the grain mite were identical, and in 1758 he described them as subspecies *scabiei* and *farinae* of *A. siro*. Western workers retain *A. siro* for the free-living mite and transfer *scabiei* to *Sarcoptes*, while some eastern European workers, e.g. Petrov *et al.* (1976) retain *A. siro* for the mange mite.

Fain (1968) made a detailed study of *Sarcoptes* mites from a wide range of hosts with the intention of defining species or subspecies on the various hosts. He found that some morphological characters were stable and others were unstable. Variation in the unstable characters occurred: (i) within the same population; (ii) between populations on different hosts; and (iii) between populations on the same host but in geographically different localities. He recognized nine different forms but concluded that these were of little taxonomic value although they might be helpful in identifying the origin, i.e. host or locality, or the degree of adaptation of a population to a host. The conclusion is that there is one species, *Sarcoptes scabiei* which infests a very wide range of mammalian hosts.

The populations of *S. scabiei* infesting different mammalian species may differ more physiologically than morphologically. Populations from one host species do not readily establish themselves on another host species. Human infections with *S. scabiei* acquired from infected dogs, horses or pigs are considered to produce mild infestations which cure spontaneously (Yunker, 1973; Muller *et al.*, 1989; Chakrabarti, 1990; Burgess, 1994). Spontaneous recovery never happens with infestations of *S. scabiei* of human origin.

Life cycle

Females are to be found at the end of burrows in the horny layer of the skin. The burrows contain faeces and relatively large eggs, which are laid singly. The egg of *S. scabiei* var. *canis* hatches in 50–53 h giving rise to a hexapod larva (Arlian and Vyszenski-Moher, 1988) which can move on the surface of the skin practically as rapidly as the adults. Larvae find shelter, and presumably also food, by entering hair follicles. In 2 to 3 days the larva moults into an octopod protonymph, which is also to be found in hair follicles. This is followed by the tritonymph which gives rise to an adult male or immature female. At this stage both adults are about 250 µm long.

Both sexes make short burrows (< 1 mm) in the skin. Pairing probably occurs on the surface of the skin, and then the female makes a permanent burrow. As the ovaries develop the female increases in size so that the mature female is 300 to 500 µm long. The female never voluntarily leaves the burrow but, if removed undamaged from a burrow, will construct another. The female takes about an hour to bury herself in the horny layer of the skin using her chelicerae and the 'elbows' of the first two pairs of legs. The rate of extension of the burrow varies from 0.5 to 5 mm per day.

The female takes 3 to 4 days to become mature and then lays one to three eggs a day during a reproductive period lasting about two months. The total length of the life cycle from egg to egg is of the order of 10 to 14 days during which there is a mortality of about 90%, i.e. 10% survival (Mellanby, 1972). Making certain assumptions these data can be used to calculate the growth of a population of *S. scabiei*. Assuming that a female lays two eggs a day during a reproductive life of 60 days, and that the development cycle from egg to ovigerous female is 12 days with a survival of 10% and a 1:1 sex ratio, there is a 17-fold increase in the ovigerous female population in two months. Arlian (1989) recorded the development time from egg to adult as 10–13 days which would produce a longer egg to egg cycle than that used in the above calculations and a lower, but still substantial, rate of increase.

Diagnosis of scabies or mange

A firm diagnosis of scabies or mange must be based on recovery of mites from the affected host. In humans this requires the recognition of the burrows of the female mite in the skin, removal of the mite, and its examination under suitable magnification. With animals a skin scraping is made of the infected area. The material removed may be examined directly, or preferably after disrupting the keratin by boiling for a few minutes in 10% caustic soda or potash. The fluid is then centrifuged and the sediment examined. In skin scrapings males are rarer than females, and this probably reflects their being shorter-lived.

In diagnosing scabies it is important to appreciate the fact that the distribution of the rash on the body bears no relation to the distribution of the mites. Nearly two-thirds of the mites are to be found on the hands and wrists with the remainder being more or less equally distributed between the elbows, feet and genital area. The rash develops bilaterally being concentrated on the axillae, waist, and inner and posterior parts of the upper thighs and buttocks. The use of contact microscopy allows the examination of a large number of lesions to be made rapidly without discomfort to the patient (Burgess, 1994).

Sarcoptes scabiei may be readily identified by its size, shape and morphology. The skin is striated but dorsally bears a central patch of raised scales which extend in lesser density posterolaterally. Also on the dorsal surface there are three pairs of lateral spines about midway along the body and six or seven pairs of spines posteromedially. Dorsally on the propodosoma, behind the gnathosoma, there is a pair of anterior vertical setae, which are of taxonomic significance (Figs 20.2 and 20.3).

In both sexes the pretarsi of legs I and II bear empodial claws and stalked pulvilli (Figs 20.2 and 20.3); the latter are sometimes referred to as suckers. In life they function as adhesive flaps which grip the substrate and give the mite purchase for movement. The epimeres of the first pair of legs are fused in the midventral line (Fig. 20.3). Legs III and IV in the female end in long setae and lack stalked pulvilli. They are located on the ventral surface and are not visible in dorsal view. In the female the oviporus is a transverse slit in the middle of the ventral surface of the body and the copulatory bursa, in which the spermatozoa are deposited, is on the dorsal side just anterior to the anus, which is terminal and slightly dorsal (Fig. 20.2).

The male is similar to, but smaller than, the female and is distinguished by the presence of stalked pulvilli on the fourth pair of legs between which is the obvious, sclerotized genital apparatus (Fig. 20.3). The nymphs are similar to the female but smaller and lack an oviporus, while the larvae resemble nymphs but have only one pair of legs posteriorly.

Scabies in humans

Survival of *S. scabiei* off the host is dependent upon a high relative humidity restricting dehydration. Survival is best at low temperatures and high relative humidities but, like other ectoparasites, *S. scabiei* is virtually immobile at temperatures below 20°C (Mellanby, 1972). Survival beyond 48 h at normal ambient temperatures is rare (Burgess, 1994). Live *S. scabiei* have been recovered from the home environment of scabietic patients indicating that transmission by fomites is possible (Arlian, 1989).

The incidence of scabies in the human population increases through the autumn, into winter and declines in spring (Burgess 1994) and there is evidence that epidemics of human scabies occur in 30 year cycles with 15 year gaps between (Arlian, 1989). Transmission of *S. scabiei* among a population is dependent upon prolonged and close personal contact. Under normal working conditions transmission of mites from one human being to another is highly unlikely. Less unlikely is transmission by fomites, but the greatest risk is in close bodily contact under warm conditions, such as in bed, when the mite will have optimal mobility. *S. scabiei* var. *hominis* exhibits a strong thermotaxis, moving to the warmest part of a temperature gradient and *S. scabiei* var. *canis* responds positively to both temperature and host odour (Arlian *et al.*, 1984). The most likely stage in the life cycle of the mite to establish a new infestation is a newly inseminated female mite whose next activity is establishing a permanent burrow.

In humans there is a period of one or more months before symptoms of scabies become manifest. This appears to be a period of sensitization during which the host does not react. Following sensitization the host reacts strongly by scratching and thus induces secondary infections, complicating the direct effect of the mite. Scratching appears to exert a certain degree of control because the average number of ovigerous female *S. scabiei* on an infected person was only 11 with most patients having 1 to 5 adult female mites, and only 3% had more than 50 ovigerous female mites on them. The highest number found on one individual was 511. People that have been cured of scabies by treatment with an acaricide such as benzyl benzoate retain their sensitivity and, when reinfected, react immediately. In theory there should ultimately be a stage when tolerance is reached, and this may be the case in Norwegian or crusted scabies in which there are huge numbers of mites in a host who does not itch or scratch (Mellanby, 1972). This condition frequently involves mites being present on the nail beds or nail plates on hands or feet. The existence of asymptomatic carriers with high mite counts has serious implications for the control of outbreaks of scabies and emphasizes the importance of treating all contacts whether symptomatic or not (Burgess, 1994).

Treatment

Benzyl benzoate and lindane, the γ isomer of benzene hexachloride, have been used successfully to cure scabies. Some lindane enters the blood stream and is stored in body fat. For that reason lindane is no longer used and has been superseded by the use of malathion and permethrin, a synthetic pyrethroid insecticide. The malathion used must be pure, since toxic reactions may occur if low grade agricultural material is used (Burgess, 1994).

Mange in animals

On animals *S. scabiei* is found more frequently on the sparsely-haired parts of the body such as the face and ears of goats, sheep and rabbits; the hock, muzzle and root of the tail in dogs and foxes; the inner surface of the thighs, under side of neck and brisket, and around the root of the tail in cattle, from which it may spread and involve the whole body surface in as little as 6 weeks; the head and neck of equines; and the trunk of pigs. The burrowing and feeding of the mites in the skin cause irritation and consequential scratching which leads to inflammation and exudations which form crusts on the skin. If left untreated the skin wrinkles and thickens with proliferation of the connective tissue. This is followed by depilation (loss of hair). Small foci of infection do not appear to affect the health of an animal adversely but under certain conditions infestation may spread all over the body and, if untreated, cause death of domestic animals. Mange is a disease associated with animals in poor condition and therefore is commoner at the end of the winter or in the early spring (Radostits *et al.*, 1994).

The spread of *S. scabiei* among animals is by close contact and this is facilitated by close herding of domestic animals, and in wild animals by living in family or social groups. Large numbers of mites can be found in the ears of normal sows and transmitted to piglets after farrowing. When a case of mange is diagnosed it is necessary to treat all animals that have been in contact with the infected animal because the early stages of infection may be clinically inapparent. The external application of organophosphorus insecticides has proved effective in controlling *S. scabiei*, as has ivermectin given subcutaneously or orally. The latter is now the preferred treatment for sarcoptic mange, a single dose being completely effective in pigs. A second treatment may be necessary for heavy infestations in goats. The length of time that *S. scabiei* can survive off its host will depend upon environmental conditions. Quarters previously occupied by infected animals should either be left in a dry state for 3 weeks or be treated with an acaricide (Radostits *et al.*,1994). Kraals previously occupied by infected goats were free from infestation when left unoccupied for 17 days (Du Toit and Bedford, 1932).

Notoedres

More than 20 species of *Notoedres* have been described, most of them being parasites of tropical bats (Fain, 1965). Three species are of interest to the veterinary entomologist, and one, *N. cati*, is important. *N. muris* occurs on rats throughout the world, including laboratory colonies (Yunker, 1973) and *N. musculi* on the house mouse in Europe.

Notoedres cati (Yunker, 1973; Muller *et al.*, 1989)

Notoedres cati is a mange mite of cats which on occasions may also infest dogs and cause a transient dermatitis in humans. A morphologically identical form occurs on

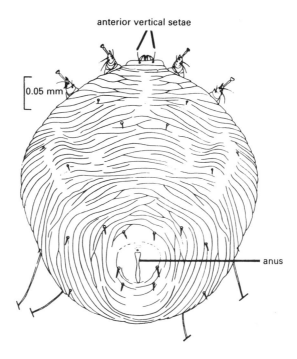

Fig. 20.4. Dorsal view of a female *Notoedres muris*. Source: from Lavoipierre M.M.J. (1964) *Journal of Medical Entomology* 1, 11. Reproduced by permission of the editor.

rabbits, but since it is difficult to transmit the cat parasite to the rabbit, two sibling species may be involved. Cats and rabbits share a predator–prey relationship in which it would be easier for the parasite to move from the prey to the predator rather than the reverse. In both hosts *N. cati* attacks the head and ears and, more rarely, in advanced cases, the legs, genitalia and perineum. In cats the infestation begins at the nape and spreads to the ears, to the head, and to the anterior region of the neck. The original lesion is the size of a pinhead but as it spreads, a crust develops and hair falls out. Notoedric mange is highly contagious and intensely pruritic among cats.

Diagnosis is by recovery of mites from skin scrapings. *N. cati* is similar to *S. scabiei* having stalked pulvilli on legs I and II in all stages, and on leg IV in the male. It is considerably smaller, the female being 225 μm and the male 150 μm. In *N. cati* the anus is located on the dorsal surface, as it is in *N. muris* (Fig. 20.4); there are no projecting scales, but mid-dorsally the striae are broken into a scale-like pattern; and stout setae replace the lanceolate spines of *S. scabiei* (Baker *et al.*, 1956).

The life cycle of an unnamed species of *Notoedres*, probably *N. muris*, was studied by Gordon *et al.* (1943). In outline, the life cycle is similar to that of *S. scabiei* but differs in that transmission from host to host is by larvae or nymphs. The female makes a burrow in the stratum corneum in which eggs are deposited. Larvae and nymphs may stay in the female's burrow or move on to the surface of the skin where they make small pits in which they moult. All moults may occur

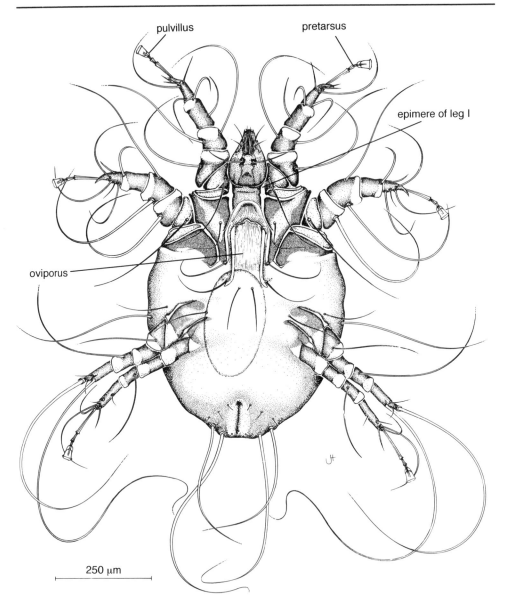

pulvillus   pretarsus

epimere of leg I

oviporus

250 μm

Fig. 20.5. Ventral view of female *Psoroptes ovis*.

in the pit made by the larva, or each stage may make a separate pit. The immature female remains in the moulting pit until she has been inseminated when she forms a permanent burrow. The cycle from egg to adult takes 17 days, and maturation and deposition of·the first egg 4 to 5 days so that the generation time is 3 weeks. The ovigerous female lays about 60 eggs in a lifetime of 2 to 3 weeks at a rate of 3 to 4 eggs per day.

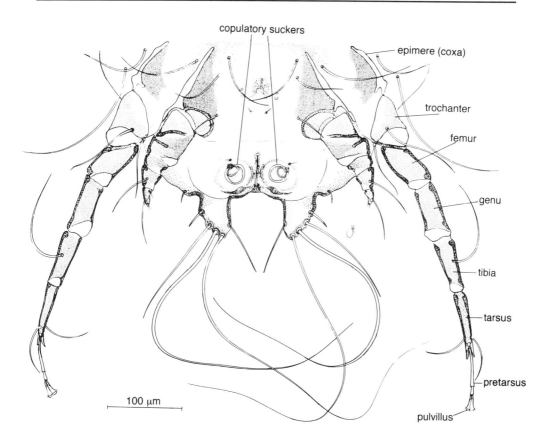

Fig. 20.6.   Ventral view of hysterosoma of male *Psoroptes ovis*.

Psoroptoidea – Psoroptidae

Members of the Psoroptidae are oval, non-burrowing mites, which are parasites on the skin of mammals. The third and fourth pairs of legs are usually visible from above, the epimeres of the first pair of legs are not fused (Fig. 20.5), and there are no vertical setae on the propodosoma. There are two nymphal stages in the life cycle. The male has prominent adanal copulatory suckers (Figs 20.6, 20.7 and 20.8) which engage with copulatory tubercles on the female tritonymph (= pubescent female) (Fig. 20.7). On the ventral surface of the ovigerous female, just posterior to the second pair of legs, there is an obvious inverted U-shaped oviporus through which the eggs are passed (Fig. 20.5). Three genera, *Psoroptes*, *Chorioptes* and *Otodectes*, are of economic importance. The status of species attributed to these genera have been subjected to critical study by Sweatman (1957, 1958a,b,c).

Psoroptes

Psoroptes is a cosmopolitan genus of obligate ectoparasites which cause a debilitating dermatitis involving hair or wool loss and pruritic scab formation. All stages of *Psoroptes* are distinguished by sucker-like pulvilli borne at the end of long, jointed pretarsi (peduncles) (Figs 20.5 and 20.6). Sweatman (1958c) recognized five species – *P. ovis*, a cosmopolitan body mite causing scab in sheep and body mange in cattle and horses; *P. equi*, a body mite of horses found in England and South Africa; *P. cuniculi*, a cosmopolitan ear mite of rabbits, goats, horses, sheep, big-horn sheep and deer; *P. natalensis*, a body mite of domestic cattle occurring mainly in the southern hemisphere (South Africa, South America and New Zealand); and *P. cervinus*, an ear mite on the American bighorn sheep and a body mite on the wapiti.

It has proved difficult to separate the first three species on morphological characters or on host specificity. The important unanswered questions are: how many species of *Psoroptes* are responsible for mange outbreaks and how host specific are they? One emerging trend is that *Psoroptes* species responsible for mange in the ears of rabbits, goats and horses do not infest the bodies of sheep or cattle and vice versa. New techniques will be required to decide how different body and ear mites really are and how many species there are in each group (Strong and Halliday, 1992). The position is further complicated by the fact that cross-mating of *P. ovis* and *P. cuniculi* produces viable fertile offspring in spite of the fact that the parent strains could be easily distinguished morphologically (Wright *et al.*, 1983).

Psoroptes ovis

The relationship between *P. ovis* on sheep and cattle is not clear. In Britain the two populations appear to be distinct because in that country *P. ovis* from cattle does not survive on sheep (Kirkwood, 1985). This separation finds support from a survey of other countries of which 12 reported having *P. ovis* on sheep but not on cattle and 16 countries reporting the reverse, *P. ovis* on cattle but not on sheep (Jensen *et al.*, 1979).

Life cycle and recognition

The ovigerous *P. ovis*, 750 μm long, lays oval, glistening white eggs, about 250 μm long. The larva which emerges has pulvilli on legs I and II, and leg III terminates in two exceptionally long setae. The larva is about 330 μm long. It moults to a protonymph with pulvilli on legs I, II and IV and long setae on leg III. Legs III and IV are shorter and less stout than legs I and II. The female protonymph and tritonymph are similar but bear copulatory tubercles posterodorsally (Fig. 20.7).

The adult male is readily recognizable by its possession of paired adanal copulatory suckers and paired posterior lobes which carry two long and three shorter setae (Fig. 20.6). Leg III is the longest, leg IV the shortest, and legs I and II markedly stouter than legs III and IV. Pulvilli are borne on legs I, II and III. Female tritonymphs attach to males and remain so until they moult to ovigerous females, when insemination occurs. The legs of the ovigerous females are more or less equal with pulvilli on all except leg III which bears two long setae. A female will live 11 to 42 days and lay between 30 and 40 eggs at a rate of one to five eggs per day, the rate being inversely related to air temperature (Baker *et al.*, 1956).

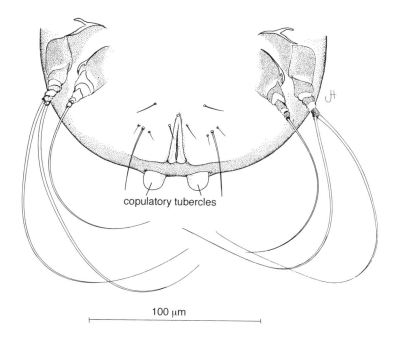

Fig. 20.7. Ventral view of hysterosoma of female tritonymph (pubescent female) of *Psoroptes ovis*.

In the developmental stages of the life cycle there is a period of active feeding, followed by a quiescent immobile phase prior to moulting. Under optimal conditions the quiescent phase lasts about a day and the active feeding phase about two days. The minimum duration of the life cycle from egg to egg is 11 days for *P. ovis* (Downing, 1936) while *P. cuniculi* and *P. equi* require about three weeks for the complete cycle (Sweatman, 1958c).

Effect on sheep and cattle

Psoroptes ovis causes a highly contagious disease of sheep and cattle. In cattle, lesions appear on the withers, neck and around the root of the tail from which, in severe cases, the condition may spread to the rest of the body. In sheep, lesions may occur on any part of the body but in badly affected animals they are most obvious on the sides. The adults puncture the epidermis and feed on lymph and tissue fluids and cause the formation of scabs. As the lesion increases a dry scab forms in the centre and is surrounded successively by zones of moist crust and moist, reddened skin. The mites are most active in the moist areas feeding at the periphery of the scab which extends rapidly (Radostits *et al.*, 1994). In sheep this leads to loss of fleece, markedly reducing wool production. The disease spreads rapidly and in 6 to 8 weeks three-quarters of the body of the sheep may be covered with crusts or be denuded of wool (Baker *et al.*, 1956).

Psoroptic mange is commoner in autumn and winter. In the summer the disease enters a latent phase during which the skin recovers and the animals appear normal, only to relapse in the following winter. Latency is due in part to the loss of mites during shearing or hair shedding, improved host nutrition and condition, and increased physical activity including grooming (Strong and Halliday, 1992). There is no diapausing stage and in summer mites can be found anywhere on the body surface and the proportions of the various stages are more or less constant throughout the year (Roberts *et al.*, 1971; Blachut *et al.*, 1973). In winter the ovigerous female lives a shorter time but produces more eggs, contributing to the build-up of a high winter population and the return of disease (Downing, 1936).

Transmission

Psoroptes ovis spreads rapidly through herds and flocks by direct contact between infested and clean animals. Under favourable conditions *P. ovis* can remain infective for 17 days after its removal from a host (Strong and Halliday, 1992). Nevertheless no natural transmission of *P. ovis* has been reported where a minimum of 10 days has elapsed between the removal of contaminated sheep and the introduction of clean ones (Wilson *et al.*, 1977).

Diagnosis and treatment

Diagnosis of psoroptic mange is made by finding the mites in skin scrapings made from the moist areas at the periphery of the scabs. Identification of *Psoroptes* to species is a matter for the specialist. In Britain sheep scab is controlled by sheep being dipped twice a year during a 4-week period in summer and a 6-week period in winter. Eggs are not killed by dipping and the active ingredient must give 4 weeks' protection to kill newly emerged larvae and nymphs. Diazinon and propetamphos are approved acaricides. As *Psoroptes* mites do not take blood, they are less susceptible to systemic insecticides such as ivermectin (Kirkwood, 1985).

Eradication

The last introduction of *P. ovis* into Australia in 1884 was eradicated by 1896 and since then Australia has remained free of sheep scab (Seddon and Albiston, 1968). Britain became free of sheep scab in 1952 but then suffered outbreaks in 1973 and 1976–1977 and compulsory dipping of sheep was introduced in 1982. In the USA the last reported outbreak of *P. ovis* in sheep was eradicated in 1970. This coincided with a decline in the number of sheep being raised and an increase in psoroptic mange on cattle (Strong and Halliday, 1992).

Chorioptes

Chorioptic mange is the commonest form of mange in horses and cattle. In horses the mites occur on the lower parts of the legs and are rarely found on other parts of the body. They are a source of irritation and reduce the performance of infected horses. In cattle *Chorioptes* mites most commonly cause lesions at the base of the tail, on the perineum, and the back of the udder. Chorioptic mange is largely a winter disease and in the summer the mites are to be found on the area above the hooves on the hind legs. In cattle the damage caused by chorioptic mange is mainly aesthetic. A single application of 0.25% solution of crotoxyphos is claimed to give effective control and leave no residue in meat or milk. Infection is passed from animal to animal by contact and possibly by grooming tools (Radostits *et al.*, 1994). In sheep the mites affect the woolless areas particularly the lower parts of the hind legs and scrotum, and cause a decrease in fertility (Rhodes, 1976).

Based on a detailed biological and morphological study of *Chorioptes* Sweatman (1957, 1958a) concluded that only two species are involved: *C. bovis* on horses, cattle, sheep, goats and llamas, and *C. texanus*, described originally from goats, has been found on cattle in Brazil (Faccini and Massard, 1976). Separation of species is a matter for the specialist. *C. bovis* and *C. texanus* are identical in all stages except for the adult male, in which there are differences in the lengths of setae on the paired posterior processes, but similar criteria have proved ineffective in separating species of *Psoroptes* (Strong and Halliday, 1992).

Unlike *Psoroptes*, *Chorioptes* mites do not pierce the skin but feed on skin debris and it has been possible, therefore, to rear them in the laboratory on epidermal material from a range of herbivores including deer, antelopes, water buffalo, zebu, zebra and donkey, from which Sweatman (1957) concluded that chorioptic mites were potentially ubiquitous. He commented that survival was less related to host species than to differences between individual hosts of the same species.

The life cycle of *C. bovis* is similar to that of *P. ovis* but at each stage *C. bovis* is considerably smaller. The ovigerous female *C. bovis* is little more than 300 μm in length. *Chorioptes* differs from *Psoroptes* in that the pretarsi are not jointed (Fig. 20.8). Male *C. bovis* have pulvilli on all four pairs of legs, and two long broad flat setae, together with three normal setae of varying lengths, on well-developed paired posterior lobes (Fig. 20.8). In the nymphal stages leg IV ends in moderately long setae. The complete life cycle takes about 3 weeks, and an egg-laying ovigerous female may live for 3 weeks, while non-laying females and adult males may live for up to 7 or 8 weeks. Ovigerous females lay a total of 3 to 17 eggs with an average of 9.5 eggs per female (*n* = 56) (Sweatman, 1957).

Otodectes cynotis

Otodectes mites generally live deep in the ear canal, near the eardrum of dogs, foxes, cats and ferrets but lesions have also been seen on the body. These mites are mainly parasites of carnivores, although they have been collected from the body of a captive white tailed deer (Sweatman, 1958b). They resemble *Chorioptes* in being of a similar size and having unjointed pretarsi. They can be reared *in vitro* on epidermal debris and hair from the inside of the ears of carnivores. As a result it has been shown that mites from the ears of dogs, red fox, cat and ferret are biologically and morphologically identical, and should be referred to the same species, *Otodectes cynotis*. In addition to the hosts already listed, the life cycle has been completed *in vitro* on ear debris from other carnivores including coyote, timber wolf and black bear.

The life cycle of *O. cynotis* is similar to that of other psoroptids. In the protonymph leg IV is greatly reduced and has no pulvillus. In the tritonymph leg IV is lacking, and copulatory tubercles occur in both sexes. The male has pulvilli

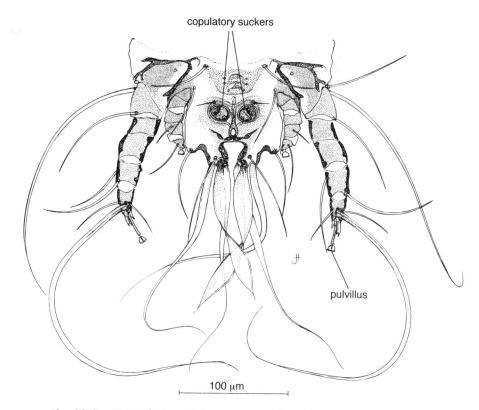

Fig. 20.8. Ventral view of hysterosoma of male *Chorioptes bovis*.

on all four pairs of legs, copulatory suckers, weakly developed posterior processes and a slightly emarginate hind margin to the body (Fig. 20.9). Males attach to tritonymph and copulation occurs as the female emerges. Females that are not attached are not inseminated at ecdysis and are infertile (Muller *et al.*, 1989). In the ovigerous female leg IV is reduced and lacks a pulvillus.

In heavily infested cats and dogs convulsions may occur (Yunker, 1973). This condition requires treatment with a suitable insecticide. In kennels and catteries, since the mites can survive for some time off their hosts, it is necessary to treat the premises weekly for 4 weeks with a residual insecticide such as diazinon or malathion. Individual animals may be treated with one or two subcutaneous injections of ivermectin or the application of Amitraz in mineral oil as ear drops twice a week (Muller *et al.*, 1989).

Analgoidea

Knemidokoptidae

Twelve species of Knemidokoptidae have been described of which three are of veterinary importance (Fain and Elsen, 1967). *Knemidokoptes mutans* and *K.*

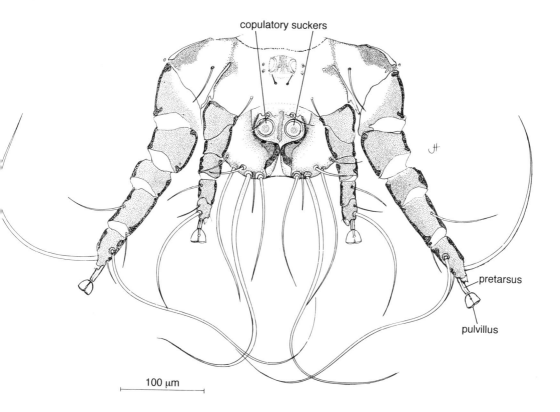

copulatory suckers

pretarsus

pulvillus

100 µm

Fig. 20.9. Ventral view of hysterosoma of male *Otodectes cynotis*.

gallinae infest poultry, and *K. pilae* is common on caged parakeets (Yunker, 1973). Female knemidokoptid mites are about 400 µm long but have neither spines, sharp pointed scales nor anterior vertical setae. Anteriorly on the mid-dorsal surface there are two sclerotized, more or less parallel, longitudinal bands, which are connected posteriorly by a less well-developed transverse band (Fig. 20.10). In the female the epimeres of the first pair of legs are concave laterally and do not meet in the midline (Fig. 20.11). In the male the epimeres of the first pair of legs fuse in the midline and have a posteromedian extension (Fig. 20.12). Stalked pulvilli are present on all legs of the male and larva, but are absent in the nymphal stages and female. The female is viviparous, and there are one larval and two nymphal stages.

Knemidokoptes mutans (Yunker, 1973; Arends, 1991)

Knemidokoptes mutans causes scaly leg in domestic poultry. At first the infestation is localized on the legs to the lower ends of the tarsus and digits, where the epidermal scales swell up and exude a whitish floury powder. This may develop into a thick, nodular, spongy crust, and in advanced cases the comb and neck may also be affected. The disease develops slowly over many months while the bird loses its appetite and wastes away. Diagnosis requires finding the adult mites on the

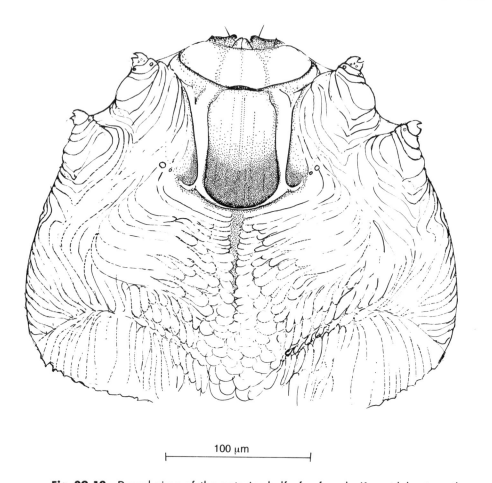

100 μm

Fig. 20.10 Dorsal view of the anterior half of a female *Knemidokoptes pilae*.

underside of the crust where ovigerous females will be found surrounded by a proliferation of epidermal cells.

Female *K. mutans* can be distinguished from *K. gallinae* by the presence, mid-
dorsally, of rounded or oval plaques resembling smooth scales (Fig. 20.10). In *K.
gallinae* the dorsal surface has only regular striations, some of which may be very
finely toothed (Fig. 20.13). Male *K. gallinae* have a pair of copulatory suckers
posteroventrally. These are absent in male *K. mutans*.

Knemidokoptes gallinae (Arends, 1991)

Knemidokoptes gallinae has been variously referred to as *Neocnemidocoptes
gallinae* (Fain, 1974) and *Mesoknemidokoptes laevis gallinae* (Krantz, 1978). *K.
gallinae* infests poultry, pheasants and geese in which it causes depluming itch,
characterized by the loss of feathers over extended areas of the body. The mite
attacks the base of the feathers on the back, head, neck, around the vent and on
the breast and thighs. The condition can be diagnosed by plucking a few feathers

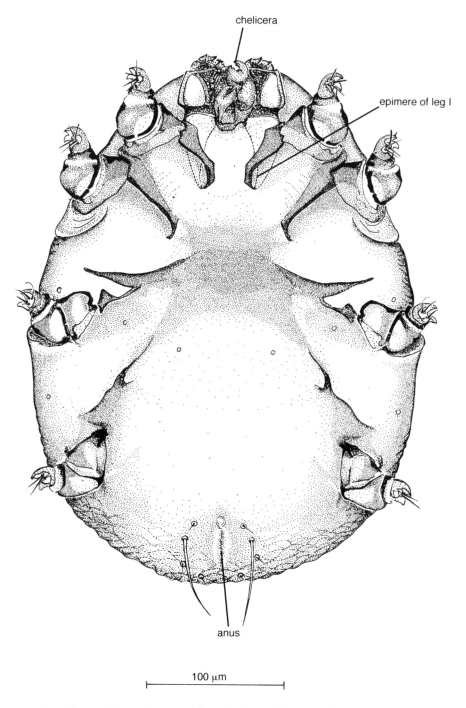

Fig. 20.11. Ventral view of female *Knemidokoptes pilae*.

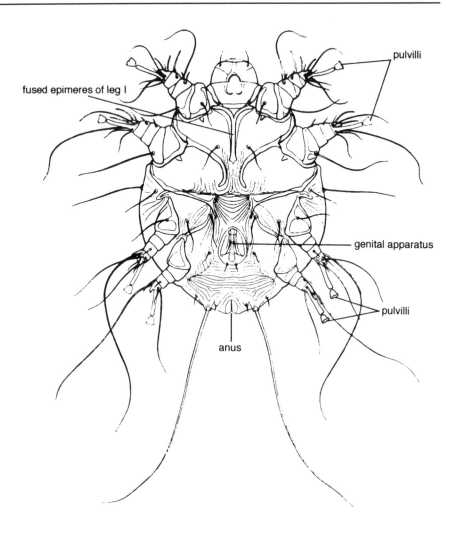

fused epimeres of leg I

pulvilli

genital apparatus

pulvilli

anus

Fig. 20.12. Ventral view of a male *Knemidokoptes mutans*. Source: from Hirst (1922).

from these areas when the mites will be found embedded in the tissue or scales at the base of the quill. This disease is more noticeable in the spring and summer.

Cytoditidae – *Cytodites nudus* (Fain, 1960; Arends, 1991)

This oval mite is usually more than 500 μm long. It occurs in the linings of the air sacs and air passages of chickens and less frequently in other poultry. Small numbers of mites have no noticeable effect on the host, but in vast numbers they may cause death. They have sometimes been found within the peritoneal and thoracic cavities. Diagnosis is only possible on clinical symptoms, or by finding the mites on post mortem examination. The mites have a smooth cuticle, largely

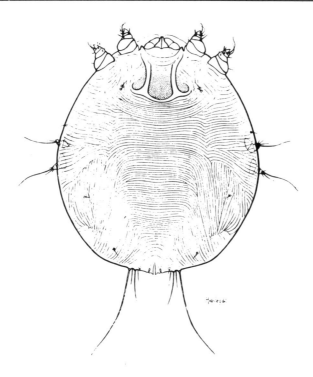

Fig. 20.13. Dorsal view of a female *Knemidokoptes gallinae*. Source: from Hirst (1922).

devoid of striations (Fig. 20.14). They have a few short setae but no anterior vertical setae. The chelicerae are absent and the pedipalps are fused to form a soft sucking organ through which fluids exuded by the host are imbibed. The epimeres of the first pair of legs are fused into Y-shaped structures. In the female all the legs have long pretarsi bearing subglobose pulvilli, while in the male there are pulvilli and the pretarsi are short.

Laminosoptidae – *Laminosoptes cysticola*

Laminosoptes cysticola is present in many parts of the world including North and South America, Europe and Australia. The mites may occur in millions in the cellular tissue of turkeys and chickens where they destroy the fibres. The mites bring about the formation of nodules which become calcified on the death of the mite, and reduce the market value of the carcass. *L. cysticola* is a small elongate mite about 250 μm long with smooth cuticle and few long setae (Fig. 20.15). The gnathosoma is recessed on the ventral surface. The epimeres of the first pair of legs are fused into a Y-shaped structure and those of the second pair of legs meet in the midventral line and then diverge posteriorly. Legs I and II bear claw-like tarsi and legs III and IV end in long spatulate pretarsi according to Hirst (1922) but Fain (1981) states that all legs end in a 'pedunculate sucker'.

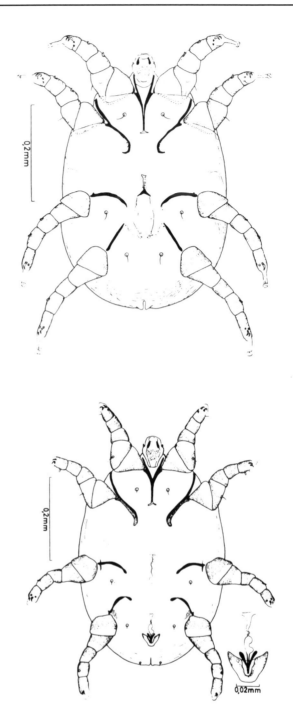

Fig. 20.14. Ventral view of *Cytodites nudus* female (top) and male (bottom). Source: from Fain (1960).

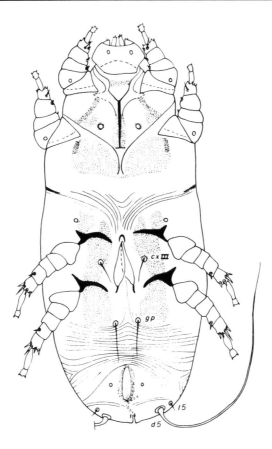

Fig. 20.15. Ventral view of female *Laminosoptes cysticola*. Source: from Fain (1981).

Astigmatic Mites and Human Allergies

Astigmatic mites have been extremely successful in exploiting patchy or ephemeral food resources, their short generation times allowing a rapid build-up in populations. Dispersal is typically accomplished by a phoretic association with an arthropod or vertebrate which carries the modified deuteronymph from habitat to habitat (OConnor, 1994). They are commonly found infesting stored food of plant or animal origin and sensitive individuals handling heavily infested produce may develop dermatitis. *Tyrophagus putrescentiae* has been associated with copra itch, *Glycyphagus domesticus* with grocer's itch, and *G. destructor* with hay itch. They are often found in very large numbers in houses and are responsible for some of the allergic material associated with house dust.

In most domestic situations two groups of mites, house dust and storage, are present. The house dust mites *Dermatophagoides pteronyssinus*, *D. farinae* and *Euroglyphus maynei* are members of the Pyroglyphidae and produce biochemically equivalent active allergens (Stewart *et al.*, 1992). Different allergens are produced

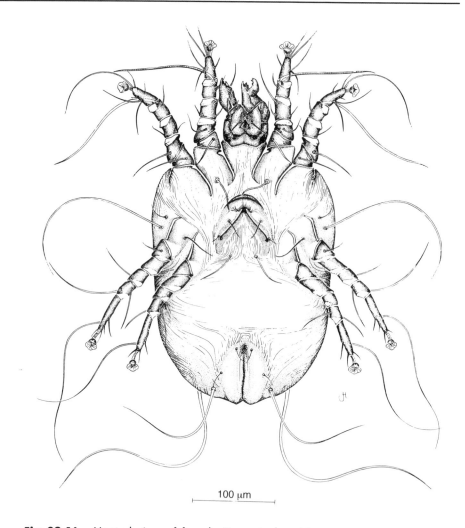

100 μm

Fig. 20.16. Ventral view of female *Dermatophagoides pteronyssinus*.

by the stored food mites *Acarus siro, T. putrescentiae, Lepidoglyphus destructor, G. destructor* and *Blomia kulagini*. Sensitization to storage mites is not restricted to occupational exposure but occurs among urban people (Hage-Hamsten and Johansson, 1992). In a study in which 87 allergic patients were tested for antibodies against house dust allergens, 66 (75%) reacted positively and all but one of these was positive for *D. pteronyssinus*. About two-thirds (42/66) also reacted to *T. putrescentiae* and 11 to *G. destructor* (Araujo-Fontaine *et al.*, 1974). The allergen associated with *D. pteronyssinus* is present not only in the mite itself, but also in its secretions and excreta (Fain, 1966; Voorhorst *et al.*, 1969).

Dermatophagoides pteronyssinus is widespread throughout the world and has been recorded from all inhabited continents (Fig. 20.16) (Fain, 1967). It was present in every one of 150 houses sampled in Holland and accounted for 61% of all the mites recovered from house dust. Densities as high as 3500 per gram of house dust were recorded in one house in Leiden but the average for 150 houses was 11; while in Hobart, Tasmania, the average was 27 mites per gram of house dust (Spieksma and Spieksma-Boezeman, 1967; Murton and Madden, 1977). House dust mites thrive in humid environments in human dwellings. At relative humidities above 65–70% they are able to extract water from air by secreting a hygroscopic solution from the supracoxal glands. This solution takes up water from the atmosphere and is then ingested by the mite. Survival during dry periods is by means of a desiccation-resistant protonymphal stage (Arlian, 1992).

Dermatophagoides pteronyssinus has been reared in the laboratory on electric razor cuttings of human beard growth. In nature the main food of *D. pteronyssinus* is considered to be human skin scales and under optimum conditions the life cycle of *D. pteronyssinus* from egg to adult can be completed in about 21 days. Laboratory populations of *D. pteronyssinus* increase most rapidly at 25°C and 75 to 80% RH (Spieksma and Spieksma-Boezeman, 1967; Murton and Madden, 1977). A female will produce 100 eggs in a lifetime of 10 weeks with 90% of them being produced in the first 5 weeks (Lebrun *et al.*, 1991).

Fur and Feather Mites

Four families of psoroptoid mites live on the hair shafts of small and medium-sized mammals (OConnor, 1994). *Lynxacarus radovsky* is the cat fur mite, endemic in Australia and Hawaii and has been reported from Florida. The mites cause little itching and are not highly contagious but occasionally they may be found all over the body when treatment with a general insecticide is needed (Muller *et al.*, 1989). Two species of fur mites occur on laboratory animals: *Myocoptes musculinus* on mice and *Chirodiscoides caviae* on guinea-pigs. In both species certain legs are adapted for clasping the hairs of the host. These adaptations are found on the third and fourth pairs of legs of female *M. musculinus* and on the third pair of legs of the male. The fourth pair of legs in the male is greatly enlarged but is not adapted for clasping hair. In *C. caviae* the first and second pairs of legs are modified for clinging to hair. These mites also feed at the base of the hairs and *M. musculinus* appears to feed on epidermal tissue but not on tissue fluids (Yunker, 1973).

Feather mites in the superfamilies Pterolichoidea and Analgoidea provide a great diversity of mite taxa occurring on the flight and tail feathers of birds. Mites of 20 families live among the feathers and those of seven other families live in the space within the feather quill (OConnor, 1994). More than 25 species are found on domestic poultry throughout the world but records of economic damage by these mites is rare. Members of the Epidermoptidae are normally encountered on the skin of birds and one species, *Epidermoptes bilobatus*, is a common skin parasite of galliform birds and can cause a scaly skin disease in chickens. Infestation may result in emaciation and death (Arends, 1991).

The most specialized developmental cycle of an astigmatic mite occurs in *Hypodectes propus*, which occurs in the nests of pigeons. The female mite produces eggs in the nest which develop into a deuteronymph while in the egg shell. The larva and protonymph are reduced to apodermas, i.e. cuticles lacking mouthparts, appendages and body setae. Deuteronymphs penetrate nestlings and increase greatly in size in the subcutaneous tissues until the host begins to develop eggs. Then the deuteronymphs pass through the skin of the bird into the nest where they rapidly develop into the adult stage, the tritonymph being reduced to an apoderma. The females do not feed but produce many eggs which develop into deuteronymphs and repeat the cycle. It was once thought that these deuteronymphs were the hypopodes of the pigeon feather mite, *Falculifer rostratus* (Evans, 1992).

Oribatida

Oribatid mites are free-living, dark coloured mites with a rigid exoskeleton from which the popular name of 'beetle' mite is derived. Apart from some rare exceptions, they possess prominent club-like sensilla (pseudostigmatic or bothridial organs), which arise from large pits on the posterolateral margins of the propodosoma. In heavily sclerotized species respiration is conducted through tracheal tubes with stigmata opening at the bases of the legs. The alternative name Cryptostigmata refers to the fact that the stigmata are poorly defined and difficult to observe. Oribatid mites are primarily fungivorous or saprophagous inhabitants of the upper layers and surface litter of soil. Their economic importance lies in their being intermediate hosts of various cestodes, especially *Moniezia expansa*, the broad tapeworm of cattle (Kates and Runkel, 1948).

References

Araujo-Fontaine, A., Miltgen, F., Rombourg, H., Molet, B., Pauli, G. and Basset, A. (1974) Contribution à l'étude du rôle allergisant des acariens de la poussière. Etude immunologique des sérums des malades et des sérums des lapins hyperimmunisés. *Revue Francaise d'Allergologie et d'Immunologie Clinique* 14, 91–96.

Arends, J.J. (1991) External parasites and poultry pests. In: Calnek B.W. (ed.) *Diseases of Poultry*. Iowa State University Press, Ames, pp. 702–730.

Arlian, L.G. (1989) Biology, host relations, and epidemiology of *Sarcoptes scabiei*. *Annual Review of Entomology* 34, 139–161.

Arlian, L.G. (1992) Water balance and humidity requirements of house dust mites. *Experimental and Applied Acarology* 16, 15–35.

Arlian, L.G. and Vyszenski-Moher, D.L. (1988) Life cycle of *Sarcoptes scabiei* var. *canis*. *Journal of Parasitology* 74, 427–430.

Arlian, L.G., Runyan, R.A., Sorlie, L.B. and Estes, S.A. (1984) Host-seeking behavior of *Sarcoptes scabiei*. *Journal of the American Academy of Dermatology* 11, 594–598.

Baker, E.W., Evans, T.M., Gould, D.J., Hull, W.B. and Keegan, H.L. (1956) *A Manual of Parasitic Mites of Medical and Economic Importance*. National Pest Control Association, New York.

Blachut, K., Roberts, I.H. and Meleney, W.P. (1973) Seasonal independence and relative frequency of motile stages of scab mites, *Psoroptes ovis*, on sheep. *Annals of the Entomological Society of America* 66, 285–287.

Burgess, I. (1994) *Sarcoptes scabiei* and scabies. *Advances in Parasitology* 33, 235–292.

Chakrabarti, A. (1990) Pig handler's itch. *International Journal of Dermatology* 29, 205–206.

Downing, W. (1936) The life history of *Psoroptes communis* var. *ovis* with particular reference to latent or suppressed scab. *Journal of Comparative Pathology and Therapeutics* 49, 63–206.

Du Toit, P.J. and Bedford, G.A.H. (1932) Goat mange – the infectivity of kraals. *18th Report of the Director of Veterinary Services and Animal Industry, Union of South Africa*, pp. 145–152.

Evans, G.O. (1992) *Principles of Acarology*. CAB International, Wallingford, Oxon, UK.

Faccini, J.L.H. and Massard, C.L. (1976) O gènero *Chorioptes* Gervais, 1895, parasita de ruminantes no Brasil (Psoroptidae: Acarina). *Revista Brasileira de Biologia* 36, 871–872.

Fain, A. (1960) Révision du genre *Cytodites* (Megnin) et description de deux espèces et un genre nouveaux dans la famille Cytoditidae Oudemans (Acarina: Sarcoptiformes). *Acarologia* 2, 238–249.

Fain, A. (1965) Notes sur le genre *Notoedres*, Railliet 1893 (Sarcoptidae: Sarcoptiformes). *Acarologia* 7, 321–342.

Fain, A. (1966) Nouvelle description de *Dermatophagoides pteronyssinus* (Trouessart, 1897) importance de cet acarien en pathologie humaine (Psoroptidae: Sarcoptiformes). *Acarologia* 8, 302–327.

Fain, A. (1967) Le genre *Dermatophagoides* Bogdanov 1864 son importance dans les allergies respiratoires et cutanées chez l'homme (Psoroptidae: Sarcoptiformes). *Acarologia* 9, 179–225.

Fain, A. (1968) Etude de la variabilité de *Sarcoptes scabiei* avec une revision des Sarcoptidae. *Acta Zoologica et Pathologica Antverpiensia* 47, 1–196.

Fain, A. (1974) Notes sur les Knemidokoptidae: avec description de taxa nouveaux. *Acarologia* 16, 182–188.

Fain, A. (1981) Notes on the genus *Laminosoptes* Megnin, 1880 (Acari: Astigmata) with description of three new species. *Systematic Parasitology* 2, 123–132.

Fain, A. and Elsen, P. (1967) Les acariens de la famille Knemidokoptidae producteurs de gale chez les oiseaux (Sarcoptiformes). *Acta Zoologica Pathologica Antverpiensia* 45, 3–145.

Gordon, R.M., Unsworth, K. and Seaton, D.R. (1943) The development and transmission of scabies as studied in rodent infestations. *Annals of Tropical Medicine and Parasitology* 37, 174–194.

Griffiths, D.A. and Bowman, C.E. (eds) (1984) *Acarology VI*. Ellis Horwood, Chichester, UK.

Hage-Hamsten, M.V. and Johansson, S.G.O. (1992) Storage mites. *Experimental and Applied Acarology* 16, 117–128.

Hirst, S. (1922) *Mites Injurious to Domestic Animals*. British Museum, Natural History, London, Economic Series No. 13.

Houck, M.A. (ed.) (1994) *Mites, Ecological and Evolutionary Analyses of Life History Patterns*. Chapman & Hall, London.

Houck, M.A. and OConnor, B.M. (1991) Ecological and evolutionary significance of phoresy in the Astigmata. *Annual Review of Entomology* 36, 611–636.

Jensen, R., Fitzhugh, H.A., Gaafar, S.M., Loomis, E.C., Matthysse, J.G., McDonald, R.P. and Wagstaff, D.J. (1979) *Psoroptic Cattle Scabies Research: An Evaluation*. National Research Council, Washington, DC.

Kates, K.C. and Runkel, C.E. (1948) Observations on oribatid mite vectors of *Moniezia expansa* on pastures, with a report of several new vectors from the United States. *Proceedings of the Helminthological Society of Washington* 15, 10–33.

Kirkwood, A.C. (1985) Some observations on the biology and control of the sheep scab mite *Psoroptes ovis* (Hering) in Britain. *Veterinary Parasitology* 18, 269–279.

Krantz, G.W. (1978) *A Manual of Acarology*. Oregon State University Book Stores, Corvallis, Oregon.

Lebrun, Ph., van Impe, C., de Saint Georges-Gridelet, D., Wauthy, G. and Andre, H.M. (1991) The life strategies of mites. In: Schuster, R. and Murphy, P.W. (eds) *The Acari, Reproduction, Development and Life History Strategies*. Chapman & Hall, London, pp. 3–22.

Mellanby, K. (1972) *Scabies*. E.W. Classey, Hampton, England.

Muller, G.H., Kirk, R.W. and Scott, D.W. (1989) *Small Animal Dermatology*. W.B. Saunders, Philadelphia.

Murton, J.J. and Madden, I.L. (1977) Observations on the biology, behaviour and ecology of the house-dust mite, *Dermatophagoides pteronyssinus* (Trouessart) (Acarina: Pyroglyphidae) in Tasmania. *Journal of the Australian Entomological Society* 16, 281–287.

OConnor, B.M. (1994) Life-history modifications in astigmatid mites. In: Houck, M.A. (ed.) *Mites, Ecological and Evolutionary Analyses of Life-History Patterns*. Chapman & Hall, New York, pp. 136–159.

Petrov, D., Milushev, I. and Monov, M. (1976) Acariasis in pigs. *Veterinarna Sbirka* 74, 35–38.

Radostits, O.M., Blood, D.C. and Gay, G.C. (1994) *Veterinary Medicine – a Textbook of the Diseases of Cattle, Sheep, Pigs and Horses*. Baillière-Tindall, London.

Rhodes, A.P. (1976) The effect of extensive chorioptic mange of the scrotum on reproductive function of the ram. *Australian Veterinary Journal* 52, 250–257.

Roberts, I.H., Blachut, K. and Meleney, W.P. (1971) Oversummering location of scab mites, *Psoroptes ovis*, on sheep in New Mexico. *Annals of the Entomological Society of America* 64, 105–108.

Schuster, R. and Murphy, P.W. (eds) (1991) *The Acari, Reproduction, Development and Life History Strategies*. Chapman & Hall, London.

Seddon, H.R. and Albiston, H.E. (1968) *Arthropod Infestations (Ticks and Mites)*. Commonwealth of Australia, Department of Health, Service Publications (Veterinary Hygiene) Number 7.

Spieksma, F.Th.M. and Spieksma-Boezeman, M.I.A. (1967) The mite fauna of house dust with particular reference to the house-dust mite *Dermatophagoides pteronyssinus* (Trouessart, 1897) (Psoroptidae: Sarcoptiformes). *Acarologia* 9, 226–241.

Stewart, G.A., Bird, C.H., Krska, K.D., Colloff, M.J. and Thompson, P.J. (1992) A comparative study of allergenic and potentially allergenic enzymes from *Dermatophagoides pteronyssinus, D. farinae* and *Euroglyphus maynei*. *Experimental and Applied Acarology* 16, 165–180.

Strong, K.L. and Halliday, R.B. (1992) Biology and host specificity of the genus *Psoroptes* Gervais (Acarina: Psoroptidae), with reference to its occurrence in Australia. *Experimental and Applied Acarology* 15, 153–169.

Sweatman, G.K. (1957) Life history, non-specificity, and revision of the genus *Chorioptes*, a parasitic mite of herbivores. *Canadian Journal of Zoology* 35, 641–689.

Sweatman, G.K. (1958a) Redescription of *Chorioptes texanus*, a parasitic mite from the ears of reindeer in the Canadian Arctic. *Canadian Journal of Zoology* 36, 525–528.

Sweatman, G.K. (1958b) Biology of *Otodectes cynotis*, the ear canker mite of carnivores. *Canadian Journal of Zoology* 36, 849–862.

Sweatman, G.K. (1958c) On the life history and validity of the species in *Psoroptes*, a genus of mange mites. *Canadian Journal of Zoology* 36, 905–929.

Voorhorst, R., Spieksma, F.T.M. and Varekamp, H. (1969) *House-dust Atopy and the House-dust Mite* Dermatophagoides pteronyssinus *(Trouessart 1897)*. Stafleu's Scientific Publishing Co., Leiden.

Wilson, G.I., Blachut, K. and Roberts, I.H. (1977) The infectivity of scabies (mange) mites,

Psoroptes ovis (Acarina: Psoroptidae), to sheep in naturally contaminated pastures. *Research in Veterinary Science* 22, 292–297.

Woolley, T.A. (1988) *Acarology, Mites and Human Welfare.* John Wiley, New York.

Wright, F.C., Riner, J.C. and Guillot, F.S. (1983) Cross-mating studies with *Psoroptes ovis* (Hering) and *Psoroptes cuniculi* Delafond (Acarina: Psoroptidae). *Journal of Parasitology* 69, 696–700.

Yunker, C. (1973) Mites. In: Flynn, R.J. (ed.) *Parasites of Laboratory Animals.* Iowa State University Press, Ames, pp. 425–492.

21

Acari – Prostigmata and Mesostigmata (Chiggers, Blood-sucking Mites)

Prostigmata

The Prostigmata are a large and complex group of mites which vary in size from 100 μm to 16 mm and have equally diverse forms. Most species live by sucking the juices of animals and plants. Predatory prostigmatid mites occur in terrestrial, freshwater and marine habitats, and include the Hydrachnidia or freshwater mites. Plant-feeding Prostigmata include the Tetranychidae or spider mites and the Eriophyoidea or bud mites, many of which are economically important pests in orchards and horticulture. Other Prostigmata are parasites of invertebrates including insects, and vertebrates including humans. It is the parasitic forms which are of medical and veterinary importance.

The typical prostigmatid mite is weakly sclerotized and, where there is an internal respiratory system, the stigmata open on the gnathosoma or anterior part of the propodosoma. It is this feature that gives rise to the name Prostigmata. The chelicerae are either blade-like as in *Trombicula* or styletiform as in *Pyemotes*; they are rarely chelate. Specialized sensilla, the trichobothria, are often present on the propodosoma, e.g. *Trombicula* larva. The coxae of the legs may be incorporated into the ventral surface of the body as in *Trombicula*, and may be united with each other as in *Demodex*.

Trombidioidea – Trombiculidae

Trombidioids are generally parasitic in the larval stage and free-living predators in the adult and nymphal stages. The Trombidioidea contains eight families including the Trombidiidae and Trombiculidae. Many species and genera have been described in these two families which, in the adult and nymphal stages, are predators on the eggs and young of other arthropods. The larval stages of both families are parasitic with trombidiids parasitizing invertebrates and trombiculids parasitizing vertebrates.

More than 1200 species of trombiculid mites have been described and about 50

414

of these have been known to attack human beings or livestock. They are widely distributed throughout the world and in many countries cause trombidiosis, a dermatitis due to the feeding of trombiculid larvae. In Europe they are known as harvest bugs, e.g. *Neotrombicula autumnalis*; in the Americas as chiggers, e.g. *Eutrombicula alfreddugesi*; and as scrub itch mites in Asia and Australia, e.g. *Eutrombicula sarcina*. In the southern States of North America *Neoschongastia americana* is a serious pest of turkeys and a minor pest of chickens, attacking free-ranging birds. Infestations are sporadic and localized (Arends, 1991).

In Japan, south-east Asia and parts of Australia larval trombiculids may also be vectors of *Rickettsia tsutsugamushi*, the causative agent of scrub typhus in humans (see Chapter 25). Two important vectors of scrub typhus are *Leptotrombidium akamushi* and *L. deliense*. Trombiculid mites normally parasitize rodents, insectivores and ground-dwelling birds, but given the opportunity will feed on people, domestic animals and free-ranging poultry.

The female trombiculid deposits its spherical eggs in damp but well-drained soil. They give rise to larvae which are about 250 µm long. They ascend grass stems to a height of 60–80 mm to await the passage of a suitable host to which they cling. This habit leads to trombiculid larvae being picked up on the faces and legs of grazing animals. Horses and cattle are most noticeably affected on the face. Infestations with *E. sarcina* can cause severe pruritus in horses and sheep (Radostits *et al.*, 1994).

The larva attaches itself by its chelicerae and feeds on the host's tissues by partially digesting them with saliva, which it pumps to and fro, leading to the formation of a feeding tube or stylostome in the host at the point of larval attachment. It feeds for several days before falling from the host and entering a quiescent phase before moulting to the protonymph. There are three nymphal stages but only one, the second, is active. The adult mite is about 1 mm long, and its body is 'waisted' producing a figure-of-eight shape.

In the warmer regions of the world trombiculid mites probably breed throughout the year but in the cooler regions the number of generations per year is limited and infestations with trombiculid mites are seasonal, e.g. larvae of *N. autumnalis* are abundant in late summer and early autumn and hence the popular name 'harvest bug'.

The prominent gnathosoma of the larva bears strong, blade-like chelicerae anteromedially (Fig. 21.1). The chelicerae are flanked by stout, segmented palps, the penultimate segment (palpal tibia) of which bears a claw. The terminal segment (palpal tarsus) opposes the claw and bears ciliated setae (Baker *et al.*, 1956). There are no true stigmata or tracheae and respiration is cutaneous. Behind the gnathosoma on the dorsal surface, there is a scutum which is very important taxonomically. The scutum bears a pair of sensilla or trichobothria, which are setaceous in many genera but inflated in some, e.g. *Schoengastia*. Typically the scutum also bears five ciliated setae, one at each corner of the more or less rectangular scutum and a single anteromedian seta (paired in *Apolonia*). Both the ventral and dorsal body surfaces carry moderately long ciliated setae, the number and distribution of which are specific characters. Nymphal and adult trombiculids and trombidiids are known as velvet mites, a term which refers to the dense covering of pilose setae which cover their legs, body and, to some

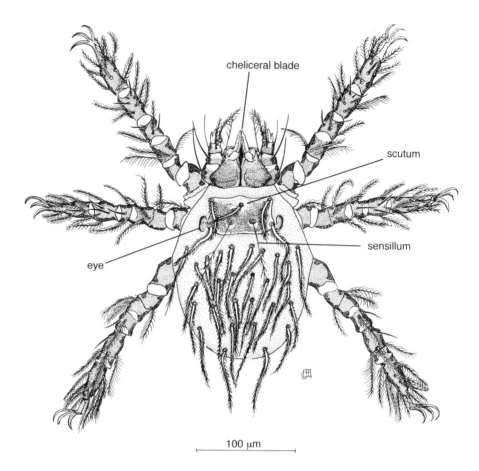

Fig. 21.1. Dorsal view of larva of *Leptotrombidium deliense*.

degree, palps. In these free-living stages the stigmata open at the base of the chelicerae.

Cheyletoidea

The Cheyletoidea are a diverse assemblage of nine families, which are mainly parasitic on arthropods, reptiles, birds and mammals. This superfamily includes: the Demodicidae, a widely distributed family of medical and veterinary importance; the Psorergatidae and Cheyletiellidae of moderate veterinary importance; and the Myobiidae of minor importance (Woolley, 1988).

Psorobia (Psorergatidae)

Two species of *Psorobia* (= *Psorergates*) have been recovered from domestic stock. *P. bos* has been found on cattle in the United States where its effect is so slight as to be almost undetectable (Roberts and Meleney, 1965). A more important

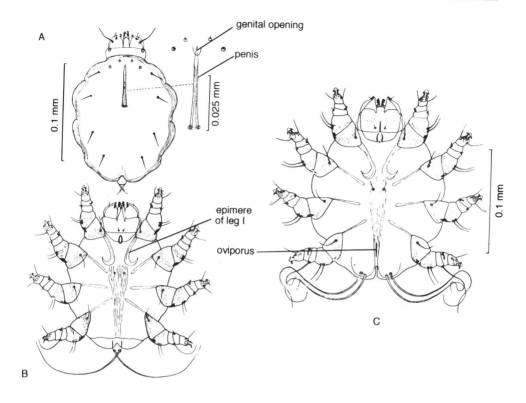

Fig. 21.2. Adult *Psorobia ovis*. (**A**) Dorsal view of male with enlargement of penis and genital opening; (**B**) ventral view of male; (**C**) ventral view of female. Source: from Fain (1961).

parasite is *P. ovis*, which occurs on sheep mainly in Australia, New Zealand, South Africa, South America and the USA (Radostits *et al.*, 1994). Adult *Psorobia* can be recognized by the fact that the legs are radially arranged around a more or less circular body (Fig. 21.2).

P. ovis Female *P. ovis* lay few eggs in their lifetime. These give rise to larvae with reduced legs. There are three nymphal stages in which the legs become progressively larger until, in the adult stage, the legs are well developed and the mites are motile (Murray, 1961). Adults of both sexes are very small measuring only 200 µm. The coxae of all the legs are sunk into the body as epimeres. Those of the first pair of legs are relatively broad and reflected laterally to be hook-shaped (Fig. 21.2B, C). The female has two pairs of long setae posteriorly, and the oviporus opens posteriorly on the ventral surface. The male is more oval than the rounded female; has only a single pair of long setae on a small posteromedian process; and the genital opening is located anterodorsally (Fig. 21.2A, B), behind which an elongated penis may be seen (Fain, 1961). The complete life cycle of *P. ovis* takes 4 to 5 weeks.

Psorobia ovis mainly affects merino sheep which respond to the irritation by

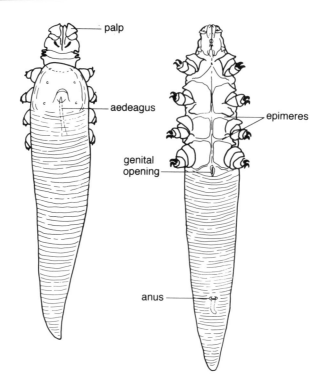

Fig. 21.3. Left, dorsal view of male *Demodex*. Right, ventral view of female *Demodex*. Source: from Krantz (1978).

chewing at the fleece, resulting in a condition known as fleece derangement and leading to the wool clip being downgraded (Sinclair, 1976). Affected sheep may become tolerant after 1 or 2 years but remain infested. *P. ovis* spreads very slowly through a flock and may affect 15% of sheep in a neglected flock (Radostits *et al.*, 1994). Only the adults are motile and they are very sensitive to desiccation, so that they survive for only a short time in the fleece and die within 24–48 h when removed from their host (Murray, 1961). Consequently the period of transmission from one host to another occurs during a brief period following shearing. Most mites occur under the stratum corneum in the superficial layers of the skin, and infestation with *P. ovis* is confirmed by finding the mites in skin scrapings. Peak numbers of *P. ovis* are reached in the spring and treatment is best applied after shearing, using two dippings with phoxim, an organophosphorus insecticide, one month apart (Radostits *et al.*, 1994).

Demodex (Demodicidae)

Members of the genus *Demodex* are minute, annulate, worm-like, parasitic mites (Fig. 21.3), which live head down in hair follicles and in the sebaceous and Meibomian glands of the skin. Mites found in lymph nodes and other internal locations are usually dead and degenerating (Muller *et al.*, 1989). In humans *D. folliculorum*

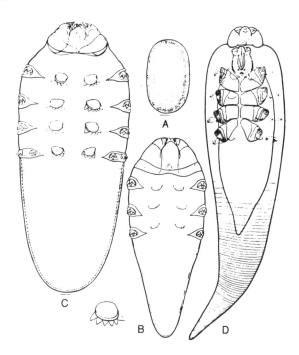

Fig. 21.4. Stages in life cycle of *Demodex*: (**A**) egg; (**B**) larva; (**C**) deutonymph of *D. muscardinus* with below an enlarged sternal scute; (**D**) deutonymph of *D. bovis* containing fully formed adult within. Source: from Hirst, S. (1922) Economic Series No. 13. *Mites Injurious to Domestic Animals*. British Museum (Natural History), London.

and *D. brevis* respectively destroy epithelial cells and sebaceous cells and *D. brevis* may penetrate the dermis (Nutting, 1976a,b). Demodicid mites occur in humans and a wide range of wild and domestic animals, including bats, insectivores, carnivores, rodents, horses and ruminants. They form a group of sibling species with different species occurring on different hosts, and more than one species may occur on the same host, e.g. *D. folliculorum* and *D. brevis* on humans (Desch and Nutting, 1972), and *D. cati* and an unnamed species on cats (Muller *et al.*, 1989). The relationship between demodicid mites on a host and clinical disease is not simple. They are to be found on both healthy and diseased hosts, in which they cause an itchless mange which is often not noticed until the hides of animals are being prepared for leather.

Life cycle of *D. canis*   This cycle is representative of all species of *Demodex*. The female lays eggs which give rise to larvae with short legs endings in a single trifid claw (Fig. 21.4B). An unusual feature of the life cycle is the production of a second hexapod form, designated the protonymph by Nutting and Desch (1978), in which each leg terminates in a pair of trifid claws. The deutonymph stage which follows is octopod. Both protonymph and deutonymph have a pair of crescent-shaped sternal scutes on the ventral surface between each pair of legs (Fig. 21.4C). The deutonymph moults into an adult in which the coxal epimeres are united to form a median longitudinal bar (Fig. 21.3). The female genital opening is on the ventral surface just posterior to the fourth pair of legs, while the male genital opening is located dorsally at the level of the second and third pair of legs, as in *Psorobia*. In *Demodex* the legs are closely associated anteriorly so that the striated opisthosoma forms at least half the total body length. Adults measure 250–300 μm long by 40 μm wide.

Demodex in domestic animals   Among domestic animals demodectic mange is of greatest importance in dogs, goats and pigs, of lesser importance in cattle and horses, rare in sheep (Radostits *et al.*, 1994) and very rare in cats (Muller *et al.*, 1989). The appearance of small nodules and pustules is a sign of demodicosis in domestic animals. In pigs *D. phylloides* is found on the face, spreading down the ventral surface to the neck and chest to the belly. In cattle (*D. bovis*) and goats (*D. caprae*) the lesions occur most commonly on the brisket, lower neck, forearm and shoulder and dorsally behind the withers. In pigs and goats pustules may develop into large abscesses and in some instances cause the death of goats (Radostits *et al.*, 1994). In East Africa infestation of cattle with *D. bovis* may take a more severe form and end fatally (Bwangamoi, 1970). Species of *Demodex* are host specific and no instance has been documented of interspecific transfer of any *Demodex* species (Nutting and Desch, 1978). Transmission of *Demodex* within a host species occurs very early in life while the young are suckling (Fisher, 1973; Muller *et al.*, 1989).

Demodex canis is acquired in the first 2 to 3 days of neonatal life. Two clinical conditions are distinguished: a localized and a generalized demodicosis. In young dogs (3–12 months) localized demodicosis involves a small number of squamous patches developing on the face, especially around the eyes and mouth. In most cases the condition cures itself without treatment, and recurrences are rare. It may continue and develop into the generalized condition, which is commoner in purebred and short-haired dogs. It involves an hereditary specific T-cell defect for *D. canis*. In dogs more than 5 years old, generalized demodicosis is a rare but serious condition indicating that the host is immuno-suppressed. Generalized demodicosis begins on the face and spreads to the head, legs and trunk. Secondary infections with bacteria follow and crusted, pyogenic, haemorrhagic lesions develop on much of the body. This condition is difficult to treat and may terminate fatally. Four to eight topical applications of Amitraz at two weekly intervals gave 86% complete recovery (Muller *et al.*, 1989).

Diagnosis of demodectic mange requires the recovery of mites from skin scrapings. They are easily identified as *Demodex* but specific identification is a matter for a specialist. Keys to species of medical and veterinary importance are given by Nutting (1976a).

Cheyletiella (Cheyletiellidae)

The Cheyletiellidae contains nine genera of mites which are parasitic on birds and small mammals (Smiley, 1978). They are characterized by having stiletiform chelicerae which are used for piercing the host and strong, curved, palpal claws for maintaining the mite in the fur or feathers of its host (Fig. 21.5). Species of

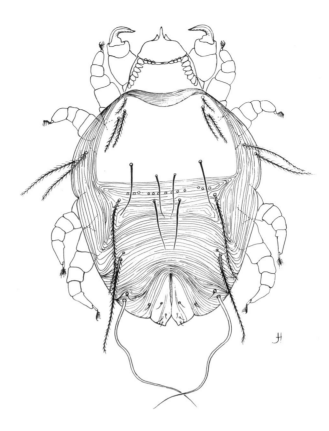

Fig. 21.5. *Cheyletiella parasitivorax* female. Leg setae omitted.

Cheyletiella are large mites (385 μm) which cause a mild, non-suppurative dermatitis in dogs, cats and rabbits, and a transitory dermatitis in humans. Other species have been described from wild animals (Bronswijk and Kreek, 1976). *Cheyletiella* is an obligate parasite which lives on the keratin layer of the epidermis, and is not associated with hair follicles but periodically may pierce the skin and engorge on a clear colourless fluid (Foxx and Ewing, 1969).

The mites move about rapidly and this behaviour gives rise to the term 'walking dandruff'. *C. yasguri* causes a highly contagious infection of puppies. This usually begins on the rump from which it may spread over the back to the head. Older dogs may be symptomless carriers with light populations of mites (Foxx and Ewing, 1969). *C. blakei* causes a mild dermatitis in cats, and *C. parasitivorax* occurs in the scapular region of rabbits. Human infestations with *Cheyletiella* are transitory and the reaction variable. Treatment involves alleviation of the symptoms, and eradication of the mites on the infested pet. Cats, dogs and rabbits may be safely treated with pyrethrins, carbaryl powders or lime-sulphur dips, and dogs with malathion or carbaryl. Three treatments are needed at weekly intervals. Two subcutaneous injections of ivermectin 2 to 3 weeks apart are also effective (Muller *et al.*, 1989). *Cheyletiella* dermatitis is diagnosed by finding the mites in the hair of the host. Keys to the three species mentioned are given by Bronswijk and Kreek (1976), but identification of species is really a matter for the specialist.

The life cycle is completed on the host. The large eggs (230 × 100 µm) are attached to the hairs of the host 2 to 3 mm above the skin; the prelarva and larva develop within the eggshell. The larva is hexapod but the prelarva possesses only rudimentary gnathosomal appendages. There are two nymphal stages before the adult stage is reached. Females can survive for approximately 10 days in a cool atmosphere, but males and immature stages die within 48 h of removal from the host. Females have been found attached to fleas and louse flies (Hippoboscidae), and phoresy could be an important mode of transfer from host to host (Bronswijk and Kreek, 1976). The ready mobility of these mites through the hair of the host makes for rapid spread among hosts which are in close contact.

Myobiidae

The Myobiidae is a cosmopolitan family of ectoparasites which occur on marsupials, rodents, bats and insectivores. In modest numbers they appear to have little effect on the health of the host. Myobiids are readily recognized by the first pair of legs being highly modified for clinging to a single hair (Fig. 21.6). *Myobia musculi* causes a mild dermatitis in mice; *Radfordia ensifera* and *R. affinis* infest rats and mice, respectively. They feed at the base of hairs ingesting extracellular tissue fluid and sometimes blood (Yunker, 1973; Krantz, 1978).

Pyemotes (Pyemotoidea, Pyemotidae)

Species of *Pyemotes* are predators of insects which attack humans and domestic animals that come in contact with infested materials. The grain or hay itch mite (*Pyemotes tritici*) is the cause of dermatitis in people in many parts of the world. On susceptible humans a vesicle develops in the centre of an erythematous weal. Rubbing and scratching burst the vesicle and introduce the possibility of secondary infection. The mites do not establish breeding populations on mammals and, if reinfestation is prevented, the condition subsides in a few days. Infestations of *Pyemotes* are mainly associated with grains, straw, hay and grasses but also occur with pulses and other crops.

The inseminated female *P. tritici* attaches herself to an insect with her chelicerae. The eggs are fertilized within the female, and all the immature stages are spent inside the female's swollen opisthosoma (Figs 21.7 and 21.8). Adult males emerge 2 days before the females and remain on the outside of the female's opisthosoma, assisting in the birth of virgin females by dragging them through the birth pore. Copulation takes place and the inseminated female moves away to find a host. *P. tritici* has haploid males and diploid females and at birth the progeny is 92% female. Males are fully potent for their first 15 matings. On average a female produces 250 offspring with the daily intrinsic rate of increase being 0.63 and the population doubling in 1.1 days (Wrensch and Bruce, 1991).

The insect hosts of *P. tritici* are larvae of the Angoumois grain moth (*Sitotroga cerealella*), the saw-toothed grain beetle (*Oryzaephilus surinamensis*), the cowpea weevil (*Callosobruchus maculatus*) and the rice weevil (*Sitophilus oryzae*) (Moser,

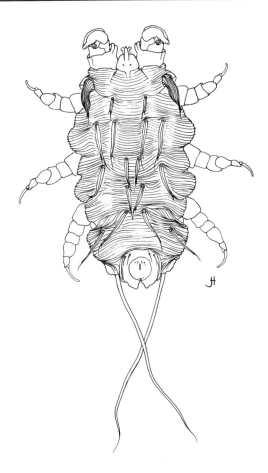

Fig. 21.6. *Myobia musculi.* Minor setae omitted.

1975). Species of *Pyemotes* vary in their toxicity to humans. *P. tritici* is highly toxic, as is *P. beckeri* (Hewitt *et al.*, 1976) while *P. scolyti* is not toxic. The venom of *P. tritici* causes immediate paralysis and eventual death of its insect host. The use of genetic engineering techniques to manufacture the venom offers the possibility of the venom becoming available as a biological-control agent (Wrensch and Bruce, 1991).

Actinotrichida

The Mesostigmata, Ixodida and Holothyrida, a small order of no medical importance, are collectively referred to as the Parasitiformes. In the Mesostigmata the hypostome is rarely toothed, the peritreme is usually well developed, the last segment of the palp bears a palpal claw or apotele, and the stigmata are located above coxae II to IV. In the Ixodida the hypostome is toothed, there is no peritreme or palpal claw, and the stigmata are posterior to the coxae of the fourth pair of legs (Ixodoidea), or above the second or third pair of legs (Argasoidea).

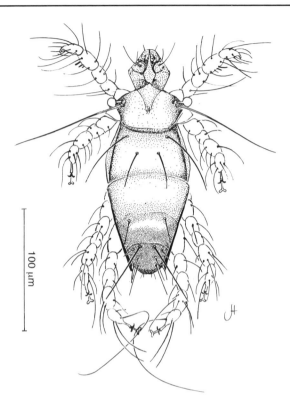

100 μm

Fig. 21.7. Dorsal view of female *Pyemotes tritici.*

Mesostigmata (Evans, 1992)

Of the 12 suborders of Mesostigmata only the Dermanyssina contains species of medical and veterinary importance, and they are all in the superfamily Dermanyssoidea. Mesostigmatic mites are comparatively large, ranging from 200 μm to 2.5 mm in length. Some dermanyssine mites have been used as biocontrol agents, e.g. *Phytoseiulus persimilis* (Phytoseiidae) against red spider mites (Tetranychidae) in glasshouses. Another dermanyssine, *Macrocheles muscae-domesticae*, is a ubiquitous inhabitant of dung and is commonly phoretic on houseflies and predacious on their eggs. In Australia *Macrocheles glaber* is phoretic on several species of native Australian dung beetles, e.g. *Onthophagus granulatus* (Scarabaeidae). Experiments in both the field and laboratory have shown that complete control of bushfly (*Musca vetustissima*) breeding can be achieved by as few as 50 *M. glaber* in a litre of dung upon which 300–400 bushfly eggs have been laid (Wallace *et al.*, 1979).

Dermanyssoidea

In the Dermanyssoidea the intermediate gradations between free-living organisms and obligatory parasites are well represented by existing species. In a single genus,

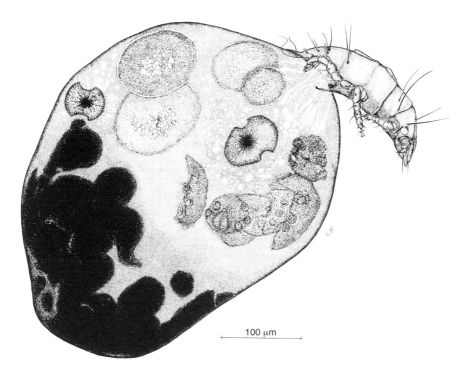

Fig. 21.8. Lateral view of gravid *Pyemotes tritici*. All of the swollen area is the opisthosoma.

e.g. *Haemogamasus*, are to be found non-parasitic, polyphagous nest-dwellers; facultative parasites; and obligatory haematophagous parasites. The early stages of this progression towards parasitism are shown in the family Laelapidae, in which one subfamily, the Hypoaspidinae, does not include any parasites of vertebrates although a number of hypoaspidines are typically found in the nests of small mammals and birds. Two other subfamilies, the Laelapinae and Haemogamasinae, are largely parasitic, but have some polyphagous, nest-dwelling species (Radovsky, 1969, 1994). The main families of importance to the medical entomologist are the blood-sucking Dermanyssidae and Macronyssidae; the Rhinonyssidae and the subfamily Halarachninae of the Halarachnidae, which infest the respiratory passages of birds and mammals, respectively; and the Raillietiinae, another halarachnid subfamily, which are found in the ears of bovids.

Structure (see Figs 21.9, 21.10 and 21.11)

In most dermanyssoids there is a narrow, elongated, anteriorly-directed peritreme associated with each stigma. On the dorsal and ventral surfaces there are a number of sclerotized shields or plates, the arrangement of which is of considerable use in classification. The base of the gnathosoma is known as the basis capituli and bears laterally a pair of palps between which the hypostome extends anteriorly on

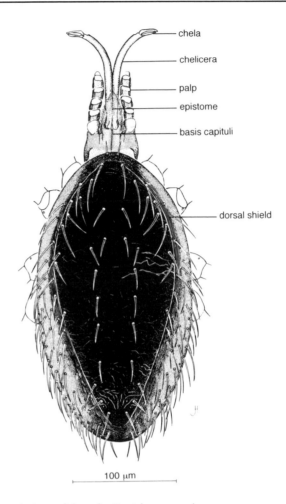

Fig. 21.9. Dorsal view of female *Ornithonyssus bursa*.

the ventral surface. The palps have five free segments, of which the terminal segment, the tarsus, bears at its base a two-tined, claw-like structure, the apotele. The chelicerae are enclosed within the basis capituli inside a system of sheaths from which they can be extruded and into which they can be withdrawn. The chelicerae are 3-segmented and consist of a short, basal segment, followed by a second segment, which may be elongate, and which bears distally the third segment or movable chela (Fig. 21.9).

The gnathosoma or capitulum is basically a tube through which fluid is carried to the oesophagus. The roof of the tube is the epistome which covers the chelicerae dorsally. In *Dermanyssus* and other obligate parasites the epistome is continued anteriorly as a forwardly-pointing, triangular or rounded structure. In *Haemogamasus* the lobate epistome is deeply serrated. Ventrally, the basis capituli is formed from the fused, expanded coxal segments of the palps. In the midventral line there is the shallow deutosternum, which is marked by several transverse rows

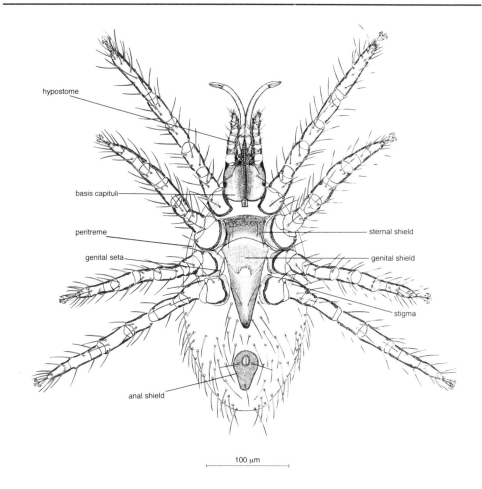

hypostome

basis capituli

peritreme

genital seta

sternal shield

genital shield

stigma

anal shield

100 μm

Fig. 21.10. Ventral view of female *Ornithonyssus bursa.*

of denticles (Fig. 21.11). In parasitic species the number of denticles per row tends to be reduced and often only one is present. Ventrally, the hypostome bears on its anterior border two pairs of structures, the corniculi and between them the internal malae. In free-living dermanyssoid mites the corniculi form obvious horn-like structures and the internal malae are free, while in obligatory parasites the hypostome is elongated and the corniculi and internal malae, both membranous, form a preoral trough. In the midventral line between the first coxae, is the tritosternum bearing two forwardly-directed laciniae, which in life engage with the denticles in the deutosternum. The tritosternum is absent in some reduced dermanyssoids.

There is usually one, large, sclerotized shield on the dorsal surface (Fig. 21.9) and a series of smaller shields in the midline on the ventral surface. The main unpaired shields are: the sternal shield at the level of the second and third pairs of legs; an epigynial or genital shield, the anterior border of which covers the genital opening; a ventral shield; and a shield surrounding the anus (Fig. 21.10).

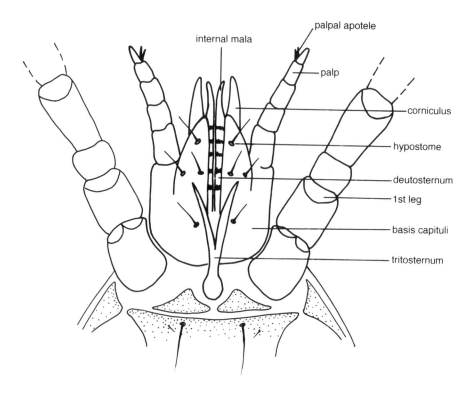

Fig. 21.11. Ventral view of the gnathosoma of a dermanyssoid mite. Redrawn from Krantz (1978).

The ventral shield is usually fused with either the genital or anal shields, and in males the shields on the ventral surface may merge to form a single holoventral shield.

The chaetotaxy, i.e. distribution of setae on the body, is useful in identification. Full details are given in Evans (1992). On the ventral surface there are three pairs of setae on the hypostome; one pair on the basis capituli; three pairs on the sternal shield; usually one pair posterior to the sternal shield; often one pair on the genital shield; and usually three setae on the anal shield, one on each side of the anus and a single seta posterior to the anus.

The pretarsus bears the ambulacral apparatus consisting of a pulvillus, often deeply incised, and a pair of well-developed claws. The anterior legs are not wholly used for walking but serve as sensory organs and are often different from the other walking legs.

Biology and medical importance

Dermanyssoids have evolved a complicated process for the transfer of sperm to the female. Sperm are passed to the chelicerae from the male genital opening and move into the hollow spermadactyl which is a specialized structure on the movable digit of the male chelicera. Sperm are then transferred to special sperm induction pores on the third or fourth coxae of the female. This phenomenon is referred to as podospermy. The pores lead by a complicated route to the spermatheca which connects through a minute lumen with the ovary. In both sexes the genital openings are on the ventral surface about one-third of the way back from the anterior end. In the female the opening is a slit posterior to the sternal plate and covered by the flexible epigynial plate. In the male the genital opening is at the anterior edge of the sterno-genital shield.

In the life cycle of an external parasitic, haematophagous dermanyssoid, the female lays a few, large, heavily yolked eggs after each blood meal. In her lifetime an individual female will lay about six batches of eggs totalling less than 100. These hatch into larvae which are followed by two nymphal stages before moulting to the adult mite. One or more of the immature stages may be inactive and non-feeding. One feature which is common to the haematophagous dermanyssoids is the speed with which the life cycle can be completed. Under optimal conditions development from egg to adult may take as little as 7 days and always less than a month. As a result, populations of these mites can build up to astronomical levels in a very short space of time.

Most of the damage caused by haematophagous dermanyssoids results from the direct effect of large numbers feeding on the host rather than from their transmitting pathogens. Various pathogens have been recovered from dermanyssoids in the field and some have been transmitted under experimental conditions in the laboratory. The results are inconclusive and it is generally considered that, with the exception of *Rickettsia akari*, blood-sucking dermanyssoids play little part in the epidemiology of blood-dwelling pathogens.

Blood-sucking dermanyssoids may be host-specific or parasitize a range of related hosts, but in certain circumstances, e.g. when birds desert an infested nest, they will attack unusual hosts. Attacks on humans are in that category. There is no blood-sucking mite for which humans are the main host. The common hosts are birds and rodents. It is this ability to feed opportunistically on other hosts which accounts for *Liponyssoides sanguineus* (=*Allodermanyssus sanguineus*) acting as a vector of *R. akari* from rodents to humans. Infestations of mites have to be dealt with positively. Merely vacating infested premises for a short period will not eliminate an infestation because adult mites are able to survive for several months in the unfed state, e.g. *Dermanyssus gallinae* can survive for 4 to 5 months, and *L. sanguineus* for 2 months (Baker *et al.*, 1956).

Dermanyssidae (Baker *et al.*, 1956; Radovsky, 1994)

Dermanyssid mites are haematophagous ectoparasites of birds and mammals. The adults are 750–1000 μm long, and in the unfed state are greyish-white becoming bright red after feeding and darker as the meal is digested. When not feeding they spend most of their time in the nest. Their eggs are deposited in the nest or in associated crevices. They have a high engorgement capacity, which enables them to withstand starvation for months. Females are most often found on the host, an action which favours dispersal of the species.

The chelicerae are chelate with minute, weakly dentate digits. In the nymphs and female the chelicerae are elongate and stylet-like with the second segment considerably lengthened. In the male the second segment is of normal length and a long, grooved spermadactyl is fused with the movable digit which is considerably longer than the fixed digit. The corniculi are membranous and there are nine or more deutosternal denticles in a single file. In the life cycle the larva is non-feeding while both the protonymph and deutonymph are actively feeding stages. Two species are of medical and veterinary importance, *Dermanyssus gallinae* and *Liponyssoides sanguineus*.

Dermanyssus gallinae

Dermanyssus gallinae, the chicken mite, is a cosmopolitan ectoparasite on poultry and a range of wild birds including pigeons, sparrows and starlings, and will attack mammals when other hosts are not available. Unfed *D. gallinae* are approximately 700 µm long × 400 µm wide but increase to more than 1 mm long when engorged. In the female the second segment of the chelicera measures about 275 µm compared with a first segment of 45 µm. In the male the corresponding lengths are 84 µm and 54 µm and the spermadactyl 105 µm long. The female has a single large dorsal shield which is truncate posteriorly. The setae on the dorsal shield are shorter than those on the adjacent body surface. There are only two pairs of setae on the sternal plate, the posterior pair being remote from the plate. There are also two pairs of pores on the sternal plate. The genito-ventral plate is rounded posteriorly and bears one pair of setae. The anal plate is large and bears three setae. In the male there is a single holoventral shield (Evans and Till, 1966).

Very large populations of *D. gallinae* can build up rapidly in poultry houses and bird nests. In North America *D. gallinae* is rare in modern commercial caged-layer operations but seen frequently in broiler breeder farms with peak numbers occurring in the summer (Arends, 1991). The mites are nocturnal, feeding on roosting birds and by day hide away in cracks and crevices. Under optimal conditions the life cycle can be completed very rapidly in the presence of hosts. The eggs hatch in 2 to 3 days, both nymphal stages moult 1 to 2 days after a blood meal, and adult females are ready to oviposit 12 to 24 h after feeding. Fed adults can survive 34 weeks without feeding, making it difficult to eliminate infestations by removing the domestic hosts (Arends, 1991). The control measures described for *Ornithonyssus sylviarum* are applicable to the control of *D. gallinae* but with the need to pay greater attention to the surroundings of the birds.

Large populations of *D. gallinae* have serious effects on domestic poultry. Egg production is reduced, hens may leave eggs that they are incubating and death may occur from exsanguination. A number of pathogens have been recorded from *D. gallinae* and it has been shown to transmit some of these experimentally, but *D. gallinae* is not considered to play a significant role in their transmission.

Liponyssoides sanguineus

Liponyssoides sanguineus is an ectoparasite of small rodents. It was originally described from Egypt where it was found on rats and has since been identified in the United States. *L. sanguineus* is very similar to *D. gallinae* but differs by having two dorsal shields, a larger tapering anterior shield and a very small posterior one, bearing one pair of setae. The sternal plate has three pairs of setae and two pairs of pores. The genital plate is slender and tapering and there is a small anal plate. Development of *L. sanguineus* is slower than that of *D. gallinae* with the life cycle from egg to adult taking 18–23 days (Baker *et al.*, 1956), compared with 7 days for *D. gallinae. L. sanguineus* is the vector of *Rickettsia akari*, which causes rickettsial pox in humans. *L. sanguineus is* normally a nest dweller, only occurring on the host when it is feeding.

Macronyssidae

Members of the Macronyssidae are haematophagous ectoparasites of mammals, birds and reptiles. It is believed that they evolved primarily on bats and secondarily have transferred to other mammals, birds and reptiles (Radovsky, 1985). In the feeding protonymph and female the chelicerae are chelate and edentate, and in the inactive, non-feeding larva and deutonymph the cheliceral digits are rudimentary. In the male the grooved spermadactyl is relatively short and incompletely fused to the movable digit and is rarely longer than the fixed digit. The deutosternal denticles are arranged in a single file.

Two genera are of interest, *Ornithonyssus*, which occurs on birds and mammals, and *Ophionyssus*, an ectoparasite of reptiles. The characters given here to separate the two genera apply to the four species dealt with (*Ornithonyssus bacoti, O. sylviarum, O. bursa* and *Ophionyssus natricis*), but are not necessarily applicable to all species in these genera. Female *Ornithonyssus* have the genital setae inserted on the genital shield and there is only a single dorsal shield, while in *Ophionyssus* there are two dorsal shields and the genital setae are inserted on the integument adjoining the genital shield. In male *Ornithonyssus* the anal shield is fused with the other ventral shields while in male *Ophionyssus* the anal shield is discrete.

In female *Ornithonyssus* the dorsal shield tapers posteriorly, and in the male it is more extensive. On the ventral surface there are three separate shields. The sternal shield bears three pairs of setae (two pairs in *O. sylviarum*); the genital shield, which tapers posteriorly, bears one pair of setae; and the anal shield is of the usual type being oval, tapering posteriorly and bearing three setae.

Ornithonyssus bacoti

Although *Ornithonyssus bacoti* is referred to as the tropical rat mite, it is cosmopolitan, occurring in both tropical and temperate areas of the world, especially in sea ports. *O. bacoti* is a pest of mice, rats, hamsters and small marsupials. It is distinguished by the setae on the dorsal shield being of similar length to those on the adjoining integument, and there being three pairs of setae on the sternal shield.

The life cycle is completed rapidly under optimal conditions with the cycle from egg to adult being completed in 11 to 16 days. As with other haematophagous mesostigmatic mites, high populations can cause the death of their host by exsanguination. *O. bacoti* is the vector of the filarial worm *Litomosoides carinii* to rodents. *O. bacoti* has transmitted a number of other pathogens experimentally but its role in nature is not considered to be important.

Ornithonyssus sylviarum (Axtell and Arends, 1990; Arends, 1991)

Ornithonyssus sylviarum, the northern fowl mite, is a serious pest of poultry and wild birds throughout the northern temperate region of Europe and North America, and is common in southern Australia. It has also been recorded from wild birds in South Africa. The setae on the dorsal shield are shorter than those on the adjoining integument, and there are only two pairs of setae on the sternal plate, the third pair being on the integument. An unusual feature of the life cycle of *O. sylviarum* is that the adults remain on the host giving heavily infested birds a greyish to blackish appearance.

Oviposition occurs on the host, primarily in the area of the vent, and egg development is completed in 1 to 2 days. The protonymph requires at least two feeds before moulting to the non-feeding deutonymph. The entire life cycle can be completed in a week, enabling populations of *O. sylviarum* to build up rapidly. In North America peak populations occur in winter. Infested poultry cause losses to commercial producers through lower egg production and higher feed costs.

Ornithonyssus sylviarum can survive for 3 to 4 weeks off the host and is introduced into new settings on contaminated egg crates, trays and flats, and even on personnel. The introduction of *O. sylviarum* into a poultry facility can be minimized by careful examination of all introduced materials to ensure that they are mite-free. *O. sylviarum* has developed resistance to malathion, carbaryl and stirofos. Permethrin is an effective insecticide providing it is applied as high pressure, high volume spray wetting the birds to the skin. A second treatment is required 5 to 7 days later and the effect lasts 9 weeks.

Ornithonyssus bursa (see Figs 21.9 and 21.10)

Ornithonyssus bursa is the tropical poultry mite, which occurs also on pigeons, sparrows and mynah birds as well as, on occasions, attacking people. Its effect on humans is irritating but temporary because *O. bursa* is unable to survive for long away from its bird hosts. In *O. bursa*, as in *O. sylviarum*, the setae on the dorsal shield are shorter than those on the integument, but it differs from the latter by having three pairs of setae on the sternal plate. *O. bursa* occurs either on the bird or in its nest. Attacks on humans result from the dispersal of mites from infested nests, deserted by the breeding birds and their nestlings.

Ophionyssus natricis (Camin, 1953)

Ophionyssus natricis is an ectoparasite of reptiles which is rare in the wild but troublesome in zoos. This species has two dorsal plates, an anterior lemon-shaped plate on the podosoma and a posterior plate which is immediately dorsal to the anal plate but considerably smaller. The posterior plate bears no setae. There are two pairs of setae and two pairs of pores on the sternal plate with the third pair of each located behind the plate on the integument. The anal plate bears the usual three setae.

The adult mite feeds under the scales on the snake and then leaves its host to oviposit in dark, moist crevices. Where there is no shortage of hosts the life cycle can be completed in 2 to 3 weeks and in their absence, some females can survive unfed for 5 to 6 weeks. Heavy infestations seriously affect the health of snakes causing a severe anaemia, which may lead to the snake's death. *O. natricis* is believed to be a mechanical vector of a haemorrhagic septicaemia of snakes caused by *Aeromonas hydrophila hydrophila (= Proteus hydrophilus)*, a motile, facultatively anaerobic bacillus (Camin, 1948). Infestations of snakes in captivity can be reduced by providing the snakes with water in which they may immerse themselves.

Halarachnidae

Two subfamilies are recognized within the Halarachnidae (Radovsky, 1969). The Halarachninae are obligatory parasites of the respiratory tract of mammals, and the Raillietiinae are obligatory parasites occurring in the external ear of mammals. As adaptations to an internal parasitic mode of life, the dorsal and ventral shields in the Halarachninae are delicate and reduced and the genital plate is much reduced. In addition the tritosternum is reduced or absent and the peritremes, associated with the stigmata, are reduced or vestigial. These mites are active within the respiratory system and consequently the ambulacral apparatus at the ends of the legs is well developed. In the Raillietiinae the genital shield is well developed, a bifid tritosternum is present and the peritremes are elongate and well developed (Furman, 1979). In both subfamilies the life cycle is compressed. The hexapod larva is active and the protonymph and deutonymph are non-feeding, non-motile, ephemeral stages with rudimentary claws.

Halarachninae

The Halarachninae includes about 35 species, which parasitize primates, terrestrial carnivores, phocid and otariid seals, rodents, hyraxes and artiodactyls. Some species are larviparous and others oviparous. The larva has well developed ambulacra and is the only stage capable of moving between hosts (Radovsky, 1994). They appear to transfer easily from one host species to another. An Asiatic macaque, which had been kept in a zoo where numerous African monkeys were present, was found to be infected with three species of lung mites, normally parasites of African monkeys. This may help to explain the wide distribution among mammals of mite species of the same genus. For example the 14 species in the genus *Pneumonyssus* are mainly found on Old World simians but species also occur on hyraxes and on a marsupial, the cuscus (*Phalanger maculatus*). Similarly the four species of *Pneumonyssoides* occur on pigs, the dog and New World monkeys (Furman, 1979). Two species are of veterinary importance: the dog parasite *Pneumonyssoides caninum* and the monkey parasite *Pneumonyssus simicola*.

Pneumonyssoides caninum   *Pneumonyssoides caninum* occurs in the sinuses and nasal passages of dogs in Australia, South Africa and the USA. Infections with *P. caninum* run a mild course and are relatively free from marked symptoms (Chandler and Ruhe, 1940; Yunker, 1973). However, in some cases, the mites penetrate tissues and migrate throughout the body. They have been found in the bronchi, the renal fat and the liver (Garlick, 1977). Adult *P. caninum* are pale

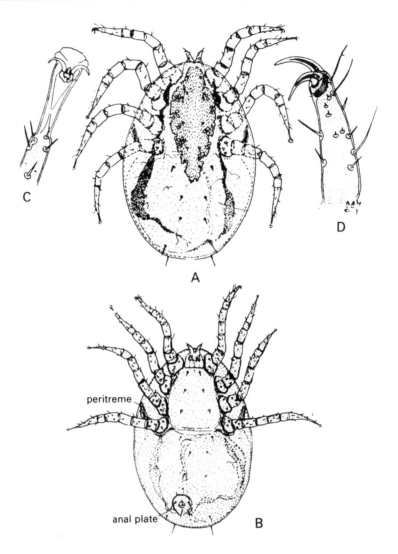

Fig. 21.12. Female *Pneumonyssoides caninum*. (**A**) Dorsal view; (**B**) ventral view; (**C**) detail of tarsus, typical of legs II, III and IV; (**D**) tarsus of leg I. Source: from Chandler and Ruhe (1940).

yellow, oval mites with few body setae (Fig. 21.12). The chelicerae are well developed with opposable digits. The dorsal plate is small, irregular in shape and covered with microscopic spines. The first pair of legs is equipped with a pair of heavily sclerotized brown claws, while the other three pairs are tipped with a long, stalked pulvillus and two slender claws.

Pneumonyssus simicola *Pneumonyssus simicola* is an extremely common parasite of the lungs of the rhesus monkey (*Macaca mulatta*) in which 100% infestation may occur. This species also occurs, but less commonly, in a range of Old

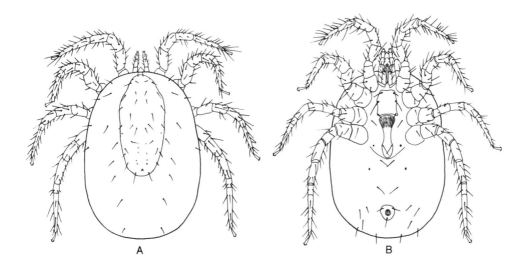

Fig. 21.13. Female *Raillietia auris:* (**A**) dorsal view; (**B**) ventral view. Redrawn from Hirst, S. (1922) *Mites Injurious to Domestic Animals.* British Museum (Natural History), London. Economic series No. 13.

World primates. Furman (1979) considers *Pneumonyssus* to be the most specialized genus on the grounds that its palps are reduced both in size and number of segments, and the denticles in the deutosternal groove are reduced to a single file. *P. simicola* is similar to *P. caninum* but smaller. The mites live and feed in the lung where they may be grouped in nodules superficially resembling tubercles. The nodules contain a characteristic golden brown to black pigment which may be faecal material resulting from the mite feeding on blood. Clinical signs are usually absent. The mite spreads readily through susceptible animals by coughing and sneezing. Rhesus monkeys taken from their mothers at birth and reared in isolation are free from infection.

Raillietiinae

The four species of *Raillietia* occur in the ears of bovids, two in the ears of East African antelopes of the genus *Kobus* (Potter and Johnston, 1978), *R. caprae* in goats and *R. auris* in domestic bovids. (*R. australis* from the common Australian wombat (*Vombatus ursinus*) is now regarded as not belonging to the Raillietiinae (Domrow, 1961; Radovsky, 1994).) *R. auris* is an oval mite about 1 mm long with a small oval dorsal plate (Fig. 21.13). The second pair of legs in the male are modified for grasping the female. The movable digit of the chelicera is entirely fused with the hypertrophied spermadactyl which collects sperm from the male genital opening and deposits them into the sperm induction pores of the female. *R. auris* occurs in the ears of cattle in North America, Europe and Australia, and of sheep in Iran (Rak and Naghshineh, 1973). The mite is considered to feed on epidermal cells and wax but not on blood. Infestations are usually benign, lacking

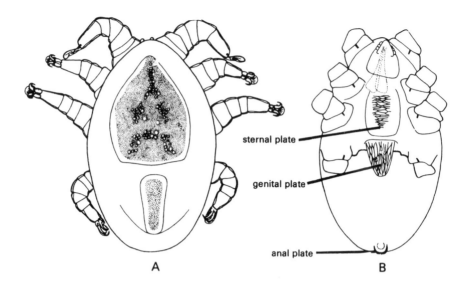

Fig. 21.14. Female *Sternostoma tracheacolum*: (**A**) dorsal view; (**B**) ventral view. Source: from Lawrence (1948).

obvious symptoms, but in northern Queensland *R. auris* has been associated with otitis media (Ladds *et al.*, 1972).

Rhinonyssidae

Most members of the Rhinonyssidae are parasites of the nasopharynx of birds. Rhinonyssids are weakly sclerotized, elongate mites with well-developed legs; peritreme reduced or absent; and the tritosternum usually absent. They are an extraordinarily successful group of parasites which are most closely related to the Macronyssidae. In both families the protonymph is a feeding stage and the deutonymph a non-feeding stage (Radovsky, 1994). One species of minor veterinary importance is the canary lung mite *Sternostoma tracheacolum*.

Sternostoma tracheacolum

Most species of *Sternostoma* are nasal mites but *S. tracheacolum* has been recorded from the respiratory tract of a range of domestic and wild birds, including canaries and budgerigars (Lawrence, 1948; Fain and Hyland, 1962; Mathey, 1967). The species is widely distributed throughout the world, occurring in Africa, North and South America, Europe, Australia and New Zealand. *S. tracheacolum* is a yellowish brown mite about 0.5 mm long with two dorsal plates, a pentagonal anterior plate and a narrower, posterior one (Fig. 21.14); distinct sternal and genital plates, and a reduced anal plate; strong, thick-set palps; large, mobile and chelate chelicerae. On legs II to IV the ambulacral apparatus consists of paired claws and a pulvillus, but on leg I it is much modified (Fig. 21.14A).

Sternostoma tracheacolum has been found in the tracheae, air sacs, bronchi, parenchyma of the lung and also on the surface of the liver, but rarely in the nasal cavities (Baker *et al.*, 1956). In canaries the mites were firmly attached to the inner walls of the tracheae from which they were imbibing blood (Lawrence, 1948). Mites were found in the lungs as well as the tracheae and air sacs of Gouldian finches in aviaries, and a wasting disease developed from which the birds died (Riffkin and McCausland, 1972). In northern Australia *S. tracheacolum* is considered to be the main factor keeping wild populations of Gouldian finches suppressed, if not the cause of their decline, with an infection rate of 62% in the wild (Tidemann *et al.*, 1992).

References

Arends, J.J. (1991) External parasites and poultry pests. In: Calnek, B.W. (ed.) *Diseases of Poultry*. Iowa State University Press, Ames, pp. 702–730.

Axtell, R.C. and Arends, J.J. (1990) Ecology and management of arthropod pests of poultry. *Annual Review of Entomology* 35, 101–126.

Baker, E.W., Evans, T.M., Gould, D.J., Hull, W.B. and Keegan, H.L. (1956). *A Manual of Parasitic Mites of Medical or Economic Importance*. National Pest Control Association, New York.

Bronswijk, J.E.M.H. van and Kreek, E.J. de (1976) *Cheyletiella* (Acari: Cheyletiellidae) of dog, cat and domesticated rabbit, a review. *Journal of Medical Entomology* 13, 315–327.

Bwangamoi, O. (1970) The pathogenesis of demodicosis in cattle in East Africa. *British Veterinary Journal* 127, 30–33.

Camin, J.H. (1948) Mite transmission of a hemorrhagic septicemia in snakes. *Journal of Parasitology* 34, 345–354.

Camin, J.H. (1953) Observations on the life history and sensory behaviour of the snake mite, *Ophionyssus natricis* (Gervais) (Acarina: Macronyssidae). *Chicago Academy of Sciences Special Publication No. 10*, 1–75.

Chandler, W.L. and Ruhe, D.S. (1940) *Pneumonyssus caninum* n.sp., a mite from the frontal sinus of the dog. *Journal of Parasitology* 26, 59–70.

Desch, C.E. and Nutting, W.B. (1972) *Demodex folliculorum* (Simon) and *D. brevis* Akbulatova of man: redescription and reevaluation. *Journal of Parasitology* 58, 169–177.

Domrow, R. (1961) New and little known Laelaptidae, Trombiculidae and Listrophoridae (Acarina) from Australian mammals. *Proceedings of the Linnean Society of New South Wales* 86, 60–95.

Evans, G.O. (1992) *Principles of Acarology*. CAB International, Wallingford, Oxon, UK.

Evans, G.O. and Till, W.M. (1966) Studies on the British Dermanyssidae (Acari: Mesostigmata). Part II. Classification. *Bulletin of the British Museum of Natural History (Zoology)* 14, 107–370.

Fain, A. (1961) Notes sur le genre *Psorergates* Tyrrell. Description de *Psorergates ovis* Womersley et d'une espèce nouvelle. *Acarologia* 3, 60–71.

Fain, A. and Hyland, K.E. (1962) The mite parasitic in the lungs of birds. The variability of *Sternostoma tracheacolum* Lawrence, 1948, in domestic and wild birds. *Parasitology* 52, 401–424.

Fisher, W.F. (1973) Natural transmission of *Demodex bovis* Stiles in cattle. *Journal of Parasitology* 59, 223–224.

Foxx, T.S. and Ewing, S.A. (1969) Morphologic features, behavior, and life history of *Cheyletiella yasguri*. *American Journal of Veterinary Research* 30, 269–284.

Furman, D.P. (1979) Specificity, adaptation and parallel evolution in the endoparasitic Mesostigmata of mammals. *Recent Advances in Acarology*, 11, 329–337.

Garlick, N.L. (1977) Canine pulmonary acariasis. *Canine Practice* 4(4), 42–47.

Hewitt, M., Barrow, G.l., Miller, D.C. and Turk, S.M. (1976) A case of *Pyemotes* dermatitis with a note on the role of these mites in skin disease. *British Journal of Dermatology* 94, 423–430.

Krantz, G.W. (1978) *A Manual of Acarology*. Oregon State University Book Stores, Corvallis, Oregon.

Ladds, P.W., Copeman, D.B., Daniels, P. and Trueman, K.F. (1972) *Raillietia auris* and otitis media in cattle in northern Queensland. *Australian Veterinary Journal* 48, 532–533.

Lawrence, R.F. (1948) Studies on some parasitic mites from Canada and South Africa. *Journal of Parasitology* 34, 364–379.

Mathey, W.J. (1967) Respiratory acariasis due to *Sternostoma tracheacolum* in the budgerigar. *Journal of the American Veterinary Medical Association* 150, 777–780.

Moser, J.C. (1975) Biosystematics of the straw itch mite with special reference to nomenclature and dermatology. *Transactions of the Royal Entomological Society of London* 127, 185–191.

Muller, G.H., Kirk, R.W. and Scott, D.W. (1989) *Small Animal Dermatology*. W.B. Saunders, Philadelphia.

Murray, M.D. (1961) The life cycle of *Psorergates ovis* Womersley, the itch mite of sheep. *Australian Journal of Agricultural Research* 12, 965–973.

Nutting, W.B. (1967a) Hair follicle mites (*Demodex* spp.) of medical and veterinary concern. *Cornell Veterinarian* 66, 214–231.

Nutting, W.B. (1967b) Hair follicle mites (Acari: Demodicidae) of man. *International Journal of Dermatology* 15, 79–98.

Nutting, W.B. and Desch, C.E. (1978) *Demodex canis* redescription and reevaluation. *Cornell Veterinarian* 6, 139–149.

Potter, D.A. and Johnston, D.E. (1978) *Raillietia whartoni* sp.n. (Acari: Mesostigmata) from the Uganda kob. *Journal of Parasitology* 64, 139–142.

Radostits, O.M., Blood, D.C. and Gay, G.C. (1994) *Veterinary Medicine – a Textbook of the Diseases of Cattle, Sheep, Pigs and Horses*. Baillière Tindall, London.

Radovsky, F.J. (1969) Adaptive radiation in the parasitic Mesostigmata. *Acarologia* 11, 450–478.

Radovsky, F.J. (1985) Coevolution of mammalian mesostigmate mites. In: Kim, K.C. (ed.) *Coevolution of Parasitic Arthropods and Mammals*. John Wiley, New York, pp. 441–504.

Radovsky, F.J. (1994) The evolution of parasitism and the distribution of some dermanyssoid mites (Mesostigmata) on vertebrate hosts. In: Houck, M.A. (ed.) *Mites, Ecological and Evolutionary Analyses of Life History Patterns*. Chapman & Hall, New York, pp. 186–217.

Rak, H. and Naghshineh, R. (1973) First report and redescription of *Raillietia auris* (Trouessart, 1902) (Acar.: Gamasidae). *Entomologist's Monthly Magazine* 109, 59.

Riffkin, G.G. and McCausland, I.P. (1972) Respiratory acariasis caused by *Sternostoma tracheacolum* in aviary finches. *New Zealand Veterinary Journal* 20, 109–112.

Roberts, I.H. and Meleney, W.P. (1965) Psorergatic acariasis in cattle. *Journal of the American Veterinary Medical Association* 146, 17–23.

Sinclair, A.N. (1976) Fleece derangement of merino sheep infested by the itch mite *Psorergates ovis*. *New Zealand Veterinary Journal* 24, 149–152.

Smiley, R.L. (1978) Further studies on the family Cheyletiellidae (Acarina). *Acarologia* 19, 225–241.

Tidemann, S.C., McOrist, S., Woinarski, J.C.Z. and Freeland, W.J. (1992) Parasitism of wild Gouldian finches (*Erythrura gouldiae*) by the air-sac mite *Sternostoma tracheacolum*. *Journal of Wildlife Diseases* 28, 80–84.

Wallace, M.M.H., Tyndale-Biscoe, M. and Holm, E. (1979) The influence of *Macrocheles glaber* on the breeding of the Australian bushfly, *Musca vetustissima* in cow dung. *Recent Advances in Acarology* 11, 217–222.

Woolley, T.A. (1988) *Acarology, Mites and Human Welfare*. John Wiley, New York.

Wrensch, D.L. and Bruce, W.A. (1991) Sex ratio, fitness and capacity for population increase in *P. tritici* (L.-F. and M.) (Pyemotidae). In: Schuster, R. and Murphy, P.W. (eds) *The Acari, Reproduction, Development and Life History Strategies*. Chapman & Hall, London, pp. 209–221.

Yunker, C. (1973) Mites. In: Flynn, R.J. (ed.) *Parasites of Laboratory Animals*. Iowa State University Press, Ames, pp. 425–492.

Ixodida – Argasidae (Soft Ticks) 22

The Ixodida or ticks are relatively large acarines, which are blood-sucking ectoparasites of vertebrates. The movable capitulum consists of the basis capituli, paired 4-segmented palps, paired chelicerae and a ventral median hypostome (Figs 22.1C, D and 22.6C, D), armed with rows of backwardly-directed teeth which securely attach the tick to its host. The genital opening and anus are both located ventrally, the genital opening being at the level of the second pair of legs, and the anus a little posterior to the fourth pair of legs (Fig. 22.3). Haller's organ, which is used in host seeking, is located on the tarsus of the first pair of legs (Fig. 22.6E).

About 800 species are included in the Ixodida and they are arranged in three families: the largest, the Ixodidae (hard ticks) has 13 genera, and 650 species; the Argasidae (soft ticks) has five genera and about 170 species; and the Nuttalliellidae has only a single species, known only from females found in the Afrotropical region (Sonenshine, 1991; Evans, 1992). The terms hard and soft refer to the possession of a dorsal scutum in the Ixodidae (Fig. 23.2), which is absent in the Argasidae. Hoogstraal and Kim (1985) give a fascinating picture of the evolution of ticks over the last 200 million years as they adapt from reptiles to the newly-evolving free-ranging, warm-blooded birds and mammals.

Argasids are tough, leathery ticks in which there is little differentiation between the sexes. In nymphs and adults the capitulum is not visible from the dorsal view, being located ventrally in a recess, the camerostome (Figs 22.3 and 22.5). The fourth segment of the palp is similar in size to the other three (Fig. 22.1D). When eyes are present they are lateral in position in folds above the legs (Fig. 22.4). The stigmata are small and placed anterior to the coxae of the fourth pair of legs. The pad-like pulvillus between the claws is either absent or rudimentary.

In the Ixodidae sexual dimorphism is well developed, the dorsal scutum being small in the female and almost covering the whole of the dorsal surface in the male (Fig. 23.2). The capitulum is terminal and always visible when the tick is viewed from above. When eyes are present they are located dorsally at the sides of the

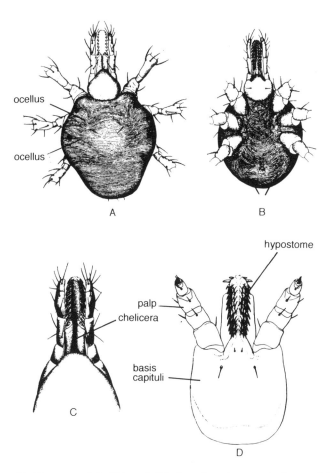

Fig. 22.1. *Otobius megnini*: (**A**) and (**B**) dorsal and ventral views of larva; (**C**) and (**D**) dorsal and ventral views of larval capitulum. Source: from Cooley and Kohls (1944).

scutum (Fig. 23.5). The fourth segment of the palp is reduced and recessed on the ventral surface of the third segment (Fig. 23.1). The stigmata are large and posterior to the coxae of the fourth pair of legs (Fig. 23.1). The pulvillus is well developed.

The two families differ in many aspects of their biology. In the Ixodidae there is a single nymphal stage. The adult female engorges, develops a very large batch of eggs, which she lays and then dies. In the Argasidae there are several nymphal stages, and the female feeds several times during her lifetime and lays several batches of eggs. Ixodid ticks feed on the host for a number of days. Argasids, with some exceptions, are nocturnal and visit the host to feed for a period of minutes. The Argasidae will be considered in this chapter and the Ixodidae in the next.

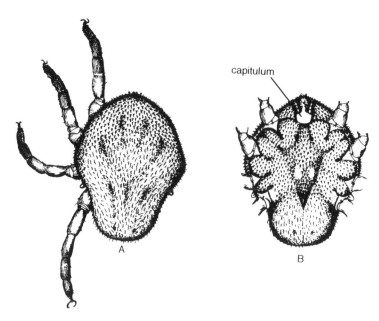

Fig. 22.2. *Otobius megnini:* (**A**) dorsal and (**B**) ventral views of partially fed nymph. Source: Cooley and Kohls (1944).

Argasidae

Three genera of argasids, *Argas, Ornithodoros* and *Otobius,* contain species of medical and/or veterinary importance, and a fourth genus, *Antricola,* parasitizes bats (Cooley and Kohls, 1944). In *Argas* the margin of the body is distinctly flattened and usually structurally different from the dorsal surface (Fig. 22.5C). The flattened margin remains distinct even when the tick is fully fed. There is usually a lateral sutural line present (Fig. 22.5D). Eyes are absent. All stages of *Argas* are found in the resting places of birds and bats which they parasitize.

In *Ornithodoros* and *Otobius* there is no lateral sutural line and no distinct margin to the body. In *Ornithodoros* the integument bears mammillae (Fig. 22.3), and in *Otobius* the integument is spiny in the nymph (Fig. 22.2) and granulated in the adult. The main species of medical importance is *Ornithodoros moubata,* and of veterinary importance *Otobius megnini* and species of *Persicargas,* a subgenus of *Argas.* In the western United States three species of *Ornithodoros* – *O. hermsi, O. parkeri* and *O. turicata* – feed readily on humans to whom they are capable of transmitting relapsing fever due to infection with *Borrelia* spp. In the western USA and Mexico *O. coriaceus* is notorious for attacking humans and causing severe toxic reactions (Sonenshine, 1993). Only two species of *Otobius* have been described, *O. megnini* from cattle, and *O. lagophilus* from cottontail and jack rabbits in western North America (Oliver, 1989).

Otobius megnini (see Figs 22.1 and 22.2)

Otobius megnini, the spinose ear tick, originated in the Americas from where it has been introduced into southern Africa and India. It is mainly a parasite of cattle and horses but has been recorded from a range of hosts in North America including donkeys, sheep, goats, dogs, cats, deer and rabbits (Cooley and Kohls, 1944). In India it has been recorded from cattle, sheep and humans (Chellappa and Alwar, 1972; Chellappa, 1973). This tick is associated with stables and animal shelters which may explain why it 'has apparently not spread to wild animals' in southern Africa (Walker *et al.*, 1978). The female tick lays her eggs in cracks and crevices several feet above the ground in the walls of animal shelters. This behaviour ensures that the emerging larva is at a height to transfer easily to the bodies of large, stabled domestic animals.

Life cycle

The eggs hatch in 11 days in summer, and 3 to 8 weeks under cooler conditions (Nuttall *et al.*, 1908; Bedford, 1925). The eggs are small, oval and reddish in colour. A hexapod larva 0.5 mm in length emerges from the egg (Fig. 22.1A, B). Its terminal capitulum is very long, accounting for more than one-third of the length of the unfed larva. There are two pairs of ocellus-like eyes present dorsally (Cooley and Kohls, 1944). The larva enters an ear of its host where it engorges and may attain a length of 4 mm (Nuttall *et al.*, 1908). Engorgement which takes 5 to 10 days is followed by a quiescent period before the larva moults to become an octopod nymph.

There are two octopod nymphal stages, characterized by having the capitulum on the ventral surface, and a spiny integument, from which the tick acquires its common name, spinose ear tick (Fig. 22.2). The nymphs re-attach to the skin lining the ear, suck blood and remain in the ear for an unusually long time. Most nymphs leave the host within 5 weeks but they may remain for several months (Bedford, 1925). The fully-fed nymph measures up to 8 mm. Nymphs drop from the host and 'creep into cracks and crevices in walls and woodwork, under stones or under the bark of trees, usually low down, where they develop into adults' (Walker *et al.*, 1978). The second stage nymph moults to an adult 1 to 4 weeks later. The body of the adult is fiddle-shaped, being constricted posterior to the fourth pair of legs. The adult does not feed, and its hypostome is poorly developed and without teeth.

Bionomics

The female can wait up to 18 months to be impregnated and then 'lays up to 1500 eggs in small batches over a period of a few weeks to several months'. This species has considerable powers of survival in the absence of hosts and it can persist in empty cattle sheds and stables for more than 2 years (Walker *et al.*, 1978). Unfed larvae usually survive for less than a month but under favourable circumstances can survive for as long as 4 months. Infestations of *O. megnini* cannot be eradicated by vacating premises except for excessively long periods.

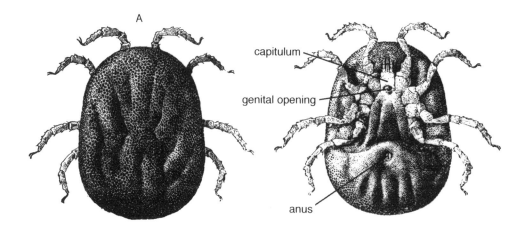

Fig. 22.3. *Ornithodoros moubata*: dorsal (left) and ventral (right) views of female. From: Castellani, A. and Chalmers, A.J. (1913) *A Manual of Tropical Medicine*. Baillière Tindall, London.

Otobius megnini favours hot, arid areas and is not present in wet areas. In South America it has been found up to 2600 m. At this altitude in Bolivia nymphs of *O. megnini* were present all the year round in the ears of dairy cattle, and during the rainy season clusters of 150 nymphs and larvae could be found under the tails of a small percentage (< 15%) of cattle (Bulman and Walker, 1979). These ticks do not transmit pathogens but do considerable damage to the ears, ear drums and auricular nerves by their feeding. The ear can be choked with ticks, wax and other debris, and the ear drum be ruptured, favouring secondary infections. Badly infested calves, sheep and goats may die and loss of condition in infested beasts is common (Bedford, 1925).

Ornithodoros

There are about 100 species of *Ornithodoros* of which three species *O. moubata*, *O. lahorensis* and *O. savignyi*, have become associated with people and/or their domestic animals. The taxonomic position of the two or more strains of *O. moubata* is not satisfactorily resolved and the term *O. moubata* will be used to cover both the hut-dwelling strain feeding on people and chickens, and a strain living in burrows and feeding on the occupants – wart hogs, antbears and porcupines. A note on this taxonomic difficulty is given in Chapter 26.

Ornithodoros moubata (see Fig. 22.3)

Ornithodoros moubata is widely distributed throughout East Africa and northern South Africa, extending into the drier parts of central Africa (Hoogstraal, 1956). It lives in cracks in walls and in the earth floors of huts. During the day it hides away in dark locations including the possessions of the occupants. Consequently

O. moubata has been spread by people with their goods and chattels as they move from one area to another.

Oviposition

Over a considerable period of time a female will lay several batches of comparatively large, spherical (0.9 mm in diameter), glistening golden yellow eggs (Nuttall *et al.*, 1908). During oviposition, the tick bends its head towards the genital opening and extrudes Gene's organ from the base of the capitulum. Each egg is 'handled' by Gene's organ which coats the egg with a waxy waterproof layer. Eggs which do not get this treatment quickly shrivel and die. The wax on the eggs has a lower melting point (50–54°C) than that of the crystalline cuticular wax of the female (65°C). The function of both waxes is to waterproof the organism, an essential requirement for survival in its natural arid environment (Lees and Beament, 1948).

Life cycle

The eggs hatch in about 8 days at 30°C and give rise to larvae which do not feed but remain motionless until they moult into nymphs 4 days later. The nymphs feed on blood taking 20–25 min to acquire a full meal. After an interval the first nymphal stage moults into the second stage which feeds and repeats the process. The number of nymphal stages is variable with adult males being produced after four nymphal stages and females after five (Jobling, 1925). Mating, which occurs after the female has fed, stimulates ovarian development with oviposition occurring 10 to 15 days later. Full digestion of the blood meal and egg development can be delayed for many months if the female is not mated. Virgin females mated 150–200 days after a blood meal produce eggs in the normal preoviposition period of 10 to 15 days (Aeschlimann and Grandjean, 1973b).

Feeding

When the tick feeds, the tips of the chelicerae press against the skin and cut it by alternate movements. The hypostome enters the host passively with the chelicerae. Feeding involves periods of active suction in which blood is stored in the midgut and its diverticula, and periods of rest when saliva, containing an effective anticoagulant, is introduced into the host (Lavoipierre and Riek, 1955; Sonenshine, 1991).

Phagocytosis is the predominant means by which blood absorption is accomplished (Sonenshine, 1991). There is no open connection between the midgut and the rectum and no faeces are passed. The Malpighian tubules open into the hindgut and their excretory products are deposited. The tick excretes a great deal of the watery component of blood via the coxal apparatus which opens just behind the coxae of the first pair of legs. Excretion of coxal fluid begins about 15 min after the start of feeding, continues during feeding and for about an hour afterwards. The coxal apparatus can excrete 30 times its own volume in 20 min (Lees, 1946b).

Effect of host on fecundity

The source of the blood meal can have a considerable effect on the fecundity of the female. Thus a laboratory strain of *O. moubata* produced nearly twice as many eggs (147 cf. 80) when fed on porcine compared with bovine blood. In both cases the ticks took up similar amounts of blood, increasing their body weight from 45 mg to 175 mg, i.e. taking up three times their body weight of blood. Nymphs reared on porcine blood developed more quickly and reached the adult stage in fewer instars than nymphs fed on bovine blood. Some nymphs fed on porcine blood became adults after four instars and 62% were adult within six instars, whereas when fed on bovine blood only 13% became adult within six instars and none took less than five instars (Mango and Galun, 1977).

In the laboratory vigorous colonies of *O. moubata* have been maintained on rabbits, a host that this tick will not have encountered in nature. So well has it become adapted that its fecundity per unit weight of blood is 30% higher than when the colony is fed on porcine blood (Mango and Galun, 1977). The same colony of *O. moubata* fed equally readily on rabbit, chicken and guinea-pig (84–92%), but 'showed considerable reluctance to feed on rat, mouse or hamster'. The blood meals taken from different hosts were comparable in size (118–125 mg), with a smaller meal (100 mg) being taken from a mouse and a larger meal (147 mg) from a guinea-pig; yet the fecundity of females fed on mouse or guinea-pig was similar, and the highest fecundity was recorded on rabbit and chicken. The last observation is of interest because this colony had been derived from ticks collected from a wart-hog burrow and reared for many generations on rabbits, and yet gave high fecundity when fed on chickens. In contrast *O. tholozani* fed equally readily on all six hosts, and produced similar numbers of eggs (Galun *et al.*, 1978).

Mating

Guanine has been shown to act as a non-specific, non-volatile, persistent assembly pheromone, which acts by contact and is responded to by argasids (*A. persicus*, *O. porcinus*) and ixodids (*Amblyomma cohaerens*, *Rhipicephalus appendiculatus*) (Otieno *et al.*, 1985). Female *Ornithodoros* secrete a non-specific pheromone in their coxal fluid which is most active 4–6 days after the female has fed and evokes courtship behaviour in sexually active males (Sonenshine, 1991). Aggregations of *O. moubata* develop in response to pheromones produced by both sexes. Males respond more readily to the female pheromone than do females to the male pheromone. The receptors are located on the fourth segment of the palps. The function of the pheromone is to bring the sexes together for mating and for food location, and hence starved ticks show an enhanced response to pheromones. The pheromones are not species specific and *O. moubata* is attracted to *O. tholozani* and *A. persicus* (Leahy, 1979).

In mating the male crawls beneath the female so that their ventral surfaces are in contact and the male uses its mouthparts to dilate the female genital opening. A bulb-shaped spermatophore appears at the genital opening of the male and is introduced into the female opening by the mouthparts of the male. The outer exospermatophore remains attached externally and the endospermatophore evaginates, entering the female genital tract and depositing developing sperm in paired capsules. In *O. tholozani*, and probably also in *O. moubata*, adlerocysts (*Adlerocystis* sp.), symbiotes of the sperm cells, are deposited in the contents of the endospermatophore. They are considered to contribute to maintaining the sperm viable. Later when the sperm are mature, the capsule ruptures and they are released into the uterus (Robinson, 1942; Hoogstraal, 1956; Feldman-Muhsam, 1967, 1991). A female may mate more than once and 80 capsules have been found in the uterus of a single female in a laboratory colony, indicating that 40 successive matings had taken place (Aeschlimann and Grandjean, 1973a).

Mating and fecundity

A female may produce up to eight egg batches but the number of eggs in each batch steadily declines from about 140 in the first batch to 33 in the seventh batch. In the absence of repeated mating the percentage of females ovipositing declines rapidly after the third batch of eggs to less than 30%. When females were mated after every blood meal, the proportion which oviposited remained very high for the first four batches and was about 70% of the surviving females in the seventh batch. On average a female will lay about 500 eggs in her lifetime. Even where the female is mated after every blood meal there is a steady decline in fertility from 93% in the first batch to 73% in the seventh batch. Mating after each blood meal ensured high fertility in the eggs subsequently laid but shortened the life of the female (Aeschlimann and Grandjean, 1973a).

Bionomics

In common with other argasid ticks *O. moubata* has considerable powers of survival against starvation and desiccation. The crystalline cuticular wax coating over the body has a melting point of 63°C and that of *O. savignyi* 75°C. It is protected by a cement layer and prevents water loss through the cuticle (Lees, 1947). It is extremely effective, and the cuticle of *O. moubata* has a very high vapour diffusion resistance, being fourth highest out of 18 terrestrial organisms, which included vertebrates, arthropods and plants (Monteith and Campbell, 1980).

Unmated and unfed adult female *O. moubata* have survived for more than 3 months when kept at 32°C over concentrated sulphuric acid in an atmosphere of 0% RH. When fed once before being starved and maintained under a more favourable humidity (85% RH), there were differences in survival between different strains of *O. moubata*. In three different populations 60% survived for 9, 18 and 56 months (Walton, 1960). Survival is favoured by the fact that *O. moubata* is able to take up water through the spiracles when exposed to humid air (95% RH) (Lees, 1946a).

Peirce (1974) studied the distribution of *O. m. porcinus* in animal burrows in East Africa and found over 40% of them infested. The estimated numbers of ticks ranged from a few up to 250,000 with a predominance of second- and third-stage nymphs accounting for more than 70% of the population. Adults formed only 6% of the population, and they and the larger nymphs were suspected of being subject to predation by insectivorous carnivores, rodents and reduviid bugs. The ticks were found on and in the soil to a depth of 5 cm and on the roof of the burrow, the latter being interpreted as a response to hunger.

The most important environmental conditions for the tick are a neutral soil, high pH and a favourable temperature with an optimum at 24°C. Vegetation around the burrow conceals the occupants from predators and favours the presence of suitable hosts for the tick. The optimum altitude for *O. m. porcinus* is 900–1500 m and the tick has not been found above 1900 m (Peirce, 1974). In the absence of hosts the tick is able to survive for long periods, even up to 5 years under suitable conditions of humidity (Walton, 1960). Such resistance to starvation enables foci of *O. moubata* to persist in the absence of hosts for a long period. Coupled with the ability to feed on alternate hosts, foci can be regarded for practical purposes as permanent.

Medical and veterinary importance

The population of *O. moubata* which lives in human habitations is of medical importance as the vector of endemic relapsing fever caused by *Borrelia duttoni* (Chapter 26). Wild suids are considered to act as reservoirs of African swine fever, and the virus has been isolated from *O. m. porcinus* taken from wart-hog burrows. Experimentally the virus passes from one generation to another by transovarian transmission but there is a problem with regard to wart hogs being reservoirs of the virus for ticks. The level of viraemia needed to infect the tick has never been detected in wart hogs, which raises the possibility of another host being involved or of special conditions being required for wart hogs to produce a sufficiently intense viraemia (Pini and Hurter, 1975).

Ornithodoros savignyi (see Fig. 22.4)

Ornithodoros savignyi is known as the eyed tampan in contrast to *O. moubata*, the eyeless tampan, on account of its possession of two pairs of simple eyes (Fig. 22.4) in the folds above coxa I and between coxae III and IV (Nuttall *et al.*, 1908). It is also known as the sand tampan because it buries itself in sandy and loose clayey soils, under trees, near wells and shady spots frequented by domestic stock – particularly camels, cattle, mules – and people, on which it feeds (Hoogstraal, 1956; Walker *et al.*, 1978). It does not occur in huts. On standing cattle it feeds on the legs just above the hooves. The numbers of *O. savignyi* in infested localities reach plague proportions. Their behaviour has been graphically described by Hoogstraal (1956, p. 197):

> At the Khartoum quarantine one may see a long, seething line of thousands of hungry tampans helplessly confined to the shade of a row of acacia trees. A few yards away, separated only by the hot, 9 o'clock sun, newly arrived cattle tied to a post fence tempt the tampans to cross the glaring strip. The next morning, in the coolness of 7 o'clock, those tampans under the trees are all blood bloated and resting comfortably in the sand, others are dragging back from their hosts across the now nonexistent barrier and the legs of the cattle are beaded with yet other podshaped ticks taking their fill of blood in a regular line just above the hoof.

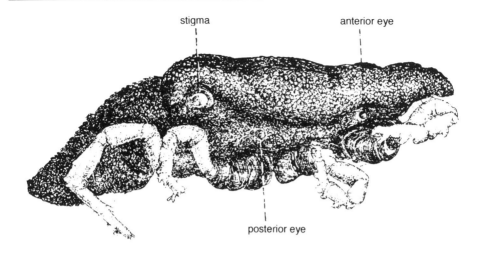

Fig. 22.4. *Ornithodoros savignyi.* Lateral view. Source: from Patton, W.S. and Evans, A.M. (1929) *Insects, Ticks, Mites and Venomous Animals. Part I: Medical.* H.R. Grubb, Croydon.

The biology and life cycle of *O. savignyi* is very similar to that of *O. moubata*, but it has a much wider geographical distribution occurring in the arid parts of Africa, Arabia, India and Sri Lanka. *O. savignyi* is not known to transmit any pathogen but 'camels and cattle suffer greatly and may even be killed by the volume of blood lost' (Hoogstraal, 1956).

Ornithodoros lahorensis

Ornithodoros lahorensis is a serious pest of sheep, cattle and camels in Asia, the southern republics of the former Soviet Union and south-east Europe from sea level to 2900 m. When larvae find a host they remain on it for 3 to 6 weeks, engorging four times and moulting three times. The engorged nymph detaches and drops to the ground where it moults into an adult. Given the opportunity the adult will feed rapidly on another host after which females will deposit batches of 300–500 eggs. This species is facultatively autogenous and can mature two batches of eggs without a blood meal. Unfed adults can live for 18 years and larvae for a year (Hoogstraal, 1985).

Argas (see Figs 22.5 and 22.6)

Fifty-six species have been described in the genus *Argas* and they are allocated to seven subgenera which are structurally and biologically distinct. Two subgenera, *Argas* and *Persicargas*, parasitize birds; other subgenera are associated with bats and a small number of other mammals; and *Argas* (*Microargas*) *transversus* is a permanent ectoparasite of the Galapagos giant tortoise (*Geochelone elephantopus*). All species of *Argas* are nocturnal except one, and with one exception all are restricted

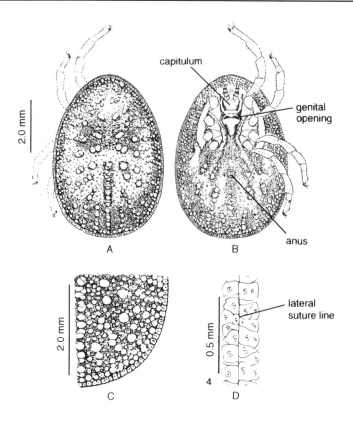

Fig. 22.5. *Argas (Persicargas) walkerae*: (**A**) dorsal and (**B**) ventral views of female. (**C**) Female dorsal integument (posterior quadrant), and (**D**) lateral integument in the same area. Source: from Kaiser and Hoogstraal (1969).

to arid habitats with long dry seasons. Where birds are present all year round, enormous populations of ticks can build up and this applies to domestic poultry. These ticks have a strongly developed positive thigmotactic response and penetrate deeply into crevices in wood or stone to a situation where both surfaces of their body are in contact with the substrate (Hoogstraal *et al.*, 1979).

Argas (Persicargas) persicus originated in the Palaearctic region where it occurs on domestic poultry and wild birds, and has been introduced by humans with their poultry into all other zoogeographical regions, with the possible exception of the Neotropical region (Hoogstraal *et al.*, 1979). It has been reported from a galah's nest in Australia (Hoogstraal *et al.*, 1975).

In southern Africa the main argasid parasitizing poultry is *A. (P.) walkerae* (Kaiser and Hoogstraal, 1969), *A. (P.) arboreus* parasitizes Ciconiformes (herons, egrets, storks, ibises) in the Afrotropical region and *A. (P.) robertsi* fills the same niche in the Australian and Oriental regions. In Queensland *A. (P.) robertsi* also parasitizes chickens, and may do so elsewhere in the region, but the commoner chicken argasid in the region is *A. (P.) persicus* (Hoogstraal *et al.*, 1975). The validity of three of these four closely related species has been confirmed by

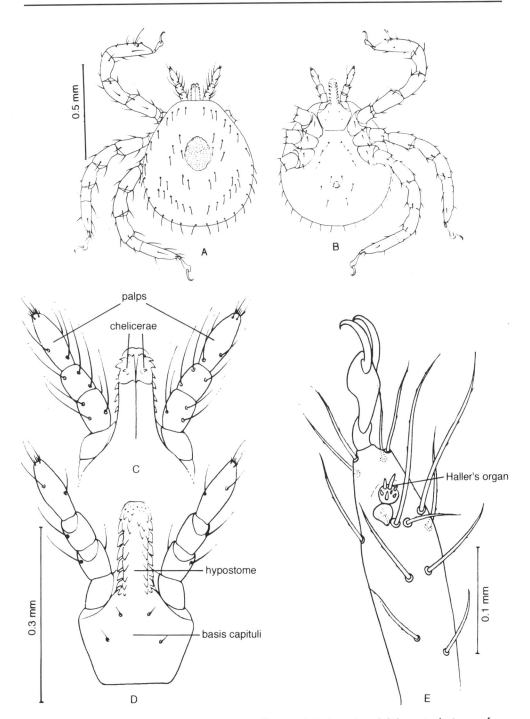

Fig. 22.6. *Argas (Persicargas) walkerae:* (**A**) dorsal and (**B**) ventral views of larvae. (**C**) Dorsal and (**D**) ventral views of larval capitulum. (**E**) Dorsal view of tarsus I. Source: from Kaiser and Hoogstraal (1969).

cross-breeding in the laboratory. Few eggs and no progeny are produced when *A.* (*P.*) *arboreus*, *A.* (*P.*) *walkerae* and *A.* (*P.*) *persicus* are cross-mated, proving that these three species are reproductively isolated (Gothe and Koop, 1974b).

Argas (P.) persicus

Argas persicus is of considerable veterinary importance as the most widespread argasid tick feeding on poultry. The unfed adult tick is pale yellow in colour, becoming darker when fed. In outline the body is oval to pear-shaped being broadest behind the legs at about the level of the anus. Females measure 7–10 mm × 5–6 mm and males are slightly smaller measuring 4–5 mm × 2.5–3 mm (Nuttall *et al.*, 1908). As in all species of *Persicargas* the quadrangular cells at the body margin are not striated (Fig. 22.5C, D), a point which distinguishes this subgenus from the subgenus *Argas* (Kaiser *et al.*, 1964). The life cycle is similar to that of *O. moubata* with two major differences: there are fewer nymphal stages and the larva feeds on the host for several days.

Life cycle

Females deposit yellowish-brown, spherical eggs in cracks and crevices of poultry houses where they hatch to produce a larva (Fig. 22.6), which has a more or less circular outline, subterminal mouthparts visible from above, and no stigmata. Larvae of *A. arboreus* have a simple respiratory system opening to the exterior through slit-like ostia above coxae I and II (Sonenshine, 1991). The larva attaches itself to its host, particularly under the wings, where it feeds for about a week. It then falls off and has a period of relative inactivity before moulting to become a nymph. There are two to four nymphal stages before the adult. Female *A. walkerae* orientate to their host by its odour and a carbon dioxide gradient, increasing towards the bird. The main stimulus is the host odour, and carbon dioxide and the host's radiant heat are secondary stimuli (Beelitz and Gothe, 1991). Digestion of blood is in three stages – haemolysis, 'rapid' digestion involving endocytosis and 'slow' digestion in which the stored blood meal is gradually consumed (Sonenshine, 1991).

Mating

The adults produce an aggregation pheromone which brings the sexes together for mating, and males of *A. persicus* were the most responsive not only to their own females, but to the pheromones produced by other species of *Argas* and *Ornithodoros* (Leahy, 1979). Usually the female feeds before mating and the coxal fluid is excreted after the tick has fed and not during feeding as in *O. moubata*. During mating the male inserts a spermatophore into the female genital opening.

Eggs are matured and laid over a period of several days after a blood meal and not in a single batch (Hoogstraal, 1956).

Quantitative studies

Quantitative studies on the life cycle of *A. persicus* have been made by Gothe and Koop (1974a) and Khalil (1979). Both sets of observations were made on material originating from the same colony. The environmental conditions were similar: 28–29°C and 75% RH and 27°C and 90–95% RH. Khalil fed her ticks on domestic pigeons and Gothe and Koop fed theirs on chickens. This may explain the differences between the two sets of results. In Khalil's study the range of variation was greater, the average time taken for each stage longer, and an extra nymphal stage occurred.

It is known that there is an interaction between host and tick. Tatchell *et al.* (1973) found that when *A. persicus* was fed on pigeons it took in only two-thirds of the amount of blood it would imbibe from poultry. The reverse was true for *A. (P.) arboreus*. Galun *et al.* (1978) obtained a different result and found *A. persicus* taking a considerably larger (×1.5) meal from pigeons compared with chickens. However, in terms of egg production, pigeon blood was less productive (×0.75) than chicken blood. This may explain the differences between the results, and since chickens are the more important host economically, the results of Gothe and Koop will be considered first, and then additional results from Khalil be presented.

Larvae of *A. persicus* attached to chickens for 4 to 7 days with the majority falling off after 5 days' attachment. There followed a premoult period of 8 days (6–10 days) before moulting to the first nymph. Following a blood meal the nymph moulted to a second nymphal stage in 12 days (10–17 days). After the first blood meal most second-stage nymphs moulted to the adult stage in 14 days (11–20 days) with only a small proportion (6%) becoming third-stage nymphs. After these had fed they moulted to the adult in 15 days and there was no fourth nymphal stage.

A fed and mated female has a preoviposition period of 8 days, before an oviposition period of 11 days during which it deposits on average about 140 eggs but with considerable variation (standard deviation = 38). Eggs were deposited at a rate of 12–18 per day from the second to ninth day of oviposition inclusively. Larval emergence took place 2 weeks after the first egg was laid and just over 3 weeks from the blood meal. Just over 60% of the nymphs become adults (Gothe and Koop, 1974a).

In view of the large variation in individual results, Khalil's data will be presented as median values. The ability of larval *A. persicus* to attach successfully was a function of age and was highest when larvae were 6 to 14 days old, when 60% of the larvae attached. Nymphs fed for varying periods up to an hour, and after each moult there was a pre-feeding period of 1 to 4 days in all stages. Adults fed for up to 2 to 3 h. In all cases the coxal fluid was excreted after detaching from the host. Rather less than half the individuals required three or more nymphal stages to become adult, and 10% required four nymphal stages. All adults produced from the fourth stage were females. Following blood meals females had four to six periods of oviposition, producing 63 and 74 eggs in the first and second ovipositions, about half the number (142) reported by Gothe and Koop (1974a) for the same species after a single blood meal.

Bionomics

The median time for survival of unfed larvae of *A. persicus* was 20 days. Nymphs survived considerably longer, especially the first nymphal stage which had a median survival period of nearly 4 months compared with 8 weeks for the second to fourth nymphal stages. The latter value was closer to the survival ability of the adult which was 9 weeks for males and 10 weeks for females. Depending upon the availability of hosts, there could be up to ten generations a year but the life cycle could extend to 2 years where hosts were only available infrequently (Khalil, 1979).

Desiccation is a major threat to the survival of unfed ticks. Experiments with *A. (P.) arboreus* showed that this species has considerable resistance to desiccation. More than 50% of adults survived unfed for 105 days when kept over concentrated sulphuric acid at 0% RH. During that time they lost about half their initial body weight, mostly due to loss of water. At 96% RH survival is almost 100% and the loss of weight a mere 4–5% after 105 days. Indeed, in the early stages the ticks actually took up water and increased their weight by 7% (Hefnawy *et al.*, 1975).

Economic importance

Three species, *A. persicus*, *A. sanchezi*, and *A. radiatus*, attack poultry in the southern USA along the Gulf of Mexico and the Mexican border. They are rarely found in commercial caged-layer operations but in simpler poultry houses. Ticks are most active in the warm dry season, causing emaciation, weakness, slow growth and a fatal anaemia from loss of blood. Unfed adults can survive for 4 years and larvae for several months. Hence control measures must concentrate on treating all possible resting places with an approved insecticide (Arends, 1991). *A. persicus*, and probably the other related chicken-feeding species, transmits two important pathogens to poultry, *Borrelia anserina* and *Aegyptianella pullorum*, which are dealt with in Chapters 25 and 26. The feeding of *A. persicus* and *A. walkeri* can cause a condition known as fowl paralysis. A similar phenomenon is caused by ixodid ticks on mammals and will be considered in the next chapter.

References

Aeschlimann, A. and Grandjean, O. (1973a) Observations on fecundity in *Ornithodorus moubata*, Murray (Ixodoidea: Argasidae). *Acarologia* 15, 206–217.
Aeschlimann, A. and Grandjean, O. (1973b) Influence of natural and 'artificial' mating on feeding, digestion, vitellogenesis and oviposition in ticks (Ixodoidea). *Folia Parasitologia* 20, 67–74.
Arends, J.J. (1991) External parasites and poultry pests. In: Calnek, B.W. (ed.) *Diseases of Poultry*. Iowa State University Press, Ames, pp. 702–730.

Bedford, G.A.H. (1925) The spinose ear-tick (*Ornithodorus megnini* Dugès). *Journal of the Department of Agriculture, Union of South Africa* 10, 147–153.

Beelitz, P. and Gothe, R. (1991) Investigations on the host seeking and finding of *Argas (Persicargas) walkerae* (Ixodidae: Argasidae). *Parasitology Research* 77, 622–628.

Bulman, G.M. and Walker, J.B. (1979) A previously unrecorded feeding site on cattle for the immature stages of the spinose ear tick, *Otobius megnini* (Dugès, 1844). *Journal of the South African Veterinary Association* 50, 107–108.

Chellappa, D.J. (1973) Notes on spinose ear tick infestations in man and domestic animals in India and its control. *Review of Applied Entomology B* 63, 2902.

Chellappa, D.J. and Alwar, V.S. (1972) On the incidence of *Otobius megnini* (Dugès, 1883) on sheep in India. *Review of Applied Entomology B* 62, 1459.

Cooley, R.A. and Kohls, G.M. (1944) *The Argasidae of North America, Central America and Cuba.* American Midland Naturalist Monograph No. 1, University Press, Notre Dame, USA.

Evans, G.O. (1992) *Principles of Acarology.* CAB International, Wallingford, Oxon, UK.

Feldman-Muhsam, B. (1967) Spermatophore formation and sperm transfer in *Ornithodoros* ticks. *Science* 156, 1252–1253.

Feldman-Muhsam, B. (1991) The role of *Adlerocystis* sp. in the reproduction of argasid ticks. In: Schuster, R. and Murphy, P.W. (eds) *The Acari, Reproduction, Development and Life History Strategies.* Chapman & Hall, London, pp. 179–190.

Galun, R., Sternberg, S. and Mango, C. (1978) Effects of host species on feeding behaviour and reproduction of soft ticks (Acari: Argasidae). *Bulletin of Entomological Research* 68, 153–157.

Gothe, R. and Koop, E. (1974a) Zur biologischer Bewertung der Validität von *Argas (Persicargas) persicus* (Oken, 1818), *Argas (Persicargas) arboreus* Kaiser, Hoogstraal und Kohls, 1964 and *Argas (Persicargas) walkerae* Kaiser und Hoogstraal, 1969. I. Untersuchungen zur Entwicklungsbiologie. *Zeitschrift für Parasitenkunde* 44, 299–317.

Gothe, R. and Koop, E. (1974b) Zur biologischen Bewertung der Validität von *Argas (Persicargas) persicus* (Oken, 1818), *Argas (Persicargas) arboreus* Kaiser, Hoogstraal und Kohls, 1964 und *Argas (Persicargas) walkerae* Kaiser und Hoogstraal, 1969. II. Kreuzungsversuche. *Zeitschrift für Parasitenkunde* 44, 319–328.

Hefnawy, T., Bishara, S.I. and Bassal, T.T.M. (1975) Biochemical and physiological studies of certain ticks (Ixodoidea): effects of relative humidity and starvation on the water balance and behaviour of adult *Argas (Persicargas) arboreus* (Argasidae). *Experimental Parasitology* 38, 14–19.

Hoogstraal, H. (1956) *African Ixodoidea. I. Ticks of the Sudan (with special reference to Equatoria Province and with Preliminary Reviews of the Genera* Boophilus, Margaropus *and* Hyalomma). Research Report NM 005 050.29.07, Bureau of Medicine and Surgery, Department of Navy, Washington.

Hoogstraal, H. (1985) Argasid and nuttelliellid ticks as parasites and vectors. *Advances in Parasitology* 24, 135–238.

Hoogstraal, H. and Kim, K.C. (1985) Tick and mammal coevolution, with emphasis on *Haemaphysalis.* In: Kim, K.C. (ed.) *Coevolution of Parasitic Arthropods and Mammals.* John Wiley, New York, pp. 505–568.

Hoogstraal, H., Kaiser, M.N. and McClure, H.E. (1975) The subgenus *Persicargas* (Ixodoidea: Argasidae: *Argas*). 20. *A. (P.) robertsi* parasitizing nesting wading birds and domestic chickens in the Australian and Oriental regions, viral infections and host migration. *Journal of Medical Entomology* 11, 513–524.

Hoogstraal, H., Clifford, C.M., Keirans, J.E. and Wassef, H.Y. (1979) Recent developments in biomedical knowledge of *Argas* ticks (Ixodoidea: Argasidae). In: Rodriguez, J.G. (ed.) *Recent Advances in Acarology.* Academic Press, New York, vol. 2, pp. 269–278.

Jobling, B. (1925) A contribution to the biology of *Ornithodorus moubata* Murray. *Bulletin of Entomological Research* 15, 271–279.

Kaiser, M.N. and Hoogstraal, H. (1969) The subgenus *Persicargas* (Ixodoidea: Argasidae: *Argas*). 7. *A. (P.) walkerae*, new species, a parasite of domestic fowl in southern Africa. *Annals of the Entomological Society of America* 62, 885–890.

Kaiser, M.N., Hoogstraal, H. and Kohls, G.M. (1964) The subgenus *Persicargas*, new subgenus (Ixodoidea: Argasidae: *Argas*). I. *A. (P.) arboreus*, new species, an Egyptian *Persicus*-like parasite of wild birds, with a redefinition of the genus *Argas*. *Annals of the Entomological Society of America* 57, 60–69.

Khalil, G.M. (1979) The subgenus *Persicargas* (Ixodoidea; Argasidae: *Argas*). 31. The life cycle of *A. (P.) persicus* in the laboratory. *Journal of Medical Entomology* 16, 200–206.

Lavoipierre, M.M.J. and Riek, R.F. (1955) Observations on the feeding habits of argasid ticks and on the effect of their bites on laboratory animals, together with a note on the production of coxal fluid by several of the species studied. *Annals of Tropical Medicine and Parasitology* 49, 96–113.

Leahy, M.G. (1979) Pheromones of argasid ticks. In: Rodriguez, J.G. (ed.) *Recent Advances in Acarology*. Academic Press, New York, vol. 2, pp. 297–308.

Lees, A.D. (1946a) The water balance in *Ixodes ricinus* L. and certain other species of ticks. *Parasitology* 37, 1–20.

Lees, A.D. (1946b) Chloride regulation and the function of the coxal glands in ticks. *Parasitology* 37, 172–184.

Lees, A.D. (1947) Transpiration and the structure of the epicuticle in ticks. *Journal of Experimental Biology* 23, 379–410.

Lees, A.D. and Beament, J.W.L. (1948) An egg-waxing organ in ticks. *Quarterly Journal of Microscopical Science* 89, 291–332.

Mango, C.K.A. and Galun, R. (1977) *Ornithodoros moubata*: breeding *in vitro*. *Experimental Parasitology* 42, 282–288.

Monteith, J.L. and Campbell, G.S. (1980) Diffusion of water vapour through integuments – potential confusion. *Journal of Thermal Biology* 5, 7–9.

Nuttall, G.H.F., Warburton, C., Cooper, W.F. and Robinson, L.E. (1908) *Ticks: A Monograph of the Ixodoidea. Part 1. The Argasidae*. Cambridge University Press, Cambridge.

Oliver, J.H. (1989) Biology and systematics of ticks (Acari: Ixodoidea). *Annual Review of Ecology and Systematics* 20, 397–430.

Otieno, D.A., Hassanall, A., Obenchain, F.D., Sternberg, A. and Galun, R. (1985) Identification of guanine as an assembly pheromone of ticks. *Insect Science and its Application* 6, 667–670.

Peirce, M.A. (1974) Distribution and ecology of *Ornithodoros moubata porcinus* Walton (Acarina) in animal burrows in East Africa. *Bulletin of Entomological Research* 64, 605–619.

Pini, A. and Hurter, L.R. (1975) African swine fever: an epizootiological review with special reference to the South African situation. *Journal of the South African Veterinary Association* 46, 227–232.

Robinson, G.G. (1942) The mechanism of insemination in the argasid tick, *Ornithodorus moubata* Murray. *Parasitology* 34, 195–198.

Sonenshine, D.E. (1991) *Biology of Ticks*, vol. 1, Oxford University Press, New York.

Sonenshine, D.E. (1993) *Biology of Ticks*, vol. 2. Oxford University Press, New York.

Tatchell, R.J., Kerr, J.D. and Boctor, F.N. (1973) Biochemical and physiological studies of certain ticks (Ixodoidea). Haemolysis rate and meal size in the interactions between

Argas (Persicargas) arboreus Kaiser, Hoogstraal and Kohls, *A. (P.) persicus* (Oken) (Argasidae) and some avian hosts. *Parasitology* 67, 41–51.

Walker, J.B., Mehlitz, D. and Jones, G.E. (1978) *Notes on the Ticks of Botswana.* German Agency for Technical Cooperation, Eschborn, West Germany.

Walton, G.A. (1960) The reaction of some variants of *Ornithodoros moubata* Murray (Argasidae: Ixodoidea) to desiccation. *Parasitology* 50, 81–88.

Ixodida – Ixodidae
(Hard Ticks)

23

The Ixodidae are large, blood-sucking Acari with a terminal capitulum in all stages; a dorsal shield or scutum showing sexual dimorphism, being small in the female and almost covering the dorsal surface in the male (Fig. 23.1A and C); reduced fourth palpal segment inserted on the ventral side of the third (Plate 23.1); and the stigmata located behind the fourth pair of legs. There is only one nymphal stage in the life cycle. Thirteen genera are recognized of which seven contain species of medical and veterinary importance.

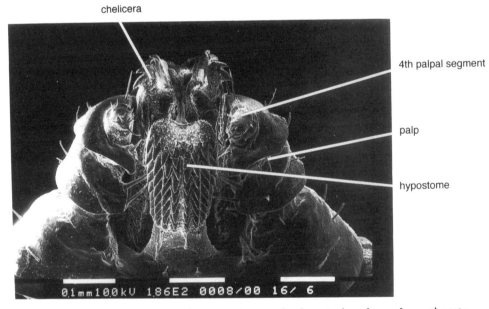

Plate 23.1. Scanning electron micrograph of ventral surface of mouthparts and palps of female *Boophilus microplus*. Note: lateral palps with 4th segment located on ventral surface of 3rd segment; median toothed hypostome; and, at the top the chelicerae protruding from their sheaths. Courtesy of T. McRae.

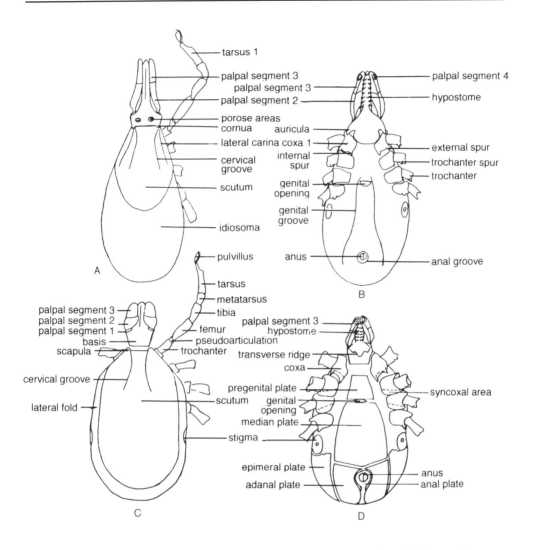

Fig. 23.1. Diagnostic features of *Ixodes* ticks: (**A**) dorsal and (**B**) ventral views of female. (**C**) Dorsal and (**D**) ventral views of male. Source: Arthur (1965).

Genera of Medical and Veterinary Importance (Hoogstraal, 1986; Varma, 1993)

Ixodes (see Fig. 23.1)

Ixodes is the largest genus in the family with more than 200 species. They are small, inornate ticks, easily overlooked when searching a host. The capitulum of the female is considerably longer than that of the male. Often the second segment of the palp is constricted at the base, creating a gap between the palp and the mouthparts. There are no eyes (Fig. 23.3) or festoons (Fig. 23.2). The anal groove passes anteriorly to the anus and *Ixodes* is said to be prostriate. In other genera the anal groove is either posterior to the anus or obsolete and they are referred to as metastriate. In the male there are seven ventral plates including a median row of three – pregenital, median, anal – a pair of adanals and a pair of epimerals. The margins of the epimerals, which are placed posterolaterally, are often indistinct (Arthur, 1965). *Ixodes* ticks are highly specialized in their habits with species parasitizing bats and sea birds. The most important group is the holarctic *ricinus–persulcatus* complex which includes the North American *I. dammini*, the European sheep tick, *I. ricinus*, and the northern palaearctic, *I persulcatus* which occurs from the Baltic Sea to Japan. All three are vectors of Lyme disease and *I. persulcatus* is also the chief vector of Russian spring–summer encephalitis. *I. rubicundus* and *I. holocyclus* cause paralysis in mammals in southern Africa and Australia, respectively.

Haemaphysalis (see Figs 23.2 and 23.9)

There are 155 easily recognizable species in the genus *Haemaphysalis*. They are small, inornate ticks with short mouthparts, i.e. brevirostrate. The basis capituli is rectangular and the base of the second palpal segment is expanded, projecting laterally beyond the basis capituli. The second and third palpal segments taper anteriorly so that the capitulum anterior to the basis capituli appears to be triangular. There are no eyes in either sex and no ventral plates in the male. Festoons are present. These are uniform, rectangular areas along the posterior margin of the body, separated by grooves. They are best seen in unfed specimens, and are lost in engorged females. This genus reaches its maximum development in the Oriental region where *H. spinigera* is the vector of the arbovirus causing Kyasanur Forest Disease (see Chapter 24). *H. leachi*, the yellow dog tick, is a widespread parasite of carnivores, especially dogs, in the tropics and subtropics. *H. punctata* parasitizes cattle in the Palaearctic region, as does *H. longicornis* in the Australian and Oriental regions (Hoogstraal *et al.*, 1968).

Boophilus (see Fig. 23.3)

The five species of *Boophilus* are small, inornate, brevirostrate ticks in which the anal groove is obsolete. The basis capituli is hexagonal dorsally, and there are simple eyes laterally on the scutum. Ventrally coxa I is bifid, and there are paired adanal and accessory adanal plates flanking the anus posteriorly. In some species the replete male develops a median tail. They are one-host ticks (see below) which parasitize large mammals, especially cattle. *B. microplus*, the pantropical cattle tick, occurs in the Neotropical, Afrotropical and Australian regions, *B. decoloratus* in tropical Africa, and *B. annulatus* in North America. They are vectors of babesiosis to cattle.

Rhipicephalus (see Fig. 23.4)

The 70 species of *Rhipicephalus* are small metastriate, brevirostrate, reddish or blackish-brown ticks which are mostly inornate. The basis capituli is hexagonal

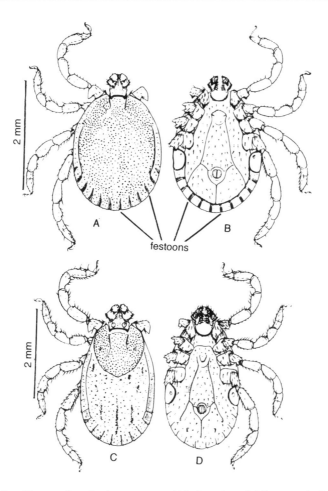

Fig. 23.2. *Haemaphysalis longicornis*: (**A**) dorsal and (**B**) ventral views of male. (**C**) Dorsal and (**D**) ventral views of female. Source: from Hoogstraal *et al.* (1968).

dorsally and eyes and festoons are present. Coxa I is bifid in both sexes. The male has adanal and accessory adanal plates on the ventral surface and when replete has a tail. The genus reaches its greatest development in the Afrotropical region where *R. appendiculatus*, the brown ear tick, is the main vector of *Theileria parva*, the causative agent of east coast fever in cattle. Other important species on domestic animals are *R. evertsi* and *R. simus*. The red dog tick, *R. sanguineus*, is the most widespread ixodid species, and although commoner in the warmer parts of the world, it survives in heated buildings in urban communities in Canada and Scandinavia.

Dermacentor (see Figs 23.5 and 23.6)

The 30 species of *Dermacentor* are medium to large, usually ornate, metastriate, brevirostrate ticks. The basis capituli is rectangular dorsally and eyes and festoons

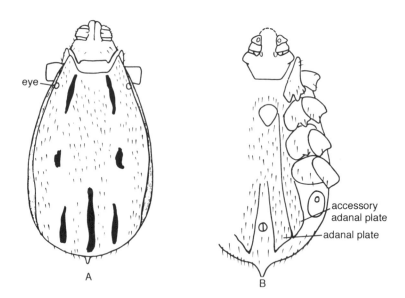

Fig. 23.3. (**A**) Dorsal and (**B**) ventral views of male *Boophilus microplus*. Source: Arthur, D.R. (1960). *Ticks: A Monograph of the Ixodoidea. Part V.* p. 209. Cambridge University Press.

are present. Coxa I is bifid in both sexes and coxa IV is greatly enlarged in the male which has no ventral plates (Fig. 23.6B). This genus has its greatest development in the New World, where it includes the wood tick *D. andersoni* (Fig 23.5), the vector of Rocky Mountain spotted fever and the cause of tick paralysis, and the American dog-tick, *D. variabilis*, which is abundant on the eastern coast of the USA. In the Palaearctic region species of *Dermacentor*, including *D. marginatus*, are vectors of Siberian tick typhus and *D. reticulatus* (Fig. 23.6) parasitizes cattle and horses.

Hyalomma (see Fig. 23.7)

The 29 species and subspecies of *Hyalomma* are medium-sized metastriate ticks with long mouthparts, i.e. longirostrate. The basis capituli is subtriangular dorsally (Fig. 23.7E) and eyes are present. Festoons and ornamentation of the scutum are variable characters which may be present or absent. The male has one pair of adanal plates and accessory adanal plates may or may not be present. Coxa I is bifid. Hyalommas are tough, hardy ticks which survive where humidity is low, climatic conditions extreme, hosts rare and hiding places sparse. They probably originated in the desert regions of Kazakhstan and Iran in the Palaearctic region (Hoogstraal, 1956). *H. anatolicum anatolicum* and species of the *H. marginatum* complex play a major part in the transmission of the arbovirus causing Crimean–Congo haemorrhagic fever (see Chapter 24).

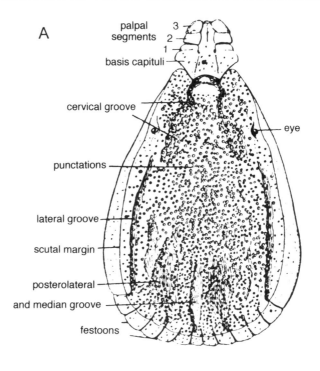

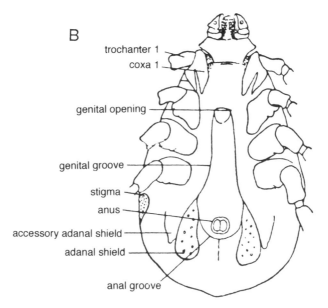

Fig. 23.4. (**A**) Dorsal and (**B**) ventral views of male *Rhipicephalus sanguineus*. Source: from Nuttall, G.H.F. (1911) *Ticks: A Monograph of the Ixodoidea. Part II.* p. 122. Cambridge University Press.

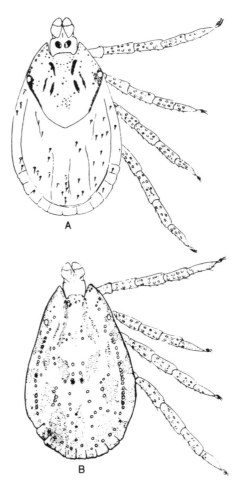

Fig. 23.5. *Dermacentor andersoni*: (**A**) dorsal view of female; (**B**) dorsal view of male. Source: from Arthur, D.R. (1962) *Ticks and Disease*. Pergamon Press, Oxford.

Amblyomma (see Fig. 23.8)

The 102 species of *Amblyomma* are large ornate, metastriate, longirostrate ticks with eyes and festoons but no adanal plates in the male. The genus is particularly well represented in the New World. In the Afrotropical region *A. hebraeum* (Fig. 23.8) and *A. variegatum* transmit *Cowdria ruminantium*, the cause of heart-water in cattle. The feeding of numbers of large longirostrate ticks can cause extensive damage to hides and skins, and their lesions provide a route for the invasion of pathogens (Yeoman and Walker, 1967).

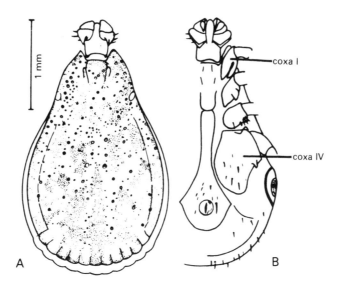

Fig. 23.6. (**A**) Dorsal and (**B**) ventral views of male *Dermacentor reticulatus*.
Source: from Arthur, D.R. (1963) *British Ticks*. Butterworth, London.

Life Cycle

Oviposition

There are four stages in the life cycle: egg, larva, nymph and adult. The female
drops off its vertebrate host and seeks a sheltered situation in which to develop
and lay a single large batch of eggs, after which she dies. Typically a batch contains
several thousand brown globular eggs, and oviposition continues for many days.
I. ricinus has been recorded as depositing an egg every 3 to 12 min (Lees and
Beament, 1948). During oviposition the tick bends its capitulum in an arc ventrally
until it is tightly oppressed to the ventral surface and the tip of the capitulum is
near the genital opening. Gené's organ is everted between the scutum and the basis
capituli. This structure has a swollen base and two short horns. During the extru-
sion of an egg the lining of the vagina prolapses through the genital opening
holding the egg which is deposited between the horns of Gené's organ. The pro-
lapse is then retracted.
    The function of Gené's organ is to apply a waterproofing wax to the egg after
which the organ is withdrawn and the egg is left on the hypostome. When the
capitulum returns to its normal forwardly directed position the egg is deposited
above the female. Consequently the egg batch is to be found above the shrivelled
body of the female. If the eggs do not receive a wax coat they shrivel and die.
In *I. ricinus* an incomplete waterproof layer is added during the egg's passage down
the vagina, and waterproofing is completed by Gené's organ which, in this species,
secretes an easily spreading and penetrating soft wax. In the first few days of

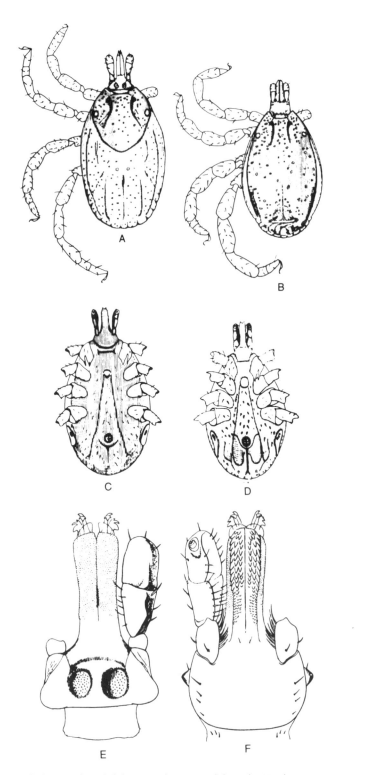

Fig. 23.7. (**A**) Dorsal and (**C**) ventral views of female *Hyalomma excavatum*. (**B**) Dorsal and (**D**) ventral views of male *H. excavatum*. (E) Dorsal and (F) ventral views of capitulum of female *H. aegyptium*. Sources: (**A**) (**B**) (**C**) (**D**) from Arthur, D.R. (1962) *Ticks and Disease*. Pergamon Press, Oxford. (**E**) and (**F**) from Nuttall, G.H.F. (1911) *Ticks: A Monograph of the Ixodoidea. Part II.* p. 122. Cambridge University Press.

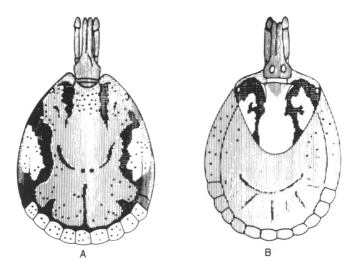

Fig. 23.8. (**A**) Dorsal view of male and (**B**) dorsal view of female *Amblyomma hebraeum*. Source: from Arthur, D.R. (1962) *Ticks and Disease*. Pergamon Press, Oxford.

incubation the wax penetrates the outer layers of the egg to reach the inner membrane. The critical temperature at which the wax ceases to waterproof the egg is 35°C in *I. ricinus*, and 44°C in *Hyalomma savignyi* (Lees and Beament, 1948). Depending upon climatic conditions, eggs hatch in 2 weeks to several months (Hoogstraal, 1956), giving rise to a hexapod larva.

Larva (see Fig. 23.9C, D)

Larvae have neither spiracles nor tracheal system and therefore water loss is solely through the cuticle. When a larva succeeds in getting on to a host it attaches and feeds for several days before dropping off engorged. All larvae are not successful in finding a host and larvae of *I. ricinus* move down to the more humid layers near the ground where, at suitable humidities, they are able to take up water from the atmosphere (Sonenshine, 1993). In this way larvae can survive for considerable periods unfed.

Nymph and adult

The larva digests its blood meal and moults to an octopod nymph (Fig. 23.9A, B), which has spiracles and a tracheal system making it more susceptible to desiccation. The nymph feeds for 4 to 8 days on the host, engorging more rapidly towards the end of the period before dropping off to find a suitable location in which to digest its meal and moult to the adult stage. When a female finds a host she attaches but does not engorge until mating has taken place. The male feeds but does not engorge. The engorged female drops off the host and hides away in a suitable location to digest the blood meal and lay an egg batch.

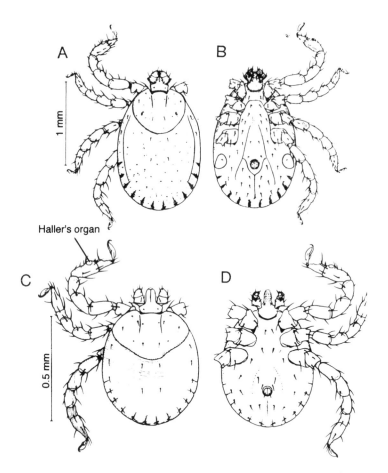

Fig. 23.9. *Haemaphysalis longicornis.* (**A**) Dorsal and (**B**) ventral views of nymph. (**C**) Dorsal and (**D**) ventral views of larva. Source: from Hoogstraal *et al.* (1968).

Variations on life cycle

The majority of ixodid ticks (620/650, Hoogstraal, 1985) have the type of life cycle just described and are referred to as three-host ticks, but two genera *Boophilus* and *Margaropus* are one-host ticks in which engorged larvae and nymphs do not drop off the host but remain attached and moult *in situ*, and the subsequent stage re-attaches on the same individual host. The genus *Margaropus* contains three species, two of which are parasites of giraffe, and *M. winthemi*, the winter horse tick of the high veld of southern Africa (Hoogstraal, 1978).

A few species in other genera are one-host ticks, including *D. (Anocentor) nitens*, a Neotropical ear-tick of horses; *D. albipictus*, the North American moose tick; and *Hy. scupense* in the Palaearctic region. A few species have adopted a two-host cycle in which the larva and nymph occur on the same individual host, the nymph then dropping off and the adult parasitizing a different individual of the same or another host species. Included in the two-host ticks, which parasitize domestic animals, are: *R. evertsi* in the Afrotropical region, and *R. bursa*, *Hy. marginatum* and *Hy. detritum* in the Palaearctic region (Hoogstraal, 1978).

Although ixodid ticks are not host-specific they are not indiscriminate in the hosts they parasitize (Hoogstraal and Aeschlimann, 1982). A few show an extremely wide range and these are often of economic importance, but most occur on a limited range of hosts which they parasitize with varying intensities. Some hosts are commonly and heavily infested and others infrequently and lightly. Each species is adapted to its hosts, concentrating on particular parts of the host's body and adjusting its seasonal and daily activity cycles to the host's behaviour and availability. To overcome adverse conditions or times when the host is not present, diapause may occur at any stage of the life cycle and both pre- and post-feeding. There may be several generations in a year, e.g. *B. microplus*, or the life cycle may extend over 3 to 4 or even 7 years in cold subarctic latitudes (Hoogstraal, 1978).

Biology and Behaviour

The structure and function of the organs and tissues of ticks has been comprehensively covered in Sonenshine (1991) and their ecology, behaviour, control and the diseases they transmit in Sonenshine (1993).

Host finding and attachment

Where there is little or no vegetational cover, ticks e.g. *Hyalomma asiaticum*, detect the presence of a host by ground vibrations and emerge from their burrows and move towards the host. In vegetated areas tick larvae seek a host by climbing vegetation and accumulate near the tips of grasses and similar plants. When a host approaches, larvae detect its presence mainly by sensory receptors in Haller's organ (Fig. 23.9C) on the tarsus of the first pair of legs and they quest for the host by waving those legs in the air. Sensory receptors in the anterior pit of Haller's organ respond to odours, especially of phenolic compounds, and other receptors respond to humidity, temperature and ammonia, while those in the posterior capsule respond to carbon dioxide, odours and temperature. Other receptors on the tarsus detect phenolic compounds, which are produced in large quantities by feeding females and act as sex pheromones, and others detect temperature gradients and respond to the host's body heat. On the host, the sensory receptors of the tarsi and the palps provide information to enable the tick to find a suitable site for attachment (Waladde, 1987).

The tick pierces the skin of its host with its chelicerae and inserts the barbed hypostome to secure it to the host initially. The cheliceral digits have mechanoreceptors and sensory receptors which respond to ATP and other haematophagostimulants (Waladde, 1987). The pattern of feeding of female *R. appendiculatus* involves the pharyngeal dilator muscles initiating fast sucking for 5 min, followed by 1 to 2 min rest when the floor of the salivarium, the fused paired salivary ducts, is actively lowered (Lösel *et al.*, 1992). When the chelicerae of a feeding *Boophilus* are exposed to the blood of a resistant host the feeding pattern deteriorates and the tick may detach (Waladde, 1987).

In brevirostrate ticks, cement secreted during the first 24 h, spreads over the surface of the skin to secure it to the host initially. During the next 96 h extra cement is secreted, which penetrates the keratinized layers of the stratum corneum, fills the lesion and attaches the tick more firmly (Moorhouse, 1969). In longirostrate ticks there is less need of cement which is reduced to a sheath around the mouthparts within the host. In *I. holocyclus* the female is longirostrate and secretes no cement (Binnington and Kemp, 1980).

Feeding and water elimination

During engorgement the body weight of a tick increases by about × 200 (Binnington and Kemp, 1980). An unfed female *I. ricinus* weighs about 2 mg and takes in 600 mg of blood while feeding, but half to two-thirds of the water contained in the ingested blood is eliminated before feeding ceases and the final weight of the engorged tick is about 240 mg (Lees, 1946b). In argasid ticks excess water is eliminated via the coxal apparatus but no such structure is present in ixodid ticks (Kaufman, 1979). Little urine is produced by the Malpighian tubules and excess fluid is eliminated by salivation, the voluminous saliva being passed back into the host.

Mammalian blood is hyposmotic to tick tissues. When *I. ricinus* feeds on its main host, sheep, which have a blood chloride concentration of 0.5%, the chloride content of its haemolymph rises from 0.72% in the unfed tick to 0.88% in the fed tick (Lees, 1946a). This can only be achieved by the secretion of hyposmotic saliva (Kaufman, 1979). The saliva contains anticoagulants, weak hydrolytic enzymes, and pharmacological agents, some of which could play important roles in the feeding process (Binnington and Kemp, 1980). The question of toxins in the saliva is considered under Tick Paralysis (p. 480).

Water balance

The only liquid imbibed by ticks is blood and hence it is vital to their survival off the host to conserve their water content and minimize losses. The cuticle of newly moulted three-host ticks is 'as permeable as it ever will be'. One-host ticks are not exposed to desiccation as they develop on the host. When engorged female *B. microplus* drop off the host their cuticular lipids are increased to provide greater protection from desiccation. There is a critical equilibrium activity (CEA) at which ticks neither lose nor gain water. For *A. americanum*, a species associated with woodlands, the CEA value is 0.88 (= 88% RH) and for the more xeric *A. cajense* it is 0.82. At 75% RH and 22–23°C 50% of female *A. cajense* survived for 9 months and males for 7 months. The comparable figures for *A. americanum* were 7 weeks and 6 weeks (Needham and Teel, 1991).

Ticks have the ability to absorb water when the moisture content of the atmosphere is above the CEA by passive and active water sorption. Active sorption is achieved by the salivary glands secreting a hygroscopic fluid which takes up water from the atmosphere and is then imbibed by the tick (Sonenshine, 1991). Passive and active sorption enabled female *A. cajense* to survive successive cycles of being dehydrated at 0% RH to a non-ambulatory status and then exposed to 96% RH for 24 h. On average they survived 4.4 such cycles with the maximum being 12 cycles in 61 days (Needham and Teel, 1991). The water content of *I dammini* increases with the stage of development, being highest in adults and lowest in larvae. This differential is maintained by the adults having the greatest resistance to water loss and the highest rate of water imbibition (Yoder and Spielman, 1992).

The critical temperature of the cuticular wax layer, which is not protected by a cement layer in the Ixodidae, varies from 32°C in *I. ricinus* to 45°C in *Hyalomma savignyi*, not as high as in argasids. *I. ricinus* inhabits a moist microclimate, and when kept at 0% RH and 25°C, unfed females lost 17% of their body weight in 24 h, while *Hy. savignyi*, an inhabitant of arid regions, lost only 0.8% under the same conditions (Lees, 1947). Loss of water also occurs through the spiracles, complex structures which can be opened and closed. Female *A. variegatum* held in an atmosphere of high carbon dioxide, when the spiracles would be kept open, lost 17 times as much water at 0% RH than at 93% RH, but the water loss of larvae, which have no spiracles, remained unchanged (Rudolph and Knulle, 1979).

Pheromones, aggregation, mating and allomones

Several different pheromones are produced by ixodid ticks. They are detected by sensory receptors on Haller's organ, on the terminal segment of the palps, and on the cheliceral digits. Assembly pheromones are interspecific, affect all stages in the life cycle but are not universally produced. They have been demonstrated in *I. holocyclus*, a paralysis tick, and *Aponomma concolor*, which parasitizes echidnas, monotreme mammals (Treverrow *et al.*, 1977). A similar pheromone is produced by the American deer tick, *I. dammini*. Males of species of *Amblyomma*, which parasitize large ungulates in North America and southern Africa, secrete an aggregation-attachment pheromone which attracts both adults and nymphs. Female *A. maculatum* and *A. hebraeum* will not attach to hosts unless feeding males are present. The pheromone is a blend of phenols and is secreted by males which have been feeding for 5 days and reaches a peak after 6 to 8 days. A similar period is required to complete spermatogenesis (Sonenshine, 1991).

In *Ixodes* mating occurs both on and off the host, but in the Metastriata mating occurs only on the host because both sexes need to feed before becoming sexually competent. Attractant sex pheromones are secreted by females and attract males which have fed for 3 to 4 days. One such pheromone, 2, 6-dichlorphenol, has been found in 14 species and five genera, including *Amblyomma*, *Rhipicephalus*, *Dermacentor* and *Hyalomma* (Sonenshine *et al.*, 1979; Sonenshine, 1991). In *D. andersoni* the pheromone is synthesized early in adult life by the foveal gland but it is not released until the female is feeding. In *R. appendiculatus* both synthesis and release of the pheromone are associated with feeding. In response to the pheromone the male detaches and orientates to the female. In *D. variabilis* males may migrate as far as 43 cm to reach females on the opposite side of the host, a dog (Sonenshine *et al.*, 1979). Males of *Hy. dromedarii* and *Hy. anatolicum excavatum* both secrete the same attractant sex pheromone but in differing amounts which enables males to find conspecific females. Discrimination in *D. variabilis* and *D. andersoni* involves two further sex pheromones, a mounting sex pheromone, which in *D. variabilis* is cholesteryl oleate, and a highly species-specific genital sex pheromone (Sonenshine, 1991).

In *I. dammini* insemination is essential for the female to engorge fully, and more blood is ingested when mating and feeding occur together, i.e. perprandial insemination (Yuval and Spielman, 1990). In *I. ricinus* mating occurs off the host in the herbage where males are readily attracted to a pheromone emitted by virgin females. This water-soluble pheromone functions as both an assembly and sex pheromone (Sonenshine, 1991). Unmated female *I. ricinus* of the spring population diapause over winter, but if they are mated in the summer they do not enter diapause but become active in the autumn (Gray, 1987).

During mating the mouthparts of the male are used to stimulate the female prior to spermatophore transfer. In *Ixodes* the hypostome and chelicerae are inserted into the genital opening of the female, and in the Metastriata only the tips of the chelicerae. Males do not usually copulate before spermatids have been formed. They are transferred to the female in a spermatophore containing only one endospermatophore. Males can copulate 20–30 times (Oliver, 1989). A maximum of five capsules have been found in a single female *D. occidentalis* indicating that it had mated five times (Oliver *et al.*, 1974). *Ha. longicornis* is unusual in that it has sexual reproduction in the greater part of its range but in the cold northern part the species is parthenogenetic (Hoogstraal, 1978).

Metastriate ticks, e.g. *A. americanum*, *D. variabilis*, are characterized by the presence of large wax glands on their dorsolateral surfaces. When the tick is disturbed these glands secrete copiously a squalene-rich substance which functions as an allomone by repelling fire ants (*Solenopsis invicta*) (Yoder *et al.*, 1993).

Two ticks, the three-host *Rhipicephalus appendiculatus* and the one-host *Boophilus microplus*, of economic importance will be considered in more detail. They are both vectors of important protozoal diseases to cattle.

Rhipicephalus appendiculatus

Distribution

Rhipicephalus appendiculatus occurs mainly in the eastern and southern parts of Africa, south of the equator. It does not occur in West Africa. It ranges from sea level at the coast up to 2100 m inland but is absent from deserts and areas lacking shrub cover (Hoogstraal, 1956; Yeoman and Walker, 1967). It is commoner in areas of tall grass in warm humid ecotypes with an annual rainfall of 700–1500 mm. The closely related *R. zambeziensis* is more xerophilic and is found in the warmer, drier valleys of the Zambezi and Limpopo rivers where the annual rainfall is 400–700 mm (Pegram and Walker, 1988). In areas where the rainfall exceeds 1000 mm heavy infestations of *R. appendiculatus* can be expected wherever cattle are present in any number. The distribution of *R. appendiculatus* is the result of a complex interaction between climate, vegetation and cattle. Yeoman and Walker (1967) do not believe that *R. appendiculatus* is present in undisturbed habitats but that it is introduced with cattle during settlement.

Biology

Daily activity

There is a rhythm in the host-seeking behaviour of *R. appendiculatus* with activity being greater by day with bimodal diurnal periodicity. In the cooler months one peak occurred before midday and the other before sunset while in the warmer months the morning peak was earlier (0800 h), and the afternoon peak reduced. Under shade conditions peak activity occurred in the late afternoon (Punyua and Newson, 1979).

Hosts

Rhipicephalus appendiculatus is primarily a parasite of cattle but also occurs on sheep and goats and to a very limited extent on wild animals. Infestations of up to nearly 2000 adult *R. appendiculatus* have been found on a single beast but infestations are usually considerably smaller. In the absence of cattle *R. appendiculatus* is not able to sustain itself on other domestic or wild animals. Smaller wild animals tend to be infested with nymphs but only in the presence of cattle (Yeoman and Walker, 1967). The site most frequented by adult *R. appendiculatus* is inside the ear flap which can support a maximum population of about 250 adults. On the same individual host larvae and nymphs of the two-host tick *R. evertsi* are present deep inside the ear, and the adults almost exclusively in the perianal region, whereas 87% of adult *R. appendiculatus* occur on the ears and only 2% in the perianal region. Adults of *R. kochi* which also parasitizes cattle, occur mainly elsewhere on the body with only small numbers on the ear or in the perianal region. Larvae and nymphs of the blue tick *B. decolaratus* are often found on the ears of cattle in a similar position to the immature stages of *R. appendiculatus*. Where low infestations of *R. appendiculatus* are present the ticks will be confined to the ear, but in heavier infestations they will occur elsewhere on the head, spreading to the neck and body (Yeoman and Walker, 1967; Newson, 1978).

Length of life cycle

Branagan (1973a) studied the effect of temperature and humidity on the duration of those stages in the life cycle of *R. appendiculatus* which occur off the host. He found that at 25°C these stages took 9 weeks (Table 23.1), and at 18°C 6 months. Assuming that hosts are readily available and that the total time spent on the host by the three feeding stages is 3 to 4 weeks, then the minimum time in which the life cycle could be completed at 25°C would be 3 months, and at 18°C 7 months. Over the range 18 to 85% RH, humidity had no effect on the speed of development. At 25°C, 40% of the egg batch is laid in the first 4 days of oviposition and 90% within 12 days, but at 18°C the respective percentages are 29% and 46%. At 25°C, hatching of the eggs is spread over 2 weeks with 50% of the larvae emerging by day five, and at 18°C hatching is prolonged to nearly 4 weeks with 50% emergence by day eight (Branagan, 1973a).

Table 23.1. Average duration (days) of various stages in life cycle of *Rhipicephalus appendiculatus*.

Stage in life cycle	25°C	18°C
Pre-oviposition	6.2	12.8
Pre-eclosion	29.3	71.8
Engorged larva to nymph	11.9	34.3
Engorged nymph to female (male)	16.5 (18.2)	56.5 (64.2)
Total	63.9	175.4

Source: Branagan (1973a).
Notes: Pre-oviposition – time to deposition of first egg. Pre-eclosion – time to emergence of first larva from eggs.

Seasonal abundance

In southern Africa there is a definite seasonal cycle with one generation per year. In the eastern Cape, which has summer rainfall, adult *R. appendiculatus* appear in November, reach a peak in January–February and then decrease in the autumn. Peak numbers of larvae on cattle occur in the autumn (April–May), and of nymphs in late winter (August–September). The activity of adult *R. appendiculatus* is regulated by day length which controls the timing of the annual cycle. The size of the *R. appendiculatus* population is dependent upon the rainfall distribution (Rechav, 1981).

A similar cycle has been recorded on the highveld of Zimbabwe with peak adult activity coinciding with the rainy season, and larval and nymphal activity with the dry season. Here, too, adult activity was controlled by day length, temperature and humidity, and there was evidence to suggest that adult *R. appendiculatus* had a quiescent period when they were present on the vegetation but not actively seeking a host (Short and Norval, 1981).

Under optimal conditions the life cycle of *R. appendiculatus* can be completed in about 4 months in Malawi, but it was found that females only engorged and oviposited in the rainy season, when the relative humidity was above 75%. Humidity itself would not limit female engorgement and oviposition and, since unengorged larvae failed to survive the cold, dry months (Wilson, 1946, 1950), Branagan (1973a) considers that *R. appendiculatus* survives the winter as engorged nymphs.

At Mwanza on Lake Victoria in Tanzania the duration of the off-the-host stages in the field was estimated to be 90 days. Allowing a month for the three feeding stages to find a host and engorge, it would be possible to have two or three cycles completed in a year, but optimal conditions are not continuously present, and the number of cycles is dependent on local conditions. There were smaller fluctuations in the numbers of adult *R. appendiculatus*, which were present all the year round, than in the immature stages, in which cycles of abundance were more apparent . There were two generations per annum with peaks of larvae and nymphs in the dry interlude in the wet season early in the year, and a second peak following the rains in the middle of the year (McCulloch *et al.*, 1968).

In the drier zone of Sukumaland, east of Lake Victoria, the seasonal pattern of *R. appendiculatus* was closer to that of southern Africa with a single generation per annum with peak adult activity during the rains (Yeoman, 1966). At a similar latitude on the Kenya coast adult *R. appendiculatus* were present all the year round, but the annual cycle was bimodal with a small peak in April at the start of the rains, and a larger second peak in September–October, when dry conditions were likely to prevail. It is suspected that much of the second generation is lost (Newson, 1978).

Survival of *R. appendiculatus* depends on temperature, humidity and activity. Exposure to 4°C continuously for 2 to 3 days is fatal to all engorged stages. In the dry season at Kedong in the Rift Valley desiccation of egg batches frequently prevented any emergence of larvae. Larvae that do emerge prolong survival by resting on grasses where transpiration is creating a humid microclimate (Branagan, 1973a,b). The survival of adult *R. appendiculatus* is inversely related to activity. Regular activation reduced survival from more than 4 months to less than 3 months (Payne and Purnell, 1975).

Veterinary importance

Rhipicephalus appendiculatus is the main vector of *Theileria parva*, a virulent pathogen causing east coast fever of cattle in eastern and central Africa. It is also a vector of *Babesia bigemina*, the cause of babesiosis in cattle, but it is not the most important vector of this disease (Hoogstraal, 1956; Yeoman and Walker, 1967) (see Chapter 28).

A number of predictive models have been developed using satellite data to forecast the distribution and seasonal abundance of *R. appendiculatus*. Randolph (1993) examines five of them – CLIMEX, BIOCLIM, NDVI, T3HOST and ECFXPERT – discusses their limitations, and proposes new perspective focusing on the larval stage. Larvae of *R. appendiculatus* do not undergo diapause but are immediately active. They are the most vulnerable stage in the life cycle, being the least resistant to desiccation, although they are able to replace lost water when humidity is high. In southern Africa peak numbers of larvae occur in March when temperature and saturation deficit are lowest, favouring water replenishment. The life cycle is geared to the seasonal cycle by the females entering a diapause, initiated by day length. They become active in the hot season (December–February), and lay eggs from January onwards. This results in larval numbers peaking in March when temperature and saturation deficit are lowest. The development of engorged larvae and nymphs is temperature-dependent, and the adults produced enter diapause.

Boophilus microplus

Distribution and hosts

Boophilus microplus is widely distributed in, but not limited to, the southern hemisphere. It occurs in Central and South America, southern Florida and Mexico, from which it may extend into Texas. It is present in the Oriental region from which it has been introduced into northern Australia, where it now extends down the east coast into northern New South Wales. From the Oriental region

B. microplus was introduced into Madagascar and from there into eastern, central and southern Africa. In all regions *B. microplus* is primarily a parasite of cattle, but heavy infestations can develop on horses and sheep; goats and deer can also be infested. In southern Florida *B. microplus* survives on white-tailed deer (*Odocoileus virginianus*). In Tanzania only light infestations occurred on sheep and goats, and in Australia the sheep-rearing areas are too arid for *B. microplus* (Hoogstraal, 1956; Seddon and Albiston, 1968; Graham and Hourrigan, 1977).

Life cycle

The female *B. microplus* deposits 2000–3000 shiny, dark yellow-brown eggs, almost spherical measuring 0.5 mm × 0.4 mm. Providing the humidity remains above 70%, they hatch in summer in about 2 weeks (Roberts, 1952). Larvae show positive phototaxis and ascend grass stems, occurring in groups of 20 or more on the tips of grasses. They do not show the ascending–descending pattern of activity described for *I. ricinus* but remain on the upper parts of the grass, avoiding direct sunlight, and aggregating on the tips of the grass particularly in the early morning. Larvae show a strong response to odours, breath and contact in their search for a host (Wilkinson, 1953).

On the host larvae attach themselves and feed for about 4 days after which there is a quiescent period of about 2 days before the larva moults to a nymph while still on the host. The nymph may wander around on the host for a while before attaching, when it feeds for nearly a week followed by a short quiescent period, and then it moults to the adult. The female attaches particularly on the neck, brisket, flank, inguinal region and escutcheon. The female feeds slowly at first and then, after mating, engorges rapidly and drops from the host about 3 weeks after attachment of the larva. Most female ticks drop from their hosts in the early morning (0600–1000 h). After a minimum preoviposition period of 2 days the female commences to lay eggs over a period of 10 or more days. The minimum time to complete the life cycle is therefore 5 weeks, and under less favourable conditions will extend to several months (Roberts, 1952; Seddon and Albiston, 1968).

Seasonal abundance

Boophilus microplus is a high rainfall species (750 mm or more) (Yeoman and Walker, 1967) and does not persist in dry areas with low humidities. In parts of the tropics where rainfall and humidity are high, *B. microplus* reproduces continuously throughout the year. In subtropical regions such as south-eastern Queensland, the tick has a pronounced seasonal cycle. The spring rise of larvae results in a first generation of adults in November–December, and the population steadily increases reaching a peak in the fourth generation at the end of the autumn (June), followed by a population crash in the winter. The rapid decline in the numbers of ticks on cattle in the winter results from the fact that female ticks which drop in April until mid-July produce virtually no progeny (Snowball, 1957). Ticks dropping from late July onwards do produce progeny. The tick 'overwinters' through larvae hatching from eggs laid by females that dropped from cattle in

March or early April. These are the source of the spring rise. Continuation of the species in a locality depends upon larval survival which is of the order of 3 to 4 months in summer and 5 to 6 months in winter but larvae, like eggs, are susceptible to low temperatures and humidities (Snowball, 1957; Seddon and Albiston, 1968; Wharton and Norris, 1980).

Population dynamics and modelling

Models have been constructed which relate the population of *B. microplus* to its host and the time of year. Such a model is a valuable tool in evaluating the effect of various management decisions, e.g. when and how frequently to dip. Models require more elaborate data than those given above with the components of the model having different values depending on the season. Thus egg production is constant over the range 16–33°C which equates to the period September to April, when a reasonable figure is 2000 eggs per female. In May, the second half of July and August it is half that, but in June and the first half of July egg production drops to about 200 eggs per female (Sutherst *et al.*, 1979a).

Egg development is prolonged at lower temperatures and survival of eggs laid from mid-April through to mid-July is virtually zero. Eggs laid in late July and later which contribute to the first generation have about 20% survival. Larvae of the spring rise and the following second generation (10% survival) can survive unfed for 4 to 6 weeks; this is slightly longer in the third generation and is 7 to 11 weeks in the fourth generation (April–September).

The obverse to larval survival is host finding and this is a function of density of hosts; having found a host, survival is a function of tick density on the host. Experiments conducted during the major active period of larvae of *B. microplus*, i.e. from the end of October until mid-April, showed that at a density of two beasts per hectare 30–70% of the larvae were picked up per week, and at five beasts per hectare that rate had increased to 50–85%. This is equivalent to each beast picking up all ticks from 0.022–0.75 ha day^{-1} with the lower values being associated with lower temperatures (Sutherst *et al.*, 1978).

Survival of larvae that have found a host depends upon a number of factors including the breed of host and time of year. Zebu are more resistant than Hereford cattle and all breeds have lower resistance to ticks in winter. The number of females completing engorgement (M) is dependent on the number of larvae attaching (L), a density dependent factor (c) and survival in the absence of any density effect (a). The relationship between these variables is: $M = aL^c$. Assuming a 1:1 sex ratio the percentage of females surviving to engorgement in winter from an initial infestation of 20,000 *B. microplus* larvae was 28% on Hereford cattle and 9% on zebu crossbreds. In spring, survival on zebu crossbreds was even lower: 1.3% (Sutherst *et al.*, 1979b).

Principles of Control of Ixodid Ticks

In theory it would appear to be easy to control boophilids. They are among the most species-specific of the ixodid ticks with the bulk of the population occurring on domestic animals which can be mustered and treated. In addition it is a one-host tick spending a minimum of 3 weeks on the host, and therefore dipping all the hosts in an area at 3 weekly intervals with an effective acaricide, for a period to exceed that for which unfed larvae can survive, should eradicate the tick. Using arsenic this technique was highly successful against *B. annulatus* which, except for the border area in Texas, was eradicated from the USA by 1940. *B. microplus* persisted in the south of Florida and was eliminated by dipping cattle and slaughtering deer, important alternative hosts (Graham and Hourrigan, 1977).

Acaricidal resistance

In other parts of the world similar programmes have been thwarted by the development of resistance in *B. microplus*. Resistance to arsenic appeared before 1940 in Australia; to DDT and other chlorinated hydrocarbons in the early 1950s; and to organophosphorus acaricides and carbamates in the late 1960s (Wharton and Norris, 1980). Synthetic pyrethroids are effective but share cross-resistance with DDT so that tick populations containing a small percentage of DDT-resistant ticks rapidly develop specific pyrethroid resistance (Nolan *et al.*, 1979). Two subcutaneous injections of ivermectin at $200 \, \mu g \, kg^{-1}$ four days apart were effective in cleansing cattle under field conditions (Nolan *et al.*, 1985).

Controlled dipping

The development of resistance can be slowed down and treatment made more efficient by concentrating the attack on the ticks at the most vulnerable points in their life system. In *B. microplus* there are three such points in the annual cycle – during the spring rise of larvae, in the autumn to kill the females that will produce the overwintering larvae, and in late winter when overwintering larvae will be on the host (Wharton and Norris, 1980).

Pasture spelling and repellent pastures

Since unfed *B. microplus* larvae can only survive for 3 to 4 months in pasture, a high degree of control can be obtained by pasture spelling involving grazing cattle for 3 to 4 months on alternate paddocks. This can be highly effective (Wilkinson, 1964; Wharton and Norris, 1980). Pasture spelling is likely to be ineffective for three-host ticks because their immature stages occur on wild hosts. Certain species of grasses are associated with low densities of ticks. Molasses grass (*Melinus minutiflora*) is repellent to *B. australis* in Latin America and against other ticks in Tanzania. Tropical legumes of the genus *Stylosanthes* trap and kill larvae of *B. microplus* (Sutherst *et al.*, 1982; Mwase *et al.*, 1990).

Tick-resistant hosts

Development of resistance to ticks by the host is not universal. Dogs do not become immune to *R. sanguineus* nor do sheep to *A. hebraeum* (Willadsen, 1980). In general, European breeds of cattle, e.g. Friesian and Hereford, have very low resistance (15% of *B. microplus* larvae survive to give engorged females). Jersey cattle are unexpectedly resistant (2% survival) and Brahman cattle 99% resistant (Wharton and Norris, 1980). Braham cross cattle with up to 50% *B. indicus* blood have the advantage in tick-infested areas of requiring fewer treatments but there are penalties involved with greater *B. indicus* content because such cattle are less productive (Blood and Radostits, 1989).

Vaccine

A commercially available vaccine TickGARD® has been developed against *B. microplus* by CSIRO and Biotech Australia, a member of the Hoechst group (Anon., 1994). The vaccine destroys the cells lining the tick's gut, producing gaps in the gut wall, which allow blood to leak into the haemocoele. It reduced tick fertility by up to 70%, females producing fewer eggs with lower hatch rates and less viable larvae. The primary injection should be given when the infestation on cattle is low and booster doses given every 6 to 10 weeks. The vaccine is against a 'concealed' antigen, the tick's midgut cells, to which cattle are never exposed, and hence the need for booster injections. For optimal response cattle should be in nutritionally good condition. The advantages of vaccination are a reduction in the frequency of insecticidal treatment and the absence of residues in milk and meat.

Repellents, bait stations and pheromones

People can obtain protection against ticks by the application of repellents such as deet (diethyl toluamide) or pyrethrins to the skin or deet or synthetic pyrethroids to clothing but there is no commercially available repellent for livestock. In small test areas, bait stations have been used to coat attracted rodents with a peanut oil/pesticide mixture. This method successfully eliminated larvae and nymphs of *D. variabilis* from their rodent hosts. The tick sex pheromone, 2,6-dichlorophenol, combined with the acaricide propoxur, has suppressed populations of *D. variabilis* on dogs (Mwase *et al.*, 1990).

Tick control and babesiosis

Two separate issues are involved in controlling *B. microplus*. There is the direct damage caused by the presence of large numbers of ticks, the so-called tick burden, and the indirect effect of the tick as a vector of babesiosis. Eradication, if it can be achieved, would be the ideal answer in both cases, but elimination of the vector tick followed by its reintroduction can be more damaging because the cattle will have lost their immunity. Cattle infected early in life develop immunity without developing clinical disease and maintain their immunity throughout adult life. Non-immune adult cattle develop clinical disease. The aim should be to maintain tick infestation at a level which confers immunity on young beasts so that there is no clinical disease and at the same time infestation is not heavy enough to cause direct damage. In Australia this result can be achieved by limiting numbers of engorged female *B. microplus* to about ten per beast per day (Wharton and Norris, 1980).

Economic Importance of Ixodid Ticks

In 1906 economic loss due to *B. annulatus* in the USA was estimated at 130 million dollars per annum which in 1976 terms would have been of the order of a billion dollars (Graham and Hourigan, 1977). Even when *B. annulatus* and *B. microplus* had been eradicated, tick losses in the cattle and sheep industries were estimated to be 65 million dollars in 1965 (Steelman, 1976). In Australia the cost to the cattle industry of tick control in 1975 was estimated at 40 million dollars, of which one-third was the cost of control and two-thirds loss in production (Sutherst *et al.*, 1979a). Heavy tick infestations damage hides, and cause a loss in liveweight gain which has been estimated at 0.6 g day^{-1} for every engorged female *B. microplus* (Sutherst *et al.*, 1983) and 4–5 g day^{-1} per female *A. variegatum* (Pegram *et al.*, 1989). Heavy infestations of *Haemaphysalis hoodi* can cause the death of poultry (Lucas, 1954).

Ixodid ticks play an important role as vectors of a wide range of pathogens to humans and domestic animals. Hoogstraal (1986) offers an overview of ticks and disease while Sonenshine (1993) provides comprehensive coverage. Ixodid ticks are involved in the transmission of arboviruses to humans including tick-borne encephalitis, Crimean–Congo haemorrhagic fever and Kyasanur forest disease, and other arboviruses to domestic animals (see Chapter 24). They transmit various tick-borne spotted fevers, e.g Rocky Mountain spotted fever, to humans and other rickettsial diseases and anaplasmosis to domestic animals (see Chapter 25). They transmit Lyme disease, a borreliosis of humans, and the bacterial agent of tularaemia to humans and sheep (see Chapter 26). Their most important role economically is the transmission of babesiosis and theileriosis to cattle (see Chapter 28).

Tick Paralysis

Paralysis in people caused by ticks has been known since 1843 in Australia, and 1912, possibly earlier, in the United States. At the onset, paralysis affects the lower limbs ascending to the torso, upper limbs and head regions within a few hours. Forty-six different species of ticks belonging to ten genera have been associated with this condition in people and/or animals. The most important paralysis-causing ticks are *I. holocyclus* in Australia, *D. andersoni* in western North America and *D. variabilis* in eastern North America (Stone, 1988; Sonenshine, 1993).

Paralysis is associated with the feeding of a female tick and first symptoms occur 5 to 7 days after attachment and 'a single female tick suffices to completely paralyse and kill an adult human' (Gothe *et al.*, 1979). Except in the case of *I. holocyclus*, removal of the attached tick terminates the condition and allows complete recovery (Sonenshine, 1993), but with *I. holocyclus* physical removal of the tick may rapidly worsen the symptoms (Stone, 1988). The toxin secreted by *I. holocyclus*, holocyclotoxin, is secreted by specific cells in the salivary glands. The toxin of *D. andersoni* is likely to be different. In Australia the antitoxin is given intravenously and time allowed for it to circulate before removing the tick. Isolation of holocyclotoxin provides the opportunity to develop a commercial vaccine.

Ixodes holocyclus is present in moist, vegetated habitats along the eastern coast of Australia. Its principal hosts are bandicoots but it will attach to a wide range of alternative hosts including livestock, domestic pets and humans. The natural hosts of *I. holocyclus* are not greatly inconvenienced by the paralysing toxin as the majority become immune to its effect. In domestic animals it has been estimated that *I. holocyclus* causes 100,000 cases of tick paralysis a year and the death of 10,000 calves, young animals being more susceptible than adults (Stone, 1988). In South Africa tick paralysis was recognized in sheep in 1890. The species involved are *I. rubicundus* and the two-host tick, *R. evertsi evertsi*. *I. rubicundus*, the Karoo paralysis tick, is estimated to cause about 30,000 deaths a year in livestock, mainly sheep (Spickett and Heyne, 1988). Tick paralysis due to *R. e. evertsi* occurs primarily in sheep and goats and its severity is related to the number of female ticks which have engorged (Sonenshine, 1993). The toxicity of *R. evertsi mimeticus* is comparable to that of *R. e. evertsi* (Gothe *et al.*, 1986).

Ticks of the subgenus *Argas* (*Persicargas*) can cause paralysis among poultry through the feeding of their larvae, which attach and feed for days, as do ixodid ticks. In a comparative study it was found that there was 'nearly direct proportional relation between the intensity of infestation and the clinical manifestation', and in order of pathogenicity the species could be arranged in descending order as *A.(P.) arboreus*, *A.(P.) walkerae* and *A.(P.) persicus* (Gothe and Verhalen, 1975). In a further set of experiments involving two other species similar results were obtained, and the order of pathogenicity was *A.(P.) radiatus*, *A.(P.) persicus* and *A.(P.) sanchezi* (Gothe and Englert, 1978).

References

Anon. (1994) *TickGARD® a New Approach to Cattle Tick Control*. Technical reference guide. Hoechst Australia Ltd, Melbourne.

Arthur, D.R. (1965) *Ticks of the Genus* Ixodes *in Africa*. Athlone Press, London.

Binnington, K.C. and Kemp, D.H. (1980) Role of tick salivary glands in feeding and disease transmission. *Advances in Parasitology* 18, 315–339.

Blood, D.C. and Radostits, O.M. (1989) *Veterinary Medicine – A Textbook of the Diseases of Cattle, Sheep, Pigs and Horses*. Baillière-Tindall, London.

Branagan, D. (1973a) The developmental periods of the ixodid tick *Rhipicephalus appendiculatus* Neum. under laboratory conditions. *Bulletin of Entomological Research* 63, 155–168.

Branagan, D. (1973b) Observations on the development and survival of the ixodid tick *Rhipicephalus appendiculatus* Neumann, 1901 under quasi-natural conditions in Kenya. *Tropical Animal Health and Production* 5, 153–165.

Gothe, R. and Englert, R. (1978) Quantitative Untersuchungen zur Toxinwirkung von Larven neoarktischer *Persicargas* spp. bei Hühnern. *Zentralblatt für Veterinärmedizin Reihe B* 25, 122–133.

Gothe, R. and Verhalen, K.H. (1975) Zur Paralyse-induzierenden Kapizität verschiedener *Persicargas*-Arten und -Populationen bei Hühnern. *Zentralblatt für Veterinärmedizin Reihe B* 22, 98–112.

Gothe, R., Kunze, K. and Hoogstraal, H. (1979) The mechanism of pathogenicity in the tick paralyses. *Journal of Medical Entomology* 16, 357–369.

Gothe, R., Gold, Y. and Bezuidenhout, J.D. (1986) Investigations into the paralysis-inducing ability of *Rhipicephalus evertsi mimeticus* and that of hybrids between this subspecies and *Rhipicephalus evertsi evertsi*. *Onderstepoort Journal of Veterinary Research* 53, 25–29.

Graham, O.H. and Hourrigan, J.L. (1977) Eradication programs for the arthropod parasites of livestock. *Journal of Medical Entomology* 13, 629–658.

Gray, J.S. (1987) Mating and behavioural diapause in *Ixodes ricinus* L. *Experimental and Applied Acarology*, 3, 61–71.

Hoogstraal, H. (1956) *African Ixodoidea. I. Ticks of the Sudan (with special reference to Equatoria Province and with Preliminary Reviews of the Genera* Boophilus, Margaropus *and* Hyalomma). Research Report NM 005 050.29.07, Bureau of Medicine and Surgery, Department of Navy, Washington.

Hoogstraal, H. (1978) Biology of ticks. In: Wilde, J.K.H. (ed.) *Tick-borne Diseases and their Vectors*. Centre for Tropical Veterinary Medicine, University of Edinburgh, pp. 3–14.

Hoogstraal, H. (1985) Argasid and nuttalliellid as parasites and vectors. *Advances in Parasitology* 24, 135–238.

Hoogstraal, H. (1986) Theobald Smith, his scientific work and impact. *Bulletin of the Entomological Society of America* 32, 23–34.

Hoogstraal, H. and Aeschlimann, A. (1982) Tick-host specificity. *Bulletin de la Société Entomologique Suisse* 55, 5–32.

Hoogstraal, H., Roberts, F.H.S., Kohls, G.M. and Tipton, V.J. (1968) Review of *Haemaphysalis (Kaiseriana) longicornis* Neumann (resurrected) of Australia, New Zealand, New Caledonia, Fiji, Japan, Korea and north-eastern China and USSR, and its parthenogenetic and bisexual populations (Ixodoidea: Ixodidae). *Journal of Parasitology* 54, 1197–1213.

Kaufman, W.R. (1979) Control of salivary fluid secretion in ixodid ticks. In: Rodriguez, J.G. (ed.) *Recent Advances in Acarology*. Academic Press, New York, vol. 1, pp. 357–363.

Lees, A.D. (1946a) Chloride regulation and the function of the coxal glands in ticks. *Parasitology* 37, 172–184.

Lees, A.D. (1946b) The water balance in *Ixodes ricinus* L. and certain other species of ticks. *Parasitology* 37, 1–20.

Lees, A.D. (1947) Transpiration and the structure of the epicuticle in ticks. *Journal of Experimental Biology* 23, 379–410.

Lees, A.D. and Beament, J.W.L. (1948) An egg-waxing organ in ticks. *Quarterly Journal of Microscopical Science* 89, 291–332.

Lösel, P.M., Guerin, P.M. and Diehl, P.A. (1992) Feeding electrogram studies on the African cattle brown ear tick *Rhipicephalus appendiculatus*: evidence for an antifeeding effect of tick resistant serum. *Physiological Entomology* 17, 342–350.

Lucas, J.M.S. (1954) Fatal anaemia in poultry caused by heavy tick infestation. *Veterinary Record* 66, 573–574.

McCulloch, B., Kalaye, W.J., Tungaraza R., Suda, B'Q.J. and Mbasha, E.M.S. (1968) A study of the life history of the tick *Rhipicephalus appendiculatus* – the main vector of east coast fever – with reference to its behaviour under field conditions and with regard to its control in Sukumaland, Tanzania. *Bulletin of Epizootic Diseases of Africa* 16, 477–500.

Moorhouse, D.E. (1969) The attachment of some ixodid ticks to their natural hosts. In: Evans, G.O. (ed.) *Proceedings of the 2nd International Congress of Acarology*, Akadémai Kiadó, Budapest, pp. 319–327.

Mwase, E.Y., Pegram, R.G. and Mather, T.N. (1990) New strategies for controlling ticks. In: Curtis, C.F. (ed.) *Appropriate Technology in Vector Control*. CRC Press, Boca Raton, Florida, USA, pp. 94–102.

Needham, G.R. and Teel, P.D. (1991) Off-host physiological ecology of ixodid ticks. *Annual Review of Entomology* 36, 659–681.

Newson, R.M. (1978) The life cycle of *Rhipicephalus appendiculatus* on the Kenyan coast. In: Wilde, J.K.H. (ed.) *Tick-borne Diseases and their Vectors*. Centre for Tropical Veterinary Medicine, University of Edinburgh, pp. 46–50.

Nolan, J., Roulston, W.J. and Schnitzerling, H.J. (1979) The potential of some synthetic pyrethroids for control of the cattle tick (*Boophilus microplus*). *Australian Veterinary Journal* 55, 463–466.

Nolan, J., Schnitzerling, H.J. and Bird, P. (1985) The use of ivermectin to cleanse tick infested cattle. *Australian Veterinary Journal* 62, 386–388.

Oliver, J.H. (1989) Biology and systematics of ticks (Acari: Ixodida). *Annual Review of Ecology and Systematics* 20, 397–430.

Oliver, J.H., Al-Ahmadi, Z. and Osburn, R.L. (1974) Reproduction in ticks (Acari: Ixodoidea). 3. Copulation in *Dermacentor occidentalis* Marx and *Haemaphysalis leporispalustris* (Packard) (Ixodidae). *Journal of Parasitology* 60, 499–506.

Payne, R.C. and Purnell, R.E. (1975) The effect of induced activity on the survival of the brown ear tick, *Rhipicephalus appendiculatus*, under laboratory conditions. *Bulletin of Animal Health and Production in Africa* 23, 297–301.

Pegram, R.G. and Walker, J.B. (1988) Clarification of the biosystematics and vector status of some *Rhipicephalus* species (Acarina: Ixodidae). In: Service, M.W. (ed.) *Biosystematics of Haematophagous Insects*. Clarendon Press, Oxford, pp. 61–76.

Pegram, R.G., Lemche, J., Chizyuka, H.G.B., Sutherst, R.W., Floyd, R.B., Kerr, J.D. and McCosker, P.J. (1989) Effect of tick control on liveweight gain of cattle in central Zambia. *Medical and Veterinary Entomology* 3, 313–320.

Punyua, D.K. and Newson, R.M. (1979) Diurnal activity behaviour of *Rhipicephalus appendiculatus* in the field. In: Rodriguez, J.G. (ed.) *Recent Advances in Acarology*. Academic Press, New York, vol. 1, pp. 441–445.

Randolph, S.E. (1993) Climate, satellite imagery and the seasonal abundance of the tick *Rhipicephalus appendiculatus* in southern Africa: a new perspective. *Medical and Veterinary Entomology* 7, 243–258.

Rechav. Y, (1981) Ecological factors affecting the seasonal activity of the brown ear tick *Rhipicephalus appendiculatus*. In: Whitehead G.B. and Gibson, J.D. (eds) *Tick Biology and Control*. Tick Research Unit, Rhodes University, Grahamstown, pp. 187–191.

Roberts, F.H.S. (1952) *Insects Affecting Livestock*. Angus and Robertson, Sydney.

Rudolph, D. and Knulle, W. (1979) Mechanisms contributing to water balance in non-feeding ticks and their ecological implications. In: Rodriguez, J.G. (ed.) *Recent Advances in Acarology*. Academic Press, New York, vol. 1, pp. 375–383.

Seddon, H.R. and Albiston, H.E. (1968) *Diseases of Domestic Animals in Australia. Part 3 Arthropod Infestations (Ticks and Mites)*. Department of Health, Commonwealth of Australia.

Short, N.J. and Norval, R.A.I. (1981) The seasonal activity of *Rhipicephalus appendiculatus* Neumann 1901 (Acarina: Ixodidae) in the highveld of Zimbabwe Rhodesia. *Journal of Parasitology* 67, 77–84.

Snowball, G.J. (1957) Ecological observations on the cattle tick, *Boophilus microplus* (Canestrini). *Australian Journal of Agricultural Research* 8 394–413.

Sonenshine, D.E. (1991) *Biology of Ticks*, vol. 1. Oxford University Press, New York.

Sonenshine, D.E. (1993) *Biology of Ticks*, vol. 2. Oxford University Press, New York.

Sonenshine, D.E., Silverstein, R.M. and Homsher, P.J. (1979) Female produced pheromones of Ixodidae. In: Rodriguez, J.G. (ed.) *Recent Advances in Acarology*. Academic Press, New York, vol. 2, pp. 347–356.

Spickett, A.M. and Heyne, H. (1988) A survey of Karoo tick paralysis in South Africa. *Onderstepoort Journal of Veterinary Research* 55, 89–92.

Steelman, C.D. (1976) Effects of external and internal arthropod parasites on domestic livestock production. *Annual Review of Entomology* 21, 155–178.

Stone, B.F. (1988) Tick paralysis, particularly involving *Ixodes holocyclus* and other *Ixodes* species. *Advances in Disease Vector Research* 5, 61–85.

Sutherst, R.W., Dallwitz, M.J., Utech, K.B.W. and Kerr, J.D. (1978) Aspects of host finding by the cattle tick, *Boophilus microplus*. *Australian Journal of Zoology* 26, 159–174.

Sutherst, R.W., Norton, G.A., Barlow, N.D., Conway, G.R., Birley, M. and Comins, H.N. (1979a) An analysis of management strategies for cattle tick (*Boophilus microplus*) control in Australia. *Journal of Applied Ecology* 16, 359–382.

Sutherst, R.W., Utech, K.B.W., Kerr, J.D. and Wharton, R.H. (1979b) Density dependent mortality of the tick, *Boophilus microplus*, on cattle – further observations. *Journal of Applied Ecology* 16, 397–403.

Sutherst, R.W., Jones, R.J. and Schnitzerling, H.J. (1982) Tropical legumes of the genus *Stylosanthes* immobilize and kill cattle ticks. *Nature* 295, 320–321.

Sutherst, R.W., Maywald, G.F., Kerr, J.D. and Stegeman, D.A. (1983) The effect of cattle tick (*Boophilus microplus*) on the growth of *Bos indicus* × *B. taurus* steers. *Australian Journal of Agricultural Research* 34, 317–327.

Treverrow, N.L., Stone, B.F. and Cowie, M. (1977) Aggregation pheromones in two Australian hard ticks, *Ixodes holocyclus* and *Aponomma concolor*. *Experentia* 33, 680–682.

Varma, M.R.G. (1993) Ticks and mites (Acari). In: Lane, R.P. and Crosskey, R.W. (eds) *Medical Insects and Arachnids*. Chapman & Hall, London, pp. 597–658.

Waladde, S.M. (1987) Receptors involved in host location and feeding in ticks. *Insect Science and its Application* 8, 643–647.

Wharton, R.H. and Norris, K.R. (1980) Control of parasitic arthropods. *Veterinary Parasitology* 6, 135–164.

Wilkinson, P.R. (1953) Observations on the sensory physiology and behaviour of larvae of the cattle tick, *Boophilus microplus* (Can.) (Ixodidae). *Australian Journal of Zoology* 1, 345–356.

Wilkinson, P.R. (1964) Pasture spelling as a control measure for cattle ticks in southern Queensland. *Australian Journal of Agricultural Research* 15, 822–840.

Willadsen, P. (1980) Immunity to ticks. *Advances in Parasitology* 18, 293–313.

Wilson, S.G. (1946) Seasonal occurrence of Ixodidae on cattle in Northern Province Nyasaland. *Parasitology* 37, 118–125.

Wilson, S.G. (1950) A check-list and host-list of Ixodoidea found in Nyasaland, with descriptions and biological notes on some of the rhipicephalids. *Bulletin of Entomological Research* 41, 415–428.

Yeoman, G.H. (1966) Field vector studies of epizootic east coast fever. 11. Seasonal studies of *R. appendiculatus* on bovine and non-bovine hosts in east coast fever enzootic, epizootic and free areas. *Bulletin of Epizootic Diseases of Africa* 14, 113–140.

Yeoman, G.H. and Walker, J.B. (1967) *The Ixodid Ticks of Tanzania*. Commonwealth Institute of Entomology, London.

Yoder, J.A. and Spielman, A. (1992) Differential capacity of larval deer ticks (*Ixodes dammini*) to imbibe water from subsaturated air. *Journal of Insect Physiology* 38, 863–869.

Yoder, J.A., Pollack, R.J. and Spielman, A. (1993) An ant-diversionary secretion of ticks: first demonstration of an acarine allomone. *Journal of Insect Physiology* 39, 429–435.

Yuval, B. and Spielman, A. (1990) Sperm precedence in the deer tick (*Ixodes dammini*). *Physiological Entomology* 15, 123–128.

DISEASES OF WHICH THE PATHOGENS ARE TRANSMITTED BY INSECTS OR ACARINES

Arboviruses 24

Viruses transmitted to vertebrates by insects and acarines are known as arboviruses. The term simply means arthropod (ar) borne (bo) viruses. Originally the term was arborviruses but this suggested associations with trees and the second 'r' was omitted. Arboviruses are defined as viruses which multiply in both their vertebrate and invertebrate hosts, and the term is therefore restricted to viruses which are transmitted biologically. It excludes viruses which are transmitted mechanically, such as the virus of myxomatosis which is spread among rabbits in England by the rabbit flea (*Spilopsyllus cuniculi*) and in Australia by mosquitoes; and the viruses of hog cholera and equine infectious anaemia which are transmitted mechanically by tabanids (Tidwell *et al.*, 1972; Hawkins *et al.* 1976).

The latest classification of viruses (Francki *et al.*, 1991) allocates 2430 viruses, including about 360 arboviruses (15%), to 73 families and groups. The criteria for a virus being an arbovirus is based on its biology and ecology and not on its morphology, consequently arboviruses are found in five families: the Bunyaviridae, Flaviviridae, Reoviridae, Rhabdoviridae and Togaviridae. The greatest number of arboviruses (about 250) are in the Bunyaviridae and the economically most important arboviruses in the Flaviviridae (62 species) and Togaviridae (27 species). The arboviruses in the Flaviviridae are assigned to the genus *Flavivirus* and in the Togaviridae to the genus *Alphavirus*. Within each genus the species are serologically related but quite distinct from species in other genera. Entomologically the most important feature is that the majority of species of *Alphavirus* are transmitted by mosquitoes while the vectors of *Flavivirus* species are mosquitoes or ticks.

The account given here of arboviruses and the diseases they cause will have an entomological bias. The medical aspects are to be found in Manson-Bahr and Bell (1987), Warren and Mahmoud (1990) and Strickland, (1991), and the veterinary aspects in Blood and Radostits (1989). Details of virus structure and taxonomy are given in Francki *et al.* (1991), and Porterfield (1989).

In common with other pathogens, there is interest in how arboviruses survive from season to season and spread from endemic areas to areas previously free from infection. Arbovirus infections were considered to be restricted to the adult insect vector and not to pass to the next generation, while in acarines arboviruses often passed from one generation to another by transovarian transmission. This view was first queried with regard to the virus of sandfly fever, where there was epidemiological evidence for transovarian transmission in *Phlebotomus papatasi* and was then undermined by it being demonstrated that La Crosse virus over-wintered in larvae of *Aedes triseriatus* (Watts *et al.*, 1975). Later experiments showed that the viruses of yellow fever, Japanese encephalitis and St Louis encephalitis, could be passed transovarially to the progeny of *Ae aegypti*, *Cx tri-taeniorhynchus* and *Cx tarsalis*, respectively (Beaty *et al.*, 1980; Rosen *et al.*, 1980; Reeves, 1982). The epidemiological significance of these findings is still being assessed.

Some arboviruses, such as bluetongue and African horsesickness, have spread widely in the last 30 years. These diseases could have been introduced into new areas by the introduction of infected hosts but there is evidence accumulating that the wind-borne carriage of infected vectors may be an important alternative route for arbovirus dissemination (Sellers, 1980).

Flaviviridae

In this family the virus particles or virions are spherical in shape with a diameter of 40–50 nm and bounded by a lipoprotein envelope which makes the particles sensitive to ether and other lipid solvents, providing a useful tool in the first-level separation of viruses. The genetic material or genome is a single molecule of a single-stranded RNA with a molecular weight of 4×10^6. The type species of *Flavivirus* is the yellow fever virus, which explains the generic name *Flavivirus*. Some other members of the genus are dengue virus and several encephalitis viruses including Murray Valley, Japanese, and St Louis. These viruses are mosquito-borne, while Kyasanur Forest disease and tick-borne encephalitis viruses are tick-borne.

The isolation and propagation of some arboviruses, e.g. dengue, have been aided by intrathoracic inoculation of material into adult *Aedes aegypti* and *Ae albopictus*. The usefulness of this technique is limited by the small size of these mosquitoes, and the danger posed by infected female mosquitoes. Fortunately both sexes of the large, non-blood-feeding *Toxorhynchites* mosquitoes have been shown to be highly suitable for the detection and propagation of dengue virus, of which hundreds of strains have been isolated using *Tx amboinensis*. In addition the viruses of St Louis encephalitis, Japanese encephalitis and some other flaviviruses replicated to very high titres in the same mosquito; as did Ross River, an alphavirus; and some bunyaviruses and rhabdoviruses (Rosen, 1981).

Yellow fever (YF) (Anon., 1986; Monath, 1990)

Yellow fever is a disease which has had a considerable impact on human social development. It was yellow fever which caused de Lesseps to abandon the first attempt to build the Panama Canal (Shaplen, 1964). It was greatly feared by the inhabitants of Western Europe and was one of three diseases (cholera, plague, yellow fever) for which ships entering British waters, had to fly a special yellow and black quarantine flag (Anon., 1911). Even in the present day travellers from areas where YF is endemic have to carry valid international certificates of vaccination against the disease. Fortunately, a fully effective vaccine (17D) is available, which probably protects for life and is currently valid for 10 years.

Classically the disease in an individual follows a rapid course which often terminates fatally within a week. Epidemics occurred regularly in urban areas of tropical and subtropical America during the seventeenth, eighteenth and nineteenth centuries (Shaplen, 1964). In the 25-year period 1958–1982 there were 17 outbreaks of YF in Africa, including Ethiopia (1960–1962) and Nigeria (1969) each with an estimated 30,000 deaths out of 100,000 cases. Since that time there have been seven further outbreaks in West Africa, the largest being in Nigeria (1986–1991) with 5600 deaths from 9800 cases (Service, 1993). It has been calculated that in a moderate sized epidemic, emergency mass immunization would be cheaper than preventive YF vaccination but the latter would reduce cases and deaths by 85%. In larger outbreaks, preventive YF vaccination is more cost effective than emergency mass immunization (Monath and Nasidi, 1993).

Yellow fever is endemic in Africa from 15°N to 10°S, especially in West Africa, and in tropical Central and South America, from which it may spread to other areas. Three epidemiological patterns are recognized: sylvatic, intermediate (rural) and interhuman. In Africa sylvatic YF occurs in rainforest, where it is enzootic among monkeys and a major vector is *Aedes africanus*. Intermediate (endemic) YF occurs in the humid savanna adjoining rainforest where the vectors are *Ae bromeliae* (formerly *Ae simpsoni*, Service, 1993), *Ae furcifer* and *Ae taylori* (the last two species being indistinguishable in the adult female). Interhuman outbreaks (epidemics) with mortalities of 20 to 50% occur on the dry savanna where the vector is the domestic, *Ae aegypti*. It was the recognition of *Ae aegypti* as the vector of YF by Walter Reed which enabled General Gorgas to institute strict control measures against that species and allow the construction of the Panama Canal to be completed.

Aedes aegypti becomes infected with the virus of yellow fever when it feeds on an infected person in the early stages of the disease, from about 6 h before the onset of clinical signs to about 4 days later. The virus undergoes a temperature-dependent cycle in the mosquito which takes 2 days at 30°C and 12 days at 18°C, during which time the virus multiplies in the cells of the midgut and salivary glands. When the cycle is complete the mosquito remains infective for the rest of its life, secreting the virus in its saliva. In a susceptible individual the virus will be incubated for 3 to 6 days before the onset of clinical symptoms of disease. Multiple feeding during a single gonotrophic cycle by *Ae aegypti* will greatly increase the possibility of it acquiring and transmitting an arbovirus (Scott *et al.*, 1993).

In South America 50 to 300 cases of YF are reported annually which will underestimate the actual number of cases. Monkeys are the principal wild vertebrate hosts and deaths of howling monkeys (*Alouatta* spp.) are an early sign of a YF epizootic in progress. Infection in African monkeys rarely results in illness or death, indicating a balanced host–parasite relationship and suggesting that YF is a relatively recent introduction into the Neotropical region, possibly with the slave trade when ships regularly crossed the Atlantic from Africa to the New World. Water storage containers on board would have provided ideal breeding sites for *Ae aegypti* which may have been introduced into the Americas at the same time.

The dispersal of yellow fever by shipping can be demonstrated by the history of YF in Jamaica. At the start of the nineteenth century YF took a heavy toll of army and navy personnel and in an attempt to reduce the incidence of this disease the main army and naval barracks were built in the Blue Mountains 1200 m above sea level, not the obvious place to build naval headquarters. Yellow fever disappeared from Jamaica about 1870 following cessation of the coastal schooner trade between South America and Trinidad, and the islands of the Greater and Lesser Antilles. A schooner with *Ae aegypti* breeding in its water containers, a susceptible crew and virus present either in a mosquito or a crew member, could disseminate YF at every port at which it called. When the schooner trade stopped, Jamaica ceased to have YF epidemics but they continued in Trinidad where there were indigenous monkeys to sustain the virus.

There are monkeys in the Indian subcontinent which are susceptible to yellow fever and *Ae aegypti* is well established there, yet YF is not present in the Oriental region. If the South American schooner trade could introduce yellow fever into Jamaica, Arab dhows, which sailed the coasts of the Indian Ocean, could similarly have introduced yellow fever into India. If they did, YF did not become established. Several explanations can be advanced, none of which is entirely satisfactory. Firstly, yellow fever is not prevalent in the coastal areas of East Africa. The coastal fringe and the main port of Mombasa are separated from the forested area of tropical Africa by a broad belt of desert. Secondly, Arab dhows would take considerably longer to reach India than South American schooners did to reach the Caribbean Islands. This would allow time for an outbreak to burn itself out with the survivors becoming immune. Lastly, there may be a maximum number of flaviviruses, offering some cross-immunity, which are able to sustain themselves in a given ecosystem. The Oriental region already has several other *Flavivirus* species, including dengue, Japanese encephalitis and Kyasanur Forest disease.

Humans become involved in sylvatic YF when they enter an enzootic area and are bitten by infected mosquitoes. Individuals incubating the virus can then introduce it into a human ecosystem infested with *Ae aegypti* and the scene is set for an outbreak of yellow fever. Interchange between the sylvatic and humid savanna ecosystems can be achieved by monkeys moving from one to the other. In western Uganda yellow fever circulates among the forest monkeys through *Ae africanus*. Monkeys are attracted into groves of cultivated bananas where they are exposed to attack by day-biting *Ae bromeliae* which breeds in water accumulating in the axils of plants including bananas. *Ae bromeliae* is susceptible to infection with yellow fever virus and readily bites both monkeys and humans, making it potentially a very important vector (Haddow, 1965). It is considered that *Ae bromeliae* was the main vector of yellow fever during the Ethiopian rural epidemic of the early 1960s (Neri, 1965).

Monkeys and humans were considered to be reservoirs of YF virus but in fact they play a temporary role as amplifiers of the virus. In humans the viraemia declines from the fourth day after the onset of clinical disease and on recovery people are immune, as are monkeys after infection. The virus survives longer in infected female mosquitoes who pass it on through transovarian transmission to a proportion of their offspring. Such transmission has been shown to occur in species of *Aedes* and *Haemagogus*. In both of these genera mature eggs can survive desiccation for several months in a state of diapause and it is possible that this is one way in which the virus can survive, especially if transovarian transmission occurs in successive generations of mosquitoes.

At least 14 species of mosquitoes are able to transmit YF in Africa and some have been captured in large numbers during epidemics. *Ae africanus* is a vector of sylvatic YF; and *Ae bromeliae*, *Ae furcifer* and *Ae taylori* are involved in the transmission of YF in rural situations, *Ae vittatus*, a rock hole breeding species, with *Ae furcifer* and *Ae taylori*, was considered to be a vector in the Sudan (1959) and *Ae luteocephalus* in Nigeria (1969). With two exceptions the species of *Aedes* mentioned as vectors of YF belong to the subgenus *Stegomyia*, a group of black mosquitoes with silvery markings. The exceptions, *Ae furcifer* and *Ae taylori*, belong to the subgenus *Diceromyia* in which the tarsi are all dark and the broad wings have pale and light scales (Service, 1993). In South America the vectors are various species of *Haemagogus* including *Hg spegazzini*, *Hg leucocelaenus*, and *Hg janthinomys*. *Sabethes chloropterus* is a relatively inefficient vector of YF but may play a role in virus survival because of it being relatively drought resistant. The recovery of yellow fever virus in Africa from the ixodid tick, *Amblyomma variegatum*, collected off cattle has added a new dimension to the survival of YF virus in nature. In this tick the virus is transmitted transovarially.

Dengue fever and dengue haemorrhagic fever (Halstead, 1990)

Dengue is present in nearly all tropical countries where, in urban situations, the major vector is *Ae aegypti* (Fig. 24.1). Other vectors are *Ae albopictus*, *Ae scutellaris* and in the islands of the Pacific *Ae polynesiensis*. There is a sylvatic cycle among monkeys in south-east Asia below 600 m above sea level. In Malaysia *Ae niveus* has been implicated as a vector (Rudnick, 1978). The incubation period in humans is 2 to 7 days with the individual becoming infective to mosquitoes 6 to 18 h before the onset of fever and during the fever period which lasts about 6 days. In the mosquito the virus replicates in the cells of the midgut epithelium and is transferred to the salivary gland by haemocytes circulating in the haemolymph, where the virus is passed with saliva during feeding. This extrinsic cycle takes a minimum of 8 days but more often takes 11–14 days. In the laboratory dengue virus can be transmitted transovarially and by sexual contact.

In classical dengue fever there is no mortality, recovery is complete but weakness and depression may be severe and last several weeks. Epidemics of dengue fever are noted for affecting a large proportion of the human community. More than a million cases occurred in Texas and Louisiana in 1922 and in Greece in 1928. The cost of an outbreak of dengue fever in Puerto Rico in 1977 was calculated to be $6.0–15.6 million in direct costs and indirect losses to the economy (Allmen *et al.*, 1979). There are four different serotypes of dengue virus which offer only cross-immunity of short duration. Three of these have been isolated from sentinel monkeys, maintained in high canopy forest in Malaysia (Rudnick, 1978).

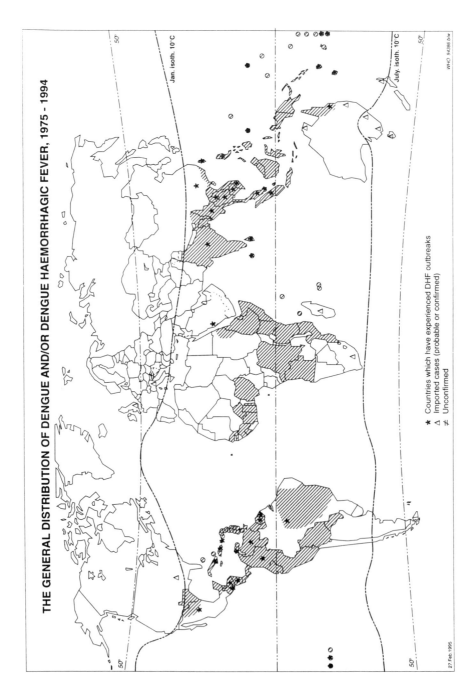

Fig. 24.1. Distribution of Dengue and/or Dengue Haemorrhagic fever, 1975–1994. Source: World Health Organization, 1995. Reproduced with permission. WHO 94386 b/w.

Dengue haemorrhagic fever (DHF) and the associated dengue shock syndrome (DSS) was first recognized in Bangkok in 1954 and is now widespread in south-east Asia from south China to Indonesia. From 1985 it has occurred annually in Puerto Rico (Gubler, 1992). DHF requires exposure to two serotypes either sequentially or during a single epidemic involving more than one serotype. It affects infants born to dengue-immune mothers during their first infection and children more than a year old, who acquire a second infection. It is rare in children above 14 years of age. The condition produces acute vascular permeability with an appreciable mortality. Between 1956 and 1986 out of 1,630,812 cases of DHF reported to the World Health Organization there were 31,178 deaths, including 159 deaths in 116,000 cases in Cuba in 1981, and 1800 deaths from 150,000 cases in Vietnam in 1983. Venezuela experienced a large epidemic of DHF/DSS in 1989–1990 (Gubler, 1992).

In the 1950s and 1960s a major effort was made to eradicate *Ae aegypti* from the Americas. It was successful in Mexico and most of Central and South America but not in the USA, Venezuela and the Caribbean. In the 1970s surveillance was reduced, most of the region was reinfested and the progress made lost. Using Puerto Rico as an example, the inter-epidemic period since 1915 has become shorter and the disease is now endemic with cases of DHS/DSS occurring annually from 1985 to 1991. Since 1977 there has been an increased incidence of dengue in the Americas with DHF occurring in Brazil, Venezuela, Mexico, Central America and Cuba (Gubler, 1992).

There is no vaccine available and control of the disease depends on controlling the vector. *Ae aegypti* is a domestic mosquito and therefore not readily exposed to ground-based ULV applications of insecticide, which in Jamaica and Venezuela achieved only a 56% reduction in female *Ae aegypti*. This makes only modest impact on the vector population which in the absence of control measures has a daily mortality of 20–30%, balanced by the emergence of a similar number of new adults (Reiter, 1992). Control therefore depends on eliminating the breeding sites of *Ae aegypti* and, since there are often small collections of water in domestic situations, requires the cooperation of the general public. This aspect dominated the 1992 Conference on dengue and *Ae aegypti* (Halstead and Gomez-Dantes, 1992).

West Nile (Manson-Bahr and Bell, 1987; Tesh, 1990)

West Nile occurs in Africa, southern Europe, the Middle East and India. It has an incubation period of 3–6 days in humans with the disease lasting another 3–6 days and occasionally ending fatally. In endemic areas it is a childhood disease with the adult population being immune. Epidemics usually occur during the summer months when populations of culicine mosquitoes are high. In Egypt, Israel and South Africa the vector is *Culex univittatus*, and in Israel and France *Cx modestus* and in India *Cx vishnui*. The virus circulates widely in birds (pigeons and crows) especially in the nesting season and they are the source of the virus which infects humans. *Cx univittatus* is usually ornithophilic but will also feed on humans. During an epidemic in an arid region of South Africa in 1974, 55% of the human population were infected by West Nile virus with isolations being made from *Cx univittatus* and *Cx theileri*. (McIntosh *et al.*, 1976).

Encephalitides (Blood and Radostits, 1989)

Four species of *Flavivirus* – Japanese, St Louis, Murray Valley and West
Nile – are associated with encephalitis in humans. They are primarily viruses of
wild hosts, often birds, and are transmitted by mosquitoes. From the human point
of view three cycles of the virus may be distinguished. There is a maintenance
cycle in a wild host, an amplifying cycle in a susceptible domestic or wild host,
and a possible cycle in the human population. In some cases the viraemia in
humans is inadequate to infect the vector. In almost all clinical cases encephalitis
viruses do irreparable damage to the brain.

Japanese encephalitis (JE) (Johnson, 1990; Igarashi, 1992, 1994)

Japanese encephalitis is the most common cause of epidemic encephalitis in the
world. It has caused epidemics in Japan since 1870 and in China it causes more
than 10,000, sometimes 20,000, cases annually and the disease has spread
southwards and westwards to India, Burma, Thailand, Indonesia, and Malaysia.
The natural cycle involves aquatic birds including herons and egrets. In Japan in
the spring there is intense virus transmission among young herons by *Culex
tritaeniorhynchus*, a ricefield breeding mosquito. This may be regarded as an amplifying cycle among the susceptible young of the maintenance hosts. The virus then
spreads to pigs, in which it causes abortion (Blood and Radostits, 1989). These
are the main amplifier hosts from which the virus is transmitted to humans and
horses. Immune mature water buffalo serve as hosts blocking multiplication of the
virus. Other vectors are *Cx gelidus*, a pig-feeder and the *Cx vishnui* group (Varma
and White, 1989). In Japan and Korea JE has been brought under control by mass
immunization. In humans mortality varies with the age group but it is always considerable (20–30%) and is 50% for people over 50 years of age.

St Louis encephalitis (SLE) (Johnson, 1990)

St Louis encephalitis is the most important arbovirus in the USA and also occurs
in northern Mexico. In the western rural areas of the United States SLE epidemics
are associated with periods of high rainfall, providing breeding sites for the vector
Culex tarsalis, and the *Culex pipiens* complex which is abundant is at best a secondary vector. In eastern urban areas SLE is associated with periods of drought and
poor drainage which provide breeding sites for *Culex quinquefasciatus*, a member
of the *Cx pipiens* complex. A contributing factor is that in more eastern parts of
the United States the strains of SLE virus produce higher viraemias in birds than
the strains present in the western USA, and *Cx tarsalis* is more readily infected
at low viraemias than are members of the *Cx pipiens* complex (Reeves, 1982). During the epidemic of 1975 more than 2000 cases and 171 deaths were reported in
the USA.

Murray Valley encephalitis (Doherty, 1977)

Murray Valley encephalitis (MVE) has caused epidemics in south-eastern
Australia, especially in the Murray–Darling Basin, but in the 1974 epidemic it

occurred in all mainland states and since then 20 out of the 31 cases reported have occurred in Western Australia (Broom *et al.*, 1993). The vector is *Culex annulirostris* which breeds in temporary breeding sites created by rainfall or flooding. Attempts to relate outbreaks of MVE with climatic conditions, particularly rainfall in the catchment areas, have not been particularly successful. In part, this may be due to the fact that by far the greater number of infections (>99%) remain subclinical. Horses, dogs and chickens develop high levels of viraemia and chickens have been used as sentinel hosts to give warning of active transmission. Chickens are convenient sentinel hosts but not the most satisfactory because *Cx annulirostris* does not readily feed on birds.

Some tick-borne flaviviruses (Hoogstraal, 1966, 1981; Manson-Bahr and Bell, 1987)

An unusual method of a tick acquiring an arboviral infection has been the demonstration that uninfected *Rhipicephalus appendiculatus* became infected by feeding on an uninfected guinea-pig at the same time as infected *R. appendiculatus*. Jones *et al.* (1991) refer to this as 'saliva-activated transmission'. The virus concerned was the unclassified Thogoto virus.

Kyasanur Forest disease

Kyasanur Forest disease virus (KFD) was detected in Karnataka (=Mysore) State in India in 1957 when an epidemic in the human population was associated with an epizootic in monkeys, with deaths occurring in both humans and monkeys. In humans the incubation period is 3–8 days with the disease sometimes having two episodes 1 to 2 weeks apart. There may be mild meningoencephalitis and some haemorrhagic manifestations with 5% mortality. Isolations of virus have been made from seven species of *Haemaphysalis* ticks of which the most important is *H. spinigera* in which trans-stadial transmission but not transovarian transmission occurs. KFD virus has been recovered from a variety of small rodents, squirrels and shrews which are probably the maintenance hosts of KFD, with monkeys (langur (*Presbytis entellus*) and bonnet macaque (*Macaca radiata*)) acting as amplifying hosts. On recovery monkeys are immune. The emergence of KFD as a human disease resulted from a rapidly increasing human population having increased contact with the forest. Cattle were grazed in and beside the forest, providing additional hosts for *H. spinigera*, and greater numbers of people visited the forest to collect firewood and other forest products.

Tick-borne encephalitis (TBE)

Tick-borne encephalitis virus is widely distributed across the northern Palaearctic region with the related louping ill virus in the British Isles and Powassan in Canada and the eastern USA. Two subtypes of TBE are recognized, a Central European subtype (CEE) and a Far Eastern subtype, the latter known originally as Russian spring–summer encephalitis (RSSE) (Francki *et al.*, 1991). The Far Eastern subtype occurs in Siberia, the southern republics of the former Soviet Union and north-eastern China; and the European subtype in Europe including

Russia west of the Ural Mountains. There are several thousand cases a year in Europe and the republics of the former USSR. Inapparent cases outnumber clinical cases by 30 to 1. RSSE virus causes a more severe disease with 20–30% mortality compared to CEE with a mortality of 1% (Johnson, 1990). In Austria 200,000 to 300,000 primary TBE vaccinations are made each year (Hoffman *et al.*, 1991).

The vectors of both subtypes are species of *Ixodes* ticks. *I. persulcatus* is the main vector of RSSE, and *I. ricinus* of CEE. Both viruses survive in the ticks by trans-stadial and transovarial transmission. RSSE was originally classified as an occupational disease of forest workers but more recently it has affected urban residents of Siberian towns, who become infected when relaxing in the adjoining countryside. In the 1960s, 65–80% of RSSE cases were contracted within 3–8 km of towns. A large range of small forest mammals and birds circulate RSSE virus and provide hosts for larvae and nymphs of *I. persulcatus*. Adult *I. persulcatus* parasitize larger wild and domestic mammals. Transmission of TBE can also occur through consuming fresh milk or cheese from infected goats or sheep (Hoogstraal, 1966, 1981).

Louping ill virus causes an acute encephalomyelitis in sheep and very rarely affects humans. It has been recorded from Scotland, northern England, and Ireland. It is enzootic in deer and grouse. The main vector is the sheep tick, *I. ricinus*, in which trans-stadial but not transovarian transmission occurs (Hoogstraal, 1966; Porterfield, 1989).

Omsk haemorrhagic fever

Omsk haemorrhagic fever (OHF) is a human disease found in western Siberia, where there are two cycles of infection. In summer the infection is tick-borne and in winter infections occur among muskrat trappers and skinners. The virus has been recovered from *Dermacentor pictus*, which transmitted the virus to laboratory animals during feeding. Infection can also be acquired by handling infected muskrat carcasses or by contact with or drinking water infected by muskrats or water voles. The disease causes fever, which is commonly bimodal, and haemorrhages with a mortality of less than 5% (Hoogstraal, 1966, 1981; McCormick and Fisher-Hoch, 1990).

Togaviridae

The virions of the Togaviridae are similar to those of the Flaviviridae being spherical in shape and bounded by a lipoprotein envelope, but slightly larger, having a diameter of 50–70 nm. The genome is a single molecule of a single-stranded RNA with a molecular weight of 4×10^6. During multiplication, lengths of RNA are formed which are less than the length of the genomic RNA, i.e. they are subgenomic. Subgenomic lengths of RNA are not formed in the Flaviviridae.

Alphaviruses were previously known as group A viruses and hence the generic name *Alphavirus*. The type species is Sindbis virus. Other species in the genus are Ross River, O'Nyong-nyong, and three equine encephalitis viruses – eastern, western and Venezuelan – which occur in the New World, causing disease in humans and horses.

Equine encephalitis viruses

The equine encephalitis viruses cause fever, followed by partial paralysis and then complete paralysis which results in death. Surviving horses are often deficient mentally (Blood and Radostits, 1989).

Eastern equine encephalitis (EEE) (Scott and Weaver, 1989)

Eastern equine encephalitis virus has been recorded from the eastern USA and Caribbean to Argentina. Two subtypes are recognized, a North American and a Central and South American. In North America the virus circulates among passerines mainly through *Culiseta melanura*, an ornithophilic, nocturnal species which breeds in forested freshwater swamps. The mosquito species that transfer the virus from the enzootic cycle to humans and horses has not been definitely identified but *Coquillettidia perturbans*, *Ae sollicitans* and other *Aedes* species have been implicated. In *Cs melanura* there are two cycles of viral multiplication in the posterior midgut cells and in the salivary glands. The virus reaches the salivary glands within 48 h of being ingested and maximum viraemia develops in 7 days. In Central America, the Caribbean and South America *Culex (Melanoconion) taeniopus* is recognized as an enzootic vector with other mosquitoes being involved. The largest outbreak of EEE among horses occurred in North America in 1947 when it was estimated that out of 14,334 infected horses 11,727 died (83%). Human cases are rarer but the mortality rate is equally high (50–90%) and humans that recover often suffer permanent brain damage. During outbreaks of EEE epizootics have occurred among exotic game birds, e.g. pheasants, causing many deaths.

Western equine encephalitis (WEE) (Hayes and Wallis, 1977)

WEE is another avian arbovirus which on occasions spills over into the human and equine populations. Human cases occur in western and central USA and western Canada. In Central and South America human cases are unknown but equine epizootics occur (Manson-Bahr and Bell, 1987). In western and central USA the primary vector is *Cx tarsalis*, populations of which show considerable variation in their ability to transmit the virus (Reeves, 1982). In the eastern USA the primary vector of WEE among its avian hosts is the ornithophilic *Cs melanura* and WEE is not a public or veterinary health problem in that area.

Venezuelan equine encephalitis (VEE)

Venezuelan equine encephalitis is more tropical in distribution being recorded from Florida, Mexico, tropical Central and South America and Trinidad. Several subtypes of VEE are enzootic in small rodents and subtype 1 infects horses and occasionally humans. In horses the viraemia is high and they act as amplifiers of the virus; consequently vaccination of horses not only protects the animal vaccinated but also eliminates a potential source of virus to the vector (Blood and Radostits, 1989). The main vectors are *Psorophora ferox*, *Culex (Melanoconion) portesi* and various species of *Aedes (Ochlerotatus)* (Varma and White, 1989). In an epidemic year up to 1000 cases can be expected to occur with a low mortality

($<3\%$) in adults and 10 to 20% mortality in children less than a year old (Johnson, 1990).

Some mosquito-borne alphaviruses

Ross River (Monath, 1991)

Ross River virus is an alphavirus which causes a syndrome with painful joints, i.e. arthralgia, rash and low grade fever in humans. It is often referred to as polyarthritis, i.e. inflamed joints and hence outbreaks are referred to as epidemic polyarthritis. The arthralgia may be prolonged for some months but no permanent damage is suffered. In Australia the virus occurs in cattle, horses, kangaroos and wallabies (Doherty *et al.*, 1966). The vectors of Ross River virus in Australia are *Aedes vigilax* and *Cx annulirostris* (Doherty, 1977), the former being restricted to coastal areas and in south-western Western Australia *Ae camptorhynchus* and elsewhere in that state other species of *Aedes* (*Ochlerotatus*) including *Ae normanensis* (Lindsay *et al.*, 1993).

In 1979 and 1980 Ross River virus spread eastwards in the Pacific, occurring in epidemic form in Fiji in the first half of 1979. The virus reached American Samoa in the second half of 1979, and Rarotonga, an island of the Cook group, early in 1980, where the vector was *Aedes polynesiensis*, from which six isolations of the virus were made, and virus transmission by bite was successfully achieved in the laboratory. For the first time many isolations of virus were made from patients by inoculating undiluted human sera into intact *Toxorhynchites amboinensis* mosquitoes (Rosen *et al.*, 1981). It caused 30,000 cases (Brès, 1988) and in some areas 90% of the population were infected and 40% developed clinical disease.

O'Nyong-nyong (Monath, 1991)

O'Nyong-nyong virus is an alphavirus which caused an epidemic in 1959–1962 in Uganda, Tanzania and Malawi affecting an estimated 2 million people. Since then there has been only one isolation of the virus but antibodies have been found more widely in the Afrotropical region. It is the only well-documented example of an anopheline-borne epidemic viral disease of humans. The vectors were *Anopheles gambiae* s.l. and *An funestus*, the main vectors of malaria in tropical Africa, and consequently outbreaks of O'Nyong-nyong coincided with outbreaks of malaria. There is an incubation period of 8 or more days followed by high fever, an unusually pruritic rash, headache and arthralgia lasting 5–7 days.

Chikungunya (Monath, 1991)

Chikungunya virus is an alphavirus found in the Afrotropical and Oriental regions. Urban epidemics are sustained by a human–mosquito–human cycle involving *Aedes aegypti* as the vector and mixed Chikungunya–Dengue and Chikungunya–Yellow Fever outbreaks have been described. In humans the disease is rarely fatal and has an incubation period of 2–6 days and an acute phase of 3–10 days characterized by fever, rash and arthralgia. In a rural epidemic in the wooded savanna of the eastern Transvaal, South Africa, the vector was *Ae furcifer/taylori* which was transmitting the virus among baboons and humans. Baboons were regarded as the primary vertebrate host from which the virus extended into the human population (McIntosh *et al.*, 1977).

Elsewhere in Africa there are similar enzootic cycles involving transmission between subhuman primates and mosquitoes. The virus has been isolated from *Ae africanus* in Uganda and the Central African Republic, and from *Ae luteocephalus* and *Ae vittatus* in Senegal. They are responsible for monkey–monkey, monkey–human and human–human transmission in rural areas. The existence of a similar forest cycle in Asia has yet to be demonstrated. In endemic areas adults are immune and outbreaks occur at 5–10 year intervals depending on the build-up of a population of susceptible children.

Sindbis (Monath, 1991)

Sindbis, an alphavirus, causes mild febrile illness with vesicular rash in humans, which in many cases goes undetected. Human cases have been recorded in Australia, Uganda and South Africa which experienced an epidemic in 1974 in which 16% of the human population had been infected. It was contemporary with an epidemic of West Nile virus (see above). Mosquitoes acquire infection from feeding on viraemic wild birds. The viraemia in humans is inadequate to infect mosquitoes. In Africa isolations were made from *Culex univittatus* and *Cx theileri* (McIntosh *et al.*, 1976). Probable vectors in other countries are *Cx annulirostris* (Australia), *Cx tritaeniorhynchus* (Malaysia), *Cx bitaeniorhynchus* (Philippines) and *Cx antennatus* (Egypt).

Bunyaviridae

Viruses of the Bunyaviridae have spherical virions, 80–100 nm diameter, surrounded by a lipid envelope with glycoprotein projections. The virions have a molecular weight of $300–400 \times 10^6$ of which 20 to 30% is due to lipids and 1–2% to the genome, consisting of three molecules of single-stranded RNA. The virus multiplies in the cytoplasm and matures by budding into smooth-surfaced vesicles. The Bunyaviridae includes three genera of arboviruses. The 166 species of *Bunyavirus* are primarily mosquito-borne, e.g. La Crosse and Californian encephalitis, but also include the *Culicoides*-borne Akabane and Oropouche viruses. The 51 species of *Phlebovirus* include the tick-borne Uukuniemi group of viruses, the mosquito-borne Rift Valley fever, and 38 species of *Phlebovirus* transmitted mainly by phlebotomine sandflies, of which about two-thirds are transmitted by *Lutzomyia* in the Neotropical region and the other third by *Phlebotomus* in the Palaearctic and Afrotropical regions (Tesh, 1988). The 33 species of *Nairovirus* primarily transmitted by ticks, include Crimean–Congo haemorrhagic fever and Nairobi sheep disease (Francki *et al.*, 1991).

California group (La Crosse) (Johnson, 1990; Porterfield, 1989)

The 14 Bunyaviruses in the California group are found in the USA, Canada, Central and South America, Finland and Central Europe. California encephalitis virus was isolated in 1943 from *Ae melanimon* and three children. The more important and commoner La Crosse virus was not isolated until 1963. It circulates in chipmunks, rabbits and squirrels among which it is transmitted by *Aedes triseriatus*, a tree-hole breeder, which overwinters in the larval stage. It was the first arbovirus to be shown to be transovarially transmitted in mosquitoes, producing infected larvae which give rise to adults, capable of transmitting the virus at their first feed. Therefore cases are likely to occur in the spring (Watts *et al.*, 1975). Amplification of the virus also occurs by venereal transmission between infected males and uninfected females. In humans La Crosse virus may produce an inapparent infection, a mild febrile illness or in children an infection involving the central nervous system. The mortality is less than 0.5% and recovery is usually complete.

Akabane and Aino viruses

Akabane and Aino viruses are credited with causing disease in cattle in Australia and Japan but Aino virus is less frequently found in cattle. Akabane occurs also in Israel, Korea and East and South Africa. Infection of a pregnant cow with Akabane virus 3 to 4 months in pregnancy results in a calf with limb deformities (arthrogryposis) and at 5 to 6 months brain deformities (hydranencephaly) (Blood and Radostits, 1989). The offspring of sheep and goats are also infected with malformation of the central nervous system, especially the brain occurring in sheep (Haughey *et al.*, 1988). The vector of Akabane virus is *Culicoides brevitarsis* which is well adapted to transmitting pathogens among cattle because its whole mode of life is dependent on them. Female *C. brevitarsis* feed readily on cattle and oviposit only in naturally-lying cattle dung, in which their larvae and pupae complete development (Kettle 1983). Allingham and Standfast (1990) could find no evidence for transovarian transmission of Akabane virus in *C. brevitarsis*.

Oropouche virus (ORO) (Pinheiro *et al.*, 1981a,b)

Oropouche virus (ORO) was first isolated in 1955 in Trinidad, and since then has caused several epidemics in Brazil. Between 1961 and 1979 there have been eight outbreaks of ORO in Pará State, Brazil, in small and large urban communities. After an incubation period of 4 to 8 days, infection with ORO causes an acute febrile illness with general aches and pains, lasting for 2 to 5 days. No deaths have been reported, although a proportion of patients become severely ill. In the 1967 outbreak in Braganca and in 1975 in Santarem over 30,000 people became infected.

ORO virus has been isolated from *Culex quinquefasciatus* and *Culicoides paraensis*, the latter proving to be the more efficient vector in the laboratory with transmission rates varying from 25 to 83% for *C. paraensis* and less than 5% for *Cx quinquefasciatus* under the same conditions. *C. paraensis* is capable of transmitting ORO 4 to 9 days after feeding on a viraemic hamster. The maximum duration of the urban cycle is apparently only 6 months and it is likely that there is a sylvatic cycle. Isolations of virus have been made from the three-toed sloth (*Bradypus tridactylus*), and antibodies against ORO have been found in several genera of monkeys. *C. paraensis* is active during the daytime reaching peak activity just before sunset, and feeds on humans both inside and outside houses (Roberts *et al.*, 1981).

Rift Valley fever (RVF)

Rift Valley fever (RVF) occurs only in Africa, where it causes an acute, febrile disease of cattle, sheep and humans, characterized by high mortality in calves and lambs, and abortion and some deaths in adult sheep and cattle (Blood and Radostits, 1989). In 1975 there was an extensive epizootic of RVF in South Africa in which thousands of livestock died, and there were many human cases with seven deaths, six of which were considered to have occurred from contact with infected animal tissues (Gear *et al.*, 1977). Transmission of RVF is normally by mosquitoes. In 1977–1978 there was a widespread epizootic and epidemic in Egypt which involved an estimated 18,000 human cases and 598 deaths (Johnson *et al.*, 1978). The most common and widespread mosquitoes were members of the *Cx pipiens* complex, from which RVF virus was isolated and laboratory transmission of RVF by *Cx pipiens* implicated it as a major vector in Egypt (Hoogstraal *et al.*, 1979). In South Africa the major vectors were *Aedes caballus* and *Culex theileri* (Gear *et al.*, 1977), in Uganda *Eretmopodites chrysogaster* (Smithburn *et al.*, 1948), and in Kenya *Aedes lineatopennis* and *Culex antennatus* (Linthicum *et al.*, 1985). Isolation of RVF virus from male *Ae lineatopennis* indicates transovarian transmission of the virus, which could be the way it survives in between outbreaks.

In enzootic areas annual vaccinations can protect livestock. Countries free from RVF but vulnerable to its introduction, should protect themselves by prohibiting not only the importation of susceptible species from Africa but also the entry of potentially infected biological materials (Blood and Radostits, 1989). It is likely that advanced warning of possible epizootics of RVF may be provided by satellite remote sensing. The normalized-difference vegetation index (NDVI) is a reliable indicator of rainfall and high NDVI values could forecast the flooding of grassland depressions, the breeding habitat of *Ae lineatopennis* and other species (Linthicum *et al.*, 1990).

Sandfly (*Phlebotomus*) fever

Eight phlebotomine-transmitted viruses are associated with human disease – four in the Old World and four in the New World. They cause a short, sharp, non-fatal fever. The vectors are *Phlebotomus papatasi*, *P. perniciosus* and *P. perfiliewi*. Five of these viruses are transmitted transovarially and have been recovered from male phlebotomines (Tesh, 1988). Transovarial transmission of Toscana virus, and probably other phleboviruses, is effected more readily in the second gonotrophic cycle after oral infection of the female than in the first cycle. The explanation appears to be that ingestion of an infected bloodmeal will initiate virus multiplication and stimulate ovarian development. The chorion of the egg is a barrier to virus entry and if the eggs have developed to that stage before the virus appears, the eggs will remain virus free. In the second ovarian cycle the virus and developing egg will be present together, favouring transovarian transmission (Maroli *et al.*, 1993).

Crimean–Congo haemorrhagic fever (CCHF) (Hoogstraal, 1979; McCormick and Fisher-Hoch, 1990)

The epidemiology of Crimean–Congo haemorrhagic fever (CCHF) was reviewed by Hoogstraal (1979). The virus is enzootic in the Palaearctic, Oriental and Afrotropical regions, chiefly in steppe, savanna, semi-desert and foothill biotopes where one or two *Hyalomma* species are the predominant ticks parasitizing domestic and wild animals. In the southern republics of the former Soviet Union 20% of human infections result in clinical illness with mortalities of 5–10%. In South Africa the rates were higher with 70% of infections resulting in clinical disease with mortalities of 35% and in Central Asia and in the Middle East 35–50% among clinical cases. The incubation period in humans is 2–5 days and clinical disease 7–10 days involving fever, severe headache and bleeding 'from virtually every orifice and venipuncture site'. Infection occurs through the bite of the infected tick or by crushing infected ticks in contact with the skin or from shearing tick-infested sheep.

CCHF virus survives trans-stadially and interseasonally in several tick species and is transmitted transovarially in members of the *Hyalomma marginatum* complex. Twenty-seven tick species and subspecies, including eleven taxa of *Hyalomma* and eight of *Rhipicephalus*, have been reported to be reservoirs or vectors of CCHF. They include one-host ticks of the genus *Boophilus*, two-host ticks of the genus *Hyalomma*, including the *H. marginatum* complex, *H. anatolicum anatolicum* and *Rhipicephalus bursa*. Three-host ticks of the genera *Haemaphysalis*, *Amblyomma*, *Dermacentor*, *Hyalomma* and *Rhipicephalus* serve chiefly to maintain enzootic foci of CCHF virus circulation between ticks and wild and domestic animals.

Ticks of the *H. marginatum* complex and *H. anatolicum* are especially important in causing epidemics and outbreaks of CCHF on account of their great numbers and their aggressiveness in seeking human hosts. Epidemics occur under a combination of favourable conditions and environmental changes which favour the survival of large numbers of hyalommas and their hosts. The severe winter of 1968–1969 in the southern Ukraine markedly reduced cases of CCHF over the next 5 years, when *H. m. marginatum* virtually disappeared but the virus was maintained in the cold-resistant *Rhipicephalus rossicus*, a tick which only rarely bites humans.

Nairobi sheep disease (Davies, 1978a,b)

Nairobi sheep disease (NSD) is a severe disease of sheep and goats in which mortality may reach 90%. The vector is the three-host tick *Rhipicephalus appendiculatus* in which transovarian transmission occurs. No virus or antibodies to NSD have been found in wild ruminants or rodents, and the virus appears to be restricted to sheep, goats and *R. appendiculatus*. Human infections with NSD occur rarely (Porterfield, 1989).

Epidemic haemorrhagic fever (EHF) (McCormick and Fisher-Hoch, 1990)

This disease is variously known as Korean HF, EHF in China and Japan, HFRS (= HF with renal syndrome) in the former Soviet Union and as NE (nephropathia epidemica) in Scandinavia. It is a *Hantavirus* in the Bunyaviridae which infects a wide range of rodents. Infected rodents pass the virus in their saliva, urine and faeces for a year and human infections occur through contact with these secretions or by inhaling infected dust particles. More recently it has been shown that the trombiculid mite *Leptotrombidium* (*L.*) *scutellare* can pass the virus transovarially and infected larvae transmit the virus while feeding (Wu *et al.*, 1992). This would account for the disease having a patchy distribution, occurring in rural areas and being associated with harvest.

Rhabdoviridae

In the Rhabdoviridae, the virions are usually bullet-shaped, with one convex and one truncated end in viruses infecting vertebrates and invertebrates, and bacilliform, with both ends convex, in those infecting plants. The virions measure 100–430 nm × 45–100 nm and are surrounded by a lipoprotein envelope with surface projections. The genome is a single molecule of single-stranded RNA with a molecular weight of $3.5–4.6 \times 10^6$, forming 1–2% of the molecular weight of the virion. Important arboviruses in this family are the vesicular stomatitis viruses and bovine ephemeral fever virus (Francki *et al.*, 1991).

Vesicular stomatitis (VS) (Porterfield, 1989)

Vesicular stomatitis, a major disease of horses, cattle and pigs in the Americas, is caused by three viruses VSI – Indiana, VSNJ – New Jersey and VSA – Alagoas. Although VSNJ is primarily an animal pathogen humans can be infected. The morbidity rate in herds is usually low and there is no mortality. This condition superficially resembles foot and mouth disease, but is much less serious. The mode of transmission during epizootics is unclear but VSV multiplies in *Lutzomyia trapidoi*, a phlebotomine sandfly, which can transmit the virus transovarially to the next generation. The infection rate obtained in the F_1 generation was 20 to 27%, and infected females transmitted VSV by bite and to the F_2 generation transovarially (Tesh and Chaniotis, 1975). Transovarian transmission also occurred in *L. ylephilator* but not in *L. sanguinaria* or *L. gomezi*. There is evidence that the epizootic in cattle in Colorado in 1982 was an extension of an enzootic cycle in elk (*Cervus elephas*) and mule deer (*Odoicoleus hemionus*) (Webb *et al.*, 1987).

Bovine ephemeral fever

Bovine ephemeral fever (BEF) is a disease of cattle, which is enzootic in Africa and Asia, and causes epizootics in Australia. In China there have been epizootics in 34 of the 43 years (1949–1991) with 11–34% morbidity and 1 to 3% mortality (Wenbin, 1993). In the 1990–1991 epizootic in Egypt morbidity was 20 to 90% and mortality 1.5 to 3% among imported dairy breeds of *Bos taurus*. The rates were much lower among unimproved local cattle and water buffalo (Davies *et al.*, 1993). There is an initial generalized inflammation and toxaemia followed by a short-term paralysis which may resolve itself suddenly or result in death. Recovered animals are considered to have a life-long sterile immunity. Economic losses are suffered from a sharp drop in milk production, deaths of dairy and beef animals and conception of the next calf being abnormally delayed (Blood and Radostits, 1989). The virus has been recovered from *Culicoides* in Africa (Davies and Walker, 1974) and from *C. brevitarsis*, *An bancrofti* and a mixed pool of culicine mosquitoes in Australia. The virus needs to be injected into the circulatory system for disease transmission. Intradermal, subcutaneous and intramuscular inoculations of BEF virus do not infect cattle. These observations favour transmission by capillary feeding mosquitoes rather than by pool feeding *Culicoides* (St George, 1993).

Reoviridae

In this family the virion is an icosahedral particle measuring 60–80 nm. The virion has no lipoprotein envelope and indeed contains no lipid, but has one or two outer protein coats. Replication occurs in the cytoplasm. Two of the eight genera recognized in this family are arboviruses. In species of *Orbivirus* the genome consists of ten pieces of double-stranded RNA with a total molecular weight of 15×10^6, which is about 20% of the molecular weight of the virion. Orbiviruses are sensitive to acidity with infectivity being lost at pH 3, and show reduced infectivity (tenfold) after exposure to lipid solvents. Three species will be considered – bluetongue virus, African horsesickness virus, and epizootic haemorrhagic disease virus. The current distribution and status of these three viruses throughout the world, their epidemiologies and their insect vectors have been reviewed in Walton and Osburn (1992). The related Colorado tick fever virus is a Colyivirus species with a genome consisting of 12 pieces of RNA (Francki *et al.*, 1991).

Bluetongue virus (BTV) (Blood and Radostits, 1989; Mellor, 1990)

Bluetongue virus (BTV) causes severe disease in sheep involving fever, inflammation of the mucous membranes of the oral cavity and nasal passages, enteritis and lameness. In cattle, African wild ruminants and North American cervids, BTV causes little clinical disease although in highly susceptible cattle a few animals may become severely affected. BTV has little clinical effect on goats which are not infected in nature. BTV was enzootic in Africa originally but in the last 50 years has become widely distributed throughout the world and now occurs from 40°N to 35°S. Outbreaks of BTV have occurred in Africa, southern Europe, the Middle East, Pakistan, India, Japan and the United States. In Australia, Brazil, Canada and the West Indies, BTV virus is present but no clinical disease has been reported.

In the USA the virus has been isolated from ruminants in all states except North Dakota and the northern New England states (Parsonson, 1979), and in a serological survey in 1977–1978 there was high prevalence of BTV antibody throughout the south-western US from California to Texas; from Nevada to western Missouri; and in north-eastern Georgia, south-western South Carolina, Florida and Puerto Rico (Metcalf *et al.*, 1981). In susceptible sheep BTV can cause morbidities of 50 to 75% and mortalities of 20 to 50%. Losses from BTV infection are both direct, i.e. mortality, and indirect through abortion of pregnant ewes, reduction in quality and quantity of the fleece, and the prolonged period of convalescence required for full recovery.

Du Toit (1944) was the first to implicate *Culicoides* in the transmission of BTV using *Culicoides imicola* (=*pallidipennis*) and his findings have been confirmed by Luedke *et al.* (1967) using a laboratory colony of *Culicoides variipennis*. In a series of experiments they transmitted BTV by the bite of *C. variipennis* from sheep to sheep, from sheep to cattle, from cattle to sheep and from cattle to cattle. Their work included the first EM photographs of an arbovirus multiplying in an insect vector (Bowne and Jones, 1966).

Four days after an infected blood meal there is a great increase in virus titre and a second marked increase between 10 and 14 days, after which the virus titre remained steady until the experiment was terminated after 5 weeks (Foster and Jones, 1979). The parent colony used by the American workers had a susceptibility rate to infection with BTV of 30%. By selected breeding, Jones and Foster (1974) were able to develop fully susceptible or highly resistant strains of *C. variipennis*. In this species susceptibility to oral infection is controlled by a single genetic locus (Tabachnik, 1991). There are four barriers to the successful development of BTV in *Culicoides*. There is entry into the midgut cells – a mesenteron infection barrier; dissemination from the midgut cells into the haemocoele – a mesenteron escape barrier; infection of the salivary glands – a salivary gland infection barrier; and finally release from the salivary glands – a salivary gland escape barrier. Infected *Culicoides* probably remain infective for the rest of their lives.

There are well over 1000 species of *Culicoides* in the world but only 17 have been connected with BTV and only six have been proven to transmit the virus with another two species likely to prove to be competent. Four of the six, *C. actoni*, *C. fulvus*, *C. imicola* and *C. wadai*, are in the subgenus *Avaritia* and *C. variipennis* and *C. nubeculosus* in the subgenus *Monoculicoides*. The two other probable vectors are *C. (Avaritia) brevitarsis* and *C. (Hoffmania) insignis*. The latter has been closely associated with BTV transmission in the Caribbean and Central America (Greiner *et al.*, 1993). The main vector in Africa, southern Europe and the Middle East is *C. imicola* and in North America *C. variipennis* of which three subspecies are recognized. Of these *C. variipennis sonorensis* is the best vector of BTV being superior to *C. v. variipennis* and *C. v. occidentalis* and there is justification for recognizing the three entities as separate species (Tabachnik, 1992). In Australia *C. actoni* and *C. fulvus* are restricted to areas where the annual summer rainfall is in excess of 1000 mm which would exclude them from the drier sheep-rearing areas. *C. wadai*, similarly restricted initially, has been extending its range and is now verging on some of the major sheep-rearing areas.

Control of BTV faces a number of practical problems. Strict quarantine measures can prevent the introduction of infected material into a country but is powerless to prevent the wind carriage of infected insects. This explanation has been advanced for the outbreaks in Portugal (1956) (Sellers *et al.*, 1978), Cyprus and Turkey (1977) (Sellers and Pedgley, 1985). It has been shown that the introduction of BTV serotype 2 into Florida in 1982 could have been by wind carriage of infected *Culicoides* from Cuba (Sellers and Maarouf, 1989). The development of a protective vaccine is handicapped by the existence of 24 different serotypes of BTV, most of which were isolated originally from South Africa (17) and Australia (4). Of the eight serotypes present in Australia three are indigenous (Gorman, 1990). Once the disease has been introduced, vaccination against the serotype involved is the only satisfactory control procedure. Control is further complicated by the existence of inapparent infections in cattle and wild ruminants. Infection rates of up to 48% have been recorded in cattle with infection lasting for up to 81 days.

Little is known concerning the maintenance of the virus during periods when there is no active transmission. A number of wild ruminants are susceptible to infection with BTV but present information is to the effect that their viraemias are comparatively short lived, i.e. 35 days (Hourrigan and Klingsporn, 1975). Sheep that recover from infection normally develop a solid immunity to the strain with which they have been infected. However, virus has been isolated from sheep 4 months after an attack, and in some cases after longer periods and latent virus in cattle has been demonstrated by recovering it from *C. variipennis* which had fed on cattle, and then been maintained for a period to allow viral multiplication (Luedke *et al.*, 1977).

African horsesickness virus (AHS) (Howell, 1968; Blood and Radostits, 1989)

African horsesickness (AHS) causes a highly fatal disease among susceptible equines. The disease is enzootic in Africa, from where, in the early years of this century, it made occasional excursions across the Red Sea and along the Nile to Palestine and Syria, e.g. the 1944 enzootic. In 1959 AHS spread eastwards into Iraq, Iran, Afghanistan, India and Pakistan and in the same year westwards to Cyprus and Turkey. In 1965–1966, the disease appeared in Spain. Following the Middle East outbreak in 1960 no clinical cases have been reported from that area in the 15-year period 1963–1978 (Howell, 1979).

African horsesickness is a disease of the vascular endothelium with three clinical expressions, all with fever. An acute or pulmonary form found in susceptible equines has an incubation period of 5 to 7 days. In enzootic areas AHS produces a more slowly developing and persistent cardiac or subacute disease and a milder horsesickness fever, which can be overlooked. The mortality rate in susceptible horses is about 90% while mules suffer a lower mortality (50%) and donkeys are even less susceptible. Nevertheless, the disease is a crippling one to mules and donkeys causing gross debility. It has been estimated conservatively that 300,000 equines died during the first phase of the 1960 epizootic in the Near East and south Asia. The spread of AHS has been attributed to the introduction of infected equines into an area but there is evidence that the disease can be spread by the wind-carriage of infected vectors from enzootic areas into previously disease-free areas (Sellers, 1980).

Evidence for AHS being transmitted by nocturnal biting flies was provided by the fact that horses, accommodated in mosquito-proof stables during the hours of darkness, were protected from infection. Du Toit (1944) incriminated *Culicoides* in the transmission of AHS in South Africa and Boorman *et al.* (1975) confirmed this in the laboratory. In the absence of a susceptible small laboratory animal they used embryonated hen eggs as infected donors and recipient hosts, and showed that *Culicoides variipennis* could transmit AHS virus from infected eggs to uninfected eggs by bite 7 days after an infective feed. There was evidence that the virus has multiplied in *C. variipennis*.

Seventeen per cent of wild-caught *Hyalomma dromedarii*, an ixodid tick, in southern Egypt harboured AHS virus which they were able to transmit to camels and horses. Experimentally the red dog tick, *Rhipicephalus sanguineus sanguineus*, transmitted AHS from sick dogs to healthy dogs and to horses. In both ticks there was trans-stadial but not transovarial transmission of the virus (Dardiri and Salama, 1988).

Nine different antigenic serotypes have been recognized between which there is no cross-immunity and there are some 42 strains within the serotypes which have antigenic differences. The vaccine used in the Middle East and India 1960 outbreak contained seven strains of attenuated virus and proved to be effective, producing solid immunity for a year. Foals born to immune mares possess passive immunity for 5 to 6 months. Horses imported into the USA from Africa, Asia and the Mediterranean countries are held in quarantine for 30 days before being released.

The only wild equine, the zebra, is highly resistant to infection. A serological survey in southern Egypt found antibodies to AHS in sheep, goats, buffalo, dogs and camels ranging from one in three sheep to one in thirty camels being positive. The virus is moderately resistant to drying and heating and can survive for 2 years in putrid blood. Dogs can become infected by eating infected meat and they develop a mild disease. Latent virus has been shown to be present in dogs after experimental infection and under repeated biting of *Culex pipiens* the virus, released into the blood stream, was capable of infecting the mosquito (Dardiri and Salama, 1988).

Epizootic haemorrhagic disease (EHD)

Epizootic haemorrhagic disease (EHD) has caused epizootics in the white-tailed deer (*Odocoileus virginianus*) in the United States and among cattle in Japan. In Australia there are five serotypes of EHD virus which infect cattle, buffalo and deer without causing clinical disease (Parsonson, 1990). Work has been done on EHD because of its similarity to BT and AHS viruses. The vectors of all three viruses are species of *Culicoides*. EHD virus was recovered from *C. variipennis* during an outbreak of the disease in Kentucky in 1971, and two strains of the virus have been transmitted from infected deer to uninfected deer by the bite of the same species (Foster *et al.*, 1977; Jones *et al.*, 1977). The virus of EHD has been shown to multiply in *Culicoides variipennis* both after oral ingestion and after intrathoracic inoculation, but in the closely related *C. nubeculosus* multiplication of the virus only occurred after intrathoracic inoculation and not after oral ingestion, indicating the existence of a mesenteron barrier (Boorman and Gibbs, 1973). This may be one factor in the development of resistance to BTV in strains of *C. variipennis*.

Colorado tick fever

Colorado tick fever is endemic in the north-western United States and western Canada where it is enzootic in wild rodents, in which it causes no apparent disease. People are susceptible to infection and encephalitis may occur, particularly in children. The vector is the three-host tick *Dermacentor andersoni* (Porterfield, 1989).

Unclassified

African swine fever

The African swine fever virus is a lipoprotein-enveloped icosahedral virion about 200 nm in diameter with a single molecule of double-stranded DNA genome. It was originally placed in the Iridoviridae but has since been removed from that family and at present is unallocated. African swine fever (ASF) is a highly fatal, highly contagious disease of pigs which produces no clinical disease in wild pigs. In domestic pigs the morbidity approaches 100%, and infections with a virulent strain are almost always fatal. ASF is enzootic to Africa but in 1957 it appeared in Portugal, probably introduced from Angola, and has become established in the Iberian Peninsular (Spain and Portugal) and Sardinia. It was also introduced into other European countries, Brazil and the island of Hispaniola (Haiti and the Dominican Republic) from all of which it has been eradicated (Mebus, 1988; Porterfield, 1989).

In Africa the reservoirs of the virus are wart hogs, bushpigs and forest hogs, among which it can be transmitted by *Ornithodoros moubata porcinus*, an argasid tick. Transmission is by bite, and virus is also excreted in the coxal fluid. Infected male *O. m. porcinus* are able to transfer virus to clean females during copulation, probably via the seminal fluid, and in one series 88% of females became infected after mating (Plowright *et al.*, 1974). Trans-stadial, sexual and transovarial transmission of ASF has been demonstrated in *O. moubata* (Hess, 1987). Hess *et al.* (1987) tested four species of New World *Ornithodoros* and found that the most efficient was *O. puertoricensis* which was readily infected and transmitted the virus trans-stadially and transovarially.

References

Allingham, P.G. and Standfast, H.A. (1990) An investigation of transovarial transmission of Akabane virus in *Culicoides brevitarsis*. *Australian Veterinary Journal*, 67, 273–274.

Allmen, S.D. von, Lopez-Correa, R.H., Woodall, J.P., Morens, D.M., Chiriboga, J. and Casta-Velez, A. (1979) Epidemic dengue fever in Puerto Rico 1977: a cost analysis. *American Journal of Tropical Medicine and Hygiene* 28, 1040–1044.

Anon. (1911) Quarantine. *Encyclopaedia Britannica*, vol. 22. Cambridge University Press, Cambridge, pp. 709–711.

Anon. (1986) *Prevention and Control of Yellow Fever in Africa*. World Health Organization, Geneva.

Beaty, B.J., Tesh, R.B. and Aitken, T.H.G. (1980) Transovarial transmission of yellow fever virus in *Stegomyia* mosquitoes. *American Journal of Tropical Medicine and Hygiene* 29, 125–132.

Blood, D.C. and Radostits, O.M. (1989) *Veterinary Medicine – a Textbook of the Diseases of Cattle, Sheep, Pigs and Horses*. Baillière-Tindall, London.

Boorman, J. and Gibbs, E.P.J. (1973) Multiplication of the virus of epizootic haemorrhagic disease of deer in *Culicoides* species (Diptera: Ceratopogonidae). *Archiv für die Gesamte Virusforschung* 41, 259–263.

Boorman, J., Mellor, P.S., Penn, M. and Jennings, M. (1975) The growth of African horsesickness virus in embryonated hen eggs and the transmission of virus by *Culicoides variipennis* Coquillett (Diptera: Ceratopogonidae). *Archives of Virology* 47, 343–349.

Bowne, J.G. and Jones, R.H. (1966) Observations on bluetongue virus in the salivary glands of an insect vector *Culicoides variipennis*. *Virology* 30, 127–133.

Brès, P. (1988) Impact of arboviruses on human and animal health. In: Monath, T.P. (ed.) *The Arboviruses: Epidemiology and Ecology* 1, 1–19.

Broom, A.K., Lindsay, M.D., Wright, A.E. and Mackenzie, J.S. (1993) Arbovirus activity in a remote community in the south-east Kimberley. In: Uren, M.F. and Kay, B.H. (eds) *Arbovirus Research in Australia, Proceedings of the Sixth Symposium*, 7–11 December 1992, Brisbane, pp. 262–266.

Dardiri, A.H. and Salama, S.A. (1988) African horse sickness: an overview. *Equine Veterinary Science* 8, 46–49.

Davies, F.G. (1978a) A survey of Nairobi sheep disease antibody in sheep and goats, wild ruminants and rodents within Kenya. *Journal of Hygiene* 81, 251–258.

Davies, F.G. (1978b) Nairobi sheep disease in Kenya. The isolation of virus from sheep and goats, ticks and possible maintenance hosts. *Journal of Hygiene* 81, 259–265.

Davies, F.G. and Walker, A.R. (1974) The isolation of ephemeral fever virus from cattle and *Culicoides* midges in Kenya. *Veterinary Record*, 20 July, 63–64.

Davies, F.G., Moussa, A. and Barsoum, G. (1993) The 1990–1991 epidemic of ephemeral fever in Egypt and the potential for spread to the Mediterranean Region. In: St George, T.D., Uren, M.F., Young, P.L. and Hoffman D. (eds) *Bovine Ephemeral Fever and Related Rhabdoviruses. Proceedings of the 1st International Symposium held in Beijing, PRC, 25–27 August 1992*, ACIAR, Canberra, pp. 54–56.

Doherty, R.L. (1977) Arthropod-borne viruses in Australia 1973–1976. *Australian Journal of Experimental Biology and Medical Science* 55, 103–130.

Doherty, R.L., Gorman, B.M., Whitehead, R.H. and Carley, J.G. (1966) Studies of arthropod-borne virus infections in Queensland. V. Survey of antibodies to group A arboviruses in man and in other animals. *Australian Journal of Experimental Biology and Medical Science* 44, 365–378.

Du Toit, R.M. (1944) The transmission of bluetongue and horse-sickness by *Culicoides*. *Onderstepoort Journal of Veterinary Science and Animal Industry* 19, 7–16.

Foster, N.M. and Jones, R.H. (1979) Multiplication rate of bluetongue virus in the vector *Culicoides variipennis* (Diptera: Ceratopogonidae) infected orally. *Journal of Medical Entomology* 15, 302–303.

Foster, N.M., Breckon, R.D., Luedke, A.J. and Jones, R.H. (1977) Transmission of two strains of epizootic hemorrhagic disease virus in deer by *Culicoides variipennis*. *Journal of Wildlife Diseases* 13, 9–16.

Francki, R.I.B., Fauquet, C.M., Knudson, D.L. and Brown, F. (1991) *Classification and Nomenclature of Viruses. Fifth Report of the International Committee on Taxonomy of Viruses. Archives of Virology Supplementum 2*, Springer-Verlag, Wien.

Gear, J.H.S., Ryan, J., Rossouw, E., Spence, I. and Kirsch, Z. (1977) Haemorrhagic fever with special reference to recent outbreaks in southern Africa. In: Gear, J.H. (ed.) *Medicine in a Tropical Environment*. A.A. Balkema, Rotterdam, pp. 350–358.

Gorman, B.M. (1990) The bluetongue viruses. *Current Topics in Microbiology and Immunology* 162, 1–19.

Greiner, E.C., Mo, C.L., Homan, E.J., Gonzalez, J., Oviedo, M.-T., Thompson, L.H. and Gibbs, E.P.J. (1993) Epidemiology of bluetongue in Central America and the Caribbean: initial entomological findings. *Medical and Veterinary Entomology* 7, 309–315.

Gubler, D.J. (1992) Dengue/dengue haemorrhagic fever in the Americas: prospects for the year 2000. In: Halstead, S.B. and Gomez-Dantes, H. (eds) *Dengue: a Worldwide Problem, a Common Strategy*. Ministry of Health, Mexico, pp. 19–27.

Haddow, A.J. (1965) Yellow fever in Central Uganda, 1964: Part 1. Historical introduction. *Transactions of the Royal Society of Tropical Medicine and Hygiene* 59, 436–440.

Halstead, S.B. (1990) Dengue. In: Warren, K.S. and Mahmoud, A.A.F. (eds) *Tropical and Geographical Medicine*. McGraw-Hill, New York, pp. 675–685.

Halstead, S.B. and Gomez-Dantes, H. (1992) *Dengue A Worldwide Problem, a Common Strategy*. Ministry of Health, Mexico.

Haughey, K.G., Hartley, W.J., Della-Porta, A.J. and Murray, M.D. (1988) Akabane disease in sheep. *Australian Veterinary Journal* 65, 136–140.

Hawkins, J.A., Adams, W.V., Wilson, B.H., Issel, C.J. and Roth, E.E. (1976) Transmission of equine infectious anaemia virus by *Tabanus fuscicostatus*. *Journal of the American Veterinary Medical Association* 168, 63–64.

Hayes, C.G. and Wallis, R.C. (1977). Ecology of western equine encephalomyelitis in the eastern United States. *Advances in Virus Research* 21, 37–83.

Hess, W.R. (1987) African swine fever virus in nature. In: Becker, Y. (ed.) *African Swine Fever*. Martinus Nijhoff, Boston, pp. 5–9.

Hess, W.R., Endris, R.G., Haslett, T.M., Monahan, M.J. and McCoy, J.P. (1987) Potential arthropod vectors of African swine fever virus in North America and the Caribbean basin. *Veterinary Parasitology* 26, 145–155.

Hoffman, H., Kunz, C. and Heinz, F.X. (1991) Laboratory diagnosis of tick-borne encephalitis. In: Calisher, C.H. (ed.) *Haemorrhagic Fever with Renal Syndrome, Tick- and Mosquito-Borne Viruses. Archives of Virology Supplementum 1*, Springer-Verlag, Wien, pp. 153–159.

Hoogstraal, H. (1966) Ticks in relation to human diseases caused by viruses. *Annual Review of Entomology* 11, 261–308.

Hoogstraal, H. (1979) The epidemiology of tick-borne Crimean–Congo hemorrhagic fever in Asia, Europe and Africa. *Journal of Medical Entomology* 15, 307–417.

Hoogstraal, H. (1981) Changing patterns of tickborne diseases in modern society. *Annual Review of Entomology* 26, 75–99.

Hoogstraal, H., Meegan, J.M., Khalil, G.M. and Adham, F.K. (1979) The Rift Valley fever epizootic in Egypt 1977–78. 2. Ecological and entomological studies. *Transactions of the Royal Society of Tropical Medicine and Hygiene* 73, 624–629.

Hourrigan, J.L. and Klingsporn, A.L. (1975) Epizootiology of bluetongue: the situation in the United States of America. *Australian Veterinary Journal* 51, 203–208.

Howell, P.G. (1968) African horsesickness. In: *Emerging Diseases of Animals*. FAO, Agricultural Studies No. 61, Rome, pp. 73–108.

Howell, P.G. (1979) The epidemiology of bluetongue in South Africa. In: St George, T.D.

and French E.L. (eds) *Proceedings 2nd Symposium Arbovirus Research in Australia.* CSIRO and QIMR, Brisbane, pp. 3–13.

Igarashi, A. (1992) Epidemiology and control of Japanese encephalitis. *World Health Statistics Quarterly* 45, 299–305.

Igarashi, A. (1994) Japanese encephalitis virus. In: Webster, R.G. and Granoff, A. (eds) *Encyclopaedia of Virology.* Academic Press, San Diego, pp. 746–751.

Johnson, B.K., Chanas, A.C., Tayeb, E. el, Abdel-Wahab, F.A. and Mohamed, A. el D. (1978) Rift Valley fever in Egypt, 1978. *Lancet* 2, 745.

Johnson, R.T. (1990) Arboviral encephalitis. In: Warren, K.S. and Mahmoud, A.A.F. (eds) *Tropical and Geographical Medicine.* McGraw-Hill, New York, pp. 691–700.

Jones, L.D., Hodgson, E. and Nuttall, P.A. (1991) Characterization of tick salivary gland factor(s) that enhance Thogoto virus transmission. In: Calisher, C.H. (ed.) *Haemorrhagic Fever with Renal Syndrome, Tick- and Mosquito-Borne Viruses. Archives of Virology Supplementum 1,* Springer-Verlag, Wien, pp. 227–234.

Jones, R.H. and Foster, N.M. (1974) Oral infection of *Culicoides variipennis* with bluetongue virus: development of susceptible and resistant lines from a colony population. *Journal of Medical Entomology* 11, 316–323.

Jones, R.H., Roughton, R.D., Foster, N.M. and Bando, B.M. (1977) *Culicoides,* the vector of epizootic hemorrhagic disease in white-tailed deer in Kentucky in 1971. *Journal of Wildlife Diseases* 13, 2–8.

Kettle, D.S. (1983) The bionomics of *Culicoides brevitarsis. Queensland Naturalist* 24, 33–39.

Lindsay, M.D., Johansen, C., Broom, A.K., D'Ercole, M., Wright, A.E., Condon, R., Smith, D.W. and Mackenzie, J.S. (1993) The epidemiology of outbreaks of Ross River virus infection in Western Australia in 1991–1992. In: Uren, M.F. and Kay, B.H. (eds) *Arbovirus Research in Australia. Proceedings Sixth Symposium,* 7–11 December 1992, Brisbane, Australia, pp. 72–76.

Linthicum, K.J., Davies, F.G. and Kairo, A. (1985) Rift Valley fever (family Bunyaviridae, genus *Phlebovirus*). Isolation from Diptera collected during an inter-epizootic period in Kenya. *Journal of Hygiene, Cambridge* 95, 197–209.

Linthicum, K.J., Bailey, C.L., Tucker, C.J., Mitchell, K.D., Logan, T.M., Davies, F.G., Kamau, C.W., Thande, P.C. and Wageteh, J.N. (1990) Application of polar-orbiting, meteorological satellite data to detect flooding of Rift Valley fever virus vector mosquito habitats in Kenya. *Medical and Veterinary Entomology* 4, 433–438.

Luedke, A.J., Jones, R.H. and Jochim, M.M. (1967) Transmission of bluetongue between sheep and cattle by *Culicoides variipennis. American Journal of Veterinary Research* 28, 457–460.

Luedke, A.J., Jones, R.H. and Walton, T.E. (1977) Overwintering mechanism for bluetongue virus: biological recovery of latent virus from a bovine by bites of *Culicoides variipennis. American Journal of Tropical Medicine and Hygiene* 26, 313–325.

Manson-Bahr, P.E.C. and Bell, D.R. (1987) *Manson's Tropical Diseases.* Baillière-Tindall, London.

Maroli, M., Ciufolini, M.G. and Verani, P. (1993) Vertical transmission of Toscana virus in the sandfly *Phlebotomus perniciosus,* via the second gonotrophic cycle. *Medical and Veterinary Entomology* 7, 283–286.

McCormick, J.B. and Fisher-Hoch, S. (1990) Viral haemorrhagic fevers. In: Warren, K.S. and Mahmoud, A.A.F. (eds) *Tropical and Geographical Medicine.* McGraw-Hill, New York, pp. 700–728.

McIntosh, B.M., Jupp, P.G., Dos Santos, I. and Meenehan, G.M. (1976) Epidemics of West Nile and Sindbis viruses in South Africa with *Culex (Culex) univittatus* Theobald as vector. *South African Journal of Science* 72, 295–300.

McIntosh, B.M., Jupp, P.G. and Dos Santos, I. (1977) Rural epidemic of Chikungunya in South Africa with involvement of *Aedes* (*Diceromyia*) *furcifer* (Edwards) and baboons. *South African Journal of Science* 73, 267–269.

Mebus, C.A. (1988) African swine fever. *Advances in Virus Research* 35, 251–269.

Mellor, P.S. (1990) The replication of bluetongue virus in *Culicoides* vectors. *Current Topics in Microbiology and Immunology* 162, 141–161.

Metcalf, H.E., Pearson, J.F. and Klingsporn, A.L. (1981). Bluetongue in cattle: a serologic survey of slaughter cattle in the United States. *American Journal of Veterinary Research* 42, 1057–1061.

Monath, T.P. (1990) Yellow fever. In: Warren K.S. and Mahmoud, A.A.F. (eds) *Tropical and Geographical Medicine*. McGraw-Hill, New York, pp. 661–674.

Monath, T.P. (1991) Viral febrile Illnesses. In: Strickland, G.T. (ed.) *Hunter's Tropical Medicine*. W.B. Saunders, Philadelphia, pp. 200–219.

Monath, T.P. and Nasidi, A. (1993) Should yellow fever vaccine be included in the expanded program of immunisation in Africa? A cost-effectiveness analysis for Nigeria. *American Journal of Tropical Medicine and Hygiene* 48, 274–299.

Neri, P. (1965) Revue taxonomique aspect écologique et biologique des diptères (Culicidae) présents dans la forêt de Manera (Province du Kaffa) Ethiopie. *Cahiers ORSTOM Entomologie Médicale* 3 and 4, 47–56.

Parsonson, I.M. (1979) Recent developments on bluetongue (BT) in the United States of America. In: St George, T.D. and French, E.L. (eds) *Proceedings 2nd Symposium Arbovirus Research in Australia*. CSIRO and QIMR, Brisbane, pp. 13–19.

Parsonson, I.M. (1990) Pathology and pathogenesis of bluetongue infections. *Current Topics in Microbiology and Immunology* 162, 119–141.

Pinheiro, F.P., Travassos da Rosa, A.P.A., Travassos da Rosa, J.F.S., Ishak, R., Freitas, R.B., Gomes, M.I.C., LeDuc, J.W. and Oliva, O.F.P. (1981a) Oropouche virus. I. A review of clinical, epidemiological and ecological findings. *American Journal of Tropical Medicine and Hygiene* 30, 149–160.

Pinheiro, F.P., Hoch, A.L., Gomes, M.L.C. and Roberts, D.R. (1981b) Oropouche virus. IV. Laboratory transmission by *Culicoides paraensis*. *American Journal of Tropical Medicine and Hygiene* 30, 172–176.

Plowright, W., Perry, C.T. and Greig, A. (1974) Sexual transmission of African swine fever virus in the tick, *Ornithodoros moubata porcinus* Walton. *Research in Veterinary Science* 17, 106–113.

Porterfield, J.S. (1989) *Andrewes' Viruses of Vertebrates*. Ballière-Tindall, London.

Reeves, W.C. (1982) Gaps in current knowledge of vector biology critical to control or to epidemiological studies of arboviruses. In: St George, T.D. and Kay, B.H. (eds) *Proceedings 3rd Symposium Arbovirus Research in Australia*. CSIRO and QIMR, Brisbane, pp. 10–15.

Reiter, P. (1992) Status of current *Aedes aegypti* control methodologies. In: Halstead, S.B. and Gomez-Dantes, H. (eds) *Dengue a Worldwide Problem, a Common Strategy*. Ministry of Health, Mexico, pp. 41–48.

Roberts, D.R., Hoch, A.L., Dixon, K.E. and Llewellyn, C.H. (1981) Oropouche virus. III. Entomological observations from three epidemics in Para, Brazil, 1975. *American Journal of Tropical Medicine and Hygiene* 30, 165–171.

Rosen, L. (1981) The use of *Toxorhynchites* mosquitoes to detect and propagate dengue and other arboviruses. *American Journal of Tropical Medicine and Hygiene* 30, 177–183.

Rosen, L., Shroyer, D.A. and Lien, J.C. (1980) Transovarial transmission of Japanese encephalitis virus by *Culex tritaeniorhynchus* mosquitoes. *American Journal of Tropical Medicine and Hygiene* 29, 711–712.

Rosen, L., Gubler, D.J. and Bennett, P.H. (1981) Epidemic polyarthritis (Ross River) virus infection in the Cook Islands. *American Journal of Tropical Medicine and Hygiene* 30, 1294–1302.

Rudnick, A. (1978) Ecology of dengue virus. *Asian Journal of Infectious Diseases* 2, 156–160.

Scott, T.W. and Weaver, S.C. (1989) Eastern equine encephalomyelitis virus: epidemiology and evolution of mosquito transmission. *Advances in Virus Research* 37, 277–328.

Scott, T.W., Clark, G.C., Lorenz, L.H., Amerasinghe, P.H., Reiter, P. and Edman, J.D. (1993) Detection of multiple blood feeding in *Aedes aegypti* (Diptera: Culicidae) during a single gonotrophic cycle using a histologic technique. *Journal of Medical Entomology* 30, 94–99.

Sellers, R.F. (1980) Weather, host and vector – their interplay in the spread of insect borne animal virus diseases. *Journal of Hygiene* 81, 65–102.

Sellers R.F. and Maarouf, A.R. (1989) Trajectory analysis and bluetongue virus serotype 2 in Florida 1982. *Canadian Journal of Veterinary Research* 53, 100–102.

Sellers R.F. and Pedgley, D.E. (1985) Possible windborne spread to western Turkey of bluetongue virus in 1977 and of Akabane virus in 1979. *Journal of Hygiene* 95, 149–158.

Sellers R.F., Pedgley, D.E. and Tucker, M.R. (1978) Possible windborne spread of bluetongue to Portugal, June–July 1956. *Journal of Hygiene* 5, 189–196.

Service, M.W. (1993) Mosquitoes (Culicidae). In: Lane, R.P. and Crosskey, R.W. (eds) *Medical Insects and Arachnids*. Chapman & Hall, London, pp. 120–240.

Shaplen, R. (1964) *Toward the Well-being of Mankind*. Doubleday, New York.

Smithburn, K.C., Haddow, A.J. and Gillett, J.D. (1948) Rift Valley fever. Isolation of the virus from wild mosquitoes. *British Journal of Experimental Pathology* 29, 107–121.

St George, T.D. (1993) The natural history of ephemeral fever of cattle. In: St George, T.D., Uren, M.F., Young, P.L. and Hoffman, D. (eds) *Bovine Ephemeral Fever and Related Rhabdoviruses. Proceedings of the 1st International Symposium held in Beijing, PRC, 25–27 August 1992*, ACIAR, Canberra, pp. 13–19.

Strickland, G.T. (1991) *Hunter's Tropical Medicine*. W.B. Saunders, Philadelphia.

Tabachnik, W.J. (1991) Genetic control of oral susceptibility to infection of *Culicoides variipennis* with bluetongue virus. *American Journal of Tropical Medicine and Hygiene* 45, 666–671.

Tabachnik, W.J. (1992) Genetic differentiation among populations of *Culicoides variipennis* (Diptera: Ceratopogonidae), the North American vector of bluetongue virus. *Annals of the Entomological Society of America* 85, 140–147.

Tesh, R.B. (1988) The genus *Phlebovirus* and its vectors. *Annual Review of Entomology* 33, 169–181.

Tesh, R.B. (1990) Undifferentiated arboviral fevers. In: Warren, K.S. and Mahmoud, A.A.F. (eds) *Tropical and Geographical Medicine*. McGraw-Hill, New York, pp. 685–691.

Tesh, R.B. and Chaniotis, B.N. (1975) Transovarial transmission of viruses by phlebotomine sandflies. *Annals of the New York Academy of Sciences* 266, 125–134.

Tidwell, M.A., Dean, W.D., Tidwell, M.A., Combs, G.P., Anderson, D.W., Cowart, W.O. and Axtell, R.C. (1972) Transmission of hog cholera virus by horseflies (Tabanidae: Diptera). *American Journal of Veterinary Research* 33, 615–622.

Varma, M.G.R. and White, G.B. (1989) Mosquito-borne virus diseases. *Geographical Distribution of Arthropod-borne Diseases and their Principal Vectors*. World Health Organization, Geneva, pp. 35–54.

Walton, T.E. and Osburn, B.I. (1992) *Bluetongue, African Horse Sickness, and Related Orbiviruses. Proceedings of the Second International Symposium*. CRC Press, Boca Raton, Florida.

Warren. K.S. and Mahmoud, A.A.F. (1990) *Tropical and Geographical Medicine*. McGraw-Hill, New York.

Watts, D.M., Pantuwatana, S., Yuill, T.M., DeFoliart, G.R., Thompson, W.H. and

Hanson, R.P. (1975) Transovarial transmission of La Crosse virus in *Aedes triseriatus*. *Annals of the New York Academy of Sciences* 266, 135–143.

Webb, P.A., McLean, R.G., Smith, G.C., Ellenberger, J.H., Francy, D.B., Walton, T.E. and Monath, T.P. (1987) Epizootic vesicular stomatitis in Colorado, 1982: some observations on the possible role of wildlife populations in an enzootic maintenance cycle. *Journal of Wildlife Diseases* 23, 192–198.

Wenbin, Bai (1993) Epidemiology and control of bovine ephemeral fever in China. In: St George, T.D., Uren, M.F., Young, P.L. and Hoffman D. (eds) *Bovine Ephemeral Fever and Related Rhabdoviruses. Proceedings of the 1st International Symposium held in Beijing, PRC, 25–27 August, 1992*, ACIAR, Canberra, pp. 13–19.

Wu, G., Zhang, Y., Zhao, X., Zhang, B., Hu, Y., Zhang, Y., Shi, J., Jiang, K., Faan, G., Zhang, J., Men, R., Zhou, Y., Gan, Y. and Qian, J. (1992) Studies on natural infection of EHFV in *Leptotrombidium* (*L.*) *scutellare* and its role in transmission of EHFV. *Proceeding of the XIXth International Congress of Entomology, Abstracts. Beijing, China, 28 June–4 July 1992*, p. 671.

Typhus and Other Rickettsial Diseases

25

This chapter and the next deal with a diverse range of pathogens which, on an evolutionary scale, would be placed above the viruses and below the Protozoa. They are included in the standard reference work, *Bergey's Manual of Systematic Bacteriology* (Krieg and Holt, 1984). Three groups of these organisms are relevant to medical entomology: the Rickettsiales, which includes the organisms responsible for typhus and the spotted fevers; among the Spirochaetales the genus *Borrelia*, which causes relapsing fever in humans; and two Gram-negative rod-shaped bacteria, *Yersinia pestis* and *Francisella tularensis*, the causative organisms, respectively, of plague and tularaemia. The Rickettsiales will be dealt with in this chapter and the other two groups in the next chapter.

Rickettsiales (Weiss and Moulder, 1984)

Most Rickettsiales are rod-shaped, coccoid and often pleomorphic microorganisms, which are Gram-negative and multiply inside host cells. All Rickettsiales are regarded as parasitic or mutualistic and are associated with arthropods which may act as vectors or primary hosts. Three families are recognized within the order Rickettsiales: the Rickettsiaceae, the Bartonellaceae and the Anaplasmataceae. The Anaplasmataceae resemble rickettsiae and are obligate parasites found within or on erythrocytes or free in the plasma of vertebrates. The Bartonellaceae are rod-shaped, bacteria-like parasites of the erythrocytes of humans and other vertebrates. Members of both families are transmitted by arthropods.

The most important family is the Rickettsiaceae which, in vertebrates, are parasites of tissue cells other than erythrocytes and are transmitted by arthropods. Three tribes are recognized within the Rickettsiaceae: the Wolbachieae which are symbionts of arthropods and do not occur in vertebrates; the Ehrlichieae, most of which are pathogenic for certain mammals with two new forms causing disease in humans; and the Rickettsieae which are capable of infecting suitable vertebrate

hosts including humans, who may be the primary host but are more often incidental hosts.

Symbiotic rickettsiae of the Wolbachieae have been described from the mosquitoes *Aedes scutellaris* and *Culex pipiens*, from the sheep ked *Melophagus ovinus*,
and from argasid and ixodid ticks. In addition similar microorganisms have been
described from the human body louse *Pediculus humanus* and *Glossina* spp. but
they are not listed by Weiss and Moulder (1984).

Rickettsieae

Within the Rickettsieae three genera are recognized; *Rickettsia, Rochalimaea* and
Coxiella. *Coxiella* grows preferentially in vacuoles of the host cells and, being
highly resistant to physical and chemical conditions in the extracellular environment, can be transmitted in the absence of an arthropod vector. *Rochalimaea* can
be cultivated in host cell-free media and grow profusely on the surface of
eukaryotic cells. Species of *Rickettsia* have not been cultivated in the absence of
host cells and they multiply in the cytoplasm or sometimes in the nucleus of certain
vertebrate and arthropod cells. They are unstable when separated from host components (Weiss and Moulder, 1984).

Species of *Rickettsia* are the most important disease-causing agents within the
Rickettsiales and indeed among all human pathogens. With the exception of louse-
borne typhus and trench fever, all these human infections are zoonoses with no
person to person or person to animal transmission occurring (Wisseman, 1991).
Three groups can be distinguished within the genus: typhus group, spotted fever
group and scrub typhus group. The last group contains only one species. Audy
(1968) believes that rickettsiae were originally associated with soil-dwelling
acarines from which they were introduced into rodents by trombiculid mites. Once
established in rodents they were taken up by ticks and blood-sucking fleas and lice.
Some rodent fleas will feed on humans as alternative hosts and in this way rickettsiae were introduced into the human population, from which they became
established in *Pediculus humanus*.

There are varying degrees of adaptation between rickettsiae and their
arthropod hosts. Their long association with acarines is shown by the ease with
which they pass the gut barrier, and disperse in the haemocoele and tissues of the
acarine, enabling both transovarian transmission and transmission by bite to occur.
In insects the rickettsiae are confined to the gut and transmission is via the faeces.
Rodent fleas have had longer exposure to rickettsiae, in an evolutionary sense, than
human lice and are apparently unaffected by infection with *R. typhi*. The longevity
of *P. humanus* is greatly reduced by infection with *R. typhi* or *R. prowazekii* but
greater adaptation has occurred to *Rochalimaea quintana* which is non-pathogenic
for *P. humanus*.

Epidemic Typhus *Rickettsia prowazekii*

The aetiological agent of classical epidemic typhus is *Rickettsia prowazekii*, named
after the American H.T. Ricketts and the Austrian S. von Prowazek, both of whom

died from typhus contracted as a result of their researches into the disease. Epidemic typhus is a severe disease with a high mortality (30–60%) in populations weakened by malnutrition (Buxton, 1950). Epidemics of typhus have changed the course of history as Zinsser (1935) has documented in his very readable book *Rats, Lice and History*. At the end of the First World War and the period immediately succeeding it (1917–1923) it is believed that 30 million cases of epidemic typhus occurred in Russia and Europe with over 300,000 deaths (Manson-Bahr and Bell, 1987). Official Soviet statistics maintain that during that period 10% of the population was affected (Zdrodovskii and Golinevich, 1960).

A little more than a century earlier typhus had played a major role in the defeat of Napoleon's armies which invaded Russia. Diseases, of which typhus was the major one, rather than military opposition defeated Napoleon. The potato famine in Ireland in the 1840s led to a major movement of the population to America. Of the 75,000 Irish who migrated in 1847, 30,000 (40%) contracted typhus of whom 20,000 (67%) died from the disease, reflecting the debilitated state of the health of the migrants (Zinsser, 1935).

The epidemic in Naples in 1943–1944 during the Second World War was the first time an epidemic of typhus had not exhausted itself but had been terminated by human action. This involved dusting fully clothed individuals with effective anti-louse powders. Initial applications were with MYL, a pyrethrum preparation or AL63, containing derris and naphthalene. These were later replaced with 10% DDT in talc. Nearly 3 million dust treatments were made to individuals from the middle of December 1943 to the beginning of April 1944 (Craufurd-Benson, 1946) and the epidemic brought to a halt.

Although *P. humanus* occurs widely in human populations, epidemic typhus is commoner in the temperate regions and in the cooler regions of the tropics above 1600 m, and is absent from the lowland tropics. It is present in mountainous regions where heavy clothing is worn continuously, favouring infestations with body lice and occurs in Mexico, Guatemala, the Andean highlands, the Himalayan region including Pakistan and Afghanistan, the highlands of Ethiopia, Burundi, Rwanda, Lesotho and North China (Wisseman, 1991). The incidence of louse-borne typhus has been steadily declining with 10,548 cases in 1975, 8065 in 1976, and 6087 in 1977 (Anon., 1978). The mortality during those three years was less than 2%. As in previous years, the majority of the cases in 1977 occurred in Africa (96.3%), with most of the remainder occurring in Peru and Ecuador (200/218 = 92%). In 1979 the number of cases increased sharply to 18,359, of which 17,476 (95.2%) were in Ethiopia, but elsewhere the decline continued (Anon., 1981).

Transmission

The head louse *P. capitis*, and the crab louse *Pthirus pubis*, can transmit *R. prowazekii* experimentally, but epidemics have always occurred in conditions where body lice *P. humanus* were particularly prevalent, and this species is the usual vector (Busvine, 1976).

Rickettsia prowazekii multiplies in the epithelial cells of the midgut, and when these burst it is passed out with the faeces of the louse (Buxton, 1950). *R. prowazekii* is pathogenic to the body louse and kills it in about 10 days. People become infected by scratching in response to the feeding of the louse, scarifying the skin, facilitating the entry of *R. prowazekii*. It is possible that *R. prowazekii* can gain entry into the human body by other routes, e.g. by inhalation of louse faeces, or by penetrating the mucosa or the conjunctiva of the eye. Fatalities among first research workers arose because they concentrated on the feeding of the louse and not on its faeces. *R. prowazekii* can survive 66 days in dry louse faeces at ambient temperatures. This means that fresh cases of typhus can be contracted for 2 months after the conclusion of a successful body louse eradication programme.

Survival of *R. prowazekii*

Louse-borne typhus can be epidemic as indicated by examples already given or endemic–epidemic as in Ethiopia or endemic with sporadic cases as in Andean villages (Wisseman, 1991). The question arises as to how *R. prowazekii* survives between outbreaks. There is no transovarial transmission; individual lice survive for only about 6 weeks, less if infected (Buxton, 1950), and *R. prowazekii* survives for only about 2 months in louse faeces. According to Weiss and Moulder (1984), humans are the primary host of *R. prowazekii* and individuals who recover from epidemic typhus often retain small numbers of organisms, presumably in their lymph nodes. They may give rise to a mild form of typhus, Brill-Zinsser disease, which occurs in the absence of body lice.

In different localities of the former USSR recrudescence has been recorded as occurring in 1 to 16% of recovered patients (Zdrodovskii and Golinevich, 1960) and may be much higher but other estimates put the recrudescent rate at 1 in 10,000 (Wisseman, 1991). Such relapses are considered to be the main reservoir of *R. prowazekii* in eastern Europe (Manson-Bahr and Bell, 1987). In Peru there is evidence for asymptomatic infections of humans with *R. prowazekii* (Zdrodovskii and Golinevich, 1960). These could be the reservoirs maintaining endemic areas of infection.

Rickettsia prowazekii has been shown to be present as natural infections of flying squirrels (*Glaucomys volans*) in Florida and Virginia in the USA (Sonenshine *et al.*, 1978). *R. prowazekii* was recovered from the blood-sucking louse *Neohaematopinus sciuropteri* and the flea *Orchopeas howardii*. The louse is host-specific but *O. howardii* has an extensive host range, which includes humans. The significance of this finding to the epidemiology of epidemic typhus is not known, but sporadic human cases have arisen in houses harbouring flying squirrels (Wisseman, 1991).

Entomological aspects of typhus epidemics (Wigglesworth, 1941; Buxton, 1950)

Epidemics of louse-borne typhus are associated with overcrowded, insanitary conditions. In England, typhus was known as gaol fever. From the sixteenth century until almost recent times epidemic typhus has been prevalent in times of war, occurring among both refugees and fighting men. *P. humanus* is well adapted to spread *R. prowazekii*. It is a permanent human ectoparasite living on the skin and in the clothing immediately adjoining it. In that situation, on a healthy person the body louse lives in a stable environment. Exposed to a temperature gradient it shows a marked preference for a temperature range of 29–30°C. When the temperature goes above or below this range the louse moves away. This has the effect that lice leave a person in a fever, e.g. suffering from typhus, and also leave a corpse.

When refugees are crowded together for warmth and shelter lice spread rapidly throughout the human population. In experiments in which two volunteers, one infested with 200 lice, shared a large bed, the louse-free individual complained of being bitten in an hour or so if the infested companion had a fever, but if the infested person had a normal body temperature the uninfested individual did not complain for 5 or 6 hours.

Pediculus humanus orientates to human odours and the excreta of its own species. Both these responses will keep body lice in the vicinity of their host. Body lice avoid moisture and they will therefore leave a sweating feverish patient; behaviour which will favour the spread of typhus. They prefer rough to smooth surfaces, i.e. woollen stockinette to cotton stockinette and both to silk (Wigglesworth, 1941). This behaviour might explain in part the fashion in earlier times for silk underwear. Although silk would be less favourable to lice it would not prevent infestation.

Pediculus humanus has great powers of reproduction. Under optimal conditions, a female louse will live for nearly 5 weeks, during which period she will produce 279 eggs of which 117 will become adults. On these figures a female louse will, in a period of 2 months, give rise to 15,000 progeny of which 10,500 will be eggs and 4500 will be nymphs and adults. However, under normal conditions an infested human being will take corrective action as the louse population builds up. In practice, 'natural populations of head and body lice commonly consist of about 10 or 20 insects (nymphs and adults) though hundreds are not very rare and populations exceeding 1000 have been recorded' (Buxton, 1950). It is easy to appreciate that, when human society is disrupted by war, famine or natural disaster, personal hygiene practices break down and louse populations flourish. These are the conditions under which louse-borne typhus becomes epidemic.

Murine typhus (*Rickettsia typhi*) (Azad, 1990)

Rickettsia typhi (= *R. mooseri*) is placed in the typhus group of the genus *Rickettsia* together with *R. prowazekii* (Weiss and Moulder, 1984). Murine typhus is worldwide in distribution and, being a disease associated with commensal rats and their fleas, has been disseminated by shipping and tends to have a coastal distribution. It occurs inland in India, Burma, Thailand, Pakistan and the southern USA. Murine typhus is the most prevalent rickettsial infection in humans. In the period 1931–1946, 42,000 cases of murine typhus occurred in the United States of which 94% were in coastal Texas. It is still regarded as a public health problem in Texas and in 1980–1984 200 cases were reported. It is considered that recorded cases may represent only one-fifth of the total number of infections.

Rickettsia typhi causes a milder disease than *R. prowazekii* but it is still a serious debilitating illness with high fever. There is an incubation period of 12 days followed by a similar period of clinical disease with a mortality of less than 5% in untreated cases and zero using appropriate antirickettsial drugs (Wisseman,

1991). Murine typhus is a household infection associated with commensal rats (*Rattus rattus* and *R. norvegicus*) and their fleas (*Xenopsylla cheopis* and *Leptopsylla segnis*) in an urban environment, but in the countries listed in the paragraph above it has spread into rural areas.

Rickettsia typhi harms neither the rat nor the flea. In *X. cheopis*, it multiplies in the cells of the midgut, from which it tends to escape after 3–5 days without damaging the cells and spreads to the entire midgut lining within 7–9 days. Fleas begin to pass infected faeces 10 days after an infective feed and continue to pass *R. typhi* for 40 days. After a flea has been infected for more than 3 weeks it is capable of transmitting *R. typhi* during feeding, but contamination with infected faeces remains the main source of infection. Transmission can also occur when an infected flea is crushed on the skin and the infective material scarified in by scratching. *R. typhi* can also be transmitted transovarially in *X. cheopis*. Although *X. cheopis* is primarily an ectoparasite of rats, it readily feeds on humans in the absence of its main host. In *X. cheopis* feeding and defecation are closely associated, and infected fleas feeding on people would deposit *R. typhi* in the human environment.

In hyperendemic foci infection rates of 3 to 10% have been found in *X. cheopis* and 9 to 16% in *L. segnis*, but the latter is a semisessile flea and has tended to be disregarded as a vector of murine typhus though it may prove to be a significant vector in parts of the world where *X. cheopis* is absent or very rare. *R. typhi* has similar cycles in *L. segnis* and *Ctenocephalides felis* and experimentally they remained infective from 10 days after an infective feed to 30 days when the experiments were terminated. Infected *Ct. felis* have been collected from opossums which were naturally infected with *R. typhi* and associated with human cases. *R. typhi* has been found in a range of rodents and also in domestic cats. It is likely that blood-sucking mites, e.g. *Ornithonyssus bacoti*, and lice, e.g *Polyplax spinulosa*, are involved in maintaining infection in the rodent population. Infection rates in commensal rats in field studies in Ethiopia, Burma and Egypt varied from 0 to 46% and in Texas in 1945 an infection rate of 63% was reported for *R. norvegicus*.

Once *R. typhi* has become established in the human population it could be transmitted by *Pulex irritans*, acquired by *Pediculus humanus* and become epidemic, but this does not appear to occur. In *P. humanus*, *R. typhi* becomes intracellular and causes the death of the louse, perhaps even more rapidly than does *R. prowazekii* (Buxton, 1950).

Spotted Fever Group (Tick- and Mite-borne Typhus)

Tick-borne spotted fevers are widely spread throughout the world, occurring in every continent and zoogeographical region. They are caused by various species of *Rickettsia* which circulate in a wide range of mammals, and sometimes birds, through the agency of ixodid ticks. The rickettsiae are transmitted trans-stadially and transovarially in the tick vector. They cause benign infections of the non-human host but serious disease in humans, which may be fatal. Hoogstraal (1967) cites the following quotation from Dr David B. Lackman: 'Whenever you have ixodid ticks biting man there is the possibility of Rocky Mountain spotted fever.' This remark could be generalized to the effect that wherever ixodid ticks bite people there is the possibility of human spotted fever.

Five species of *Rickettsia* are involved in tick-borne typhus: *R. rickettsii*, *R. sibirica*, *R. conorii*, *R. australis* and the recently described *R. japonica* (Uchida *et al.*, 1992; Uchida, 1993). Most detailed information is available on the epidemiology of the first two species and they will be considered at greater length than the others.

Rickettsia rickettsii

The disease associated with *R. rickettsii* is commonly referred to as Rocky Mountain spotted fever, but the organism is widely distributed outside the Rocky Mountains occurring in 46 mainland States of the USA with most cases originating in the eastern States. It is also found in Mexico, Central America (Panama and Costa Rica) and South America (Brazil and Colombia), being given different names in each country, e.g. Sao Paulo fever in Brazil. In the USA the number of cases built up to more than 1000 around 1980 and since then has declined to 600 in 1987. There is an incubation period of about a week with 2 to 3 weeks of disease characterized by chills, headache, fever, rash and photophobia. This disease has a normal mortality of about 20% and prompt medication is essential to reduce that but even with early treatment there may be 5% mortality (Wisseman, 1991).

In the eastern United States the vector is *Dermacentor variabilis* which in the immature stages occurs on rodents, and in the adult occurs on larger mammals including humans and dogs. Infestation of dogs with *D. variabilis* brings infection indoors and exposes individuals to infection in a domestic situation. In the western United States, the disease is associated with field workers and the vector is *D. andersoni*, of which the immature stages are indiscriminate parasites of small mammals, and the adults parasitize hares, larger wild and domestic animals, and humans (Hoogstraal, 1967).

Among wild hosts infections with *R. rickettsii* are transitory, but serological tests have demonstrated antibodies to *R. rickettsii* in 18 species of birds and 31 species of mammals, including 18 species of rodents, 5 species of leporids and 5 carnivores. In the laboratory, infected female *D. andersoni* transmit the rickettsia to 100% of their daughters and from them to their progeny. Nevertheless, in the field, infections of *D. andersoni* with *R. rickettsii* are much lower, less than 14%. Trans-stadial and transovarian transmission of the rickettsia does not reduce its virulence. Infection results from the bite of the tick and the faeces of infected *D. andersoni* do not appear to be important in transmission (Hoogstraal, 1967).

Argasid ticks have been found infected in nature but their role in transmission is not clear. They do not appear to be involved in transmitting *R. rickettsii* to humans. The main human vectors are *Amblyomma cajennense* in Central and South America; *Rhipicephalus sanguineus* in Mexico; *A. americanum* in Texas and Oklahoma and *D. andersoni* and *D. variabilis* in the USA and probably also in Canada. *R. sanguineus* has established itself in North America and a serious situation will develop should that tick become infected with *R. rickettsii*.

Control of the disease involves avoidance of likely tick-infested locations and the wearing of boots and protective clothing, especially on the lower limbs which are most likely to pick up ixodid ticks. The protection offered by clothing can be increased by the application of a suitable repellent, e.g. permethrin or *N-N*-butylacetanilide, followed by daily body examinations to remove any attached ticks. Early removal is considered to reduce the likelihood of infection (Wisseman, 1991).

Rickettsia sibirica (Hoogstraal, 1967)

Rickettsia sibirica, sometimes spelt *siberica*, is closely related to *R. rickettsii*. The two species are geographically distinct with *R. rickettsii* being confined to the western hemisphere, and *R. sibirica* to the northern Palaearctic region from Armenia in the west to north China and eastern Siberia but the geographical boundaries of this rickettsia are still not known (Weiss, 1988). *R. sibirica* is less virulent than *R. rickettsii* and came into prominence during the development of 'The Virgin Lands' of the Soviet Union in the 1930s: 200 to 600 cases of Siberian tick typhus are recorded annually at Krasnoyarsk.

Strains of *R. sibirica* have been recovered from at least 18 species of mammals, mostly rodents. Birds are considered to play a secondary role as reservoirs of *R. sibirica*. Nine species of ixodid ticks act as reservoirs and vectors: four species of *Dermacentor*, three of *Haemaphysalis; Rhipicephalus sanguineus*; and *Hyalomma asiaticum*. In *Dermacentor marginatus*, naturally or experimentally infected with *R. sibirica*, the rickettsia survives for at least 5 years or through four generations. The immature stages of *Dermacentor* occur on small mammals, particularly rodents, hedgehogs, hares and small carnivores, and are rare on birds. This may explain the secondary role of birds as reservoirs of *R. sibirica*. Adult *Dermacentor* parasitize medium to large wild and domestic mammals.

Ticks of the genus *Dermacentor* are common and widely distributed in Eurasia, with species occupying different ecological zones. *Dermacentor marginatus* occurs in lowland and alpine steppes of western Eurasia and further west into central Europe. *D. silvarum* has an eastern distribution, extending from western Siberia to the Pacific Ocean. It infests the taiga and shrub-wormwood steppes. In the Far East in shrub and fern marshes adjoining taiga, *Haemaphysalis concinna* is the chief vector. It gives way to *Hyalomma asiaticum* in the semi-desert steppe. In central and eastern Siberia *Dermacentor nuttalli* occurs in cisalpine, alpine, forested and desert steppes.

Rickettsia slovaca was recovered from *Dermacentor marginatus* from Central Slovakia but is considered to be a serotype of *R. sibirica* (Weiss and Moulder, 1984). It is widely distributed in Slovakia and probably in most parts of Europe. All stages of *D. marginatus* transmit *R. slovaca* by bite and in their faeces and it is passed trans-stadially and transovarially. Antibodies to *R. slovaca* have been found in wild rodents which are considered to be major reservoir hosts. Nevertheless, the importance of *R. slovaca* in human and animal pathology has not yet been proven, but there is evidence that this *Rickettsia* can be a human pathogen (Rehacek, 1979).

Rickettsia conorii, *Rickettsia australis*, *Rickettsia japonica*

Rickettsia conorii is the most ubiquitous rickettsia of the spotted fever group occurring in southern Europe, Africa, India and the Oriental region. The popular names for the disease caused by *R. conorii* are often formed from the geographical location plus tick typhus, e.g. Kenya tick typhus, but it is also known as Marseilles fever and *fièvre boutonneuse* (pimply fever). *R. conorii* causes acute suffering but few deaths. There have been no extensive epidemiological studies on *R. conorii* comparable to those on *R. rickettsii* and *R. sibirica*. In Zimbabwe during anti-guerilla warfare, *R. conorii* infections among black, African troops reached 'mission aborting' levels (Hoogstraal, 1986).

The chief vectors of South African tick bite fever on the veldt appear to be *Amblyomma hebraeum* and *Rhipicephalus appendiculatus*, the immature stages of both species being common human parasites. Human infections with *R. conorii* in South Africa also occur in urban settings where the vectors are *Haemaphysalis leachi* and *Rhipicephalus sanguineus*, the common dog ticks of tropical and subtropical areas. The normal route of transmission of *R. conorii* to humans is through the bite of the tick but infection may also result from contamination of eye and nasal mucosa from crushed ticks or tick faeces, particularly when dogs are being de-ticked (Rehacek, 1979).

In South Africa the dog is not regarded as a reservoir host of *R. conorii* (Hoogstraal, 1967), but in many countries of southern Europe the dog is suspected of being a major reservoir. In the Crimea 15 to 71% of dogs carried antibodies to *R. conorii*; nevertheless Rehacek (1979) concluded that the role of the dog as a reservoir host has not yet been proved conclusively. In South Africa the striped mouse (*Rhabdomys pumilio*) and the vlei rat (*Otomys irroratis*) are as often infected in nature as the commensal black rat *Rattus rattus* (Gear, 1988), but their role in the epidemiology of the disease is not fully understood. A drop in human infections in Europe following an epizootic of myxomatosis suggests that rabbits may have been maintaining an infected tick population (Weiss and Moulder, 1984).

Rickettsia australis is the cause of tick typhus in Queensland, Australia. Several species of marsupials, including bandicoots and possums, are considered to be reservoir hosts and, on circumstantial evidence, the probable vector is *Ixodes holocyclus*. *I. holocyclus* is a widely distributed, unusually indiscriminate feeder, attacking almost any bird or mammal (Hoogstraal, 1967). There is evidence that *Haemaphysalis longicornis* could be both a reservoir and vector of *R. japonica*, the causative agent of Oriental spotted fever (Uchida *et al.*, 1995).

Rickettsial pox (*Rickettsia akari*) (Rehacek, 1979)

Rickettsia akari is included with the agents of tick typhus in the spotted fever group within the genus *Rickettsia*. Rickettsial pox was first recognized as a new human disease in the mid-1940s when it was described from Boston and New York in the USA, and a few years later from the former USSR. In the mid-1940s about 180 cases were reported annually in the United States (Weiss and Moulder, 1984). In 1949–1950 it occurred in epidemic form in the Ukraine causing approximately 1000 cases in the Donets basin. The vector is the gamasid mite *Liponyssoides sanguineus*, an ectoparasite of the house mouse, *Mus musculus*. Rickettsial pox has since been identified in South Korea, South Africa and French Equatorial Africa. It is also suspected of occurring in Yugoslavia and Italy (Sicily). In *L. sanguineus*, *R. akari* is transmitted transovarially to the next generation of mites. In the laboratory the tropical rat mite *Ornithonyssus bacoti* can maintain *R. akari* and transmit it transovarially. There is little information on the existence of a cycle of *R. akari* in wild rodents but in Korea it has been recovered from *Microtus fortis pellicieus* (Weiss and Moulder, 1984).

Scrub typhus (*Rickettsia tsutsugamushi*) (Audy, 1968; Wisseman, 1991)

Rickettsia tsutsugamushi (= *orientalis*) occurs in an area bounded by Pakistan and Tadzhikistan in the west to Japan and south-east Siberia in the east, and to Indonesia and tropical North Queensland in the south. It is also present on certain islands in the Pacific and Indian Oceans, including Diego Garcia, midway between Madagascar and the Oriental region, where it is considered that infected vectors were introduced by birds or flying foxes. *R. tsutsugamushi* is the aetiological agent of scrub typhus, an acute febrile disease, transmitted by trombiculid mites. When an infected mite feeds on a susceptible human, an ulcer-like eschar commonly forms at the site of the mite's attachment. The disease involves fever, severe headache, rash and lymphadenopathy lasting for 2 to 3 weeks, with a mortality which may exceed 30% in untreated cases. Prompt treatment with tetracycline drugs can reduce the mortality to zero. There are multiple serotypes of *R. tsutsugamushi* which give transient cross-immunity to humans and allow multiple sequential infections.

The association between scrub typhus and mites was reported in a sixteenth-century Chinese work on natural history and has been known for at least 200 years by the Japanese who named the disease tsutsugamushi, meaning dangerous mite. More than 1000 species of trombiculid mites have been described but only about 10 species in the genus *Leptotrombidium* are vectors of *R. tsutsugamushi* to humans. They include *L. akamushi*, *L. deliense*, *L. pallidum*, and *L. scutellare*. Species of *Ascoschoengastia* are involved in transmitting *R. tsutsugamushi* among rodents.

Scrub typhus is a zoonosis and humans become involved when they enter an enzootic focus, where the pathogen, *Leptotrombidium* spp., and wild rodents, especially those in the subgenus *Rattus (Rattus)*, are present. Such foci tend to be characterized by the presence of transitional vegetation, a habitat much favoured by *Rattus (Rattus)*. *L. akamushi* and *L. deliense* have a wide geographical distribution within which they are patchily distributed, being abundant in ecologically favourable 'mite islands' and absent from other habitats. In Japan, classical tsutsugamushi was present in limited areas and associated with high mortality. Tsutsugamushi is, in fact, more widely distributed throughout Japan where, as in other parts of south-east Asia, it causes a milder but still severe disease.

Only trombiculid mites which parasitize rodents will have access to *R. tsutsugamushi* while species of *Eutrombicula* and *Schoengastia*, which parasitize birds and reptiles, will have little chance of becoming infected. About 15 species of these last two genera cause scrub itch but not scrub typhus in humans. This explains why the distributions of scrub typhus and scrub itch do not necessarily coincide.

The rodent hosts of *R. tsutsugamushi* are various species of rats which are abundant in the region with more than 500 named forms being described from the Malaysian subregion alone.

The limited distribution of scrub typhus within a country is well demonstrated by groups undertaking a 4-day exercise in Sri Lanka during the Second World War. At that time *R. tsutsugamushi* was not known to be present in that country yet the division engaged in the exercise developed 750 cases of scrub typhus. These were traced to a 'typhus island' at Embilipitiya where slash-and-burn cultivation in virgin forest had produced a mixture of grassy areas and fields of millet. Field rodents established themselves in the grassy areas, fed in the millet fields and built up high populations of their trombiculid parasites.

Scrub typhus is not associated with clearing virgin rainforest or among cultivation where weeds are kept to a minimum. However, when such cultivated areas are neglected, scrub typhus is a common infection among the workers who undertake clearance of lapsed cultivation. The environment in which scrub typhus occurs is essentially man-made. In Malaya, Sumatra, New Guinea and tropical Queensland scrub typhus is associated with a coarse, rasping, fire-resistant grass, *Imperata cylindrica*, known in some areas as kunai grass. It provides a suitable habitat for field rodents and its dominance is maintained by the use of fire, which prevents the establishment of shrubs and trees, whose shade would control the grass. Kunai grass represents a pyrophytic subclimax in a succession which should have continued to forest. In areas where the forest is re-established, scrub typhus is either absent or at a negligible level. There is also an ecotone effect with scrub typhus being more likely to occur at the fringes of the grassland habitat.

In the life cycle of a trombiculid mite only the larval stage is parasitic; the active nymph and adult are predators on other arthropods. The larva feeds on only one individual host and therefore has the potential to acquire or to transmit an infection, but not to do both. Therefore trombiculid mites can only be vectors of *R. tsutsugamushi* if the rickettsia is transmitted transovarially. Transovarian transmission has been shown to occur in various species of *Leptotrombidium*. Both sexes of *L. pallidum* can be infected with *R. tsutsugamushi* but, while transovarian transmission occurred, the spermatophores of infected males were free of rickettsia (Takahashi *et al.*, 1988). *R. tsutsugamushi* was found in almost all organs and tissues of larval and adult *L. pallidum*, especially the salivary glands, epidermal cells and reproductive organs. Infected mites are the main reservoirs of *R. tsutsugamushi* because they will maintain infections longer by transovarian transmission than will the rodent host.

In the coastal areas of south-east China scrub typhus is a summer–autumn disease, but in Fujian Province there is a winter form of the disease of which the suspected vector is *L. scutellare*, which has been shown to be the vector of the winter form in Japan (Dun-Qing, 1988). An outbreak of scrub typhus in residential areas of Nagano Prefecture in Japan was associated with *L. pallidum*, which was present in the gardens of housing lots close to patients' houses (Uchikawa *et al.*, 1988). In a suburban Bangkok focus 96% of *Rattus rattus* showed serological signs of exposure to *R. tsutsugamushi*; nearly 2% of *L. deliense* and *Ascoschoengastia* spp. were positive for the pathogen and 30% of the human population were serologically positive (Tanskul *et al.*, 1992).

Preventive measures involve treatment of clothing and exposed skin with repellents such as diethyltoluamide (Deet), benzyl benzoate or dibutyl phthalate, paying particular attention to the lower legs, ankles and feet. They are the parts of the body likely to encounter larval mites first as the mites rarely ascend higher than a few centimetres when seeking a rodent host. Rodent control may not be the most effective short-term measure to reduce scrub typhus. The effect of removing many rodent hosts from the ecosystem is to leave large numbers of trombiculid larvae seeking alternative hosts. On deserted Jarak Island, six out of eight scientific investigators who indulged in rodent control contracted scrub typhus, while another group who visited the island later were equally overrun by rats but carried out no control, and none of the ten scientists developed scrub typhus.

Rochalimaea quintana

The two species of *Rochalimaea* are *R. quintana* which causes trench or 5-day fever in humans and *R. vinsonii* which is considered to be a recent mutation of *R. quintana*. Epidemics of trench fever occurred in the First World War (1914–1918) inflicting at least 1 million military personnel (Weiss and Moulder, 1984). It was virtually unheard of during the inter-war period but reappeared in Germany in 1941–1942 and had become widespread by 1943 (Zdrodovskii and Golinevich, 1960). The disease is only rarely fatal and nearly half the convalescents recovering from trench fever were carriers of the pathogen for months or even years to form a source of further cases.

Rochalimaea quintana is spread among human populations by the body louse, *Pediculus humanus*, which acquires the pathogen when feeding on the blood of an infected person. *R. quintana* multiplies in the lumen of the midgut in the cuticular margin of the midgut epithelial cells (Wisseman, 1991). After 6 to 10 days *R. quintana* appears in the faeces of the louse and infection is caused either by the faeces being scarified into the skin or possibly by inhalation. The longevity of the louse is not affected by the presence of *R. quintana*, and remains infective for the rest of its life which, in an adult louse, would not exceed 5 weeks (Buxton, 1950). There is no transovarian transmission so that newly emerged lice are free from infection. However, since transmission of the pathogen is by the faeces of the louse, it is possible for new cases to arise for some time after elimination of the louse population. *R. quintana* remains viable in dry louse faeces for many months and possibly in excess of a year (Zdrodovskii and Golinevich, 1960).

Coxiella burnetii

Coxiella is a monotypic genus of which the sole representative is *C. burnetii*. Weiss *et al.* (1991) consider that it is not related to other Rickettsieae but is closer to *Wolbachia persica*, a symbiont of *Argas arboreus*, and *Legionella*. Described originally from Australia, *C. burnetii* has a worldwide distribution. In Queensland *C. burnetii* is circulated among bandicoots by the ticks, *Haemaphysalis humerosa* and *Ixodes holocyclus*, and among kangaroos by *Amblyomma triguttatum* (Fenner, 1990).

Human infection is by inhalation and is most common in people associated with domestic animals. It causes Q-fever, an acute, self-limiting disease of 3 to 6 days' duration with fever, severe headache and pneumonia occurring in about 60% of cases and hepatitis in a third. Most cases are inapparent or unrecognized and infection produces a solid, but non-sterile immunity and may produce endocarditis, a serious condition, 2 to 20 years later (Wisseman, 1991).

Coxiella burnetii is enzootic in cattle, sheep and goats, which shed large numbers of rickettsiae at parturition in the placenta and fetal fluids. It is normally regarded as being non-pathogenic for domestic livestock but there are records of very heavy infections causing abortion in sheep and goats (Radostits *et al.*, 1994). In 1978–1979 a vaccine was developed to protect abattoir workers who were succumbing to Q-fever when slaughtering infected feral goats in South Australia. Over the period 1981–1989 6000 abattoir workers were given complete, long-lasting protection (Fenner, 1990).

Coxiella burnetii is widely disseminated among wild mammals and birds and has been isolated from domesticated camels, buffaloes and horses. In invertebrates natural infections have been found in 32 species from 8 genera of Ixodidae; 5 species from 3 genera of Argasidae; 3 species of gamasid mites and the body louse, *Pediculus humanus* (Waag *et al.*, 1988). Ixodid ticks play a role in transmitting *C. burnetii* among wild and domestic animals.

Ehrlichieae (Ristic and Huxsoll, 1984)

Ehrlichieae are minute rickettsia-like organisms which are pathogenic for certain mammals, including humans. In the vertebrate, ehrlichiae grow in the cytoplasm of reticuloendothelial cells and not in erythrocytes. *Ehrlichia* spp. are found in leucocytes in the circulating blood and the vectors, where known, are ixodid ticks. *Cowdria ruminantium* is found in the vascular endothelial cells and is transmitted by species of *Amblyomma*. In the tick the ehrlichiae are transmitted trans-stadially but not transovarially.

Ehrlichia canis

Canine ehrlichiosis is a worldwide disease of dogs which is usually mild but in certain breeds, e.g. German shepherds, it can produce a severe haemorrhagic condition known as tropical canine pancytopenia which killed 200 to 300 military dogs in Vietnam and has been reported elsewhere in the tropics, Malaysia, Puerto Rico and Florida. *E. canis* can last for up to 5 years in dogs without clinical expression. The vector is the red dog tick, *Rhipicephalus sanguineus* (Huxsoll, 1990).

E. (= Cytoecetes) phagocytophila

This is transmitted by *Ixodes ricinus* and causes a mild disease in cattle and sheep in the United Kingdom, Ireland, Scandinavia, Sweden and Spain but, more importantly, it increases the susceptibility of lambs to staphylococcal infection and louping-ill (Radostits *et al.*, 1994). In Norway, infected lambs had lower liveweights in the autumn (Stuen *et al.*, 1992). *E. phagocytophila* may persist in the blood stream of sheep for life, forming a major source of infection in an enzootic area. The mortality rate is negligible in cattle, the main losses being due to abortion when pregnant cows become infected. Related organisms occur in Africa and India (Radostits *et al.*, 1994).

Other *Ehrlichia* spp.

Horses are infected by *E. equi* in the USA and Switzerland and by *E. risticii*, the causative agent of Potomac horse fever, in the USA, Canada and France (Ristic, 1990). The vectors are not known. *E. sennetsu* causes human ehrlichiosis in Japan and Malaysia. One hundred cases of a tick-associated human ehrlichiosis have been identified in 15 states in the USA but the agent has not been formally named (Fishbein, 1990).

Cowdria ruminantium

Cowdria ruminantium causes heartwater in cattle, sheep and goats in southern Africa where it is regarded as the most important tick-borne disease in the region. It also occurs in wild ruminants, including blesbok (*Damaliscus albifrons*), wildebeest (*Connochaetes gnu*) and springbok (*Antidorcas marsupialis*) (Provost and Bezuidenhout, 1987), which may become symptomless carriers. Peracute cases show only high fever, prostration and convulsions ending in death. In acute cases the disease lasts about 6 days with 50 to 90% mortality. *C. ruminantium* is found in the Afrotropical region and has been introduced into some islands in the West Indies. Twelve species of *Amblyomma* are known to be capable of transmitting *C. ruminantium*, of which *A. variegatum* is the most important and widely distributed vector in Africa. *A. maculatum* and *A. cajennense* have transmitted *C. ruminantium* experimentally but have not been implicated in outbreaks of disease (Walker and Olwage, 1987). There is evidence that in the vector *C. ruminantium* only invades the salivary glands, when the tick is feeding (Kocan and Bezuidenhout, 1987).

Bartonellaceae (Ristic and Kreier, 1984)

Members of the Bartonellaceae are polymorphic, often rod-shaped microorganisms which are distinguished from the Anaplasmataceae by cultural and structural characteristics. The Bartonellaceae have cell walls, as do the Rickettsiaceae, but Bartonellaceae can be grown on non-living media. The Bartonellaceae include two genera, *Bartonella* and *Grahamella*. *Grahamella* is found within the erythrocytes of voles and deer mice. Only one species of *Bartonella* has been described, *B. bacilliformis*, which is found on or in the erythrocytes and within the cytoplasm of endothelial cells of humans to whom it is highly pathogenic. It is also found in the vector, a phlebotomine sandfly, *Lutzomyia verrucarum*.

Bartonella bacilliformis (Schultz, 1968)

Bartonella bacilliformis is restricted to certain high mountain valleys in the western and central Cordilleras of the Andes in South America. *B. bacilliformis* is unusual in that it produces two strikingly different human diseases: a progressive anaemia with high mortality (40%), referred to as Oroya fever, and a benign cutaneous eruption known as Verruga peruana. These two conditions are referred to jointly as Carrion's disease in memory of Carrion, who in 1885 inoculated himself with the organisms of Verruga peruana, and developed and died from Oroya fever, thus tragically and dramatically proving that the same organism caused both diseases. During the building of the Trans-Andean railway in 1870, Oroya fever was responsible for 7000 deaths.

Carrion's disease is an anthroponosis. No animal reservoir host is known. Transmission is via the phlebotomine sandfly *Lutzomyia verrucarum*. *B. bacilliformis* is in the circulating blood and is picked up by *L. verrucarum* when it feeds on an infected human. There is no known cycle of development in the sandfly and the pathogen has been found in its gut and on its mouthparts. *L. verrucarum* is nocturnal in habit and the incidence of Carrion's disease among railway workers was greatly reduced by removing them from the high-level valleys before nightfall. The disease is endemic in certain mountain valleys between 750 and 2750 m in altitude where it has been known to persist for 300 years. Above that height the night temperatures are too low for *L. verrucarum* to be active and below 750 m the climate is too arid for the survival of *L. verrucarum*. However, that does not explain the absence of *L. verrucarum* and Carrion's disease from well-watered riverine habitats below 750 m.

It is worth noting that the essential features of the epidemiology of Carrion's disease, except for the role of the insect vector, were known by the end of the nineteenth century. It is instructive to read the following paragraph from Weinman and Kreier (1977):

> The history of bartonellosis is, in part, a scientific documentation for what seemed one of the most improbable of the New World marvels. The excellent monograph of Odriozola (1898) drew attention to a disease believed to occur only in Western South America. There it was restricted to certain mountain valleys and contracted only at night. This unique disease was said to exist in two forms, clinically distinct and apparently unrelated. One, an anaemia, occurring at times in epidemics, could be fulminant and kill in a few days. The other form was benign and distinguished by a skin eruption, the like of which was unknown outside the Andean valleys. To the surprise of the scientific community all of these facts have proven to be substantially correct.

Anaplasmataceae (Ristic and Kreier, 1984; Radostits *et al.*, 1994)

The Anaplasmataceae are very small, rickettsia-like particles occurring in or on the erythrocytes of vertebrates and transmitted by arthropods. They are obligate parasites, which multiply intracellularly by binary fission. They differ from the Bartonellaceae by having no cell wall and by not multiplying on non-living media.

Two of the four genera, *Anaplasma* and *Aegyptianella*, are important pathogens of domestic stock but not of humans.

Anaplasma

Species of *Anaplasma* are very small (0.3 to 1.0 μm in diameter) parasites of the erythrocytes of ruminants, especially bovids and cervids. *Anaplasma marginale* is a pathogen of cattle in the tropics and subtropics occurring in South Africa, Australia, Asia, South America, the former Soviet Union and the USA. It causes severe debility, anaemia, jaundice and abortion in adult cattle. Young animals are relatively resistant to infection but in cattle more than 3 years old a peracute condition may develop with death occurring within 24 h. When susceptible animals move into an infected area or the vector population expands into a previously disease-free area the morbidity rate can be very high and the mortality may be 50% or more. Recovered animals require prolonged convalescence and will continue to be infected for the rest of their lives.

Anaplasma is transmitted biologically by ticks and mechanically by blood-sucking flies, especially tabanids. In the USA three regions may be distinguished: West Coast where the vector is *Dermacentor occidentalis*; Intermountain West where *D. andersoni* is the vector; and the south-eastern where the vectors are tabanid flies. Trans-stadial transmission of *A. marginale* has been demonstrated in *Rhipicephalus sanguineus* ticks (Parker, 1982) and colonies of *A. marginale* were found in the gut tissues of engorged nymphs of *Dermacentor andersoni*, which gave rise to infected adults of both sexes (Kocan *et al.*, 1983).

In an area of Tanzania where the cattle tick, *Boophilus decoloratus*, was absent 44% of the cases of anaplasmosis were attributed to *Tabanus taeniola*. Although *T. taeniola* and *T. fraternus* were the most abundant of 12 species of *Tabanus*, cases of anaplasmosis correlated with *T. taeniola* and not with *T. fraternus* (Wiessenhutter, 1975). This difference may reflect a difference in behaviour if, for example, *T. fraternus* did not immediately seek another host after having its blood-feeding interrupted.

Anaplasma marginale has a wide range of hosts including zebu cattle, water buffalo, African antelopes, American deer and camels. The African buffalo (*Syncerus caffer*) is refractory to infection. In cattle *Anaplasma centrale* causes mild, inapparent disease. *A. ovis* occurs in Africa, the Mediterranean region, the former USSR and the USA where it usually produces subclinical infections in sheep and goats although a severe anaemia may develop especially in goats suffering from a concurrent disease.

Aegyptianella pullorum (Gothe and Kreier, 1977, 1984)

Aegyptianella pullorum formerly included in the Babesidae (Protozoa: Sporozoa) is now included in the Anaplasmataceae. *A. pullorum* infects a wide range of wild and domestic birds in the warmer parts of the world and has been recorded from Africa, Asia and southern Europe but is probably more widely distributed. In the host's erythrocytes, *A. pullorum* has a diameter of 0.3 to 4.0 μm.

Aegyptianella pullorum infects chickens, geese, ducks and quail, and the ostrich has been found infected naturally. Pigeons and turkeys are refractory to infection and guinea fowl of uncertain susceptibility. A range of wild birds have been infected experimentally but natural infections of wild birds need to be examined more closely before deciding that these are infections with *A. pullorum*. The effect of infection varies with the age of the bird. Fatalities occur in chickens up to the age of 4 weeks with mortality declining rapidly over that period. Poultry infected after the age of 12 weeks develop a low persistent parasitaemia. Fowls that have recovered clinically may remain infective to argasid ticks for up to 18 months.

The vectors of *A. pullorum* are ticks of the subgenus *Persicargas* of the genus *Argas*, in particular *A. (P.) persicus* and *A. (P.) walkerae*. Transmission in the argasid is both trans-stadial and transovarial, although only a small percentage of larvae are infected by the transovarial route which is regarded as being of little epizootological importance.

In *A. (P.) walkerae*, *A. pullorum* develops in the intestinal epithelium, haemocytes and salivary glands. In each of these three separate locations intensive multiplication of the parasite takes place and 30 days are required to complete the cycle. After an infected feed the parasites are to be found in the intestinal epithelium after 24 h and at the end of 14 days the intestinal cells are heavily parasitized. After 2 to 3 weeks the parasites appear in the haemocytes and multiply rapidly until the end of the fourth week, when the parasites appear in the salivary glands. This development can take place in all stages of the tick, larva, nymph and adult. Once infected, the tick remains infected for life, with infections even persisting for 2 years in starved nymphs and female ticks. *A. pullorum* is introduced into its bird host with the saliva of the feeding tick.

Eperythrozoon and *Haemobartonella* (Gothe and Kreier, 1977; Kreier and Ristic, 1984)

Haemobartonella are coccoid or rod-shaped organisms located on or within the erythrocytes while *Eperythrozoon* appears as rings or cocci on the erythrocytes or free in the plasma. They are worldwide in their distributions and have been found in domestic and wild mammals. *Eperythrozoon suis* causes icteroanaemia of swine, a disease of some economic importance in the United States. In cattle clinical disease due to *E. wenyonii* is uncommon (Radostits *et al.*, 1994). In sheep *E. ovis* causes anaemia which is more severe in young sheep and may be the principal cause of ill-thrift in lambs. Experimentally *E. ovis* has been transmitted by *Aedes camptorhynchus* and *Culex annulirostris* (Kabay *et al.*, 1991). *Haemobartonella muris* is spread by the rat louse, *Polyplax spinulosa*, both mechanically and biologically, and it is also transmitted by the rat flea, *Xenopsylla cheopis*. *Haemobartonella felis* is an exception in that it spreads readily by the oral route and during cat fights.

References

Anon. (1978) Louse-borne typhus in 1977: the decline continues. *WHO Chronicle* 32, 401–402.

Anon. (1981) Louse-borne typhus in 1979. *WHO Chronicle* 35, 188–189.

Audy, J.R. (1968) *Red Mites and Typhus.* Athlone Press, London.

Azad, A.F. (1990) Epidemiology of murine typhus. *Annual Review of Entomology* 35, 553–569.

Busvine, J.R. (1976) *Insects, Hygiene and History.* Athlone Press, London.

Buxton, P.A. (1950) *The Louse.* Edward Arnold, London.

Craufurd-Benson, H.J. (1946) Naples typhus epidemic 1943–4. *British Medical Journal* 1, 579–580.

Dun-Qing, W. (1988) Biosystematic problems in relation to the vectors of scrub typhus in China. In: Service, M.W. (ed.) *Biosystematics of Haematophagous Insects.* Clarendon Press, Oxford, pp. 177–190.

Fenner, F. (1990) *History of Microbiology in Australia.* Brolga Press, Canberra.

Fishbein, D.B. (1990) Human ehrlichiosis in the United States. In: Williams, J.C. and Kakoma, I. (eds) *Ehrlichiosis: a Vector-borne Disease of Animals and Humans.* Kluwer Academic, Dordrecht, pp. 100–111.

Gear, J.H.S. (1988) Other spotted fever group rickettsioses: clinical signs, symptoms and pathophysiology. In: Walker, D.H. (ed.) *Biology of Rickettsial Diseases.* CRC Press, Boca Raton, Florida, USA, pp. 101–114.

Gothe, R. and Kreier. J.P. (1977) *Aegyptianella, Eperythrozoon* and *Haemobartonella.* In: Kreier, J.P. (ed.) *Parasitic Protozoa,* vol. IV. Academic Press, New York, pp. 251–294.

Gothe, R. and Kreier. J.P. (1984) Genus II. *Aegyptianella* Carpano 1929, 12. In: Krieg, N.R. and Holt, J.G. (eds) *Bergey's Manual of Systematic Bacteriology.* Williams and Wilkins, Baltimore, pp. 722–723.

Hoogstraal, H. (1967) Ticks in relation to human diseases caused by *Rickettsia* species. *Annual Review of Entomology* 12, 377–420.

Hoogstraal, H. (1986) Theobald Smith: his scientific work and impact. *Bulletin of the Entomological Society of America* 32, 23–35.

Huxsoll, D.L. (1990) The historical background and global importance of ehrlichiosis. In: Williams, J.C. and Kakoma, I. (eds) *Ehrlichiosis a Vector-borne Disease of Animals and Humans.* Kluwer Academic, Dordrecht, pp. 1–8.

Kabay, M.J., Richards, R.B. and Ellis, T.E. (1991) A cross-sectional study to show *Eperythrozoon ovis* infection is prevalent in Western Australian sheep farms. *Australian Veterinary Journal* 68, 170–173.

Kocan K.M. and Bezuidenhout, J.D. (1987) Morphology and development of *Cowdria ruminantium* in *Amblyomma* ticks. *Onderstepoort Journal of Veterinary Research* 54, 177–182.

Kocan K.M., Holbert, D., Ewing, S.A., Hair, J.A. and Barron, S.J. (1983) Development of colonies of *Anaplasma marginale* in the gut of incubated *Dermacentor andersoni. American Journal of Veterinary Research,* 44, 1617–1620.

Kreier, J.P. and Ristic, M. (1984) Genus III. *Haemobartonella* Tyzzer and Weinman 1939, 143 and Genus IV. *Eperythrozoon* Schilling 1928, 1854. In: Krieg, N.R. and Holt, J.G. (eds) *Bergey's Manual of Systematic Bacteriology.* Williams and Wilkins, Baltimore, pp. 724–729.

Krieg, N.R. and Holt, J.G. (eds) (1984) *Bergey's Manual of Systematic Bacteriology.* Williams and Wilkins, Baltimore.

Manson-Bahr, P.E.C. and Bell, D.R. (1987) *Manson's Tropical Diseases.* Baillière Tindall, London.

Parker, R.J. (1982) The Australian brown dog tick *Rhipicephalus sanguineus* as an experimental parasite of cattle and vector of *Anaplasma marginale. Australian Veterinary Journal* 58, 47–50.

Provost, A. and Bezuidenhout, J.D. (1987) The historical background and global importance of heartwater. *Onderstepoort Journal of Veterinary Research* 54, 165–169.

Radostits, O.M., Blood, D.C. and Gay, C.C. (1994) *Veterinary Medicine – a Textbook of the Diseases of Cattle, Sheep, Pigs and Horses.* Baillière Tindall, London.

Rehacek, J. (1979) Spotted fever group rickettsiae in Europe. In: Rodriguez, J.G. (ed.) *Recent Advances in Acarology* II. Academic Press, New York, pp. 245–255.

Ristic, M. (1990) Current strategies in research on ehrlichiosis. In: Williams, J.C. and Kakoma, I. (eds) *Ehrlichiosis a Vector-borne Disease of Animals and Humans.* Kluwer Academic, Dordrecht, pp. 136–153.

Ristic, M. and Huxsoll, D.L. (1984) Tribe II, Ehrlichieae Philip 1957, 948. In: Krieg, N.R. and Holt, J.G. (eds) *Bergey's Manual of Systematic Bacteriology.* Williams and Wilkins, Baltimore, pp. 704–711.

Ristic, M. and Kreier, J.P. (1984) Bartonellaceae Gieszczykiewicz 1939, 25. In: Krieg, N.R. and Holt, J.G. (eds) *Bergey's Manual of Systematic Bacteriology.* Williams and Wilkins, Baltimore, pp. 717–719.

Schultz, M.G. (1968). A history of bartonellosis (Carrion's disease). *American Journal of Tropical Medicine and Hygiene* 17, 503–515.

Sonenshine, D.E., Bozeman, F.M., Williams, M.S., Masiello, S.A., Chadwick, D.P., Stocks, N.I., Lauer, D.M. and Elisberg, B.L. (1978) Epizootiology of epidemic typhus (*Rickettsia prowazekii*) in flying squirrels. *American Journal of Tropical Medicine and Hygiene* 27, 339–349.

Stuen, S., Hardeng, F. and Larsen, H.J. (1992) Resistance to tick-borne fever in young lambs. *Research in Veterinary Science* 52, 211–216.

Takahashi, M., Murata, M., Nogami, S., Hori, E., Kawamura, A. and Tanaka, H. (1988) Transovarial transmission of *Rickettsia tsutsugamushi* in *Leptotrombidium pallidum* successively reared in the laboratory. *Japanese Journal of Experimental Medicine* 58, 213–218.

Tanskul, P., Strickman, D., Watcharapichat, P., Kelly, D., Inlao, I. and Futrakul, A. (1992) Longitudinal study of *Rickettsia tsutsugamushi* (scrub typhus) in a suburban Bangkok community. *Abstracts XIIIth International Congress for Tropical Medicine and Malaria, Jomtien, Pattaya, Thailand, 29 November–4 December 1992*, vol. 2, p. 10.

Uchida, T. (1993) *Rickettsia japonica*, the etiologic agent of Oriental spotted fever. *Microbiology and Immunology* 37, 91–102.

Uchida, T., Uchiyama, T., Kumano, K. and Walker, D.H. (1992) *Rickettsia japonica* sp. nov., the etiological agent of spotted fever group rickettsiosis in Japan. *International Journal of Systematic Bacteriology* 42, 303–305.

Uchida, T., Yan, Y. and Kitaoka, S. (1995) Detection of *Rickettsia japonica* in *Haemaphysalis longicornis* ticks by restriction fragment length polymorphism of PCR product. *Journal of Clinical Microbiology* 33, 824–828.

Uchikawa, K., Kumada, N. and Yamada, Y. (1988) Studies on tsutsugamushi by Tullgren's funnel method 3. Occurrence of vector trombiculid, *Leptotrombidium pallidum* in the neighbourhood of patients' houses. *Japanese Journal of Sanitary Zoology* 39, 13–17.

Waag, D.M., Williams, J.C., Peacock, M.G. and Raoult, D. (1991) Methods of isolation, amplification and purification of *Coxiella burnetii*. In: Williams, J.C. and Thompson, H.A. (eds) *Q fever: The Biology of Coxiella burnetii.* CRC Press, Florida, pp. 73–115.

Walker, J.B. and Olwage, A. (1987) The tick vectors of *Cowdria ruminantium* (Ixodoidea, Ixodidae, genus *Amblyomma*) and their distribution. *Onderstepoort Journal of Veterinary Research* 54, 353–379.

Weinman, D. and Kreier, J.P. (1977) *Bartonella* and *Grahamella*. In: Kreier, J.P. (ed.) *Parasitic Protozoa*, vol. IV. Academic Press, New York, pp. 197–233.

Weiss, E. (1988) History of rickettsiology. In: Walker, D.H. (ed.) *Biology of Rickettsial Diseases*, vol. 1. CRC Press, Florida, pp. 15–32.

Weiss, E. and Moulder, J.W. (1984) Order 1. Rickettsiales Gieszckiewicz 1939, 25. In: Krieg, N.R. and Holt, J.G. (eds) *Bergey's Manual of Systematic Bacteriology*. Williams and Wilkins, Baltimore, pp. 687–704.

Weiss, E., Williams, J.C. and Thompson, H.A. (1991) The place of *Coxiella burnetii* in the microbial world. In: Williams, J.C. and Thompson, H.A. (eds) *Q Fever: The Biology of Coxiella burnetii*. CRC Press, Boca Raton, Florida, USA pp. 1–17.

Wiesenhutter, E. (1975) Research into the relative importance of Tabanidae (Diptera) in mechanical disease transmission. III. The epidemiology of anaplasmosis in a Dar-es-Salaam dairy farm. *Tropical Animal Health and Production* 7, 15–22.

Wigglesworth, V.B. (1941) The sensory physiology of the human louse *Pediculus humanus corporis* de Geer (Anoplura). *Parasitology* 33, 67–109.

Wisseman, C.L. (1991) Rickettsial infections. In: Strickland, G.T. (ed.) *Hunter's Tropical Medicine*, W.B. Saunders, Philadelphia, pp. 256–286.

Zdrodovskii, P.K. and Golinevich, H.M. (1960) *The Rickettsiial Diseases*. Pergamon Press, London.

Zinsser, H. (1935) *Rats, Lice and History*. Atlantic Monthly Press, Boston.

Relapsing Fevers, Borrelioses, Plague and Tularaemia

26

Relapsing Fevers (*Borrelia* spp.) (Kelly, 1984)

The Borrelias are loosely helically coiled, Gram-negative, motile, parasitic spirochaetes 3 to 20 μm long by 0.2 to 0.5 μm in diameter. They are the causative agents of tick-borne and louse-borne relapsing fever in humans. *Borrelia recurrentis* is the species transmitted by *Pediculus humanus*. Eighteen of the other 19 are transmitted by ticks, and the vector of one species is not known. Sixteen species are transmitted by argasid ticks. *B. anserina*, the type species and cause of avian spirochaetosis, is transmitted by species of *Argas* (*Persicargas*) and the other 15 by species of *Ornithodoros*. Two species, *B. theileri* and *B. burgdorferi*, are transmitted by ixodid ticks. *B. burgdorferi* is the causative agent of Lyme disease in humans.

Lyme Disease (*Borrelia burgdorferi*) (Lane *et al.*, 1991; Paleologo, 1991)

Lyme disease was first recognized in the United States in 1975, the main vector on the Atlantic coast, *I. dammini** identified in 1979, and the causative agent, *Borrelia burgdorferi*, isolated in 1981. The spirochaetes of *B. burgdorferi* are the longest and narrowest in the genus measuring 20–30 μm and 0.2–0.3 μm in diameter. Three stages of disease have been recognized: localized with flu-like symptoms and a characteristic skin rash occurring 1 to 3 weeks after infection; disseminated disease with neurological and/or cardiac symptoms usually weeks to months later; and a persistent stage with arthritic manifestations weeks to years later. Lyme disease has been recorded in 20 countries in North America, Europe, Asia and Australia. In Europe it is the most common arthropod-borne infection in humans. In the USA it occurs in 43 States in the north-east, mid-west and west with 80%

---

*Oliver *et al.* (1993) consider *I. dammini* to be conspecific with *I. scapularis*. In this chapter the name used by author(s) will be retained.

of the cases being in the north-east. In 1984–1986 1149 cases were reported in Connecticut. Asymptomatic infections in the USA are estimated at 10% and up to 50% in Europe.

The primary vectors of *B. burgdorferi* are members of the *Ixodes ricinus* complex: *I. dammini* in the north-east and upper mid-west of the USA; *I. pacificus* in the far west; *I. ricinus* in Europe and *I. persulcatus* in Asia. Ixodid ticks of four other genera (*Amblyomma, Dermacentor, Haemaphysalis, Rhipicephalus*) have been found infected with *B. burgdorferi* in nature but only in New Jersey has *A. americanum* been implicated with Lyme disease in humans. The other genera may be involved in maintaining the zoonosis. In the laboratory two strains of *B. burgdorferi* from Florida were transmitted by *I. scapularis* (= *I. dammini*) but not by *A. americanum* or *D. variabilis* (Sanders and Oliver, 1995).

In the virtual absence of transovarial transmission of *B. burgdorferi* in the tick, trans-stadial transmission is essential for a species to be a vector. This has been shown to occur in *I. hexagonus* which has a wide range of hosts but does not feed on humans (Gern *et al.*, 1991). The Australia *I. holocyclus* feeds readily on humans and in the laboratory ingested spirochaetes of a North American isolate, but there was no trans-stadial transmission. A different result may be obtained with an Australian isolate (Piesman and Stone, 1991). In eastern Germany the natural infection rate of *I. ricinus* (18%) was significantly higher than that found in *D. reticulatus* (9%) in the same locality and no infections were found in *H. concinna* (Kahl *et al.*, 1992).

Ixodes dammini and Lyme disease in the north-east USA

In north-east USA Lyme disease is associated with white-tailed deer (*Odoicoleus virginianus*), the white-footed mouse (*Peromyscus leucopus*) and *I. dammini*. *P. leucopus* is a major reservoir of *B. burgdorferi* and the main host for the immature stages of *I. dammini*, and *O. virginianus* the principal host for adult *I. dammini*. The life cycle of *I. dammini* takes 2 to 4 years. Larvae are active from August to October when they become infected by feeding on *P. leucopus*. Peak feeding of nymphs with an infection rate of 25% occurs from May to June, producing human cases which peak in July and augmenting the infection rate of *P. leucopus*, prior to their being hosts to uninfected larvae. Adults actively seek hosts in October to December, and if unsuccessful, renew the search in March to May with oviposition occurring in the summer.

A tick needs to feed for more than 48 h to transmit *B. burgdorferi* successfully. Adult *I. dammini* have infection rates of 50 to 100% but on humans they are likely to be found and removed within 48 h of attachment before achieving transmission. Nymphs are more likely to be overlooked and are considered to be the main source of human infections. In most *I. dammini*, ingested *B. burgdorferi* persist in the midgut and only in a small number (4 to 5%) of ticks does the infection spread to the haemocoele. It is possible that invasion of the haemocoele may not occur until the tick is feeding. If it does then transmission with the saliva would be possible. The alternative is transmission by regurgitation of *B. burgdorferi* from the midgut into the host (Burgdorfer and Hayes, 1990).

Ixodes dammini requires an environment with high humidity and is found in dense brush on islands off the north-east coast and in heavily forested areas inland. In Wisconsin it occurs in the ecotone between the northern coniferous-hardwood and the southern deciduous forests. Changing land use has provided ideal conditions for an explosion of deer, mice, ticks and Lyme disease by creating a mosaic of woods, ornamental plantings and lawns. Islands without deer had no *I. dammini*. Removal of deer from an island over 2 years reduced the tick population over the next 3 years.

Ixodes pacificus **and Lyme disease in western North America**

In western North America the primary vector of *B. burgdorferi* is *Ixodes pacificus*, which occurs from California to British Columbia and eastwards to Nevada, Utah and Idaho. It ranges from sea level to 2150 m and has been found on about 80 host species. Larvae and nymphs are active from March to June attaching to lizards, rodents and birds, and adults are found on large mammals such as bears, deer, dogs, horses and humans from November to May. The infection rate of adult *I. pacificus* with *B. burgdorferi* is low (1.5%). There are three reasons for this. A common host for the immature ticks is the western fence lizard (*Sceloporus occidentalis*) which is not susceptible to *B. burgdorferi*. Peak larval feeding occurs before that of nymphs, reducing the opportunity for larvae to feed on hosts recently infected by nymphs, and by nymphs feeding more frequently on lizards than on competent reservoir species such as the deer mouse (*Peromyscus maniculatus*) and the piñon mouse (*P. truei*).

Very high serum antibodies to *B. burgdorferi* or other borreliae have been found in brush rabbits (*Sylvilagus bachmani*) and jack rabbits (*Lepus californicus*) but the ticks associated with them, *D. perumapertus, H. leporispalustris* and *I. neotomae*, do not feed on humans. In Texas *B. burgdorferi* has been isolated from the cat flea (*Ctenocephalides felis*) and antibodies to it have been found in 31 out of 85 domestic cats.

Ixodes ricinus **and Lyme disease in Europe** (Gray, 1991)

Details of the transmission of Lyme disease in Europe are less well known than those operating in north-east America. The vector *Ixodes ricinus* occurs in a broad band of Europe eastwards to the Ural Mountains and south to the Mediterranean. Like *I. dammini*, non-parasitic stages of *I. ricinus* are found in deciduous woodland where there is high humidity (>80% RH) or in meadows and moorland with high rainfall. Its life cycle takes 3 years which may be extended to 6 years. All stages become actively questing in the spring and, if unsuccessful in finding a host, resume activity in the autumn, avoiding unfavourable summer conditions. The length of time over which activity can be sustained is greatest in highly sheltered locations and shortest in exposed meadows, where the tick's need to maintain its water balance will more rapidly deplete the tick's energy reserves. Behavioural diapause is shown by inactivity of unfed stages and by delayed engorgement of ticks on the host. Morphogenetic diapause occurs in egg development, in delayed ecdysis of the immature stages and slow maturation of the ovaries. The main hosts of larval *I. ricinus* are rodents but unlike *I. dammini* fewer nymphs fed on rodents,

the greater proportion feeding on other mammals, and the peak feeding activities
of larvae and nymphs are not well separated in time, reducing the opportunity for
larvae to feed on recently infected rodents.

Disease control

Avoidance of Lyme disease involves the same anti-tick precautions for protection
against *Rickettsia rickettsii*, i.e. avoidance of tick habitats, wearing of protective
clothing treated with a tick repellent. A simple method of reducing numbers of
I. dammini has been the provision of cardboard tubes filled with cotton wool
impregnated with permethrin. *P. leucopus* uses the cotton wool to line its nest,
where it kills larval and nymphal *I. dammini*. This technique reduced infected ticks
by 72% but it does not work where the tick host is the vole (*Microtus pennsylvanicus*) because voles do not use cotton wool as nesting material.

Epidemic Relapsing Fever (*Borrelia recurrentis*) (Buxton, 1950;
Geigy, 1968; Butler, 1991a)

Between 1910 and 1945 there were seven major epidemics of louse-borne relapsing
fever in Africa, eastern Europe and Russia with an estimated 15 million cases and
5 million deaths. From 1960 to 1979 louse-borne relapsing fever flourished in the
Sudan and Ethiopia, the latter country having the highest prevalence with an
estimated 10,000 cases per annum. The high incidence of the disease in Ethiopia
is reminiscent of the survival of the other louse-borne human disease, epidemic
typhus, in that country (see p. 519). The first attack of relapsing fever lasts about
6 days and the first relapse about 2 days with a 9-day period of remission in
between. In untreated cases mortality may be up to 40%. Treatment with tretra-
cycline or erythromycin can produce in most patients a short-lived, non-fatal severe
reaction involving a sharp rise in body temperature and a fall in blood pressure
while the borreliae are being cleared from the blood stream.

Borrelia recurrentis occurs in the circulating blood of an infected person from
where it can be acquired by *Pediculus humanus* when it feeds on that person. Most
of the spirochaetes die in the midgut of the louse but a few survive to penetrate
the midgut and reach the haemocoele, in which they multiply. From the sixth day
after an infective feed spirochaetes become increasingly abundant in the
haemolymph of the louse. They invade the neural ganglia and the muscles of the
head and thorax but have never been observed in other organs, e.g. salivary glands
or ovaries (Burgdorfer and Hayes, 1990). Their absence from those organs means
that *B. recurrentis* can neither be transmitted by the bite of the louse nor by transovarian transmission.

Few *B. recurrentis* are passed in the faeces of the louse and most of them are
moribund so that transmission via the faeces is highly unlikely. Transmission of
B. recurrentis occurs when a louse is crushed and the infected haemolymph released
on to the skin. The borreliae may be scratched into the skin but there is also
evidence that *B. recurrentis* can penetrate unbroken skin. This means that transmission involves the death of the louse and an individual louse can only infect one

person. Therefore louse-borne relapsing fever depends upon high louse populations to become epidemic in the human population. The conditions for this are
similar to those associated with epidemic typhus (see p. 520). Humans are the only
known reservoir for *B. recurrentis*. This raises the problem of the survival of *B.
recurrentis* during non-epidemic periods. The life of a louse is less than 2 months
and in the absence of transovarian transmission *B. recurrentis* cannot survive in
the louse population.

Endemic Relapsing Fever (*Borrelia duttoni*) (Geigy, 1968; Burgdorfer and Hayes, 1990)

Borrelia duttoni is the aetiological agent of endemic relapsing fever found in
eastern, central and southern Africa. It is named after J.E. Dutton, who died from
this disease. It is an anthroponosis, with humans being the only known host. The
vector is usually referred to as *Ornithodoros moubata* or one of its subspecies and
a note on the *O. moubata* complex is given below. Some populations of *O. moubata*
feed almost exclusively on wart hogs (*Phacochoerus*) or chickens and such populations are not infected with *B. duttoni*.

In the 1940s thousands of cases of endemic relapsing fever occurred in east
and central Africa. In 1946 there were 8200 cases admitted to hospital in East
Africa alone and many more thousands of cases remained in the villages and were
not reported (Walton, 1962). Of recent years the number of cases has fallen, partly
as a result of the use of synthetic insecticides and improved standards of living,
reducing human–tick contact.

When a tick feeds on a person infected with *B. duttoni*, it takes in borreliae
with the blood into the midgut. The borreliae then penetrate the midgut and enter
the haemocoele where they multiply and invade various organs including the central nerve mass, the coxal glands, and ovaries. Only in nymphal ticks do the
salivary glands become heavily infected and therefore only nymphal ticks can
transmit *B. duttoni* by bite. In adult ticks the salivary glands are virtually free of
spirochaetes and they transmit *B. duttoni* via infected coxal fluid. Borreliae are able
to enter a new host through skin abrasions or by direct penetration of the unbroken
skin.

In *O. moubata*, *B. duttoni* is passed both trans-stadially and transovarially with
filial infection rates being as high as 90%. *B. duttoni* can be transmitted transovarially through at least five generations, and hence centres of endemic relapsing
fever tend to persist for long periods of time. Transmission of *B. duttoni* to humans
is associated with populations of *O. moubata*, which are synanthropic and have
habits comparable to those of bedbugs. The ticks live and multiply in cracks and
crevices of human habitations from which they emerge at night to feed upon the
sleeping occupants. Survival of the disease in a location is aided by the long time
for which individual ticks may live. Burgdorfer (1976) refers to this ability of
Ornithodoros to survive for long periods and states that he has kept them alive for
15 years without feeding, but the species of *Ornithodoros* was not stated.

A Note on the *Ornithodoros moubata* Complex (Walton, 1962, 1979; Van der Merwe, 1968)

Walton studied *Ornithodoros moubata* in East Africa for many years and concluded that several different forms had been referred to the species *O. moubata*. He recognized four species with two subspecies: *O. compactus*, a parasite of tortoises in Cape Province of South Africa; *O. apertus*, a rare species associated with porcupines in one locality of Kenya; *O. moubata*, a species frequenting huts and the lairs of wild animals in arid conditions, defined as an annual rainfall of 50 mm or less; and a new species, *O. porcinus*, with two subspecies *O. p. porcinus* and *O. p. domesticus*.

O. p. porcinus is described as being exceedingly abundant in lairs of wild animals, particularly wart hog (*Phacochoerus*), and occurring on the humid Central African plateau between altitudes of 900 and 1500 m; *O. p. domesticus* is described as being completely interfertile with *O. p. porcinus*, but morphologically and physiologically distinct. Later Walton (1979) added *O. p. avivora*, a chicken-feeding subspecies present on the East African coast.

Van der Merwe (1968) examined a range of material, including some of Walton's, but came to different conclusions. Her studies were entirely morphological and she found difficulty in differentiating between the various forms. She concluded that *O. compactus* was a valid species, but that the other forms were within the single species *O. moubata* in which she recognized three subspecies: *O. m. moubata*, *O. m. apertus* and *O. m. porcinus*. Van der Merwe's *O. m. porcinus* included Walton's *O. porcinus porcinus* and his *O. p. domesticus*. Both workers recognized wild and domestic populations within their taxa: Van der Merwe in her *O. m. moubata* and *O. m. porcinus*, and Walton in his *O. moubata*.

From the medical point of view neither of these two classifications is particularly helpful in that wild and domestic populations are included within the one taxon. Since only domestic populations will play any role in the transmission of *B. duttoni* it would have been preferable, were it taxonomically possible, to have distinguished between wild and domestic populations, i.e. between vectors and non-vectors. Unfortunately that is not possible.

Other *Ornithodoros*-borne Relapsing Fevers (*Borrelia* spp.)

The quotation regarding tick-borne spotted fevers (see p. 522) can be adapted to cover relapsing fevers carried by *Ornithodoros* spp., to the effect that 'Whenever you have species of *Ornithodoros* biting humans there is the possibility of relapsing fever'. Species of *Ornithodoros*, known to be reservoirs or vectors of *Borrelia* species, occur in each major zoogeographical region with the exception of Australia, New Zealand and Oceania (Butler, 1991a). Ten species of *Borrelia*, transmitted among rodents and other small mammals by *Ornithodoros* ticks, can cause relapsing fever in humans. In North America the three main species are *B. hermsii*, *B. parkeri* and *B. turicatae*. Typical human infections involve three to seven attacks at 4- to 7-day intervals.

The generation time for *B. hermsii* in mice is approximately 6 h. Twenty-five different serotypes have been isolated within which there is no antigenic cross-reactivity and all can arise from a single variant (Barbour, 1990). There is considerable species specificity in *Borrelia–Ornithodoros* relationships (Hoogstraal, 1986). Borreliae from different vectors have been given specific status but further biochemical characterization and nucleic acid hybridization studies are needed to determine their validity (Kelly, 1984).

Most foci of *Ornithodoros*-borne borreliae exist in nature in restricted tick biotypes such as burrows, nests or caves. People become exposed to infection when visiting such areas or using caves as shelter. Such infections are accidental and play no role in the population dynamics of the zoonosis among the small mammal population. Two outbreaks of relapsing fever caused by *B. hermsii* in the western United States could be traced to people occupying log cabins infested with rodents and *O. hermsi* – 10 out of 20 boy scouts and scoutmasters who spent at least one night in an infested cabin developed relapsing fever. In 1973, 62 visitors and park service employees contracted relapsing fever caused by *B. hermsii* after spending one or more nights in log cabins in the North Rim Park area of the Grand Canyon in Arizona.

Borrelia hermsii is a zoonosis of chipmunks (*Eutamias* spp.) and pine squirrels (*Tamiasciuris* spp.). Transmission of rodent borreliae is either via the saliva or the coxal fluid. *Ornithodoros hermsi* secretes too little coxal fluid for it to be the route of infection. Some species of *Ornithodoros* do not secrete coxal fluid until they have left the host and consequently coxal fluid cannot be the route of infection. *O. hermsi* and *O. turicata* are highly efficient at transmitting borreliae through their saliva, and infection may result from a tick feeding for less than a minute.

Although transovarian transmission of borreliae is common among species of *Ornithodoros* it does not occur in all species, e.g. *O. rudis* does not transmit *B. venezuelensis* transovarially.

Avian borreliosis (*Borrelia anserina*) (Hoogstraal *et al.*, 1979)

Borrelia anserina is pathogenic for geese, ducks, turkeys, pheasants, canaries, chickens and grouse. Avian borreliosis occurs in southern Europe, the Middle East, Africa, India, Australia, South America and the western United States. The vectors of *B. anserina* are species of *Argas* (*Persicargas*), including *A.* (*P.*) *persicus* and *A.* (*P.*) *arboreus*. *A.* (*P.*) *persicus* has been conveyed by humans with domestic chickens to many parts of the tropics and subtropics.

The development of *B. anserina* in argasid ticks involves the borreliae penetrating the midgut and appearing in the haemocoele, in which they multiply and invade certain tissues including the central nerve mass, salivary glands and gonads. Heavy infections of *B. anserina* develop in *A.* (*P.*) *persicus* and *A.* (*P.*) *arboreus*, in which both trans-stadial and transovarial transmission occur (Diab and Soliman, 1977). Two other common Egyptian bird-parasitizing argasids, *A.* (*P.*) *streptopelia* and *A.* (*Argas*) *hermanni*, resist infection. They show only limited trans-stadial and no transovarial transmission (Zaher *et al.*, 1977).

Borrelia theileri (Hoogstraal, 1979; Burgdorfer and Hayes, 1990)

Borrelia theileri occurs in Africa, India, Indonesia, Australia and South America. It causes a mild disease in cattle and a febrile disease in horses. The vectors are species of *Boophilus* and in Africa it is also associated with *Rhipicephalus evertsi*. *Boophilus* spp. are one-host ticks and can only be vectors if there is transovarian transmission. *B. theileri* invades the haemocoele of the tick and infects the central ganglion and ovary from which rates of transovarian transmission as high as 80% have been recorded, but larval ticks seem to be incapable of transmitting *B. theileri*. In feeding nymphs and adults there is massive multiplication of *B. theileri* in the haemocytes and release into the haemolymph.

Plague (*Yersinia pestis*) (Twigg, 1978; Bercovier and Mollaret, 1984)

Plague is a disease of rodents caused by *Yersinia pestis*, to which humans are susceptible. Three biovars of *Y. pestis* have been defined on the basis of their geographical distributions and biochemical properties. They are biovar *antiqua* present in Central Asia and Central Africa; biovar *mediaevalis* present in Iran and the former Soviet Union and biovar *orientalis* which has a worldwide distribution. They are individually regarded as having been responsible for separate pandemics in the last 1500 years. A pandemic is an epidemic which is prevalent over a whole country, continent or widespread throughout the world at the same time.

Characteristics of *Yersinia pestis* (Butler, 1991b)

Yersinia pestis is a Gram-negative, facultatively anaerobic, non-motile micro-organism, which varies in shape from straight rods to coccobacilli. The characteristics expressed by *Y. pestis* are temperature-dependent being different at 37°C (human body temperature) compared to 28°C the optimum temperature for growth. *Y. pestis* produces a large number of antigens and toxins with its virulence depending on plasmid-mediated V and W antigens. The capsular envelope surrounding the organism produces fraction-1 antigen readily at 37°C but not at 28°C. This antigen has antiphagocytic activity. A powerful endotoxin causes fever, leucopenia, intravascular coagulation and tissue damage. Murine toxin, an exotoxin, is cardiotoxic to animals but its role in human disease is unclear. *Y. pestis* produces a coagulant enzyme at 28°C but not at 37°C. It has the effect of causing ingested blood to clot in the proventriculus of the flea producing blocked fleas, a prerequisite for the transmission of *Y. pestis* by fleas. The fact that the coagulant enzyme is not produced at 37°C is part of the explanation for the declining incidence of plague during the hot season.

Plague in humans (Butler, 1991b)

The most common clinical form of plague is an acute lymphadenitis, called bubonic plague. Less common forms include septicaemic, pneumonic and meningeal plague. During the incubation period *Y. pestis* proliferates in the regional lymph nodes for 2 to 8 days. Infected persons then suffer fever, headache and painful swellings (buboes) in the lymph nodes of groin, axilla or neck. This condition can proceed to septicaemic plague when there is massive growth of *Y. pestis* in the circulating blood, where it is readily available to blood-sucking fleas. To infect fleas successfully a high level of bacteraemia is required. Pneumonic plague is highly contagious by airborne transmission, in which insect vectors play no part. Mortality in untreated cases is estimated to exceed 50%. Early treatment with streptomycin can reduce this to less than 5%. Protection can be provided to persons at risk by a vaccine involving two shots at an interval of 1 to 3 months followed by 6-monthly booster shots.

The pandemics

Plague is considered to have originated in Central Asia from which it has spread virtually throughout the world. It is believed that plague had spread to Asia Minor, Egypt and Africa in pre-Christian times. In the first pandemic, plague spread from Arabia in AD 542 into the Byzantine empire of Justinian, throughout Europe and North Africa, and possibly also reached East Africa. The organism responsible is considered to be *Y. p. antiqua*.

The second pandemic, attributed to *Y. p. mediaevalis*, was the 'Black Death' of the Middle Ages which ravaged Europe from the fourteenth to the seventeenth centuries. [Butler (1991b) recognizes two pandemics during this period. One in the fourteenth century and another in the fifteenth to eighteenth centuries.] It originated in Central Asia, probably near Alma Ata, the modern capital of Kazakhstan, and spread eastwards to China, south to India and westwards to the Crimea, which it reached in 1346. From there it spread rapidly throughout Europe reaching France and England in 1348, and in 18 months had spread throughout the whole of England and southern Scotland. In two and a half years the human population had been reduced by a third. It took 200 years for the population of Europe to recover numerically from the ravages of the 'Black Death'.

The third pandemic, associated with *Y. p. orientalis*, is currently coming to an end. It originated in Yunnan Province of south-west China in 1892 and 2 years later reached Canton and Hong Kong, from which it spread rapidly throughout the world through the agency of ocean-going liners. It had spread to Calcutta by 1895, to the Transvaal of South Africa and Santos in Brazil in 1899, and by 1900 plague had reached San Francisco in North America, and Sydney in Australia (Pollitzer, 1954; Twigg 1978).

The third pandemic continues to produce cases of human plague in many parts of the world. In the 1960s Vietnam experienced 10,000 deaths a year from plague and in the 1970s more than a hundred cases in a year were reported from Brazil, Burma, Kenya and Sudan sometime during the decade. In 1980 to 1986 17 countries reported cases of human plague to WHO and nine of these reported more than a hundred, ranging from 148 cases in the USA to 927 in Tanzania. The current distribution throughout the world is shown in Fig. 26.1. It should be noted that the former USSR has not notified WHO of its cases of plague for 50 years, but it is unlikely to have been plague free (Velimirovic, 1990).

Plague took a heavy toll among overcrowded, underfed, poorly housed urban

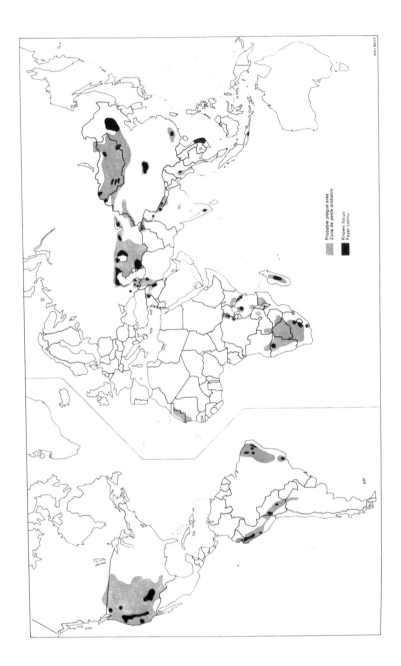

Fig. 26.1. Known and probable foci and areas of plague, 1959–1979. Reproduced, by permission, from *World Epidemiological Record*, 55(32) : 243 (1980).

populations of the East. In India, during the period 1898 to 1908, there were more than half a million deaths per annum from plague and in the succeeding decade the annual mortality from it was 420,000 (Pollitzer, 1954). This aroused the compassion of the international community and the British Plague Commission worked for many years combating this disease. At this time there were five national Plague Commissions in Bombay alone (Busvine, 1976).

Epidemiology of plague in India

The Plague Commission (early 1900s)

The essential features of the epidemiology of plague were established by the British Plague Commission. In nature plague spread among rats by the agency of rat fleas, and in the absence of fleas there was no epizootic among rats (Anon., 1910). The first sign of an outbreak of plague was an epizootic among the peridomestic brown rat *Rattus norvegicus*, with numerous deaths. This was followed in about 10 days by an epizootic among the domestic black rat *Rattus rattus* (Anon., 1907). After an interval of about 2 weeks human cases of plague appeared (Anon., 1910). In the winter of 1905/6 in Bombay peak mortality among *R. norvegicus* was reached on 18 February, peak mortality among *R. rattus* on 17 March, and human deaths reached a maximum on 28 April (Twigg, 1978).

The explanation of these observations was that infected rat fleas left the bodies of dead *R. norvegicus* and some transferred to *R. rattus*, infecting them with plague. The rat flea *Xenopsylla cheopis* readily bites people in the absence of its main host and infected *X. cheopis* would transfer from *R. rattus* to humans (Anon., 1910).

It was shown that epidemics of plague were more likely when the temperature was between 10 and 29°C (Anon., 1908) and that they declined rapidly at the onset of hot weather. Humidity also played a leading role in the epidemiology of plague. Between 1901 and 1910 there were four winters in which plague was severe and five in which only mild plague occurred. The average temperatures in mild and severe plague years did not differ but the humidities in severe plague years were considerably higher than in the years when there was less plague (Gloster *et al.*, 1917).

Brooks (1917) showed that the limiting factor was saturation deficit, which had a critical value of 10.2 mbar. Epidemic plague ceases when the temperature exceeds 26.7°C and the saturation deficit is more than 10.2 mbar; and rapidly wanes when the temperature is below 26.7°C but the saturation deficit exceeds 10.2 mbar. Conversely, plague epidemics may commence when the temperature is well above 26.7°C, providing the saturation deficit is less than 10.2 mbar.

Rogers (1928) extended these findings to take into account the climatic factors of the previous year, in particular temperature and saturation deficit during the hot weather and monsoon seasons. Hot, dry conditions act directly on the transmission of plague by reducing the life expectancy of infected fleas, and indirectly by reducing the flea population by causing higher mortality among larvae, and reducing the fecundity and survival of adult fleas.

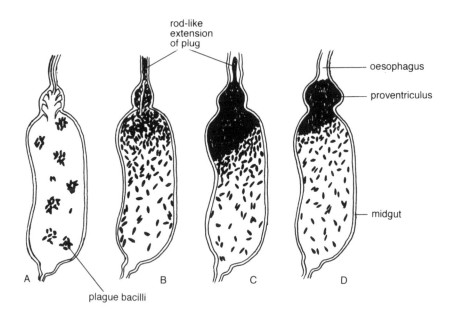

Fig. 26.2. Development of blockage in the alimentary tract in fleas. (**A**) Fed flea with midgut full of blood containing plague bacilli; (**B**) partially blocked flea; (**C**) and (**D**) formation of the plug. Source: from Bibikova, V.A. and Klassevskii, L.N. (1974) *Transmission of Plague by Fleas* (In Russian), Meditsina, Moscow.

More recent status of plague in Calcutta and Bombay

Calcutta and Bombay became free from plague in 1925 and 1935, respectively. Later both cities had epidemics in 1948 and 1949, and Bombay again in 1952. Investigations at this time showed that in the previous 50 years there had been a number of changes in the rodent and flea populations. By 1960 the lesser bandicoot rat (*Bandicota bengalensis*) had become the dominant rodent in both cities, forming 80% of the rodent population in Calcutta and 49% in Bombay. *B. bengalensis* is fairly susceptible to plague but carried fewer *X. cheopis* than *Rattus rattus*, and more *X. astia* which feeds on people less readily. In Calcutta the flea population was 66% *X. astia* and 34% *X. cheopis*. In Bombay, although *X. cheopis* was the dominant flea (76%) on *R. rattus* and *R. norvegicus*, both rats were highly resistant to plague (Seal, 1960b).

Development of *Y. pestis* in fleas

When a flea feeds, blood is propelled along the oesophagus to the midgut by the pharyngeal pump, with the proventriculus acting as a valve to prevent regurgitation. If the blood is infected, *Y. pestis* multiplies in the midgut and in the interstices of the spine-like epithelial cells of the proventriculus. The culture of *Y. pestis* in the flea's midgut forms a cohesive, gelatinous body which fills the midgut and proventriculus, effectively occluding the lumen of the latter. Such a flea is referred to as 'blocked' (Fig. 26.2).

When a blocked flea attempts to feed, the pharyngeal pump is unable to force blood through the proventriculus into the midgut and its pumping distends the oesophagus. When the pharyngeal pump ceases to work, the stretched oesophagus recoils and forces blood, contaminated with fragments of the gelatinous bacterial culture, into the host on which the flea has been attempting to feed (Bacot and Martin, 1914). Being unable to feed successfully a blocked flea will repeatedly attempt to feed and in so doing has the potential to infect many hosts. Being unable to take in liquid food a blocked flea is susceptible to desiccation and survives only a short time under hot, dry conditions.

Blockage of the gut is not necessarily permanent or fatal to the flea. A passage may be re-established through the plug, giving rise to a partially blocked flea. Such a flea is more dangerous than a fully blocked flea because not only is it able to feed and therefore live longer, but its proventriculus is unable to function as an effective valve, and infective material is regurgitated from the midgut into the flea's host (Bacot and Martin, 1914; Bacot, 1915).

The process of blockage in the flea is a complicated one and in only a percentage of infected fleas does it develop. Burroughs (1947) studied blockage in ten species of fleas and found that the percentage infected varied with the species from 0 to 50%, the latter being in *Xenopsylla cheopis*. Bibikova (1977) also working with *X. cheopis*, but having a rather larger sample, obtained 33 blocked fleas out of 214 fleas exposed (15%). Many factors influence the development of *Y. pestis* in a flea including: the strain of *Y. pestis*; the temperature at which the flea is held; the frequency and duration of feeding; and the specificity of the host (Gubareva *et al.*, 1976; Bibikova 1977).

From his experimental work Burroughs (1947) concluded that: a single flea was unlikely to transmit plague mechanically; and transmission by the ingestion of infected fleas, or by crushing such fleas or their faeces, into the skin was a rare occurrence of minor importance. When a flea feeds it deposits viscous faeces on the hair of its host, and any plague bacilli would be retained in the pellet and not come into contact with the host's skin. Under moderate temperatures *Y. pestis* would remain viable on the mouthparts of a flea for only 3 h which, for most species of fleas, is less than the interval between feeds, making mechanical transmission unlikely (Bibikova, 1977).

Rat fleas as vectors of plague (Burroughs, 1947; Pollitzer, 1954, 1960; Bahmanyar and Cavanaugh, 1976)

Xenopsylla cheopis has a worldwide distribution between 35°S and 35°N where it is a common ectoparasite of rats in cities, ports and rural situations. It is undoubtedly the most important vector of human plague throughout the world. In a comparison of ten species of fleas *X. cheopis* showed: the highest rate of blockage (58%); the highest ratio of transmissions (35 transmissions from 53 fleas); and the lowest rate of eliminating *Y. pestis* from its body to become plague free.

Xenopsylla brasiliensis is an ectoparasite of *Rattus* spp. in rural situations in Africa, South America and India. It is as efficient as *X. cheopis* as a vector of plague and is considered to be the major vector of plague to humans in rural situations in Africa and in the hilly, woody tracts of Bombay State in India.

Xenopsylla astia is a parasite of *Rattus* spp. in south-east Asia, where it occurs in fields, villages and ports. It is regarded as a mediocre vector of plague. It was the dominant flea of rats in Madras where in 1931 it formed 94% of the rodent flea population, and Madras was considered to be relatively plague free. (The other 6% were *X. cheopis*.) *X. astia* feeds more readily on rats than on humans, and will play a greater role in maintaining plague among the *Rattus* population than in transmitting it to people (Twigg, 1978).

The role of the human flea (*Pulex irritans*) is more debatable. It is reluctant to feed on rats, and when it becomes infected the gut rarely blocks. In one series of experiments only 1 out of 57 *P. irritans* became blocked (Burroughs, 1947). *P. irritans* is worldwide in distribution, and it is to be expected that there will be differences between populations in different parts of the world. In Brazil *P. irritans* was implicated in an epidemic of plague when 20 to 70% of *P. irritans* fed on infected *Cercomys* sp., became infected and transmitted *Y. pestis* to healthy *Cercomys* (Karimi *et al.*, 1974).

Ctenocephalides felis and *Ct. canis*, the dog and cat fleas, are weak vectors of *Y. pestis*. *Nosopsyllus fasciatus* is widely distributed throughout the world in cool, temperate areas and can maintain plague among *Rattus* spp. It is, however, reluctant to feed on people and human cases are rare in epizootics when *N. fasciatus* is the vector (Bahmanyar and Cavanaugh, 1976). *Leptopsylla segnis* is a cosmopolitan ectoparasite of mice, which also occurs on rats. It is less readily infected with *Y. pestis* than *N. fasciatus* and may play a minor role in plague epizootics but feeds only reluctantly on humans and is considered to play a negligible role in the transmission of plague to humans.

The role of *Echidnophaga gallinacea* is uncertain. It is a slightly better vector than *N. fasciatus* but considerably less efficient than *X. cheopis*. Although *E. gallinacea* is known as the sticktight flea of poultry it is found on a large range of mammals and birds, and is capable of transmitting *Y. pestis*. Nevertheless, its habit of attaching itself permanently to the head of its host and not detaching for some time after the host's death would militate against it being a significant vector. In addition mice, and presumably rats, rapidly clean fleas off their heads and devour them, which would reduce the probability of *E. gallinacea* being an effective vector of plague among rodents (Burroughs, 1947).

Strains within a species may differ in their ability to act as vectors of plague. Divergent results have been obtained with *Diamanus montanus*, a parasite of ground squirrels, which Wheeler and Douglas (1945) found to be a better vector of plague than *X. cheopis*, while Burroughs (1947) found it to be a very poor vector. There was also considerable difference in the vector efficiency of male and female *D. montanus* with females being very much more efficient (Wheeler and Douglas, 1945).

Sylvatic Plague (Twigg, 1978) (see Fig. 26.1)

A disturbing feature of the third pandemic has been the way in which plague has spread from *Rattus* spp. in relatively close association with humans to wild rodents in the field. Twigg (1978) refers to the first phase when plague is in *Rattus* spp. as the murine phase, and to the second phase in wild rodents as sylvatic plague, a term in common use. Sylvatic plague is now established in all continents except Australia. Some measure of the complexity of the epidemiology of plague can be given by the numbers of small mammals and their flea parasites which have been found infected. Excluding the cosmopolitan species of rats and commensal mice, 209 taxa (179 species and 30 subspecies) of Rodentia and 16 taxa (15 species and 1 subspecies) of Lagomorpha have either been found infected in nature or their ectoparasites have been positive for plague. Ninety-nine taxa (95 species and 4 subspecies) of fleas, associated with these small mammals, have been found infected with *Y. pestis* in nature (Pollitzer, 1960). In addition, in Vietnam and Kampuchea the major carrier of plague is a shrew, *Suncus murinus*, which is parasitized by *X. cheopis* (Twigg, 1978).

Plague in the USA (Craven *et al.*, 1993)

Within 40 years of plague being introduced into San Francisco on the west coast of the USA, it had crossed the continental divide and become firmly established east of the Rocky Mountains. This rapid spread was facilitated by the existence of continuous rodent populations. Plague is now enzootic in the USA west of 100°W where it has been recorded in 12 states. In the 22-year period 1970 to 1991 inclusive, 295 cases of plague were reported of which 166 (56%) were in New Mexico, 75 (25%) in Arizona and Colorado and a further 9% in California.

In New Mexico and Arizona human cases were associated with the rock squirrel (*Spermophilus variegatus*) and the antelope ground squirrel (*Ammospermophilus leucurus*) and in California with the ground squirrel (*Spermophilus beechyi*). The associated vector fleas were *Oropsylla* (*Diamanus*) *montana* on *S. variegatus* and *S. beechyi* and *Oropsylla* (*Thrassis*) *bacchi* on *A. leucurus*. In addition to flea transmission, domestic cats becoming infected by feeding on plague-infected rodents can transmit *Y. pestis* to humans by biting, scratching and coughing.

Sylvatic plague was originally a low risk disease resulting from chance exposure but with the human population extending into traditional enzootic areas there is now peridomestic transmission of plague. To minimize human infections populations of plague-prone rodents are kept under surveillance; effective flea and rodent control measures are being developed; and concerned citizens are being enrolled to monitor signs of peridomestic plague.

Southern Africa

In southern Africa plague is now enzootic in wild rodents over a large part of the country where the rainfall is less than 625 mm per annum. The main reservoir of plague is the gerbil, *Tatera brantsi*, which is host to three species of fleas of which *Xenopsylla philoxera* is the most important. Gerbil colonies experiencing an epizootic of plague decline over a period of many months and recovery of the population is even slower, with total recovery being long delayed. There are two routes through which plague may spread from gerbils to humans. The first is the more direct route from gerbils to domestic *Rattus rattus* and on to humans, and the other involves the peridomestic multimammate mouse *Mastomys natalensis* (= *M. coucha*) as an intermediate host between gerbils and *R. rattus*. Both *R. rattus* and *M. natalensis* are parasitized by *Xenopsylla brasiliensis*, an extremely efficient vector of plague (Davis, 1953).

Palaearctic region

In western Europe plague has not become established in wild rodents and sylvatic plague is unknown. In south-eastern Europe there are plague foci in the Caspian lowlands at the northern end of the Caspian Sea. Here the main reservoir for plague is the little sisel (*Citellus pygmaeus*), a relative of the ground squirrels of the USA. An extensive campaign, carried out to eradicate plague by controlling the little sisel, has involved the large-scale destruction of colonies of the rodent so that surviving colonies are isolated, and should plague break out in one colony it is unlikely to spread to others. These control measures have been greatly aided by changes in land usage, breaking the originally continuous suitable habitat into isolated small pockets. The campaign has been highly successful and it is confidently anticipated that enzootic plague can be completely eradicated from the area (Fenyuk, 1960).

In the adjoining region of the Middle East there is the Kurdistan focus which extends for 1000 km through the mountainous area of southern Turkey and northern Syria, Iraq and Iran. Here a close examination of the gerbil population revealed the existence of two species, *Meriones libycus* and *M. persicus*, which were resistant to plague, and two other species, *M. tristrami* and *M. vinogradovi*, which were very susceptible. The maintenance of plague in Kurdistan results from a balance between resistant and susceptible species (Baltazard *et al.*, 1960).

Plague is transmitted among the gerbils by *Xenopsylla buxtoni*. In this area there are no domestic rats and there is a break in the normal sequence of wild rodent to domestic rat to humans. It is considered that epidemics of plague result from rare cases of human plague contracted in the field, infecting *Pulex irritans*, which is abundant in the living quarters of the human population (Baltazard *et al.*, 1960). A similar plague focus based on gerbils is present in Transcaucasia, and there are reasons for regarding the Kurdistan and south-eastern European Russian foci as a single unit, the Kurdo-Caspian focus (Baltazard and Seydian, 1960).

There was a plague focus in Asiatic Russia, east of Lake Baikal in Transbaikalia, where the reservoir host was the Siberian marmot (*Marmota sibirica*). As a result of control measures between 1939 and 1955 against the marmot no plague has been recorded since 1946. The area is considered to be free from plague but vulnerable to its reintroduction from the adjoining Mongolian People's Republic where foci still exist (Baltazard and Seydian, 1960).

Oriental region

In the Oriental region sylvatic plague exists in India and in some countries of south-east Asia. In India the main reservoir host is the Indian gerbil (*Tatera indica*), which is not truly sedentary or sufficiently resistant to plague to form permanent pockets of infection. Baltazard and Bahmanyar (1960b) concluded that the lack of stubborn foci should ensure the disappearance from India of sylvatic plague. This appears to be taking place. In 1951 sylvatic plague was present in foci in northern, central and southern India but by 1969 central India was free, and in southern India only one of three foci was still active (Baltazard and Bahmanyar, 1960b; Seal, 1960a). Plague continues in the foothills of the Himalayas in Nepal and northern

India. In Burma and Vietnam plague has continued to be active since it was introduced 17 years ago.

After a silent period of 7 years human plague reappeared in Java in Indonesia, where the situation is unusual, in that the wild reservoir host is another species of *Rattus*, *R. exulans* (Velimirovic, 1972). In mobility and susceptibility to plague *R. exulans* is comparable to *T. indica* and unlikely to create permanent foci of infection. The domestic rat in Java is *R. rattus diardi* which is highly susceptible to plague (Baltazard and Bahmanyar, 1960a). A similar situation exists in Hawaii where sylvatic plague is maintained in *Rattus hawaiiensis* by its parasite *Xenopsylla hawaiiensis* (Twigg, 1978).

South America

In South America foci of sylvatic plague are present in Brazil, Argentina, Bolivia, Paraguay, Peru and Ecuador (Anon., 1970).

Survival of *Y. pestis* (Anon., 1970)

Plague can survive the hot dry season in India by persisting in aestivating Indian gerbils (Baltazard and Bahmanyar, 1960b). It can also overwinter in hibernating rodents, and latent infections in rodents can relapse, later become active and initiate an epizootic. It is possible for flea larvae to ingest *Y. pestis* when feeding on the faeces of infected adult fleas. When larvae of *N. fasciatus*, which had been exposed to infected flea faeces, were examined few were positive, and the few bacilli they contained showed no signs of multiplication (Bacot, 1914).

The survival of blocked fleas is dependent on external conditions and at 27°C blocked *X. cheopis* survived an average of 10 days (Pollitzer, 1954). Infected fleas, not necessarily blocked, can survive for long periods in the favourable microclimate of a rodent burrow. Davis (1953) reported finding infected *X. philoxera* in gerbil burrows which had been deserted for up to 4 months. There is evidence that infected fleas may live for at least a year, and some species for as long as 4 years (Anon., 1970), but the latter claim is refuted by Kir'yakova (1973). In addition plague bacilli can survive, and indeed multiply, in the soil layers of a rodent burrow where microclimatic and other conditions are favourable. Bacilli in soil have infected healthy rodents re-occupying burrows which had been vacant for at least 11 months (Anon., 1970).

Tularaemia (*Francisella tularensis*) (Eigelsbach and McGann, 1984; Sandford, 1991)

Francisella tularensis is a Gram-negative, obligately aerobic, pleomorphic, non-motile coccobacillus which causes tularaemia in humans and many other warm-blooded animals including sheep, horses, pigs, cattle and birds. Tularaemia is ubiquitous in the northern hemisphere between latitudes 30° and 71° north, occurring in Japan, Russia, Canada, Mexico and all the mainland states of the USA but not on the Iberian Peninsula or in Great Britain. Two serologically identical biovars are recognized: *F. tularensis tularensis* (= *nearctica*) and *F. t. palaearctica* (= *holarctica*). *F. t. tularensis* is the more virulent biovar and is found only in North America, while *F. t. palaearctica*, which causes a milder disease, occurs in both the Palaearctic and Nearctic regions.

Francisella tularensis can be transmitted from host to host by a variety of routes including blood-sucking arthropods, water, food and inhalation. In addition the organism possesses the potent property of being able to penetrate the unbroken skin, and therefore can be transmitted by contact with infected material. It usually produces a marked reaction at the site of entry, which in 70 to 80% of cases is an ulcer.

Some 100 species of wild mammals, 25 species of birds and more than 50 species of arthropods have been found naturally infected. In North America *F. t. tularensis* occurs particularly in lagomorphs and wild rodents and is transmitted by the bites of ticks (*Amblyomma americanum*, *Dermacentor andersoni*, *D. variabilis*, *Haemaphysalis leporispalustris*) and tabanid flies. *F. t. palaearctica* is associated with lemmings in Sweden, jackrabbits in Utah and hares and voles in Russia, where *D. nuttalli* is a vector (Thorpe *et al.*, 1965; Kloch *et al.*, 1973; Antisiferov *et al.*, 1976).

The ability of *F. tularensis* to survive away from its vertebrate host favours mechanical transmission by blood-sucking flies, especially tabanids. *Chrysops discalis* is recognized as being a major mechanical vector of *F. tularensis* in the USA. Half (19/39) of the human cases of tularaemia in Utah in the summer of 1971 could be attributed to the biting of *C. discalis* and another quarter (9) were suspected of being due to *C. discalis* (Kloch *et al.*, 1973).

Infections of *F. tularensis* in ticks are long-lasting. The organism multiplies in the midgut epithelium and haemolymph of the tick and is transferred when the tick is feeding (Ol'sufyev, 1978, personal communication). Transovarian transmission has been demonstrated in *D. andersoni* and other ixodid ticks in North America (Guerrant *et al.*, 1976).

Tularaemia is most severe in sheep in which the morbidity rate in North America may be as high as 40%, with a mortality of 50% especially in young animals. In horses there is fever and foals are more seriously affected than older animals, and in swine tularaemia causes fever in piglets but is latent in adult pigs (Radostits *et al.*, 1994).

Treatment with streptomycin has reduced the mortality rate of 7% in humans to virtually zero. Human tularaemia is an acute, febrile infectious zoonotic disease with an incubation period of 3 to 4 days. The introduction of as few as ten organisms subcutaneously can cause disease in non-vaccinated individuals. In the former USSR there were 10,000 cases a year before vaccination was introduced, which reduced the incidence to less than 200 per annum (Ol'sufyev, 1978, personal communication).

References

Anon. (1907) Reports on plague investigations in India. 22. Epidemiological observations made by the Commission in Bombay City. *Journal of Hygiene* 7, 724–798.

Anon. (1908) Reports on plague investigations in India. 31. On the seasonal prevalence of plague in India. *Journal of Hygiene* 8, 266–301.

Anon. (1910) Reports on plague investigations in India. 39. Interim report of the advisory committee for plague investigations in India. *Journal of Hygiene* 10, 566–568.

Anon. (1970) WHO expert committee on plague – fourth report. WHO *Technical Reports Series* 447, 1–25.

Antisiferov, M.I., Zykina, N.A., Sizykh, L.V., Charnaya, T.G., Zharov, V.R., El'shanskaya, N.I. and Evdokimov, A.V. (1976) Establishment of natural nidality of tularaemia in the Tuva ASSR. *Zhurnal Microbiologii, Epidemiologii i Immunobologii* 2, 5–8.

Bacot, A.W. (1914) On the survival of bacteria in the alimentary canal of fleas during metamorphosis from larvae to adult. *Journal of Hygiene* 13, *Plague Supplement* III, 655–664.

Bacot, A.W. (1915) Further notes on the mechanism of the transmission of plague by fleas. *Journal of Hygiene* 13, *Plague Supplement* IV, 774–776.

Bacot, A.W. and Martin, C.J. (1914) Observations on the mechanism of the transmission of plague by fleas. *Journal of Hygiene* 13, *Plague Supplement* III, 423–439.

Bahmanyar, M. and Cavanaugh, D.C. (1976) *Plague Manual.* World Health Organization, Geneva.

Baltazard, M. and Bahmanyar, M. (1960a) Recherches sur la peste à Java, *Bulletin of the World Health Organization* 23, 217–246.

Baltazard, M. and Bahmanyar, M. (1960b) Recherches sur la peste en Inde. *Bulletin of the World Health Organization* 23, 169–215.

Baltazard, M. and Seydian, B. (1960) Enquête sur les conditions de la peste au Moyen-Orient. *Bulletin of the World Health Organization* 23, 157–167.

Baltazard, M., Bahmanyar, M., Mostachfi, P., Eftekhari, M. and Mofidi, Ch. (1960) Recherches sur la peste en Iran. *Bulletin of the World Health Organization* 23, 141–155.

Barbour A.G. (1990) Multiphasic antigenic variation in the bacterium that causes relapsing fever. In: Van Der Ploeg, L.H.T., Cantor, C.R. and Vogel, H.J. (eds) *Immune Recognition and Evasion: Molecular Aspects of Host–Parasite Interaction.* Academic Press, pp. 183–199.

Bercovier, H. and Mollaret, H.H. (1984) Genus XIV. *Yersinia* Van Loghem 1944, 15. In: Krieg, N.R. and Holt, J.G. (eds) *Bergey's Manual of Systematic Bacteriology.* Williams and Wilkins, Baltimore, pp. 498–506.

Bibikova, V.A. (1977) Contemporary views on the interrelationships between fleas and the pathogens of human and animal diseases. *Annual Review of Entomology* 22, 23–32.

Brooks, R. St J. (1917) The influence of saturation deficiency and of temperature on the course of epidemic plague. *Journal of Hygiene* 15, *Plague Supplement* V, 881–899.

Burgdorfer, W. (1976) The epidemiology of the relapsing fevers. In: Johnson R.C. (ed.) *The Biology of Parasitic Spirochetes.* Academic Press, New York, pp. 191–200.

Burgdorfer, W. and Hayes, S.F. (1990) Vector–spirochaete relationships in louse-borne and tick-borne borrelioses with emphasis on Lyme disease. *Advances in Disease Vector Research* 6, 127–150.

Burroughs, A.L. (1947) Sylvatic plague studies. The vector efficiency of nine species of fleas compared with *Xenopsylla cheopis. Journal of Hygiene* 45, 371–396.

Busvine, J.R. (1976) *Insects, Hygiene and History.* Athlone Press, London.

Butler, T. (1991a) Relapsing fever. In: Strickland, G.T. (ed.) *Hunter's Tropical Medicine.* Saunders, Philadelphia, pp. 312–317.

Butler, T. (1991b) Plague. In: Strickland, G.T. (ed.) *Hunter's Tropical Medicine.* Saunders, Philadelphia, pp. 408–416.

Buxton, P.A. (1950) *The Louse.* Edward Arnold, London.

Craven, R.B., Maupin, G.O., Beard, M.L., Quan, T.J. and Barnes, A.M. (1993) Reported cases of human plague infections in the United States, 1970–1991. *Journal of Medical Entomology* 30, 758–761.

Davis, D.H.S. (1953) Plague in South Africa: a study of the epizootic cycle in gerbils (*Tatera brantsi*) in the northern Orange Free State. *Journal of Hygiene* 51, 427–449.

Diab, F.M. and Soliman, Z.R. (1977) An experimental study of *Borrelia anserina* in four species of *Argas* ticks. 1. Spirochete localisation and densities. *Zeitschrift für Parasitenkunde* 53, 201–212.

Eigelsbach, H.T. and McGann, V.G. (1984) Genus *Francisella* Dorofe'ev 1947, 176. In: Krieg, N.R. and Holt, J.G. (eds) *Bergey's Manual of Systematic Bacteriology*. Williams and Wilkins, Baltimore, pp. 394–399.

Fenyuk, B.K. (1960) Experience in the eradication of enzootic plague in the north-west part of Caspian region of the USSR. *Bulletin of the World Health Organization* 23, 263–273.

Geigy, R. (1968) Relapsing fevers. In: Weinman, D. and Ristic, M. (eds) *Infectious Blood Diseases of Man and Animals*. Academic Press, New York, vol. II, pp. 175–216.

Gern, L., Toutoungi, L.N., Chgang Min Hu and Aeschlimann, A. (1991) *Ixodes* (*Pholeoixodes*) *hexagonus*, an efficient vector of *Borrelia burgdorferi* in the laboratory. *Medical and Veterinary Entomology* 5, 431–435.

Gloster, T.H., White, F.N., Mukhari, A.N., Chaudhuri, J.S.R., Mitra, C.C., Mandal, G.C. and Ram, M. (1917) Epidemiological observations in the United Provinces of Agra and Oudh, 1911–1912. *Journal of Hygiene* 15, *Plague Supplement* V, 793–880.

Gray, J.S. (1991) The development and seasonal activity of the tick *Ixodes ricinus*: a vector of Lyme borreliosis. *Review of Medical and Veterinary Entomology* 79, 323–333.

Gubareva, N.P., Akiev, A.K., Zemel'man, B.M. and Abdulrakhmanov, G.A. (1976) The effect of some factors on block-formation in the fleas *Ceratophyllus tesquorum* Wagn. 1898 and *Neopsylla setosa setosa* Wagn., 1898. *Parazitologiya* 10, 315–319.

Guerrant, R.L., Humphries, M.K., Butler, J.E. and Jackson, R.S. (1976) Tickborne oculoglandular tularaemia. *Archives of Internal Medicine* 136, 811–813.

Hoogstraal, H. (1979) Ticks and spirochetes. *Acta Tropica* 36, 133–136.

Hoogstraal, H. (1986) Theobald Smith his scientific work and impact. *Bulletin of the Entomological Society of America*, 32, 23–35.

Hoogstraal, H., Clifford, C.M., Keirans, J.E. and Wassef, H.Y. (1979) Recent developments in biomedical knowledge of *Argas* ticks (Ixodoidea: Argasidae). In: Rodriguez J.G. (ed.) *Recent Advances in Acarology*. Academic Press, New York, vol. II, pp. 269–278.

Kahl, O., Janetzki, C., Gray, J.S., Stein, J. and Bauch, R.J. (1992) Tick infection rates with *Borrelia*: *Ixodes ricinus* versus *Haemaphysalis concinna* and *Dermacentor reticulatus* in two locations in eastern Germany. *Medical and Veterinary Entomology* 6, 363–366.

Karimi, Y., Eftekhari, M. and Almeida, C.R. (1974) Sur l'écologie des puces impliquées dans l'épidémiologie de la peste et le rôle éventuel de certains insectes hématophages dans son processus au nord-est du Brésil. *Bulletin de la Société de Pathologie Exotique* 67, 583–591.

Kelly, R.T. (1984) Genus IV. *Borrelia* Swellengrebel 1907, 582. In: Krieg, N.R. and Holt, J.G. (eds) *Bergey's Manual of Systematic Bacteriology*. Williams and Wilkins, Baltimore, pp. 57–62.

Kir'yakova, A.N. (1973) On the length of life of fleas in burrows. *Parazitologiya* 7, 261–263.

Kloch, L.E., Olsen, P.F. and Fukushima, T. (1973) Tularaemia epidemic associated with the deerfly. *Journal of American Medical Association* 226, 149–152.

Lane, R.S., Piesman, J. and Burgdorfer, W. (1991) Lyme borreliosis: relation of its causative agent to its vectors and hosts in North America and Europe. *Annual Review of Entomology* 36, 587–609.

Oliver, J.H., Owsley, M.R., Hutcheson, H.J., James, A.M., Chen, C., Irby, W.S., Dotson, E.M. and McLain, D.K. (1993) Conspecificity of the ticks *Ixodes scapularis* and *I. dammini* (Acari: Ixodidae). *Journal of Medical Entomology* 30, 54–63.

Paleologo, F.P. (1991) Lyme disease. In: Strickland, G.T. (ed.) *Hunter's Tropical Medicine*. Saunders, Philadelphia, pp. 324–331.

Piesman, J. and Stone, B.F. (1991) Vector competence of the Australian paralysis tick, *Ixodes holocyclus*, for the Lyme disease spirochete *Borrelia burgdorferi*. *International Journal for Parasitology* 21, 109–111.

Pollitzer, R. (1954) *Plague*. WHO, Geneva.

Pollitzer, R. (1960) Review of recent literature on plague. *Bulletin of the World Health Organization* 23, 313–400.

Radostits, O.M., Blood, D.C. and Gay, G.C. (1994) *Veterinary Medicine – a Textbook of the Diseases of Cattle, Sheep, Pigs and Horses*. Baillière Tindall, London.

Rogers, Sir Leonard (1928) The yearly variations in plague in India in relation to climate: forecasting epidemics. *Proceedings of the Royal Society of London B* 103, 42–72.

Sanders, F.H. and Oliver, J.H. (1995) Evaluation of *Ixodes scapularis, Amblyomma americanum* and *Dermacentor variabilis* (Acari: Ixodidae) from Georgia as vectors of a Florida strain of the Lyme disease spirochete, *Borrelia burgdorferi*. *Journal of Medical Entomology* 32 (in press).

Sanford, J.P. (1991) Tularaemia. In: Strickland, G.T. (ed.) *Hunter's Tropical Medicine*. Saunders, Philadelphia, pp. 416–417.

Seal, S.C. (1960a) Epidemiological studies of plague in India. 1. The present position. *Bulletin of the World Health Organization* 23, 283–292.

Seal, S.C. (1960b) Epidemiological studies of plague in India. 2. The changing pattern of rodents and fleas in Calcutta and other cities. *Bulletin of the World Health Organization* 23, 293–300.

Thorpe, B.D., Sidwell, R.W., Johnson, D.E., Smart, K.L. and Parker, D.D. (1965) Tularaemia in the wildlife and livestock of the Great Salt Lake Desert Region, 1951 through 1964. *American Journal of Tropical Medicine and Hygiene* 14, 622–637.

Twigg, G.l. (1978) The role of rodents in plague dissemination: a worldwide review. *Mammal Review* 8, 77–110.

Van der Merwe, S. (1968) Some remarks on the 'tampans' of the *Ornithodoros moubata* complex in southern Africa. *Zoologischer Anzeiger* 181, 280–289.

Velimirovic, B. (1972) Plague in South-East Asia. *Transactions of the Royal Society of Tropical Medicine and Hygiene* 66, 479–504.

Velimirovic, B. (1990) Plague and Glasnost. First information about human cases in the USSR in 1989 and 1990. *Infection* 18, 388–393.

Walton, G.A. (1962) The *Ornithodoros moubata* superspecies problem in relation to human relapsing fever epidemiology. *Symposia of the Zoological Society of London*, 6, 83–156.

Walton, G.A. (1979) A taxonomic review of the *Ornithodoros moubata* (Murray) 1877 (*sensu* Walton, 1962) species group in Africa In: Rodriguez J.G. (ed.) *Recent Advances in Acarology*. Academic Press, New York, vol. 11, pp. 491–500.

Wheeler, C.M. and Douglas, J.R. (1945) Sylvatic plague studies. V. The determination of vector efficiency. *Journal of Infectious Diseases* 77, 1–12.

Zaher, M.A., Soliman, Z.R. and Diab, F.M. (1977) An experimental study of *Borrelia anserina* in four species of *Argas* ticks. 2. Transstadial survival and transovarial transmission. *Zeitschrift für Parasitenkunde* 53, 213–223.

Malaria (*Plasmodium*) and Other Haemospororina (Sporozoa)

27

The Apicomplexa are parasitic Protozoa which lack cilia and flagella, except in some male gametes. Two subclasses, the Coccidiasina and Piroplasmasina, are of interest to the medical entomologist. The Coccidiasina has one suborder, the Haemospororina, members of which undergo asexual reproduction (merogony) in vertebrate erythrocytes and sexual reproduction (gametogony) and sporozoite formation (sporogony) in blood-sucking insects (Diptera) (Cox, 1991; Peters and Gilles, 1995).

The Haemospororina contains three families: the Plasmodiidae, the Haemoproteidae and the Leucocytozoidae (Garnham, 1966). The most important of these is the Plasmodiidae, which contains the genus *Plasmodium* and includes the four species responsible for human malaria. In *Plasmodium*, schizogony occurs in the blood; gametocytes develop in mature erythrocytes; the end product of the digestion of haemoglobin is a dark pigment, haemozoin; and the vectors are mosquitoes. In the other two families, schizogony does not occur in the blood, and the vectors are Diptera other than mosquitoes.

In the Haemoproteidae, haemozoin is produced, and the gametocytes develop in mature erythrocytes. In the Leucocytozoidae no haemozoin is produced, and the gametocytes are not found in mature erythrocytes. Two genera of Haemoproteidae, *Haemoproteus* and *Hepatocystis*, and one of Leucocytozoidae, *Leucocytozoon*, are of minor veterinary importance.

In the Piroplasmasina pigment is absent; merogony occurs in vertebrate erythrocytes or lymphocytes; and sexual development and sporozoite formation in ticks. Two families, the Babesiidae and Theileriidae, are of considerable veterinary importance, with species of *Babesia* and *Theileria* causing severe disease in livestock (Cox, 1991).

The Haemospororina will be dealt with in this chapter and the Piroplasmasina in the next.

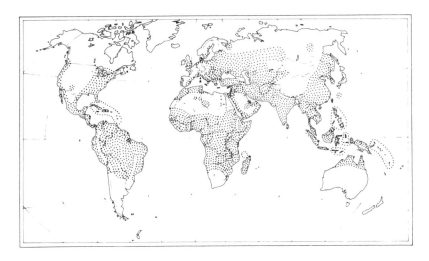

Fig. 27.1. Geographical distribution of malaria in the mid-nineteenth century. Malarious areas stippled. Source: from Wernsdorfer (1980).

Distribution and Incidence of Malaria in the World
(López-Antuñano and Schmunis, 1993)

Malaria is the most widespread and persistent disease which affects human populations throughout the world. It is estimated that out of a world population of 5 billion people, 110 million (2%) develop clinical disease each year and 280 million (5.6%) carry the parasite. Malaria is an ancient disease, recognized by Hippocrates about 400 BC, who described the three characteristic stages of an attack – chilly rigor, high fever and profuse sweating.

In the nineteenth century malaria was widely distributed throughout the world and even in 1950, 64% of the world's population (excluding China) was at risk of contracting malaria, which was widespread throughout the Afrotropical, Oriental and Palaearctic regions. It was present in northern Australia, and in western Europe, extended as far north as Arkhangelsk (64°N). In the Americas the endemic area embraced South America north of 32°S; Central America; a narrow belt west of the Rocky Mountains extending to Vancouver in Canada; and a broader belt on the eastern coast of North America extending into south-east Canada (Fig. 27.1).

Through the strenuous efforts of the World Health Organization, considerable progress has been made in reducing the incidence and distribution of malaria (Fig. 27.2). It is estimated that 27% of the world's population live in areas in which malaria has never existed or from which it disappeared without intervention; 30% in areas rendered malaria free; 34% in areas of reduced incidence and the other 9%, mainly in the Afrotropical region, live in areas where the level of endemic malaria has remained unchanged. It is estimated that the 515 million people in the Afrotropical region suffered 90 million cases of malaria in 1990, a case rate of 17%.

Excluding Africa, 4.7 million cases of malaria were reported to WHO in 1990

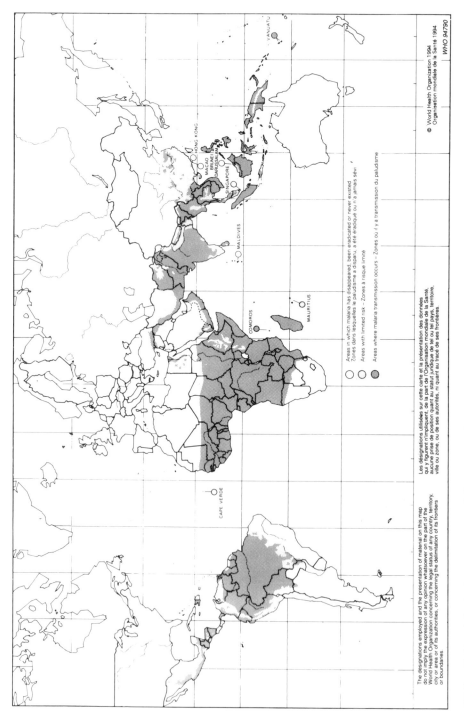

Fig. 27.2. Epidemiological assessment of the status of malaria, 1992. Reproduced by permission, from *World Malaria Situation, 1992. Parts I and II. Weekly Epidemiological Record. 69* (42–43): 309–314; 317–321 (1994).

of which 38% were in India and 18.5% in other countries of the Oriental region. In South America Brazil reported 560,000 cases (just over 52%) of those reported for the whole region. Papua New Guinea reported more than 100,000 cases and Afghanistan more than 300,000.

In endemic areas infants have passive immunity from antibodies acquired from their mothers across the placenta and in breast milk, which protect them for about the first 3 months of life after which they suffer repeated bouts of malaria, which may prove fatal if untreated, especially when associated with measles, severe gastroenteritis or malnutrition (Strickland, 1991). In later life the mortality rate from malaria declines from around 1% for children 1 to 4 years old, to 0.1% in adolescence (10 to 14 years old) and to 0.03% in adults (Wernsdorfer and Wernsdorfer, 1988).

Life Cycle of *Plasmodium*

History

Laveran (1880) was the first person to recognize the malaria parasite within the red blood cells and later he observed the process of exflagellation in fresh preparations. It was the demonstration of exflagellation which convinced Pasteur in 1884 that the object being observed was indeed an independent organism, and not the result of degenerative processes. The significance of exflagellation was shown by MacCullum (1897) when he observed the formation of microgametes by exflagellation, the fertilization of the macrogamete, and the subsequent development of a motile ookinete in a related genus *Haemoproteus*. MacCullum referred to the anterior tip of the ookinete puncturing erythrocytes with which it made contact. Indeed the pointed end of the ookinete of *Plasmodium* is used to penetrate the cells of the gut of mosquitoes (Garnham, 1966).

At this stage it was not known in what host exflagellation and formation of the ookinete occurred. In 1897 Ross, working in India, observed oocysts of *P. falciparum* on the midgut of an *Anopheles* mosquito. Ross was then moved to Calcutta where he was unable to continue to work on human malaria and turned his attention to bird malaria. In 1898 he obtained the complete life cycle of *P. relictum* (= *P. danilevskyi*) in *Culex quinquefasciatus* (= *fatigans*), including the presence of sporozoites in the salivary glands, and the transmission of *P. relictum* from infected to healthy sparrows by the bite of *Cx quinquefasciatus* (Ross, 1911).

At this stage, Ross's work was again interrupted and he was not able to return to human malaria until late 1899 in West Africa. Of this work he wrote 'In a few weeks I was able to show that the parasites of quartan, tertian and malignant fever all develop in *Pyretophorus costalis* [now *An gambiae*] or *Myzomyia funesta* [now *An funestus*], precisely as the *Proteosoma* [now *Plasmodium*] of birds develops in *Culex fatigans*.' In the meantime Grassi *et al.* (1899) had followed the development of *P. falciparum* in *An claviger*. The final seal on their work was the sending of *An maculipennis*, infected with *P. vivax*, from Italy to London where P. Manson fed them on his son, P.T. Manson, who developed malaria (Ross, 1911).

There remained one problem. Schaudinn had illustrated and described the sporozoites of *Plasmodium* directly invading erythrocytes. If that were so then it

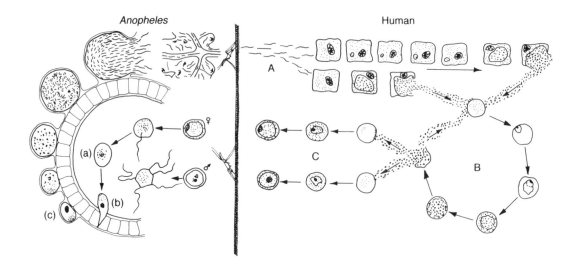

Fig. 27.3. The life cycle of malaria parasites in the *Anopheles* mosquito and the human host, according to present views on exo-erythrocytic schizogony. A, sporozoites injected into a human by an *Anopheles* female develop in the liver either into latent hypnozoites (above) which sometime later undergo schizogony to cause relapses or undergo immediate schizogony (below). Both release merozoites which enter red blood cells. B, erythrocytic schizogony involving release of merozoites. C, some merozoites develop into male or female gametocyctes, which only develop further when ingested by a an *Anopheles*. D, male gametocytes undergo exflagellation to produce male gametes one of which will fuse with a female gamete to form a zygote (a). The zygote becomes a motile ookinete (b) which pass between the cells of the midgut to form an oocyst (c). E, the oocyst enlarges. There is much nuclear division, ending in the formation of motile sporozoites which invade the haemocoele and penetrate the salivary glands, from which they are passed into the host with the saliva when the *Anopheles* next feeds. Source: Redrawn from Giles and Warrell (1993).

should be possible to transmit malaria by blood transfusion soon after the inoculation of sporozoites. This was not the case. Following inoculation of sporozoites there was a period of several days during which it was impossible to transmit infection by blood transfusion. The answer was that the sporozoites did not parasitize red cells but other tissues in the body. In 1936 Raffaele showed that the first cycle of development of the bird parasites, *P. elongatum* and *P. relictum*, was not in the erythrocytes but in the cells of the reticulo-endothelial system (Raffaele, 1936a,b). Later Shortt and Garnham (1948) showed that the first cycle of development of *P. vivax* occurred in the parenchyma cells of the liver and this was confirmed for *P. falciparum* by Shortt *et al.* (1951).

Cycle of *Plasmodium* in humans and *Anopheles* (see Fig. 27.3 and Table 27.1)

When an infected *Anopheles* mosquito feeds, it injects sporozoites with its saliva. The sporozoites disappear from the circulating blood in 30 min and invade the parenchyma cells of the liver where a cycle of schizogony takes place. This is the exoerythrocytic or pre-erythrocytic cycle of the parasite. A large unpigmented schizont is formed, containing several thousand merozoites (Shortt and Garnham, 1948). The merozoites are released into the circulation and invade the erythrocytes. Their release marks the end of the prepatent period. In *P. vivax* and *P. ovale* some parasites may delay development and remain as hypnozoites for variable periods (Garnham, 1988). Depending on the intensity of the initial infection, there may be one or two erythrocytic schizogonic cycles before clinical symptoms are produced and end the incubation period.

The merozoite attaches to an erythrocyte and is invaginated into the red cell within a parasitophorous vacuole, where it feeds and deposits a pigment, haemozoin. The ingested merozoite becomes a feeding trophozoite and, in the early stages of an infection, the fully grown trophozoite becomes a schizont, producing a small number of new merozoites (6–16) (Garnham, 1966). Release of the merozoites from the erythrocytes brings on an attack of malaria, and the interval between attacks is the length of the schizogonic cycle. The released merozoites repeat the cycle and invade other erythrocytes.

After a number of cycles of schizogony some of the trophozoites do not divide but become gametocytes which develop no further in humans. Gametocytes are mature in 4 days in *P. vivax* and in 8 days in *P. falciparum* (Wernsdorfer, 1980). When taken up by a female *Anopheles*, gametocytes shed the remains of the erythrocyte and become free in the midgut. It should be noted that 'Individuals with very low-density gametocytemias are often infectious: and conversely, individuals with very high-density gametocytemias often are not infectious' (Nedelman, 1990). Microgametocytes undergo exflagellation producing eight microgametes; the process taking 10–15 min. The microgametes move away to find and enter macrogametes to form zygotes. The zygote is at first immobile and then becomes an active ookinete which enters the midgut epithelium to form an oocyst under the basal laminar.

The oocyst may be recognized by the presence of pigment derived from the gametocyte. As it grows it disrupts the basal laminar and projects into the haemocoele (Sinden, 1975). Considerable nuclear division goes on within the oocyst and sporozoites develop around a number of germinal centres and in the case of *P. falciparum*, contain around 10,000 sporozoites (Pringle, 1965). The time taken to complete sporogony is temperature-dependent and times at 28°C are given in Table 27.1. Sporogony is not completed at temperatures above 33°C or below 16°C (Rieckmann and Silverman, 1977). This explains the association between the distribution of malaria and the summer isotherm of 16°C for *P. vivax* and 20°C for *P. falciparum* (Wernsdorfer, 1980). When sporogony is complete, the sporozoites escape into the haemocoele and accumulate in the salivary glands. In *P. yoelii nigeriensis* the sporozoites escape from the oocyst through small holes and tears (Sinden, 1975).

The salivary glands of anopheline mosquitoes are paired with each gland consisting of three lobes, two lateral and one median. The central duct in the proximal portion of the lateral lobes is described as being 'a thick chitinized cuticular duct'. Sporozoites that enter cells of the salivary gland in that region have difficulty in penetrating the duct wall. Sporozoites of *P. berghei* in *An stephensi* accumulate in

Table 27.1. Quantitative data on the four species of *Plasmodium* which cause malaria in humans.

Plasmodium species	Exoerythrocytic schizont			Disease in humans				Sporogonic cycle at 28°C duration (days)
	Size (μm)	No. of merozoites	Prepatent (days)	Incubation (days)	Periodicity of paroxysms (h)	Average persistence (years)	Maximum persistence (years)	
P. falciparum	60	30,000–40,000	9–10	12	Note 1	1 to 2	4	9 to 10
P. vivax	45	10,000–20,000	11–13	13	48	1.5 to 4	8	8 to 10
P. ovale	70	15,000	10–14	17	48–50	1.5 to 4	5	12 to 14
P. malariae	45	2,000	15	28	72	3+	53	14 to 16

Source: Data abstracted from Garnham (1980, 1988), Wernsdorfer (1980), Harinasuta and Bunnag (1988), Strickland (1991) and López-Antuñano and Schmunis (1993).

Note 1. Duration of paroxysms 16–36 hours at irregular intervals cf. 8–12 h for other species (24, 36 or 48 h).

the cells of the distal portion of the lobes where they have easy access into the lumen of the duct (Sterling *et al.*, 1973). When next the mosquito feeds, sporozoites will be passed with the saliva into the host, and if the host is susceptible the cycle is repeated. Once infective, anopheline mosquitoes remain so for up to 12 weeks (Rieckmann and Silverman, 1977). For all practical purposes that means that mosquitoes are infected for life.

Sporogony (Vanderberg and Gwadz, 1980)

At 30°C sporogony of *P. vivax* is completed in 7 to 8 days and this time doubles to 15–16 days at 20–21°C. At 16°C sporogony takes 55 days (Wernsdorfer, 1980). When an infective *Anopheles* female feeds, only a small percentage of the sporozoites in the salivary gland are injected into its host, with some being deposited directly into capillaries and others into subcutaneous tissues. Although Shortt *et al.* (1951) found that relatively few sporozoites of *P. falciparum* developed into schizonts in the liver, yet 'as few as ten sporozoites (of *P. vivax*) is ordinarily sufficient to infect'.

Observations on field-caught *An gambiae* s.l. and *An funestus* by Beier *et al.* (1991a,b) have estimated the geometrical mean of the number of sporozoites in the salivary glands as 962 and 812 respectively, with a maximum of 41,830 in *An funestus* and 117,544 in *An gambiae* s.l. When feeding, 98% of infected females transmitted less than 25 sporozoites, about 3% of the number in the salivary glands, suggesting that only sporozoites free in the salivary duct at the time of feeding were actually transmitted.

The Human Malaria Parasites (*Plasmodium*) (Garnham, 1980; Wernsdorfer, 1980)

More than a hundred species of *Plasmodium* have been described from vertebrates. Four species occur in humans, about 20 species in other primates, a similar number in other mammals, and about 40 each in birds and reptiles. The vectors of the mammalian species of *Plasmodium* are invariably species of *Anopheles* mosquitoes and those of bird plasmodia are most often culicine mosquitoes (Garnham, 1988).

Four species of *Plasmodium* cause malaria in humans. Although they produce a similar illness with paroxysms recurring at 48 h or 72 h intervals, and are grouped under the one heading 'malaria', there are in fact four different diseases each with its own specific pattern. The species and the diseases they cause are: *P. falciparum* – malignant tertian malaria; *P. vivax* – benign tertian malaria; *P. malariae* – quartan malaria; and *P. ovale*. The terms tertian and quartan derive from the Roman way of counting, in which a fever which recurs at 48 h intervals is said to be tertian (third day) because there is fever on day one, normal temperature on day two, and fever again on day three. By the same line of argument, *P. malariae* with a 72 h periodicity is said to be quartan (fourth day). Some comparative data on the four species are given in Table 27.1.

The subgenus *Laverania*, which includes *P. falciparum*, is characterized by the possession of crescent-shaped gametocytes whose production takes up to six times that required for the asexual cycle and which show a wave-like invasion of the blood stream. The other three species are in the subgenus *Plasmodium* in which the gametocytes are spherical and develop at a similar rate to the asexual forms (Nedelman, 1990). In its ultrastructure and DNA composition *P. falciparum* is closer to rodent and avian malarias than to the other primate malarias (Garnham, 1988).

Plasmodium falciparum

Malignant tertian malaria is the most severe form of the disease and in the absence of treatment may kill up to 25% of non-immune adults within 2 weeks. Merozoites of *P. falciparum* parasitize erythrocytes of all ages. Schizogony occurs in the capillaries of the internal organs, interrupting their blood supply. When infected erythrocytes attach to the vascular epithelia of vital organs infection may be confused with other febrile illnesses such as gastroenteritis, hepatitis, typhoid fever, endocarditis and pyelonephritis. Cerebral malaria is the most dangerous and a frequent cause of mortality in children and non-immune adults. After the initial series of attacks have passed malignant tertian malaria may recur from the activation of latent erythrocytic forms. These are recrudescences. In contrast, relapses, which occur in *P. vivax* and *P. ovale* infections, arise from hypnozoites, dormant tissue forms from the initial exoerythrocytic cycle (Garnham, 1988; Strickland, 1991).

 Plasmodium falciparum is associated with two severe reactions: blackwater fever, and haemolytic anaemia in children. In blackwater fever the patient passes a urine dark with methaemoglobin, a condition associated with irregular and inadequate treatment of malignant tertian malaria with quinine. With the replacement of quinine by other drugs, blackwater fever is now rarely encountered. Mortality from blackwater fever could be up to 50%. Haemolytic anaemia is a very rapidly fatal condition of African infants in areas with holoendemic malaria (Garnham, 1980).

 Plasmodium falciparum has a higher temperature threshold for development than *P. vivax* and is commoner in the warmer areas of the world, being limited by a summer isotherm of 20°C. Infections of *P. falciparum* develop faster in humans than *P. vivax*. Their exoerythrocytic cycle is the shortest and produces several times as many merozoites (Strickland, 1991).

Plasmodium vivax

Plasmodium vivax causes a milder disease, but is more persistent than *P. falciparum*. It is widely distributed throughout the world, being limited by the 16°C summer isotherm and was therefore often the only species present in the cooler temperate regions. Merozoites of *P. vivax* invade young erythrocytes, i.e. reticulocytes. Not all the *P. vivax* sporozoites which enter the liver develop immediately but some remain as hypnozoites which may persist for up to 4 years, causing relapses at intervals as short as 2 months. It is absent from West Africa because the indigenous people lack the Duffy factor on their erythrocytes, which is essential if *P. vivax* merozoites are to enter erythrocytes. Some strains of *P. vivax* have a prepatent period of 250–350 days (Garnham, 1980, 1988).

Plasmodium ovale

Plasmodium ovale is the rarest of the human malaria parasites and was not described until 1922. It produces hypnozoites which may cause relapses at 3-monthly intervals for a period of up to 4 years. Its merozoites invade young erythrocytes. It is the least pathogenic, and produces a tertian fever after a longer incubation period than either *P. vivax* or *P. falciparum* (Garnham, 1980; Strickland, 1991) (Table 27.1). In West Africa it replaces *P. vivax* among the indigenous people. It is also present in Papua New Guinea, Thailand, Kampuchea and Vietnam (Wernsdorfer, 1980).

Plasmodium malariae

Plasmodium malariae is a slow-growing parasite which has a worldwide but patchy distribution. Although widespread it is usually less common than either *P. falciparum* or *P. vivax*. However, it is next to *P. falciparum* in pathogenicity, with death resulting from kidney failure. *P. malariae* does not produce hypnozoites and its merozoites invade ageing erythrocytes. It has remarkable powers of persistence with recrudescences occurring for up to 53 years (Table 27.1). *P. malariae* is limited by the 16°C summer isotherm. The West African chimpanzee parasite *P. rodhaini* is identical with *P. malariae* and it is likely that the quartan parasite (*P. brasilianum*) of New World monkeys is another form of *P. malariae* (Garnham, 1988). The epidemiologies of these three parasites is such that there is virtually no interchange and the simian parasites present no obstacle to the control of human malaria (Wernsdorfer, 1980). Very rarely, infections with simian parasites occur in humans. Three cases of human infection with *P. knowlesi* in Malaya and one of *P. simium* in Brazil have been reported. *P. knowlesi* produces a daily paroxysm, i.e. its periodicity is quotidian, and *P. simium* has a tertian periodicity (Garnham, 1988). *P. cynamolgi bastianellii* has also been found in humans (Garnham, 1966).

Features of malaria in humans

Clinical symptoms are associated with the bursting of infected erythrocytes releasing merozoites, malarial pigment and other debris into the blood stream. A typical paroxysm has an abrupt onset with chill which turns within an hour into profuse sweating with headache and high temperature lasting for 2 to 6 h, after which the temperature falls rapidly to normal and the patient may feel 'well' (Strickland, 1991). Paroxysms recur at intervals of 48 h or 72 h with increasing parasitaemia to a maximum after which the parasitaemia declines rapidly. This maximum is reached with *P. falciparum* about 10–14 days after the prepatent period (Garnham, 1966). The result of an attack of malaria is anaemia with the possibility of oxygen deprivation to the tissues. As part of the immune response the spleen is enlarged and the spleen rate in children aged 2 to 9 years of age has been used by malariologists as an index of malarial endemicity.

Certain mutations confer a degree of protection against *P. falciparum* and are prevalent in areas where that parasite is dominant. In the heterozygous state, genes for sickle cell anaemia and thalassaemia provide protection against *P. falciparum* but are lethal in the homozygous state (Allison, 1961; Luzzatto, 1974). *P. falciparum* cannot develop in haemoglobin-S-containing cells which have reduced oxygen tension as encountered in venous blood. Thalassaemia heterozygotes may be protected against *P. falciparum* during the critical first year of life when passive immunity is waning and active immunity not yet established. There is excellent epidemiological evidence that glucose-6-phosphate dehydrogenase (G-6-PD) deficiency protects against malaria although the mechanism has yet to be determined (Weatherall, 1988).

Epidemiology of Malaria (Wernsdorfer, 1980)

Endemicity and stability

Malaria is said to be endemic where there is 'a constant measurable incidence of natural transmission over a succession of years'. Four grades are recognized based on the frequency of enlarged spleens in the susceptible 2- to 9-year-old age group of children. In holoendemic areas their spleen rate is above 75% and the adult spleen rate low. In hyperendemic areas the adult spleen rate is high and the spleen rate in children above 50%. In mesoendemic and hypoendemic areas the spleen rate is between 11 and 49% and below 10%, respectively. Malaria is said to be epidemic when the 'incidence of cases in an area rises rapidly and markedly above its usual level or when the infection occurs in an area where it was not present previously'.

Malaria is also said to be stable or unstable or in an intermediate state. Stable malaria is associated with holoendemic and hyperendemic areas. Characteristically transmission is perennial with little change in the incidence of malaria from season to season; the vector is both strongly anthropophilic, feeding almost exclusively on the human population, and long-lived. Stable malaria is associated with the warmer areas of the world, favouring rapid sporogony and with the main parasite being *P. falciparum*. Unstable malaria is associated with sudden, very intense epidemics, a short-lived vector, and a limited transmission period. The vector is not strongly anthropophilic and its importance is derived from it being present in high density. Sporogony is not rapid and the parasite usually involved is *P. vivax*.

In temperate regions, malaria transmission is seasonal and often bimodal with peaks in late spring to early summer and late summer to early autumn with a decline in midsummer. There are two reasons for the midsummer decline. High temperatures, particularly if associated with low humidities, will markedly reduce the longevity of *Anopheles*, and if high temperatures are maintained, they could exceed the threshold (33°C) for sporogony. Infections with *P. vivax* predominate in the spring peak and *P. falciparum* in the autumn peak. The reasons are: higher minimum temperature (20°C) for sporogony of *P. falciparum*; the long periods between infection and production of gametocytes (minimum of 10 days after patency) (Short and Garnham, 1948); and the lower parasite reservoir in the population as a result of recoveries during the winter. The average duration of infection with *P. falciparum* is 80 days.

Anopheles vectors of malaria (Table 27.2; Fig. 27.4)

Some 422 species of *Anopheles* have been described (Table 7.1) of which 68 have been associated with malaria, 40 as main vectors and 28 as subsidiary vectors in one or more epidemiological zones (Service, 1993). Some species are important vectors wherever they occur, e.g. *An albimanus*, *An aquasalis*, and *An darlingi* in the Central and South American zones; *An fluviatilis* in the Indo-Iranian and Indo-Chinese Hills zones; and *An dirus* in the Indo-Chinese Hills and Malaysian zones. Some species, e.g. *An melas* in the Afrotropical region, are important local vectors but are not classified as main vectors because of their limited geographical distribution. Wernsdorfer (1980) includes *An melas* as one of eight species responsible for stable malaria. Of the 40 major species, 15 are main vectors in one zone and subsidiary vectors in another zone; *An fluviatilis* is a main vector in zones 8 and 9 (Indo-Iranian; Indo-Chinese Hills) and a subsidiary vector in zone 6 (Afro-Arabian); and *An minimus* is a main vector in zone 10 (Malaysian) and a subsidiary vector in zones 8 and 9 (Indo-Iranian; Indo-Chinese Hills).

Four of the eight species associated with stable malaria occur in the Afrotropical region, where three (*An arabiensis*, *An funestus*, *An gambiae*) are widespread; two (*An minimus*, *An fluviatilis*) occur in the Oriental region; and two (*An labranchiae*, *An sacharovi*) in the Palaearctic region (Table 27.2). Another eight species are associated with intermediate malaria, including *An quadrimaculatus* and *An darlingi* in the Nearctic and Neotropical regions, respectively; *An farauti* in the Australian region; three species (*An balabacensis*, *An sinensis*, *An sundaicus*) in the Oriental region; and two (*An atroparvus*, *An sergenti*) in the Palaearctic region. Thirteen species are associated with unstable (epidemic) malaria, and of these seven are in the Oriental region; two each in the Neotropical and Palaearctic regions; and one in each of the Afrotropical and Australian regions.

Macdonald (1957) attempted a 'tentative classification of some notorious vectors' of malaria, using survival and degree of anthropophily as his criteria. Two problems arise in making such a classification. Firstly, data on these aspects of anopheline bionomics are not available for all vector species, and those that are, are often unavoidably biased, e.g. anthropophilic index being determined on catches made in houses. The second arises from the assumption that survival and anthropophily remain constant throughout the species range. Survival will depend on local microclimatic conditions, and the degree of anthropophily on the extent to which a species is an opportunistic feeder, likely to be diverted from people in the presence of abundant domestic animals. In Italy, *An messeae* was zoophilic and not associated with malaria, but in eastern Europe and the Soviet Union after the Second World War *An messeae* was an important vector. Its increased anthropophily was attributed to drastically reduced numbers of domestic stock as a result of the war.

Host preference and gonotrophic cycle

The main factor governing the ability of an *Anopheles* species to act as a vector of malaria is the frequency with which it feeds on humans. The vectors associated with stable malaria are those which are strongly anthropophilic, often feeding on

Table 27.2. Main *Anopheles* vectors of malaria in 12 epidemiological zones. Zone numbers and geographical naming follow Macdonald (1957). North America is included for completeness but it is no longer a malarious area. Species which are main vectors in a zone are in italics and where the same species is a subsidiary vector in another zone it is in ordinary type. (Data from Service, 1993; and Foley *et al.*, 1994.) The association of species with stable, unstable or intermediate malaria is taken from Wernsdorfer (1980).

Zone 1. North American
- *An. freeborni*
- An quadrimaculatus — intermediate
- An albimanus

Zone 2. Central American
- An albimanus — unstable
- An aquasalis — unstable
- An argyritarsis
- An darlingi — intermediate
- An albitarsis
- An pseudopunctipennis

Zone 3. South American
- An albimanus — unstable
- An albitarsis
- An aquasalis — unstable
- An darlingi
- An pseudopunctipennis
- An punctimacula
- An argyritarsis

Zone 4. North Eurasia
- An atroparvus — intermediate
- An sacharovi
- An sinensis

Zone 5. Mediterranean
- An atroparvus — intermediate
- An labranchiae — stable
- An sacharovi — stable
- An superpictus — unstable

Zone 6. Afro-Arabian
- An pharoensis — unstable
- An sergentii
- An culicifacies
- An fluviatilis

Zone 7. Afrotropical
- *An arabiensis* — stable
- *An funestus* — stable
- *An gambiae s.s* — stable
- An pharoenis

Zone 8. Indo-Iranian
- An culicifacies — unstable
- An fluviatilis
- An minimus
- An sacharovi
- An sundaicus
- An superpictus

Zone 9. Indo-Chinese Hills
- An dirus
- An fluviatilis — stable
- An culicifacies
- An maculatus
- An minimus
- An nigerrimus

Zone 10. Malaysian
- An aconitus — unstable
- An balabacensis — intermediate
- An campestris
- An dirus
- An donaldi
- An flavirostris — unstable
- An letifer
- An leucosphyrus
- An maculatus — unstable
- An minimus — stable
- An nigerrimus
- An subpictus
- An sundaicus — intermediate

Zone 11. Chinese
- An anthropophagus
- An sinensis
- An balabacensis

Zone 12. Australian
- An farauti type 1 — intermediate
- An koliensis
- An punctulatus — unstable
- An subpictus

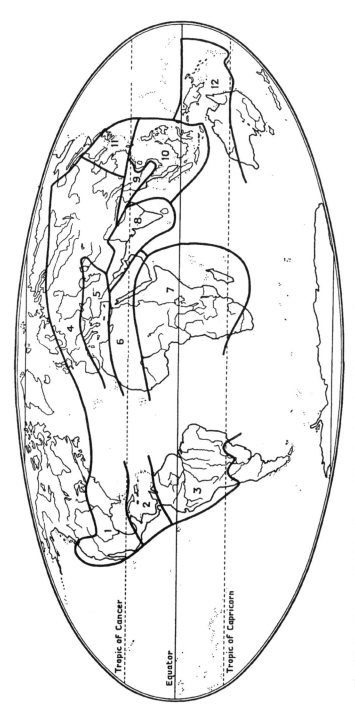

Fig. 27.4. Geographical locations of Macdonald's 12 epidemiological zones of malaria: **1**, North American; **2**, Central American; **3**, South American; **4**, North Eurasia; **5**, Mediterranean; **6**, Afro-Arabian; **7**, Afrotropical; **8**, Indo-Iranian; **9**, Indo-Chinese Hills; **10**, Malaysian; **11** Chinese; **12**, Australasian. Redrawn from Macdonald (1957).

humans to the exclusion of other hosts. To transmit malaria an individual *Anopheles* has to feed on humans at least twice: the first time to acquire an infection and the second to transmit the parasite. This means that the ability of a species to transmit malaria is related to the product of the two probabilities of an individual feeding on humans twice and not simply to the proportion feeding on humans. This is important in comparing the potential of two species.

If the probability of an *Anopheles* feeding on humans (p) is 1.0, i.e. the species feeds exclusively on humans, then the probability of it feeding on humans twice is p^2, which is also 1.0. If a species takes only half its feeds from people, then $p = 0.5$ and the probability of it feeding twice on a human is 0.25. When a species feeds only occasionally on humans, e.g. $p = 0.1$, then the probability of that species acting as a vector is 0.01, and when a species feeds infrequently on humans, e.g. 1% of feeds, the probability of an individual feeding on a human twice is 1 to 10,000. This explains how species with a high rate of anthropophily can transmit malaria when present in low density, while species with low anthropophilic rates only act as vectors when present in high density and are associated with unstable malaria.

Anopheline mosquitoes show gonotrophic concordancy in which blood-feeding and oviposition alternate. In some species, e.g. *An funestus, An gambiae*, some females, especially those newly emerged, do not develop eggs after a blood meal (Gillies, 1955). The taking of multiple blood meals early in life will increase the probability of their being infected with *Plasmodium* at an early age. A blood meal is required for ovarian development. The gonotrophic cycle begins with a blood meal, and as this is digested the ovaries develop, leading to a stage when the meal has been fully digested and the eggs are ready for oviposition. Under optimal conditions this cycle can be completed in 2 days.

It is possible for a female to oviposit and feed again during the same night, and blood-feeding to occur at 2-day intervals, but it is more likely that the next blood meal will be taken on the night following oviposition, giving an interval of 3 days. In the tropics female *Anopheles* collected in the morning can be classified as blood-fed; half gravid; gravid, with fully developed eggs; and empty, which includes newly-emerged nullipars and recently oviposited parous females. At lower temperatures, both blood digestion and ovarian development will take longer.

Longevity

An *Anopheles* mosquito cannot transmit malaria until the sporogonic cycle of the *Plasmodium* has been completed, and this takes a minimum of 8 days (Table 27.1). Only mosquitoes that live at least that long can act as vectors of malaria. A species could be very abundant and feeding mainly on humans, but if its mortality was such that few survived long enough to become infective then that species would play only a minor role, if any, in transmission. It is important, therefore, to be able to determine the age structure of an *Anopheles* population.

A method of determining the physiological age of an *Anopheles* mosquito, developed by Polovodova (1949) has been applied widely by Detinova (1962). The method relies on the fact that in each ovarian cycle each ovariole matures only one egg. When this is deposited, the stretched tissues of the ovariole contract to form a dilatation at the position of the former follicle in which the egg developed. A separate dilatation is formed at each ovarian cycle so that the number of dilatations indicates the number of ovarian cycles that have been completed, and the number of blood meals which have been taken. Knowledge of the durations of the gonotrophic and sporogonic cycles under local conditions enables calculation of the proportion of the Anopheles population which has lived long enough to be infective.

Gillies and Wilkes (1965) used this technique to study the longevity and infectivity of populations of An gambiae in Tanzania. At Muheza on the coast, survival from one ovarian cycle to the next was 0.62 which, on a 3-day ovarian cycle, represented a daily survival of 0.854. Mosquitoes will have to complete a minimum of three ovarian cycles, i.e. be 3-parous, before being infective. The sporozoite rate in 3-parous An gambiae was 4.1% and increased to 32% among females which were 7-parous or more. The heavier infection rate in older An gambiae was countered by their small numbers – 20% of the population survived to complete three ovarian cycles or more but only 1% completed seven cycles. The greatest contribution to transmission of P. falciparum came from female An gambiae which had completed four, five and six cycles. These mosquitoes formed only 16% of the population but 73% of those infective.

The same workers made parallel observations on the An gambiae population at Gonja, inland from Muheza, where the dry season lasted 5 months. At Muheza only An gambiae s.s. was present, and at Gonja, both An gambiae s.s. and An arabiensis, and this may have had some influence on the findings. At Gonja, the daily survival rate of An gambiae s.l. was lower than at Muheza (0.791 cf. 0.854). Correspondingly, only 14% (cf. 20%) completed three or more ovarian cycles, and only 0.3% (cf. 1%) completed seven cycles. The maximum parity recorded at Gonja was a female that had completed eight cycles and at Muheza 12 cycles. These differences can be attributed to the effect of differences in aridity on survival, and are in agreement with Wernsdorfer (1980) who recorded daily survival rates of 0.95, 0.90 and 0.85 at relative humidities of 65%, 55% and 50%, respectively, at mean temperatures of 27–30°C. The survival rate at Muheza corresponds with that found for 50% RH.

Density of Anopheles – Effects of temperature and rainfall

Two main factors which influence the abundance of Anopheles are temperature and availability of suitable breeding sites. In the tropics the abundance of a species may show very little variation throughout the year. At Muheza the population of An funestus, which breeds in permanent bodies of water, showed little change throughout the year and had a stable physiological age structure. An gambiae breeds in temporary collections of water and its populations show greater variation throughout the year, being more dependent on rainfall. Nevertheless for much of the year (December–July) the density of An gambiae at Muheza was high and the physiological age structure almost constant (Detinova, 1968).

The speed of development of Anopheles is temperature-dependent and can be determined more accurately using non-feeding stages such as eggs or pupae. In members of the An maculipennis complex the length of the egg stage increased from

2 days at 22–24°C to 10 to 12 days at 10-12°C (Kettle and Sellick, 1947). Muirhead-Thomson (1951) found a fourfold increase in the durations of the egg and pupal stages of *An minimus* with a decrease in temperature from 30°C to 16°C. When, in the summer in Assam, water temperature fluctuated from 26.5 to 30.6°C, the larval stage of *An minimus* took 7 days. Ribbands (1949) considered that under these conditions the complete life cycle from egg to egg would take 14 days – egg to adult 11 days and oviposition of the first egg batch 3 days. At 16°C, the egg to adult cycle could be expected to take about 6 weeks.

In warmer temperate regions, e.g. North Africa, breeding will continue all the year, but the life cycle would be greatly extended during the cooler months. In Egypt the life cycle of *An pharoensis* lasts several months in the winter but adults continue to emerge. In more severe climates, breeding ceases and the adult females hibernate. In the Moscow region *An messeae* hibernates for 6 months (October–March) and there are only two generations in a year: an overwintering generation which feeds and oviposits in the spring and early summer, and a summer generation whose progeny go into hibernation (Detinova, 1968).

Species of *Anopheles* have the potential to increase their populations very rapidly when, under optimal conditions, batches of 100–200 eggs can be laid at intervals of 2 to 3 days by females which can complete several cycles in their lifetime. Development from egg to adult may take as little as 10 days and it is possible to have an increase of $\times 50$ to $\times 100$ in the *Anopheles* population within 2 weeks. One limiting factor on this exponential rate of increase is the availability of breeding sites, which are dependent on rainfall but this relationship is complex, depending on the species of *Anopheles* involved and its particular breeding sites.

Heavy rains will flush out water courses and reduce the population of species breeding in pools in river beds. Such autumnal rains in Algeria normally brought the malaria transmission season to an end by flushing out the breeding sites of *An labranchiae*. In 1934 the rains began with steady gentle rain, in place of the more usual violent storms, and increased the breeding sites for *An labranchiae*, resulting in a severe late-season epidemic of malaria.

In Ceylon in 1934/5, it was a drought which provided the conditions for a dramatic increase in the population of *An culicifacies*. *An culicifacies* breeds in still water in pools and its breeding sites were greatly increased when a severe drought led to the rivers ceasing to flow, becoming a series of discontinuous pools with no water movement. This was ideal for *An culicifacies* and led to a severe epidemic of malaria. The effect of rainfall will depend upon local conditions, and in northern India, *An culicifacies*, which was favoured by drought in Ceylon, has been responsible for severe epidemics in years of excessive rainfall with widespread pool formation (Hackett, 1937).

Models of Malaria Transmission

Once the intimate relationship between malaria and the *Anopheles* mosquito was understood, efforts were made to express it in quantitative terms. In the first decade of this century, Ross (1911) was already modelling the transmission of malaria. The appropriateness of a model needs to be evaluated against reliable quantitative field data. An efficient model should be able to predict the likely effects of different strategies, enabling limited facilities (staff, money) to be used most effectively. The tradition of modelling malaria was continued by Macdonald (1957), when he was, appropriately, Director of the Ross Institute of Tropical Hygiene in London. More recent models (Dietz, 1988) differ in their treatment of superinfection and immunity in the vertebrate host. Various quantitative entomological aspects of malaria transmission are considered below.

Measure of stability of malaria

One critical piece of information is the probability of a mosquito surviving long enough for sporogony to be completed. If the daily survival of a female *Anopheles* is p and the extrinsic incubation period takes n days, then the probability of an individual mosquito surviving n days is p^n. The life expectancy of the mosquito at emergence is $1 - \ln p$, and the index of stability is the product of the life expectancy times the average number of humans bitten by one mosquito in one day (a) and is therefore $a - \ln p$. Stability of malaria is therefore a function of the longevity of the vector and its degree of anthropophily. Stable malaria is associated with an index exceeding 2.5, unstable malaria with an index less than 0.5 and intermediate malaria with an index of 0.5–2.5 (Macdonald, 1957; Wernsdorfer, 1980).

Basic reproduction rate and vectorial capacity

Basic reproduction rate

The reproduction rate (R) is the average number of new infections which will be produced from a single existing infection. The basic reproduction rate (R_0) is the average number of new infections produced by a single infection when an infected individual, possessing no immunity, is introduced into a community where neither the mosquitoes nor the inhabitants are or have been previously infected. The basic reproduction rate is given by the expression:

$$R_0 = ma^2bp^n - r\ln p$$

where m is the number of female *Anopheles* per person; a is the daily biting rate of an individual female on humans; b is the proportion of mosquitoes with sporozoites which are actually infective, and in a non-immune population b will approach 1; r is the recovery rate which for *P. falciparum* has been taken as 0.0125, i.e. the average duration of an infection being 80 days (Macdonald, 1957; Wernsdorfer, 1980). The reproduction rate is a function of a^2, the probability of a mosquito feeding twice on a human being. One important application of this expression is that when R_0 falls below 1.0 the incidence of malaria is declining and if other conditions remain constant, the disease should disappear.

Wernsdorfer (1980) has calculated various values of R_0 using selected, reasonable values for the other parameters. The values chosen were $m = 10$; $a = 0.4$, i.e. the mosquito is 100% anthropophilic feeding at 2- or 3-day intervals; $p = 0.90$; $n = 8$; $b = 1.0$ and $r = 0.0125$. When these are substituted in the above expression the calculated value of R_0 is 525.

Any change in m produces a similar proportional change in the value of R_0 while the effect of a change in a produces an effect proportional to the square of the change. Thus when the density of *Anopheles* (m) is reduced by one-fifth, from ten to two, the value of R_0 is reduced by the same factor from 525 to 105, but when the *Anopheles* feeds less frequently or there is a change in its degree of anthropophily then the change in R_0 is enhanced. Thus if a changes from 0.4 to 0.16, i.e. feeding at 3-day intervals and taking only half of its meals from humans, R_0 changes from 525 to 84. A trebling of the daily mortality, reducing survival (p) from 0.90 to 0.70, has a marked effect on R_0, reducing it from 525 to 21, i.e. to 4% of the initial value.

Vectorial capacity

Garrett-Jones (1964) highlighted the entomological components of Macdonald's equation, setting them apart as the vectorial capacity (C), the daily rate at which future inoculations arise from a currently infective case. It is defined as:

$$C = ma^2p^n/-\ln p$$

and differs from the basic reproduction rate by the omission of terms b (infectivity of sporozoites) and r (recovery rate).

Dye (1992) points out that exceptional quantities of data will be required to give reasonable values to the variables involved in the calculations of R_0 and C and that measurements on blood-sucking insect populations have relatively little absolute value. A number of assumptions underlie the calculation of vectorial capacity some of which are given below:

1. Vector competence is unity, i.e. all uninfected vectors biting an infectious host acquire a viable infection.
2. Female *Anopheles* feed equally on all members of the human population. Should they feed selectively, the values of R_0 and C will be increased.
3. Survival is independent of age and the parasite has no effect on the vector.
4. The relationship between temperature and the length of the extrinsic incubation period (EIP) needs to be known with considerable accuracy.
5. The possibility of subpopulations of the vector having different feeding preferences. In Natal, the human blood index of *An arabiensis* was in excess of 90% for indoor collections in unsprayed areas compared to 31% in sprayed areas and 66% in exit traps in sprayed areas (Sharp and le Sueur, 1991).

Nájera (1974) would be sympathetic to these sources of error. He found the model in use in northern Nigeria predicted different results from those found and attributed these discrepancies to inadequate estimation of the basic variables required for the model. He criticized the preparation not the model, but the model may not have adequately represented conditions in the field.

Sporozoite rate

Macdonald (1957) defined the sporozoite rate (s) as: $s = p^n ax/(ax - \ln p)$ where x is the infectivity of the human population to the *Anopheles*, i.e. the gametocyte rate. Gillies and Wilkes (1965) tested this expression at Muheza for *An gambiae*, where daily survival of the mosquito (p) was 0.854; the average number of feeds on a human by a mosquito in one day (a) was taken as 0.33, i.e. the gonotrophic cycle was 3 days; and the EIP (n) was 13 days. The sporozoite rate observed was 2.5% ($s = 0.025$). Substituting the numerical estimates into the expression gave x a value of 0.114, i.e. at each blood meal 11.4% of *An gambiae* became infected.

Gillies and Wilkes (1965) obtained their sporozoite rate by carefully dissecting out the salivary glands of female *Anopheles* and examining them under high magnification. The development of species-specific anticircumsporozoite protein monoclonal antibodies (MAb) enables the rapid determination of the presence or absence of sporozoites and the species of *Plasmodium* involved, providing the target epitope of MAb is conserved in the study area. Such MAbs can be used in immunofluorescent assays, immunohistochemical assays and enzyme-linked immunosorbent assays (ELISA) (Wirtz and Burkot, 1991).

Cyclic feeding model

Saul *et al.* (1990) reformulated the transmission model of Macdonald (1957) by substituting a cyclic feeding model in place of one involving daily rates of feeding and survival. A feature of the cyclic model is the ease with which non-uniform conditions can be included. Graves *et al.* (1990) used this model to interpret malaria transmission in three villages (Butelgut, Mebat, Maraga) in Papua New Guinea from indoor and outdoor catches of *An farauti*, *An koliensis* and *An punctulatus*. Survival from one feeding cycle to another was similar for all three species (0.580 to 0.608; Table 27.3). Survival over the EIP ranged from 0.151 for *An farauti* to 0.294 for *An punctulatus*. The vectorial capacity, i.e. the ratio of infectious bites per day per host to the proportion of bites on infectious hosts was

Table 27.3. Estimates of indices of the intensity of malaria transmission by *An punctulatus*, *An farauti* and *An koliensis* in three villages (Butelgut, Mebat and Maraga) in Papua New Guinea.

	An punctulatus	*An farauti*	*An koliensis*	
Survival through feeding cycle	0.580	0.608	0.595	
Survival through extrinsic incubation period	0.294	0.151	0.171	
			Mebat	Butelgut
Vectorial capacity	n.a.	0.027	0.252	0.154
Total vectorial capacity	n.a.	9.8	29.2	2.2
Mosquito infection probability	n.a.	0.074	0.049	0.174

Source: From Graves *et al.* (1990).
n.a.: estimates not available.

Table 27.4. Sporozoite rates in *An farauti, An koliensis* and *An punctulatus* collected in three villages (Butelgut, Mebat and Maraga) in Papua New Guinea.

A. Sporozoite rate by species and site of collection

	Indoor	Outdoor	Combined
An farauti	0.23 (14,374)	0.23 (16,757)	0.23 (31,131)
An koliensis	0.74 (7,014)	0.84 (7,719)	0.79 (14,733)
An punctulatus	2.21 (2,081)	2.80 (2,500)	2.53 (4,581)

B. Sporozoite rate by species and village

	Butelgut	Mebat	Maraga
An farauti	1.59 (63)	1.43 628)	0.20 (30,440)
An koliensis	2.32 (862)	1.04 (6,839)	0.37 (7,032)
An punctulatus	3.27 (3,177)	0.95 (1,064)	0.42 (240)

significantly lower for *An farauti* at Maraga (0.027) than for *An koliensis* in the other two villages (0.154, 0.252). When the number of bites per person per night is included to provide the total vectorial capacity, the value for *An farauti* (9.8) in Maraga was comparable to that of *An koliensis* in Butelgut (2.2) and Mebat (29.2). The mosquito infection probability, i.e the probability of a mosquito feeding on a host and becoming infected, ranged from 0.049 to 0.174 (Table 27.3).

The sporozoite rates of females of the same species caught indoors and outdoors were not significantly different (Table 27.4A). *An punctulatus* appears to be more heavily infected than the other two species but this is due to disproportionate numbers being taken in the different villages. In the same village, the sporozoite rates of the three species are similar (Table 27.4B).

Malaria Control Strategy

Malaria is the most prevalent and devastating disease in the tropics with 40% of the world's population being at risk of infection. The disease is becoming more difficult to control due to the spread of resistance to antimalarial drugs among parasites and an increase in the number of foci of intense malarial transmission due to changing environmental condition (Anon., 1993a). The WHO Global Strategy has four elements:

1. The early diagnosis and treatment of infections. Such disease management is inadequate in most of subSaharan Africa.

2. To implement preventive measures including vector control, which should be part of a country's general health programme, aiming at reduced morbidity and mortality by lowering the level of transmission.

3. Early detection and response to epidemics arising from a resurgence of malaria in areas where preventive measures have substantially decreased or even interrupted transmission.

4. To strengthen local capacities for regular assessment of the country's malaria situation, facilitating an appropriate response. For example indoor spraying of residual insecticides should not become an open-ended operation but be justified by current epidemiological information.

Epidemiological types of malaria

Throughout the world, seven different epidemiological types of malaria are currently recognized (Anon., 1993b). Their names, geographical distributions and some characteristics are given below:

1. *Savanna malaria* – present in subSaharan Africa and Papua New Guinea; characterized by perennial transmission; *P. falciparum* predominating; morbidity and mortality mainly in young children and pregnant women; expanding drug resistance.

2. *Malaria of plains and valleys* – present in Central America, China and the Indian subcontinent with variable transmission; *P. vivax* may predominate; strong seasonal variation and the risk of epidemics; drug resistance well established; large-scale insecticide programmes often ineffective.

3. *Highland and desert fringe malaria* – occurring in African and south-east Asian highlands, the Sahel, southern Africa and south-west Pacific – risks of epidemics due to climatic aberrations, changing agricultural practice or migration; insecticide spraying can often curb transmission and sometimes restore the previous malaria-free status.

4. *Agricultural development projects* – in Africa, Asia and South America; transmission increased due to irrigation in certain circumstances; risk of seasonal malaria outbreaks due to non-immune labourers; insecticide resistance frequent in cotton-growing areas; control by larvivorous fish in certain rice-growing areas and house-spraying.

5. *Urban and periurban malaria* – in Africa, South America and South Asia; transmission and population immunity highly variable over short distances; epidemics caused by specially adapted vectors in south Asia; breeding sites identifiable and larval control is practical.

6. *Malaria of forests and forest fringes* – in south-east Asia and South America; locally intense transmission; disease often an occupational risk; severe multi-drug resistance; benefits of house-spraying and larval control questionable; personal protection essential.

7. *War-zone malaria* – due to displacement of parasite-carrying or non-immune populations coupled with environmental degradation allowing increased mosquito breeding; prevention by personal protection and chemoprophylaxis; space-spraying in emergency situations and environmental measures in refugee camps.

Personal protection

In areas where ineffective or no control measures are in operation the individual must rely on personal protection, using screening of living accommodation, especially bedrooms; sleeping under nets; wearing protective clothing; and using repellents. The use of bednets impregnated with a residual insecticide, such as a synthetic pyrethroid, has been an outstanding success, reducing mortality of infants by 20 to 40% in endemic areas (Walgate, 1994). Such measures can be effective but require a high level of self-discipline. Chemoprophylaxis is recommended for travellers from non-endemic areas and as a short-term measure for workers serving in highly endemic areas (see Treatment and Prevention below).

Principles of control

Any attack on malaria must have the full support of the local community. Measures may be directed against either the parasite or the vector, and in the case of the latter, against the immature stages in the breeding sites or against emerged adults in houses and animal shelters. Anti-parasite measures would be directed to the detection and treatment of all gametocyte carriers in the community. In theory, if there are no gametocyte carriers then no malaria transmission will take place regardless of the habits and density of the *Anopheles* population. The reverse can also occur where gametocyte carriers move into an area previously rendered malaria free (Kondrashin and Orlov, 1989).

Adult control, involving the indoor spraying of residual insecticides, will be preferred where the vector is endophilic and endophagic. It is ineffective against an exophilic vector such as *An balabacensis balabacensis*. The aim of anti-adult measures is to increase the daily mortality of female *Anopheles* so that few live long enough to become infective. Larval control is only practical where the breeding sites are well defined and limited. Breeding sites can be eliminated by filling (a permanent solution) or by installing a drainage system. The latter will require regular maintenance to remain free of breeding. The opposite to drainage is irrigation, which often results in increased malaria transmission, but this can be minimized with cooperation from the irrigation authorities (Reuben, 1989). The use of chemicals to control breeding of *Anopheles* is rarely feasible in rural areas but may be practical in urban and periurban situations.

Before the advent of synthetic insecticides oil and Paris green (a copper–arsenic compound) were the main larvicides, while pyrethrum sprays were virtually the only adulticide available. Rigorously applied, such simple materials could be highly effective. Gorgas used them to control both yellow fever, transmitted by *Ae aegypti*, and malaria, at first in Cuba and subsequently in the Panama Canal Zone. An outstanding achievement using these methods was the elimination of *An gambiae* from Brazil. It had been introduced from West Africa, but its presence was not detected until it had established itself over a considerable area. Success was achieved by first containing *An gambiae* in the area it already occupied, and then steadily reducing that area by eliminating the invader from the periphery inwards. This splendid achievement of Soper and Wilson (1943) was published when attention of much of the world was elsewhere and it did not receive the acclaim it deserved.

The discovery in the 1940s of the insecticidal properties of DDT and other chlorinated hydrocarbons dramatically changed the approach to malaria control by concentrating the attack against adult *Anopheles*. The outstanding property of

DDT and related compounds was their persistence, and deposits applied to resting places of mosquitoes in houses and animal shelters remained insecticidally active for up to 6 months (Wernsdorfer, 1980). Previously, adult control measures had to be carried out daily during the transmission season and the extension of this interval to 3 months opened up the prospect of not just reducing the incidence of malaria to a level at which it ceased to be a major public health problem, but complete eradication became a distinct possibility. This ambitious aim has now been recognized as unrealistic in most countries (Anon., 1993b).

Technical problems have hindered the control of malaria: the development of drug-resistant strains of *Plasmodium*; insecticidal resistance in the vector so that DDT is no longer effective and has had to be replaced by organophosphorus compounds such as malathion, and the carbamates, which have shorter periods of activity, e.g. malathion less than 1 month on mud walls (Carnevale *et al.*, 1992). Technical problems are solvable but where a government lacks the will to implement control measures, no amount of technical expertise can succeed.

Treatment and Prevention (Drugs and Vaccines)

Drugs and vaccines are being produced to attack malaria parasites at particular stages of their development and to make *Anopheles* resistant to infection.

Drugs (Strickland, 1991)

Chloroquine is the drug of choice for treatment and prevention of infection with *P. vivax, P. malariae, P. ovale* and strains of *P. falciparum* sensitive to the drug. Unfortunately, resistance to chloroquine has been developed by *P. falciparum* in most parts of the world including South America, the Indian subcontinent, southeast Asia, China, New Guinea, the Philippines and subSaharan Africa. Quinine is effective against chloroquine-resistant *P. falciparum* and quinidine against both chloroquine- and quinine-resistant *P. falciparum*. Artemisinin (qinghaosu), a drug extracted from the herb, *Artemisia annua* (sweet wormwood), is effective against chloroquine-resistant *P. falciparum* and it is the hope of the malaria specialists of the World Health Organization that artemesinin will be restricted to the treatment of multi-drug-resistant *P. falciparum* (Anon., 1994). Other drugs effective against multi-drug-resistant *P. falciparum* are halofantrine and mefloquine. Primaquine is an excellent gametocytocidal and sporontocidal drug which is used to prevent relapses of *P. vivax* and *P. ovale* by its action against exoerythrocytic hypnozoites. Pyrimethamine and proguanil act slowly against blood schizonts but are effective against the exoerythrocytic schizonts of *P. falciparum*. Resistance to both compounds is extensive, but pyrimethamine together with sulfadoxine is now widely used for treatment They are less active against the latent exoerythrocytic stages of *P. vivax* and *P. ovale*. Proguanil in conjunction with chloroquine was extensively used for prophylaxis against chloroquine-resistant *P. falciparum* in Africa, but is currently being replaced in many areas by mefloquine (Peters, 1995, personal communication).

Vaccines (Saul, 1992)

Plasmodium species exist as antigenically distinct strains so that infected humans develop an effective immunity against the infecting strain but only a weaker immunity to heterologous strains. Therefore an effective vaccine will necessarily incorporate critical antigens which cannot readily be dispensed with or altered by the parasite (Wellems, 1990). One vaccine (SPf66) is being tested in the field and five other promising vaccines are being prepared for human trials, which should take place over the next few years (Anon., 1994). SPf66 is a combination of three *P. falciparum* merozoite proteins. In field trials in southern Tanzania it proved to be safe, immunogenic and reduced the risk of clinical malaria among children exposed to intense *P. falciparum* transmission. Its efficacy was estimated to be 31% (Alonso *et al.*, 1994).

The other five vaccines active against blood forms are as follows:

1. Apical membrane antigen 1 (AMA-1), a protein found in an apical structure in merozoites within the erythrocyte and which appears on the surface of free merozoites.

2. Serine-rich protein (SERA), a soluble protein secreted into the parasitophorous vacuole containing the parasite and released when the schizont ruptures, immune sera react with and trap emerging merozoites.

3. Merozoite surface antigen (protein) 1 (MSA-1 or MSP-1), a large protein anchored to the merozoites' surface membrane through a glycolipid tail.

4. Erythrocyte binding antigen (EBA-175), the ligand on a *P. falciparum* merozoite which binds with sialic acid (*N*-acetylneuraminic acid) on the erythrocyte membrane (Sim *et al.*, 1994).

5. The vaccine Pfs25, a protein found on the surface of zygotes and ookinetes, is the only mosquito stage antigen to have reached the stage of field testing. Antibodies which target this antigen should block parasite development in *Anopheles*. Work on an anti-sporozoite vaccine has focused on the major circumsporozoite protein of *P. falciparum*. It has not reached the stage of being tested in the field.

The use of vaccines to control blood-sucking insects or to protect a potential vector from becoming infective is in its infancy. Lehane (1994) considers that potential vaccine targets for the control of blood-sucking insects were their lipid-digesting enzymes, their haemolysins and, where present, their symbionts. The malaria ookinete has to penetrate the peritrophic membrane and the midgut epithelium before developing into an oocyst adjacent to the basal lamina. Billingsley (1994) considers that the peritrophic membrane is of lesser importance as a target than previously thought and suggests that the basal laminar and the glycosylated molecules on the microvilli of the midgut epithelium should be regarded as potential vaccine targets.

Other Haemosporina

Avian plasmodia (Garnham, 1980)

According to Garnham (1980), 'over 30 valid species of malaria parasites have been described from about 500 species of birds'. They are mostly infections of wild birds and where transmissible to domestic poultry, cause little disease in local breeds of birds but can cause severe epizootics in introduced poultry. Three species, *P. gallinaceum*, *P. juxtanucleare* and *P. durae*, are of minor veterinary importance. Species of *Plasmodium* in birds have two pre-erythrocytic cycles before invading the erythrocytes. The pre-erythrocytic schizonts are smaller than those in mammals, producing less than 100 merozoites, but two successive cycles offer the possibility of releasing more than 1000 merozoites to invade the erythrocytes.

Plasmodium gallinaceum is native to Sri Lanka, India and Malaysia. Its pre-erythrocytic schizonts grow in the cells lining the capillaries and block them causing cerebral damage. Infections develop rapidly in 1-day-old chicks but adult birds are only weakly susceptible. *P. juxtanucleare* is more widely distributed in tropical Asia and Japan and has been introduced into Latin America, where it has caused more severe epizootics than in its native Asia. Both these species infect jungle fowl, partridges and other wild birds, which form a reservoir of infection. *P. durae* has caused epizootics among domestic turkeys in Kenya and West Africa, and a very similar species, *P. hermani*, has been found in wild turkeys in Florida. *P. durae* causes considerable mortality in domestic turkeys.

Haemoproteidae

Hepatocystis (Garnham, 1980)

In *Hepatocystis*, the pre-erythrocytic cycle occurs in the liver of mammals and only gametocytes occur in the erythrocytes. Thirteen species of *Hepatocystis* have been described. They are mostly parasites of arboreal, tropical mammals – lower monkeys, bats and squirrels. There is also one species which occurs in mouse deer (*Tragulus* spp.), and another in the hippopotamus. Most work has been done on *H. kochi*, a parasite of monkeys in Africa, for which the vectors are species of *Culicoides*, including *C. adersi* on the East African coast, and *C. fulvithorax*, and probably other species of *Culicoides*, in both inland and coastal areas.

The sporogonic cycle in *Culicoides* follows the usual pattern with rapid exflagellation of the microgametocyte with the formation of eight microgametes. In *H. kochi*, the ookinete penetrates the basement membrane and enters the haemocoele. Oocysts are free in the haemocoele and accumulate anteriorly in the head, particularly near the eyes and supraoesophageal ganglia. Oocysts measure about 40 μm in diameter, have several germinal centres and produce hundreds of slender sporozoites measuring 11–13 μm. Oocysts mature in 5 days at 27°C. Sporozoites are rarely seen in the salivary glands and they may invade the mouthparts to be transmitted without being introduced with the saliva.

In monkeys the sporozoites invade the hepatic parenchyma cells and develop into slowly growing schizonts (merocysts) which may take 1 to 2 months to reach a size of 2 mm and more. Most of the released merozoites invade erythrocytes to form gametocytes, but some re-invade the liver and repeat schizogony. In the erythrocyte, the merozoite produces haemozoin pigment. The presence of free oocysts in *H. kochi* may not be typical of the genus because oocysts of *H. brayi* occur in the usual location on the midgut of *Culicoides* (*Monoculicoides*) *nubeculosus* and *C.* (*M.*) *variipennis* (Miltgen *et al.*, 1976). In some areas, infections of *H. kochi* in monkeys can be 100% but its pathogenicity is doubtful. An unusual feature is that parasitaemia increases with the age of the host and there appears to be little or no immunity to *H. kochi*.

Haemoproteus and *Parahaemoproteus* (Fallis and Desser, 1977)

More than 80 species of *Haemoproteus* have been named. Schizogony occurs in the tissues and only gametocytes are in the circulating blood. The parasite in the erythrocyte encircles the nucleus of the host cell, in which pigment is deposited. Species of *Haemoproteus* are parasites mainly of birds, but also of lizards and turtles. *H. columbae* is a parasite of pigeons and is transmitted by the hippoboscid *Lynchia maura*. *H. metchnikovi* occurs in turtles and is transmitted by the tabanid *Chrysops callidus* (DeGiusti *et al.*, 1973). This is the first record of a tabanid acting as the vector of a species of Haemospororina, and undermines the basis for establishing the genus *Parahaemoproteus* on the grounds that the vectors are not hippoboscids but ceratopogonids. *P. nettionis*, a parasite of ducks, is transmitted by *Culicoides downesi* and other species of *Culicoides*.

In the vector the microgamete undergoes exflagellation. The fertilized macrogamete becomes a motile ookinete about 25 μm long, and forms an oocyst up to 30 μm in diameter, in which sporozoites are produced from several germinal centres. Sporozoites of *H. columbae* are slender bodies with one end blunt and the other tapered. They are produced in thousands, and those of *H. columbae* are released when the oocyst bursts. The minimum time for the sporogonic cycle in *H. columbae* is 10 to 12 days. The sporozoites accumulate in the salivary gland and are passed with the saliva when the vector is feeding.

The cycle of *P. nettionis* in *Culicoides* shows several differences. The oocyst grows very little and there is only a single germinal centre. Fewer sporozoites are produced and they escape gradually from the oocyst. The minimum time for the sporogonic cycle is 7 to 10 days.

In the vertebrate host schizonts occur in many tissues but are most frequent in the lungs of birds. Those of *H. metchnikovi* occur in the spleens of turtles. Merozoites released from schizonts develop into mature gametocytes in 4 to 6 days in birds, and considerably longer (3 months) for *H. metchnikovi* in turtles. Infections of *Haemoproteus* are often heavier than those of *Leucocytozoon* but their pathogenicity is uncertain. As with *Leucocytozoon*, there is increased parasitaemia in late winter and early spring, favouring infection of the vector and transmission to nestlings.

Leucocytozoidae

Leucocytozoon (Fallis *et al.*, 1974; Fallis and Desser, 1977)

About 70 species of *Leucocytozoon* have been named and they are all parasites of birds. Schizogony occurs in the tissues and only the gametocytes appear in the peripheral circulation. Gametocytes occur in both leucocytes and erythrocytes but no pigment is produced in the latter. Economically-important species include *L. simondi*, a parasite of domestic and wild ducks and geese in Europe, North America and south-east Asia; *L. smithi*, a parasite of turkeys in Europe and North America; and probably the most important, *L. caulleryi*, which parasitizes chickens in south-east Asia and Africa. *L. caulleryi* is placed in the subgenus *Akiba* which is sometimes raised to generic rank. Species of *Akiba* are characterized by the disappearance of the nucleus of the parasitized cell as the gametocyte matures, and by being transmitted by biting midges of the genus *Culicoides*. The vectors of most species of *Leucocytozoon* are species of Simuliidae of various genera, including *Simulium*, *Prosimulium*, *Eusimulium* and *Cnephia*.

The life cycle of *L. simondi* has been studied in considerable detail. An infective simuliid injects sporozoites when feeding, and in a susceptible host they develop in the parenchyma of the liver, growing to a size of 20–40 μm in 4 to 5 days. These hepatic schizonts contain several thousand merozoites which invade erythrocytes and erythroblasts to become rounded gametocytes in 48 h. It is not known whether these merozoites are able to invade the liver and produce a second cycle of schizogony. The hepatic schizonts also produce multinucleated syncytia, which are phagocytized by macrophages and grow into megaloschizonts, measuring up to 200 μm, in cells of the reticuloendothelial system, especially of the spleen and lymph nodes. Megaloschizonts produce a million or more merozoites which either invade the liver to start another cycle of schizogony or enter leucocytes to form elongated gametocytes. Maximum parasitaemia is reached after 10 to 12 days after which the infection can remain chronic for 2 or more years in ducks. The host cell in which a megaloschizont develops becomes hypertrophied.

The density of gametocytes in the peripheral circulation shows a diurnal periodicity with peak numbers being reached in the daytime when simuliids are active and hence favours gametocytes being taken up by feeding simuliids. Both microgametocytes and macrogametocytes escape through a small opening in the pellicle of the host cell. Rounded gametocytes escape more readily than elongated ones. The microgametocytes undergo exflagellation, producing eight free microgametes. When a microgamete penetrates a macrogamete, the two nuclei fuse and the zygote becomes a motile ookinete measuring 30 μm × 4 μm.

The ookinete penetrates between the cells of the midgut and forms an oocyst beneath the basement membrane. Some ookinetes remain in the midgut for 3 to 4 days until the peritrophic membrane disrupts and they are able to reach the midgut epithelium. A mature oocyst is 10–14 μm in diameter and produces about 50 sporozoites from a single germinal centre. The oocyst does not rupture; sporozoites escape gradually from the cyst and move to the salivary glands which they penetrate.

Sporogony occurs at a variable rate even under identical conditions. In the same vector species, sporogony can take 6–18 days at 18–20°C. Exflagellation occurs within 1 to 3 min of ingestion, with the stimulus being changes in oxygen and carbon dioxide tensions rather than temperature. It takes 6–12 h for the rounded zygote to become an elongated ookinete, and about 48 h to develop from ookinete to oocyst.

The minimum time from introduction of sporozoites to production of mature gametocytes is 6 days. Gametocytes will continue to circulate for weeks but their viability declines with time. *L. caulleryi* has a similar cycle in *Culicoides arakawae* in which sporozoites are produced in 3 days at 25°C and 6 days at 15°C. They remain infective for 3 to 5 weeks. The cycle of *L. smithi* in turkeys is slightly different. The primary cycle occurs in the liver, and merozoites released from that cycle either form rounded gametocytes, or invade the liver or kidney producing a second cycle of schizogony; but the schizont and its host cell are not hypertrophied and no megaloschizont is formed.

There is no evidence of a specific relationship between vector and parasite. *L. caulleryi* develops equally well in its major vector *C. arakawae*; in *C. odibilis*, a minor vector which feeds on chickens; and also in *C. schultzei* which feeds on cattle. Similar observations have been made on other species of *Leucocytozoon*, which parasitize only birds and yet they develop apparently with equal ease in mammal-feeding simuliids. Consequently, simuliid–*Leucocytozoon* relationships are more dependent on the simuliid's feeding preferences than on the specificity of the parasite for a particular simuliid.

During winter, the parasitaemia in the avian host is low and a small increase occurs in early spring in response to the host's developing reproductive cycle. This ensures that gametocytes will be available in the peripheral blood when the vectors appear in the spring, and when susceptible nestlings will be available for infection. Transmission occurs on the birds' nesting grounds.

Species, such as *L. caulleryi* and *L. simondi*, which include a megaloschizont stage in the reticuloendothelial system in their cycle are more pathogenic than those, such as *L. smithi*, in which schizogony occurs in the liver and kidneys. Ducks infected with *L. simondi* show some or all of the following symptoms: lethargy, loss of appetite, diarrhoea, convulsions and anaemia. The condition may be fatal. The anaemia cannot be accounted for by simple parasitization of the erythrocytes but involves intravascular haemolysis. Similar symptoms occur in chickens infected with *L. caulleryi*. There is conflicting evidence as to the pathogenicity of *L. smithi* in turkeys. It may be more important in the presence of other diseases. In general, *Leucocytozoon* infections are more severe in domestic than wild species, and deaths are commoner in young birds.

References

Allison. A.C. (1961) Genetic factors in resistance to malaria. *Annals of the New York Academy of Sciences* 91, 710–729.

Alonso, P.L., Smith, T., Schellenberg, J.R.M.A., Masanja, H., Mwankusye, S., Urassa, H., Bastos de Azevedo, I., Chongela, J., Kobero, S., Menendez, C., Hurt, N., Thomas, M.C., Lyimo, E., Weiss, N.A., Hayes, R., Kitua, A.Y., Lopez, M.C., Kilama, W.L.,

Teuscher, T. and Tanner, M. (1994) Randomised trial of efficacy of SPf66 vaccine against *Plasmodium falciparum* malaria in children in southern Tanzania. *The Lancet* 344, 1175–1181.

Anon. (1993a) Implementation of the Global Malaria Control Strategy. *WHO Technical Report Series* 839.

Anon. (1993b) *A Global Strategy for Malaria Control*, World Health Organization, Geneva.

Anon. (1994) TDR calls for global alliance to vanquish 'microscopic murderer'. *TDR News* 44, 1–2.

Beier, J.C., Onyango, F.K., Ramadhan, M., Koros, J.K., Asiago, C.M., Wirtz, R.A., Koech, D.K. and Roberts, C.R. (1991a) Quantitation of malaria sporozoites in the salivary glands of wild Afrotropical *Anopheles*. *Medical and Veterinary Entomology* 5, 63–70.

Beier, J.C., Onyango, F.K., Koros, J.K., Ramadhan, M., Ogwang, R., Wirtz, R.A., Koech, D.K. and Roberts, C.R. (1991b) Quantitation of malaria sporozoites transmitted *in vitro* during salivation by wild Afrotropical *Anopheles*. *Medical and Veterinary Entomology* 5, 71–79.

Billingsley, P.F. (1994) Vector–parasite interactions for vaccine development. *International Journal for Parasitology* 24, 53–58.

Carnevale, P., Robert, V. and Mouchet, J. (1992) *The Biology of Vectors and Malaria Control*. Laveran Journals No. 1, Les Pensières, Veyrier-du-Lac, France.

Cox, F.E.G. (1991) Systematics of parasitic Protozoa. In: Krier, J.P. and Baker, J.R. (eds) *Pasitic Protozoa*. Academic Press, San Diego, vol. I, pp. 55–80.

DeGiusti, D.L., Sterling, C.R. and Dobrzechowski, D. (1973) Transmission of the chelonian haemoproteid *Haemoproteus metchnikovi* by a tabanid fly *Chrysops callidus*. *Nature, London* 242, 50–51.

Detinova, T.S. (1962) *Age-grouping Methods in Diptera of Medical Importance*. World Health Organization, Geneva.

Detinova, T.S. (1968) Age structure of insect populations of medical importance. *Annual Review of Entomology* 13, 427–450.

Dietz, K. (1988) Mathematical models for transmission and control of malaria. In: Wernsdorfer, W.H. and McGregor, I. (eds) *Malaria Principles and Practice of Malariology*. Churchill Livingstone, Edinburgh, vol. 2, pp. 1091–1133.

Dye, C. (1992) The analysis of parasite transmission by bloodsucking insects. *Annual Review of Entomology* 37, 1–19.

Fallis, A.M. and Desser, S.S. (1977) On species of *Leucocytozoon*, *Haemoproteus* and *Hepatocystis*. In: Kreier, J.P. (ed.) *Parasitic Protozoa*. Academic Press, New York, vol. 3, pp. 239–266.

Fallis, A.M., Desser, S.S. and Khan, R.A. (1974) On species of *Leucocytozoon*. *Advances in Parasitology* 12, 1–67.

Foley, D.H., Meek, S. and Bryan, J.H. (1994) The *Anopheles punctulatus* group of mosquitoes in the Solomon Island and Vanuatu surveyed by allozyme electrophoresis. *Medical and Veterinary Entomology* 8, 340–350.

Garnham, P.C.C. (1966) *Malaria Parasites and Other Haemosporidia*. Blackwell Scientific Publications, Oxford.

Garnham, P.C.C. (1980) Malaria in its various vertebrate hosts. In: Kreier, J.P. (ed.) *Malaria*. Academic Press, New York, vol. 1, pp. 95–144.

Garnham, P.C.C. (1988) Malaria parasites of man: life-cycles and morphology (excluding ultrastructure). In: Wernsdorfer, W.H. and McGregor, I. (eds) *Malaria Principles and Practice of Malariology*. Churchill Livingstone, Edinburgh, vol. 1, pp. 61–96.

Garrett-Jones, C. (1964) The human blood index of malaria vectors in relation to epidemiological assessment. *Bulletin of the World Health Organization* 30, 241–261.

Giles, H.M. and Warrell, D.A. (1993) *Bruce-Chwatt's Essential Malariology*. Edward Arnold, London.

Gillies, M.T. (1955) The pregravid phase of ovarian development in *Anopheles funestus*. *Annals of Tropical Medicine and Parasitology* 49, 320–325.

Gillies, M.T. and Wilkes, T.J. (1965) A study of the age-composition of populations of *Anopheles gambiae* Giles and *A. funestus* Giles in north-eastern Tanzania. *Bulletin of Entomological Research* 56, 237–262.

Grassi, B., Bignami, A. and Bastianelli, G. (1899) Ciclo evolutivo delle semilune nell' *Anopheles claviger* ed altri studi sulla malaria dall' ottobre 1898 al maggio 1899. *Atti della Societá per gli studi della Malaria* 1, 14–27.

Graves, P.M., Burkot, T.R., Saul, A.J., Hayes, R.J. and Carter, R. (1990) Estimation of anopheline survival rate, vectorial capacity and mosquito infection probability from malaria vector infection rates in villages near Madang, Papua New Guinea. *Journal of Applied Ecology* 27, 134–147.

Hackett, L.W. (1937) *Malaria in Europe*. Oxford University Press, London.

Harinasuta, T. and Bunnag, D. (1988) The clinical features of malaria. In: Wernsdorfer, W.H. and McGregor, I. (eds) *Malaria Principles and Practice of Malariology*. Churchill Livingstone, Edinburgh, vol. 1, pp. 709–734.

Kettle, D.S. and Sellick, G. (1947) The duration of the egg stage in the races of *Anopheles maculipennis* Meigen (Diptera: Culicidae). *Journal of Animal Ecology* 16, 38–43.

Kondrashin, A.V. and Orlov, V.S. (1989) Migration and malaria. In: Service, M.W. (ed.) *Demography and Vector-borne Diseases*. CRC Press, Boca Raton, Florida, USA, pp. 353–365.

Laveran, A. (1880) Note sur un nouveau parasite trouvé dans le sang de plusieurs malades atteints de fièvre palustre. *Bulletin de l'Académie de Médicine Paris Series* 2, 9, 1235–1236.

Lehane, M.J. (1994) Digestive enzymes, haemolysins and symbionts in the search for vaccines against blood-sucking insects. *International Journal for Parasitology* 24, 27–32.

López-Antuñano F.J. and Schmunis, G.A. (1993) Plasmodia in humans. In: Kreier, J.P. (ed.) *Parasitic Protozoa*. Academic Press, San Diego, vol. 5, pp. 135–266.

Luzzatto. I. (1974) Genetic factors in malaria. *Bulletin of the World Health Organization* 50, 195–202.

MacCullum, W.G. (1897) On the flagellated form of the malarial parasite. *Lancet* 2, 1240–1241.

Macdonald, G. (1957) *The Epidemiology and Control of Malaria*. Oxford University Press, London.

Miltgen, F., Landau, I., Canning, E.U., Boorman, J. and Kremer, M. (1976) *Hepatocystis* de Malaise.III. Développement d'*Hepatocystis brayi* chez *Culicoides nubeculosus* et *C. variipennis*. *Annales de Parasitologie Humaine et Comparée* 51, 299–302.

Muirhead-Thomson, R.C. (1951) *Mosquito Behaviour in Relation to Malaria Transmission and Control in the Tropics*. Edward Arnold, London.

Nájera, J.A. (1974) A critical review of the field application of a mathematical model of malaria eradication. *Bulletin of the World Health Organization* 50, 449–457.

Nedelman, J. (1990) Gametocytemia and infectiousness in falciparum malaria: observations and models. *Advances in Disease Vector Research* 6, 59–89.

Peters, W. and Gilles, H.M. (1995) *A Colour Atlas of Tropical Medicine and Parasitology*. Mosby-Wolfe, London.

Polovodova, V.P. (1949) Determination of the physiological age of female *Anopheles*. *Meditsinskaya Parazitologiya i Parazitarnye Bolezni* 18, 352–355.

Pringle, G. (1965) A count of the sporozoites in an oocyst of *Plasmodium falciparum*. *Transactions of The Royal Society of Tropical Medicine and Hygiene* 59, 289–290.

Raffaele, G. (1936a) Il doppio ciclo schizogonico di *Plasmodium elongatum. Rivista di Malariologia* 15, 309–317.

Raffaele, G. (1936b) Presumibili forme iniziali di evoluzione di *Plasmodium relictum. Rivista di Malariologia* 15, 318–324.

Reuben, R. (1989) Obstacles to malaria control in India – the human factor. In: Service, M.W. (ed.) *Demography and Vector-borne Diseases.* CRC Press, Boca Raton, Florida, USA, pp. 144–154.

Ribbands, C.A. (1949) The duration of the aquatic stages of *Anopheles minimus* Theo., determined by a new method. *Bulletin of Entomological Research* 40, 371–377.

Rieckmann, K.H. and Silverman, P.H. (1977) Plasmodia of man. In: Kreier J.P. (ed.) *Parasitic Protozoa.* Academic Press, New York, vol. 3, pp. 493–527.

Ross, R. (1911) *The Prevention of Malaria.* John Murray, London.

Saul, A. (1992) Towards a malaria vaccine: riding the roller-coaster between unrealistic optimism and lethal pessimism. *SouthEast Asian Journal of Tropical Medicine and Public Health* 23, 656–671.

Saul, A.J., Graves, P.M. and Kay, B.H. (1990) A cyclical feeding model for pathogen transmission and its application to determine vectorial capacity from vector infection rates. *Journal of Applied Ecology* 27, 123–133.

Service, M.W. (1993) Mosquitoes (Culicidae). In: Lane, R.P. and Crosskey, R.W. (eds) *Medical Insects and Arachnids.* Chapman & Hall, London, pp. 120–240.

Sharp, B.L. and le Sueur, D. (1991) Behavioural variation of *Anopheles arabiensis* (Diptera; Culicidae) populations in Natal, South Africa. *Bulletin of Entomological Research* 81, 107–110.

Shortt, H.E. and Garnham, P.C.C. (1948) The pre-erythrocytic development of *Plasmodium cynomolgi* and *Plasmodium vivax. Transactions of the Royal Society of Tropical Medicine and Hygiene* 41, 785–795.

Shortt, H.E., Fairley, N.H., Covell. G., Shute, P.G. and Garnham, P.C.C. (1951) The pre-erythrocytic stage of *Plasmodium falciparum. Transactions of the Royal Society of Tropical Medicine and Hygiene* 44, 405–419.

Sim, K.L., Chitnis, C.E., Wasniowska, K., Hadley, T.J. and Miller, L.H. (1994) Receptor and ligand domains for invasion of erythrocytes by *Plasmodium falciparum. Science* 264, 1941–1944.

Sinden, R.E. (1975) The sporogonic cycle of *Plasmodium yoelii nigeriensis*: a scanning electron microscope study. *Protistologica* 11, 31–39.

Soper, F.L. and Wilson, D.B. (1943) Anopheles gambiae *in Brazil 1930 to 1940.* Rockefeller Foundation, New York.

Sterling, C.R., Aikawa, M. and Vanderberg, J.P. (1973) The passage of *Plasmodium berghei* sporozoites through the salivary glands of *Anopheles stephensi*: an electron microscope study. *Journal of Parasitology* 59, 593–605.

Strickland, G.T. (1991) Malaria In: Strickland, G.T. (ed.) *Hunter's Tropical Medicine.* Saunders, Philadelphia, pp. 586–615.

Vanderberg, J.P. and Gwadz, R.W. (1980) The transmission by mosquitoes of plasmodia in the laboratory. In: Kreier, J.P. (ed.) *Malaria.* Academic Press, New York, vol. 2, pp. 153–234.

Walgate, R. (1994) What has TDR been up to over the past two years? *TDR News* 45, 1, 4.

Weatherall, D.J. (1988) The anaemia of malaria. In: Wernsdorfer, W.H. and McGregor, I. (eds) *Malaria Principles and Practice of Malariology.* Churchill Livingstone, Edinburgh, vol. 1, pp. 735–751.

Wellems, T.E. (1990) Mechanisms of antigenic variation in *Plasmodium.* In: van der Ploeg L.H.T., Cantor, C.R. and Vogel, H.J. (eds) *Immune Recognition and Evasion, Molecular*

Aspects of Host–Parasite Interaction. Academic Press, New York, pp. 201–224.

Wernsdorfer, W.H. (1980) The importance of malaria in the world. In: Kreier, J.P. (ed.) *Malaria.* Academic Press, New York, vol. 1, pp. 1–93.

Wernsdorfer, G. and Wernsdorfer, W.H. (1988) Social and economic aspects of malaria and its control. In: Wernsdorfer, W.H. and McGregor, I. (eds) *Malaria Principles and Practice of Malariology.* Churchill Livingstone, Edinburgh, vol. 2, pp. 1421–1471.

Wirtz, R.A. and Burkot, T.R. (1991) Detection of malarial parasites in mosquitoes. *Advances in Disease Vector Research* 8, 77–106.

Babesiosis and Theileriosis $\underline{\underline{28}}$

This chapter will deal with babesiosis and theileriosis, clinical diseases produced in domestic animals by species of *Babesia* and *Theileria*, respectively. Two similar terms are used in describing infections with *Babesia*; babesiasis refers to the presence of *Babesia* in the vertebrate host; and babesiosis to clinical disease caused by infection with *Babesia* (Joyner and Donnelly, 1979). The comparable terms for *Theileria* are theileriasis and theileriosis. Both *Babesia* and *Theileria* have an intracellular stage in erythrocytes which is pear-shaped and hence referred to as a piroplasm. In general the piroplasms of *Theileria* are smaller than 1–2 μm and those of *Babesia* larger (Mehlhorn and Schein, 1984).

Babesioses*

Ninety-nine species of *Babesia* have been described from nine orders of mammals with the greatest numbers being found in rodents (32), carnivores (26) and ruminants (21) – most of the last being in domestic animals (Telford *et al.*, 1993). Their greatest importance is as agents of disease among cattle, to which the greater part of the 1.2×10^9 cattle in the world are exposed (McCosker, 1981). The discovery by Smith and Kilborne (1893) of the transmission of *B. bigemina* by the ixodid tick, *Boophilus annulatus*, was the first record of a protozoan being transmitted by an arthropod. As *Bo. annulatus* is a one-host tick the cycle involved transovarian transmission of the *Babesia* by the female tick to the next generation.

    Babesia species are informally divided into two groups: large and small. Small species have piroplasms measuring less than 2.5 μm (Telford *et al.*, 1993) or 3 μm (Mehlhorn and Schein, 1984) in length and, when budding, form four daughter cells. This group includes *B. bovis*, the pyriform bodies of which measure 2.5×1.5 μm. Large species measure 2.5–5.0 μm and form two daughter cells.

---

*Kakoma and Mehlhorn (1994) was unavailable when this chapter was revised.

This group includes *B. bigemina* which measures 5 × 2 µm. The division cycle of *B. bigemina* is completed in about 8 h, and that of *B. bovis* in a similar time, multiplying tenfold in 24 h (Hoyte, 1961). The parasite ingests the cytoplasm of the erythrocyte by pinocytosis, and the haemoglobin is completely digested without the production of pigment (Mahoney, 1977).

Economic importance

Of the 171 countries providing information on diseases of veterinary importance for inclusion in the *Animal Health Yearbook*, 120 (70%) listed bovine babesiosis and 19 (11%) reported equine babesiosis. This emphasis is reflected in the level of epidemiological study of babesiosis throughout the world in which attention has been mainly concentrated on *B. bovis* and its boophilid tick vectors; secondly to other species infecting cattle including *B. bigemina*, *B. divergens* and *B. major*; less information is available on equine (*B. caballi*, *B. equi*) and canine (*B. gibsoni*, *B. canis*) babesioses (Smith, 1984). Susceptible cattle exposed to babesial infection can suffer a morbidity rate of 90% and an almost equally high mortality rate (Radostits *et al.*, 1994). Smith (1984) estimates that 70% of the 250 million cattle in Central and South America are in tick-infested regions and that ticks and tick-borne diseases represent a loss of US$870 million a year throughout Latin America. The losses include increased labour cost (36%); losses in beef and dairy production (36%); cost of acaricides, hide damage, loss by death and increased drought loss (28%).

Babesiosis in cattle (Radostits *et al.*, 1994)

After an incubation period of 2 to 3 weeks babesial infections cause fever (41°C), profound anaemia, haemoglobinuria (hence 'redwater' being a common name for the disease) and death may either occur within 24 h or the disease last for 3 weeks with survivors being carriers for a variable period of time, usually about 6 months, followed by a further 6 months of sterile immunity (NB the periods mentioned are subject to significant variation due to the different responses of races of cattle and species of *Babesia*. They give a generalized picture). Repeated infections can make the immunity permanent, but in the absence of further infections the beast becomes susceptible again after a year. All races of cattle are equally susceptible to *B. bigemina* but zebu and Afrikaner cattle have a higher resistance to *B. bovis* than European breeds. Indian (*Bos indicus*) and zebu-type cattle are relatively free from the disease because of their resistance to heavy tick infestations (Joyner and Donnelly, 1979). In enzootic areas calves receive passive immunity from maternal antibodies in the colostrum which protect for about 11 weeks. The greatest infection rate is in animals 6 to 12 months of age and uncommon in animals more than 5 years old, but the severity of the disease increases with age. Cattle that have recovered from infections are immune to developing disease in response to an homologous challenge but may suffer subclinical superinfection. Such animals may develop clinical disease in response to a heterologous challenge (Mahoney, 1977).

Species involved

In addition to the four species of *Babesia* already mentioned, three other species are involved – *B. jakimovi* in Russia (Nikol'ski *et al.*, 1977), *B. ovata* in Japan, and *B. ocultans* in South Africa (Smith, 1984). The most widespread species are *B. bovis* and *B. bigemina* which are widely distributed between 32°S and 40°N where they are responsible for serious losses in Latin America and Asia (McCosker, 1981). In the USA the main parasite was *B. bigemina* which was brought under control by the eradication of the vector *Bo. annulatus* (See Chapter 23). In Australia *B. bovis* is more important than *B. bigemina* although both species occur and are transmitted by the same vector, *Bo. microplus*. In 73% of cases of redwater in Australia the parasite involved was *B. bovis* and in only 6% of cases was it *B. bigemina*. The other 21% being due to *Anaplasma marginale* (see Chapter 25) (Radostits *et al.*, 1994).

Control of babesiosis (Radostits *et al.*, 1994)

Effective drugs are available to treat the disease in cattle but early treatment is essential to prevent the animal dying from anaemia even though its blood is sterile. Effective drugs are diminazene, imidocarb and amicarbalide and the anti-theilerial drugs, parvaquone and buparvaquone have proved promising in preliminary trials. Vaccination with living parasites and imidocarb has to be carefully controlled to ensure that the drug does not prevent the parasite eliciting an immune response from the host, leaving the animal unprotected. An attenuated vaccine has given excellent results and has the advantage of being less virulent and non-infective to ticks. The use of irradiated infected blood has given excellent results in protecting beasts against *B. major* and *B. divergens* in small trials.

Controlling babesiosis by controlling the tick vector must either be complete and proceed to tick eradication, a condition which must be maintained, or the level of tick infestation must be reduced to one which minimizes tick damage but allows all animals to become infected early in life when they are least susceptible and to be reinfected throughout their lives to maintain immunity. Too successful tick control can allow some beasts to escape infection, risking a more damaging outbreak later.

Babesia bovis (= *B. argentina*, *B. berbera*, *B. colchica*)

Babesia bovis is widely distributed throughout the world occurring in Central and South America, Africa, Asia, Europe and Australia. In tropical and subtropical areas the vectors are species of *Boophilus*, and in southern Europe *Rhipicephalus bursa* (Mehlhorn and Schein, 1984). *Bo. microplus* is widely distributed in the tropics and is the only boophilid in Australia. In Africa *Bo. decoloratus* is an important vector. Infection is acquired by the female tick, which transmits the parasite transovarially to its larvae which are the infective stage. The larval tick injects sporozoites which invade the erythrocytes, and the severity of the reaction is a function of the parasitaemia reached in the host. Maximum parasitaemias of 15,000 parasites mm^{-3} (= 0.2% parasitaemia) are fatal; non-fatal, severe infections have

5000 parasites mm^{-3}; and mild infections less than 1000 ($= 0.01\%$ parasitaemia) (Purnell, 1981). Infected erythrocytes clump together and block capillaries causing brain damage and anoxia of internal organs.

The duration of a single infection of *B. bovis* is of the order of 18 months to 4 years. Such an infection shows a fluctuating parasitaemia with cycles of 3 to 8 weeks. The fluctuations are due to change in antigenic type of which more than 100 have been recognized. On passage through the tick the strain of *B. bovis* reverts to its basic antigen but this also is variable. In the more natural situation where the infected animal is exposed to repeated infections, i.e. superinfections, the parasitaemia rises smoothly from zero to a maximum in 1–2 years and then declines (Mahoney, 1977).

Babesia bigemina

Babesia bigemina has a similar distribution to that of *B. bovis* and is also found in the southern former USSR (Purnell, 1981). One important vector of *B. bigemina* is *Bo. microplus* which acquires the parasite when the adult female is feeding, and transmits it in the succeeding nymphal stage (Hoyte, 1961) and adult *Rhipicephalus evertsi* acquire infection in the adult and transmit the parasite in the succeeding nymphal stage (Smith, 1984). *B. bigemina* can be transmitted transovarially through several generations of *Bo. microplus* (Radostits *et al.*, 1994). Clinical disease is produced when the parasitaemia exceeds 1% but there is no clumping of infected red cells and therefore no blockage of capillaries in organs, such as the brain (Mahoney, 1977). The duration of a single infection of *B. bigemina* is of the order of 6 to 12 months. Because *Bo. microplus* is a vector of several pathogens of cattle, the clinical picture can be complicated by synergistic pathogenicity among the various pathogens, such as *B. bigemina*, *B. bovis* and *Anaplasma marginale* (Purnell, 1981). Strains of *B. bigemina* exist, with the African strain being highly pathogenic and the Australian strain considerably less so (Mahoney, 1977).

Babesia divergens (= B. caucasica, B. occidentalis, B. karelica)

Babesia divergens is a small parasite which causes disease and death in cattle in northern Europe. It can produce high parasitaemias, exceeding 10% and as high as 24%, but the infected erythrocytes do not clump and therefore there is no cerebral involvement (Purnell, 1981). In western and central Europe the vector is *Ixodes ricinus* (McCosker, 1981). The parasite is acquired by the adult tick and transmitted transovarially to the next generation, of which all stages, but especially larvae, can transmit *B. divergens*. Infection may occasionally continue into the next (F$_2$) generation (Joyner and Donnelly, 1979). There is no such carry-over with either *B. bovis* or *B. bigemina* (Riek, 1966). Since the life cycle of *I. ricinus* extends for 3 years the parasite may survive up to 4 years in the tick (Joyner and Donnelly, 1979).

Babesia major

Babesia major is a large species found in cattle in Europe and the Middle East (McCosker, 1981). The vector is *Haemaphysalis punctata* in which the parasite acquired by the female tick is transmitted transovarially and probably passed in the succeeding adult stage (Smith, 1984). Although in infections the erythrocytes come together, few fatalities occur in cattle. Infections in American bison, introduced into south-east England, were very severe with 50% mortality before drugs were used to cure the survivors (Purnell, 1981).

Babesia jakimovi

Babesia jakimovi is a large *Babesia*, whose natural host is the Siberian roe deer (*Capreolus capreolus*). *B. jakimovi* causes a severe and often fatal disease of cattle. It also infects reindeer and elk. Species of *Ixodes* are vectors in the field, and in the laboratory *I. ricinus* can transmit *B. jakimovi* transovarially. The parasite is acquired by the adult female and transmitted by adults of the succeeding generation (Nikol'skii *et al.*, 1977).

Species of *Babesia* parasitizing other domestic animals (Purnell, 1981)

Sheep

Two species of *Babesia* occur in sheep: a large species, *B. motasi*, and a small one, *B. ovis*. Some strains of *B. motasi* are only infective for sheep, and others for both sheep and goats. *B. motasi* is a significant pathogen on its own but participates in synergistic pathogenicity with *Theileria* and *Ehrlichia*. The distribution of *B. motasi* coincides with that of various species of *Haemaphysalis* in Europe, the Mediterranean Basin and north-western Africa and that of *B. ovis* with that of *Rhipicephalus bursa* and *R. evertsi* in the Mediterranean Basin, central-western Asia and equatorial Africa (Morel, 1989). In the former USSR the vector is *R. bursa*, in which transovarian transmission has been reported to occur through 44 generations over 20 years (Markov and Abramov, 1970). When adult sheep become infected they can act as carriers for more than 2 years (Habela *et al.*, 1990). An attenuated vaccine has produced solid immunity in sheep (Radostits *et al.*, 1994).

Horses (De Waal, 1992)

Both a large species, *B. caballi*, and a small one, *B. equi*, parasitize equines. They cause equine piroplasmosis, also known as biliary fever. The disease occurs in southern Europe, France, Belgium, Russia, Africa and South and Central America. It was introduced into the USA in 1958 but confined by quarantine. It is absent from Great Britain, Germany, Austria, Switzerland and Japan. There is a constant danger of introducing the disease into non-infested areas. *B. equi* is more widespread than *B. caballi*. Fourteen species of three genera of ixodid ticks, *Dermacentor*, *Hyalomma* and *Rhipicephalus*, have been identified as vectors. *B. equi* has been transmitted trans-stadially by nine species and *B. caballi* trans-stadially by four species and transovarially by nine species.

Infections with *B. equi* develop into clinical cases after an incubation period of 12–14 days and intrauterine infection of the fetus is a complication, which causes serious losses. The incubation period for *B. caballi* is 10–30 days with infections being often inapparent and intrauterine infection rare. The symptoms are fever, anorexia, malaise and weight loss and in neglected horses, severe anaemia. *B. equi* produces a parasitaemia of 1 to 5% and occasionally to more than 20%, while in *B. caballi* infections parasitaemia is less than 0.1%. Infections can be diagnosed using DNA probes which are sensitive to parasitaemias of 0.002%. Treatment with diminazene is effective in eliminating *B. caballi* but not *B. equi*. Chemical control of ticks is probably feasible under highly intensive management systems. In free-ranging systems in endemic areas strategic tick control should expose foals to natural infection, allowing them to develop immunity without clinical disease while protected by natural non-specific resistance. This will result in an endemic stable disease situation.

Pigs

Pigs support a large, *B. trautmanni*, and a small, *B. perroncitoi*, species of *Babesia*. *B. trautmanni* is present in southern Europe, former USSR and parts of Africa. It may cause a severe condition with 60–65% of the erythrocytes being parasitized. Infections are commoner in pigs aged 4 to 6 months. Wild pigs, wart hog and bush-pig are likely significant reservoirs of this parasite. *B. perroncitoi* has a very limited distribution in the Sudan, Sardinia and Italy, and may be a significant pathogen of pigs. The vectors of both species are suspected rather than known (Morel, 1989).

Dogs

The large species in dogs is *B. canis*, and the small species *B. gibsoni*. Both parasites are widely distributed but *B. gibsoni* has a more limited distribution. Parasitaemias of 40–45% have been reported in fatal cases where death has been due to anoxia, and parasitaemias of 2 to 14% in non-fatal cases (Mahoney, 1977). Foxes and jackals act as reservoirs of *B. gibsoni* In south-east Asia the vector is *Haemaphysalis longicornis*.

In warmer countries *B. canis* is transmitted by *Rhipicephalus sanguineus* and *Haemaphysalis leachi* and in temperate regions of Europe by *Dermacentor marginatus*. In France infections with *B. canis* have two peaks in the year: a spring–summer peak among domestic dogs with *R. sanguineus* as the vector, and an autumn–winter peak among hunting dogs when *Dermacentor spp.* are the vectors (Joyner and Donnelly, 1979). *B. canis* shows synergistic pathogenicity with *Ehrlichia canis*, with *B. canis* destroying the erythrocytes and *E. canis* impeding red cell production. Infected red cells clump together and block capillaries in the brain, leading to death. Young puppies are highly susceptible to *B. canis* (Mahoney, 1977). In France, where there are more than 400,000 cases of canine babesiosis a year, a commercial vaccine, Pirodog®, is available. In trials with 740 dogs it gave 70–100% protection with the disease in vaccinated dogs being generally mild (Moreau *et al.*, 1988). Transovarian transmission of *B. canis* has been observed through five generations (Joyner and Donnelly, 1979).

Cycle of *Babesia* in tick vector (see Fig. 28.1)

Ixodid ticks acquire infections with *Babesia* by the alimentary route when they feed on an infective host. Infections may also be acquired transovarially via the female parent. Strictly speaking transovarian transmission is vertical transmission but it is useful to limit that term to situations where the parasite is passed from generation to generation of the tick without the intake of additional parasites. Such vertical transmission occurs with *B. ovis* in *R. bursa* (Markov and Abramov, 1970) while *B. bovis* and *B. bigemina* have only transovarial transmission in *Bo. microplus* (Friedhoff and Smith, 1981). Infections within the tick can be transmitted transstadially, and this is the only mode of transmission for *B. microti*. Morel (1989) points out that reinfection can occur in one-host ticks and in two-host ticks which are monotropic, i.e. all stages feeding on the same host species.

General cycle (Mehlhorn and Schein, 1984)

Multiplication of *Babesia spp.* involves schizogony in the vertebrate host, gamogony in the intestinal cells of the tick with fusion of gametes followed by sporogony in the salivary glands and other organs. All species invade the erythrocytes of their vertebrate host but in *B. equi* and *B. microti* sporozoites introduced by the tick initially invade the lymphocytes, in which schizogony occurs. The infected lymphocytes burst, releasing motile merozoites which invade erythrocytes. The merozoites undergo rapid reproduction and in the process destroy the erythrocytes in which they develop, causing severe anaemia in the host. Some merozoites do not divide but develop into ovoid or spherical gamonts, which develop no further until ingested by the tick. In the tick the gamonts become free and develop into uninucleate ray-bodies, the Strahlenkörper of Koch, 4–7 μm long. Two to four days after ingestion the ray-bodies pair off and fuse to form zygotes 7–8 μm. They develop into motile kinetes which leave the intestine and enter the haemolymph, from which they invade the cells of the Malpighian tubules, muscle fibres, oocytes and haemocytes. In the host cell the kinete subdivides forming single-membraned uninucleate cytomeres which differentiate into new kinetes. They may leave and invade other cells. On the second day of attachment of the tick the kinetes invade the cells of the salivary gland and multiply. By the fifth day of attachment thousands of sporozoites lie in each enlarged host cell. Sporozoites vary in size being 1.5–2.1 μm in *B. bovis*, 2.5 μm in *B. bigemina* and 3–3.4 μm in *B. equi*. They are pyriform in shape with a broad apex and a pointed posterior pole.

Babesia bigemina and *Babesia bovis*

Riek (1964, 1966) has studied the development of *B. bigemina* and *B. bovis* in *Bo. microplus*. He considers that most of the parasites ingested by the tick are destroyed; and that after 24–36 h kinetes appear in the gut and invade the epithelial cells. Multiple fission occurs in these cells and 72 h after detachment of the tick kinetes appear in the haemolymph. The slightly larger kinetes of *B. bovis* (16 × 3 μm) invade the ovaries, while those of *B. bigemina* (11 × 2.5 μm) have a second cycle of multiple fission in the Malpighian tubules and the cells of the haemolymph, where they form more kinetes which invade the ovary.

A further cycle of multiple fission occurs in the cells of the gut of the developing larva, giving rise to kinetes which appear in the salivary glands of the larva 2 to 3 days after attachment. Further multiple fission occurs in the cells of the

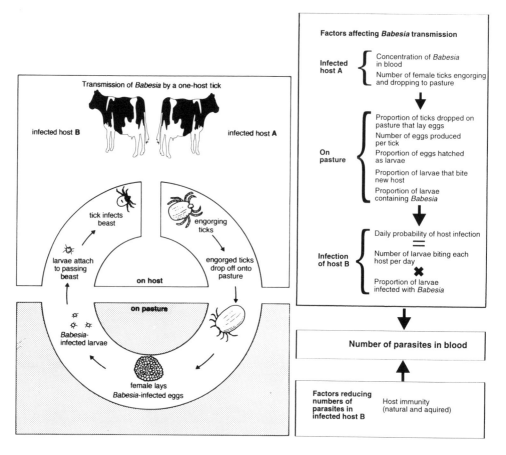

Fig 28.1

Fig. 28.1. Diagrammatical representation of the transmission of a *Babesia* parasite by a one-host tick, accompanied by the factors which determine the level of babesial infection in the environment. Source: from Mahoney (1977).

salivary glands, leading to the production of sporozoites. Development of *B. bigemina* occurs in cells *a* of acinus II and *d* of acinus III, while those of *B. bovis* appeared to be in type *e* cells of acinus III and absent from acini I and 11 (Binnington and Kemp, 1980). In the case of *B. bigemina*, sporozoites are not formed until 8 to 10 days after attachment when the tick is in the nymphal stage. It is of interest that heavy parasitaemia in the bovine host (more than 5%) causes mortality among *Bo. microplus* (Riek, 1966). In *Bo. microplus*, *B. bigemina* does not invade the ovaries until 16–32 h after oviposition has begun. Consequently eggs deposited early in the oviposition cycle, amounting to 13–53% of the egg batch, are uninfected (Friedhoff and Smith, 1981).

Babesia canis, Babesia ovis and other species of Babesia

Infections are usually acquired by the adult female tick, although *R. sanguineus* can transmit *B. canis* in the adult stage having become infected in the preceding nymphal stage (Mahoney, 1977). The three-host ticks *I. ricinus, H. punctata* and *R. sanguineus* acquire, respectively, *B. divergens, B. major* and *B. canis* in the adult stage, and can transmit the parasite in all stages of the next generation, although larvae of *R. sanguineus* only transmit when present in large numbers. Male *R. sanguineus* and *Bo. microplus* can transmit *B. canis* and *B. bigemina*, respectively. *R. bursa* acquires infection with *B. ovis* in the last 4 h before detachment, and in this two-host tick transmission is by the adult stage of the succeeding generation. In *R. bursa*, where both alimentary and vertical infections with *B. ovis* occur, alimentary infections produce higher infection rates and grades of infection than from vertical transmission, but the onset of ovarian infection is retarded compared with vertical transmission (Friedhoff and Smith, 1981).

In the early stages of alimentary infection of *B. ovis* in *R. bursa*, Friedhoff (1981) has described five phases of development involving ray-bodies. Ray-bodies persist in the gut of the tick up to 120 h after detachment, but some give rise to multiple fission bodies in the epithelial cells of the gut, and developing kinetes become discernible 92 h after detachment. Similar bodies have been described in four species of *Babesia* and three species of *Theileria* (Friedhoff, 1981).

The development of sporozoites of *B. canis* in *D. reticulatus* has been studied by Schein *et al.* (1979). The sporozoites are formed in the salivary glands by binary divisions and not by multiple fission. Development of the sporozoites took 2 to 3 days, being completed 4 to 5 days after the adult female had attached and transmission occurred at engorgement.

Epizootiology

The treatment of epizootiology is essentially quantitative. Many of the processes involved change steadily with time, e.g. the protection conferred on calves by the colostrum and innate immunity protects them for some months before it wanes. This is not a sudden change but a steady loss of protection with time. It is convenient for the purposes of calculation to subdivide smoothly changing responses into convenient sections but it must be appreciated that the limits of the sections are arbitrary.

Epizootics occur in three situations: when infected ticks are introduced into a clean area; when susceptible animals are moved into an infected area; and when a temporary reduction in the vector by control or climate leads to animals escaping early infection, and remaining susceptible as adults (Mahoney, 1977). In an enzootically stable situation calves become immunized by infection during the period (about 9 months) they are protected by colostral and innate immunity, when they suffer minimal clinical disease. In a stable enzootic situation the inoculation rate is high. In an enzootically unstable situation the inoculation rate is not high enough to infect all the calves before they become adult, leading to the presence of susceptible adult animals, vulnerable to developing severe clinical disease on infection. When the inoculation rate becomes very low there is the possibility of the disease disappearing (McCosker, 1981).

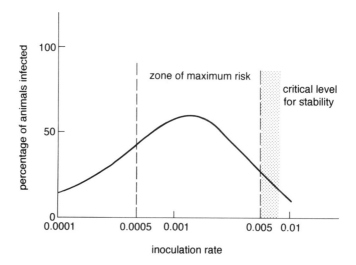

Fig. 28.2. Percentage of animals that become infected with B. bovis after passively acquired immunity wanes at average inoculation rates in the range of 0.0001 and 0.01. Source: Mahoney (1977).

Inoculation rate

The inoculation rate (h), an important parameter in the epizootiology of babesiosis, has been defined as:

$$h = mab$$

when m = bites by vector per beast per day; a = proportion of vectors infected; and b = proportion of infective bites that successfully infect a host (Fig. 28.1). With *Bos taurus* a single larval tick infective with *B. bovis* can cause disease, i.e. $b = 1.0$, but b may have a lower value for resistant cattle, e.g. *Bos indicus* (Mahoney, 1977). In an enzootic situation the inoculation rate is more easily calculated from the proportion of animals infected (I) at age t when

$$I = 1\text{-}e^{-ht}$$

Mahoney (1977) has plotted the 'proportion of cattle in a herd that become infective with *B. bovis* within four years after passive immunity wanes for a wide range of inoculation rates' (Fig. 28.2). These are the infections which are likely to result in severe clinical disease.

As the inoculation rate increases, the proportion of animals infected rises to a maximum of over 50% and then declines. The critical level for enzootic stability is an inoculation rate of 0.005, which Mahoney (1977) equates with 'at least twelve *Boophilus microplus* larvae being required to bite each cow (*Bos taurus*) daily'. Higher rates of infestation with ticks will ensure greater stability and minimal disease, but introduce the complication of other damaging effects of high tick burdens. The zone of maximum risk of babesiosis in this model and that of Smith (1983) is when the inoculation rate (*h*) lies between 0.0005 and 0.005, which Smith equates to two to eight engorged ticks dropping per day.

Basic reproduction rate

Another important parameter is the 'basic reproduction rate (*z*) which is a hypothetical number representing the number of secondary cases of a disease disseminated by a single primary case in a nonimmune individual in an environment in which neither host nor vector populations were previously infected' (Mahoney, 1977). The basic reproduction rate is given by the expression:

$$z = 2 \; dna$$

where d = duration of infectivity in days; n = number of engorged female ticks dropped on the pasture per day; and a = the average infection rate in the larval progeny (Mahoney, 1977). When the basic reproduction rate is below 1.0 then the parasite will disappear. Mahoney and Ross (1972) suggested values of 230–350 days for d for *B. bovis* in *Bos taurus* and 0.04% for a which give a critical value of 3.6 to 5.4 ticks/head/day. Smith (1983) finds that the critical value for the maintenance of babesiae in nature ($z = 1$) is approximately two engorged female ticks/head/day. The actual values are of less importance than the general conclusion that there is a level of transmission below which, other factors remaining constant, the parasite, and therefore the disease, will disappear.

Epizootiology of *Babesia divergens*

There is no proof of a vertebrate reservoir host for a bovine babesia, although non-bovine hosts may harbour *B. bigemina* for a period, and *B. capreoli* in red deer is very close to and may be identical with, *B. divergens* in cattle. Joyner and Donnelly (1979) describe the epizootiology of *B. divergens* and its vector *I. ricinus* in the United Kingdom. Both the vector and the disease have a bimodal seasonal pattern. The autumn rise of *I. ricinus* is composed of new emergences, and the spring rise represents renewed activity of ticks that failed to feed in the autumn. If these are unsuccessful they die off in the late spring. Disease incidence was positively correlated with temperature 14 days earlier, with temperature acting by increasing tick activity. The increase in disease incidence in spring ($\times 2.5$) compared to autumn was reflected in a similar difference in the inoculation rates, which in spring was $\times 2.25$ that in autumn by increasing tick activity.

Human babesiosis (Telford *et al.*, 1993)

The first convincingly demonstrated case of human babesial infection was in a spleenectomized resident of Yugoslavia who died in 1957 after an illness involving fever, anaemia and haemoglobinuria. Since then another 18 cases have been reported in Europe of which half ended fatally – 16 of those 19 patients (84%) were spleenectomized. Another 5 asymptomatic cases occurred in individuals with intact spleens. In 14 of the 19 clinical cases the parasite was identified as *B. divergens*, a parasite of cattle. The probable vector was *Ixodes ricinus*. This disease is more severe than the comparable one in the USA, and requires blood exchange transfusion to reduce the proportion of infected erythrocytes and drug therapy.

The first American case was in 1966, involving an aspleenic Californian resident and the fifth, in 1969, a patient with an intact spleen living on Nantucket Island, off the Massachusetts coast. The pathogen was identified as *B. microti*, a rodent piroplasm, described originally from a vole in Portugal though Uilenberg (1986) cautions that the American parasite may not be the same species. In the next 10 years 43 cases occurred among residents on islands off the New England coast. There are now more than 100 documented human infections by *B. microti* in Massachusetts and New York (Spielman *et al.*, 1985). Human infections with *B. microti* are self-limiting with parasitaemia and symptoms due to red cell destruction persisting for several months (Ruebush, 1991). The severity of the disease varies with the age of the patient and is often asymptomatic, as demonstrated by parasites being found in 101 asympotomatic residents of Shelter Island in New York State. Treatment, where necessary, involves oral quinine and intravenous clindamycin. The epidemiology of human babesiosis is similar to that of Lyme disease (see Chapter 26). *B. microti*, a parasite of the common white-footed mouse *Peromyscus leucopus*, is transmitted by *Ixodes dammini*. The larval tick acquires *B. microti* by feeding on infected *P. leucopus* and transmits the parasite in the nymphal stage. On Nantucket Island 5% of *I. dammini* nymphs were infected with *B. microti* (Spielman *et al.*, 1985).

Theilerioses*

Four genera, *Theileria, Gonderia, Hematoxenus* and *Cytauxzoon*, have been recognized within the Theileriidae. The differences between these four genera seem less absolute today than they did earlier and only a single genus *Theileria* is now recognized (Morel, 1989).

Economic importance of *Theileria* (Morel, 1989; Radostits *et al.*, 1994)

Five species of *Theileria* are found in cattle in various parts of the world. The most important ones are *T. parva* which can cause 90 to 100% morbidity and mortality in *Bos taurus* and *T. annulata* which is highly virulent to European breeds in which it can cause 40 to 80% mortality in enzootic areas. These two species will be considered in more detail later.

Theileria orientalis (= *T. sergenti, T. buffeli*) has a worldwide distribution. It causes a mild disease which may involve severe anaemia in heavily parasitized beasts, especially in the more susceptible European breeds. In eastern Asia and Australia the vector is *Haemaphysalis longicornis* and in western Eurasia and north-western Africa, *H. punctata*. For a long time *T. orientalis* was confused with *T. mutans* but the latter is now regarded as a strictly African species (Uilenberg, 1986).

---

* Mehlhorn *et al.* (1994) was not available when this chapter was revised.

Theileria mutans is widely distributed in the Afrotropical region, where it is a parasite of cattle which is rarely pathogenic but in East Africa there is a pathogenic strain, *T. mutans* (Aitong) (Barnett, 1977). In cattle 45% of the erythrocytes can be parasitized, producing marked anaemia (Purnell, 1977; Young *et al.*, 1978). In tropicoequatorial Africa, Madagascar, Mauritius, Reunion and Guadeloupe the vector is *Amblyomma variegatum*. The infection rate of *T. mutans* in *A. variegatum* is lower than that of *T. p. parva* in *R. appendiculatus* and fewer parasites are produced (Purnell *et al.*, 1975). *T. mutans* (Aitong) has also been transmitted from buffaloes to cattle by *A. cohaerens* (Barnett, 1977; Purnell, 1977).

Theileria velifera causes a mild disease in cattle in tropical Africa. Its distribution coincides with that of *A. variegatum* in Africa and Madagascar; *A. lepidum* in East Africa; and *A. hebraeum* in southern Africa. *T. taurotragi* is a parasite of eland (*Taurotragus oryx*) which is normally non-pathogenic to cattle but in southern Africa it has been associated with turning sickness, when parasitized lymphoblasts accumulate in the central nervous system.

Theileria hirci (= *T. lestoquardi*) causes malignant theileriosis in sheep and goats in a region extending from North Africa through the Middle East to India. In the Sudan 22% of the sheep involved in an outbreak died within 3 to 6 days. It is transmitted by *Hyalomma anatolicum* and *Rhipicephalus* species.

Theileria ovis and *T. separata* cause mild theilerioses in small ruminants and sheep respectively. In tropicoequatorial Africa they are transmitted by *R. evertsi* and *A. variegatum* and in the Mediterranean basin and central-western Asia the distribution of *T. ovis* coincides with that of the vectors *R. bursa* and *R. evertsi*.

Development of *Theileria* species (Mehlhorn and Schein, 1984)

In the vertebrate host

Sporozoites, about 1 μm in diameter, are introduced by an infective tick after it has been feeding for 3 to 5 days during which time kinetes have invaded the salivary glands of the tick and produced infective sporozoites. Within 10 min the sporozoites have invaded lymphoid cells of the host. After 72 h they are 2 μm in diameter and begin to multiply by binary fission producing large schizonts 10 to 15 μm in diameter, containing 13 to 50 nuclei. These macroschizonts are 'Koch's blue bodies'. They stimulate the infective lymphocytes to divide, producing infected daughter cells. (Uninfected lymphocytes do not divide.) When only one or two schizonts are present in a lymphocyte they become spherical and relatively large (6–10 μm in diam.). These microschizonts multiply to produce merozoites. The merozoites are set free and are to be found in erythrocytes beginning 8 days after infection (*T. annulata*) and 13 days (*T. parva*). Up to 90% of the erythrocytes may be infected with *T. annulata*. Commonly two different forms of merozoites are seen – comma-shaped and spherical. Comma-shaped merozoites measuring 1 to 1.5 μm (*T. parva*) and up to 2.5 μm (*T. annulata*) divide by binary fission, leading to the destruction of the erythrocytes, causing anaemia in the host. Spherical merozoites measuring 0.5 to 0.6 μm are probably gamonts which do not develop further until ingested by a tick.

The pathogenicity of a *Theileria* species is related to the density of schizonts in lymphocytes and piroplasms in erythrocytes. *T. parva*, *T. annulata* and *T. hirci* produce numerous schizonts and piroplasms and are very pathogenic. *T.*

orientalis, T. ovis and *T. mutans* produce few schizonts but may cause varying degrees of anaemia when piroplasms are abundant. Schizonts of the benign parasites *T. velifera* and *T. separata* have not been described and only scanty infections of the erythrocytes are found (Radostits *et al.*, 1994).

In the tick vector

In the tick, sexual stages develop from spherical merozoites 2 to 4 days after cessation of feeding. Ray-bodies (8 to 12 μm by 0.8 μm) are found together with spherical 'macrogametes' (4 to 5 μm diam.). The ray-bodies are considered to be microgamonts which develop into uninucleated gamete-like stages. Syngamy of gametes occurs 6 days after feeding giving rise to a zygote – 6 to 24 days later the spherical zygote becomes a motile kinete which is to be found in the intestinal cells of the tick. After the tick has moulted and attached to a host the motile kinetes leave the intestinal cells and invade the cells of the salivary glands but not those of other organs. They are to be found in cell types *d* and *e* of the type III acinus. The parasite becomes polymorphic with nuclear division producing thousands of small cytomeres. Five days after attachment cytomere production is complete and they develop into ovoid sporozoites (1 μm in diam.). It has been estimated that there are 50,000 sporozoites per host cell.

Theileria parva

Three subspecies of *T. parva* are recognized. Classic east coast fever (ECF) of East and Central Africa is caused by *T. parva parva*. Corridor disease of east and southern Africa is caused by *T. parva lawrencei* and January disease or Zimbabwean theileriosis by *T. parva bovis*. These are probably not sound taxa because *T. p. lawrencei* can be transformed into *T. p. parva*, and *T. p. bovis* and *T. p. parva* are serologically inseparable, but they are useful, if not precise, terms to have available (Irvin *et al.*, 1989; Norval et al., 1991a).

In susceptible hosts, imported cattle or previously unexposed indigenous stock, *T. p. parva* and *T. p. lawrencei* can cause 90 to 100% mortality and even zebu cattle, which have a natural immunity to the disease, suffer a 5% mortality in calves in enzootic areas (Dolan, 1987). The incubation period is 1 to 3 weeks. The disease is characterized by high fever, catastrophic lymphoblastosis which destroys vital organs and impairs the immune system, emaciation and death in 7 to 10 days (Yeoman, 1991; Radostits *et al.*, 1994). It has been estimated that ECF killed 1.1 million cattle in 1989, causing losses of US$168 million through loss of milk and meat production, loss of livestock and the cost of tick control (Mukhebi, 1992).

Theileria parva bovis

In Zimbabwe *T. p. bovis* is transmitted by *R. appendiculatus*. It produces a low mortality (15%), and surviving cattle remain infective to ticks for a long period, during which piroplasms are present in the circulating blood (Barnett, 1977).

Theileria parva parva

Theileria parva parva occurs in eastern tropical Africa from the southern Sudan to Malawi and west to Ruanda and eastern Zaire, involving 11 countries in eastern, central and southern Africa. It is lethal to European (*Bos taurus*) and zebu (*Bos indicus*) cattle, and the water buffalo (*Bubalus bubalis*). The main vector is the brown ear tick *Rhipicephalus appendiculatus*, and in the more arid regions of eastern Africa *R. zambeziensis* (Pegram and Walker, 1988). In Angola where *R. appendiculatus* is absent, and in Zaire the vector is *R. duttoni* (Lessard *et al.*, 1990).

In cattle, fever commences when the density of schizonts in the host reaches 7×10^9 and its time of onset is dependent on the number of sporozoites injected. The growth rate of the parasite is dependent on the size of the initial inoculation with a tenfold increase occurring in 4.9 days and 1.4 days following inoculations of 2 and 2000 ticks, respectively. Formation of merozoites is time dependent, and at all dosages piroplasms appear in erythrocytes in 13 days, i.e. the prepatent period is independent of dosage (Barnett, 1977). At death 50% of the erythrocytes may contain piroplasms (Purnell, 1977). With a virulent strain of *T. parva*, single tick infections can kill over 90% of susceptible cattle (Barnett, 1977). Cattle that survive have a solid immunity in which piroplasms are not visible but Young *et al.* (1981) found that in their area of Kenya 'it is possible that the carrier state of *T. p. parva* in cattle is widespread and extends through all age groups'.

Theileria parva lawrencei

Theileria parva lawrencei is a benign parasite of the African buffalo in which it causes a mild disease, but it is lethal to cattle and the water buffalo. *T. p. lawrencei* occurs in East and Central Africa where the vector is *R. appendiculatus*. Infections of *T. p. lawrencei* in cattle are self-limiting because the cattle die before being able to infect ticks. When *T. p. lawrencei* is experimentally transmitted from cattle to cattle four or five times, *T. p. lawrencei* becomes indistinguishable from *T. p. parva*. The process being referred to as transformation (Norval *et al.*, 1991a). There is cross-immunity between *T. p. parva* and *T. p. lawrencei* (Barnett, 1977).

Development of *Theileria parva* in *Rhipicephalus appendiculatus*

Transmission of *T. parva* is trans-stadial. There is no transovarial transmission. Infections acquired by the larva or nymph are transmitted in the succeeding nymph or adult stage (Barnett, 1977). There is considerable variation in the proportion of feeding ticks which become infected and in the intensity of infection in individual ticks. One factor that influences this is the parasitaemia of the host. In one series of experiments the overall infection rate among ticks was 35%; but a significantly higher percentage (61%) were infected when the parasitaemia of the host was 41–50% than at parasitaemias of 6–40% and <5%, when the respective infection rates were 33 and 27%.

The mean number of infected acini per tick was 6.0–6.4 at parasitaemias of 1 to 40%, and somewhat higher (8.6) when the host's parasitaemia was 41–50% (Purnell *et al.*, 1974). There was no close relationship between the intensity of infection in the salivary glands of the tick and the level of parasitaemia in the host animal. It was considered that a random factor, such as juxtaposition of infected gut cells and developing salivary glands during the nymphal moult, was involved (Purnell *et al.*, 1974). When nymphal and larval *R. appendiculatus* were fed on the same host the infection rate in the resulting adults was higher (45%) than in the nymphs (35%). In addition there were more infected acini in adults than in nymphs, 2.25 compared to 0.7, but sporozoites were produced more rapidly in nymphs than in adults, appearing in 2 days compared to 4 days in adult *R. appendiculatus* (Purnell *et al.*, 1971). Infectivity with *T. parva* is lost after 11 months even though the tick may survive for 1 to 2 years (Barnett, 1977).

Development of *T. p. lawrencei* in *R. appendiculatus*

The infection rate among adult *R. appendiculatus* fed in the nymphal stage on infective buffalo is much lower (5.9%) than that obtained with *T. parva* in cattle. Even lower infection rates (2.1%) are obtained when nymphal *R. appendiculatus* feed on infected cattle. The lower infectivity may be related to the presence of fewer piroplasms in erythrocytes. Commonly in cattle less than 0.1% of erythrocytes are infected, and the rate is always below 1% (Young and Purnell, 1973). Buffalo may act as carriers of *T. lawrencei* for more than 26 months but recovered cattle are not carriers. Continued passage of *T. lawrencei* through cattle increases its infectivity to *R. appendiculatus* by increasing the density of piroplasms in the circulating blood and the infection can become indistinguishable from that of *T. parva* (Young and Purnell, 1973; Purnell, 1977).

Control of bovine theileriosis

Halofuginone given orally or intramuscularly and intramuscular parvaquone are effective treatments. Cattle can be immunized by the 'infection-treatment method', which involves injecting *T. parva* sporozoites into a beast and controlling the resulting infection by the administration of parvaquone (Radostits *et al.*, 1994). Irvin *et al.* (1989) used a mild stock of *T. p. bovis* to vaccinate cattle which were challenged by virulent strains of *T. p. parva* and *T. p. lawrencei*. Only 1 beast out of 33 died on challenge and 30 (90%) had either a mild or no reaction. Controlling the disease by controlling the vector is an expensive and demanding operation. It is necessary to treat cattle weekly and the cost of acaricides to Kenya's livestock industry has been estimated at US$6 million to US$10 million per annum (Tatchell, 1987). Barnett (1977) suggests that immune cattle should be kept in a fenced paddock and dipped at 7- or preferably 5-day intervals for a period of 15 months to exhaust the pasture of infected ticks. Susceptible stock may now be introduced but dipping should still be carried out weekly, and if there are buffalo in the area, even this regime will not work.

Epizootiology of *T. p. parva* and *T. p. lawrencei*

Rhipicephalus zambeziensis occurs in seven countries of southern and eastern Africa from South Africa to Tanzania and west to Zambia and Namibia. The distribution of *R. appendiculatus* is much wider being found in 17 countries with an area of more than 3 million km², about 11% of the African continent, stretching from South Africa north to Kenya and the southern Sudan and west to Angola, Zaire and the Central African Republic (Lessard *et al.*, 1990). It is found in savanna and woodland savanna where there is a mixture of grassland and tree cover, but it is absent from open plains and densely forested areas, and its abundance in heavily overgrazed areas is reduced (Perry *et al.*, 1990).

At present *R. appendiculatus* is absent from the horn of Africa. Using CLIMEX to calculate an ecoclimatic index of suitability for survival and development of *R. appendiculatus*, Norval *et al.* (1991b) found that the Ethiopian highlands were climatically suitable and the normalized difference vegetation index (NVDI) showed similarity in vegetation cover between the Ethiopian highlands and highland areas of neighbouring countries where *R. appendiculatus* occurs. Deforestation, which is occurring in the Ethiopian highlands, will transform an area partially suitable for *R. appendiculatus* to one that is highly suitable. *R. appendiculatus* is now widespread in the southern Sudan and could be introduced into Ethiopia. Such an introduction has already occurred with *Bo. annulatus*, which has been present in southern Sudan for many years, and has recently been found in an adjoining region of Ethiopia. The implication is that it has been introduced from the Sudan.

Diapause in adult *R. appendiculatus* appears to occur approximately south of 12°S which coincides closely with the northern boundary of high, dry stress index values (Perry *et al.*, 1990). ECF was introduced into South Africa in 1901 and by 1960, as a result of vigorous control measures, the disease had disappeared. Norval *et al.* (1991a) believe that *T. parva* was introduced into southern Africa with non-diapausing *R. appendiculatus* and that its elimination was due to the eradication of non-diapausing ticks. They argue that diapausing *R. appendiculatus*, which have a seasonal cycle involving little overlap between adults and larvae and none between adults and nymphs, would militate against their being effective vectors of *T. parva*.

The only wild reservoir host for *T. parva* and *T. lawrencei* is the African buffalo (*Syncerus caffer*), and the only effective vector of both organisms is *R. appendiculatus* (Barnett, 1977) [NB *R. zambeziensis* was included in *R. appendiculatus* until 1981 (Pegram and Walker, 1988)]. Seven other species of *Rhipicephalus* and three of *Hyalomma* can transmit *T. parva* experimentally but they are not considered to be able to sustain the parasite on their own (Cunningham, 1974).

Yeoman (1966, 1991) showed that there is a relationship between tick burden and the intensity of ECF which was meaningful at low infestation rates. An average of five adult ticks per beast (two or three per ear) will sustain enzooticity; one to four per head will invite epizooticity in which heavy losses of adult cattle can occur; while an average of less than one per beast can allow sporadic outbreaks; and in ECF-free areas infestations of adult *R. appendiculatus* average one adult for every five or more beasts.

Theileria annulata (Radostits *et al.*, 1994)

Theileria annulata is the agent causing tropical theileriosis or Mediterranean coast fever (not particularly appropriate terms) in a region extending from Morocco and

Portugal in the west via the Middle East to China, an area in which it is estimated that 200 million cattle are at risk of contracting the disease. Although less pathogenic than *T. parva*, *T. annulata* may be economically the more important species because of its wider distribution in the world (Barnett, 1977). Its presence is an obstacle to livestock improvement. In enzootic areas *Bos taurus* cattle may suffer mortalities of up to 40 to 80% and even *Bos indicus* cattle, which are much more resistant to *Theileria* than imported cattle breeds, may still suffer 15 to 20% mortality, with deaths occurring mainly in calves. The disease involves hypertrophy of the lymphoid tissue, fever, weight loss and haemolytic anaemia with more than 30% of the erythrocytes being parasitized. Recovered cattle have a good, persistent immunity of the premunity type in which there is persistent parasitaemia (Hashemi-Fesharki, 1988).

The vectors of *T. annulata* are various species of *Hyalomma*, most of which have only one generation a year. Cattle which have recovered from infection with *T. annulata* continue to harbour parasites in the circulating blood and ticks easily become infected with infection rates of up to 64% being recorded for field-collected *H. anatolicum*. *T. annulata* is transmitted by three-host ticks, such as *H. anatolicum anatolicum*, two-host ticks such as *H. detritum*, and even by *H. scupense*, a one-host tick in which the adult remains on the host over the winter period and transmits the disease by moving from one host to another when cattle are in close contact (Barnett, 1977). According to Barnett (1977), *H. anatolicum excavatum* is a two-host tick in which the early stages occur on rodents, and since transmission is transstadial, the tick cannot be a vector in the field, although it is readily infected experimentally. In other areas *H. a. excavatum* is described as a three-host tick and a vector of *T. annulata* (Samish and Pipano, 1978).

The development of *T. annulata* in *H. a. excavatum* has been described by Schein *et al.* (1975) and Schein and Friedhoff (1978). They described the formation of microgametes and macrogametes, four microgametes being formed from each microgamont. These are formed in the first 96 h, and from day 5 after repletion zygotes appear in the epithelial cells of the gut and grow steadily to day 12. The zygotes transform into kinetes which at first move within the epithelial cells, and then from day 17 kinetes, measuring about 18 μm, are to be found in the haemolymph.

They reach the salivary glands 18 days after repletion and transform into fission bodies about 10 μm in diameter in type III and less frequently in type II acini. Infected host cells become greatly enlarged, growing from 15 to 110 μm. The parasite divides several times before sporozoites are formed and released into the saliva. This takes about 2 days in young ticks; 5 to 7 days in ticks which have been starved for 6 months; and when ticks have been starved for 6 to 9 months no 'sporozoites' may develop during their feeding period. When ticks feed on hosts with parasitaemias exceeding 40%, they suffer disease and mortality (Schein and Friedhoff, 1978).

An attenuated live schizont tissue culture vaccine has been used in Iran to protect 100,000 *Bos taurus* cattle of eight breeds and it has been so successful that it is planned to extend the programme to *Bos indicus* cattle. Mortality among vaccinated cattle was 0.04% compared to 80% among non-vaccinated beasts. Similar results have been obtained in the former USSR (Hashemi-Fesharki, 1988). It is hoped that a better understanding of the nature of attenuation will reduce or eliminate the need for virulence tests and speed up the process of attenuation (Hall and Baylis, 1993).

References

Barnett, S.F. (1977) *Theileria*. In: Kreier J.P. (ed.) *Parasitic Protozoa*. Academic Press, New York, vol. IV, pp. 77–113.

Binnington, K.C. and Kemp, D.H. (1980) Role of tick salivary glands in feeding and disease transmission. *Advances in Parasitology* 18, 315–339.

Cunningham, M.C. (1974) East coast fever – ECF – cycle in host and vector. In: *East Coast Fever and Related Tick-Borne Diseases*. Food and Agricultural Organization, Rome (1980), pp. 1–2.

De Waal, D.T. (1992) Equine piroplasmosis: a review. *British Veterinary Journal* 148, 6–13.

Dolan, T.T. (1987) Immunization to control East coast fever. *Parasitology Today* 3, 4–6.

Friedhoff, K.T. (1981) Morphologic aspects of *Babesia* in the tick. In: Ristic, M. and Kreier, J.P. (eds) *Babesiosis*. Academic Press, New York, pp. 143–169.

Friedhoff, K.T. and Smith, R.D. (1981) Transmission of *Babesia* by tick. In: Ristic, M. and Kreier J.P. (eds) *Babesiosis*. Academic Press, New York, pp. 267–321.

Habela, M., Reina, D., Nieto, C. and Navarrete, I. (1990) Antibody response and duration of latent infection in sheep following experimental infection with *Babesia ovis*. *Veterinary Parasitology* 35, 1–10.

Hall, R. and Baylis, H.A. (1993) Tropical theileriosis. *Parasitology Today* 9, 310–312.

Hashemi-Fesharki, R. (1988) Control of *Theileria annulata* in Iran. *Parasitology Today* 4, 36–40.

Hoyte, H.M.D. (1961) Initial development of infections with *Babesia bigemina*. *Journal of Protozoology* 8, 462–466.

Irvin, A.D., Morzaria, S.P., Munatswa, F.C. and Norval, R.A.J. (1989) Immunization of cattle with a *Theileria parva bovis* stock from Zimbabwe protects against challenge with virulent *T. p. parva* and *T. p. lawrencei* stocks from Kenya. *Veterinary Parasitology* 32, 271–278.

Joyner, L.P. and Donnelly, J. (1979) The epidemiology of babesial infections. *Advances in Parasitology* 17, 115–140.

Kakoma, I. and Mehlhorn, H. (1994) *Babesia* of domestic animals. In: Kreier, J.P. (ed.) *Parasitic Protozoa*, 2nd edn. Academic Press, San Diego, vol. 7, pp. 141–216.

Lessard, P., L'Eplattenier, R., Norval, R.A.I., Kundert, K., Dolan, T.T., Croze, H., Walker, J.B., Irvin, A.D. and Perry, B.D. (1990) Geographical information systems for studying the epidemiology of cattle diseases caused by *Theileria parva*. *The Veterinary Record* 126, 255–262.

Mahoney, D.F. (1977) *Babesia* of domestic animals. In: Kreier, J.P. (ed.) *Parasitic Protozoa*. Academic Press, New York, vol. IV, pp. 1–52.

Mahoney, D.F. and Ross, D.R. (1972) Epizootiological factors in the control of bovine babesiosis. *Australian Veterinary Journal* 48, 292–298.

Markov, A.A. and Abramov, I.V. (1970) Results of twenty years' observations on repeated life cycles of *Babesia ovis* in 44 generations of *Rhipicephalus bursa*. *Veterinary Bulletin* 42, 1911(1972).

McCosker, P.J. (1981) The global importance of babesiosis. In: Ristic, M. and Kreier J.P. (eds) *Babesiosis*. Academic Press, New York, pp. 1–24.

Mehlhorn, H. and Schein, E. (1984) The piroplasms: life cycle and sexual stages. *Advances in Parasitology* 23, 37–103.

Mehlhorn, H., Schein, E. and Ahmed, J.S. (1994) *Theileria*. In: Kreier, J.P. (ed.) *Parasitic Protozoa*, 2nd edn. Academic Press. San Diego, vol. 7, pp. 217–304.

Moreau, Y., Martinod, S. and Fayet, G. (1988) Epidemiologic and immunoprophylactic aspects of canine babesiosis in France. In: Ristic, M. (ed.) *Babesiosis of Domestic Animals and Man*. CRC Press, Boca Raton, Florida, USA, pp. 191–196.

Morel, P. (1989) Tick-borne diseases of livestock in Africa. In: Shah-Fischer, M. and Say, R.D. (eds) *Manual of Tropical Veterinary Parasitology*. CAB International, Wallingford, Oxon, UK, pp. 301–463.

Mukhebi, A.W. (1992) Economic impact of theileriosis and its control in Africa. In: Norval, R.A.I., Perry, B.D. and Young, A.S. *The Epidemiology of Theileriosis in Africa*. Academic Press, London, pp. 379–403.

Nikol'skii, S.M., Nikiforenko, V.I. and Pozov, S.A. (1977) Epizootiology of piroplasmosis in Siberia. *Veterinariya* 4, 71–75.

Norval, R.A.I., Lawrence, J.A., Young, A.S., Perry, B.D., Dolan, T.T. and Scott, J.B. (1991a) *Theileria parva*: influence of vector, parasite and host relationships on the epidemiology of theileriosis in southern Africa. *Parasitology* 102, 347–356.

Norval, R.A.I., Perry, B.D., Gebreab, F. and Lessard, P. (1991b) East Coast fever a problem of the future for the horn of Africa? *Preventive Veterinary Medicine* 10, 163–172.

Pegram, R.G. and Walker, J.B. (1988) Classification of the biosystematics and vector status of some African *Rhipicephalus* species (Acarina: Ixodidae). In: Service, M.W. (ed.) *Biosystematics of Haematophagous Insects*. Clarendon Press, Oxford, pp. 61–76.

Perry, B.D., Lessard, P., Norval, R.A.I., Kundert, K. and Kruska, R. (1990) Climate, vegetation and the distribution of *Rhipicephalus appendiculatus* in Africa. *Parasitology Today* 6, 100–104.

Purnell, R.E. (1977) East coast fever: some recent research in East Africa. *Advances in Parasitology* 15, 83–132.

Purnell, R.E. (1981) Babesiosis in various hosts. In: Ristic, M. and Kreier J.P. (eds) *Babesiosis*. Academic Press, New York, pp. 25–63.

Purnell, R.E., Boarer, C.D.H. and Peirce, M.A. (1971) *Theileria parva*: comparative infection rates of adult and nymphal *Rhipicephalus appendiculatus*. *Parasitology* 62, 349–353.

Purnell, R.E., Ledger, M.A., Omwoyo, P.L., Payne, R.C. and Peirce, M.A. (1974) *Theileria parva*: variation in the infection rate of the vector tick *Rhipicephalus appendiculatus*. *International Journal of Parasitology* 4, 513–517.

Purnell, R.E., Young, A.S., Payne, R.C. and Mwangi, J.M. (1975) Development of *Theileria mutans* (Aitong) in the tick *Amblyomma variegatum* compared to that of *T. parva* (Muguga) in *Rhipicephalus appendiculatus*. *Journal of Parasitology* 61, 725–729.

Radostits, O.M., Blood, D.C. and Gay, G.C. (1994) *Veterinary Medicine. A Textbook of the Diseases of Cattle, Sheep, Pigs, Goats and Horses*. Baillière Tindall, London.

Riek, R.F. (1964) The life cycle of *Babesia bigemina* (Smith and Kilborne, 1893) in the tick vector *Boophilus microplus* (Canestrini). *Australian Journal of Agricultural Research* 15, 802–821.

Riek, R.F. (1966) The life cycle of *Babesia argentina* (Lignières, 1903) (Sporozoa: Piroplasmidea) in the tick vector *Boophilus microplus* (Canestrini). *Australian Journal of Agricultural Research* 17, 247–254.

Ruebush, T.K. (1991) Babesiosis. In: Strickland, G.T. (ed.) *Hunter's Tropical Medicine*. Saunders, Philadelphia, pp. 655–658.

Samish, M. and Pipano, E. (1978) Transmission of *Theileria annulata* by two and three host ticks of the genus *Hyalomma* (Ixodidae). In: Wilde, J.K.H. (ed.) *Tick-borne Diseases and Their Vectors*. Centre of Tropical Veterinary Medicine, University of Edinburgh, pp. 371–372.

Schein, E. and Friedhoff, K.T. (1978) Lichtmikroskopische Untersuchungen über die

Entwicklung von *Theileria annulata* (Dschunkowsky and Luhs, 1904) in *Hyalomma anatolicum excavatum* (Koch, 1844). II. Die Entwicklung in Hämolymphe und Speicheldrüsen. *Zeitschrift für Parasitenkunde* 56, 287–303.

Schein, E., Buscher, G. and Friedhoff, K.T. (1975) Lichtmikroskopische Untersuchungen über die Entwicklung von *Theileria annulata* (Dschunkowsky und Luhs, 1904) in *Hyalomma anatolicum excavatum* (Koch, 1844).1. Die Entwicklung im Darm vollgesogener Nymphen. *Zeitschrift für Parasitenkunde* 48, 123–136.

Schein, E., Mehlhorn, H. and Voigt, W.P. (1979) Electron microscopical studies on the development of *Babesia canis* (Sporozoa) in the salivary glands of the vector tick *Dermacentor reticulatus*. Acta Tropica 36, 229–241.

Smith, R.D. (1983) *Babesia bovis*: computer simulation of the relationship between the tick vector, parasite and bovine host. *Experimental Parasitology* 56, 27–40.

Smith, R.D. (1984) Epidemiology of babesiosis. In: Ristic, M., Ambroise-Thomas, P. and Kreier, J.P. (eds) *Malaria and Babesiosis*. Martinus Nijhoff Publishers, Dordrecht, pp. 207–232.

Smith, T. and Kilborne, F.L. (1893) Investigations into the nature, causation and prevention of Texas or southern cattle fever. *United States Department of Agriculture, Bureau of Animal Industry Bulletin* 1, 1–301.

Spielman, A., Wilson, M.L., Levine, J.F. and Piesman, J. (1985) Ecology of *Ixodes dammini*-borne human babesiosis and Lyme disease. *Annual Review of Entomology* 30, 439–460.

Tatchell, R.J. (1987) Tick control in the context of ECF immunization. *Parasitology Today* 3, 6–10.

Telford, S.R., Gorenflot, A., Brasseur, P. and Spielman, A. (1993) Babesial infections in humans and wildlife. In: Kreier, J.P. (ed.) *Parasitic Protozoa*. Academic Press, San Diego, vol. 5, pp. 1–47.

Uilenberg, G. (1986) Highlights in recent research on tick-borne diseases of domestic animals. *The Journal of Parasitology* 72, 485–491.

Yeoman, G.H. (1966) Field vector studies of epizootic east coast fever. I. A quantitative relationship between *R. appendiculatus* and the epizooticity of east coast fever. *Bulletin of Epizootic Diseases of Africa* 14, 5–27.

Yeoman, G.H. (1991) East coast fever: Africa's stubborn problem. *The Veterinary Record* 129, 414–415.

Young, A.S. and Purnell, R.E. (1973) Transmission of *Theileria lawrencei* (Serengeti) by the ixodid tick, *Rhipicephalus appendiculatus*. *Tropical Animal Health and Production* 5, 146–152.

Young, A.S., Purnell, R.E., Payne, R.C., Brown, C.G.D. and Kanhai, G.K. (1978) Studies on the transmission and course of infection of a Kenyan strain of *Theileria mutans*. *Parasitology* 76, 99–115.

Young, A.S., Leitch, B.L. and Newson, R.M. (1981) The occurrence of a *Theileria parva* carrier state in cattle from an east coast fever endemic area of Kenya. In: Irvin, A.D., Cunningham, M.P. and Young A.S. (eds) *Advances in the Control of Theileriosis*. Martinus Nijhoff, The Hague, pp. 60–62.

Trypanosomiases and Leishmaniases

29

The trypanosomiases are diseases of humans and livestock, and the leishmaniases largely diseases of humans. They are caused by parasitic flagellate Protozoa of the order Kinetoplastida (Phylum Sarcomastigophora). Members of this order possess a single mitochondrion, Golgi; a relatively small and compact kinetoplast, a DNA-containing particle, close to the flagellar pocket; and a single flagellum (Vickerman, 1976; Seed and Hall, 1992).

Species of two genera, *Trypanosoma* and *Leishmania*, are of economic importance, causing, respectively, trypanosomiasis and leishmaniasis. At different stages of the developmental cycle the parasite takes various forms, five of which will be defined briefly. In the amastigote the flagellar base and kinetoplast are anterior to the nucleus and there is no free flagellum; while in the promastigote the flagellar base is also anterior to the nucleus and there is a flagellum which emerges from the anterior end of the body. In the opisthomastigote the flagellar base is behind the nucleus and there is a long flagellar pocket leading to the anterior end. In the epimastigote the flagellar base is anterior to the nucleus and the flagellum emerges laterally to form an undulating membrane as it runs along the body to the anterior end; while in the trypomastigote the flagellar base is behind the nucleus, the flagellum emerges laterally and there is a long, undulating membrane (Vickerman, 1976).

Species of *Trypanosoma* occur as blood parasites in a wide range of vertebrates, from fish to mammals. Trypomastigote and epimastigote stages are common to nearly all trypanosome life cycles. *Leishmania* species are characterized by intracellular amastigotes in the mammalian host and extracellular promastigotes in the gut lumen of phlebotomine sandflies (Vickerman, 1976).

Trypanosomiases

The economically important and other representative trypanosomes transmitted by insects are listed in Table 29.1. The genus *Trypanosoma* is divided into two

Table 29.1. List of species of *Trypanosoma* transmitted by insects.

Subgenus/species/subspecies	Vector	Susceptible host
Section Salivaria		
Trypanozoon evansi	Tabanids, *Stomoxys*	Livestock
Trypanozoon brucei brucei	*Glossina*	Livestock (horse)
Trypanozoon brucei gambiense	*Glossina*	Humans
Trypanozoon brucei rhodesiense	*Glossina*	Humans
Nannomonas congolense	*Glossina*	Livestock
Nannomonas simiae	*Glossina*	Pigs
Duttonella vivax	*Glossina*	Livestock
Duttonella vivax viennei	Tabanids, *Stomoxys*	Livestock
Duttonella uniforme	*Glossina*	Cattle
Pycnomonas suis	*Glossina*	Pigs
Section Stercoraria		
Schizotrypanum cruzi	Triatomine bugs	Humans
Megatrypanum theileri	Tabanids	Cattle
Megatrypanum melophagium	*Melophagus ovinus*	Sheep
Herpetosoma lewisi	Rat fleas	Rats
Herpetosoma rangeli	Triatomine bugs	Humans, dog
*(unresolved) grayi**	*Glossina palpalis*	Crocodile

*Note: Dr J.R. Baker informs me that subgenera have only been erected for mammalian trypanosomes and hence no subgenus can be given for *T. grayi*, a parasite of crocodiles.

sections, the Salivaria and the Stercoraria, which differ in their modes of transmission. Salivarian trypanosomes either undergo cyclic development in the insect before being transmitted with the saliva, or are transmitted mechanically. Stercorarian trypanosomes also undergo development in the insect but, with the exception of *T. rangeli*, the infective forms are deposited in the faeces of the vector. Infection rates of salivarian trypanosomes in the vector are often low, even when the vector has been fed on infective hosts in the laboratory. In contrast vectors of stercorarian trypanosomes fed on an infective host have a very high infection rate, approaching 100%.

A characteristic feature of trypanosome infections in animals is that their abundance follows a wave-like pattern with time. In each wave of parasitaemia one variant will dominate and then be cleared as the host develops the appropriate antibody to the surface-coat protein (VSA = variant specific antigen) of the parasite. This is a glycoprotein of approximately 60,000 kDa. Following clearance, another variant will dominate but the switching from one to another is not due to exposure to the antibody because it occurs in immunosuppressed animals. More than 60 different antigenic types (VAT) have been detected in a cloned strain of *T. b. gambiense*. The VSA of metacyclic forms is similar and the injected population may contain as many as 27 different VATs (Turner *et al.*, 1988). The predominant VAT in the metacyclics will be the one that dominated in the infective blood meal. In *Glossina* the procyclic trypomastigotes have no VSA coating but are uniformly coated with a new antigen, procyclin (Seed and Hall, 1992).

In the following account of the trypanosomiases, the salivarian species transmitted by *Glossina* will be considered first, other insect-borne salivarian trypanosomes next, and finally the stercorarian trypanosomes.

Life Cycles of Salivarian Trypanosomiases

Salivarian trypanosomes, with two exceptions, are transmitted by species of *Glossina*. The two exceptions, *T. evansi* and *T. vivax viennei*, are mechanically transmitted by tabanids and *Stomoxys* outside the range of *Glossina*. Four subgenera are recognized within the salivarian species of *Trypanosoma*. They are: *Trypanozoon, Nannomonas, Duttonella* and *Pycnomonas* (Logan-Henfrey *et al.*, 1992).

Life cycle of *T. brucei* (Fig. 29.1) (Seed and Hall, 1992)

In the vertebrate host, infective metacyclic forms injected by a feeding *Glossina* develop into rapidly dividing long slender (LS) trypomastigotes (20–40 × 1 μm). At a later stage there is a switch to an intermediate (I) form, which is rarely seen to divide and then to a non-dividing short stumpy (SS) form (15–25 × 3.5 μm) which lacks a free flagellum. It has been suggested that the SS forms are not infective to the mammalian host and that they and the I forms are infective for *Glossina*.

Trypanosomes pass with the ingested blood into the crop from which they are passed via the proventriculus into the midgut where they are contained within the peritrophic membrane, i.e. they are in the endoperitrophic space. In the midgut the I and SS forms differentiate into procyclic trypomastigotes which develop and divide extensively. It has been shown that a single ingested trypanosome can infect a tsetse fly. The trypanosomes move into the ectoperitrophic space via the open free end of the membrane or by penetrating the membrane. This process takes about 2 weeks after an infective feed.

In the ectoperitrophic space the trypanosomes move forward to the proventriculus where they pass through the soft newly secreted peritrophic membrane to become free in the oesophagus and move down the food channel to the opening of the salivary duct in the hypopharynx. They pass along the salivary duct to reach the salivary glands, where they develop into epimastigotes which attach to the epithelial cells of the salivary glands by means of a junction (hemidesmosome) between the flagellum and the microvilli. Extensive division takes place and finally metacyclic forms detach from the microvilli and develop a surface coat like that found in trypanosomes in the vertebrate host. Metacyclic forms injected into a susceptible vertebrate host develop into rapidly multiplying long slender trypomastigotes.

There is evidence to suggest that genetic exchange may occur in the vector but it is not obligatory (Tait and Turner, 1990). Five different genotypes were obtained by Gibson (1989) by cotransmitting two clones of *T. b. rhodesiense* through *G. morsitans*; Schweizer and Jenni (1991) recovered four hybrids from eight salivary gland isolates and one from 34 midgut isolations.

The probability of transmission occurring is enhanced by the fact that the first drop of saliva produced by the feeding fly is the largest and will presumably carry

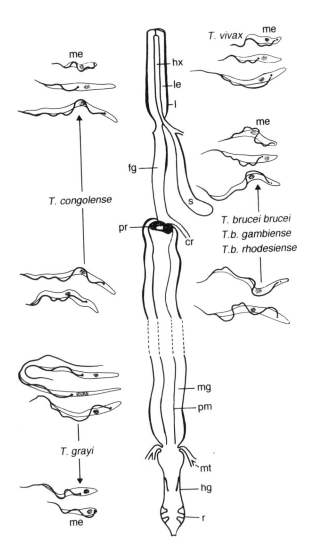

Fig. 29.1. Developmental patterns of trypanosomes in *Glossina*. Top left, *T. congolense*. Bottom left, *T. grayi*. Top right, *T. vivax*. Middle right, *T. brucei brucei*, *T. b. gambiense* and *T. b. rhodesiense*. *cr*, crop duct; *fg*, foregut; *hg*, hindgut; *hx*, hypopharynx; *l*, labium; *le*, labrum-epipharynx; *me*, metacyclic trypanosomes; *mg*, midgut; *mt*, Malpighian tubes; *pm*, peritrophic membrane; *pr*, proventriculus; *r*, rectum; *s*, salivary gland. Redrawn from Hoare (1972).

with it most metacyclics. If feeding is interrupted another large drop of saliva is produced when the tsetse attempts to feed again (Youdeowei, 1975). This implies that a probing fly is almost as likely to transmit as one that feeds to repletion.

Evans and Ellis (1983) have reviewed observations which indicate that the above description may be incomplete. Trypanosomes have been found in the haemolymph of *Glossina*, and penetrating the cells of the midgut. Undamaged parasites have been seen between the basement membrane of the midgut cells and the haemocoele membrane. The significance of these observations is unclear. It is claimed that trypanosomes that reach the haemocoele infect the salivary glands, enabling normal transmission to occur when the tsetse fly feeds. Certainly such a route would be considerably shorter than that involved in moving to the salivary gland via the salivary duct.

Life cycles of *T. congolense* and *T. vivax* (Hoare, 1972; Logan-Henfrey *et al.*, 1992)

The development cycle of *T. congolense* in *Glossina* is similar in part to that of *T. brucei*. In the midgut procyclic trypomastigotes develop with a surface coat of procyclin. They pass through the same stages in the endo- and ecto-peritrophic spaces and move forward through the proventriculus to the pharynx and food channel, where they differentiate into epimastigotes and anchor themselves to the walls of those structures. Here they divide and eventually differentiate into metacyclic trypomastigotes. Uncoated premetacyclics attach to the wall of the hypopharynx before becoming free, coated, metacyclic trypomastigotes capable of infecting a susceptible host.

In *Glossina morsitans morsitans T. congolense* penetrates the peritrophic membrane in the central region of the midgut and 7 days after an infective feed heavy infections were found in both the endoperitrophic and ectoperitrophic spaces; by 21 days epimastigotes were to be found attached to the food canal by their flagella and a week later free forms were found in the hypopharynx. *T. congolense* was found to penetrate the midgut in folds between the midgut cells and not by penetrating the cells (Evans *et al.*, 1979).

Trypanosoma congolense forms large clusters of organisms on the labrum of *Glossina*, and it has been postulated that they interfere with the sensory receptors of the fly, favouring increased probing and therefore facilitating transmission of the parasite. It has been observed that *Glossina* infected with *T. brucei* or *T. congolense* probe more than uninfected flies (Molyneux *et al.*, 1979b).

Ingested trypomastigotes of *T. vivax* attach to the walls of the food channel, pharynx and oesophagus by means of hemidesmosomes formed between their flagella and the chitinous walls of the surrounding structures. Attached trypomastigotes multiply and differentiate into epimastigotes and finally into premetacyclic forms which detach and migrate to the hypopharynx where they mature into coated infective metacyclics. Trypomastigotes of *T. vivax* that pass into the midgut of the tsetse fly perish.

Implications of trypanosome cycle in *Glossina*

As would be expected from the above descriptions of the developmental cycles of trypanosomes in *Glossina*, the cycle of *T. vivax* is completed in a much shorter time (5–13 days) than those of *T. congolense* (2 to 3 weeks) and *T. brucei* (3 to 5 weeks). The cycle of *T. (Pycnomonas) suis* in *Glossina* takes a similar time to that of *Trypanozoon*, i.e. about 4 weeks. This relatively long development time is consistent with the longevity of *Glossina*, which is about 6 weeks for males and 15 weeks for females. Infections with *Trypanozoon* and *Nannomonas* in *Glossina* last for the life of the fly, while those of *Duttonella* persist in the proboscis for up to 8 weeks (Raadt and Seed, 1977; Soltys and Woo, 1977).

In natural populations of *Glossina* there is considerable variation in their infection rates with *Trypanosoma*. In the data tabulated by Jordan (1974), infection rates varied from 0.2% in *G. fuscipes martinii* in Zambia to 76.6% in *G. morsitans submorsitans* in Nigeria. The highest infection rates were of *T. vivax*, and the lowest of *T. brucei*, with infections of *T. congolense* being intermediate in frequency. As a reasonable approximation it can be taken that there is an order of magnitude difference in the percentage infectivity of the three subgenera of *Trypanosoma* in *Glossina*, being of the order of 20% for *Duttonella*, 2% for *Nannomonas*, and 0.2% for *Trypanozoon* (Ford, 1971; Hoare, 1972; Jordan, 1974).

Susceptibility to infection with *T. brucei* is a function of age of the vector. Unless a *Glossina* feeds on an infected host very soon after eclosion it is unlikely to become infective. Highest percentages of infection are established in flies that feed on an infected host within 48 h of emergence and preferably within 24 h; that is at the first feed. Even then infectivity is unlikely to exceed 10%. This relationship between early feeding and infectivity gave rise to the observation that flies reared from puparia kept at high temperatures were more readily infected. At eclosion such flies would be deficient in food reserves and feed early.

The dependence of infectivity on age is related to the development of the peritrophic membrane, which is not secreted until after eclosion. Flies that feed early in adult life have shorter membranes, and the trypomastigotes would more easily reach the ectoperitrophic space. When the peritrophic membrane is fully developed the trypomastigotes would have to travel further down the midgut to find the end of the membrane, and this could carry them into the zone where the pH is lethal (Freeman, 1973) and account for resistance to infection. The peritrophic membrane is secreted continuously, but particularly after a blood meal.

Three stages (initial, established, mature) can be recognized in the development of an infection of *T. brucei* in *Glossina*. When *G. morsitans* was fed on infected blood, 50% of the flies developed initial infections in the midgut 3 days later but only 9% developed established infections in the ectoperitrophic space and foregut. Established infections were detectable 5 to 30 days after the infective feed. However, the number of mature infections, i.e. flies able to transmit metacyclics, was very low. Only 39 flies out of many hundreds examined successfully transmitted *T. brucei* by bite and in only 15 of those were infections observed in the salivary glands (Dipeolu and Adam, 1974).

Human Trypanosomiasis (Sleeping Sickness) (Bales, 1991; Kuzoe, 1993)

In Africa, human trypanosomiasis is present from 14°N to 29°S involving 36 countries and some 200 foci of the disease. At the turn of the century human trypanosomiasis caused half a million deaths in Zaire and half that number around Lake Victoria, East Africa. By the early 1950s the disease was under control but civil and political unrest plus the cost to the health budget of AIDS has resulted in epidemics with about 25,000 new cases being reported annually. In 1976–1983 in Busoga, Uganda 20,000 new cases resulted from the failure of the fly control barrier to prevent the spread of *G. f. fuscipes* (Abaru, 1985).

Two forms of human sleeping sickness are recognized. Gambian sleeping sickness is caused by *T. brucei gambiense*, and Rhodesian sleeping sickness by *T. brucei rhodesiense*. The organisms concerned are morphologically indistinguishable from the animal pathogen *T. brucei brucei*. Trypanosomes of *T. b. brucei* are lysed by a non-specific trypanocidal factor found in human serum while those of *T. b. gambiense* and *T. b. rhodesiense* are unaffected. Taxonomically there is little justification for differentiating between *T. b. gambiense* and *T. b. rhodesiense*, but epidemiologically there is value in making the distinction. *T. b. gambiense* and *T. b. rhodesiense* are geographically and ecologically separated and produce clinically different human diseases.

A chancre (ulcer) commonly develops at the site of an infective bite after 5 to 15 days. Within 1 to 3 weeks of being infected trypanosomes can be found in the haemolymphatic system producing lymphadenopathy and splenomegaly. This is followed by the trypanosomes insidiously invading the central nervous system, a development which occurs in weeks to months after infection by *T. b. rhodesiense* or in months to years with *T. b. gambiense*. The victim shows increasing indifference, lassitude and daytime somnolence leading to death in 9 months (*T. b. rhodesiense*) or 4 years (*T. b. gambiense*) (Seed and Hall, 1992).

Early infections with *T. b. gambiense* are treated with intramuscular injections of pentamidine, a drug which acts slowly but is remarkably effective combined with intravenous suranim, a drug which causes rapid elimination of trypanosomes of *T. b. gambiense* and *T. b. rhodesiense*. Both drugs have poor penetration of the central nervous system and are ineffective against late stage infections. Melarsoprol (Mel B) has to be given intravenously. Its poor penetration of the central nervous system (CNS) is overcome by it being highly trypanocidal and it is effective against late infections and relapses of both subspecies. In 1990 eflornithine (DMFO) became the first new drug to become available for 40 years. It is given orally; shows excellent penetration of the CNS and is very effective against late-stage *T. b. gambiense* infections but it is expensive (Pépin and Milord, 1994).

Trypanosoma brucei gambiense occurs in West and in western and northern Central Africa and *T. b. rhodesiense* in Central and East Africa. *T. b. gambiense* is associated with the *Nemorhina* subgenus of *Glossina*, and its vectors are the riverine and lacustrine tsetse *G. palpalis*, *G. fuscipes* and *G. tachinoides*, whereas the vectors of *T. b. rhodesiense* are the savanna species of *Glossina* (*Glossina*), *G. morsitans*, *G. pallidipes* and *G. swynnertoni*. Genetic exchange can occur between *T. b. brucei* and both *T. b. rhodesiense* and *T. b. gambiense*. There are indications that the *brucei–rhodesiense* population is evolving rapidly while *gambiense* trypanosomes are more stable. *Gambiense* stocks can be distinguished from non-*gambiense* stocks by their isoenzyme characteristics (Seed and Hall, 1992).

Gambiense sleeping sickness

Gambiense sleeping sickness had been regarded as an anthroponosis, involving transmission from one person to another, but evidence is accumulating that domestic pigs may play an active role as reservoirs of *T. b. gambiense* (Molyneux, 1980). In some forest and forest/savanna villages in West Africa where domestic pigs are kept, peridomestic populations of *G. palpalis* and *G. tachinoides* have become established, feeding on pigs and humans, and with the pigs acting as hosts of *T. gambiense*, the villages become foci of human trypanosomiasis (Baldry, 1980).

Species of *Glossina (Nemorhina)* are opportunist feeders, feeding equally on reptiles, bovids and humans and to a lesser extent on suids. Weitz (1970) gives percentages of feeds for the respective hosts as 9–34%; 30–38%; 8–40%; and 2–6%. Gambiense sleeping sickness is transmitted when there is close association between tsetse and humans as occurs, for example, in the dry season when people and *G. palpalis* concentrate around waterholes (Nash, 1978). Other hazardous situations occur when palm cutters work in raffia beds (*Raphia sudanica*), the haunt of *G. tachinoides*, or fishermen land on a lake shore frequented by *G. fuscipes*.

Gambiense sleeping sickness survives in relatively modest-sized, more or less separated foci. One problem has been the survival of the disease at low endemic levels. One explanation would be the existence of an animal reservoir, but there is an alternative explanation. Molyneux *et al.* (1979a) have presented evidence that tsetse flies can be dispersed by wind over long distances in West Africa. They draw attention to the fact that foci tend to be distributed in south-west to north-east series, an arrangement which could be explained by *Glossina* being carried by winds associated with the Inter Tropical Convergence Zone. The foci would then represent particularly favourable areas for dispersed flies to settle, feed and transmit infection. Wind carriage of insects over long distances in Africa has been well documented for the desert locust (*Schistocerca gregaria*) and *Simulium damnosum* (see p. 667).

Rhodesiense sleeping sickness

This disease is an anthropozoonosis in which the natural cycle is from animal to animal and humans are an unimportant, accidental intrusion. *T. rhodesiense* is readily infective to humans, laboratory animals and wild and domestic animals. Strains of *T. rhodesiense* retain their infectivity to humans, and the strain used in the Tinde experiment has continued to be infective to humans over 20 years, during which it has been cyclically transmitted through antelopes and sheep. Strains pathogenic to humans have been isolated from bushbuck (Heisch *et al.*, 1958), cattle (Hoeve *et al.*, 1967) and some other animals including donkeys. People become infected when their activities, as hunter, game warden, entomologist or tourist, take them into the habitat of game animals and *Glossina (Glossina)* tsetse flies. Humans are not their preferred host, 80–90% of the feeds of *G. pallidipes* being made from bushbuck, 65% of those of *G. swynnertoni* on suids, and *G. morsitans* 30–35% on suids and 25–40% on bovids (Weitz, 1970).

Trypanosomiases of Domestic Animals Transmitted by
Glossina (Soltys and Woo, 1977; Logan-Henfrey *et al.*, 1992; Radostits *et al.*, 1994)

In Africa tsetse flies infest approximately 11 million km^2; involving 37 countries in which they occupy about half of the available arable land. They cause an estimated annual loss of US$5 billion. Virtually all domestic animals suffer from trypanosomiasis caused by one or more species of *Trypanosoma*. The common name for trypanosomiases transmitted by *Glossina* is nagana, the Zulu word for the infection in cattle and horses, and surra for the disease in camels and horses caused by *T. evansi*, transmitted mechanically by other blood-sucking flies. The three main pathogens are *T. brucei*, *T. congolense* and *T. vivax*. All three species exhibit some degree of pleomorphism but this is most marked in *T. brucei* in which long slender, intermediate and short stumpy trypomastigotes are found in the blood. Economically the most important trypanosome infection of domestic animals is the one in cattle.

Species infecting livestock

Trypanosoma vivax

This is a very motile trypanosome (21–25 µm) with a free flagellum. It multiplies rapidly in the blood. A smaller form (14–16 µm), originally described as *T. uniforme*, is now regarded as a form of *T. vivax*. *T. vivax* is the main cause of trypanosomiasis in domestic animals, especially cattle, to which it maintains high virulence, particularly in West Africa. It is also pathogenic to sheep, goats, horses and camels but not to cats or dogs. The main vectors are species of the subgenus *Glossina*.

Trypanosoma congolense

This is a sluggish trypanosome (12–18 µm long) which lacks a free flagellum. It tends to aggregate in small blood vessels and capillaries with fewer trypanosomes entering the circulation. It produces severe disease in cattle, sheep and goats, and chronic infections in them and camels and horses. It is transmitted by members of the subgenus *Glossina* and *G. (A.) brevipalpis* and *G. (A.) longipennis*.

Trypanosoma brucei

This trypanosome is particularly pathogenic to equines, and causes severe disease in sheep, goats, cats and dogs. It produces only a benign infection in cattle. Indeed cattle and game animals are the main reservoirs of *T. brucei*. Its pathogenicity to horses led Ford (1971) to write '*T. brucei* more than any other trypanosome has protected African people from invasion and African wildlife from destruction' (p. 64). *T. brucei* is transmitted especially by the subgenus *Glossina* and other species.

Trypanosoma simiae

Trypanosoma simiae (18 μm long) occurs naturally in wart hogs in East and Central Africa, where it is highly pathogenic to domestic pigs, with death occurring within hours of the parasites first appearing in the circulating blood. In view of the high feeding rates of *Glossina* species on suids, this trypanosome is transmitted by five species of *Austenina*, four *Glossina* and two *Nemorhina*. Cattle are resistant.

Trypanosoma suis

This is a stumpy (14 μm), monomorphic trypanosome with a free flagellum, which is rarely reported as a pathogen of swine. Most domestic animals seem to be refractory to *T. suis*. Natural hosts are wart hogs, bushpigs and forest hogs. It is transmitted by *G. (A.) brevipalpis* and *G. (A.) vanhoofi*.

Trypanosomiasis in cattle

The most important species in cattle are *T. vivax* and *T. congolense*. In enzootic areas infection rates can be over 60% and in outbreaks infections with *T. vivax* may reach 70%. After an incubation period of 8–20 days acute infections can cause death within several weeks and in chronic cases infection persists for months or years with the infected animals becoming carriers. Infected animals suffer intermittent fever, anaemia, are emaciated but have good appetites, and are reproductively impaired.

Treatment and control

Hormidium has been widely used in the treatment of trypanosome-infected ruminants but drug resistance has reduced the usefulness of hormidium and quinapyramine. Currently diminazene and isometamidium are most widely used because they have no cross-resistance. Equines and camels are treated with quinapyramine. The use of drugs as prophylaxis against trypanosomiasis runs the risk of producing drug-resistant strains of trypanosomes. Some native breeds of cattle are trypanotolerant becoming infected but not developing anaemia. The West African N'Dama and Muturu cattle are more resistant than West African zebu (Murray *et al.*, 1982). Prevention of disease is achieved by maintaining livestock in tsetse-free areas or by instituting control of *Glossina* (see Chapter 12).

Trypanosomiasis in African wildlife (Ford, 1971)

The response of wildlife to infection with trypanosomes differs from species to species. Some, such as baboons, are totally resistant to infection. Others, such as the suids and buffalo, develop a scanty parasitaemia and are tolerant of infection. This suggests a long relationship between parasite, *Glossina* and host, as suggested when considering the evolution of *Glossina* (p. 228). Some bovids are tolerant of infection but develop high parasitaemias which could make them important reservoir hosts. This group includes eland, reedbuck, bushbuck and impala, all of which are browsers on vegetation at the forest edge, and in thickets where close contact would be made with *Glossina*. High parasitaemias suggest that their association with *Glossina* and *Trypanosoma* is, in evolutionary terms, of recent origin. Other bovids and families of mammals are intolerant of infection and develop fatal parasitaemias. These include the plains-dwelling gazelles which would rarely have contact with *Glossina* in nature, and hyrax on which tsetse do not commonly feed.

Surveys of infection rates of game animals with *T. vivax*, *T. congolense* and *T. brucei* give overall values of 6%, 10% and 4%, respectively, which are quite different from those found in *Glossina* where the infection rates are closer to 20%, 2% and 0.2%. This difference highlights the effect of selective feeding by *Glossina* and its varying susceptibility to infection.

Some data concerning trypanosome infections in wildlife are presented in Table 29.2. Infection rates in buffalo, zebra and suids are low. (Zebra are plains animals with little contact with tsetse.) There is considerable difference between the specific infections found in different mammals. High infection rates occur in the Reduncinae, Tragelaphinae of the Bovidae, and in the Giraffidae. Infections with *T. vivax* were common in giraffe, waterbuck and reedbuck. *T. congolense* infections were particularly abundant in the Tragelaphinae, reedbuck and giraffe. Infections with *T. brucei* were 5% or more in hartebeeste, waterbuck, reedbuck, eland and bushbuck. The last named is particularly important because an isolation of *T. b. rhodesiense* has been made from it (Heisch *et al.*, 1958), and it is a major host of tsetse (Table 12.2).

General observation on *Glossina*-borne trypanosomiases

It should be clear that transmission of salivarian trypanosomes to humans and livestock is very complex, depending on the distribution of the reservoir host and the species of *Glossina* involved. Even when reservoir host and tsetse coincide in the same habitat their interaction will be influenced by the feeding preferences of the fly, and susceptibility of both host and fly to the strain of trypanosome. In general, species of *Glossina* (*Nemorhina*) are relatively poor vectors of *Trypanosoma* (*Nannomonas*) species (Jordan, 1974). Finally, the onset of fresh cases of trypanosomiasis will depend on contact being established between susceptible hosts and infective *Glossina*.

Salivarian Trypanosomes not Transmitted by *Glossina*
(Gardiner and Mahmoud, 1992)

Trypanosoma (T.) evansi

This is a monomorphic trypanosome (24 µm long) with a free flagellum. It is found in Africa north of the Sahara, Sudan, northern Kenya, Middle East, through India to south-east Asia and southern China, and south to Indonesia. It has been introduced into South America and to Mauritius and Reunion, islands in the Indian Ocean. It is transmitted mechanically by tabanids and *Stomoxys*, and in South America by vampire bats. In the Sudan there is a definite correlation

Table 29.2. Infections in game animals, expressed as percentages of animals examined.

Family	Subfamily	Genus	Common name	n	T.v.	Percentage infected T.c.	T.b.	All
Suidae	Suinae	Phacochoerus	Wart hog	154	1	6	2	10
		Potamochoerus	Bushpig	26	12	0	0	12
Giraffidae		Giraffa	Giraffe	180	2	5	2	10
				68	16	28	1	37
Bovidae	Alcelophinae	Alcelophus	Hartebeeste	76	1	3	9	13
		Connochaetes	Wildebeeste	38	0	0	0	0
		Damaliscus	Topi	30	3	3	3	13*
				144	1	2	6	10
	Reduncinae	Adenota	Cob	70	3	0	1	4
		Kobus	Waterbuck	110	33	9	24	52*
		Redunca	Reedbuck	39	25	15	8	43
				219	22	7	14	35
	Aepycerotinae	Aepyceros	Impala	151	1	9	1	11
	Antilopinae	Gazella	Gazelles	20	0	0	0	0
	Tragelaphinae	Strepsiceros	Kudu	40	3	45	0	45
		Taurotragus	Eland	63	5	21	5	29
		Tragelaphus	Bushbuck	55	2	25	5	31
				158	3	28	4	34
	Bovinae	Syncerus	Buffalo	87	0	3	3	7
Equidae		Equus	Zebra	109	2	3	0	6*

Source: From Lumsden (1962).
T.v., T.c. and T.b. indicate, respectively, *T. vivax*, *T. congolense* and *T. brucei*.
*One unidentified infection included in total.

between outbreaks of *T. evansi* infections and the increase in number of tabanids in the rainy season (Mahmoud and Gray, 1980). In Somalia the tabanids, *Philoliche zonata* and *P. magretti*, have been incrimated as major vectors of *T. evansi* among camels (Dirie *et al.*, 1989).

In bats the trypanosomes penetrate the buccal mucosa, enter the blood stream and are transmitted when they subsequently feed. *T. evansi* causes a severe disease in horses with mortality approaching 100%. In camels the disease may be acute and end fatally, or be chronic and infected animals survive 3 to 4 years. Dogs and cats are highly susceptible to infection but it is only slightly pathogenic to sheep, goats and pigs. In cattle and buffalo it is usually a stable enzootic condition with low mortality but subject to occasional severe outbreaks such as that in Vietnam in 1978 which killed 20,000 buffalo. *T. evansi* is considered to have evolved from *T. brucei* but its mitochondrion is now incapable of supporting the development of procyclic forms. In Central and South America, a variant of *T. evansi*, *T. equinum*, causes a chronic disease of horse known as mal de caderas. In *T. equinum* the kinetoplast DNA is scattered throughout the mitochondrion.

Trypanosoma (D.) vivax viennei

This trypanosome is thought to have been introduced into the Caribbean Islands of Martinique and Dominique from West Africa in imported cattle about 1830. It is pathogenic to cattle in which the infection rates in some areas of South America may be as high 15–54%. It causes only mild symptoms in horses and a range of ruminants, which presumably act as reservoir hosts. Dogs are refractory to infection. *T. v. viennei* is transmitted mechanically by tabanids and biting muscids and is no longer competent to develop in *Glossina*.

Stercorarian Trypanosomiasis

The only stercorarian trypanosome of economic importance is *Trypanosoma (Schizotrypanum) cruzi*, the causative organism of Chagas' disease.

Chagas' Disease (*Trypanosoma (Schizotrypanum) cruzi*)
(Dusanic, 1991; Garcia-Zapata *et al.*, 1991)

Chagas' disease is most prevalent in Argentina, Brazil and Venezuela but also occurs widely throughout Central and South America. It is estimated that there are 24 million seropositive people in Latin America of which 10 to 30% will develop clinical disease, resulting in 70,000 deaths per annum. *T. cruzi* has a broad host range being found in more than 100 species of mammals of 24 families including marsupials and six orders of eutherian mammals. It is found in wild animals in the USA but human cases are rare north of Mexico. Although *T. cruzi* is maintained in urban communities disease is most common in rural areas. The vectors of *T. cruzi* are haematophagous triatomine bugs (Reduviidae, Hemiptera; see Chapter 19).

Cycle of *T. cruzi*

In the vertebrate host slender and stumpy trypomastigotes (16–22 × 1–6 µm) are to be found in the circulating blood. It is believed that it is the slender forms which penetrate the host's cells where they develop into oval amastigotes (2.4 to 6.5 µm) and multiply forming pseudocysts which after 4 to 5 days burst, releasing trypomastigotes into the blood stream from which they invade other cells. When a triatomine bug ingests an infected blood meal the trypomastigotes differentiate into epimastigotes in the midgut and later adhere to the walls of the rectum where they develop into metacyclic trypomastigotes infective to a vertebrate host. The cycle in the bug takes about 20 days after which the bug remains infective for life and may transmit *T. cruzi* for several years. Transmission to the vertebrate is effected by metacyclics being deposited in excretory material on the skin of the host. The initial faecal deposit may contain fewer metacyclics than the subsequent drops of urine.

The acute phase of Chagas' disease is often asymptomatic with clinical disease usually occurring in children. This phase lasts 4 to 8 weeks followed by an indeterminate phase, in which the infected individual shows no clinical signs. This may continue indefinitely but about 10 to 30% of infections pass to the chronic phase, which may take place up to 30 years or longer after the initial infection. In the chronic phase patients may develop myocarditis or marked dilation of the oesophagus or colon. Treatment with nifurtimox and benznidazole will alleviate, but not always cure acute or chronic Chagas' disease. A new, less toxic drug, allopurinol, is currently being assessed for the treatment of asymptomatic Chagas' disease (Anon., 1991).

Triatomine vectors of *T. cruzi* (Wilton and Cedillos, 1978; Zeledon and Rabinovitch, 1981)

The vectors of *T. cruzi* differ in their ecological requirements and geographical distributions. *Triatoma dimidiata* is found in dry areas where the climate is not too warm, and occurs in Mexico, Central America, Colombia and Ecuador. *Triatoma infestans* occurs in warmer but equally dry habitats in Argentina, Paraguay, Bolivia, Chile, and parts of Peru and Brazil. It is the most domestic of the triatomines and extends widely, being found at 3682 m in Argentina and as far south as 45°S. *Panstrongylus megistus* occurs in the moister areas of Brazil and Peru. It requires a high humidity for breeding (above 60% RH), and is found as wild, peridomestic and domestic populations in rural situations, and even as urban populations in the city of Salvador. By contrast *Triatoma braziliensis* flourishes under dry conditions in north-east Brazil where *P. megistus* cannot survive.

Rhodnius prolixus occurs north of the equator in the mainland countries bordering the Caribbean, i.e. Mexico, Central America (except Panama and Costa Rica), Colombia, Venezuela and the Guianas. It is an ancient human pest, well adapted to poor households. In Venezuela it occurs as high as 1500 m and in Colombia it has been found at 2600 m. In El Salvador it is replaced by *T. dimidiata* above 340 m.

Epidemiology and epizootiology of Chagas' disease

The main wild hosts of *T. cruzi* are opossums, marsupials of the genus *Didelphys*, which are widely distributed from Argentina to the USA. They are parasitized by triatomine bugs, and have a high incidence of infection with *T. cruzi*. In sylvatic situations the disease is enzootic, but when opossums become established near houses they infect peridomiciliary triatomines. Human infections result when these triatomines enter houses and infect the residents directly, or by infecting dogs and cats which become sources of infection for resident domiciliary triatomines, which in turn will feed and infect the human occupants of the house.

Trypanosoma cruzi is spread by faecal contamination of the host by the bug. Since bugs only visit hosts to feed, it follows that they will only be potential vectors if they defecate while feeding. Some bugs, e.g. *Triatoma infestans*, do so and are vectors, while *Triatoma protracta* defecates off the host and is not a vector.

In infected houses in Bahia, Brazil, Minter (1978) found that the infection rate of *T. cruzi* in *P. megistus* increased steadily with age. In 1st instar nymphs the infection rate was 6.5%, and in adults 70%. In one house he found that 82% of *P. megistus*, found in the bedroom, had fed on humans and 56% were infected with *T. cruzi*. In the same house bugs in the living room had fed almost exclusively (97%) on chickens which roosted along the walls in which the bugs were found, and only 10% were infected. In laboratory experiments, Miles *et al.* (1975) found that both *P. megistus* and *T. infestans* were more readily infected than *R. prolixus*, and this applied to both the percentage of bugs which became infected and the degree of infection established, but there was no correlation between the size of the blood meal and the development of infection.

Control

Argentine, Brazil and Venezuela have instituted national programmes to control Chagas' disease. In the preparatory phase the incidence of *T. cruzi* infections and domestic bug infestations are assessed. This is followed by an attack phase in which insecticides are used against the vector and repeated until less than 5% of the houses are infested. In the following vigilance phase the community is responsible for detecting bug-infested houses. In Brazil this has been highly successful with the seroprevalence of school children in many endemic areas being reduced from 20–40% to 0–2% (Dias, 1987; Garcia-Zapata *et al.*, 1991).

Other Stercorarian Trypanosomes

Trypanosoma (Herpetosoma) rangeli (D'Alessandro-Bacigalupo and Saravia, 1992)

Trypanosoma rangeli resembles *T. cruzi* in being transmitted by triatomine bugs, particularly *R. prolixus*. It occurs in Central and South America where it has been recovered from humans and 23 species of marsupials, carnivores, edentates, primates and rodents. *T. rangeli* is not pathogenic to vertebrates but in humans it must be differentiated from *T. cruzi*. Flagellates ingested with the blood meal multiply in the midgut of the bug and penetrate between the cells lining the posterior midgut to enter the haemocoele and multiply and form epimastigotes. These move through the haemocoele to penetrate the salivary glands, in which they produce metacyclic forms, which are passed with the saliva when the bug feeds. In the haemocoele, some flagellates parasitize haemocytes and form dividing amastigotes which give rise to trypomastigotes. They are released when the infected cell bursts but their subsequent fate is not known. Some flagellates will proceed down the midgut to the rectum and be deposited with the faeces but there is no attachment to the wall of the rectum. Posterior station transmission is considered to occur infrequently, if at all. Six species of *Rhodnius* have been found with salivary infections in nature and two more *Rhodnius* species and five species of *Triatoma* have produced salivary gland infections experimentally. *T. rangeli* infections are pathogenic to *R. prolixus* causing the formation of excessive haemolymph and adversely affecting moulting.

Other non-pathogenic trypanosomes (D'Alessandro and Behr, 1991)

Trypanosoma (Herpetosoma) lewisi is a much studied, benign parasite of rats (*Rattus* spp.), transmitted via the faeces of the rat fleas *Nosopsyllus fasciatus* in temperate regions and *Xenopsylla cheopis* in the warmer parts of the world. *T. (Megatrypanum) theileri* is a large trypanosome which produces a low density, benign infection of cattle. In the tabanid *Haematopota pluvialis*, trypomastigotes give rise by multiplication and transformation to epimastigotes within 24 h of ingestion. From day five trypanosomes were found only in the hindgut and from day six small metacyclic forms, which were presumed to be infective to cattle, were present (Böse and Heister, 1993). *T. (M.) melophagium* is a large trypanosome which occurs in sheep. It completes its cycle in the sheep ked (*Melophagus ovinus*), a hippoboscid fly and sheep become infected by ingesting infected keds. *T. grayi* is a parasite of crocodiles, transmitted by *Glossina palpalis* by faecal contamination of the mucosa of the reptiles' mouth while the tsetse is feeding. Epimastigotes and metacyclic trypomastigotes occur in the hindgut of infected *Glossina* (Vickerman, 1976).

Leishmaniasis

Leishmaniasis is a disease of humans caused by infection with *Leishmania*, in which the parasites are intracellular amastigotes (2–5 μm) in the reticuloendothelial cells. They are transmitted from host to host by the bites of phlebotomine sandflies in which the parasites are motile and extracellular. In the Old World the vectors are species of *Phlebotomus* and, in the New World, species of *Lutzomyia* (*Lu.*). Clinical leishmaniasis takes three main forms – visceral, cutaneous and mucocutaneous. In the New World the diseases are zoonoses involving reservoir hosts, while in the Old World some important diseases are anthroponoses with no animal reservoir (Table 29.3).

Table 29.3. List of species of *Leishmania* transmitted by phlebotomine sandflies, together with their distributions, reservoir hosts and the type of human disease that they cause.

Leishmania	Vector	Geographical distribution	Human disease	Reservoir host
Tropica complex				
L. tropica	*P. sergenti*	Morocco, Saudi Arabia, former USSR	ACL	None
	P. rossi	Namibia	CL	Hyrax
Major complex				
L. major	*P. duboscqui*	Kenya, Senegal	ZCL	Rodents (*Avicanthis, Mastomys, Tatera*)
	P. papatasi	Iran, Israel, Jordan, Morocco, Saudi Arabia, Tunisia, former USSR	ZCL	*Meriones, Rhombomys, Psammomys*
	P. salehi	India	ZCL	*Meriones hurrianae*
Aethiopica complex				
L. aethiopica	*P. longipes*	Ethiopia	ZCL	Hyrax
	P. pedifer	Ethiopia, Kenya	ZCL	Hyrax
Donovani complex				
L. donovani	*P. argentipes*	India	AVL	None
	P. martini	Kenya	AVL	None
L. infantum	*P. alexandri*	China	ZVL	Dog
	P. ariasi	France, Spain	ZVL + CL	Dog, fox
	P. chinensis	China	VL	?
	P. longiductus	Former USSR	ZVL	Badger (*Meles*), dog, fox (*Vulpes*), jackal
	P. major neglectus	Greece	ZVL	Dog
	P. perfiliewi	Italy	CL	?
	P. perniciosus	Algeria, France, Italy, Malta, Spain	ZVL + CL	Dog, fox, *Rattus rattus*
L. chagasi	*Lu. longipalpis*	Bolivia, Brazil, Colombia	ZVL	Dog, fox

Mexicana complex				
L. amazonensis	Lu. flaviscutellata	Brazil, Colombia, French Guiana	ZCL	Rodents (Dasyprocta, Heteromys, Nectomys, Oryzomys, Proechimys), marsupials (Caluromys, Marmosa, Metachirus) ?
L. mexicana	Lu. olmeca olmeca	Belize, southern Mexico	ZCL	?
Braziliensis complex				
L. braziliensis	Lu. carrerai carrerai	Bolivia	ZCL	?
	Lu. llanos martinsi	Bolivia	ZCL	
	Lu. yucumensis	Bolivia	ZCL	
	Lu. spinicrassa	Colombia	ZCL	
	Lu. wellcomei	Brazil		
	Lu. whitmani	Brazil		
Guyanensis complex				
L. guyanensis	Lu. anduzei	Brazil	ZCL	⎱ Anteater (Tamandua), opossum (Didelphis)
	Lu. umbratilis	North Brazil, Colombia, French Guiana	ZCL	⎰ Sloth (Choloepus)
	Lu. whitmani	Brazil	ZCL	
L. panamensis	Lu. trapidoi	Panama, Costa Rica, Colombia	ZCL	Sloth (Bradypus, Choloepus), primate (Aotus), Rodent (Heteromys)
	Lu. ylephilator	Panama, Costa Rica	ZCL	
	Lu. panamensis	Panama	ZCL	
	Lu. gomezi	Panama	ZCL	

Source: WHO Technical Report Series 793 (1990).
ACL = anthroponotic cutaneous leishmaniasis; AVL = anthroponotic visceral leishmaniasis; ZCL = zoonotic cutaneous leishmaniasis; ZVL = zoonotic visceral leishmaniasis. Only proven vectors and reservoir hosts have been listed. Where more than one country is cited reservoir hosts will not necessarily have been proven in all the countries listed.

Visceral leishmaniasis (Oster and Chulay, 1991)

Visceral leishmaniasis, known also as kala-azar, causes hepatosplenomegaly, weight loss and anaemia, which ends fatally in untreated, established cases. In the southern Sudan it is estimated that 25% of the population are infected and that there have been 40,000 deaths over a 5-year period (Anon., 1993). Response to infection is highly variable. In Brazil, out of 86 children with positive antibodies, 20 showed no signs of disease for 5 years (end of observations); 38 had prolonged subclinical illness which resolved itself in 3 years without treatment; the other 28 developed classical visceral leishmaniasis within 15 months. Treatment is with pentavalent antimonials and diamidines. There are reports of resistance to antimonials in Bihar State with 5 to 10% mortality occurring in antimonial-treated patients (Anon., 1990). Up to 30% of 'cured' patients may relapse.

Cutaneous and mucocutaneous leishmaniasis (Chulay, 1991)

Cutaneous leishmaniasis involves nodular and ulcerative skin lesions which, in the Old World, may heal completely in a few months to a few years. In the New World such infections are more dangerous leading in 3 to 5% of cases to mucocutaneous leishmaniasis (espundia) in which there are destructive nasopharyngeal lesions, which can result in death. This development usually occurs within 2 years but can occur up to 30 years after the primary infection. Other cutaneous manifestations of infection are leishmaniasis recidivans, a chronic, drug-resistant infection, found mainly in Iraq and Iran, which may persist for 20 to 40 years, and in India, post kala-azar dermal leishmaniasis (PKDL) in people cured of kala-azar 2 to 10 years earlier.

Taxonomy of *Leishmania* (Lainson and Shaw, 1987; WHO, 1990)

A range of criteria are used to separate species of *Leishmania*, which are morphologically similar. Use is made of immunological, biological, geographical, clinical, behavioural and morphological characteristics to define and separate species. Two subgenera are recognized – *Leishmania* and *Viannia* – on the basis of where they develop in the gut of the phlebotomine vector. Development of *L. (Leishmania)* species is suprapylarian, being confined to the midgut and foregut. In *L. (Viannia)* species development is peripylarian, with a prolific and prolonged phase of development occurring in the hindgut (pylorus, ileum), followed by migration to the midgut and foregut. Using biochemical techniques and a cladistic approach, Thomaz-Soccol *et al.* (1993) have further refined the species complexes listed in Table 29.3, proposing separate complexes for *L. amazonensis* and *L. mexicana* and for *L. donovani* and *L. infantum*. They also sink *L. chagasi* as a synonym of *L. infantum*. The New World and Old World species complexes are distinct except for the *L. donovani* complex which occurs in both. It is possible that the New World representative (*L. chagasi*) was introduced into the New World with dogs of the conquistadors (Oster and Chulay, 1991).

Development and transmission of *Leishmania* in phlebotomine sandflies (Killick-Kendrick, 1990)

When a phlebotomine feeds on an infected host, it ingests amastigotes with the blood meal, some of which will divide before becoming promastigotes. Promastigotes quickly become the predominant form in the midgut. Two forms of promastigotes are recognized: a long, slender, unattached, electron-dense nectomonad promastigote and a broad, electron-lucid haptomonad promastigote with flagella which attach to the cuticular lining of the stomodeal (proventricular) valve by hemidesmosomes. In the foregut the commonest form is the oval, less motile paramastigote which attaches to the lining by hemidesmosomes in the tip of the flagellum. Highly motile metacyclic promastigotes are found in the proboscis and more posterior parts of the gut. (Those of *L. major* in *P. papatasi* have a body length of 10 μm and a flagellum length of 20 μm.) They are adapted for life in the vertebrate host, having an enzyme profile more comparable to that of amastigotes than to that of multiplying promastigotes. Grimm and Jenni (1993) found that promastigotes, formed during the period of mass multiplication in the midgut, were sensitive to normal human serum but later (days 7 to 11 of the infection) small highly motile promastigotes (presumably metacyclic promastigotes) were formed which had partial resistance to normal human serum.

When an infected phlebotomine feeds, transmission occurs by the introduction of metacyclic forms from the proboscis and by regurgitation of metacyclics from more posterior parts. Transmission of leishmaniae from infected phlebotomines to a vertebrate host has not proved easy. It has long been known that sugar feeding enhances transmission and it has been suggested that this is achieved by inhibiting attachment and thereby increasing the number of free-swimming promastigotes able to colonize the proboscis. Carbohydrates are inhibitors of lectin-mediated agglutination reactions of leishmanial promastigotes (Molyneux and Killick-Kendrick, 1987). The saliva of *Lu. longipalpis* substantially enhances the infectivity of *L. major*, a parasite with which it is not associated in nature (Titus and Ribeiro, 1988). In *L. braziliensis* paramastigotes were the principal form found dividing and attached in the pylorus and ileum of *Lu. longipalpis*. There is evidence for two pathways of development: one involving the production of metacyclic promastigotes and the other involving the development of non-dividing promastigotes which attach to the cuticle of the stomodeal valve and foregut and block the intestine (Ashford, 1991).

Infected phlebotomines have difficulty in engorging and tend to probe repeatedly. It has been postulated that this may be due to parasites blocking the cibarial sensilla, which are few in number. The absence of sensory information from the cibarium would restrict or inhibit ingestion, but not affect probing. Increased probing would be beneficial to the parasites in ensuring that some were deposited in sites suitable for development (Molyneux and Killick-Kendrick, 1987).

Infection rates in wild-caught phlebotomines have varied from 0 to 15.4%, the latter being found in *Lu. trapidoi*. These infections will include both pathogenic and non-pathogenic leishmaniae. Dissection of substantial numbers of *P. orientalis* in the Sudan and *P. longipes* in Ethiopia gave infection rates of 2.4

and 3.1%, respectively. Infection rates in *P. papatasi* have ranged from 0.2 to 8.7%, and in *P. caucasicus* from 2.6 to 10.5%. In 11 out of 17 species of *Lutzomyia* infection rates were low, mostly below 1%, while in the other six species they were much higher (Williams and Coelho, 1978).

Vectors and reservoir hosts of *Leishmania* (WHO, 1990)

Visceral leishmaniasis

Vectors of *L. donovani* include *P. argentipes* in India and *P. martini* in Kenya. There is no known animal host. In the Palaearctic region *L. infantum* is a zoonosis of canines which principally infects children. Three weeks after feeding on a dog infected with *L. infantum*, *P. ariasi* transmitted the parasite to a healthy dog (Rioux *et al.*, 1979b). In the Cévennes region of southern France the greatest risk of infection with *L. infantum* is in the late summer when the proportion of parous female *P. ariasi* is high (Guilvard *et al.*, 1980). In the same area the parasite is dispersed by movement of infected dogs and by *P. ariasi* which has been found to disperse 750 m (Rioux *et al.*, 1979a). In the Neotropical region *L. chagasi* is a zoonosis of dogs and foxes with the vector being *Lu. longipalpis*.

Old World cutaneous leishmaniasis

Leishmania tropica is associated with dry cutaneous leishmaniasis, an anthroponosis of urban areas, while *L. major* is associated with wet cutaneous leishmaniasis, a zoonosis of rural areas. *L. tropica* is found predominantly in densely populated settlements from Greece through Turkey and the Middle East to India. Person to person transmission is maintained by *P. sergenti*. The Namibian parasite is considered to be separate from both *L. tropica* and *L. major* (Lainson and Shaw, 1987). *L. major* occurs along the North African coast; in the arid inland region of West Africa; in the Middle East and Arabian Peninsular; and east and south of the Caspian and Aral Seas. *L. major* parasitizes ground-dwelling rodents with the major reservoir host being the giant gerbil, *Rhombomys opimus*, among which the main vector is the strongly zoophilic *P. caucasicus*. The disease is spread to humans by *P. papatasi*, which is markedly but not exclusively anthropophilic. Control of *L. major* has been achieved in Central Asia by applying insecticides and rodenticides to gerbil burrows to control both the reservoir hosts and the vectors.

   Leishmania aethiopica is a parasite of hyraxes (*Procavia, Heterohyrax, Dendrohyrax*) transmitted by *P. longipes* and *P. pedifer*. These sandflies feed equally easily on hyraxes and cattle, and humans are bitten and become infected when they associate closely with the sandflies' main hosts. It has been calculated that in compounds where human disease was present the average biting rate was as low as 21 bites/person/month (Foster *et al.*, 1972).

New World cutaneous leishmaniasis

In humans the suprapylarian leishmaniae, *L. mexicana* and *L. amazonensis*, produce mild cutaneous lesions. They are parasites of forest rodents and opossums and are transmitted by species of the *Lu. flaviscutellata* group. *L. mexicana* infects the log cutters and collectors of chicle (chewing gum latex) who work in the forest for periods of about 6 months during the rainy season. The main host is the rodent *Ototylomys phyllotis*, and the vector *Lu. olmeca olmeca*, which is not especially attracted to humans, except when its daytime resting places in the forest floor leaf litter are disturbed. It is particularly liable to bite in the hour or so after dawn (Williams, 1970). By contrast *L. amazonensis*, whose main hosts are species of *Proechimys*, rarely causes human infections because the vector *Lu. flaviscutellata* is not very anthropophilic (Shaw and Lainson, 1987).

Human infections with *L. braziliensis* are associated with jungle activities, particularly land clearing, when the risk of infection may be very high (70–80%). It also occurs in forest remnants close to human populations. The primary cutaneous lesions may be followed by mucocutaneous disease while the primary lesion is still active or may occur many years after the primary lesion has disappeared. There is no proven reservoir host. The vectors are listed in Table 29.3. *L. peruviana* is a member of the *L. braziliensis* complex which causes self-healing skin lesions in humans in the high (1200–3000 m) Peruvian and Ecuadorean Andes. The condition is known as 'uta'. Dogs are probably the most likely peridomestic reservoir but infections also occur in wild rodents (*Phyllotis andinum*, *Akodon* spp.). The most likely vector is *Lu. peruensis* which has been found infected naturally.

Two other leishmanial infections associated with humans involved in forest activities are caused by *L. guyanensis* and *L. panamensis*. Infection rates of 60 to 90% have been recorded in small exposed populations. *L. guyanensis* is rarely involved in mucocutaneous leishmaniasis but it has a tendency to develop multiple ulcers by secondary spread via the lymphatic system from a single lesion. Infections of *L. panamensis* may produce a chain of enlarged nodes along efferent lymph channels. Its cutaneous lesions are slow to heal and may persist for more than 10 years. Infections can proceed to mucocutaneous leishmaniasis.

Control (WHO, 1990)

Depending on the particular situation, measures to reduce human leishmaniasis can be directed against the vector and/or the reservoir. In the absence of such control measures, reliance must be placed on personal protection. Where transmission occurs in a domestic setting, the spraying of houses with a residual insecticide such as DDT, as part of an anti-malaria campaign, can have the additional benefit of reducing human leishmaniasis. It was noted in India during the anti-malaria campaign that there were fewer peridomestic phlebotomines and fewer cases of leishmaniasis. When spraying ceased there was a resurgence of cases, but it should be noted that no detailed observations on phlebotomine numbers are available to support this thesis. There are recent reports that *P. argentipes* has become resistant to DDT (Mukhopadhyay *et al.*, 1990). Where dogs are the reservoir host, strays and feral animals should be eliminated. The possibility of reducing human leishmaniasis by controlling the rodent reservoir needs to be carefully evaluated. Spectacular results were obtained in the former USSR by the complete destruction of *Rhombomys* colonies, but where fat sand rats (*Psammomys* spp.) are the reservoir hosts, successful control has proven to be difficult and it might be more practical to locate new human settlements away from *Psammomys* colonies. In forest situations, reliance must be on personal protection involving the use of repellents, wearing of long-sleeved shirts and trousers, sleeping under bednets and avoiding infested areas at the time when the vectors are active. Phlebotomines do not disperse widely and avoidance of infested localities can be a practical measure.

References

Abaru, D.E. (1985) Sleeping sickness in Busoga, Uganda, 1976–1983. *Tropical Medicine and Parasitology* 36, 72–76.

Anon. (1990) Antimonials: large-scale failure in leishmaniasis 'alarming'. *TDR News* 34, 1,7.

Anon. (1991) Chagas products in action. *TDR News* 37, 4.

Anon. (1993) 'Killing disease' spreads in Sudan. *TDR News* 43(1) 4.

Ashford, R.W. (1991) Intravectorial cycle of *Leishmania* in sandflies. *Annales de Parasitologie Humaine et Comparée* 66, Suppl. 1, 71–74.

Baldry, D.A.T. (1980) Local distribution and ecology of *Glossina palpalis* and *G. tachinoides* in forest foci of West African human trypanosomiasis, with special reference to associations between peri-domestic tsetse and their hosts. *Insect Science and its Application* 1, 85–93.

Bales, J.D. (1991) African trypanosomiasis. In: Strickland, G.T. (ed.) *Hunter's Tropical Medicine*. Saunders, Philadelphia, pp. 617–628.

Böse, R. and Heister, N.C. (1993) Development of *Trypanosoma (M.) theileri* in tabanids. *Journal of Eukaryotic Microbiology* 40, 788–792.

Chulay, J.D. (1991) Leishmaniasis. In: Strickland, G.T. (ed.) *Hunter's Tropical Medicine*. Saunders, Philadelphia, pp. 638–641, 648–655.

D'Alesandro. A. and Behr, M.A. (1991) *Trypanosoma lewisi* and its relatives. In: Kreier, J.P. and Baker, J.R. (eds) *Parasitic Protozoa*. Academic Press, San Diego, vol. I, pp. 225–263.

D'Alessandro-Bacigalupo A. and Saravia, N.G. (1992) *Trypanosoma rangeli*. In: Kreier, J.P. and Baker, J.R. (eds) *Parasitic Protozoa*. Academic Press, San Diego, vol. II, pp. 1–54.

Dias, J.C.P. (1987) Control of Chagas' disease in Brazil. *Parasitology Today*, 3, 336–341.

Dipeolu, O.O. and Adam, K.M.G. (1974) On the use of membrane feeding to study the development of *Trypanosoma brucei* in *Glossina*. *Acta Tropica* 31, 185–201.

Dirie, M.F., Wallbanks, K.R., Aden, A.A., Bornstein, S. and Ibrahim, M.D. (1989) Camel trypanosomiasis and its vectors in Somalia. *Veterinary Parasitology* 32, 285–291.

Dusanic, D.G. (1991) *Trypanosoma (Schizotrypanum) cruzi*. In: Kreier, J.P. and Baker, J.R. (eds) *Parasitic Protozoa*. Academic Press, San Diego, vol. I, pp. 137–194.

Evans, D.A. and Ellis, D.S. (1983) Recent observations on the behaviour of certain trypanosomes within their insect hosts. *Advances in Parasitology* 22, 1–42.

Evans, D.A., Ellis, D.S. and Stamford, S. (1979) Ultrastructure studies of certain aspects of the development of *Trypanosoma congolense* in *Glossina morsitans morsitans*. *Journal of Protozoology* 26, 557–563.

Ford, J. (1971) *The Role of the Trypanosomiases in African Ecology*. Clarendon Press, Oxford.

Foster, W.A., Boreham, P.F.L. and Tempelis, C.H. (1972) Studies on leishmaniasis in Ethiopia. IV. Feeding behaviour of *Phlebotomus longipes* (Diptera: Psychodidae). *Annals of Tropical Medicine and Parasitology* 66, 433–443.

Freeman J.C. (1973) The penetration of the peritrophic membrane of the tsetse flies by trypanosomes. *Acta Tropica* 30, 347–355.

Garcia-Zapata, M.T.A., McGreevey, P.B. and Marsden, P.D. (1991) American trypano-somiasis. In: Strickland, G.T. (ed.) *Hunter's Tropical Medicine.* Saunders, Philadelphia, pp. 628–637.

Gardiner, P.R. and Mahmoud, M.M. (1992) Salivarian trypanosomes causing diseases in livestock outside sub-Saharan Africa. In: Kreier, J.P. and Baker, J.R. (eds) *Parasitic Protozoa.* Academic Press, San Diego, vol. II, pp. 277–314.

Gibson, W.C. (1989) Analysis of a genetic cross between *Trypanosoma brucei rhodesiense* and *T. b. brucei. Parasitology* 99, 391–402.

Grimm, F. and Jenni, L. (1993) Human serum resistant promastigotes of *Leishmania infantum* in the midgut of *Phlebotomus perniciosus. Acta Tropica* 52, 267–273.

Guilvard, E., Wilkes, T.J., Killick-Kendrick, R. and Rioux, J.A. (1980) Ecologie des leishmanioses dans le sud de la France. 15. Déroulement des cycles gonotrophiques chez *Phlebotomus ariasi* Tonnoir, 1921 et *Phlebotomus mascittii* Grassi, 1908 en Cévennes. Corollaire épidémiologique. *Annales de Parasitologie Humaine et Comparée* 55, 659–664.

Heisch, R.B., McMahon, J.P. and Manson-Bahr, P.E.C. (1958) The isolation of *Trypanosoma rhodesiense* from a bushbuck. *British Medical Journal* 2, 1203–1204.

Hoare, C.A. (1972) *The Trypanosomes of Mammals.* Blackwell Scientific Publications, Oxford.

Hoeve, K. van, Onyango, R.J., Harley, J.M.B. and Raadt, P. de (1967) The epidemiology of *Trypanosoma rhodesiense* sleeping sickness in Alego Location, Central Nyanza, Kenya. II. The cyclical transmission of *Trypanosoma rhodesiense* isolated from cattle to a man, a cow and to sheep. *Transactions of the Royal Society of Tropical Medicine and Hygiene* 61, 684–687.

Jordan, A.M. (1974) Recent developments in the ecology and methods of control of tsetse flies (*Glossina* spp.) (Dipt., Glossinidae) – a review. *Bulletin of Entomological Research* 63, 361–399.

Killick-Kendrick, R. (1990) The life cycle of *Leishmania* in the sandfly with special reference to the form infective to the vertebrate host. *Annales de Parasitologie Humaine et Comparée* 65, Suppl. 1, 37–42.

Kuzoe, F. A.S. (1993) Current situation of African trypanosomiasis. *Acta Tropica* 54, 153–162.

Lainson, R. and Shaw, J.J. (1987) Evolution, classification and geographical distribution. In: Peters, W. and Killick-Kendrick, R. (eds) *The Leishmaniases in Biology and Medicine.* Academic Press, vol. I, pp. 1–120.

Logan-Henfrey, L.L., Gardiner, P.R. and Mahmoud, M.M. (1992) Animal trypanosomiasis in sub-Saharan Africa. In: Kreier, J.P. and Baker, J.R. (eds) *Parasitic Protozoa.* Academic Press, San Diego, vol. II, pp. 157–276.

Lumsden W.H.R. (1962) Trypanosomiasis in African Wildlife. *Proceedings of the First International Conference on Wildlife Diseases, New York,* pp. 66–95.

Mahmoud, M.M. and Gray, A.R, (1980) Trypanosomiasis due to *Trypanosoma evansi* (Steel, 1885) Balbiani, 1888. A review of recent research. *Tropical Animal Health and Production* 12, 35–47.

Miles, M.A., Patterson, J.W., Marsden, P.D. and Minter, D.M. (1975) A comparison of *Rhodnius prolixus, Triatoma infestans* and *Panstrongylus megistus* in the xenodiagnosis of a chronic *Trypanosoma (Schizotrypanum) cruzi* infection in a rhesus monkey (*Macaca mullatta*). *Transactions of the Royal Society of Tropical Medicine and Hygiene* 69, 377–382.

Minter, D.M. (1978) Triatomine bugs and the household ecology of Chagas's disease. In: Willmott, S. (ed.) *Medical Entomology Centenary Symposium Proceedings.* Royal Society of Tropical Medicine and Hygiene, London, pp. 85–93.

Molyneux, D.H. (1980) Animal reservoirs and residual 'foci' of *Trypanosoma brucei gambiense* sleeping sickness in West Africa. *Insect Science and its Application* 1, 59–63.

Molyneux, D.H. and Killick-Kendrick, R. (1987) Morphology, ultrastructure and life-cycles. In: Peters, W. and Killick-Kendrick, R. (eds) *The Leishmaniases in Biology and Medicine*. Academic Press, London, vol. I, pp. 121–176.

Molyneux, D.H., Baldry, D.A.T. and Fairhurst, C. (1979a) Tsetse movement in wind fields: possible epidemiological and entomological implications for trypanosomiasis and its control. *Acta Tropica* 36, 53–65.

Molyneux, D.H., Lavin, D.R. and Elce, B. (1979b) A possible relationship between salivarian trypanosomes and *Glossina* labrum mechano-receptors. *Annals of Tropical Medicine and Parasitology* 73, 287–290.

Mukhopadhyay, A.K., Saxena, N.B.L. and Narasimham, M.V.V.L. (1990) Susceptibility status of *Phlebotomus argentipes* to DDT in some kala-azar endemic areas of Bihar (India). *Indian Journal of Medical Research. Section A, Infectious Diseases* 91, 458–460.

Murray, M., Morrison, W.I. and Whitelaw, D.D. (1982) Host susceptibility to African trypanosomiasis: trypanotolerance. *Advances in Parasitology* 21, 1–68.

Nash, T.A.M. (1978) A review of mainly entomological research which has aided the understanding of human trypanosomiasis and its control. In: Willmott, S. (ed.) *Medical Entomology Centenary Symposium Proceedings*. Royal Society of Tropical Medicine and Hygiene, London, pp. 39–47.

Oster, C.N. and Chulay, J.D. (1991) Visceral leishmaniasis (kala-azar). In: Strickland, G.T. (ed.) *Hunter's Tropical Medicine*. Saunders, Philadelphia, pp. 642–648.

Pépin, J. and Milord, F. (1994) The treatment of human African trypanosomiasis. *Advances in Parasitology* 33, 1–47.

Raadt, P. de and Seed, J.R. (1977) Trypanosomes causing disease in man in Africa. In: Kreier, J.P. (ed.) *Parasitic Protozoa*. Academic Press, New York, vol. I, pp. 175–237.

Radostits, O.M., Blood, D.C. and Gay, G.C. (1994) *Veterinary Medicine. A Textbook of the Diseases of Cattle, Sheep, Pigs, Goats and Horses*. Baillière Tindall, London.

Rioux, J.A., Killick-Kendrick, R., Leaney, A.J., Turner, D.P., Bailly, M. and Young, C.J. (1979a) Ecologie des leishmanioses dans le sud de la France. 12. Dispersion horizontale de *Phlebotomus ariasi* Tonnoir, 1921. Expériences préliminaires. *Annales de Parasitologie Humaine et Comparée* 54, 673–682.

Rioux, J.A., Killick-Kendrick, R., Leaney, A.J., Young, C.J., Turner, D.P., Lanotte, G. and Bailly, M. (1979b) Ecologie des leishmanioses dans le sud de la France. 11. La leishmaniose viscerale canine: succès de la transmission expérimentale 'chien – phlebotome – chien' par la piqûre de *Phlebotomus ariasi* Tonnoir, 1921. *Annales de Parasitologie Humaine et Comparée* 54, 401–407.

Schweizer, J. and Jenni, L. (1991) Hybrid formation in the life cycle of *Trypanosoma (T.) brucei*: detection of hybrid trypanosomes in a midgut-derived isolate. *Acta Tropica* 48, 319–322.

Seed, J.R. and Hall, J.E. (1992) Trypanosomes causing disease in man in Africa. In: Kreier, J.P. and Baker, J.R. (eds) *Parasitic Protozoa*. Academic Press, San Diego, vol. II, pp. 85–155.

Shaw, J.J. and Lainson, R. (1987) Ecology and epidemiology: New World. In: Peters, W. and Killick-Kendrick, R. (eds) *The Leishmaniases in Biology and Medicine*. Academic Press, London, vol. I, pp. 291–363.

Soltys, M.A. and Woo, P.T.K. (1977) Trypanosomes producing disease in livestock in Africa. In: Kreier, J.P. (ed.) *Parasitic Protozoa*. Academic Press, New York, vol. I, pp. 239–268.

Tait, A. and Turner, C.M.R. (1990) Genetic exchange in *Trypanosoma brucei*. *Parasitology Today* 6, 70–75.

Thomaz-Soccol, V., Lanotte, G., Rioux, J.A., Patlong, F., Martini-Dumas, A. and Serres, E. (1993) Monophyletic origin of the genus *Leishmania* Ross, 1903. *Annales de Parasitologie Humaine et Comparée* 68, 107–108.

Titus, R.G. and Ribeiro, J.M.C. (1988) Salivary gland lysates from the sand fly *Lutzomyia longipalpis* enhance *Leishmania* infectivity. *Science* 239, 1306–1308.

Turner, C.M.R., Barry, J.D., Maudlin, I. and Vickerman, K. (1988) An estimate of the size of the metacyclic variable antigen repertoire of *Trypanosoma brucei rhodesiense*. *Parasitology* 97, 269–276.

Vickerman, K. (1976) The diversity of the kinetoplastid flagellates. In: Lumsden, W.H.R. and Evans, D.A. (eds) *Biology of the Kinetoplastida*. Academic Press, New York, vol. I, pp. 1–34.

Weitz, B.G.F. (1970) Hosts of *Glossina*. In: Mulligan H.W. (ed.) *The African Trypanosomiases*. John Wiley, New York, pp. 317–326.

WHO (1990) Control of the leishmaniases. *WHO Technical Report Series*, 793.

Williams, P. (1970) Phlebotomine sandflies and leishmaniasis in British Honduras (Belize). *Transactions of the Royal Society of Tropical Medicine and Hygiene* 64, 317–364.

Williams, P. and Coelho, M. de V. (1978) Taxonomy and transmission of *Leishmania*. *Advances in Parasitology* 16, 1–42.

Wilton, D.P. and Cedillos, R.A. (1978) Domestic triatomines (Reduviidae) and insect trypanosome infections in El Salvador, C.A. *Bulletin of the Pan American Health Organization* 12, 116–123.

Youdeowei, A. (1975) Salivary secretion in three species of tsetse flies (Glossinidae). *Acta Tropica* 32, 166–171.

Zeledon, R. and Rabinovich, J.E. (1981) Chagas' Disease: an ecological appraisal with special emphasis on its insect vectors. *Annual Review of Entomology* 26, 101–133.

Lymphatic Filariasis (*Wuchereria bancrofti, Brugia malayi, B. timori*)

30

The most important group of helminths, transmitted by insects, belongs to the superfamily Filarioidea, order Spirurida, class Nematoda. The classification used here is that of Anderson *et al.* (1974). Members of this superfamily, often referred to as filarial worms, are responsible for serious, widespread diseases of humans and, with the exception of *Dirofilaria immitis*, comparatively minor conditions in domestic animals. The vectors of the economically important filarial worms are various species of blood-sucking Diptera, mostly Nematocera, particularly mosquitoes. Other Spirurida have a wide range of intermediate hosts, including insects of many different orders, and some species are of minor veterinary importance (*Habronema, Thelazia*). A few species of tapeworms (Cestoda) use insects as their intermediate hosts and are of minor veterinary importance.

In the vertebrate host the Filarioidea are parasites of the blood or lymphatic system; muscles or connective tissue; or of the serous cavities of their host. The Filarioidea contains two families, the Onchocercidae and the Filariidae. All the medically important species are in the Onchocercidae, including *Wuchereria bancrofti, Brugia malayi* and *B. timori*, the causative organisms of bancroftian and brugian filariasis; *Onchocerca volvulus*, the cause of onchocerciasis or river blindness. Minor filarial pathogens are *Loa loa* and *Mansonella streptocerca* which respectively cause Calabar swellings and skin lesions. *M. perstans* and *M. ozzardi* are parasites to which humans have developed a remarkable tolerance (Buck, 1991).

Parasites of veterinary importance in the Onchocercidae include species of *Onchocerca; D. immitis*, the heartworm of dogs; and species of *Setaria* and *Elaeophora*. In the Filariidae, species of *Parafilaria* and *Stephanofilaria* are of minor veterinary importance.

The diseases caused by these parasites, with emphasis on the role played by the insect vectors, will be dealt with as follows: bancroftian and brugian lymphatic filariasis in this chapter, onchocerciasis in the next, and other helminths transmitted by insects in Chapter 32.

Bancroftian and Brugian Filariasis

The World Health Organization (WHO, 1992) estimated that in 1990 750 million people were living in areas endemic for lymphatic filariasis and nearly 80 million were infected, 73 million with bancroftian and 6 million with brugian filariasis. In Pondicherry and Shertallai, two Indian towns of similar size, 10% of their working populations were infected with *W. bancrofti* and *B. malayi* respectively. Working days lost per infected person were twice as high in Pondicherry than in Shertallai (17.4 cf. 8.9) due to *B. malayi* causing fewer acute attacks per year (4.5 cf. 2.2).

Wuchereria bancrofti is widely distributed throughout tropical Africa and the Indo-Pacific region. In the latter it occurs in India, Bangladesh, Myanmar, Vietnam, New Guinea and Polynesia. Its incidence has been much reduced in China, Indonesia, Malaysia, Sri Lanka and Thailand, and eradicated from the Solomon Islands. In Africa *W. bancrofti* is endemic in the Nile Delta of Egypt and in much of the Afrotropical region. In the latter its northern limit is a line from Senegal to Somalia (Mogadishu) and its southern limit a line from Angola (Benguela) to Mozambique (Beira). There is high prevalence of *W. bancrofti* in the coastal regions of East Africa, Madagascar and the islands off the East African coast and in the Gulf of Guinea. The distribution of *W. bancrofti* in tropical America is much reduced, being present in the Guianas and areas around Belem and Recife in coastal north-east Brazil. *B. malayi* occurs in Malaysia, Indonesia and Mindanao in the Philippines. *B. timori* occurs on Timor, Flores and other islands of the Savu Sea.

The dynamics of human lymphatic filariasis is such that mass migration of infected humans is required to introduce the disease to new areas. Laurence (1989) postulates that *W. bancrofti* was a parasite of nomadic humans in south-east Asia (Indonesia), a sea-going people who introduced the parasite to Polynesia in prehistoric times and across the Indian Ocean to Madagascar in the first five centuries AD. The slave trade of the seventeenth and eighteenth centuries introduced *W. bancrofti* to the New World.

Lymphatic filariasis – the disease (Buck, 1991; WHO, 1992)

The adult worms inhabit the lymphatic vessels and nodes of the human host causing local inflammation of the lymphatic vessels (lymphangitis), swelling of the lymphatic nodes (lymphadenitis) and destruction of the lymphatics. The females release microfilariae into the circulation. Four states of infection can be recognized. Symptomless, healthy individuals with filarial antigens, but no microfilariae in their blood. Other asymptomatic individuals have circulating microfilariae and, although these individuals have no oedema, lymphoscintigraphy has revealed abnormalities in their lymphatic systems (Dreyer *et al.*, 1994). Initial overt disease involves acute episodic adenolymphangitis and fever, which untreated can lead to a reversible lymphoedema of the extremities and on to irreversible elephantiasis. Elephantiasis is more commonly found in the legs and scrotum than in the arms, breasts and labia. Other complications include chyluria due to the rupture of the lymphatics into the urinary tract and, in males, hydrocoele and lymph scrotum, chronic epididymitis and inflammatory swelling of the spermatic cord, some of which can be relieved by surgical treatment.

The disease develops slowly with recurrent episodes of fever and adenolym-phangitis in the first decade; lymphoedema of the extremities and genital lesions in the second decade; these symptoms becoming more frequent and severe in the third and fourth decades, after which symptoms may remain steady or decline. Genital lesions are rare in brugian filariasis and manifestations of the disease are unusually severe in infections with *B. timori*, in which in some communities elephantiasis, which is usually below 5%, can be present in 35% of the adult population.

Diethylcarbamazine (DEC) causes rapid disappearance of microfilariae from the circulation and also kills adult worms, but some microfilariae and some adult worms survive repeated treatments. DEC is given as a course of treatment spread over several days. A single dose of ivermectin cleared microfilariae from the blood more rapidly that did a 12-day course of DEC, but ivermectin may not be effective against adult worms. The reactions produced by both treatments were similar, being related to the density of microfilariae. Reactions to DEC are more severe in brugian than in bancroftian infections.

The Parasites

Discovery of transmission of *W. bancrofti* (Service, 1978)

The discovery of the insect vector of *W. bancrofti* is of particular significance to medical entomologists because it is considered to be the birth of medical entomo-logy. It was the first time that an insect had been incriminated as the vector of an agent of any human or animal disease. Microfilariae of the parasite were found by Wucherer in the urine of a patient suffering from chyluria in Bahia, Brazil, in 1866. (Nelson, 1981, states that they were seen earlier in 1863 by Demarquay.) Interest in filariasis was worldwide. Microfilariae were seen in the blood of patients in India by Lewis in 1872 and by Manson and Bancroft in China and Australia, respectively, in 1876. In the same year Bancroft found the adult worms and sent them to Cobbold in England, who named the worm after him as *Filaria bancrofti* in 1877.

The existence of microfilariae in the blood suggested the possibility that a blood-sucking insect might be involved in transmission. In 1877, at Amoy in China, Manson was able to follow the development of the microfilariae in *Cx quin-quefasciatus* to the infective third stage. At the time it was thought that mosquitoes only fed once and therefore Manson considered that transmission occurred through drinking water infected with worms which had escaped from mosquitoes, trapped in the surface film. This view persisted until Bancroft in 1899 showed that transmission occurred during the feeding of an infective mosquito.

Life cycle of parasite (Sasa, 1976)

The female worm is viviparous, liberating microfilariae into the lymphatic system in large numbers (50,000/female/day) and appear in the peripheral blood. The microfilariae are enclosed in a membrane and are said to be sheathed. They are long and slender, those of *W. bancrofti* measuring 250–300 × 7–9 µm and those of *B. malayi* are somewhat shorter measuring 200–275 × 4–7 µm (Buck, 1991). The microfilariae are taken up with the blood ingested by the vector. In the midgut they shed their sheaths and penetrate the epithelium to reach the haemocoele through which they migrate to the thoracic flight muscles. Here they develop into thicker, shorter 'sausage' forms, which undergo two moults before developing into elongate, snake-like mature infective larvae (L_3) measuring about 1.5 × 0.02 mm. Mature third-stage larvae leave the thoracic musculature and enter the haemocoele in which they move around actively and accumulate in the head (Wharton, 1957a). When the mosquito is feeding they enter the labium and escape by rupturing the labella. They are deposited in a drop of haemolymph, and enter the host through the puncture made by the feeding mosquito (McGreevy *et al.*, 1974). In the vertebrate host they pass through more stages before becoming adult (Buck, 1991).

The number of adult worms in a host cannot exceed the number of infective larvae introduced. A year after a leaf monkey (*Presbytis melalophos*) had been subcutaneously inoculated with 471 infective larvae of *W. bancrofti*, it died, and 77 adult worms were recovered. The sexes are separate and the female has to pair before producing microfilariae. In the experimentally infected leaf monkey microfilariae first appeared 287 days after inoculation (Sucharit *et al.*, 1982). In humans the prepatent period for *W. bancrofti* is 7–8 months, considerably longer than that of *B. malayi* (2 months) (WHO, 1992). Microfilariae of *W. bancrofti* are seldom found in children less than 3 years of age, but in areas where there is a high level of transmission of *B. malayi* microfilariae can be found in the blood at an earlier age. In East Pahang, Malaysia, microfilariae were present in 8% of children under 1 year, and in 60% of children aged 5 to 9 years (Edeson and Wilson, 1964).

The adult worms are long and slender with male and female *W. bancrofti* measuring 40 × 0.1 mm and 90 × 0.7 mm, respectively, and being about twice the size of adult *B. malayi* (Sasa, 1976; Denham and McGreevy, 1977). Webber (1977) estimated the reproductive expectancy of *W. bancrofti* in the Solomon Islands to be about 8 years. In Samoa it was estimated to be 2 to 4 years (Buck, 1991), and in India 5.4 years (Vanamail *et al.*, 1989).

Periodicity of microfilariae (Hawking, 1975)

Populations of microfilariae show variation in their abundance in the circulating blood at different times of the day. Manson recognized a marked nocturnal periodicity in China in 1877 (Service, 1978), and Thorpe in 1896 found that microfilariae were more abundant in the circulating blood during the day than at night in Fiji and Tonga. These two populations are now referred to as the nocturnal periodic and the diurnal subperiodic forms of *W. bancrofti* and geographically they are separated in the Pacific islands at the 170° east longitude. Malaria and periodic filariasis occur to the west of this line of demarcation and islands to the east have the diurnal subperiodic form and no malaria (Nelson, 1978). This distinction reflects the adaptation of the parasite to its mosquito vector and it should be expected that in different ecosystems populations of parasites will have arisen which are adapted to the biting habits of the local vector. Nocturnal periodic and nocturnal subperiodic forms of *B. malayi* occur (Sasa, 1976). *B. timori* has only a nocturnal periodic form which equates with it having only one known vector, *An barbirostris* (WHO, 1992).

Hawking (1975) pointed out the need to indicate at what time in the 24 h cycle the peak is reached, diurnal or nocturnal, and to give quantitative expression to the fluctuations throughout the 24 h period. For the latter he uses a periodicity index (PI) which is the standard deviation expressed as a percentage of the mean. During the time when microfilariae are absent from the peripheral blood they accumulate in the lungs in the terminal arterioles before the capillaries. In the nocturnal periodic *W. bancrofti* and *B. malayi* this accumulation is in response to a greater difference in oxygen tension between the alveolar and arterial blood by day than by night (55 mmHg cf. 45 mm). There is no simple explanation for the less marked accumulation of microfilariae of the diurnal subperiodic form in the lungs at night.

Both the nocturnal periodic *W. bancrofti* and *B. malayi* have similar periodicities with peak numbers of microfilariae being in the circulating blood from 2300–0300 h and a PI of 90–110%. The diurnal subperiodic form of *W. bancrofti* peaks at 1400–1700 h and has a PI of 20–25%. In east Thailand there is a nocturnal subperiodic form of *W. bancrofti* which peaks at 2100 h with a PI of 50%. The nocturnal subperiodic form of *B. malayi* is associated with dense swamp forest in south-east Asia and its microfilariae have a peak in the early evening and a PI of 30% (Sasa, 1976). There is a localized, diurnally subperiodic form of *B. malayi* in west Malaysia, which has a periodicity similar to that of the diurnal subperiodic *W. bancrofti* (Denham and McGreevy, 1977).

Nocturnal subperiodic *B. malayi* is common in wild monkeys and wild and domestic carnivores (dogs and cats). The disease it causes is therefore a zoonosis. The closely related *B. pahangi* is sympatric with this form of *B. malayi* and occurs in a similar range of hosts. *B. pahangi* is primarily a parasite of carnivores with primates and other vertebrates as incidental hosts whereas subperiodic *B. malayi* is primarily a parasite of humans and leaf monkeys (*Presbytis* spp.) with wild and domestic carnivores as incidental hosts (Wharton, 1963). There has been no proven natural human infection with *B. pahangi* (WHO, 1992).

Fate of deposited infective larvae

Infective larvae of the related filarioid worm *D. immitis* escape from the tip of the labella of *Ae aegypti* in a drop of haemolymph which, under laboratory conditions, evaporated in 4 min (McGreevy *et al.*, 1974). Larvae that fail to enter the host in that time will perish on the surface of the skin. This finding is in accord with observations on the fate of larvae of *B. pahangi* deposited from *Ae togoi*, of which 62–90% die on the skin surface (Ewert and Ho, 1967; Ho and Ewert, 1967). Conflicting results have been obtained with larvae of *B. pahangi* deposited from

Ae aegypti on to cats in which it was found that 90% of the deposited larvae penetrated and that penetration was independent of humidity (20% or 80% RH). The only difference was that at the lower humidity fewer larvae escaped on to the skin (Denham and McGreevy, 1977).

Larvae escape rapidly from the labium, the main stimulus being the bending of the labium (McGreevy *et al.*, 1974). The nature of the host on which the mosquito is feeding also plays a role. When *Ae togoi* probed a cat, 57% of the infective larvae of *B. pahangi* escaped in the first 5 s; at the end of a minute this value had risen to 77%; and to 91% on engorgement. The corresponding figures when infected *Ae togoi* fed on a mouse were 28% in 5 s, 44% in 1 min, and 74% at engorgement (Ho and Lavoipierre, 1975).

The Vector

Mosquito vectors of bancroftian and brugian filariasis (Table 30.1; WHO, 1992)

The most widespread cause of lymphatic filariasis is the nocturnal periodic form of *W. bancrofti* of which the main vector in urban areas is *Cx quinquefasciatus*, a highly anthropophilic species which feeds readily both indoors and outdoors and has its peak biting period between midnight and 0300 h, coinciding with the peak microfilarial abundance in the peripheral blood (Chow, 1973). In rural areas the vectors are species of *Anopheles*. Fourteen species are vectors of periodic *W. bancrofti*; four of periodic *B. malayi*; and one of *B. timori*. Many of these species are also vectors of malaria, and measures taken to control malaria also reduce filarial transmission.

The vector of nocturnal periodic *W. bancrofti* in a focus on the Pahang River in Malaysia was *An whartoni* (= *An letifer* auct.), which had a low infection in nature, but when fed on a carrier nearly 40% became infective, while *Cx quinquefasciatus* was a comparatively poor vector with only 5% becoming infective. *An whartoni* is exophilic, exophagic, and nocturnal with maximum activity in the 2 h after sunset. It breeds in shaded swamp forest, open grass swamp and clear seepage pools (Wharton, 1960).

In the Philippines an important vector of nocturnal periodic *W. bancrofti* is *Ae poecilius* which breeds in the axils of the abaca (*Musa textilis*) and banana. It is an exophilic, endophagic, nocturnal species which becomes active 3 h after sunset and ceases before dawn (WHO, 1992). In west Thailand the vector of nocturnal subperiodic *W. bancrofti* is *Ae niveus* (Sasa, 1976).

Sixteen species of *Aedes* are listed as vectors of diurnal subperiodic *W. bancrofti* of which ten are in the subgenus *Stegomyia* (nine in the *Ae scutellaris* group) and six in the subgenus *Finlaya*. The main vector is the day-biting, exophilic *Ae polynesiensis* which breeds in a wide range of small water containers, including coconut shells, tins, tyres, drums, tree holes, crab holes, canoes and the axils of *Pandanus*, and has a flight range of 400 m (WHO, 1992). *Ae polynesiensis* has a minor peak of feeding at 08.00 h and a major one just before sunset at 1700–1800 h, which more or less corresponds with the time of

Table 30.1. Vectors of lymphatic filariasis in various parts of the world.

Geographical region	Wuchereria bancrofti		Brugia malayi	
	Periodic	Subperiodic	Periodic	Subperiodic
Americas	Cx quinquefasciatus			
Eastern Mediterranean				
Egypt	Cx antennatus, Cx molestus,			
Sudan	An funestus, An gambiae			
Afrotropical region	Cx quinquefasciatus, An arabiensis, An funestus, An gambiae, An melas, An merus			
South-East Asia region *	Cx quinquefasciatus, An jamesi, An subpictus, Anopheles spp.	Ae niveus group	Ma annulifera, Ma indiana, Ma uniformis, An barbirostris, Mansonia spp.	Co crassipes, Mansonia spp.

Region			
Western Pacific China, Korea	Cx quinquefasciatus, Cx pallens, Ae togoi, An dirus	An anthropophagus, An sinensis, Ae kiangensis, Ae kwaiagensis. Ae togoi	
Papua New Guinea	Cx quinquefasciatus, An farauti, An punctulatus, Ae kochi		
Malaysia, Vietnam, Philippines	Cx quinquefasciatus, An balabacensis, An donaldi, An flavirostris, An maculatus, Ae poecilius	An campestris, Ma annulifera, Ma uniformis, Mansonia spp.	Co crassipes, Ma annulata, Ma bonneae, Ma dives, Ma uniformis
Pacific Islands	Ae polynesiensis, Ae aegypti, Ae cooki, Ae fijiensis, Ae kesseli, Ae kochi, Ae oceanicus, Ae pseudoscutellaris, Ae rotumae, Ae samoanus, Ae scutellaris, Ae tabu, Ae tongae, Ae tutuilae, Ae upolensis		

Source: WHO Technical Report Series 821 (1992).
* The vector of *B. timori* which occurs only in Indonesia is *An barbirostris*.

microfilarial maximum abundance at 1600 h (Chow, 1973; Hawking, 1975).

Near the coast of west Malaysia the main vectors of nocturnal subperiodic *B. malayi* are four species of *Mansonia* (*Mansonioides*): *Ma dives*, *Ma bonneae*, *Ma annulata* and *Ma uniformis*. The first two species breed in dense swamp forest where the larvae attach to the pneumatophores of trees. The adults are exophilic and exophagic and largely zoophilic. At ground level in swamp forest biting occurred all day and night with a peak after sunset. In more open areas around houses there was a sharp peak in biting after sunset, and within houses the peak of biting occurred after midnight. These species act as vectors because of their large numbers and the catholicity of their feeding responses. Although largely exophilic and zoophagic they will enter houses and feed on humans. Around houses 10% of *Ma dives* and *Ma bonneae* were found to have fed on humans (Wharton, 1962).

On the coastal rice-plains in west Malaysia, nocturnal periodic *B. malayi* is largely transmitted by *An campestris*, an anthropophilic, endophilic, endophagic species which breeds in ditches, wells and 'burrow-pits' under semi-shade (Chow, 1973). *Anopheles* species are poor hosts for the nocturnal subperiodic form of *B. malayi*. The reverse response is shown by the *Mansonia* vectors of nocturnal subperiodic *B. malayi*. No larvae of the periodic form developed in *Ma bonneae* and few in *Ma dives* (Wharton, 1962). In China and Korea the vector of nocturnal periodic *B. malayi* is *Ae togoi* which breeds in brackish water in rock holes and also in rain-filled artificial containers. The adults are endophilic with the peak biting rate occurring after sunset (Chow, 1973).

Genetics of susceptibility of mosquitoes to infection (Macdonald, 1976; Wakelin, 1978)

Susceptibility of *Ae aegypti* to *B. malayi* was increased from 17 to 90% in one generation, and subsequently retained at that level for 15 generations. Susceptibility to *B. malayi*, *B. pahangi* and *W. bancrofti* is controlled by a sex-linked, recessive gene designated f^m. This gene has no effect on susceptibility to *D. immitis* and *D. repens*, which is controlled by another sex-linked, recessive gene, f^t. It is considered that these genes act through the tissues in which the worms develop. Development of *Dirofilaria* takes place in the Malpighian tubes and of *Brugia* and *W. bancrofti* in the thoracic musculature.

Even in a susceptible strain of *Ae aegypti* 75–80% of *B. malayi* die in the course of development with the peak number of deaths occurring 2 to 3 days after the infecting meal. In the same strain of *Ae aegypti*, *B. pahangi* suffers 25% mortality in the course of development in the thoracic muscles. This probably occurs early in development because third-stage larvae introduced into the thorax survive equally well in susceptible and refractory hosts (Beckett and Macdonald, 1971).

Aedes polynesiensis and *Ae pseudoscutellaris* are both vectors of diurnal subperiodic *W. bancrofti* and they are also susceptible to infection with urban nocturnal periodic *W. bancrofti* and *B. pahangi*. *Ae malayensis* is not considered to be a vector of subperiodic *W. bancrofti* in the field and in the laboratory it is refractory to both urban periodic *W. bancrofti* and *B. pahangi*. Cross-mating of *Ae polynesiensis* and *Ae malayensis* and testing the progeny for susceptibility to *B. pahangi* shows that susceptibility is a recessive character because the hybrids are refractory.

All populations of *Cx quinquefasciatus* tested were very susceptible to urban periodic *W. bancrofti* but had only a low susceptibility to rural periodic *W. bancrofti* (5% infective cf. 87%) (Wharton, 1960); and were refractory or with low susceptibility to *B. pahangi*. Susceptibility to *B. pahangi* is a sex-liked recessive gene, which has no effect on susceptibility to *W. bancrofti*.

Uptake of microfilariae and their development in the vector

Nelson (1978) states that a temperature of 25–30°C and a relative humidity greater than 70% are required for development of the worm in the mosquito, and that no development occurs if the relative humidity is below 50%. The number of microfilariae taken up by feeding mosquitoes is greater than would be expected from the size of the blood meal. For example, although *Ma dives* takes up the number of microfilariae found in 5 mm³ of blood, it ingests only 3 mm³ of blood (Wharton, 1957a). An even greater discrepancy is found with *Cx quinquefasciatus* which ingests 4 mm³ of blood and the number of microfilariae found in three times that volume (Wharton, 1960). Both mosquitoes excrete a clear fluid either while feeding or shortly afterwards, so that the blood volume could have been underestimated, but it is possible that microfilariae respond to the feeding mosquito and accumulate at the site of feeding. This would be more likely to occur with pool feeding than with capillary feeding.

Microfilariae of *B. malayi* reached the thorax of *Ma dives* in 12 h, and half of them were mature after 11 days. The number of larvae matured was directly related to the density of microfilariae in the blood of the host, and to the numbers taken up. However, when large numbers of microfilariae were ingested (50 or more) there was a high mortality among the infected mosquitoes after 7 days. When the density of microfilariae in the blood exceeded two per mm³ there was 100% infection. Wharton (1957a) proposed an index of experimental infection based on survival of the mosquitoes, proportion infected and density of the infection. It was equivalent to:

$$\frac{\text{Total no. of mature larvae}}{\text{No. of mosquitoes fed}}$$

When this expression was applied to *B. malayi* and *Ma dives*, the index increased to a peak between three and eleven microfilariae per mm³, and then decreased due to the lethal effects of developing larvae (Wharton, 1957a,b).

Bryan and Southgate (1976) reported that *Ae polynesiensis* ingested more microfilariae of diurnal subperiodic *W. bancrofti* than expected from their concentration in the circulating blood. They found no effect of parasite load on mosquito survival but they were working with very low infections (less than 0.01 microfilariae mm⁻³). Rosen (1955), working with much heavier human infections (8–10 microfilariae mm⁻³), found that the longevity of *Ae polynesiensis* was adversely affected by large numbers of maturing larvae, with deaths occurring 13 days after an infective feed.

Cibarial and pharyngeal armatures

Wharton (1957b) found that there was no appreciable loss of larvae of *B. malayi* in *Ma dives*, and Bryan and Southgate (1976) that *Ae polynesiensis* was easily infected with *W. bancrofti* at low microfilarial densities. The ease of development of microfilariae is in part due to the absence of cibarial armature in these mosquitoes. Denham and McGreevy (1977) state that none of the vectors of *B. malayi* has cibarial armature, and the effect of their pharyngeal armatures is unknown.

In some *Anopheles* vectors of *W. bancrofti*, e.g. *An gambiae*, *An arabiensis*, and *An farauti* No. 1, the cibarial armature is well developed and damages microfilariae during feeding. *An gambiae* damaged 49% of ingested microfilariae of *W. bancrofti* and rendered 36% amotile. Cibarial armatures were absent in four species of *Anopheles* (*Anopheles*) and five species of *Aedes*, including *Ae polynesiensis*, *Ae togoi* and *Ae aegypti* and only weakly developed in *Cx quinquefasciatus*. *Cx quinquefasciatus* damaged and rendered amotile only 6% of ingested microfilariae of *W. bancrofti*, while the corresponding results for ingested microfilariae of *B. pahangi* were 9% damaged and 6% amotile in *Ae aegypti* and 22% and 11% in *Ae togoi* (McGreevy *et al.*, 1978). An efficient cibarial armature can play a substantial role in reducing infections in the mosquito host.

Proportionality, facilitation and limitation (Brengues and Bain, 1972; Bain and Chabaud, 1975; WHO, 1992)

In proportionality, the number of maturing larvae is proportional to the number of microfilariae ingested and occurs with subperiodic *B. malayi* in *Ma dives*. When *An gambiae* fed on a heavily infected (1.25 microfilariae mm^{-3}) volunteer, 97% of them ingested microfilariae. The average number ingested was 26.5, but the distribution was not normal. More than half the females ingested less than 20 microfilariae, and a small number (2%) ingested more than 100. Passage of microfilariae through the midgut epithelium was rapid with more than 60% of the mosquitoes having microfilariae in the haemocoele 2 h after engorging, but the numbers of microfilariae entering the haemocoele became proportionately larger the greater the number of microfilariae ingested. Thus where one to ten microfilariae were ingested, only 20% had entered the haemocoele 12–18 h after engorging compared with 52% when more than 50 microfilariae had been ingested. This increasing proportion with numbers ingested has been called facilitation.

When *Ae aegypti* was fed on the same carrier of *W. bancrofti* it ingested, on the average, rather more microfilariae (36.8) but the opposite response occurred. When a few microfilariae were ingested (1–10), 58% successfully penetrated the gut epithelium and were free in the haemolymph; but as the number of microfilariae ingested increased, the percentage reaching the haemocoele decreased; and when more than 70 microfilariae were taken in, only 8% successfully reached the haemocoele. This response was called limitation. Limitation occurs with subperiodic *W. bancrofti* in *Ae polynesiensis* in Samoa and Tahiti, with periodic *W. bancrofti* in *Cx quinquefasciatus* in Sri Lanka and Tanzania, and with periodic *W. bancrofti* in *Cx molestus* in Egypt.

Limitation and facilitation are associated with different responses of the midgut epithelium to penetration. In *An gambiae* microfilariae enter the haemocoele by penetrating the columnar cells at the anterior and posterior ends of the midgut. Following penetration of the first microfilariae the neighbouring cells enlarge and protrude into the midgut, facilitating subsequent penetration by other microfilariae. The reaction of the midgut cells of *Ae aegypti* is quite different and they undergo severe lysis, which impedes passage of microfilariae.

Peritrophic membrane formation and blood clotting (Denham and McGreevy, 1977)

Escape of microfilariae from the blood meal into the haemocoele of the mosquito is likely to be affected by the formation of the peritrophic membrane, and by the speed with which the blood clots. Blood rapidly coagulates in the midgut of *Ae aegypti*, and only 30% of ingested *D. immitis* microfilariae reached the haemocoele, but this proportion increased to 80% when an anticoagulant was added to the ingested blood. Most microfilariae have escaped from the midgut before formation of the peritrophic membrane, and it is not known whether microfilariae can penetrate the membrane, once it has been formed.

Epidemiology

The diseases caused by *W. bancrofti* and the periodic form of *B. malayi* are anthroponoses and the nocturnal subperiodic form of *B. malayi* is an anthropozoonosis. Subperiodic *B. malayi* is common in wild monkeys and is limited to foci in swamp forests in south-east Asia where people and domestic animals are surrounded by virgin forest with wild animals and mosquitoes. The relative importance of domestic and wild animals is not known but it is likely that transmission to humans would occur with greater frequency from domestic animals (Denham and McGreevy, 1977). Wharton (1962) considered that the catholic feeding habits of *Mansonia*, combined with their occurrence in large numbers in houses, kampong (village) and forest, and their ability to transmit subperiodic *B. malayi*, were sufficient to explain the high human infection rates, particularly in children, and maintain the parasite in its animal reservoir.

Transmission requires the presence of microfilariae for the vector to take up when it feeds. In *W. bancrofti* and periodic *B. malayi*, the proportion of the population with microfilariae increases with age whereas with subperiodic *B. malayi* the highest infection rates are found in children below the age of 5 years and there is a decrease in infection with age (Wharton, 1963). Abundant microfilariae are produced during the inflammatory and early obstructive stage while in advanced cases of elephantiasis, microfilariae are rarely present and the patients are non-infective (Denham and McGreevy, 1977). Nevertheless, infected individuals remain infective for 5 to 10 years (Buck, 1991) and some mosquitoes, e.g. *Ae polynesiensis*, can acquire infections when the density of microfilariae in the host is very low (Bryan and Southgate, 1976).

Limiting factors involving the vector

In the vector microfilariae undergo a temperature-dependent cycle of development and in western Samoa infective larvae of *W. bancrofti* were present in the proboscis of *Ae polynesiensis* in 12 days (Bryan and Southgate, 1976). Sasa (1976) cites the results of Feng (1936) who found that *B. malayi* completed its development in *An sinensis* in 6–6.5 days at 29–32°C. To become infective the mosquito must survive long enough for the worm to complete its development. Daily survival rates of 0.80–0.85 were obtained for *Ma dives* and *Ma bonneae* in Malaysia (Wharton, 1962) and of 0.80 per day for *Cx quinquefasciatus* throughout the year in Rangoon (De Meillon *et al.*, 1967a). A survival rate of 0.80 gives 7% survival after 12 days.

De Meillon *et al.* (1967a) found that the infection rate of *Cx quinquefasciatus* with *W. bancrofti* in Rangoon was more or less constant throughout the year with 4.8% of the mosquito population infected and 0.36% infective. In west Malaysia infection rates of *B. malayi/B. pahangi* in *Ma dives* and *Ma bonneae* around houses were 1.1–1.4% infected and 0.5–0.6% infective. Two miles away in swamp forest the rates were somewhat lower, 0.7% infected and 0.4% infective. Unlike malaria where an infected *Anopheles* mosquito remains infective for the rest of her life, filarial infections are limited and can be exhausted. Infected *Ma dives* and *Ma bonneae* contained on average five mature larvae and a similar number of immature larvae, but the distribution was skewed with 60% of the infective mosquitoes containing one to four larvae and a similar distribution occurred with the immature stages (Wharton, 1962).

Transmission rate

The transmission rate depends not only on the proportion of the vectors that are infective but also on vector density. In Rangoon the infection rate of *Cx quinquefasciatus* with *W. bancrofti* remained unchanged throughout the year, but the risk of infection was markedly seasonal, being lowest in the monsoon season when the breeding sites of *Cx quinquefasciatus* were flushed away and highest at the end of the dry season. In addition the risk of infection was greater late in the evening, towards midnight, than in the period following sunset and also greater outdoors than indoors (De Meillon *et al.*, 1967a). In the west Malaysian *Brugia/Mansonia* system, the intensity of transmission was highest in the swamp forest on account of the larger number of vectors, lower around houses and lowest at night within houses where the rate was 31 infective bites per annum (Wharton, 1962).

Three comparable investigations of the transmission of nocturnal periodic *W. bancrofti* by *Cx quinquefasciatus* have been carried out in Rangoon (Hairston and De Meillon, 1968), Calcutta (Gubler and Bhattacharya, 1974) and Jakarta (Self *et al.*, 1978). The average annual biting rate per person was of the order of 100,000, of which 0.3 to 1.6% of the mosquitoes were infective, carrying 3.0 to 4.5 infective larvae per infective mosquito. De Meillon *et al.* (1967b) calculated that during blood-feeding 41.4% of the infective larvae were deposited on the skin. Using this correction, it was found that the average number of infective larvae deposited on a person each year in the three cities ranged from 560 in Rangoon to nearly 2500 in Calcutta. When the same correction is made to the data collected in Pondicherry for 48 consecutive months the number of infective larvae deposited is estimated to be 2896 (Ramaiah and Das, 1992), comparable to that found in Calcutta.

To assess the value of integrated vector management (IVM) in controlling lymphatic filariasis through controlling *Cx quinquefasciatus*, half the town of Pondicherry in South India was subjected to IVM with a major emphasis on environmental modification to permanently remove or reduce the breeding sites of the vector. The other half of the town continued to use routine application of larvicides. In the 5 years of the experiment (1981–85), vector populations were reduced by around 80% and by the end of the period the microfilarial rate in the IVM area was significantly lower in all age groups up to 24 years. The difference was most marked in the 0–5 year class which had been born during the experiment. Their microfilarial rate had declined from 2.4% in 1981 to 0.2% in 1986. It showed that sustained control of the breeding of *Cx quinquefasciatus* could be achieved in a complex urban environment (Subramanian *et al.*, 1989).

Transmission was not uniform throughout the year – the maximum transmission occurred in the cooler months from November to February. Early larvae of *W. bancrofti* were found in the vector all the year but no infective (L_3) larvae were found in May to August, the hottest months, when the temperature was above 30°C. Development of the parasite in the vector takes 8 to 12 days and hence only females that have completed a minimum of two oviposition cycles would have the potential to be vectors (Ramaiah and Das, 1992).

Bryan and Southgate (1976) calculated the input of infective larvae into a population from the expression:

$$y = Mmap^n ib$$

where *m, a, p, n* refer to mean density of vector in relation to people, average number of humans bitten by vector in one day, daily survival rate and duration of the extrinsic cycle, respectively, as in similar expressions concerned with the transmission of malaria. The new factors are: *M*, the number of carriers in the population; *i*, the proportion of vectors infective; and *b* the number of infective larvae per infected mosquito. Applying this to *W. bancrofti* and *Ae polynesiensis*, the human population was being exposed to an annual injection of 1.37 million larvae per annum.

Comparison with malaria

Unlike malaria, where a single inoculation of sporozoites will produce clinical malaria, filariasis requires repeated infection. Webber (1977) made a comparison between the transmission of *Plasmodium falciparum* and *W. bancrofti* by *An farauti* in the Solomon Islands and found that the critical density of mosquitoes to maintain transmission was of the order of ten times greater for filariasis than for malaria. In other words malaria transmission would continue after transmission of filariasis had ceased, but this comparison is dependent upon accurate knowledge of the recovery rates from malaria and filariasis, and the number of bites required to produce a case of malaria or filariasis. In Rangoon it was calculated that 15,500 infective bites were needed to produce microfilaraemia, and at a rate of 300 infective bites per annum, on average 50 years would be required to achieve that level (Hairston and De Meillon, 1968).

Principles of Control (WHO, 1992; see also Chapter 7)

The worms are long lived and produce microfilariae over a period of up to 10 years, so that control measures against filariasis must be maintained at an appropriate level over many years. Measures can be directed against the parasite or against the vector or to minimizing the human–vector contact, but it is desirable that a control programme should include several lines of attack concentrating on the approach which offers the maximum benefit for effort expended. Mass treatment of populations with DEC interrupted transmission of *B. malayi* by *An sinensis* in China. While in Samoa it reduced the prevalence of *W. bancrofti* from 19% to 0.2% but low levels of microfilaraemia continued, enabling *Ae polynesiensis* and *Ae samoanus* to continue transmission. In India and China the distribution of DEC medicated salt has decreased transmission.

House spraying for malaria control has eradicated *Anopheles*-transmitted *W. bancrofti* in the Solomon Islands. Control measures are usually directed against the immature stages of the vectors of lymphatic filariasis. As *Cx quinquefasciatus* has become widely resistant to organochlorine compounds preference is given to organophosphate larvicides. Floating bags of temephos have given good control of four vector species of *Mansonia* in southern Thailand, and temephos sprayed on *Pandanus* plants has given control of *Ae samoanus* for 6 weeks.

Bacillus sphaericus has been used to control *Mansonia* breeding in ponds in Kerala, India. As *B. sphaericus* can persist and recycle in polluted waters and, in general, *Culex* larvae are highly susceptible, this organism holds out promise for the biological control of *Cx quinquefasciatus*. Physical control by using polystyrene beads to cover stagnant, confined waters, as found in cesspits, has been widely used and been highly effective, persisting for 3 years in the absence of flooding. Impregnated bednets offer individual protection against being bitten by nocturnal, endophagic vectors, e.g. *Cx quinquefasciatus*. Raising living standards, in particular by the introduction of piped water supply with accompanying drainage and sewerage, removes breeding sites of *Cx quinquefasciatus*, and has given control of filariasis in the southern USA, Puerto Rico and the Mediterranean (Nelson, 1978).

When the vector is exophilic and diurnal, control is more difficult. Mosquito-proofing of houses offers little protection, neither do mosquito nets. This is the situation with the *Ae scutellaris* group which are vectors of the diurnal subperiodic *W. bancrofti* in the Pacific. They breed in a wide range of small containers, which are too numerous to locate and deal with individually so that larval control is not practical (Macdonald, 1976). In that situation mass chemotherapy offers the best prospect of control.

Control of subperiodic *B. malayi* transmitted by *Mansonia* species is a formidable task. Although residual insecticides had little effect on the overall vector population, house-spraying with dieldrin halved the transmission rate (Wharton, 1962). Wilson (1969) has shown that supervised administration of DEC to a high proportion of the human population can bring about a long-lasting reduction in infection rates of both periodic and subperiodic *B. malayi*. Another possible approach to the control of subperiodic *B. malayi* involves eliminating the reservoir in domestic animals. This might make a significant contribution to the reduction of human filariasis (Denham and McGreevy, 1977).

References

Anderson, R.C., Chabaud, A.G. and Willmott, S. (eds) (1974) *CIH Keys to the Nematode Parasites of Vertebrates.* Commonwealth Agricultural Bureau, Slough, UK.

Bain, O. and Chabaud, A.G. (1975) Le mécanisme assurant la régulation de la traversée de la paroi stomacale du vecteur par les microfilaires (*Dipetalonema dessetae-Aedes aegypti*). *Comptes Rendus Hebdomadaire des Séances de l'Académie Scientifique, Paris, Série D* 281, 1199–1202.

Beckett, E.W. and Macdonald, W.W. (1971) The survival and development of subperiodic *Brugia malayi* and *B. pahangi* larvae in a selected strain of *Aedes aegypti*. *Transactions of the Royal Society of Tropical Medicine and Hygiene* 65, 339–346.

Brengues, J. and Bain, O. (1972) Passage des microfilaires de l'estomac vers l'hémocèle du vecteur, dans les couples *Wuchereria bancrofti–Anopheles gambiae* A, *W. bancrofti–Aedes aegypti* et *Setaria labiatopapillosa–Aedes aegypti Cahiers ORSTOM, Entomologie Médicale et Parasitologie* 10, 235–249.

Bryan, J.H. and Southgate, B.A. (1976) Some observations on filariasis in Western Samoa after mass administration of diethylcarbamazine. *Transactions of the Royal Society of Tropical Medicine and Hygiene* 70, 39–48.

Buck, A.A. (1991) Filarial infections. General principles. Filariasis. In: Strickland, G.T. (ed.) *Hunter's Tropical Medicine.* Saunders, Philadelphia, pp. 711–727.

Chow, C.Y. (1973) Filariasis vectors in the Western Pacific Region. *Zeitschrift für Tropenmedizin und Parasitologie* 24, 404–418.

De Meillon, B., Grab, B. and Sebastian, A. (1967a) Evaluation of *Wuchereria bancrofti* infection in *Culex pipiens fatigans* in Rangoon, Burma. *Bulletin of the World Health Organization* 36, 91–100.

De Meillon, B., Hayashi, S. and Sebastian, A. (1967b) Infection and reinfection of *Culex pipiens fatigans* with *Wuchereria bancrofti* and the loss of mature larvae in blood-feeding. *Bulletin of the World Health Organization* 36, 81–90.

Denham, D.A. and McGreevy, P.B. (1977) Brugian filariasis: epidemiological and experimental studies. *Advances in Parasitology* 15, 243–309.

Dreyer, G., Figueredo-Silva, J., Andrade, L.D., Marchetti, F. and Coutinho, A. (1994) Bancroftian filariasis: new approaches to the diagnostic and treatment. *Parasite* 1 (1S), 21–22.

Edeson, J.F.B. and Wilson, T. (1964) The epidemiology of filariasis due to *Wuchereria bancrofti* and *Brugia malayi*. *Annual Review of Entomology* 9, 245–268.

Ewert, A. and Ho, B.C. (1967) The fate of *Brugia pahangi* larvae immediately after feeding by infective vector mosquitoes. *Transactions of the Royal Society of Tropical Medicine and Hygiene* 61, 659–662.

Gubler, D.J. and Bhattacharya, N.C. (1974) A quantitative approach to the study of bancroftian filariasis. *American Journal of Tropical Medicine and Hygiene* 23, 1027–1036.

Hairston, N.G. and De Meillon, B. (1968) On the inefficiency of transmission of *Wuchereria bancrofti* from mosquito to human host. *Bulletin of the World Health Organization* 38, 935–941.

Hawking, F. (1975) Circadian and other rhythms of parasites. *Advances in Parasitology* 13, 123–182.

Ho, B.C. and Ewert, A. (1967) Experimental transmission of filarial larvae in relation to feeding behaviour of the mosquito vectors. *Transactions of the Royal Society of Tropical Medicine and Hygiene* 61, 663–666.

Ho, B.C. and Lavoipierre, M.M.J. (1975) Studies on filariasis. IV. The rate of escape of the third-stage larvae of *Brugia pahangi* from the mouthparts of *Aedes togoi* during the blood meal. *Journal of Helminthology* 49, 65–72.

Laurence, B.R. (1989) The global dispersal of bancroftian filariasis. *Parasitology Today* 5, 260–264.

Macdonald, W.W. (1976) Mosquito genetics in relation to filarial infections. *Symposia of the British Society of Parasitology* 14, 1–24.

McGreevy, P.B., Theis, J.H., Lavoipierre, M.M.J. and Clark, J. (1974) Studies on filariasis. III. *Dirofilaria immitis*: emergence of infective larvae from the mouthparts of *Aedes aegypti. Journal of Helminthology* 48, 221–228.

McGreevy, P.B., Bryan, J.H., Oothuman, P. and Kolstrup, N. (1978) The lethal effects of cibarial and pharyngeal armatures of mosquitoes on microfilariae. *Transactions of the Royal Society of Tropical Medicine and Hygiene* 72, 361–368.

Nelson, G.S. (1978) Mosquito-borne filariasis. In: Willmott, S. (ed.) *Medical Entomology Centenary Symposium Proceedings*. Royal Society of Tropical Medicine and Hygiene, London, pp. 15–25.

Nelson, G.S. (1981) Issues in filariasis – a century of enquiry and a century of failure. *Acta Tropica* 38, 197–204.

Ramaiah, K.D. and Das, P.K. (1992) Seasonality of adult *Culex quinquefasciatus* and transmission of Bancroftian filariasis in Pondicherry, South India. *Acta Tropica* 50, 275–283.

Rosen, L. (1955) Observations on the epidemiology of human filariasis in French Oceania. *American Journal of Hygiene* 61, 219–248.

Sasa, M. (1976) *Human Filariasis*. University of Tokyo Press, Tokyo.

Self, L.S., Usman, S., Sajidiman, H., Partono, F., Nelson, M.J., Pant, C.P., Suzuki, T. and Mechfudin, H. (1978) A multidisciplinary study on Bancroftian filariasis in Jakarta. *Transactions of the Royal Society of Tropical Medicine and Hygiene* 72, 581–587.

Service, M.W. (1978) Patrick Manson and the story of Bancroftian filariasis. In: Willmott, S. (ed.) *Medical Entomology Centenary Symposium Proceedings*. Royal Society of Tropical Medicine and Hygiene, London, pp. 11–14.

Subramanian, S., Pani, S.P., Das, P.K. and Rajagopalan, P.K. (1989) Bancroftian filariasis in Pondicherry, South India: 2. Epidemiological evaluation of the effect of vector control. *Epidemiology and Infection* 103, 693–702.

Sucharit, S., Harinasuta, C. and Choochote, W. (1982) Experimental transmission of subperiodic *Wuchereria bancrofti* to the leaf monkey (*Presbytis melalophos*), and its periodicity. *American Journal of Tropical Medicine and Hygiene* 31, 599–601.

Vanamail, P., Subramanian, S., Das, P.K., Pani, S.P., Rajagopalan, P.K., Bundy, D.A.P. and Grenfell, B.T. (1989) Estimation of age-specific rates of acquisition and loss of *Wuchereria bancrofti* infection. *Transactions of the Royal Society of Tropical Medicine and Hygiene* 83, 689–693.

Wakelin, D. (1978) Genetic control of susceptibility and resistance to parasitic infections. *Advances in Parasitology* 16, 219–308.

Webber, R.H. (1977) The natural decline of *Wuchereria bancrofti* infection in a vector control situation in the Solomon Islands. *Transactions of the Royal Society of Tropical Medicine and Hygiene* 71, 396–400.

Wharton, R.H. (1957a) Studies in filariasis in Malaya: observations on the development of *Wuchereria malayi* in *Mansonia* (*Mansonioides*) *longipalpis*. *Annals of Tropical Medicine and Parasitology* 51, 278–296.

Wharton, R.H. (1957b) Studies in filariasis in Malaya: the efficiency of *Mansonia longipalpis* as an experimental vector of *Wuchereria malayi*. *Annals of Tropical Medicine and Parasitology* 51, 422–439.

Wharton, R.H. (1960) Studies on filariasis in Malaya: field and laboratory investigations of the vectors of a rural strain of *Wuchereria bancrofti*. *Annals of Tropical Medicine and Parasitology* 54, 78–91.

Wharton, R.H. (1962) *The Biology of Mansonia Mosquitoes in Relation to the Transmission of Filariasis in Malaya*. Bulletin No. 11, Institute for Medical Research, Federation of Malaya.

Wharton, R.H. (1963) Adaptation of *Wuchereria* and *Brugia* to mosquitoes and vertebrate hosts in relation to the distribution of filarial parasites. *Zoonoses Research* 2, 1–12.

WHO (1992) Lymphatic filariasis the disease and its control. *WHO Technical Report Series* 821.

Wilson, T. (1969) An example of filariasis control from west Malaysia. *Bulletin of the World Health Organization* 41, 324–326.

Human Onchocerciasis *(Onchocerca volvulus)* **31**

Distribution in the World (WHO, 1987)

Human onchocerciasis is an economically important disease, which is not directly fatal but causes untold misery in certain areas of tropical Africa, Latin America and the eastern Mediterranean (Sudan and Yemen). It is caused by infection with *Onchocerca volvulus* and transmitted by blackflies of the genus *Simulium*. It places an intolerable burden on whole communities, and denies vast fertile areas to human settlement and agricultural development. Thus, although the valleys of the White Volta and Red Volta in Burkina Faso have the most fertile and best irrigated soils in the Republic, they have, until recently, been devoid of humans. The vast Onchocerciasis Control Programme (OCP) in West Africa was initially named 'OCP in the Volta River basin'.

It is estimated that there are 86 million people at risk of infection with 18 million infected and 336,400 blind. The greatest incidence is in tropical Africa (92% of total); Latin America has 6% and the Sudan and Yemen 2%. In tropical Africa the disease occurs throughout the northern Sudano-Guinean savanna of West Africa east to the Ethiopian highlands and Uganda. It extends south through the rainforest of West Africa and Zaire to Angola and eastwards to Tanzania and Malawi. In a recent survey (Akogun and Renz, 1994), eye lesions and blindness were present in 27% and 20% respectively of the inhabitants of the upper Taraba river valley in Nigeria.

In Yemen it is present in a few permanent wadis at altitudes of 300 to 1200 m and in the southern Sudan where, in the Raga area 25% of the adult population is blind. In Latin America onchocerciasis is present in adjoining areas of Mexico/Guatemala, Colombia/Ecuador, Brazil/Venezuela, and in northern Venezuela. Infections with *O. volvulus* are cryptic and new foci are found by chance, e.g. the one in the Yemen was not discovered until the mid-1950s, and Nelson (1970) considers that onchocerciasis 'should be looked for in all parts of the world where simuliids bite man'.

The Disease and the Parasite (Nelson, 1970; Duke and Taylor, 1991)

Onchocerciasis has three manifestations: an unsightly and irritating dermatitis; subcutaneous nodules; and eye lesions which result in blindness.

The long, slender, long-lived adult worms are to be found free in the subcutaneous tissue or more commonly encapsulated in nodules which may be in clusters 10 cm across. The larger female worm measures 23 to 50 cm in length and 250 to 450 μm in diameter and males 16 to 42 cm long and 125 to 200 μm in diameter. Apart from the unsightly nodules the adult worms are innocuous. It is the unsheathed migrating microfilariae, measuring 220 to 360 μm by 5 to 9 μm, which cause damage and provoke inflammatory lesions when they degenerate. The density of microfilariae is greatest in the vicinity of adult worms. The distribution of nodules and microfilariae differs with the strain of parasite, being most abundant on the lower part of the body in West Africa; mainly around the buttocks and upper thigh in East Africa; and on the torso in Central America (Nelson, 1970). This distribution is correlated with, but not necessarily determined by, the biting habits of the vectors. In Central America *S. ochraceum* attacks the upper part of the body and in West Africa *S. damnosum* the lower parts of the body.

The presence of large numbers of active microfilariae in the skin causes intense itching and scratching leading to loss of pigment in patches in the affected area. These contrast strikingly with the normal dark skin and are an obvious sign of infection. In the later stages there is thickening of the skin and a loss of elasticity, giving the sufferer a prematurely aged appearance. Scarring in the lymph node can cause regional lymphoedema and result in a condition known as 'hanging groin'.

Onchocerciasis develops slowly, with blindness being rare in people under 20 years of age and rising to more than 50% in those more than 50 years old (Akogun and Renz, 1994). Living microfilariae may be found in many parts of the eye and cause little damage but when they die they induce lesions. Snowflake ocular opacities in the cornea are temporary but sclerosing keratitis of the cornea is progressive and a cause of blindness. Profound chorioretinal atrophy may develop and optic neuritis and optic atrophy are frequent complications of onchocerciasis. One puzzle is the lower incidence of blindness in the forest zone of West Africa compared to the savanna zone (less than 1% cf. 10%), in spite of the higher transmission rate of *O. volvulus* in the forest (Nelson, 1970).

Onchocerciasis is commonly diagnosed by a skin snip in which a small piece of skin is removed and examined for the presence of living microfilariae. In Guatemala and Mexico where nodules on the head are common, their surgical removal is popular and this may have reduced the incidence of ocular onchocerciasis. In Ecuador the more drastic removal of all palpable nodules improves the clinical condition of the patient. Nodulectomy is not widely practised in Africa.

Ivermectin is a microfilaricide which is less toxic than diethylcarbamazine and safe enough for large-scale use. It is a microfilarial suppressant having no action on the adults, but preventing the escape of microfilariae from the female for 6 to 12 months. Microfilariae are not expelled by the female but leave by their own propulsion (Schulz-Key and Soboslay, 1994). Hence when the microfilariae are affected they remain in the uterus. Suramin kills adult worms but is toxic. A new drug, Amocarzine, taken after a meal, has onchocercidal effects against microfilariae and adult worms (Poltera, 1994).

Development of *O. volvulus* in *Simulium*

Blacklock (1926a,b), working in Sierra Leone, showed that *O. volvulus* was ingested by *S. damnosum* when it fed on a host with microfilariae in the skin, and that these developed in the thoracic muscles, giving rise to infective (3rd stage) larvae which escaped from the labium of the fly during feeding. The worm undergoes a very similar cycle of development to that of *W. bancrofti* in its mosquito vector. In the human host infective larvae moult to the 4th stage in 3 to 7 days and to juvenile worms several weeks later. There follows a premature period of 9 to 12 months after which mated females begin to produce microfilariae. Female worms are sessile but males regularly leave nodules (Schulz-Key and Soboslay, 1994).

A feeding, female simuliid takes up microfilariae as it scrapes its way through the skin, and hence the uptake of microfilariae is more likely to be related to the distribution of microfilariae in the skin and the time taken to penetrate it than to the size of the blood meal. Wegesa (quoted by Nelson, 1970) and Philippon (1977) relate the number of microfilariae ingested to the duration of feeding, which may reflect the time spent penetrating the skin.

There is a great deal of variation in the uptake of microfilariae by flies feeding on the same host and it is preferable to work with geometric rather than arithmetic means. The intake of microfilariae varied with their concentration in the skin up to a maximum of 150 microfilariae per mg of skin (Duke, 1962). The intake by flies that fed on the same host ranged from 0 to 171 with 55% of *S. damnosum* ingesting ten or fewer microfilariae, and 15% ingesting more than 50 (Laurence, 1966). In this series the arithmetic mean was 26.2 and the modified geometric mean 10.1. In Central America the intake of microfilariae by *S. ochraceum* and *S. metallicum* is higher than that of *S. damnosum*, possibly because their saliva attracts microfilariae to the feeding site (Shelley, 1988).

Microfilariae must avoid being trapped within the peritrophic membrane, which is secreted around the blood meal. In *S. ochraceum* and *S. metallicum* most microfilariae have escaped before the membrane hardens and becomes impenetrable. In *S. neavei* the thin, delicate membrane is a less formidable barrier but in *S. damnosum* the membrane is a major source of worm mortality (Crosskey, 1990). In two series of experiments, the proportions of microfilariae which avoided such imprisonment were 0.44 (Duke and Lewis, 1964) and 0.75 (Laurence, 1966) but in individual flies the proportion varied from none to all microfilariae. There was no correlation between the proportion of microfilariae which escaped into the haemocoele and the number ingested, i.e. there was no evidence of facilitation or limitation as observed with *W. bancrofti* and different mosquito hosts (Duke and Lewis, 1964).

In the simuliid the first microfilariae reached the thorax in 20 min and invaded the flight muscles in 2 h. The numbers of microfilariae in the thorax increased steadily from 30 min to 6 h after feeding, when 62% of the ingested microfilariae had reached the thorax (Laurence, 1966).

Once microfilariae have escaped from the midgut, a high proportion (0.91) completed development to become infective larvae (third stage). The success rate for ingested microfilariae developing into infective forms was around 35–40% and took 7 to 8 days (Duke, 1962; Duke and Lewis, 1964). A similar time (6–7 days) was found for the development of *O. volvulus* in *S. neavei* in Uganda at 21°C and 75% RH (Nelson, 1970). Wegesa found the threshold for development of *O. volvulus* in *S. woodi* to be 18°C, and the optimum temperature 24°C. In *S. ochraceum* development of *O. volvulus* is at least 8 days at 25°C, and 4 days at 30°C (Ogata, 1981).

Between oviposition and the next blood meal, female *S. damnosum* s.l. take a meal of nectar during which some infective larvae of *O. volvulus* are lost (Wenk, 1981). This proportion may be as high as one-third of the infective larvae (Philippon, 1977). Further loss of infective potential occurs when only about 80% of the infective larvae escape while the female is taking a blood meal (Wenk, 1981).

Development of *O. volvulus* may not damage the simuliid host as much as might be expected. Blackflies invading the OCP in West Africa have come from localities several hundred kilometres away, and yet they have a high infection rate. Female *S. metallicum* and *S. exiguum* which have fed on heavily infected carriers of *O. volvulus*, suffer high mortality as a result of physical damage. *S. ochraceum* does not suffer this damage because a high proportion of the ingested microfilariae are damaged by its buccopharyngeal armature (Shelley, 1988), recalling similar destruction wrought on microfilariae of *W. bancrofti* by the comparable armature of *An gambiae*.

Strains of *Onchocerca volvulus*

The strains of *O. volvulus* in Central America and Africa behave quite differently, partly in adaptation to the vectors. In Central America microfilariae are seven to ten times as abundant in the face, neck and arms than in the legs, making them more accessible to the vector, *S. ochraceum*, which preferentially feeds on the upper parts of the body. In contrast, in Africa microfilariae are more abundant in the legs and lower parts of the body where 98% of the biting of *S. damnosum* occurs (De Leon and Duke, 1966).

When fed on carriers with comparable densities of microfilariae in the skin, *S. ochraceum* ingests 20–25 times more of the Guatemalan strain of *O. volvulus* than of either the forest or Sudan savanna strains from West Africa (De Leon and Duke, 1966). When either forest or Sudan savanna *S. damnosum* are fed on a host containing the Guatemalan strain of *O. volvulus* there is no concentration of microfilariae. Indeed forest *S. damnosum* ingested 2 to 4.5 times as many micro-filariae of the forest strain of *O. volvulus* than of the Guatemalan strain, and both the forest and Sudan savanna forms of *S. damnosum* rapidly eliminated microfilariae of the Guatemalan strain (Duke *et al.*, 1967). The success rate of microfilariae of the forest strain *O. volvulus* in *S. damnosum* was 47% compared with 2% for microfilariae of the same strain in *S. ochraceum*, although slightly more microfilariae were ingested by *S. ochraceum* (De Leon and Duke, 1966).

Duke *et al.* (1966) recognized two parasite–vector complexes, one existing in the forest and Guinea savanna and the other in the Sudan savanna. Microfilariae of *O. volvulus* from the forest area of Cameroon developed well in *S. damnosum* of the forest and Guinea savanna zones, but not in *S. damnosum* from the Sudan savanna; and, conversely, microfilariae from the Sudan savanna developed well in *S. damnosum* of the same zone but achieved little or no development in *S. damnosum* from the forest or Guinea savanna. This difference is considered to reside in the peritrophic membrane because when microfilariae of the savanna and forest strains are injected into the thorax, they develop equally well in both savanna (*S. damnosum, S. sirbanum*) and forest (*S. squamosum, S. mengense*) vectors (Eichner *et al.*, 1991).

The differences between vectors can be related to sibling species of the *S. damnosum* complex. No comparable separation is available for the recognition of forms of *O. volvulus* but there is evidence for the existence of two strains of *O. volvulus* in West Africa differing in their pathogenicities, biochemical structure and vectors. The savanna strain is associated with *S. damnosum* s.s. and *S. sirbanum*, and causes severe disease involving serious eye lesions and blindness, urinary excretion of microfilariae, and depressed immunity in the host. The forest strain of *O. volvulus* has low human pathogenicity; is poorly transmitted by *S. damnosum* and *S. sirbanum*, but is well adapted to local vectors; and shows certain characteristic biochemical features (Prost *et al.*, 1980).

Simuliid Vectors of *Onchocerca volvulus*

The vectors of onchocerciasis will be considered under three headings: the *S. damnosum* group of species, the *S. neavei* group, and the vectors in Latin America.

Simulium damnosum complex (WHO, 1987; Crosskey, 1993)

Simulium damnosum was considered to be a single species widely distributed throughout the Afrotropical region. Examination of the polytene chromosomes of different populations led to the recognition of four different forms in 1966, which by 1987 had increased to 42. Nine of these are formally named species; another nine are defined as 'incautiously described morphospecies of uncertain validity from the Zaire basin'; 22 taxa from East Africa, which are probably valid; and two others. The East African species are mainly zoophilic (Dunbar and Vajime, 1981).

Thirteen taxa – nine named species and four siblings – in the *S. damnosum* complex are vectors of *O. volvulus*. They include the widespread *S. damnosum* s.s. which occurs from West Africa eastwards to the southern Sudan and Uganda, and *S. sirbanum* which occurs from Senegal to the southern Sudan and northwards along the Nile. They are long-range migrants which regularly recolonize large, open seasonal rivers. Other vectors in West Africa are *S. sanctipauli* and *S. soubrense* in forest areas; *S. squamosum* and *S. mengense* in the forest–savanna mosaic; and *S. yahense* in upland forest. *S. sanctipauli* and *S. squamosum* breed in smaller, permanently flowing, shaded rivers from which they disperse little. *S. kilibanum* is a vector in some foci in eastern Africa and *S. rasyani* which breeds in sun-warmed, west-flowing wadis of variable water flow is the vector in Yemen. Of the four siblings two are vectors in Tanzania, one in eastern Zaire and the other in Ethiopia.

The distribution of the anthropophilic species in West Africa is given in map form by Crosskey (1981). The distributions of these species can be related to ecological zones. With increasing distance from the coast, zones of increasing aridity can be recognized. Except for the Dahomey Gap the moist forest zone occupies a coastal belt of varying width. Moving inland this is succeeded by the forest–savanna mosaic zone; an undifferentiated, relatively moist Guinea or woodland savanna; a relatively dry Sudan or woodland savanna; the Sahel savanna; subdesert; and finally desert (Duke *et al.*, 1966). Onchocerciasis extends into the Sudan savanna but not further inland.

Simulium sirbanum is predominant in the Sudan savanna, spreads into the Guinea savanna and is almost absent from the forest zone. *S. damnosum* s.s. is abundant in the Guinea savanna, well represented in the Sudan savanna and present in smaller numbers in the forest. *S. squamosum* is mainly a forest species being present in heavily shaded or forested areas in the Guinea savanna but almost absent from the Sudan savanna. The other three species are mainly found in the forest zone although *S. soubrense* also occurs in the Guinea savanna.

Cytotaxonomic methods can only be applied to larvae, and a lot of work has been devoted to developing sound methods of separating the six important West African species in the adult stage. Peterson and Dang (1981) have tabulated comparisons between these species for 31 characters, from which they have derived a simplified pictorial key (Dang and Peterson, 1980). Additional detailed information on the antennae, and the maxillae are given by Quillévéré *et al.* (1977) and Meredith and Townson (1981) found two enzyme systems useful to separate *S. squamosum* and *S. yahense* from each other, and from the other four West African anthropophilic species.

Simulium neavei group and S. albivirgulatum

Three species of the *S. neavei* group are vectors of *O. volvulus* in eastern Africa. *S. neavei* is a localized vector in Uganda and Zaire and was the vector in Kenya before it was eradicated from that country. *S. woodi* is the vector in the Usambara and Uluguru mountains of eastern Tanzania and *S. ethiopiense* in south-western Ethiopia. The larvae and pupae of species of the *S. neavei* group have a phoretic association with crabs. Those of *S. neavei* are found on the carapace of *Potamonautes* (*Potamon*) *niloticus*, which lives in the rockier parts of rivers near cascades (Van Someren and McMahon, 1950; McMahon *et al.*, 1958). The eggs of *S. neavei* are deposited on vegetation in clusters near cascades (De Meillon, 1957). In western Uganda 7% of *S. neavei* contained infective larvae giving a rate of nearly 300 infective larvae per 1000 parous females (Garms *et al.*, 1992).

Simulium woodi has a bimodal cycle of biting activity with morning and afternoon peaks in which nulliparous flies are commoner in the morning and parous flies in the afternoon. When *S. woodi* feeds on humans it attacks, almost exclusively, the legs where the greatest concentration of microfilariae is to be found. In one survey 17% of the parous flies were infected and 3% were infective (Raybould, 1967). *S. albivirgulatum*, which breeds in slow to swift but smooth-flowing rivers lined by gallery forest, is a vector in the Zaire river basin (Crosskey, 1993).

Vectors in Latin America (Shelley, 1988)

The vectors of onchocerciasis in Latin America are designated as primary or secondary vectors depending on whether or not they are considered capable of sustaining transmission on their own. Primary vectors can do so and secondary vectors cannot. The position is complicated by the existence of cytotypes within species. Four cytotypes have been recognized in *S. exiguum*; three in *S. ochraceum*; two in *S. oyapokense*; and eleven in *S. metallicum*, but they have not been studied in the same detail as the African vectors. In Mexico/Guatemala the primary vector is *S. ochraceum* s.l. and the secondary vector *S. metallicum* s.l.; in northern Venezuela *S. metallicum* s.l. is the primary vector and *S. exiguum* s.l. the secondary; in Amazonia the primary vectors are *S. guianense* in the highlands and *S. oyapokense* in the lowlands where *S. guianense* is a secondary vector; in Colombia/Ecuador the vector is *S. exiguum*. (Davies and Crosskey, 1991, give *S. quadrivittatum* as the primary vector in Ecuador and restrict *S. exiguum* to Colombia.)

Simulium ochraceum s.l. is a predominantly anthropophilic species occurring in high rainfall areas between 500 and 1500 m above sea level, where it breeds in trickles and streams less than 0.5 m in width under deep vegetation. It is diurnal with peak activity occurring mid-morning when it attacks selectively the upper region of the body where the microfilariae of *O. volvulus* are most abundant. High biting rates, up to 1000 per person/per hour, were associated with high humidities and sunny skies, conditions found in the coffee-growing regions where onchocerciasis is rife. *S. ochraceum* can disperse 10 km from the breeding site.

In northern Venezuela *S. metallicum* s.l. maintains foci of onchocerciasis up to 1000 m above sea level (a.s.l.). It is largely an anthropophilic species maintaining human biting rates of 200 per hour throughout the day, feeding below the waist where microfilarial densities are high. It can have an infectivity rate of 2%. It breeds in small streams less than 5 m in width.

In Amazonia *S. guianense* maintains hyperendemic onchocerciasis above 250 m where it is the most common human biter with rates of 150 per day and up to 2% being infective. In lowland Amazonia *S. oyakopense* is considered to be a poor vector maintaining onchocerciasis by its high biting rate supplemented by the movement of infected humans from the highlands.

In Colombia/Ecuador a single endemic focus of onchocerciasis is present in lowland coastal forest up to a height of 200 m a.s.l., where the vector, *S. exiguum*, breeds in rivers more than 10 m wide. Transmission occurs at the end of the rainy season when low infectivity rates (1%) in *S. exiguum* are countered by high human-biting rates (430 to more than 2000 per person per day). An outbreak of onchocerciasis in Esmeraldas Province, Ecuador is causing considerable concern (Walgate, 1991).

Epidemiology of Onchocerciasis

Onchocerciasis is not a zoonosis, although natural infections have been found in a spider monkey (*Ageles geoffroyi*) in Guatemala and a gorilla in the Congo; and chimpanzees can be infected in the laboratory (Nelson, 1970). The disease is passed from person to person by the bites of *Simulium* and can only rarely be passed before the third blood meal providing the simuliid becomes infected at its first blood meal. The gonotrophic cycle of *S. damnosum s.s.* and *S. sirbanum* is normally 3 to 4 days, made up of a period of less than 24 h between oviposition and feeding; ovarian development of 48 h; and a variable time (less than 12 h) between the eggs being fully mature and oviposition (Bellec and Hébrard, 1980).

One hypothesis put forward to explain the difference in severity of onchocerciasis between the Sudan savanna and the forest and Guinea savanna, was that in the Sudan savanna the flies were long lived, and had more limited dispersal so that a higher rate of transmission existed. In contrast it was postulated that in the forest the life expectancy of the vector was shorter, and it dispersed more widely spreading infection more evenly over a greater area (Le Berre *et al.*, 1964).

Observation showed that indeed *S. damnosum* was longer lived in the Sudan savanna with 61% of the population being parous compared with 40% in the Guinea savanna and forest. This higher parity rate resulted in a higher proportion of the infected flies being infective, with this proportion reaching nearly 50% in the Sudan savanna, compared to 20–36% in the Guinea savanna and forest. However, these calculations do not take into account the different sizes of the human-biting population of *S. damnosum* in the different regions, and the density of microfilariae in the human population available to the vectors. Simuliid populations in the Sudan savanna were only 12% of those in the Guinea savanna, and less than 10% of those in the forest (Duke, 1968).

Using the number of infective larvae per infective fly as an estimate of the microfilarial reservoir in the human population, that in the Sudan savanna was 40% of that in the forest. Putting all these factors together, Duke (1968) calculated that the infective biting density (the number of infective bites per person per day) was 5 in the Sudan savanna, 14 in the Guinea savanna, and depending upon the season, ranged from 18 to 83 in the forest. The transmission potential (number of infective larvae per person per day) was 12 in the Sudan savanna, 42 in the Guinea savanna, and 99–556 in the forest. These figures confound the hypothesis. The severity of onchocerciasis in the Sudan savanna is not related to the transmission rate which is actually lower than in the forest (Duke, 1968).

Control of Onchocerciasis (Davies, 1994)

The problems of controlling onchocerciasis are formidable on account of the longevity of the parasite and the habits of the vector which is both exophilic and exophagic. This pattern of behaviour renders ineffective most methods of personal protection, but some protection can be achieved with repellents, and the wearing of suitable clothing. Adult simuliids disperse widely, and the population is most concentrated in the immature stages which occur in running water. Control measures have therefore been directed against the immature stages with success.

Control requires either the eradication of the vector or for the vector population to be kept at a level at which transmission does not occur for 15 years, the maximum life of the adult worm. Four stages in the development of control measures have been recognized: the pre DDT era 1932–1944; the chlorinated hydrocarbon period 1944–1972; the temephos period 1972–1992; and integrated control involving the use of the microfilaricide ivermectin and the use of a range of larvicides.

Fifty control schemes have been attempted – 6 in Latin America; 22 in West Africa from Senegal to Cameroon and Chad including the OCP; and 22 in eastern and central Africa. Thirty-five of these schemes have been against members of the *S. damnosum* complex and nine against *S. neavei* with eradication being successful on seven occasions, all in East Africa – five against *S. neavei* and two against *S. damnosum* s.l.

Foci of onchocerciasis sustained by *S. neavei* tend to be compact and isolated, favouring eradication as in the inland hilly region of western Kenya. Four foci were discovered with onchocerciasis rates varying from 21 to 72%, and eye lesions from 1.6 to 10.5%. An area of 12,000 km^2 was freed from onchocerciasis by eradicating *S. neavei* from 323 rivers by treating them with DDT at a dosage which killed the larvae of *S. neavei* but had no effect on fish or on the crabs with which *S. neavei* has a phoretic association. Treatments were repeated at 10-day intervals which was less than a minimum period for larval development. Adult *S. neavei* live for up to 2 months and by maintaining control for nine successive cycles, eradication of the vector was achieved (McMahon *et al.*, 1958).

The eradication of *S. neavei* provided the opportunity to obtain information on the longevity of infections. Eleven years after the successful campaign no new infections had occurred, but 60% of the older people were still infected and those that had mild eye lesions earlier were now completely blind. The maximum life of *O. volvulus* is probably 16 years with microfilariae persisting in the skin for 30 months (Nelson, 1970).

The two early successes against *S. damnosum* s.l. were in Uganda. One involved its eradication along the Nile from Lake Victoria to Lake Kyoga by taking advantage of the controlled water outlet from the Owen Falls hydroelectric station to introduce DDT into the released water and control breeding of *S. damnosum* for 80 km below the dam, freeing 4000 km^2 from *S. damnosum* and making it available for more intensive settlement and agriculture without the fear of onchocerciasis (Nelson, 1970). In the Ruwenzori focus, the anthropophilic sibling was replaced by a zoophilic one as happened in the OCP when *S. damnosum* was eradicated and *S. adersi* and *S. griseicolle* survived, possibly due to the larvae having different feeding habits (Walsh *et al.*, 1981).

Onchocerciasis Control Programme in West Africa (WHO, 1987) (Fig. 31.1)

This ambitious programme began in 1975 with the aim of controlling onchocerciasis in a selected area of Sudan and Guinea savanna covering 654,000 km^2

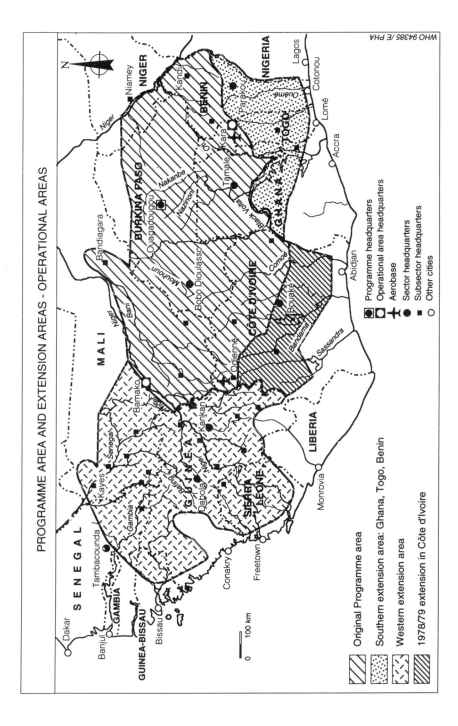

Fig. 31.1. Map showing the location of the Onchocerciasis Control Programme in West Africa. Reproduced, by permission, from Samba, E.M. *The Onchocerciasis Control Programme in West Africa: an Example of Effective Public Health Management.* Geneva, World Health Organization, 1994 (Public Health in Action, No. 1) Figure 2.

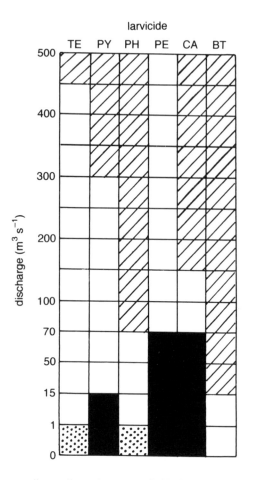

Fig. 31.2. The generally preferred insecticide(s) (open squares) for different discharges. Low cost efficiency (hatched squares), potential environmental damage (closed squares), or lack of accuracy in application (dotted) make each insecticide unusable under certain conditions. TE; temephos; PY, pyraclofos; PH, phoxim; PE, permethrin; CA carbosulfan; BT, *Bacillus thuringiensis* H-14.

extending across seven countries – Benin, Burkina Faso, Ghana, Ivory Coast, Mali, Niger, Togo – where a million people suffered from onchocerciasis and 70,000 were blind or had seriously impaired vision (Walsh *et al.*, 1979). The control area was based on the Volta River system and involved the weekly treatment of 14,500 km of rivers. In 1978 the area was increased by the addition of another 110,000 km^2 in the Ivory Coast bringing the area covered to 764,000 km^2 and involving treatment in the wet season of 18,000 km of rivers.

In 1988 the area was extended to include another 8 million people in Guinea, Guinea Bissau, Senegal and Sierra Leone bringing the area now covered to 1,300,000 km^2 and involving treatment of 46,000 km of rivers. The scheme has been outstandingly successful. Transmission is now zero in 90% of the original central area and the prevalence of onchocerciasis has fallen from 70% to 3%. After

7 to 8 years of control, the *O. volvulus* population in the OCP was ageing and dying with the productivity index of the parasite in ten villages being reduced by 55 to 97% (Karam *et al.*, 1987). In the seven countries included in the scheme from its onset 16.5 million people are no longer at risk of the disease.

From the start of the control programme it was noticed that there was an increase in the density of *Simulium* during the rainy season and this was attributed to the wind-carriage of adults along the track of the monsoon winds during the northwards movement of the Inter Tropical Convergence Zone (ITCZ). This invasion takes place along a SW–NE track bringing flies 300 km or more, even up to 500 km (Garms *et al.*, 1979). The flies concerned are mainly older parous *S. damnosum* s.s. and *S. sirbanum*, many of which are infective (15%). The average speed of the invasion in 1977 and 1978 was 7–35 km per day (Johnson *et al.*, 1985) and the average age of invading flies has been estimated by pteridine analysis to be 29 days. Extension of the control area in 1978 was to prevent this invasion.

In the Ivory Coast *S. sanctipauli* developed resistance to temephos in 14 months and to its replacement chlorphoxim in 5 months. Resistance than appeared in *S. damnosum* s.s. and *S. sirbanum* and spread to the whole of the OCP by 1989. To minimize the development of resistance the OCP uses six insecticides in a rotation scheme designed to maximize larval control. Other criteria are cost and environmental impact on non-target organisms. The six include three organophosphorus compounds (temephos, phoxim, pyraclofos), permethrin, carbosulfan and *Bacillus thuringiensis* serovar H14. The choice is influenced by the rate of river discharge (Fig. 31.2). Below 1 m^3 s^{-1} *B. thuringiensis* is used; between 1 and 15 m^3 s^{-1} *B. thuringiensis*, temephos or phoxim; from 15 to 70 m^3 s^{-1} one of the three organophosphorus compounds; from 70 to 150 m^3 s^{-1} temephos, pyraclofos, carbosulfan or permethrin; at 150–300 m^3 s^{-1} temephos, pyraclofos or permethrin; between 300 and 450 m^3 s^{-1} the choice is between temephos and permethrin; and above 450 m^3 s^{-1} permethrin is used. Recently etofenprox, a pseudo-pyrethrinoid, is showing promise as an alternative for use between 15 and 70 m^3 s^{-1} (Guillet *et al.*, 1990; Hougard *et al.*, 1993).

References

Akogun, O.B. and Renz, A. (1994) Further observations on hyperendemic onchocerciasis in the upper Taraba river valley, Nigeria. *Parasite* 1(1S), 13.

Bellec, C. and Hébrard, G. (1980) La durée du cycle gonotrophique des femelles du complexe *Simulium damnosum* en zone préforestière de Côte d'Ivoire. *Cahiers ORSTOM Entomologie Médicale et Parasitologie* 18, 347–358.

Blacklock, D.B. (1926a) The development of *Onchocerca volvulus* in *Simulium damnosum*. *Annals of Tropical Medicine and Parasitology* 20, 1–48.

Blacklock, D.B. (1926b) The further development of *Onchocerca volvulus* Leukart in *Simulium damnosum* Theob. *Annals of Tropical Medicine and Parasitology* 20, 203–218.

Crosskey, R.W. (1981) Geographical distribution of Simuliidae. In: Marshall Laird (ed.) *Blackflies*. Academic Press, New York, pp. 57–68.

Crosskey, R.W. (1990) *The Natural History of Blackflies*. John Wiley, Chichester, UK.

Crosskey, R.W. (1993) Blackflies (Simuliidae). In: Lane, R.P. and Crosskey, R.W. (eds) *Medical Insects and Arachnids*. Chapman & Hall, pp. 241–287.

Dang, P.T. and Peterson, B.V. (1980) Pictorial keys to the main species and species groups within the *Simulium damnosum* Theobald complex occurring in West Africa (Diptera: Simuliidae). *Tropenmedizin und Parasitologie* 31, 117–120.

Davies, J.B. (1994) Sixty years of onchocerciasis vector control: a chronological summary with comments on eradication, reinvasion, and insecticide resistance. *Annual Review of Entomology* 39, 23–45.

Davies, J.B. and Crosskey, R.W. (1991) *Simulium – Vectors of Onchocerciasis.* WHO/VBC/91.992.

De Leon, J.R. and Duke, B.O.L. (1966) Experimental studies on the transmission of Guatemalan and West African strains of *Onchocerca volvulus* by *Simulium ochraceum, S. metallicum* and *S. callidum. Transactions of the Royal Society of Tropical Medicine and Hygiene* 60, 735–752.

De Meillon, B. (1957) The bionomics of the vectors of onchocerciasis in the Ethiopian geographical region. *Bulletin of the World Health Organization* 16, 509–522.

Duke, B.O.L. (1962) Studies on factors influencing the transmission of onchocerciasis. II. The intake of *Onchocerca volvulus* microfilariae by *Simulium damnosum* and the survival of the parasite in the fly under laboratory conditions. *Annals of Tropical Medicine and Parasitology* 56, 255–263.

Duke, B.O.L. (1968) Studies on factors influencing the transmission of onchocerciasis. VI. The infective biting potential of *Simulium damnosum* in different bioclimatic zones and its influence on the transmission potential. *Annals of Tropical Medicine and Parasitology* 62, 164–170.

Duke, B.O.L. and Lewis, D.J. (1964) Studies on factors influencing the transmission of onchocerciasis. III. Observations on the effect of the peritrophic membrane in limiting the development of *Onchocerca volvulus* microfilariae in *Simulium damnosum. Annals of Tropical Medicine and Parasitology* 58, 83–88.

Duke, B.O.L. and Taylor, H.R. (1991) Onchocerciasis. In: Strickland, G.T. (ed.) *Hunter's Tropical Medicine.* Saunders, Philadelphia, pp. 729–744.

Duke, B.O.L., Lewis, D.J. and Moore, P.J. (1966) *Onchocerca–Simulium* complexes. I. Transmission of forest and Sudan-savanna strains of *Onchocerca volvulus*, from Cameroon, by *Simulium damnosum* from various West African bioclimatic zones. *Annals of Tropical Medicine and Parasitology* 60, 318–336.

Duke, B.O.L. Moore, P.J. and De Leon, J.R. (1967) *Onchocerca–Simulium* complexes. V. The intake and subsequent fate of microfilariae of a Guatemalan strain of *Onchocerca volvulus* in forest and Sudan-savanna forms of West African *Simulium damnosum. Annals of Tropical Medicine and Parasitology* 61, 332–337.

Dunbar, R.W. and Vajime, Ch. G. (1981) Cytotaxonomy of the *Simulium damnosum* complex. In: Marshall Laird (ed.) *Blackflies.* Academic Press, New York, pp. 31–43.

Eichner, M., Renz, A., Wahl, G. and Enyong, P. (1991) Development of *Onchocerca volvulus* microfilariae injected into *Simulium* species from Cameroon. *Medical and Veterinary Entomology* 5, 293–297.

Garms, R., Walsh, J.F. and Davies, J.B. (1979) Studies on the reinvasion of the Onchocerciasis Control Programme in the Volta River basin by *Simulium damnosum s.l.* with emphasis on the south-western areas. *Tropenmedizin und Parasitologie* 30, 345–362.

Garms, R., Kerner, H., Yocha, J. and Kipp, W. (1992) Transmission of onchocerciasis in western Uganda, East Africa. *Abstracts of the XIIIth International Congress for Tropical Medicine and Malaria*, Jomtien, Pattaya, Thailand, 29 November–4 December 1992, vol. 2, p. 326.

Guillet, P., Kutak, D.C., Philippon, B. and Meyer, R. (1990) Use of *Bacillus thuringiensis israelensis* for onchocerciasis control in West Africa. In: Barjac, H. de and Sutherland, D.J. (eds) *Bacterial Control of Mosquitoes and Black Flies. Biochemistry, Genetics and*

Applications of Bacillus thuringiensis israelensis *and* Bacillus sphaericus. Rutgers University Press, New Brunswick, pp. 187–201.

Hougard, J.M., Poudiougo, P., Guillet, P., Back, C., Akpoboua, L.K.B. and Quillévéré, D. (1993) Criteria for the selection of larvicides by the Onchocerciasis Control Programme in West Africa. *Annals of Tropical Medicine and Parasitology* 87, 435–442.

Johnson, C.G., Walsh, J.F., Davies, J.B., Clark, S.J. and Perry, J.N. (1985) The pattern and speed of displacement of females of *Simulium damnosum* Theobald *s.l.* (Diptera: Simuliidae) across the Onchocerciasis Control Programme area of West Africa in 1977 and 1978. *Bulletin of Entomological Research* 75, 73–92.

Karam, M., Schulz-Key, H. and Remme, J. (1987) Population dynamics of *Onchocerca volvulus* after 7 to 8 years of vector control in West Africa. *Acta Tropica* 44, 445–457.

Laurence, B.R. (1966) Intake and migration of the microfilariae of *Onchocerca volvulus* (Leuckart) in *Simulium damnosum* Theobald. *Journal of Helminthology* 40, 337–342.

Le Berre, R., Balay, G. Brengues, J. and Coz, J. (1964) Biologie et ecologie de la femelle de *Simulium damnosum* Theobald, 1903, en fonction des zones bioclimatiques d'Afrique occidentale. *Bulletin of the World Health Organization* 31, 843–855.

McMahon, J.P., Highton, R.B. and Goiny, H. (1958) The eradication of *Simulium neavei* from Kenya. *Bulletin of the World Health Organization* 19, 75–107.

Meredith, S.E.O. and Townson, H. (1981) Enzymes for species identification in the *Simulium damnosum* complex from West Africa. *Tropenmedizin und Parasitologie* 32, 123–129.

Nelson, G.S. (1970) Onchocerciasis. *Advances in Parasitology* 8, 173–224.

Ogata, K. (1981) Preliminary report of Japan–Guatemala onchocerciasis control pilot project. In: Marshall Laird (ed.) *Blackflies*. Academic Press, New York, pp. 105–115.

Peterson, B.V. and Dang, P.T. (1981) Morphological means of separating siblings of the *Simulium damnosum* complex (Diptera: Simuliidae). In: Marshall Laird (ed.) *Blackflies*. Academic Press, New York, pp. 45–56.

Philippon, B. (1977) Etude de la transmission d'*Onchocerca volvulus* (Leuckart, 1893) (Nematoda: Onchocercidae) par *Simulium damnosum* Theobald, 1903 (Diptera: Simuliidae) en Afrique tropicale. *Travaux et Documents de l'ORSTOM* 63, 1–308.

Poltera, A. (1994) From the laboratory to the home of the patient infected with *Onchocerca volvulus* (O.v.): clinical experience with Amocarzine (preliminary results). *Parasite* 1(1S), 25–27.

Prost, A., Rougemont, A. and Omar, M.S. (1980) Caracteres épidémiologiques cliniques et biologiques des onchocercoses de savane et de forêt en Afrique occidentale, Revue critique et éléments nouveaux. *Annales de Parasitologie Humaine et Comparée* 55, 347–355.

Quillévéré, L., Sechan, Y. and Pendriez, B. (1977) Etude du complexe *Simulium damnosum* en Afrique de l'Ouest. V. Identification morphologiques des femelles en Côte d'Ivoire. *Tropenmedizin und Parasitologie* 28, 244–253.

Raybould, J.N. (1967) A study of anthropophilic female Simuliidae (Diptera) at Amani, Tanzania: the feeding behaviour of *Simulium woodi* and the transmission of onchocerciasis. *Annals of Tropical Medicine and Parasitology* 61, 76–88.

Schulz-Key, H. and Soboslay, P.T. (1994) Reproductive biology and population dynamics of *Onchocerca volvulus* in the vertebrate host. *Parasite* 1(1S), 53–55.

Shelley, A.J. (1988) Vector aspects of the epidemiology of onchocerciasis in Latin America. *Annual Review of Entomology* 30, 337–366.

Van Someren, V.D. and McMahon, J. (1950) Phoretic association between *Afronurus* and *Simulium* species, and the discovery of the early stages of *Simulium neavei* on freshwater crabs. *Nature*, London 166, 350–351.

Walgate, R. (1991) 'Explosion' of onchocerciasis alarms Ecuadorean health officials. *TDR News* 36(1), 10.

Walsh, J.F., Davies, J.B. and Le Berre, R. (1979) Entomological aspects of the first five years of the Onchocerciasis Control Programme in the Volta River basin. *Tropenmedizin und Parasitologie* 30, 328–344.

Wenk, P. (1981) Bionomics of adult blackflies. In: Marshall Laird (ed.) *Blackflies*. Academic Press, New York, pp. 259–279.

WHO (1987) WHO expert committee on onchocerciasis. *WHO Technical Report Series* No. 752.

Other Helminths Transmitted by Insects

32

This chapter will deal with a miscellaneous collection of helminths transmitted by insects to humans and domestic animals. The human parasites will include *Loa loa* and three species of *Mansonella*. Animal parasites will include species of *Onchocerca* and other filarioid parasites of domestic animals: *Dirofilaria, Elaeophora, Habronema, Parafilaria, Setaria, Stephanofilaria, Thelazia*; and two cestodes, *Dipylidium* and *Hymenolepis*.

Loiasis

Parasite and disease (Duke, 1991)

Loiasis is a human filarial disease caused by infection with *Loa loa*. It occurs in the rainforest of tropical Africa, extending roughly from 10°N to 10°S and from 0° to 30°E, ranging from Nigeria to south-west Sudan and south to Zaire and north-west Angola. The disease is marked by recurrent, temporary subcutaneous swellings mostly in the region of the wrists and forearms. These Calabar or fugitive swellings are a response to antigenic material released by migrating worms. The thin, transparent adult worms measure 50–70 by 0.5 mm in the female and 30–35 by 0.3 mm in the male.

The adult worms live a nomadic life moving through loose connective tissue, and at times can be seen moving under the skin causing minimal local reaction. The most disquieting demonstration of their mobility is when they move across the eye under the conjunctiva at a speed of 1 cm min^{-1}. No permanent damage is done to the eye which becomes oedematous. As in other filarial infections there is a pronounced eosinophilia with, at times, the eosinophil count exceeding 70% of the white blood cells. The sheathed microfilariae, measuring 230–300 by 6–8 μm, occur in the circulating blood. Loiasis is a relatively benign disease which can be treated with the microfilaricide diethylcarbamazine (DEC). Care is needed in treating patients with high microfilaraemia when DEC can cause meningoencephalitis, a serious condition which can be fatal. Ivermectin is effective as a single dose repeated after 1 month. It is safe, well tolerated and markedly reduces the eosinophil count but its long-term effects have to be evaluated (Hovette *et al.*, 1994).

Loa loa in the vertebrate host (Duke, 1972, 1991)

Two strains of *L. loa* are recognized. Microfilariae of the human strain have a diurnal periodicity in the circulating blood, and develop in day-biting *Chrysops*. The microfilariae of the strain which parasitizes monkeys, especially the drill (*Mandrillus leucophaus*), have a nocturnal periodicity, and develop in night-biting *Chrysops*. Hybrids formed from the two strains have a characteristic periodicity, differing markedly from that of either parent, or from a 50:50 mixture of the two periodicities.

In monkeys the worms become mature and microfilariae appear in the circulating blood 4 to 5 months after the introduction of infective larvae, while in humans maturation of the worm takes 6 to 12 months before microfilariae appear in the peripheral blood stream. In monkeys, and probably also in humans, the female passes microfilariae into the connective tissue, from which they enter the vascular system and accumulate in the pulmonary blood before entering the peripheral circulation about 3 weeks after being released. The long-lived adult worms are considered to live for 4 to 17 years.

Development of *Loa loa* in *Chrysops*

Connal and Connal (1922) were the first to demonstrate the development of *L. loa* in *Chrysops silacea* and *C. dimidiata*. Later Williams (1981) carried out a detailed study into its development in *C. silacea* and found that the worm developed in the fat-body of the fly where it underwent two moults before reaching the infective stage. At 28–30°C and 92% RH, microfilariae developed to the infective stage in 7 days, and in the process increased in length from 275 µm to more than 2 mm. Most microfilariae developed in the fat-body of the abdomen, and a smaller number in the fat-body of the thorax and head. In the initial stages of development the parasite is intracellular but later it becomes free. Infective larvae move to the head where they accumulate in the subcibarial haemocoelic space and escape, when the fly is feeding, by rupturing the delicate labiohypopharyngeal membrane (Lavoipierre, 1958).

When *C. silacea* and *C. dimidiata* feed they take in about twice their body weight of blood with the heavier species, *C. silacea*, taking in about 20–25% more blood. *C. silacea* ingests only about half the number of microfilariae that would be expected from their density in the circulating blood and the size of the blood meal. After ingestion there is little mortality of microfilariae during development (Kershaw *et al.*, 1956). Duke (1972) found in the field that the numbers of *L. loa* in *C. silacea* and *C. dimidiata* were almost identical, 79 and 81 respectively, and, in view of the smaller blood meal taken by *C. dimidiata*, this suggests that *C. dimidiata* takes in proportionally more microfilariae than *C. silacea*.

Ecology of the vector and epidemiology of loiasis

Connal and Connal (1922) incriminated *C. silacea* and *C. dimidiata* as vectors of *L. loa*, and Woodman and Bokhari (1941) and Woodman (1949) showed that *L. loa* would develop in *C. distinctipennis* and *C. longicornis* in the southern Sudan. Duke (1955a) considers: (i) that *C. silacea* is the most important vector, and that *C. dimidiata* is an equally efficient vector, but is usually less numerous; (ii) that *C. distinctipennis* and *C. zahrai* are less effective local or subsidiary vectors; and (iii) that *C. langi* and *C. centurionis* are zoophilic and responsible for the transmission of the monkey strain. Duke (1972) concluded that transmission of *L. loa* from human to monkey could occur rarely but that the reverse transmission from monkey to human was most unlikely.

The biting cycle of *C. silacea* and *C. dimidiata* in the forest canopy was bimodal with morning and afternoon peaks in which nulliparous females were more numerous in the morning and parous females in the afternoon (Duke, 1960). *C. langi* and *C. centurionis* were crepuscular, biting from about 1700 h to 2100 h (Duke, 1958a). It is unlikely that *C. silacea* and *C. dimidiata* would acquire infections by feeding on monkeys, which during the daytime would quickly catch tabanids attempting to feed. Even if some fed successfully, there would be virtually no microfilariae in the monkeys' peripheral blood at that time. At night *C. langi* and *C. centurionis* would find it easier to feed on sleeping monkeys at a time when the microfilarial density in their circulating blood would be high, favouring infection.

In a Congo rainforest where *C. silacea* was three times more abundant than *C. dimidiata*, both species were markedly anthropophilic with 90% of their blood meals being taken from humans (Noireau and Gouteux, 1989). In the presence of a fire the catch of *C. silacea* increased by eightfold at ground level and fivefold in the canopy while that of *C. dimidiata* remained unchanged (Caubere and Noireau, 1991). In both species 0.6% of the flies were infective, carrying on average ten infective third stage larvae. The annual transmission potential was about 140 for the more numerous *C. silacea* and 95 for *C. dimidiata* (Noireau *et al.*, 1990a).

In a mixed farmland–forest region tabanid larvae and pupae were more numerous in 'small patches of mud in deep valleys in deep forest', and diminished with reduction of the forest to reach their lowest level in mud in open ground. The distribution of immature *Chrysops* followed that of the biting densities of *C. silacea* and *C. dimidiata*, which formed 60–70% of all *Chrysops* bred to the adult. This suggests that these species do not readily cross open ground (Williams, 1962). If this applies to hungry as well as gravid females, then the risk of infection with *L. loa* should decrease with distance from forest. Certainly the biting intensity decreases with distance with the rate of decline depending on the degree of cover available. Thus where there are only saplings 0.5 m high in the clearing, the biting density was reduced to one-tenth of its original value at 90 m from the forest, but where the saplings were taller (3–3.5 m), the same reduction was not reached until 500 m (Duke, 1955b). Even in the tropical rainforest the biting density of *C. silacea* and *C. dimidiata* are not constant throughout the year but shows seasonal cycles, with *C. silacea* being abundant from April to December and *C. dimidiata* having two peaks of abundance, November to January and March to May (Duke, 1959).

Other Human Filarial Infections

Three species of *Mansonella* infect humans, to whom they are usually no more than mildly pathogenic. Their unsheathed microfilariae are found in the circulating blood (*M. perstans*); in the dermis (*M. streptocerca*) or in both the circulating blood and the subcutaneous tissues (*M. ozzardi*) (Nelson, 1991). *Mansonella* microfilariae in the skin need to be differentiated from the sheathed microfilariae of *Onchocerca volvulus* which occur in the skin of humans in the same parts of the world. The three species will be considered separately.

Mansonella ozzardi

Mansonella ozzardi is found only in the Neotropical region where it occurs in Central America, the north coast of South America, Brazil, Colombia, the northern province of Argentina and certain Caribbean Islands. No particular symptoms are associated with *M. ozzardi*, although in Trinidad a chronic arthritis appears to be associated with infection but the detailed relationship is unknown. Diethylcarbamazine is considered to be ineffective against *M. ozzardi*. Adult female worms measure 49 by 0.15 mm and males 26 by 0.07 mm and microfilariae 220 by 3 to 4 µm (Nelson, 1991). In Brazil microfilariae were found with almost equal frequency in blood (350/701) and in bloodless skin snips (295/701) (Moraes, 1976).

Buckley (1934) followed the development of *M. ozzardi* in *Culicoides furens* on the Caribbean island of St Vincent, where a high proportion of the human population (37.5%) was infected. When *C. furens* fed on a carrier, microfilariae reached the thorax in 24 h and infective larvae were present in the head after 7 to 8 days. These laboratory observations were supported by finding 5% of *C. furens* infected naturally. In Haiti *C. barbosai* is considered to be as important a vector as *C. furens* (Lowrie and Raccurt, 1984). In northern Argentina the vector is *C. paraensis* (Romana and Wygodzinsky, cited in Wirth, 1977).

The vector of *M. ozzardi* on Trinidad, is *Culicoides phlebotomus*. The proportion of *C. phlebotomus* females infected was very low, being 0.8% and 1.3% in two large samples. The proportion infective was even lower being 0.07% and 0.13% of all females, but nearly 4% of parous females, i.e. those which had had the opportunity to become infected. Transmission of *M. ozzardi* is assured by the large numbers of *C. phlebotomus* biting humans, which can easily exceed 100 h^{-1} in the early morning. Indeed, using simple assumptions, it has been calculated that a person spending 1 h day^{-1} on the beach in the early morning would receive 38 infective bites in the course of a year (Nathan, 1981).

In Amazonas in Venezuela *Simulium sanchezi* has been shown to be a good vector of *M. ozzardi*, with synchronous development of ingested microfilariae to the infective third stage in 7 to 8 days at 23 to 27°C (Yarzabal et al., 1985). In the adjoining area of north Brazil *Simulium oyapokense* s.l. is capable of supporting the full development of microfilariae of *M. ozzardi* but it is not considered to be an efficient vector. Only 21% of *S. oyapokense* became infected after feeding on a carrier and they only matured one or two infective larvae (Moraes *et al.*, 1985).

Mansonella perstans

Mansonella perstans (= *Dipetalonema* or *Acanthocheilonema perstans*) is the most widespread of the three *Mansonella* species, occurring in the tropical rainforests of West and Central Africa and extending south to sylvatic foci in Zimbabwe. It is also found in limited foci among rainforest-dwelling Amerindian communities in Central and South America, where the infection rate may exceed 50%. Infections are usually asymptomatic but where it is suspected that the infection is responsible for clinical symptoms, treatment with diethylcarbamazine is effective. In chimpanzees and presumably in humans, adult worms have been found in serous cavities of the chest and abdomen. The female measures 60–80 × 0.1–0.15 mm and the male 35–45 × 0.05–0.07 mm. The microfilariae measure 200 × 4–5 µm (Nelson, 1991).

When microfilariae of *M. perstans* were ingested by *Culicoides austeni*, they escaped from the midgut into the haemocoele in 6 h and in 20–30 h had reached the thorax. After 7 days infective larvae were present in the head, and emerged from the membranous end of the labium 8 to 10 days after the infected blood meal. This role of *C. austeni* as a vector of *M. perstans* was confirmed by finding that 7% of wild-caught flies were infected (Sharp, 1928). In the laboratory Hopkins and Nicholas (1952) found that 40% of *C. austeni* became infected after feeding on a carrier of *M. perstans*. *C. austeni* took up more microfilariae (about × 2) than expected from their concentration in the circulating blood (Nicholas and Kershaw, 1954). Allowing for difficulties in measuring the size of a blood meal of a *Culicoides*, it is clear that there is no barrier to the uptake of *M. perstans* microfilariae by *C. austeni*.

The role of *C. grahamii* is less clear cut. In Zaire Henrard and Peel (1949) found that *C. grahamii* did not ingest microfilariae of *M. perstans* but workers in the Cameroon showed that *C. grahamii* did take them up at a rate comparable with their density in the circulating blood (Nicholas *et al.*, 1952; Nicholas and Kershaw, 1954). In an area of the southern Congo where 86% of the forest-dwelling pygmies were infected with *M. perstans*, 0.8% of the day-feeding *C. grahamii* were found to be infected and *C. grahamii* was considered to be the vector (Noireau *et al.*, 1990b). In Cameroon when *C. grahamii* and *C. inornatipennis* fed on the same carrier, similar proportions, 77% and 76%, respectively, took up microfilariae, but 6 to 9 days later few *C. grahamii* (6%) were infected compared to 41% of *C. inornatipennis* (Duke, 1956).

Mansonella streptocerca

Mansonella streptocerca (= *Dipetalonema* or *Acanthocheilonema streptocerca*) is limited to the rainforest areas of West and Central Africa. In limited areas of Zaire the infection rate in the human population may be up to 90%. The adult worms are found in the dermis of the upper trunk and shoulder girdle. Females measure 27 × 0.07 mm; males 17 × 0.05 mm and microfilariae 180–240 × 2.5–5.0 µm.

Infection results in a chronic itching dermatitis which responds to treatment with diethylcarbamazine. The drug kills both adult worms and microfilariae. After treatment papules form around dead adult worms. Axillary and inguinal lymphadenopathy is common (Meyers and Neafie, 1991).

Chardome and Peel (1949) and Henrard and Peel (1949) described the development of *M. streptocerca* in *C. grahamii* with infective larvae being produced in 7 to 8 days and 1.2% wild-caught *C. grahamii* being naturally infected. Duke (1954, 1958b) showed that *C. grahamii* readily took up microfilariae of *M. streptocerca* with 15% and 39% *C. grahamii* taking up microfilariae in two separate experiments. During development there was some loss of infection, and 7 to 9 days after feeding on a carrier the percentage of infected flies had fallen from 39 to 19%, and the number of larvae per infected *C. grahamii* from 2.1 to 1.4. *C. milnei* (= *C. austeni*) takes in very few microfilariae, less than one-tenth of those taken in by *C. grahamii*, and must be considered to be no more than a poor vector of *M. streptocerca* (Duke, 1958b).

Filarioid Parasites of Domestic Animals Transmitted by Insects

Onchocerca species

Radostits *et al.* (1994) consider seven species of *Onchocerca* to be of veterinary importance – five in cattle (*O. gibsoni, O. gutturosa, O. lienalis, O. ochengi* and *O. armillata*) and two (*O. cervicalis* and *O. reticulata*) in horses. Although species of *Onchocerca* are often referred to as parasites of cattle or of equines, Ottley and Moorhouse (1978b) have shown that species are not necessarily host specific. In a group of 20 horses, 11 were infected with *O. cervicalis* and 12 with the cattle parasite, *O. gutturosa* and *O. gibsoni*, another cattle parasite, was found in sheep.

Economic importance and vectors

Onchocerciasis of cattle, water buffaloes and horses occurs widely throughout the world wherever these domestic animals have been introduced and there are suitable vectors. Eichler and Nelson (1971) regarded *O. lienalis* (*O. gutturosa* auct.) as an unobtrusive parasite which produced no marked pathological changes or evidence of clinical disease. Losses in the beef industry arise because 'free worms cause aesthetically displeasing blemishes' to the carcass, and encapsulated adult worms in nodules have to be trimmed from the carcass (Ottley and Moorhouse, 1978a). Microfilariae occur in the skin and subcutaneous lymph, with the exception of those of *O. armillata* which occur in the blood (Nelson, 1970). Microfilariae are ingested by blood-sucking flies and develop in various species of *Simulium, Culicoides and Lasiohelea* (Steward, 1933, 1937; Ottley and Moorhouse, 1980).

Onchocerca species in domestic bovines

Adult *O. lienalis* occur in the gastrosplenic ligament and its microfilariae are concentrated in the region of the umbilicus, while adult *O. gutturosa* are found in the ligamentum nuchae and its microfilariae in the skin of the head, neck and back (Bain *et al.*, 1978). Bain and Beveridge (1979) classify the nodule-forming onchocercas in terms of their geographical distribution and the location of the adults in the bovine host. In Africa, adults of *O. dukei* are subcutaneous, and those of *O. ochengi* are dermal parasites. In Asia, adults of *O. gibsoni* and possibly those of *O. indica* occur in the subcutaneous tissues; *O. cebei* and possibly *O. sweetae* in the dermis; and *O. armillata* in the wall of the thoracic aorta. *O. ochengi* causes a dermatitis on the scrotum and udder resembling mange or pox. Nodules of *O. gibsoni*, about 3 cm in diameter, are most prevalent on the brisket and also occur on the stifle and thigh. To reduce infection in a herd it is recommended that cattle with many nodules should be culled (Radostits *et al.*, 1994).

Transmission of bovine onchocercas

Steward (1937) showed that microfilariae of a bovine onchocerca (*O. lienalis* and/or *O. gutturosa*) could develop in *Simulium ornatum* but not in two other species of *Simulium* or in *Culicoides nubeculosus*, a vector of *O. cervicalis* to horses. Microfilariae occurred at a depth of about 1 mm from the skin surface, and 42% of *S. ornatum* ingested microfilariae when feeding. Development was relatively slow with the 'sausage' stage, measuring 200 × 20 μm, being reached in 10 days, and infective forms being present in the head 19 days after the infective feed.

S. *ornatum* is well adapted to transmitting *O. lienalis* (*O. gutturosa* auct.) because it feeds preferentially in the umbilical regions where the microfilariae of *O. lienalis* are concentrated (Eichler and Nelson, 1971). The number of microfilariae ingested is a function of the period of time spent feeding, and not of the volume of blood (about 3 mg) ingested. On average, about 15 microfilariae were ingested in 3 min and 30 in 6 min (Eichler, 1971). Within 1 h 25% of the microfilariae had reached the haemocoele and most were in the thorax in 6 h, some even entering the thoracic musculature within 1 h of feeding. The ingested blood is quickly surrounded by a peritrophic membrane which becomes progressively thicker with time, making the escape of microfilariae from the midgut more difficult (Eichler, 1973).

In Malaysia, *Culicoides pungens* ingested microfilariae of *O. gibsoni* when feeding on infected cattle (Buckley, 1938). Microfilariae of *O. gibsoni* have their maximum concentration at a depth of 50–200 μm from the surface of the skin. The infection rate in *C. pungens* was less than 1%, but this is compensated for by the very large numbers biting cattle. The microfilariae of *O. gibsoni* completed their development in another ceratopogonid, *Lasiohelea townsvillensis*, in 6 days at 30°C and 85% RH (Ottley and Moorhouse, 1980).

Ingested microfilariae of *O. gutturosa* developed to infective third stage both when ingested and when introduced by intrathoracic inoculation into *C. nubeculosus*. They did not develop in *S. ornatum* whereas microfilariae of *O. lienalis* developed in *S. ornatum* but not in *C. nubeculosus* (Dohnal *et al.*, 1990). Although microfilariae of *O. gutturosa* occur mainly on the dorsal surface of cattle, particularly in the withers, *Culicoides brevitarsis*, which attacks preferentially the dorsal surface of cattle, did not ingest any microfilariae (Ottley and Moorhouse, 1980).

Microfilariae of *O. ochengi* are located in the umbilical area and legs of cattle. They developed normally in females of the *Simulium damnosum* complex, probably *S. sanctipauli* with infective larvae of *O. ochengi* being present 6 days after an infective feed. Although microfilariae of *O. gutturosa* and *O. dukei* were also ingested by *S. damnosum s.l.*, they did not develop (Omar *et al.*, 1979). In North Cameroon *Simulium bovis* is an efficient vector of *O. dukei* but not of *O. ochengi*. On average a female *S. bovis* ingested 13 microfilariae of *O. dukei* and only rarely ingested any of *O. ochengi*. After 6 to 9 days about a quarter of the ingested microfilariae had reached the infective stage (Wahl and Renz, 1991). When injected intrathoracically, microfilariae of *O. dukei* and *O. ochengi* developed to the infective stage in *Simulium hargreavesi*. *O. ochengi* developed equally well in *S. damnosum* s.l. in which few (1%) of *O. dukei* developed (Wahl *et al.* 1991).

Equine onchocercas

New infections with *O. reticulata* may cause swelling of the suspensory ligament, making the affected animals temporarily lame. After the swelling subsides the ligament remains thickened. *O. cervicalis* causes fibrotic, calcified lesions in the ligamentum nuchae without visible clinical signs. Hypersensitivity to the microfilariae of *O. cervicalis* can cause alopecia, scaliness and pruritis along the ventral abdomen, which may become more extensive (Radostits *et al.*, 1994).

After feeding on an infected host, infective larvae of *O. cervicalis* developed in *Culicoides nubeculosus, C. obsoletus, C. parroti* and *C. variipennis* but not in *C. pulicaris* or in *Simulium* (Steward, 1933; Mellor, 1975). At 23°C infective forms were produced in 14–15 days. When *C. nubeculosus* fed on an infected horse, 17% of the flies ingested microfilariae with an average intake of 1.9 microfilariae per infected fly (Mellor, 1975). The ingestion of microfilariae of *O. cervicalis* by *C. variipennis sonorensis* was independent of the time spent feeding and the amount of blood ingested (engorged weight). Microfilariae were ingested by 20% of *C. v. sonorensis* and of these 45% had ingested only one microfilaria (Higgins *et al.*, 1988).

Microfilariae of *O. cervicalis* escape from the midgut of *C. nubeculosus* within 5 min of the fly finishing feeding, and 60% of the microfilariae reach the haemocoele within 1 h. About 40% of the microfilariae fail to escape from the midgut. Most of those which enter the haemocoele have reached the thorax in 16–36 h. *C. variipennis* (subspecies unknown) is a less efficient vector with about two-thirds of the microfilariae being retained within the midgut. Early death of the midge can occur when large numbers of microfilariae penetrate the gut wall (Mellor, 1975).

Microfilariae of *O. cervicalis* are predominantly (95%) present in the skin along the abdominal midline of the host which brings them into close contact with *C. nubeculosus*, 85% of which feed on the ventral midline of the horse from the front legs to the mammae or sheath (Mellor, 1974). While the numbers of microfilariae in the whole skin remain unchanged over the year, during the active season of

C. nubeculosus (June to September) microfilariae were most abundant just under the epidermis, favouring their ingestion by blood-sucking insects. During the cooler months of the year, October to February, the microfilariae were deeper (1–2 mm) in the skin. A similar seasonal movement of microfilariae of *O. gutturosa* in the skin of cattle has been shown to coincide with the period of activity of the vector *S. ornatum* (Hawking, 1975).

Other Filarioid Parasites of Domestic Animals

Dirofilaria immitis in the dog

Dirofilaria immitis, the heartworm of dogs, occurs mainly in the tropics and sub-tropics where it infests dogs, other canids, and rarely cats or humans. *D. immitis* appears to be spreading into more temperate regions along the east coast of Australia and into the northern states of the USA and into parts of Canada. Adult worms, measuring 12–20 cm in the male and 25–31 cm in the female, are found in the right ventricle of the heart and in the pulmonary artery. They restrict the circulation, leading to a loss of exercise tolerance, chronic cardiac insufficiency and heart failure. Dogs living in infected areas can be protected by daily doses of diethylcarbamazine or monthly doses of ivermectin (Heyneman, 1973; Jacobs and Fox, 1991). Wharton (1963) found that *D. immitis* was common in domestic and forest carnivores in Malaya.

The unsheathed microfilariae of *D. immitis* show a nocturnal periodicity in the circulating blood. When they are ingested by mosquitoes the microfilariae escape from the midgut into the haemocoele and develop in the Malpighian tubes, in which development is completed in 15–16 days in temperate regions, and in 8–10 days in tropical regions. Infective larvae move into the head and enter the labium from which they escape when the mosquito is feeding. Mature worms reach the heart in 3–4 months, and microfilariae are produced in 6–8 months (Heyneman, 1973).

When fed on the same infected host *Aedes notoscriptus* matured 58% of the microfilariae it ingested compared to only 2% maturing in *Culex annulirostris*. Nevertheless *Cx annulirostris* is probably more important as a vector than *Ae notoscriptus* because of its greater abundance (Russell, 1985; Russell and Geary, 1992). Infection with *D. immitis* reduced the survival of *Aedes albopictus* but this mortality was minimal when less than 20 juveniles developed in a female (Mori and Oda, 1990). There can be a synergistic effect between two infections. When *Ae albopictus* ingests both microfilariae of *D. immitis* and the Chikungunya virus, a small percentage transmit this arbovirus to the next generation by transovarian transmission. This does not occur when only the virus is ingested (Zytoon *et al.*, 1993). The parasite can have an adverse effect on its host. The net reproductive rate of susceptible *Aedes aegypti* infected with *D. immitis* is significantly reduced. Infections with *D. immitis* cause mortality among the refractory strain of *Ae aegypti* by damaging their Malpighian tubules (Mahmood and Nayar, 1989).

Intermill (1973) found that 96% of *Ae triseriatus* ingested microfilariae of *D. immitis*, but in only a little over 50% of the mosquitoes did microfilariae produce infective larvae. On average 24 microfilariae were ingested, of which 11% reached the infective stage. At 21–27°C and 70–80% RH infective larvae reached the labium in 13 days.

Microfilariae have a nocturnal periodicity which coincides with the feeding cycle of the vector and in northern temperate regions there is a seasonal cycle with a five- to tenfold increase in microfilariae in the circulating blood in August and September, when mosquitoes are most abundant (Hawking, 1975). Nayar and Sauerman (1975) found that when the mosquito's saliva contained an anticoagulant, as in *Ae sollicitans*, ingested microfilariae readily escaped into the haemocoele, but in its absence, as in *Ae aegypti*, the microfilariae were trapped in the rapidly clotting blood in the midgut.

In *Ae aegypti* the gene f^1 controls susceptibility to infection with *D. immitis*. It is considered to act via the site of development of the parasite, i.e. the Malpighian tubes, because the development of *D. immitis* is not affected by the gene f^m which controls susceptibility to infection with *Brugia malayi*, which develops in the thoracic musculature (Wakelin, 1978).

Human dirofilariasis (Neafie and Meyers, 1991)

Dirofilariasis is a rare human condition but infections of *D. immitis* have been reported from Japan, USA and Australia. They are asymptomatic with the worm never becoming mature. Subcutaneous dirofilariasis has been reported from the Americas, Africa, Asia and Europe. The mosquito-borne parasites involved are *D. tenuis*, a parasite of the racoon, and *D. repens*, a parasite of dogs and cats. In humans no microfilariae appear in the circulation but lesions occur on the conjunctiva, eyelid, arm, leg, breast and scrotum.

Elaeophora schneideri in sheep

Elaeophora schneideri occurs in North America and Italy in sheep which have been grazed at high altitudes (1800 m) in the summer months. It has an incidence of about 1% and is commoner in 4- to 6-year-old sheep. Sheep that survive the establishment of the worm show a severe dermatitis on the head and feet (Radostits *et al.*, 1994). *E. schneideri* is a benign parasite of the mule deer *Odocoileus hemionus*, which causes clinical disease in abnormal hosts such as elk (*Cervus canadensis*) and moose (*Alces alces*). Lesions caused by the nematodes include occlusion of the cephalic and other arteries which can result in severe disease and death. Hypersensitivity to the microfilariae of *E. schneideri* can produce a severe dermatitis in the head of domestic sheep (*Ovis aries*) and Barbary sheep (*Ammotragus lervia*). At least 16 species of *Hybomitra* and *Tabanus* (Tabanidae) are biological vectors of *E. schneideri* (Pence, 1991).

In Gila National Park, New Mexico, the vector of *E. schneideri* is *Hybomitra laticornis*, of which in one survey, 16% were infected with an average of 25 developing worms (Clark and Hibler, 1973). Ingested microfilariae escape from the midgut into the haemocoele and enter the fat-body for the initial stage of development. Older larvae leave the fat-body, and develop in the abdomen to infective larvae, measuring 4.5 mm by 50 μm, before moving to the head and mouthparts from which they escape when the fly is feeding (Hibler and Metzger, 1974).

Stephanofilariasis in cattle

Stephanofilariasis is widely distributed throughout the warmer parts of the world occurring in the East Indies, Malaya, India, Australia, Japan and the USA. Several species are involved of which *Stephanofilaria stilesi* and *S. assamensis* have been most studied (Radostits *et al.*, 1994).

Stephanofilaria stilesi

This is found in cattle in North America, Hawaii and the Soviet Union, and may occur more widely in the world. It causes lesions in the skin along the midventral line between the brisket and navel, which remain raw and bloody for several years. The adult worms and the sheathed, very small microfilariae (52 × 3 μm) occur in the lesions from which they are ingested by feeding flies. In New Mexico the vector is the blood-sucking muscid, *Haematobia irritans*. An infection rate of 12% was found in field-collected female flies and a similar infection rate was obtained when laboratory-reared females were fed on infected lesions, but infections in male flies, both field-collected and laboratory-reared, were very low (less than 0.5%). The infection rate in cattle is variable, and was 98% in beef cattle but lower (25%) in drylots where the manure was removed or scattered, reducing the populations of *H. irritans*. Infection rates in *H. irritans* were highest in spring and autumn when the maximum temperature was 21–27°C, and low in July and August when the temperature was 32–38°C (Hibler, 1966).

Stephanofilaria assamensis

This is a parasite mainly of cattle but occurs in other ungulates, infesting the sub-cutaneous layer and skin of the ears and back (Kabilov, 1980). In India the vector is *Musca conducens*, which has well developed prestomal teeth and interdental armature on its proboscis, which enable it to scratch and rasp the skin (Srivastava and Dutt, 1963). *S. assamensis* produces eggs which are ingested by the fly and hatch in the midgut. Second-stage larvae occur in the abdomen and thorax, while third-stage larvae migrate to the proboscis, and are deposited on to the skin of cattle when the fly feeds (Shamsul, 1971).

The overall infection rate in cattle was 30% but it varied according to locality being as high as 95% in swampy areas. Transmission occurred all the year round but had a marked maximum in July and August. Natural infection rates of *M. conducens* were low (2%) with infected flies containing one to seven larvae (Shamsul, 1971). In Uzbekistan the vector is *Haematobia thirouxi titillans* (*Lyperosia titillans* auct.), in which the natural infection rate was 0.7%, and the infection rate in cattle was only 1%. At 26–32°C, development of *S. assamensis* in *H. t. titillans* took 21–24 days (Kabilov, 1980).

Parafilariasis in cattle and equines

Parafilaria bovicola

Parafilaria bovicola has been recorded from cattle in Europe, North Africa, South Africa, Japan, India, the Philippines and introduced into Canada from France (Radostits *et al.*, 1994). The adult worm causes slimy, bruise-like, subcutaneous lesions which reduce the value of carcasses. The ovipositing female worm perforates the skin and deposits egg and/or microfilariae into the blood which trickles down over the surface of the skin. These blood spots occur in the hottest part of the day and attract muscid flies (Nevill, 1975). In South Africa the vectors of *P. bovicola*, in descending order of importance, are *Musca lusoria*, *M. nevilli* and *M. xanthomelas* (Nevill, 1985). In Sweden the vector is *M. autumnalis* (Chirico, 1994).

Infective larvae escape from the mouthparts when the fly feeds on warm (38–40°C) citrated ox blood, but not when it feeds on blood at 22°C, or warm saline or 15% sucrose solution. The stimulus to emergence of infective larvae would appear to be a high temperature and the presence of blood proteins. *P. bovicola* was successfully transmitted when infected *M. lusoria* fed at a skin incision (Nevill, 1979). Treatment of cattle with ivermectin reduced the number of blood spots but failed to interrupt transmission. Dipping cattle every 2 weeks in deltamethrin was highly infective in reducing the number of vectors and after 9 months had reduced transmission from 50% to less than 2% (Nevill *et al.*, 1987).

Parafilaria multipapillosa

Parafilaria multipapillosa is a relatively benign parasite of equines in Europe, North Africa, China and South America, producing subcutaneous nodules which ulcerate, heal and disappear (Radostits *et al.*, 1994). In the former USSR the vector is a blood-sucking muscid (*Haematobia atripalpis*), in which infective larvae develop only in female flies (Gnedina and Osipov, 1960).

Setaria species in domestic animals

Several species of *Setaria* occur in the peritoneal cavity of cattle, horses and pigs. *S. labiato-papillosa* (= *S. digitata*) occurs in Israel, Japan, China, Korea, India and Sri Lanka (Radostits *et al.*, 1994) and *S. cervi* is worldwide in distribution. In India just under 50% of *Bos indicus* and *Bos bubalis* were infected with *Setaria* with the infection in *B. indicus* being predominantly (97%) *S. labiato-papillosa*, and in *B. bubalis* 89% *S. cervi* (Varma *et al.*, 1971). The adult worms are 5–10 cm long and of little pathological significance in their normal host, but can cause severe infections in unnatural hosts, in which the worm invades the central nervous system, causing cerebrospinal nematodiasis. Systemic diethylcarbamazine has given encouraging results (Radostits *et al.*, 1994). The cattle parasite *S. labiato-papillosa* causes serious economic losses in sheep and goats in Japan (Hagiwara *et al.*, 1992), and in horses and goats in China (Han *et al.*, 1990).

Sheathed microfilariae of *S. labiato-papillosa* in the circulating blood are ingested by mosquitoes in which they have the usual filarioid cycle in the thoracic musculature. Third-stage, infective larvae, measuring 2–2.5 mm × 40–50 µm, reach the proboscis in 11 to 13 days at laboratory temperature. Vectors of *S. labiato-papillosa* include *An sinensis, Armigeres obturbans, Ae vittatus* and *Ae togoi* (Varma *et al.*, 1971).

Other Helminths

Habronemiasis in horses (Greenberg, 1973; Radostits *et al.*, 1994)

Habronema muscae, H. majus (= *microstoma*) and *Draschia megastoma*, are the cause of gastric habronemiasis in horses. The adult *Habronema* measure 1–2.5 cm, and *D. megastoma* rarely exceeds 1.25 cm. *D. megastoma* occurs in gastric 'tumours', which may cause pyloric obstruction. The other two *Habronema* species can cause a catarrhal gastritis and may penetrate the stomach glands and cause ulceration. When larvae of *D. megastoma* are deposited in wounds, they give rise to cutaneous habronemiasis causing a condition known as summer sores or swamp cancer in which lesions may increase to 30 cm in diameter in a few months. The sores are unsightly and cause some irritation. Smaller lesions (1–5 mm) may develop on the nictitating membrane, causing conjunctival habronemiasis. Habronemiasis has a worldwide distribution being commoner in warmer, wetter areas, and also in adult horses. A single treatment of ivermectin is highly effective against adult and immature *Habronema* and *D. megastoma*.

Larvae are passed out in the faeces of the horse and ingested by muscid larvae, in which they penetrate into the haemocoele where *D. megastoma* develops in the Malpighian tubes and *H. muscae* in the fat body. The infective stage of the worm is reached in the pupal stage of the fly so that newly emerged adults are capable of transmitting the worm. Horses become infected when they ingest parasitized flies or deposited larvae. Larvae emerge from the fly's proboscis when it is feeding on the lips, nostrils or wounds. Larvae of *H. muscae* readily escape when flies are feeding on horse blood, but not on horse saliva or other media.

Habronema muscae and *D. megastoma* develop in larvae of *Musca domestica* and other species of muscid flies may be involved in transmission, including *Haematobia exigua* and *Sarcophaga melanura*. The development of larvae of *H. majus* in *Stomoxys calcitrans* modifies the fly's behaviour and it feeds on the moist surfaces of the horse and no longer pierces the skin to feed on blood. This behaviour favours deposition of larvae of *H. majus* on the skin of the horse where they can develop further.

Thelaziasis

Species of *Thelazia*, the eyeworms of mammals, have a worldwide distribution. Greenberg (1973) lists five species which inhabit the conjunctival sac of animals, and occasionally occur in humans. Over a period of 5 years ocular infections of *T. callipaeda* were found in 32 humans in Hubei Province, China, of whom 29 were less than 4 years old. The source of the infections was considered to be heavily infected dogs of which 37/39 were infected with an average of 52 worms per infected dog (Shi *et al.*, 1988). Infections, which are commoner in cattle than horses, can produce corneal ulceration and abscesses on the eyelids. Levamisole given orally or applied as an eye lotion has proved to be highly effective (Radostits *et al.*, 1994).

When ingested by *Amiota variegata* (Drosophilidae), the larvae of *T. callipaeda* shed their sheaths and penetrate the midgut to reach the haemocoele where in males they enter the testes and in females the wall of the haemocoele. Larvae develop encapsulated in the host's tissue. When development is complete infective larvae migrate through the haemocoele to the head and proboscis. At 26 to 32°C, the infective stage was reached in 17 days (Wang and Yang, 1993). Infective larvae escape from the labella when the fly is feeding on eye secretions.

Musca autumnalis was the most abundant of 23 cattle-visiting muscid species in south-central Sweden. It formed 82% of the 15,000 muscids collected and was the only one infected with nematodes including *Thelazia* spp. (Chirico, 1994). In north-east Zaire 49% of cattle being processed at an abattoir were infected with *T. balayi* and/or *T. rhodesi* with an average of 12 worms per infected beast (Chartier and Eboma, 1988).

Cestodes with insect intermediate hosts

Dipylidium caninum (Voge, 1973; Jacobs and Fox, 1991)

This is the commonest parasite of dogs and cats with 72% of racing greyhounds being found infected on post mortem examination. It has a worldwide distribution, occurs in some wild carnivores, and is rarely found in humans. *D. caninum* occurs in the small intestine causing anal irritation, digestive disturbances and ill thrift, but occasionally heavy infestations produce more severe symptoms with vomiting, convulsions and chronic enteritis.

Proglottides of the tapeworm either crawl out of the anus of the host or are passed with the faeces. They move about vigorously expelling egg capsules containing 8 to 15 eggs each. These are ingested by the intermediate host, usually larvae of dog or cat fleas, *Ctenocephalides canis* and *C. felis*. In the midgut of the flea the eggs give rise to oncospheres with three pairs of hooks, which penetrate to the haemocoele and develop to the infective metacestode stage. Further development occurs when the infected flea is swallowed by a suitable host. Development can also occur in the human flea, *Pulex irritans*, and the dog louse, *Trichodectes canis*.

Development of *D. caninum* in *C. felis felis* is independent of the development of the flea and solely dependent on temperature. At 30–32°C, infected newly emerged adult fleas contain fully developed metacestodes. At lower temperatures, *D. caninum* did not complete their development until their flea hosts had spent a time on a mammalian host. This was a temperature effect independent of the flea's feeding (Pugh, 1987).

Hymenolepis nana and *H. diminuata*

Both *H. nana* and *H. diminuata* have worldwide distributions, occurring predominantly in rodents but also in some monkeys and humans. *H. nana* occurs in humans, particularly in the tropics and subtropics, and is more pathogenic than *H. diminuata*, the effects of which are mild or inapparent. Intermediate hosts of these cestodes are fleas and flour beetles such as *Tribolium confusum*, but eggs of *H. nana* can also develop in another definitive vertebrate host (Voge, 1973).

Eggs of *Hymenolepis* ingested by *T. confusum* hatch in the midgut releasing oncospheres which penetrate into the haemocoele where they develop to the infective cysticercoid stage. At 30°C development of *H. nana* is complete in 5 days and of *H. diminuata* in 8 days (Voge and Heyneman, 1957). Infection occurs when the definitive host ingests an infected intermediate host. The fecundity of *T. confusum* was reduced exponentially with increasing parasite burden, an effect which was particularly marked in young beetles (Maema, 1986).

References

Bain, O. and Beveridge, I. (1979) Redescription d'*Onchocerca gibsoni* C. et J., 1910. *Annales de Parasitologie Humaine et Comparée* 54, 69–80.

Bain, O. Petit, G. and Poulain, B. (1978) Validité des deux espèces *Onchocerca lienalis* et *O. gutturosa*, chez les bovins. *Annales de Parasitologie Humaine et Comparée* 53, 421–430.

Beveridge, I., Kummerow, E.L., Wilkinson, P. and Copeman, D.B. (1981) An investigation of biting midges in relation to their potential as vectors of bovine onchocerciasis in north Queensland. *Journal of the Australian Entomological Society* 20, 39–45.

Buckley, J.J.C. (1934) On the development, in *Culicoides furens* Poey, of *Filaria* (= *Mansonella*) *ozzardi* Manson, 1897. *Journal of Helminthology* 12, 99–118.

Buckley, J.J.C. (1938) On *Culicoides* as a vector of *Onchocerca gibsoni* (Cleland and Johnston, 1910). *Journal of Helminthology* 16, 121–158.

Caubere, P. and Noireau, F. (1991) Effect of attraction factors on the sampling of *Chrysops silacea* and *C. dimidiata* (Diptera: Tabanidae), vectors of *Loa loa* (Filaroidea: Onchocercidae) filariasis. *Journal of Medical Entomology* 28, 263–265.

Chardome, M. and Peel, E. (1949) La répartition des filaires dans la région de Coquilhatville et la transmission de *Dipetalonema streptocerca* par *Culicoides grahamii*. *Annales de la Société Belge Médecine Tropicale* 29, 99–119.

Chartier, C. and Eboma, K.E. (1988) La thélaziose oculaire des bovins en Ituri (Haut-Zaïre): épidémiologie et clinique. *Revue de Médecine Vétérinaire* 139, 1053–1058.

Chirico, J. (1994) A comparison of sampling methods with respect to cattle-visiting Muscidae and their nematode infections. *Medical and Veterinary Entomology* 8, 214–218.

Clark, G.C. and Hibler, C.P. (1973) Horse flies and *Elaeophora schneideri* in the Gila National Forest, New Mexico. *Journal of Wildlife Diseases* 9, 21–25.

Connal, A. and Connal, S.L.M. (1922) The development of *Loa loa* (Guyot) in *Chrysops silacea* (Austen) and in *Chrysops dimidiata* (van der Wulp). *Transactions of the Royal Society of Tropical Medicine and Hygiene* 16, 64–89.

Dohnal, J., Blinn, J., Wahl, G. and Schulz-Key, H. (1990) Distribution of microfilariae of *Onchocerca lienalis* and *Onchocerca gutturosa* in the skin of cattle in Germany and their development in *Simulium ornatum* and *Culicoides nubeculosus* following artificial infestation. *Veterinary Parasitology*, 36, 325–332.

Duke, B.O.L. (1954) The uptake of the microfilariae of *Acanthocheilonema streptocerca* by

Culicoides grahamii and their subsequent development. *Annals of Tropical Medicine and Parasitology* 48, 416–420.

Duke, B.O.L. (1955a) The development of *Loa* in flies of the genus *Chrysops* and the probable significance of the different species in the transmission of loiasis. *Transactions of the Royal Society of Tropical Medicine and Hygiene* 49, 115–121.

Duke, B.O.L. (1955b) Studies on the biting habits of *Chrysops*. IV. The dispersal of *Chrysops silacea* over cleared areas from the rain-forest at Kumba, British Cameroons. *Annals of Tropical Medicine and Parasitology* 49, 368–375.

Duke, B.O.L. (1956) The intake of microfilariae of *Acanthocheilonema perstans* by *Culicoides grahamii* and *C. inornatipennis* and their subsequent development. *Annals of Tropical Medicine and Parasitology* 50, 32–38.

Duke, B.O.L. (1958a) Studies of the biting habits of *Chrysops*. V. The biting-cycles and infection rates of *C. silacea*, *C. dimidiata*, *C. langi* and *C. centurionis* at canopy level in the rain-forest at Bombe, British Cameroons. *Annals of Tropical Medicine and Parasitology* 52, 24–35.

Duke, B.O.L. (1958b) The intake of the microfilariae of *Acanthocheilonema streptocerca* by *Culicoides milnei* with some observations on the potentialities of the fly as a vector. *Annals of Tropical Medicine and Parasitology* 52, 123–128.

Duke, B.O.L. (1959) Studies on the biting habits of *Chrysops*. VI. A comparison of the biting habits, monthly biting densities and infection rates of *C. silacea* and *C. dimidiata* (Bombe form) in the rain-forest at Kumba, Southern Cameroons, U.U.K.A. *Annals of Tropical Medicine and Parasitology* 53, 203–214.

Duke, B.O.L. (1960) Studies on the biting habits of *Chrysops*. VII. The biting-cycles of nulliparous and parous *C. silacea* and *C. dimidiata* (Bombe form). *Annals of Tropical Medicine and Parasitology* 54, 147–155.

Duke, B.O.L. (1972) Behavioural aspects of life cycle of *Loa*. In: Canning, E.U. and Wright, C.A. (eds) *Behavioural Aspects of Parasite Transmission*. Academic Press, London, pp. 97–107.

Duke B.O.L. (1991) Loiasis. In: Strickland, G.T. (ed.) *Hunter's Tropical Medicine*. Saunders, Philadelphia, pp. 727–729.

Eichler, D.A. (1971) Studies of *Onchocerca gutturosa* (Neumann, 1910) and its development in *Simulium ornatum* (Meigen, 1818). II. Behaviour of *S. ornatum* in relation to the transmission of *O. gutturosa*. *Journal of Helminthology* 45, 259–270.

Eichler, D.A. (1973) Studies on *Onchocerca gutturosa* (Neumann, 1910) and its development in *Simulium* (Meigen, 1818). 3. Factors affecting the development of the parasite in its vector. *Journal of Helminthology* 47, 73–88.

Eichler, D.A. and Nelson, G.S. (1971) Studies on *Onchocerca gutturosa* (Neumann, 1910) and its development in *Simulium ornatum* (Meigen, 1818). I. Observations on *O. gutturosa* in cattle in south-east England. *Journal of Helminthology* 45, 245–258.

Gnedina, M. and Osipov, A. (1960) Contribution to the biology of *Parafilaria multipapillosa* (Condamine and Drouilly, 1878) parasitic in the horse. *Helminthologia* 2, 13–16.

Greenberg, B. (1973) *Flies and Disease*. II. *Biology and Disease Transmission*. Princeton University Press, Princeton, NJ.

Hagiwara, S., Suzuki, M., Shirasaka, S. and Kurihara, F. (1992) A survey of the vector mosquitoes of *Setaria digitata* in Ibaraki Prefecture, Central Japan. *Japanese Journal of Sanitary Zoology* 43, 291–295.

Han, G.C., Huang, G.Q., Zhu, Q.W. and Zhang, H.C. (1990) Study of cerebrospinal filariasis in horses and goats. *Chinese Journal of Veterinary Medicine* 16, 7–9.

Hawking, F. (1975) Circadian and other rhythms in parasites. *Advances in Parasitology* 13, 123–182.

Henrard, C. and Peel, E. (1949) *Culicoides grahamii* Austen. Vecteur de *Dipetalonema strep-tocerca* et non de *Acanthocheilonema perstans. Annales de la Société Belge Médecine Tropicale* 29, 127–143.

Heyneman, D. (1973) Nematodes. In: Flynn, R.J. (ed.) *Parasites of Laboratory Animals.* Iowa State University Press, Ames, pp. 203–320.

Hibler, C.P. (1966) Development of *Stephanofilaria stilesi* in the horn fly. *Journal of Parasitology* 52, 890–898.

Hibler, C.P. and Metzger, C.J. (1974) Morphology of the larval stages of *Elaeophora schneideri* in the intermediate and definitive hosts with some observations on their pathogenesis in abnormal definitive hosts. *Journal of Wildlife Diseases* 10, 361–369.

Higgins, J.A., Klei, T.R. and Foil, L.D. (1988) Factors influencing the ingestion of *Onchocerca cervicalis* microfilariae by *Culicoides variipennis* (Diptera: Ceratopogonidae). *Journal of the American Mosquito Control Association* 4, 242–247.

Hopkins, C.A. and Nicholas, W.L. (1952) *Culicoides austeni*, the vector of *Acanthocheilonema perstans. Annals of Tropical Medicine and Parasitology* 46, 276–283.

Hovette, P., Debonne, J.M., Gaxotte, P., Touze, J.E., Imbert, P., Fourcade, L. and Laroche, R. (1994) Efficacy cf Ivermectin treatment of *Loa loa* filariasis patients without microfilaraemias. *Annals of Tropical Medicine and Parasitology* 88, 93–94.

Intermill, R.W. (1973) Development of *Dirofilaria immitis* in *Aedes triseriatus* Say. *Mosquito News* 33, 176–181.

Jacobs, D.E. and Fox, M.T. (1991) Endoparasites. In: Chandler, E.A., Thompson, D.J., Sutton, J.B. and Price, C.J. (eds) *Canine Medicine and Therapeutics.* Blackwell Scientific Publications, Oxford, pp. 699–722.

Kabilov, T.K. (1980) New data on biology of *Stephanofilaria assamensis* (Nematoda: Filariata). *Helminthologia* 17, 191–196.

Kershaw, W.E., Deegan, T., Moore, P.J. and Williams, P. (1956) Studies on the intake of microfilariae by their insect vectors, their survival and their effect on the survival of their vectors. VIII. The size and pattern of the blood-meals taken in by groups of *Chrysops silacea* and *C. dimidiata* when feeding to repletion in natural conditions on a rubber estate in the Niger Delta. *Annals of Tropical Medicine and Parasitology* 50, 95–99.

Lavoipierre, M.M.J. (1958) Studies on the host–parasite relationships of filarial nematodes and their arthropod hosts. I. The sites of development and the migration of *Loa loa* in *Chrysops silacea*, the escape of the infective forms from the head of the fly, and the effect of the worm on its insect host. *Annals of Tropical Medicine and Parasitology* 52, 103–121.

Lowrie, R.C. and Raccurt, C.P. (1984) Assessment of *Culicoides barbosai* as a vector of *Mansonella ozzardi* in Haiti. *American Journal of Tropical Medicine and Hygiene* 33, 1275–1277.

Maema, M. (1986) Experimental infection of *Tribolium confusum* (Coleoptera) by *Hymenolepis diminuata* (Cestoda): host fecundity during infection. *Parasitology* 92, 405–412.

Mahmood, F. and Nayar, J.K. (1989) Effects of *Dirofilaria immitis* (Nematoda: Filarioidea) infection on life table characteristics of susceptible and refractory strains of *Aedes aegypti* (Vero Beach) (Diptera: Culicidae). *Florida Entomologist* 72, 567–578.

Mellor, P.S. (1974) Studies on *Onchocerca cervicalis* Railliet and Henry 1910: IV. Behaviour of the vector *Culicoides nubeculosus* in relation to the transmission of *Onchocerca cervicalis. Journal of Helminthology* 48, 283–288.

Mellor, P.S. (1975) Studies on *Onchocerca cervicalis* Railliet and Henry 1910: V. The development of *Onchocerca cervicalis* larvae in the vectors. *Journal of Helminthology* 49, 33–42.

Meyers, W.M. and Neafie, R.C. (1991) Streptocerciasis. In: Strickland, G.T. (ed.) *Hunter's Tropical Medicine*. Saunders, Philadelphia, pp. 746–748.

Moraes, M.A.P. (1976) *Mansonella ozzardi* microfilariae in skin snips. *Transactions of the Royal Society of Tropical Medicine and Hygiene* 70, 16.

Moraes, M.A.P., Shelley, A.J. and Dias, A.P.A.L. (1985) *Mansonella ozzardi* no Territorio de Roraima, Brasil. Distribuição e achado de um novo vetor na area do rio Surumu. *Memorias do Instituto Oswaldo Cruz* 80, 395–400.

Mori, A. and Oda, T. (1990) Effects of the microfilarial density of *Dirofilaria immitis* on the longevity of infected *Aedes albopictus*. *Japanese Journal of Sanitary Zoology* 41, 369–374.

Nathan, M.B. (1981) Transmission of the human filarial parasite *Mansonella ozzardi* by *Culicoides phlebotomus* (Williston) (Diptera: Ceratopogonidae) in coastal north Trinidad. *Bulletin of Entomological Research* 71, 97–105.

Nayar, J.K. and Sauerman, D.M. (1975) Physiological basis of host susceptibility of Florida mosquitoes to *Dirofilaria immitis*. *Journal of Insect Physiology* 21, 1965–1975.

Neafie, R.C. and Meyers, W.M. (1991) Dirofilariasis. In: Strickland, G.T. (ed.) *Hunter's Tropical Medicine*. Saunders, Philadelphia, pp. 748–749.

Nelson, G.S. (1970) Onchocerciasis. *Advances in Parasitology* 8, 173–224.

Nelson, G.S. (1991) Miscellaneous filarial infections – General principles – *Mansonella ozzardi* infection – *Mansonella perstans* infection. In: Strickland, G.T. (ed.) *Hunter's Tropical Medicine*. Saunders, Philadelphia, pp. 744–746.

Nevill, E.M. (1975) Preliminary report on the transmission of *Parafilaria bovicola* in South Africa. *Onderstepoort Journal of Veterinary Research* 42, 41–48.

Nevill, E.M. (1979) The experimental transmission of *Parafilaria bovicola* to cattle in South Africa using *Musca* species (subgenus *Eumusca*) as intermediate hosts. *Onderstepoort Journal of Veterinary Research* 46, 51–57.

Nevill, E.M. (1985) The epidemiology of *Parafilaria bovicola* in the Transvaal bushveld of South Africa. *Onderstepoort Journal of Veterinary Research* 52, 261–267.

Nevill, E.M., Wilkins, C.A. and Zakrisson, G. (1987) The control of *Parafilaria bovicola* transmission in South Africa. *Onderstepoort Journal of Veterinary Research* 54, 547–550.

Nicholas, W.L. and Kershaw, W.E. (1954) Studies on the intake of microfilariae by their insect vectors, their survival and their effect on the survival of their vectors. III. The intake of microfilariae of *Acanthocheilonema perstans* by *Culicoides austeni* and *C. grahamii*. *Annals of Tropical Medicine and Parasitology* 48, 201–206.

Nicholas, W.L., Gordon, R.M. and Kershaw, W.E. (1952) The taking up of microfilariae in the blood by *Culicoides* spp. *Transactions of the Royal Society of Tropical Medicine and Hygiene* 46, 377–378.

Noireau, F. and Gouteux, J.P. (1989) Current considerations on *Loa loa* simian reservoir in the Congo. *Acta Tropica* 46, 69–70.

Noireau, F., Nzoulani, A., Sinda, D. and Itoua, A. (1990a) Transmission indices of *Loa loa* in the Chaillu Mountains, Congo. *American Journal of Tropical Medicine and Hygiene* 43, 282–288.

Noireau, F., Itoua, A. and Carme, B. (1990b) Epidemiology of *Mansonella perstans* filariasis in the forest region of south Congo. *Annals of Tropical Medicine and Parasitology* 84, 251–254.

Omar, M.S., Denke, A.M. and Raybould, J.N. (1979) The development of *Onchocerca ochengi* (Nematoda: Filarioidea) to the infective stage in *Simulium damnosum s.l.* with a note on the histochemical staining of the parasite. *Tropenmedizin und Parasitologie* 30, 157–162.

Ottley, M.L. and Moorhouse, D.E. (1978a) Bovine onchocerciasis: aspects of carcase infection. *Australian Veterinary Journal* 54, 528–530.

Ottley, M.L. and Moorhouse D.E. (1978b) Equine onchocerciasis. *Australian Veterinary Journal* 54, 545.

Ottley, M.L. and Moorhouse, D.E. (1980) Laboratory transmission of *Onchocerca gibsoni* by *Forcipomyia* (*Lasiohelea*) *townsvillensis*. *Australian Veterinary Journal* 56, 559–560.

Pence, D.B. (1991) Elaeophorosis in wild ruminants. *Bulletin of the Society for Vector Ecology* 16, 149–160.

Pugh, R.E. (1987) Effects on the development of *Dipylidium caninum* and on the host reaction to this parasite in the adult flea (*Ctenocephalides felis felis*). *Parasitology Research* 73, 171–177.

Radostits, O.M., Blood, D.C. and Gay, G.C. (1994) *Veterinary Medicine – a Textbook of the Diseases of Cattle, Sheep, Pigs and Horses*. Baillière Tindall, London.

Russell, R.C. (1985) Report on a field study on mosquitoes (Diptera: Culicidae) vectors of dog heartworm, *Dirofilaria immitis* Leidy (Spirurida: Onchocercidae) near Sydney, NSW, and the implications for veterinary and public health concern. *Australian Journal of Zoology* 33, 461–472.

Russell, R.C. and Geary, M.J. (1992) The susceptibility of the mosquitoes *Aedes notoscriptus* and *Culex annulirostris* to infection with dog heartworm *Dirofilaria immitis* and their vector efficiency. *Medical and Veterinary Entomology* 6, 154–158.

Shamsul, A.V.M. (1971) Biology of *Stephanofilaria assamensis* and the epizootiology of stephanofilariasis of zebu cattle. *Veterinariya* 3, 112–113.

Sharp, N.A.D. (1928) *Filaria perstans*: its development in *Culicoides austeni*. *Transactions of the Royal Society of Tropical Medicine and Hygiene* 21, 371–396.

Shi, Y.E., Han, J.J., Yang, W.Y. and Wei, D.X. (1988) *Thelazia callipaeda* (Nematoda: Spirurida): transmission by flies from dogs to children in Hubei, China. *Transactions of the Royal Society of Tropical Medicine and Hygiene* 82, 627.

Srivastava, H.D. and Dutt, S.C. (1963) Studies on the life history of *Stephanofilaria assamensis*, the causative parasite of 'humpsore' of Indian cattle. *Indian Journal of Veterinary Science* 33, 173–177.

Steward, J.S. (1933) *Onchocerca cervicalis* (Railliet and Henry 1910) and its development in *Culicoides nubeculosus* Mg. *Review of Applied Entomology B* 22, 58–59 (1934).

Steward, J.S. (1937) The occurrence of *Onchocerca gutturosa* Neumann in cattle in England, with an account of its life history and development in *Simulium ornatum* Mg. *Parasitology* 29, 212–219.

Varma, A.K., Sahai, B.N., Singh, S.P., Lakra, P. and Shrivastava, V.K. (1971) On *Setaria digitata*; its specific characters, incidence and development in *Aedes vittatus* and *Armigeres obturbans* in India with a note on its ectopic occurrence. *Zeitschrift für Parasitenkunde* 36, 62–72.

Voge, M. (1973) Cestodes. In: Flynn, R.J. (ed.) *Parasites of Laboratory Animals*. Iowa State University Press, Ames, pp. 155–202.

Voge, M. and Heyneman, D. (1957) Development of *Hymenolepis nana* and *Hymenolepis diminuata* (Cestoda: Hymenolepididae) in the intermediate host *Tribolium confusum*. *University of California Publications in Zoology* 59, 549–580.

Wahl, G. and Renz, A. (1991) Transmission of *Onchocerca dukei* by *Simulium bovis* in north Cameroon. *Tropical Medicine and Parasitology* 42, 368–370.

Wahl, G., Ekale, D., Enyong, P. and Renz, A. (1991) The development of *Onchocerca dukei* and *O. ochengi* microfilariae to infective-stage larvae in *Simulium damnosum s.l.* and in members of the *S. medusaeforme* group following intrathoracic injection. *Annals of Tropical Medicine and Parasitology*, 85, 329–337.

Wakelin, D. (1978) Genetic control of susceptibility and resistance to parasitic infections. *Advances in Parasitology* 16, 219–308.

Wang, Z.X. and Yang, Z.X. (1993) Studies on the development of *Thelazia callipaeda* larvae in the intermediate host *Amiota variegata* in China. *Chinese Journal of Zoology* 28, 4–8.

Wharton, R.H. (1963) Adaptations of *Wuchereria* and *Brugia* to mosquitoes and vertebrate hosts in relation to the distribution of filarial parasites. *Zoonoses Research* 2, 1–12.

Williams, P. (1981) Studies on Ethiopian *Chrysops* as possible vectors of loiasis. II. *Chrysops silacea* Austen and human loiasis. *Annals of Tropical Medicine and Parasitology* 55, 1–17.

Williams, P. (1982) The bionomics of the tabanid fauna of streams in the rain-forest of the Southern Cameroons. III. The distribution of immature tabanids at Kumba. *Annals of Tropical Medicine and Parasitology* 56, 149–160.

Wirth, W.W. (1977) A review of the pathogens and parasites of the biting midges (Diptera: Ceratopogonidae). *Journal of the Washington Academy of Science* 67, 60–75.

Woodman H.M. (1949) *Filaria* in the Anglo-Egyptian Sudan. *Transactions of the Royal Society of Tropical Medicine and Hygiene* 42, 543–558.

Woodman H.M. and Bokhari, A. (1941) Studies on *Loa loa* and the first report of *Wuchereria bancrofti* in the Sudan. *Transactions of the Royal Society of Tropical Medicine and Hygiene* 35, 77–92.

Yarzábal, L., Basánez, M.G., Ramirez Perez, J., Ramirez, A., Bottto, C. and Yarzábal, A. (1985) Experimental and natural infections of *Simulium sanchezi* by *Mansonella ozzardi* in the Middle Orinoco region of Venezuela. *Transactions of the Royal Society of Tropical Medicine and Hygiene* 79, 29–33.

Zytoon, E.M., El-Belbasi, H.I. and Matsumura, T. (1993) Transovarial transmission of chikungunya virus by *Aedes albopictus* mosquitoes ingesting microfilariae of *Dirofilaria immitis* under laboratory conditions. *Microbiology and Immunology* 37, 419–421.

Index

Note: numbers in italic type represent pages on which illustrations appear.

Aardvark 280
Abate, *see* Temephos
Abdomen 27
Acalypterae 50
Acanthocheilonema, see Mansonella
Acari 20, 383–485
    main reference works 383
Acaridia 386
Acaridida, *see* Astigmata
Acariformes, *see* Actinotrichida
Acarus siro 388–9, 408
Accessory glands 84
Acrostichal bristles 41, *43*
Actinedida, *see* Prostigmata
Actinotrichida 383
    life cycle 286
Acyophora 35
Aedeomyia 113, 130
Aedeomyiini 113
Aedes 113, 116
    dormant eggs 115–6
    kairomones 70
Aedes aegypti 5, 7, *100*, 102, 123, 125–7,
    490–5 passim, 580
    and pathogens
      Chikungunya 500
      dengue virus 493, 495
      Dirofilaria immitis 679–80
      filarias 642–9 passim
      yellow fever virus 491–2

    biology and bionomics 99–101, 131–142
      passim
      mating 99
      proboscis TS 57
      response to adenosine nucleotides 71
    taxonomic status 101
      variation in 101
Aedes aegypti aegypti 101
Aedes aegypti formosus 101
Aedes africanus 5, 135, 141
    Chikungunya 501
    yellow fever 491–3
Aedes albopictus 101, 141, 144, 490
    dengue fever 493
    Dirofilaria immitis 679
Aedes bromeliae and yellow fever 491–3
Aedes caballus and Rift Valley fever 503
Aedes camptorhynchus
    Eperythrozoon ovis 533
    Ross River virus 500
Aedes caspius 141
Aedes communis 140
Aedes detritus 141
Aedes dorsalis 139
Aedes furcifer/taylori
    Chikungunya virus 501
    yellow fever 491–3
Aedes hexodontus 132
Aedes ingrami 136
Aedes lineatopennis and Rift Valley fever 503

Aedes luteocephalus
    Chikungunya 501
    yellow fever 493
Aedes melanimon
    Californian encephalitis 502
Aedes niveus
    dengue fever 493
    Wuchereria bancrofti 643
Aedes normanensis and Ross River virus 500
Aedes poecilius and *Wuchereria bancrofti* 643
Aedes polynesiensis 126, 135
    dengue fever 493
    Ross river virus 500
    Wuchereria bancrofti 643–652 passim
Aedes pseudoscutellaris and *Wuchereria
    bancrofti* 646
Aedes samoanus and lymphatic filariasis 652
Aedes scutellaris 135, 518
    species complex 98
    dengue fever 493
    Wuchereria bancrofti 652
Aedes simpsoni 5, 141, *see also Aedes bromeliae*
Aedes sollicitans 141, 143
    Dirofilaria immitis 680
    eastern equine encephalitis 499
Aedes taeniorhynchus 131–143 passim
    kairomones 70
Aedes taylori, *see Aedes furcifer/taylori*
Aedes togoi
    Setaria digitata 683
    Wuchereria bancrofti 642–8 passim
Aedes triseriatus 144–5, 490
    Dirofilaria immitis 679
    La Crosse virus 502
Aedes vexans 134
Aedes vigilax 141, 143
    Ross River virus 500
Aedes vittatus
    Chikungunya 501
    Setaria digitata 683
    yellow fever 493
Aedini 113
Aegyptianella pullorum 454, 532–3
Aeromonas hydrophila hydrophila 433
African horse sickness 508–9
African swine fever 510
Afrotropical region 9
Ageing 86
Ageles geoffroyi 663
AIDS and tropical diseases 11
Aino virus 502

Air sacs *76*, 81
Akabane virus 168, 582
Akiba 585
Alces alces 680
Allergens 407–9
Allodermanyssus sanguineus, *see Liponyssoides
    sanguineus*
Alouatta 491
Alphavirus 489, 498
Alula 44
Amblycera 361, 372–6 passim, *372*, *373*
    mouthparts of *56*
Amblyomma 464, 538
Amblyomma americanum 470, 472, 554
    Rickettsia rickettsii 523
Amblyomma cajennense 470
    Rickettsia rickettsii 523
Amblyomma cohaerens and *Theileria
    mutans* 603
Amblyomma hebraeum 464, 466, 471, 478
    Cowdria ruminantium 530
    Rickettsia conorii 525
    Theileria velifera 603
Amblyomma lepidum and *Theileria velifera* 603
Amblyomma maculatum 274, 471
Amblyomma triguttatum and *Coxiella
    burnetii* 528
Amblyomma variegatum 464, 471, 480
    Cowdria ruminantium 530
    Theileria mutans 603
Amblypygi 21, 22
Ambulacral combs 109, 111
Ambulacrum 385
Amiota variegata and *Thelazia callipaeda* 684
Ammotragus lervia 680
Ammospermophilus leucurus 551
Amphipneustic 252
Anactinotrichida 484
Anal vein 44
Analgoidea 385, 400, 409
Anaplasmataceae 517, 531–3
Anaplasma marginale 221, 532, 593–4
Anaplasma centrale 532
Anaplasma ovis 532
Anepisternum 41, *43*
Anocentor nitens, *see Dermacentor nitens*
Anopheles 4, 113, 561–80 passim
    adult characters 127–131
    feeding on plasma 70
    host preferences 569–72
    kairomones 70

Anopheles – *continued*
   longevity 86, 572–3
   lymphatic filariasis 614–5 passim
   malaria transmission
      sporozoite rate 576–7
      vectorial capacity 576
      vectors of malaria 569–70
   wing *110*
Anopheles aconitus and malaria 570
Anopheles aquasalis and malaria 569–70
Anopheles albimanus and malaria 569
Anopheles albitarsis and malaria 570
Anopheles annulipes 112
Anopheles anthropophagus 98
   malaria 570
Anopheles arabiensis 131, 137, 143
   malaria 569–70, 573, 576
   polytene chromosomes of 95
   Wuchereria bancrofti 648
Anopheles argyritarsis and malaria 570
Anopheles atropos 134
Anopheles atroparvus 90–3, *91*, 133, 144
   malaria 569
Anopheles aztecus 93
Anopheles balabacensis 127, 131
   malaria 569
Anopheles bancroftii and bovine ephemeral
      fever 506
Anopheles barberi 144
Anopheles barbirostris
   Brugia malayi 642
   species complex 98
Anopheles beklemishevi 92
Anopheles bellator 141
Anopheles bwambae 96
Anopheles campestris
   Brugia malayi 646
   malaria 570
Anopheles claviger 59, 132, 141–2, 561
Anopheles culicifacies 127, 132, 140
   malaria 570, 574
Anopheles darlingi 137
   malaria 569
Anopheles dirus
   malaria 569
   species complex 98
Anopheles donaldi and malaria 570
Anopheles earlei 93
Anopheles farauti 135, 137
   malaria 569, 577–8
   Wuchereria bancrofti 648, 651

Anopheles flavirostris 137
   malaria 570
Anopheles fluviatilis 140
   malaria 569
Anopheles freeborni 93, 137, 143–5
   malaria 570
Anopheles funestus 136, 140–1
   malaria 561, 565, 569–73 passim
   O'Nyongnyong 500
   Wuchereria bancrofti 644
Anopheles gambiae s.l.
   in Brazil 123
   malaria 561–80 passim,
   O'Nyong-nyong 500
   species complex 93–6
Anopheles gambiae s.s. 94, 132–42 passim
   kairomones 70
   malaria 561–80 passim
   polytene chromosomes of 95
   Wuchereria bancrofti 648–9
Anopheles hilli 125
Anopheles hyrcanus complex 98
Anopheles implexus 135
Anopheles indiensis 132
Anopheles koliensis and malaria 570, 577–8
Anopheles labranchiae 90–3, *91*
   malaria 569, 574
Anopheles letifer and malaria 570
Anopheles leucosphyrus and malaria 570
Anopheles maculatus 131–3 passim, 140
   malaria 570
   species complex 98
Anopheles maculatus willmori 124
Anopheles maculipennis s.l. 8, 59, 561, 573
   pupa 123
   species complex 88–93
Anopheles maculipennis s.s. *91*
Anopheles martini 92
Anopheles melanoon melanoon 91–3, *91*, 137
Anopheles melanoon subalpinus 91–3, *91*
Anopheles melas 94, 132, 134
   malaria 569
   Wuchereria bancrofti 644
Anopheles merus 94
Anopheles messeae 90–3, *91*
   malaria 569, 574
Anopheles minimus 131, 135
   breeding sites of 139
   malaria 569, 574
   species complex 98
Anopheles multicolor 139

Anopheles nigerrimus 570
Anopheles occidentalis 93
Anopheles pharoensis 140
    dispersal 142
    malaria 570, 574
Anopheles philippinensis 132
Anopheles plumbeus 141
Anopheles pseudopunctipennis and malaria 570
Anopheles punctimacula and malaria 570
Anopheles punctulatus
    malaria 570, 577-8
    species complex 98
Anopheles quadriannulatus 94
Anopheles quadrimaculatus 93, 144
    malaria 569, 570
Anopheles sacharovi 91-3, *91*, 143-5
    dispersal 143
    malaria 569
Anopheles sergenti 139
    growth of larva 123
    malaria 569
Anopheles sicaulti 92
Anopheles sinensis
    Brugia malayi 450, 452
    malaria 569
    Setaria digitata 683
Anopheles stephensi 133, 138, 141-2
Anopheles subpictus and malaria 570
Anopheles sundaicus
    breeding sites 140-1
    malaria 570
Anopheles superpictus 140, 144
    malaria 570
Anopheles umbrosus 140
Anopheles vagus 137
Anopheles walkerae 144
Anopheles wellcomei 109
Anopheles whartoni and *Wuchereria bancrofti* 643
Anophelinae 111, *113*
    adult characters 127-31, *128-31*
Anoplura 30-1, 361-71, 373
    bionomics 365-71 passim
    external morphology 362-3
    internal structure 363-5
    mouthparts 68, *69*
    symbiotic organisms 80
Apheloria corrugata 15
Antennae 23
Antennata 14
Anterior vertical setae 390

Anthrax 221
Anthropophily 88
Antivenom
    Atrax robustus 19
    Latrodectus hasselti 19
Antricola 442
Apicomplexa 558
Apodemus agrarius 337
Apodemus sylvaticus 337
Apolonia 415
Apolysis 75
Aponomma concolor 471
Appendix dorsalis 27
Apterygota 27
Arachnida 16
Araneae 19
    Araneamorphae 19
    Mygalomorphae 20
Arboviruses 489-516
    transovarian transmission 490
Argas 442, 449-50
Argas arboreus 450-1, 454, 543
    Borrelia anserina 543
    Coxiella burnetii 528
    paralysis caused by 481
Argas hermanni and *Borrelia anserina* 543
Argas persicus 450-4
    Aegyptianella pullorum 534
    bionomics 454
    Borrelia anserina 543
    economic importance 454
    life cycle 451-2
    mating 452-3
    paralysis caused by 454, 481
    quantitative studies 453-4
Argas radiatus, paralysis caused by 454, 481
Argas robertsi 450
Argas sanchezi, paralysis caused by 454, 481
Argas streptopelia and *Borrelia anserina* 543
Argas transversus 449
Argas walkerae 450-52
    paralysis caused by 481
Argasidae 442-454
Argasoidea 423
Arista 42
Armigeres 113, 115, 117
Armigeres dolichocephalus 141
Armigeres obturbans and *Setaria digitata* 683
Arsenicals, *see* Paris green
Artemisia annua 581
Arthropoda, classification of 13-40

Arthropoda, classification of—*continued*
    venoms 13
Aschiza 42, *50*
Ascoschoengastia and scrub typhus 526-7
Astigmata 383-419 passim
Atrax robustus 18
    antivenom 19
Atrichopogon 152
Auchmeromyia 269
Auchmeromyia luteola see A. senegalensis
Auchmeromyia senegalensis 279-82
    female 280
    hosts of 280-2
    larva *281*
Auchmeromyia bequaerti 292
Austenina 225, 228, 239-41, *see also* Glossina
Australian region 9
Austroconops 152, 155
Austrosimulium 192-3, 195
Austrosimulium bancrofti 196, 199, 202-3
Austrosimulium pestilens 196-204 passim
Autogeny 133
Avaritia 166

Babesia 558, *594*, *596-7*
    cycle in tick 396-8
    in dogs 596
    in horses 595-6
    in pigs 596
    in sheep 595
    transmission 599-601
Babesia argentina, *see Babesia bovis*
Babesia berbera, *see Babesia bovis*
Babesia bigemina 591-601 passim
    cycle in *Boophilus microplus* 597-8
Babesia bovis 591-4, 597-601 passim
    cycle in *Boophilus microplus* 597-8
Babesia caballi 592, 595-6
Babesia canis 592, 596, 599
    development in tick 599
Babesia capreoli 601
Babesia caucasicus see Babesia divergens
Babesia colchica see Babesia bovis
Babesia divergens 592-4, 601-2
    epizootiology of 601
Babesia equi 592, 595-7
Babesia gibsoni 592, 596
Babesia jakimovi 592, 595
Babesia karelica see Babesia divergens
Babesia major 592-5

Babesia microti 597, 602
Babesia motasi 595
Babesia occidentalis see Babesia divergens
Babesia ocultans 593
Babesia ovata 592
Babesia ovis 595-9 passim
    in *Rhipicephalus bursa* 599
Babesia perroncitoi 596
Babesia trautmanni 596
Babesiidae 558
Babesioses 591-602
    basic reproduction rate 601
    in cattle 592
    control 593
    economic importance 592
    epizootiology 599-600
    in humans 601-2
    immunity against 592
    inoculation rate 600
    species involved 592-3
Bacillus anthracis 221
Bacillus sphaericus 127, 652
Bacillus thuringiensis 127
Bacillus thuringiensis israelensis 122-3
Bandicota bengalensis 548
Bartonellaceae 517, 530-1
Bartonella bacilliformis 531
Basis capituli 426
Basitarsus 25
Bedbug see Cimex
Bembix bequaerti dira 220
Benzyl benzoate 391
Bironella 111, 113
Blackflies *see* Simuliidae
Blatella germanica 29
Blatta orientalis 29
    oxygen debt in 82
Blattodea 29
    and pathogens 29
Blister beetle *see* Meloidae
Blomia kulagini 408
Bluetongue 506-8
Boophilus 461, 467, 544
    Borrelia theileri 544
Boophilus australis 478
Boophilus annulatus 461, 477, 480
Boophilus decoloratus 461, 593
    anaplasmosis 533
    Borrelia theileri
Boophilus microplus 461, 469-72, 475-80
    Babesia 593-4, 596-600

Boophilus microplus – *continued*
    Babesia bigemina 597–8
    Babesia bovis 597–8
    bionomics
       hosts 475
       life cycle 476,
       seasonal abundance 476
    modelling of 477
       and babesiosis 479
    population dynamics 477
Boopidae 373
Borrelia 517, 537–44
Borrelia anserina 454, 537, 543
Borrelia burgdorferi 537–40
Borrelia duttoni 541
Borrelia hermsi 542–3
Borrelia recurrentis 537, 540–1
Borrelia theileri 537, 544
Borrelia venezuelensis 543
Borrelioses, *see* Relapsing fevers
Bos indicus, *see* Zebu
Bos taurus, *see* Cattle
Bovine ephemeral fever 505–6
Brachycera 46, 50–3
Bradypus tridactylus 503
Brahman, *see* Zebu
Brain
Brugia malayi 638, 680, *see also* lymphatic
      filariasis
    forms of 644–5
Brugia pahangi 642–50 passim
Brugia timori 638–42 passim
Brumptomyia 177–8
Bubalus bubalis, *see* buffalo, Asian
Buccal cavity 77
Buffalo, African
    Anaplasma 532, 607
    Theileria parva 607
    Theileria lawrencei 607
Buffalo, Asian (water)
    Theileria parva 605
    Setaria spp. 682
    Wohlfahrtia magnifica 285
Bunyaviridae 489, 501–5
Bunyavirus 501
Bushbuck 623
Bushpig 623
Buthidae 18

Calabar swelling 671

California encephalitis 502
Calliphora 38, 269, 275, *276*
    digestive enzymes 78
Calliphora augur 278
Calliphora erythrocephala 287
Calliphora dubia 278
Calliphora nociva, *see* *Calliphora dubia*
Calliphora stygia 269, 276
Calliphoridae 53, 268–83
Calliphorinae 268–9
    metallic 275–9
    testaceous 279–83
Callistemon viminalis 200
Callitroga americana, *see* *Cochliomyia*
      hominivorax
Callosobruchus maculatus 422
Calyptra eustrigata 36
Calypter, *see* squama 44
Calypteratae 50
Calyptra 36
Capitulum 384
Capreolus capreolus 595
Capreolus capreolus capreolus 302
Carcinops 264
Cat (*Felis domesticus*), parasites of
    Cheyletiella blakei 421–2
    Dipylidium caninum 684
    Felicola subrostrata 375, 376–7
    Notoedres cati 392–4
    Otodectes cynotis 399–400
    Trypanosoma brucei 620
Cattle (*Bos taurus*)
    arbovirus diseases of
       Aino 502
       Akabane 502
       bluetongue 506–8
       bovine ephemeral fever 505–6
       Rift Valley fever 503
       vesicular stomatitis 505
    arthropod parasites of
       Chorioptes spp. 398–9, *400*
       Damalinia bovis 377–8
       Demodex bovis 420
       Haematopinus spp. 368–9
       Linognathus vituli 370–1
       Ornithodoros lahorensis 449
       Ornithodoros savignyi 448–9
       Otobius megnini 443–4
       Psorobia bos 416–7
       Psoroptes bovis 396
       Psoroptes natalensis 396

Cattle (*Bos taurus*)—*continued*
    Raillietia auris 435-6
    Sarcoptes scabiei 392
    Solenopotes capillatus 371
    trombiculid larvae 415
    Borrelia theileri in 544
    filarioid parasites of
        Onchocerca spp. 677-8
        Parafilaria spp. 682
        Setaria spp. 682-3
        Stephanofilaria spp. 681
    myiasis agents of
        Chrysomya bezziana 274-5
        Cochliomyia hominivorax 271-4
        Dermatobia hominis 309-11
        Hypoderma spp. 297-303
    protozoal diseases of, with arthropod
        vectors
        babesiosis 592-5 passim
        theileriosis 602-9
        trypanosomiasis 620-1
    rickettsial diseases of, with arthropod
        vectors
        anaplasmosis 532
        ehrlichiosis 529-30
        heartwater 530
        Q fever 528-9
    thelaziasis 684
Cement layer 73
Centipedes, *see Chilopoda*
Centruroides limpidus 117
Centruroides pantherinus 18
Centruroides sculpturatus 17
Centruroides vittatus 18
Cephalothorax 124
Cephalopina titillator 292, 296-7
Cephenemyia 295
Ceratophyllidae 323
Ceratophyllus 324, 327, 331
Ceratophyllus gallinae 325, 332-4, 337-9
    passim
Ceratophyllus styx 332-3
Ceratopogonidae 48, 152-76
    wing *45*
Ceratopogoninae 152
Cercomys 550
Cercus plural cerci 27
Cervus canadensis 680
Cervus elephas 505
Cervus elephas hippelaphus 302
Cestoda

Dipylidium caninum 684
Hymenolepis diminuata 684-5
Hymenolepis nana 684-5
Chagas' disease, *see Trypanosoma cruzi*
Chagasia 111, 113
Chaoboridae 109
    wing 110
Chelicera 18, 384-5, *385*
Chelicerata 14
Chelifer cancroides 23
Chemoreceptors 70-1
Cheyletiella 420-1
Cheyletiella blakei 421
Cheyletiella parasitivorax 421
Cheyletiella yasguri 421
Cheyletiellidae 416, 420-2
Cheyletoidea 384, 416-22
Chikungunya virus 500-1
Chilopoda 14
Chinius 177-8
Chipmunk 543
Chirodiscoides caviae 409
Chironomidae *46, 47*
    haemoglobin antigens 47
Chironomus tentans 47
Chironomus thummi thummi
Chitin 74
Chloropidae *51*
Cholesterol 80
Chorion 84
Chorioptes 395, 398-9
Chorioptes bovis 399, *400*
Chorioptes texanus 399
Chrysomya 268-9, 276
Chrysomya albiceps 275
Chrysomya bezziana 271-5, *270, 272*
    and myiasis 274-5
Chrysomya marginalis 275
Chrysomya megacephala 275
Chrysomya putoria 275
Chrysomya rufifacies 271, 275, 278
Chrysomyinae 268, 271-5
Chrysopsinae 214
Chrysops 211, *213*, 214
    digestive enzymes 78
Chrysops atlanticus 219-20
Chrysops australis 214
Chrysops caecutiens 220
Chrysops callidus vector of *Haemoproteus*
    metchnikovi 584
Chrysops centurionis and loiasis 673

Chrysops dimidiata 219
  and loiasis 672–3
Chrysops discalis 217–8, 220
  and tularaemia 554
Chrysops distinctipennis and loiasis 673
Chrysops fuliginosus 218–20
Chrysops langi 219
  and loiasis 673
Chrysops longicornis and loiasis 673
Chrysops relictus 218
Chrysops silacea 219–20
  and loiasis 672–3
Chrysops zahrai and loiasis 673
Cibarial pump 70, 74
  resilin in 74
Cibarium 68
Cicada, *see Thopha saccata*
Cimex 344, 348–9
  behaviour 348–9
    mating 348
  life cycle 347–9
    development rate 349
    egg 347
    fecundity 349
    nymph 347–8
  medical importance 349–50
  structure 345–7
    abdomen 346
    female 345
    head 345
    male terminalia *346*
    mouthparts 67–8
    mycetome 347
    thorax 346
  survival 349
  symbionts in 80
Cimex hemipterus 344–6, 349–50
  hepatitis B virus in 350
Cimex lectularius 344–50 passim
  female *345*
  hepatitis B virus in 350
  HIV in 350
  male *346*
  symbionts in 347
Cimicidae 344–50
Circulatory system 82–3
Citellus pygmaeus, plague reservoir 552
Citellophilus tesquorum 98, 335
Cladotanytarsus lewisi 47
Cleg, *see Haematopota*
Clypeus 24

Cnephia 193, 195
  Leucocytozoon 585
Cnephia pecuarum 204–5
Coccidiasina 558
Cochliomyia 268–9
Cochliomyia hominivorax 270, 271–4
  ageing 86
  in Libya 274
  and myiasis
Cochliomyia macellaria 271, 275
Cockroach, *see* Blattodea
Cocoa, pollination of 182
Coleoptera 32, *33, 34*
Collembola 23
Colorado tick fever 510
Conjunctivitis 51–2
Connochaetes gnu 530
Coquillettidia 114–5
  egg rafts of *117*
  larva 119–20
  pupa 124–5
Coquillettidia aurites 125
Coquillettidia perturbans 134
  eastern equine encephalitis 499
Cordylobia 269
Cordylobia anthropophaga 271, 279, 282–3
  biology and bionomics 282
  female *280*
  hosts of 282
  larva *283*
Cordylobia rodhaini 283
Corpora allata 84
Corpus cardiacum 84
Costa *44*
Cowdria ruminantium 530
Coxa 25, *395*
Coxiella 518
Coxiella burnetii 528–9
  vaccine against 529
Crab hole mosquitoes, *see Deinocerites*
Cremastogaster, feeding female *Malaya* 115
Crimean-Congo haemorrhagic fever 504
Crop *77–8*
Cross veins 45
Crustacea 14
Cryptostigmata, *see* Oribatida
Ctenocephalides 327
  and plague 550
Ctenocephalides canis 334, 339
  and *Dipylidium caninum* 684
  and plague 550

Ctenocephalides felis 323, 325, 328–39 passim
  biology and control 338
  life cycle *330*
  pathogens
    Borrelia burgdorferi 539
    Dipylidium caninum 684
    plague 550
    Rickettsia typhi 522
Ctenophthalmidae 323
Cubitus 44
Cuclotogaster heterographus 374, 376–80 passim
Culex 113, 117
Culex annulirostris 141
    Eperythrozoon ovis 533
    Murray Valley encephalitis 497
    Ross River virus 500
    Sindbis virus 501
Culex antennatus
    Rift Valley fever 503
    Sindbis virus 501
Culex bitaeniorhynchus and Sindbis virus 501
Culex decens 134
Culex fatigans, see Culex quinquefasciatus
Culex gelidus and Japanese encephalitis 496
Culex modestus and West Nile virus 495
Culex molestus 133
Culex morsitans 144
Culex nigripalpus 134
Culex pipiens 133–44 passim, 164
    Rift Valley fever 503
    St Louis encephalitis 497
    symbionts 80
Culex portesi and eastern equine
      encephalitis 99
Culex quinquefasciatus 123, 134–42 passim,
      561
    control 125–7
    lymphatic filariasis 640–652 passim
    St Louis encephalitis 496
    Oropouche virus 502
Culex tarsalis 138, 144, 490
    St Louis encephalitis 497
    western equine encephalitis 499
Culex taeniopus and eastern equine encephalitis
Culex territans 134
Culex theileri 144
    Rift Valley fever 503
    Sindbis virus 501
    West Nile virus 495
Culex tritaeniorhynchus 134, 140, 491
    Japanese encephalitis 496

    Sindbis virus 501
Culex univittatus
    Sindbis 501
    West Nile virus 495
Culex vishnui
    Japanese encephalitis 496
    west Nile virus 495
Culicidae 47, 109–51
    behaviour
      biting cycle and height 135–6
      host finding 133–5
      mating 131–3
      oviposition 138
    bionomics 130–45
      aestivation 143–5
      breeding sites 138–42
      dispersal 142–3
      hibernation 143–5
      seasonal cycles 143–5
    control of immatures 122–3, 125–7
    and disease 115
    life cycle
      development times 125
      egg *116*, 117
      emergence 127
      larva 117–23, *118*, *119*
      larval feeding and respiration 119–22
      ovarian development 136–7
      pupa 124–5, 49
    myxomatosis virus 489
    structure
      antenna *42*
      mouthparts *57*, *58*
      peritrophic membrane 77
      spermathecae 130
      wing *45*, *110*
Culicinae 112
    adult characters 127, *128*, *129*, *130*, *131*
    ecological classification of 115
    life cycle
      duration 125
      egg stage *116*, 117
      larva 119–23, *120*, *121*
      pupa 124
Culicini 113
Culicoides 154
    arboviruses
      African horse sickness 509
      bluetongue 507
      bovine ephemeral fever 506
    bionomics

Culicoides — continued
  bionomics — continued
    biting behaviour 168
    breeding sites 165-7
    feeding 168-9
    flight range 167-8
    mating 161-2
    seasonal changes 169-70
    survival 168-9
  female *153*
  life cycle 156-60
    duration of 159-60
    larva *48, 157, 158*
    larval epipharynges *159, 160*
    pupa *49, 157*
  medical and veterinary importance 170
  onchocerciasis
    bovine 677
    equine 678
  protozoa transmitted by 583-4
Culicoides achrayi 166
Culicoides actoni and bluetongue virus 507
Culicoides adersi and *Hepatocystis kochi* 583
Culicoides albicans 166
Culicoides algecirensis 162
Culicoides angularis 157, 158, *166*
Culicoides anophelis 163
Culicoides arakawae 163-8 passim
    Leucocytozoon caulleryi 586
Culicoides austeni 169
    Mansonella spp. 59, 675-6
Culicoides austropalpalis 159, 166
Culicoides barbosai 164-5, 168-9
    Mansonella ozzardi 674
Culicoides brevitarsis 156
    Culicoides hypersensitivity 170
    epipharynges *160*
    pathogens
      Akabane virus 168, 502
      bluetongue virus 507
      bovine ephemeral fever 506
      Onchocerca gutturosa 678
Culicoides bundyensis 166
*Culicoides bunroensis*166
Culicoides chiopterus 166
Culicoides circumscriptus 156, 160, *165*
Culicoides copiosus 166
Culicoides cordiger 161
Culicoides cornutus 158
    bluetongue virus
Culicoides denningi 166

Culicoides dewulfi 166
Culicoides downesi and *Parahaemoproteus*
    nettionis 584
Culicoides duddingstoni 166
Culicoides fagineus 166
Culicoides fascipennis 166
Culicoides fulvithorax and *Hepatocystis*
    kochi 584
Culicoides fulvus and bluetongue virus 507
Culicoides furens 159, 162-9 passim
    kairomones 70
    Mansonella ozzardi 674
Culicoides grahamii and *Mansonella* spp. 59,
    675-6
Culicoides grisescens 156
Culicoides guttipennis 166
Culicoides halophilus 166
Culicoides heliconiae 166
Culicoides heliophilus 164
Culicoides hoffmani 166
Culicoides hollensis 165
Culicoides imicola 164, *166*, 169
    bluetongue virus 507
Culicoides impunctatus 160-70 passim
    flight range 167
Culicoides inornatipennis and *Mansonella*
    perstans 675
Culicoides insignis 170
    bluetongue virus 507
Culicoides longior 166, 169
Culicoides loughnani 166
Culicoides maritimus 166
Culicoides marksi 166-7
Culicoides marmoratus 165-9
    epipharynges *160*
    larval anal segment *158*
    wing *154*
Culicoides melleus 161, 165
Culicoides milnei and *Mansonella* spp. 676
Culicoides mississippiennis 167, 169
Culicoides molestus 161, 165, 167
Culicoides nipponensis 169
Culicoides nubeculosus 157, *158, 159*, 157-65
    passim
    anal papillae *158*
    epipharynges *159*
    pupa *157*
    pathogens
      bluetongue virus 507
      epizootic haemorrhagic disease of
        deer 509

Culicoides nubeculosus—*continued*
    pathogens—*continued*
       Hepatocystis brayi 584
       Onchocerca spp. 677–8
Culicoides obsoletus 168–70
    Onchocerca cervicalis 678
Culicoides odibilis 163, 166
    Leucocytozoon caulleryi 586
Culicoides pallidicornis 166
Culicoides pallidipennis 164
Culicoides paraensis 166
    Oropouche virus 502
Culicoides parroti and *Onchocerca cervicalis* 678
Culicoides peleliouensis *169*
Culicoides phlebotomus 164, 169–70
    Mansonella ozzardi 169, 674
Culicoides pulicaris 166
Culicoides pungens and *Onchocerca gibsoni* 677
Culicoides sanguisuga 166
Culicoides schultzei and *Leucocytozoon*
    caulleryi 586
Culicoides scoticus 166
Culicoides sphagnumensis 166
Culicoides subimmaculatus 161–9 passim
    epipharynges *160*
Culicoides truncorum 166
Culicoides variipennis 158, 160, 164–9
    pathogens
       African horse sickness 509
       bluetongue virus 507
       epizootic haemorrhagic disease of
       deer 510
       Hepatocystis brayi 584
       Onchocerca cervicalis 678
    subspecies 166
Culicoides variipennis sonorensis and bluetongue
    virus 507
Culicoides variipennis variipennis and
    bluetongue virus 507
Culicoides vexans 156, 161, 166
Culicoides victoriae 169
Culicoides wadai and bluetongue virus 507
Culiseta 113, 117
Culiseta inornata 134, 137–8, 143
Culiseta melanura 134, 140, 144
    eastern equine encephalitis 499
    western equine encephalitis 499
Culisetini 113
Cuscus 433
Cuterebrinae 53, 292–4
    hosts of 293

Cuticle 73–5
Cuticular hydrocarbons 98
Cuticular lipids 98
Cuticulin 73
Cyclops 14
Cyclorrhapha 50
    lapping mouth parts 61
Cylindroiulus sp. 16
Cytauxzoon, see *Theileria*
Cytodites, *nudus* 404–6, *406*
Cytoecetes, see *Ehrlichia*

Damalinia bovis 362, 377–80 passim
Damalinia caprae 362, 378
Damalinia crassipes 362
Damalinia equi 362, 378–9
Damalinia limbata 362
Damalinia ovis 362, 375, 376–80
Damaliscus albifrons 530
Dasyheleinae 152, 154, 160
Deer fly, see *Chrysops*
Deer, white-tailed, see *Odocoileus virginianus*
Deinocerites 113, 115
Deinocerites cancer 133, 142
Deinocerites dyari 132
Demodex 414, 418–20
    in domestic animals 420
Demodex bovis *419*, 420
Demodex brevis 419
Demodex canis 419–20
    life cycle *419*, 420
Demodex caprae 420
Demodex cati 419
Demodex folliculorum 418–9
Demodex muscardinus *419*
Demodex phylloides 420
Demodicidae 416
Dendrohyrax 632
Dengue fever 493
    distribution *494*
Dengue haemorrhagic fever (DHF) 493, 495
Dengue shock syndrome (DSS) 495
Dermacentor 463, 534, 595–6
Dermacentor albipictus 468
Dermacentor andersoni *463*, 471, 480
    Anaplasma 532
    Colorado tick fever 510
    paralysis caused by 480
    Rickettsia rickettsii 523
    tularaemia 554

Dermacentor (Anocentor) nitens 468
Dermacentor marginatus 463
   Babesia canis 596
   Rickettsia sibirica 524
   Rickettsia slovaca 524
Dermacentor nuttalli
   Rickettsia sibirica 524
   tularaemia 554
Dermacentor occidentalis 472
   Anaplasma 533
Dermacentor perumapertus and *Borrelia*
   burgdorferi 539
Dermacentor pictus and Omsk haemorrhagic
   fever 498
Dermacentor reticulatus 463, *464*
   Babesia canis 599
   Borrelia burgdorferi 538
Dermacentor silvarum and *Rickettsia*
   sibirica 524
Dermacentor variabilis 463, 471–2, 479
   Rickettsia rickettsii 523
   tularaemia 554
Dermanyssidae 425, 429–31
Dermanyssoidea 384, 424–37
   biology of 428–9
   medical importance 428–9
   structure of 425–8, *428*
Dermanyssus 426
Dermanyssus gallinae 429–31
Dermatobia hominis 293, 309–11
   eggs on carrier *310*
   female *309*
   hosts of 293
   larva *311*
   medical and veterinary
     importance 310–11
Dermatophagoides farinae 407
Dermatophagoides pteronyssinus 407–9
   female *408*
Deutonymph 386
Diamanus montanus and plague 550
Dichoptic 23
Diceromyia 493
Dicranura vinula 36
Didelphys 626
Digestive system 76–9
Diplopoda 14
Diptera 25, *37*
   classification of 46–53
   structure of 41–5
     antennae *42*

   peritrophic membrane *77*
   thorax *43*
   wing *45*
   energy sources 78
Dipylidium caninum 684
Dirofilaria immitis 638, 642, 646, 469, 479–80
Dirofilaria repens 646, 680
Dirofilaria roemeri 221
Discal cell 44
Disease transmission
   biological 4
   mechanical 4
Dixa brevis larva 111
Dixidae 109
   wing *110*
DNA probes 98
Dog (*Canis familiaris*)
   arthropod parasites of
     Cheyletiella yasguri 421
     Demodex canis 419
     Heterodoxus spiniger
     Hippobosca longipennis 320
     Linognathus setosus 370
     Otodectes cynotis 399
     Pneumonyssoides caninum 433
     Sarcoptes scabiei 392
     Trichodectes canis 375, 376
   helminths transmitted by arthropods
     Dipylidium caninum 684
     Dirofilaria immitis 679–80
   Protozoa transmitted by arthropods
     Babesia spp. 596
     Leishmania spp. 628, 632
     Trypanosoma spp. 613, 620
Dorsocentral bristles 41, *43*
Dracunculus medinensis (Guinea worm) 14
Draschia megastoma 683
Duttonella, see *Trypanosoma*
Dyar's law 123

East coast fever, see *Theileria parva*
Eastern equine encephalitis 499
Ecdysial suture 75
Ecdysis 75
Ecdyson 84
Echidnophaga 336
Echidnophaga gallinacea 323–4, 329, 352–9
     passim
   female *326*
   Yersinia pestis 550
Ectognathous 23

Ehrlichia canis 529, 596
Ehrlichia equi 530
Ehrlichia phagocytophila 529–30
Ehrlichia risticii 530
Ehrlichia sennetsu 530
Ehrlichieae 517–29
Ejaculatory duct 84
Elaeophora 638
Elaeophora schneideri 221, 680–1
Eland, *see Taurotragus oryx*
Elk, *see Cervus canadensis*
Empodium 25, 385
Encephalitis 496–7
    eastern equine 499
    Japanese 496
    Murray Valley 496
    St Louis 496
    tick-borne 497–8
    Venezuelan equine 499–500
    western equine 499
Endocrine glands 83
Endocuticle *74*
Endophagy 131
Endophily 131
Endopterygota 28
Entognathous 23
Entomophobia 11
Eperythrozoon 533
Eperythrozoon ovis 533
Eperythrozoon suis 533
Eperythrozoon wenyonii 533
Ephemeral fever, *see* Bovine ephemeral fever
Ephemeroptera 28
Epicuticle 73, *74*
Epidemic haemorrhagic fever 505
Epidermis *74*, 75
Epidermoptes bilobatus 409
Epigynium 386
Epipharynx 55
Epizootic haemorrhagic disease of deer 509–10
Equine infectious anaemia 221
Eretmapodites 113, 115, 141
Eretmapodites chrysogaster and Rift Valley
       fever 503
Eriocheir japonicus 14
Eriophyoidea 414
Euproctis edwardsi 36
Euproctis similis 36
Euroglyphus maynei 407
Eusimulium 585
Eusynanthropy 250

Eutamias and relapsing fever 543
Eutrombicula and *Rickettsia tsutsugamushi* 526
Eutrombicula afreddugesi 415
Eutrombicula sarcina 415
Exocuticle *74*
Exophagy 131
Exophily 131
Exopterygota 27
Exuviae 76
Eyeflies, *see* Chloropidae

Falculifer rostratus 410
Family 8
Fannia 263–4
Fannia canicularis 249–51, *251*, 263, *264*
Fannia femoralis 250, 263
Fannia scalaris 249–50, *264*
Fanniidae 52, 249–50, 263–5
Felicola subrostrata 362, 376, 379
    female *375*
Femur 25, 385, *395*
Ficalbia 114–5
Ficalbiini 114
Filariasis, *see* lymphatic filariasis and
      onchocerciasis
Filariidae 638
Filarioidea 638
Filth flies in poultry houses 263–4
Flaviviridae 489
Flavivirus 489
    tick-borne
Flea, *see* Siphonaptera
Food canal
Forcipomyia 152
Forcipomyiinae 152
Foregut 76, *78*
Forensic entomology 286–7
Fowl paralysis 454
Fox, reservoir of leishmaniasis
Francisella tularensis 221, 517, 553–4
    subspecies of 554
Frons 25
Frontal ganglion 77
Frontal suture *42*
Fuscuropoda vegetans 264

Galeodes arabs 22
Galindomyia 113
Gamasida, *see* Mesostigmata

Ganglia 83
Gasterophilinae 53, 292, 303–9
Gasterophilus 292, 303–9
    biology and life cycle 304
        eggs 304–5, *305*
        larvae 305–6
        pupal duration 306
        bionomics of adult 306–7
        veterinary importance 307–9
Gasterophilus haemorrhoidalis 305–9 passim,
        305, 306
Gasterophilus inermis 303–5
Gasterophilus intestinalis 303–9, *303, 305, 306,*
        307
Gasterophilus nasalis 303–9, *305, 306*
Gasterophilus nigricornis 303–4
Gasterophilus pecorum 303–9 passim
Gavia immer (Common loon) 203
Gazelles 623
Gena 25
Gene's organ 445
Genu 385, 395
Genus, plural genera 7
Geochelone elephantopus 449
Gerbil, great, *see Rhombomys opimus*
Gigantodax 193, 195
Giraffe and trypanosomes 623
Glaucomys volans 520
Gliricola porcelli 373
Glossina 225, 227–8, 238–40, 315
    ageing of 86
    bionomics of 235–42
        activity cycle 235
        dispersal 242
        feeding 70
        habitat 240–1
        host finding 70, 235–7
        host selection 238
        mating 232–3
        resting places 241–2
        sampling adults 237
    control 242–3
    digestion of blood meal 78, 233–4
    distribution 225–8
    energy production 79, 234
    feeding 233–4
        preferences, evolution of 238–9
        sensory receptors 70–1
    life cycle 228
        adult emergence 231–2
        egg 229

    larva 229–30
    pupa 230, *231*
    medical and veterinary importance 243
    mycetome 80
    reproductive cycle 228–9
    structure of adult
        antenna *42, 226*
        female reproductive system *229*
        mouthparts 65, *66*
        peritrophic membrane *615*
        wing *45, 228*
    subgenera 225
    symbionts in 230, 518
    trypanosome development in 614–6
    and trypanosomiasis 243, 612–22
Glossina austeni 227–8, 232, 239
    adenosine nucleotides, response to 71
    ageing 86
Glossina brevipalpis 228, 235, 239–40
    trypanosomes 620–1
Glossina fusca 239
Glossina fuscipes 228, 235, 238, 241–2, *226*
    and trypanosomes 617–8
Glossina fuscipes fuscipes 235, 238, 241–2
Glossina longipalpis 228, 235
Glossina longipennis 228, 232, 239–40
    and trypanosomes 620
Glossina morsitans 225–43 passim, *231*
    kairomones 70
    trypanosomes 617–9
Glossina morsitans centralis 225, 228, 230,
        240–1
Glossina morsitans morsitans 228, 232, 235–43
        passim
Glossina morsitans submorsitans 228, 240–2
Glossina nigrofusca 226
Glossina pallidipes 227–8, 232–43 passim
    and trypanosomes 618–9
Glossina palpalis 225, 226, 228, 232–3, 237,
        240–3
    and trypanosomes 618–9
    response to adenosine nucleotides 71
Glossina palpalis gambiensis 225, 240
Glossina palpalis palpalis 232, 237, 240, 243
Glossina pallicera 228
Glossina swynnertoni 227–8, 233, 238–40
    and trypanosomes 618–9
Glossina tabaniformis 239
Glossina tachinoides 228, 237, 241–2
    and trypanosomes 618–9
Glossina vanhoofi and trypanosomes 621

Glossinidae 52, 225–247
   classification 225
Glucose-6-phosphate dehydrogenase and
   malaria 568
Glycyphagus destructor 407–8
Glycyphagus domesticus 407
Gnathosoma 384, 385
Goat (*Capra hircus*)
   arthropod parasites of
      Chorioptes spp. 399
      Damalinia spp. 378
      Linognathus spp. 370
      Lipoptena capreoli 320
      Oestrus ovis 295
      Raillietia caprae 435
      Sarcoptes scabiei 392
      Wohlfahrtia magnifica 285
   diseases transmitted by arthropods
      heartwater 530
      Nairobi sheep disease 504
      Q fever 529
      theileriasis 603
      trypanosomiasis 620
Gonderia, see Theileria
Goniocotes gallinae 374, 376
Goniodes dissimilis 374, 376
Goniodes gigas 376, 379
Gonopore 27
Gonotrophic concordancy 90, 572
Gouldian finch 437
Grahamella 530
Growth in insects 75–6
Gymnopais 198
Gymnopais dichopticus 204
Gynandromorph 84
Gyropidae 373
Gyropus ovalis 373

Habronema 638
Habronema majus 683
Habronema muscae 683
Hadrurus arizonensis 17
Haemagogus 113, 115
   eggs, drought resistant 117
Haemagogus spegazzini and yellow fever 493
Haemaphysalis 460
   Crimean-Congo haemorrhagic fever 504
   Rickettsia sibirica 524
   Babesia motasi 595
Haemaphysalis asiaticum and *Rickettsia*
   sibirica 524

Haemaphysalis concinna
   Borrelia burgdorferi 538
   Rickettsia sibirica 524
Haemaphysalis hoodi 480
Haemaphysalis leachi 460
   Babesia canis 596
   Rickettsia conorii 525
Haemaphysalis leporispalustris and *Borrelia*
   burgdorferi 539
Haemaphysalis longicornis 460, 472
   Babesia gibsoni 596
   stages:
      female *460*
      larva *468*
      male *460*
      nymph *468*
   Theileria orientalis 602
Haemaphysalis punctata 460
   Babesia major 594, 599
   Theileria orientalis 602
Haemaphysalis spinigera 460
   Kyasanur Forest disease 497
Haematobia 261
   bibliography 251
Haematobia atripalpis and *Parafilaria*
   multipapillosa 682
Haematobia exigua 249–50, 259, 261
   habronemiasis 683
Haematobia irritans 248, 259
   biology and bionomics 261–2, 380
   economic importance 262
   Stephanofilaria stilesi 681
Haematobia thirouxi titillans and
   Stephanofilaria assamensis 681
Haematomyzus elephantis 361
Haematomyzus hopkinsi 361
Haematophagy 59
Haematopinus 362, 365, 368–9
Haematopinus asini 362, 368, *369*, 379
Haematopinus eurysternus 362, 368–70, 379
Haematopinus quadripertusus 362, 368–9, 379
Haematopinus suis 362, 368–9, 378–80
Haematopinus tuberculatus 362, 368, 379–80
Haematopota 211, *212*, 214–5, *215*
Haematopota pluvialis 216, 218
   Trypanosoma theileri 627
Haematopota tristis 220
Haematoxenus, see Theileria
Haemobartonella felis 533
Haemobartonella muris 533
Haemocoele 82

Haemogamasinae 425
Haemogamasus 425-6
Haemolymph 82
Haemoproteidae 558, 583-4
Haemoproteus 558, 584
Haemoproteus columbae 584
Haemoproteus metchnikovi 221, 584
Haemospororina 558, 583
Halacaroidea 383
Halarachnidae 425, 433
Halarachninae 425, 433
Haller's organ *452*
Haltere 37
Hantavirus 505
Hartebeeste 623
Haustellum *61*, 63
Head, structure of *25*
Heart 82
Heartwater, *see Cowdria ruminantium*
Heartworm, *see Dirofilaria immitis*
Heizmannia 113
Hematoxenus, see Theileria
Hemimetabola 27
Hemiptera *32*, 344-60
  blood-sucking
    mouthparts 67, *68*
Hemisynanthropy 250
Hepatitis B virus, in *Cimex* 350
Hepatocystis 558, 583-4
Hepatocystis brayi 584
Hepatocystis kochi 583
Heterocampa mantea 36
Heterodoxus spiniger 362, 373
Heterohyrax 632
Hexapoda 23
Hindgut 76, *77*, *78*
  water absorption 78
Hippelates collusor 52
Hippelates pusio 52
Hippobosca 317, 320
  bionomics 321
  control 321
  geographical distribution 320
  hosts of 315, 317
Hippobosca camelina 320
Hippobosca equina 316, 320-1
Hippobosca longipennis 317, 320-1
Hippobosca macula 321
Hippobosca rufipes 320
Hippobosca variegata 320
Hippoboscidae 53, 315-25, 422

bacteriome 315
  symbiotic organisms 80
Hippoboscinae 315
HIV mechanical transmission 62, 350
Hodgesia 114
Hodgesiini 114
Holarctic region 9
Holoconops 152, 155, *see also Leptoconops*
Holometabola 28
Holoptic 23
Holothyrida 423
Homonym 8
Hoplopleuridae 362
Horse (*Equus caballus*)
  arbovirus diseases in
    African horse sickness 508
    equine encephalitides 499
    vesicular stomatitis 505
  bacterial diseases transmitted by
    arthropods
    Ehrlichia spp. 530
    tularaemia 553
  helminths transmitted by insects
    habronemiasis 683
    onchocerciasis 678-9
    parafilariasis 682
    Setaria spp. 682-3
  lice on
    Damalinia equi 362
    Haematopinus asini 362
  mites infesting
    Chorioptes bovis 399
    Psoroptes spp. 396
    Sarcoptes scabiei 392
    trombiculid larvae 415
  myiasis agents affecting
    Gasterophilus spp. 305
    Rhinoestrus purpureus 296
    Wohlfahrtia magnifica 285
  protozoal diseases transmitted by
    arthropods
    babesiosis 595
    trypanosomiasis 620
Horsefly, *see Tabanus*
Housefly, *see Musca domestica*
Humans
  allergies to astigmatic mites 407
  arboviruses
    Chikungunya 500
    Crimean-Congo haemorrhagic fever 504
    dengue fever 493

Humans—*continued*
  arboviruses—*continued*
    encephalitides 496
    Kyasanur Forest disease 497
    O'Nyong-nyong 500
    Oropouche virus 502
    Ross River virus 500
    sandfly fever 503
    Sindbis virus 501
    West Nile virus 495
    yellow fever 490
  bacterial diseases transmitted by
    arthropods
    Lyme disease 537
    plague 544
    relapsing fevers 541
    tularaemia 553
  ectoparasites
    Auchmeromyia senegalensis 280
    Demodex spp. 419
    Ornithodoros moubata 444
    Pediculus spp. 366
    Pthirus pubis 67
    Pyemotes tritici 422
    Sarcoptes scabiei 387
    trombiculid larvae 415
  filarial diseases
    loiasis 671–4
    lymphatic filariasis 639–40
    minor filarial worms 674–6, 680, 683–4
    onchocerciasis 656–67
  myiasis in
    Chrysomya bezziana 274
    Cochliomyia hominivorax 271
    Cordylobia anthropophaga 282
    Dermatobia hominis 309
    Oestrus ovis 296
    Wohlfahrtia spp. 285
  protozoal diseases
    Chagas' disease 624
    leishmaniasis 628–9
    malaria 559
    trypanosomiasis 617
  rickettsial diseases
    Carrion's disease 531
    epidemic typhus 518
    fièvre boutonneuse 525
    murine typhus 521
    Q fever 528
    rickettsial pox 525
    Rocky Mountain spotted fever 523
    scrub typhus 526
    tick-borne typhus 524–5
    trench fever 528
Humeral cross-vein 44
Humeral pits 156
Hyalomma 464, 607
  Babesia equi 595
  Crimean-Congo haemorrhagic fever 504
  Theileria annulata 608
Hyalomma aegyptium mouthparts *465*
Hyalomma anatolicum
  Crimean-Congo haemorrhagic fever 504
  Theileria hirci 603
Hyalomma anatolicum anatolicum 464
  Crimean-Congo haemorrhagic fever 504
  Theileria annulata 608
Hyalomma anatolicum excavatum *465, 471*
  Theileria annulata 608
Hyalomma detritum 408
  Theileria annulata 608
Hyalomma dromedarii 471
  African horse sickness 509
Hyalomma marginatum 464, 468
  Crimean-Congo haemorrhagic fever 504
Hyalomma savignyi 467, 471
Hyalomma scupense 468
  Theileria annulata 608
Hybomitra *217*, 218, 680
Hybomitra laticornis and *Elaeophora*
  schneideri 680
Hydrachnidia 414
Hydrachnellae 383
Hydrotaea irritans 250, 262–3
Hymenolepis diminuata 684–5
Hymenolepis nana 684–5
Hymenoptera 34, *35*
Hypoaspidinae 425
Hypodectes propus 409
Hypoderma 297–303
  control 302–3
Hypoderma bovis 292, 297–302
  biology and life cycle 298–300
  larva *300*
  seasonal cycle *301*
  veterinary and medical importance 301–2
Hypoderma diana 297, 302
Hypoderma lineatum 292, 297–301
  biology and life cycle 298–300
  female *298*
  larva *300*
  seasonal cycle *301*

Hypoderma bovis—*continued*
   veterinary and medical importance 301–2
Hypodermatinae 297–303
   hosts of 296
Hypognathous 23
Hypopharynx 55
Hypopleuron *see* meron 41, *43*
Hypopus, plural hypopodes 386
Hypostome 440, *441*
Hyrax 628
Hysterosoma 384, *385*

Idiosoma 384, *385*
Impala 623
Imperata cylindrica 527
Indian cattle, *see* zebu
Insecta 23–38
   classification 27–38
   external structure *24*, 23–7
     abdomen 27
     head 23, *25*
     leg *26*
     thorax 25
     wing *26*
   growth 75
   internal structure
     circulatory system *82–3*
     digestive system 76–9
     female reproductive system 84, *85*
     male reproductive system 84, *85*
     nervous system *83*, 84
     respiratory system 80, *81*
     symbionts 80
     vitamin requirements 80
Instar 75
Interbifid space 62
Ischnocera 361, 372, 376–9
Ixodes 459–60, *459*, 595
Ixodes dammini 458, 470–1, *see also Ixodes*
   scapularis
   Babesia microti 602
   Borrelia burgdorferi 537–40
Ixodes hexagonus 538
Ixodes holocyclus 459, 469, 471
   Borrelia burgdorferi 538
   Coxiella burnetii 528
   paralysis caused by 480–1
   Rickettsia australis 525
Ixodes neotomae 539
Ixodes pacificus 538–9

Ixodes persulcatus 458
   Borrelia burgdorferi 538
   tick-borne encephalitis 498
Ixodes ricinus 458, 466–76 passim
   Babesia divergens 594, 599, 601–2
   Babesia jakimovi 595
   Borrelia burgdorferi 538–9
   Coxiella burnetii
   louping ill 498
   tick-borne encephalitis 498
Ixodes rubicundus, paralysis caused by 481
Ixodes scapularis 537
Ixodida 423, 440
Ixodoidea 423
Ixodidae 440–1, 458–85
   biology and behaviour
     aggregation 471–2
     feeding 470
     mating 471–2
     pheromones 471–2
     water balance 75, 470–1
     water elimination 470
   economic importance 475, 480
   genera of 458–64
   life cycle
     adult 467
     larva 467
     nymph 467
     oviposition 465–7
     variations on 467–9
   paralysis caused by 480–1
   principles of control 477
     bait stations 479
     dipping 478
     pasture spelling 478
     pheromones 479
     repellents 479
     resistance to acaricides 478
     resistant hosts 478–9
     vaccine 479
Ixodides, *see* Ixodida
Ixodoidea, *see* Ixodida

Japanese encephalitis 496
Johnbelkina 114

Kairomones and host finding
   carbon dioxide 70
   octenol (1-octen-3-ol) 70

Kala-azar 630
Katepisternum 41, *43*
Kibwezi group of *Simulium damnosum* 96, *97*
Kieferulus longilobus 47
Kinetoplastida 612
Knemidokoptes gallinae 401–2
Knemidokoptes mutans 400–2
    male *404*
Knemidokoptes pilae 401–3
    female *402, 403*
Knemidokoptidae 386, 400–4
Kobus 435
Kudu 623
Kyasanur Forest disease 497

Labellum 57
Labial salivary gland 61
Labium 23
Labrum 24
Lacinia 66, *67*
La Crosse virus 502
Laelapidae 425
Laelapinae 425
Lagenidium giganteum 126
Laminosoptidae 385, 405–7
Laminosioptes cysticola 405, *407*
Lasiohelea 154
    larva *161*
    pupa *162*
    wing *155*
Lasiohelea anabaena 160
Lasiohelea cornuta 161, *162*
Lasiohelea sibirica 155, 168, 170
Lasiohelea taiwana 160, 163–4
Lasiohelea townsvillensis and *Onchocerca*
    gibsoni 677
Latrodectus hasselti 19
    antivenom 19
Latrodectus indistinctus 19
Latrodectus mactans complex 19
Latrodectus mactans 19
Latrodectus tridecimguttatus 19
Laverania 565
Leishmania 612, 627, 630
    cycle in phlebotomine 631–2
    forms 612
    reservoir hosts 632–3
    taxonomy 628–30
    transmission 631–2
    vectors 632–3, 628–9

Leishmania aethiopica complex 628, 632
Leishmania amazonensis 629–30, 632–3
Leishmania braziliensis complex 629, 631, 633
Leishmania chagasi 628, 630, 632
Leishmania donovani complex 628, 630, 632
Leishmania guyanensis complex 629, 633
Leishmania infantum 628, 630, 632
Leishmania major complex 628, 631–2
Leishmania mexicana complex 629–33 passim
Leishmania panamensis 629, 633
Leishmania peruviana 633
Leishmania tropica complex 628, 632
Leishmaniasis 627–34
    cutaneous 630, 632
    mucocutaneous 630, 632–3
    visceral 630, 632
Lepidoglyphus destructor 408
Lepidoptera *36*
Lepisma saccharina 27
Lepismodes inquilinus 27
Leptocimex boueti 344
Leptoconopinae 152
Leptoconops 152, 154
    larva *163*
    pupa *163*
    wing *155*
Leptoconops albiventris 163
Leptoconops australiensis 155
Leptoconops becquaerti 155, 163–9 passim
Leptoconops lucidus 163
Leptoconops mediterraneus 163
Leptoconops specialis 163
Leptoconops spinosifrons 155, 165, 169
    larva *163*
    pupa *163*
Leptoconops torrens 155, 167
Leptopsyllidae 323
Leptopsylla segnis 326, 337
    plague 250
    Rickettsia typhi 522
Leptotrombidium 526
Leptotrombidium akamushi 415, 526
    Rickettsia tsutsugamushi 526
Leptotrombidium deliense 415, 416
    Rickettsia tsutsugamushi 526–7
    larva *416*
Leptotrombidium pallidum and *Rickettsia*
    tsutsugamushi 526–7
Leptotrombidium scutellare
    Epidemic haemorrhagic fever 505
    Rickettsia tsutsugamushi 526–7

Lepus californicus 539
Leucocytozoidae 558, 585–6
Leucocytozoon 558, 584–6
Leucocytozoon caulleryi 585–6
Leucocytozoon simondi 585–6
Leucocytozoon smithi 585–6
Lewisellum 199
Limatus 14
Linnaeus 7
Linognathidae 370–1
Linognathus 362, 370
   cattle 370–1
   sheep 370
Linognathus africanus 362, 370, 380
Linognathus ovillus 362, 370
   female *371*
Linognathus pedalis 362, 370, 378–80
Linognathus setosus 362, 370, 379
Linognathus stenopsis 362, 370, 380
Linognathus vituli 31, 362, 370, 380
   female *31*
Linshcosteus 350
Lipeurus caponis *374, 376, 379*
Liponyssoides sanguineus 429–31
   and rickettsial pox 431, 526
Lipoptena 317
Lipoptena capreoli 320
Lipoptena cervi 321
Lipopteninae, *see* Melophaginae
Literature, introduction to 9–10
Loa loa 221, 638
   development in *Chrysops* 672
   ecology of vectors 673–4
   strains 672
   in vertebrate host 672
Loiasis 671–4
   disease 671–2
   epidemiology 673–4
Lonomia archilous 36
Louping ill 498
Loxosceles laeta 20
Loxosceles reclusa 20
Lucilia 269, 275–6
   air sacs *81*
Lucilia cuprina 271, 276–9
   biology and bionomics 277–8
   myiasis in sheep 276–9
   structure of adult:
     abdominal air sacs 76
     labellum *63*
     tarsal claws *26*

   wing *270*
Lucilia sericata 250, 256, 276–7, 287
Lutzia 122
Lutzomyia 177–8, 185, 627, 632
   Bunyaviridae 501
   feeding on plasma 70
Lutzomyia anduzei and *Leishmania*
   guyanensis 629
Lutzomyia betrani 186
Lutzomyia californica 186
Lutzomyia carrerai and *Leishmania*
   brasiliensis 629
Lutzomyia flaviscutellata 185, 187
   Leishmania mexicana complex 629, 633
Lutzomyia gomezi 186
   Leishmania panamensis 629
   vesicular stomatitis 505
Lutzomyia llanos martinsi and *Leishmania*
   brasiliensis 629
Lutzomyia longipalpis 178, 182–3, 185–7
   and leishmaniasis 628, 631–2
Lutzomyia olmeca 186, 633
   and *Leishmania mexicana* 629
Lutzomyia panamensis and *Leishmania*
   panamensis 629
Lutzomyia peruensis and *Leishmania*
   peruviana 633
Lutzomyia pessoana 187
Lutzomyia sanguinaria and *vesicular*
   stomatitis 505
Lutzomyia spinicrassa and *Leishmania*
   brasiliensis 629
Lutzomyia trapidoi 186–7
   Leishmania panamensis 629, 631
   vesicular stomatitis 505
Lutzomyia trinidadensis 186
Lutzomyia umbratilis
   Leishmania brasiliensis 629
   Leishmania guyanensis 629
Lutzomyia verrucarum 633
   Bartonella bacilliformis 530–1
   Leishmania peruviana 633
Lutzomyia vespertilionis 182
Lutzomyia vexator 178, 180–1, 183
Lutzomyia wellcomei 185
   Leishmania brasiliensis 629
Lutzomyia whitmani
   Leishmania brasiliensis 629
   Leishmania guyanensis 629
Lutzomyia ylephilator
   Leishmania panamensis 629

Lutzomyia ylephilator – continued
    vesicular stomatitis 505
Lutzomyia yucumensis and *Leishmania*
    brasiliensis 629
Lycosa 20
Lyme disease 537–40
    control 540
    in Europe 539–40
    in north eastern USA 538–9
    in west North America 539
Lymphatic filariasis 638–53
    epidemiology 649–52
    comparison with malaria 651–2
    infection rate 650
    mosquitoes and parasites 647–9
        cibarial and pharyngeal armatures 648
        facilitation 648–9
        limitation 648–50
        susceptibility to parasites 646–7
    nature of the disease 639–40
    parasites involved 639–43
        geographical distribution of 639, 644–5
        infective larvae 642–3
        life cycle 641
    principles of control 652–3
    transmission, discovery of 640–1
    transmission rate 650–1
        Calcutta 650
        Jakarta 650
        Rangoon 650
    vectors 643–9
Lynchia maura 584
Lynxacarus radovsky 409
Lytta vesicatoria 32, *33*

Macaca mulatta 434
Macaca radiata 497
Macrocheles glaber 424
Macrocheles muscaedomesticae 264, 424
Macronyssidae 425, 431–3
Malaria 4, 559–82
    control strategy
        ecological types 579
        personal protection 579–80
        principles of control 580–1
    distribution and current incidence 559–60
        in the 19th century *589*
        in 1992 *560*
    epidemiology 568–74
        density of *Anopheles* 573–4

host preferences of *Anopheles* 569, 572
longevity of *Anopheles* 572–3
sporogony and temperature 563
vectors of stable malaria 569, *570*, *571*
in Europe 88–9
in Holland 90
in humans
in non-human primates 567
malaria parasites 565–7
model of transmission 574–8
    Anopheles vectorial capacity 578
    basic reproduction rate 575–6
    measure of stability 575
    sporozoite rate 576–7
treatment and prevention 581–2
    drugs 581
    vaccines 582
Malaya 114
    fed by ants 115
Mallophaga 30, *31*, 361, 371–80
    mouthparts of 55, *56*
Malpighian tubes 78, *79*
Mandible 23
Mange
    chorioptic 398
    demodectic 418
    diagnosis of 390
    notoedric 393
    otodectic 399
    psoroptic 395
    sarcoptic 392
Mansonella ozzardi 169, 638, 674–5
Mansonella perstans 638
    Culicoides austeni 59, 675
    Culicoides grahamii 59, 675–6
Mansonella streptocerca 638
    Culicoides austeni 59, 675
    Culicoides grahamii 59, 675–6
Mansonia 114–5
    Brugia malayi 644
    egg 117
    larva
    pupa *49*
Mansonia africana 121, 125, 135
Mansonia annulata and lymphatic filariasis 646
Mansonia bonneae and lymphatic filariasis 646, 650
Mansonia dives and lymphatic filariasis 646–50
    passim
Mansonia uniformis 121, 125 135
    larval siphon *126*

Mansonia uniformis — continued
    lymphatic filariasis 646
    pupal trumpet *126*
Mansoniini 114
Mansonioides and lymphatic filariasis 646
Maorigoeldia 114
Margaropus winthemi 467
Marmota sibirica 552
Mastomys coucha 551, *see also Mastomys*
    natalensis
Mastomys natalensis 337, 551
Maxilla 23
Media vein 44
Melinus minutiflora 478
Meloidae 32
Melophaginae 315
Melophagus 317
Melophagus ovinus *316*, 315–21, 518
    adult *316*
    bionomics 318–9
    control 32
    life cycle 317–8
    Trypanosoma melophagium 627
    economic importance 319–20
Menacanthus stramineus 372, 373, 376, 379–80
Menoponidae 372–3
Menopon gallinae 372, 373, 380
Meriones erythrourus 187
Meriones libycus 552
Meriones persicus 552
Meriones tristrami 552
Meriones vinogradovi 552
Meron 41, *43*
Mesocyclops 114
Mesocyclops aspericornis 126
Mesoknemidokoptes laevis gallinae 402
Mesopleuron, *see* anepisternum 41, *43*
Mesostigmata 384–5, 423–37
Mesothorax 25
Metacnephia 193, 195
Metapneustic 123
Metapodosoma 354
Metastigmata, *see* Ixodida
Metatarsus 25
Metathorax 25
Micropyle 84
Microtus fortis pellicieus 526
Microtus pennsylvanicus 540
Mictyris livingstonei 168
Midgut 76, *77, 78*
    enzymes 78

Millipede 14
Mimosa nigra 241
Mimomyia 114
Mites, *see* Acari
Moniezia expansa 410
Monoculicoides 167
Moose 680
Mosquitoes, *see* Culicidae
Moulting fluid 75
Mouthparts
    anopluran 68
    cyclorrhaphan, blood-sucking 64
    cyclorrhaphan, lapping 61
    flea 66
    hemipteran, blood-sucking 67
    mallophagan 55
    nematoceran, blood-sucking 56
    tabanid 59
Mucidus 122
Mules' operation 279
Murray Valley encephalitis 496–7
Mus musculus 526
Musca
    bibliography 251
    energy sources 79
Musca autumnalis
    Parafilaria bovicola 682
    Parafilaria sp. 684
    Thelazia 684
Musca conducens and *Stephanofilaria*
    assamensis 681
Musca crassirostris 263
Musca domestica 7, 248–57, 263–4, 271
    adult *249*, 253–4
    control 264–5
    Habronema species 683
    life cycle
        duration 254
        egg 251
        larva 252
        pupa 253
    medical importance 254–6
    mouthparts *61*
    myiasis 255
    oviposition 253–4
    population growth 254
    wing *44*, *250*
Musca domestica domestica 251
Musca domestica calleva 251
Musca domestica curviforceps 251
Musca lusoria and *Parafilaria bovicola* 682

Musca nevilli and *Parafilaria bovicola* 682
Musca sorbens complex 248, 256–9
Musca textilis 643
Musca vetustissima *257*, 256–9, 424
    biological control 259
    bionomics 257–9
    life cycle 257–8
    phenology 258–9
    survival 258
Musca xanthomelas and *Parafilaria*
    bovicola 682
Muscidae 52, 248, 251–263
Muscina 287
Muscina stabulans 263–4
Muscinae 248
Muscoidea 50, 52
Mycetome 80
Mygalomorphae 18, *19*, 20
Myiasis 268
Myobia musculi 422, *423*
Myobiidae 416, 422
Myocoptes musculinus 409
Myxomatosis 489

Nairobi sheep disease 504
Nairovirus 501
Nannomonas 613–4, 617, *see also*
    Trypanosoma
Nearctic region 9
Nematocera 47–50
    antennae *42*
    larvae *48*
    mouthparts *57*
    pupae *49*
    wings *45*
Nematoda, *see Brugia* spp., *Dirofilaria* spp.,
    Mansonella spp. *Onchocerca* spp.,
    Parafilaria spp. *Setaria* spp., *Thelazia*
    rhodesi, *Wuchereria bancrofti*
Nemorhina 225, 228, 238–40, *see also Glossina*
Neocnemidocoptes gallinae 402
Neohaematopinus sciuropteri and *Rickettsia*
    prowazekii 520
Neoptera 28
Neoschongastia americana 415
Neotenin 84
Neotoma 356
Neotrombicula autumnalis 415
Neotropical region 9
Nerve cord 83

Nervous system *83*, 84
Nile group of *Simulium damnosum* 97
Noctuidae 35
Nomenclature, taxonomic 7
Nosopsyllus 327
Nosopsyllus fasciatus *325*, 329–32, 338
    and plague 550, 553
    Trypanosoma lewisi 627
Notoedres 392–4
    life cycle 393
Notoedres cati 392–4
Notoedres muris *393*, 392–3
Notoedres musculi 392
Notopleuron 41, *43*
Nuttalliellidae 440
Nycteribiidae 53

Occipital foramen 23
Ocellar triangle *42*
Ocellus 23
Odocoileus hemionus 505, 680
Odocoileus virginianus 475, 509, 538
Odonata 28
Oedemagena tarandi, *see Hypoderma tarandi*
Oesophagus 77
Oestridae 53, 292–314
    hosts of 292
Oestrinae 292, 295–7
Oestrus ovis 292–6
    bionomics of 295–6
    female *293*
    hosts of 292
    larva *294*
    life cycle 295
    veterinary and medical importance 296
    wing *293*
Ommatidia 192
Omsk haemorrhagic fever 498
Onchocerca spp. parasitizing domestic animals
    bovine onchocercas 677–8
    economic importance 676
    equine onchocercas 678–9
    species of 676–9
    vectors 676–9
Onchocerca armillata 676–7
Onchocerca cebei 677
Onchocerca cervicalis 676–8
Onchocerca dukei 677–8
Onchocerca gibsoni 676–7
Onchocerca gutturosa 676–9 passim

Onchocerca indica 677
Onchocerca lienalis 676–7
Onchocerca ochengi 676–8
Onchocerca reticulata 676–8
Onchocerca sweetae 677
Onchocerca volvulus 638, 656–67
 control 663–4
  chlorinated hydrocarbon era 664
  integrated control 664
  pre DDT 664
  temephos period 664
 development in *Simulium* 658–9
 disease 657
 distribution in world 656
 epidemiology 663
  annual transmission rate 663
  dispersal of infected simuliids 660, 667
 Onchocerciasis Control Pro-
  gramme 664–7, *665, 666*
  insecticide selection *666*
  map of area *665*
 simuliid vectors
  in Central America 662
  Simulium damnosum complex 59, 660–1
  Simulium neavei group 661–2
Onchocerciasis, domestic animals 676–9
 human 656–67
Onchocercidae 638
Onthophagus binotis 259
Onthophagus ferox 259
Onthophagus gazella 259
Onthophagus granulatus 424
O'Nyong-nyong virus 500
Oocyte 84
Ootheca 29
Ophionyssus natricis 431–3
Ophyra spp. 263–4
Ophyra aenescens 264
 introduced into Europe 264
Opifex 113
Opifex fuscus 127, 133
Opisthosoma 18, 384, *385*
Orbivirus 502
Orchopeas howardii and *Rickettsia
  prowazekii* 520
Order 8
Oribatida 383–6 passim, 410
Oriental region 9
Ornithodoros 442, 444, 537
 Borrelia 541–3
Ornithodoros apertus 542

Ornithodoros compactus 542
Ornithodoros coriaceus 421
Ornithodoros hermsi 442, 543
Ornithodoros lahorensis 449
Ornithodoros moubata 444–8, 541–2
 African swine fever 448, 510
 bionomics 447–8
 Borrelia duttoni in 448, 541–2
 fecundity 447
 feeding 445
 female *444*
 mating 446–7
 life cycle 445
 medical and veterinary importance 448
 oviposition 445
 species complex 542
Ornithodoros moubata apertus 542
Ornithodoros moubata moubata 542
Ornithodoros moubata porcinus 542
 African swine fever 510
Ornithodoros porcinus avivora
Ornithodoros porcinus domesticus 542
Ornithodoros porcinus porcinus 542
Ornithodoros puertoricensis and African swine
  fever 510
Ornithodoros rudis 543
Ornithodoros savignyi 444, 448–9
 adult *449*
Ornithodoros tholozani 446
Ornithodoros turicata 442, 543
Ornithomyinae 315
Ornithonyssus bacoti 431
 Rickettsia akari 522, 526
Ornithonyssus bursa 431–2
 female *426, 427*
Ornithonyssus sylviarum 430–2
Oropouche virus 502
Oropsylla (Diamanus) montana 551
Oropsylla Thrassis bacchi 551
Orthopodomyia 114
Orthopodomyiini 114
Orthorrhapha 50
Oryzaephilus surinamensis 422
Oryzomys 185
Ostia 82
Otobius 442
Otobius lagophilus 442
Otobius megnini 441–4
 bionomics 443–4
 larva *441*
 life cycle 443

Otobius megnini—continued
   nymph *442*
Otodectes 395, 399
Otodectes cynotis 399
   male *401*
Otomys irroratis and *Rickettsia conorii* 525
Ototylomys phyllotis 633
Ovariole 84, *85*
Ovary 84, *85*
Oviduct 84, *85*
Oviparity 84
Oviporus 386
Ovoviparity 84
Ovum 84
Owen Falls 664
Oxygen debt 82

Pachydesmus crassicutis 15
Paddles, pupal 124
Paederus 32
Paederus cruenticollis *33*
Paederus sabaeus *33*
Palaearctic Region 9
Palaeoptera 28
Palmate hairs 118, *119*
Pangonia 212
Pangoniinae 214
Panstrongylus 357
Panstrongylus megistus 351–8 passim
   female *351*
     Trypanosoma cruzi 625–6
Papilio canopus 36
Parafilaria 638
Parafilaria bovicola 682
Parafilaria multipapillosa 682
Paragonimus ringeri 14
Parahaemoproteus 584
Parahaemoproteus nettionis 584
Paraheterodoxus insignis 31
Paralysis, caused by ticks 480–1
Paraneoptera 28
Parasitiformes, *see* Anactinotrichida 423
Paris green 123
Pectine 16, *17*
Pedicinus 363
Pediculus 362–5 passim
   control of on humans 368
   egg 366
   fecundity 366
   longevity 366

   medical importance 369
   nymph 366
   population dynamics 367
   symbionts of 80
Pediculus capitis 3, *363*, 366, 368
   Rickettsia prowazekii 519
Pediculus humanus 3, 365–8, 379, 540
   Borrelia recurrentis 540
   Coxiella burnetii 529
   egg *365*
   epidemic typhus 518–22
   female terminalia *364*
   male terminalia *364*
   trench fever 528
Pedipalps 18
Pericardial sinus 82
Periplaneta americana 29
Periplaneta australasiae 29, *30*
Perispiracular glands 123
Peritrophic membrane 77, *78*
   immunological attack
Peromyscus leucopus 538, 540, 602
Peromyscus maniculatus 539
Peromyscus truei 539
Persicargas, see also Argas
   paralysis caused by 454, 481
Phacochoerus aethiopicus, see warthog
Phaenicia, see Lucilia
Phalanger maculatus 433
Pharate stage 75
Pharyngeal pump 57
Pharyngobolus africanus 292
Pharynx 77
Philaematomyia insignis, see Musca crassirostris
Philoliche magretti 624
Philoliche zonata 624
Philopteridae 372, 376
Phlebotominae 177
   behaviour
     feeding 183
     mating 182–3
     resting places 186
   bionomics 185–7
   genera in 183–5
   life cycle 178–82
     duration of 181–2
     egg *179*
     larva 48, *180*
     pupa *181*
   medical and veterinary importance 188

Phlebotominae—continued
  structure
    adult 178
    cibarium 184
    peritrophic membrane 77
    wing 45, 179
Phlebotomus 177-8
  Bunyaviridae 501
  female 178
  male terminalia 182
  phlebotomus fever 503
  wing 179
Phlebotomus andrejevi 186
Phlebotomus argentipes 182, 185
  leishmaniasis 628, 632-3
Phlebotomus ariasi 183, 185, 187
  leishmaniasis 628, 632
Phlebotomus caucasicus 182, 185-7
  and leishmaniasis 632
Phlebotomus chinensis and leishmaniasis 628
Phlebotomus guggisbergi 185
Phlebotomus langeroni 187
Phlebotomus longipes 178, 181-6 passim
  and leishmaniasis 628, 631-2
Phlebotomus major and leishmaniasis 628
Phlebotomus martini and leishmaniasis 628,
    632
Phlebotomus mongolensis 186
Phlebotomus orientalis 182, 185-6
  and leishmaniasis 631
Phlebotomus papatasi 182-8 passim, 490
  cibarium 184
  DDT resistance 188
  leishmaniasis 628, 631-2
  sandfly fever 503
Phlebotomus pedifer 185
  leishmaniasis 628, 632
Phlebotomus perfiliewi
  leishmaniasis 628
  sandfly fever 503
Phlebotomus sergenti 185, 628, 632
Phlebovirus 503
Phoniomyia 114
Phoretomyia 199
Phormia regina 269, 279
Phthiraptera 30-1, 361-82
  effect on domestic animals 378-80
  host specificity 380
  life cycle length 379
  on domestic animals 362
  peritrophic membrane 78

Phytoseiidae 424
Phytoseiulus persimilis 424
Pig (Sus scrofa domesticus)
  arthropod-borne diseases
    African swine fever 510
    Babesiasis 596
    filariasis 582
    trypanosomiasis 613
    tularaemia 553
    vesicular stomatitis 305
  ectoparasites of
    Demodex phylloides 420
    Haematopinus suis 369
    Sarcoptes scabiei 392
    Wohlfahrtia magnifica 285
Pirodog® 596
Piroplasmasina 558
Plague, see Yersinia pestis
Plasmodiidae 558
Plasmodium 558, 561-2, 572, 577, 580, see
    also malaria
  avian malaria 583
  basic reproduction rate 575
  life cycle 561-5
    in Anopheles 562-5
    diagram of 562
    history of 561-2
    human parasites 565-7
    in humans 562-5
    sporogony 565
Plasmodium berghei 563
Plasmodium durae 583
Plasmodium elongatum 562
Plasmodium falciparum 561-8, 573, 582-3,
    651
Plasmodium gallinaceum 583
Plasmodium hermani 583
Plasmodium juxtanucleare 583
Plasmodium knowlesi 567
Plasmodium malariae 564-7, 581
Plasmodium ovale 563-7, 581
Plasmodium relictum 88, 561-2
Plasmodium rodhaini 567
Plasmodium simium 567
Plasmodium vivax 561-8, 581
Pleuron 13
Pneumonyssoides caninum 433-5
  female 434
Pneumonyssus simicola 433-5
Podosoma 384, 385
Podospermy 385

Poecilochirus 264
Polyctenidae 344
Polymerase chain reaction (PCR) 98
Polyplacidae 362
Polyplax spinulosa
    Haemobartonella muris 533
    Rickettsia typhi 522
Polytene chromosomes 94, *95*, *96*, *97*
Pore canal *74*, *75*
Potamonautes pseudoperlatus 202
Potamonautes (Potamon) niloticus 661
Poultry
    arthropod-borne diseases
        Aegyptianella pullorum 532
        avian borelliosis 543
        avian plasmodia 583
        Leucocytozoon infections 585
    ectoparasites
        Argas spp. 450
        Dermanyssus gallinae 430
        Knemidokoptes mutans 401
        Mallophaga 373, 376
        Knemidokoptes gallinae 402
        Ornithonyssus spp. 432
    internal arthropod parasites
        Cytodites nudus 404
        Laminosioptes cysticola 405
Prementum 63
Preoral cavity 23
Presbytis entellus 497
Presbytis melalophos 641
Prescutum 68
Prestomal teeth *63*, 64, *69*
Prestomum 64
Pretarsal sclerite 369
Pretarsus 25, 385, *395*
Procavia 632
Proctiger 27
Proechimys 185
Prognathous 23
Proline, as an energy source 79
Propneustic 124
Propodosoma 384, *385*
 Prosimulium 193, 195
    and *Leucocytozoon* 585
Prosimulium fuscum 204
Prosimulium mixtum 198, 203–4
Prosimulium mysticum 204
Prosimulium ursinum 204
Prosoma 384
Prostigmata 383, 386, 414–23

Protease 78
Proteus hydrophilus, see Aeromonas hydrophila
Proteus mirabilis
Proterosoma 384
Prothoracic notched organ 118
Prothorax 25
Protonymph 386
Protophormia terraenovae 269, 279
Proventriculus 77
Providencia rettgeri 272
Psammolestes 354
Psammomys 633–4
Pseudolynchia canariensis 317
Pseudopod 109
Pseudoscorpion *22*, *23*
Pseudotracheae *61–3*
Psilopa petrolei 123
Psocoptera 29
Psorergates see Psorobia
Psorobia 416–7, 420
Psorobia bos 416
Psorobia ovis 416–8, *417*
Psorergatidae 416
Psorophora 113
    dormant eggs 115
Psorophora ferox and Venezuelan equine
    encephalitis 499
Psoroptes cervinus 396
Psoroptes cuniculi 396
Psoroptes equi 396
Psoroptes natalensis 396
Psoroptes ovis 396–9
    female *394*
    male *395*
    life cycle 396–7
    nymph 397
    on cattle 397–8
    sheep scab 397–8
        eradication 398
        recognition 396–7
        transmission 398
        treatment 398
Psoroptidae 385–6
Psoroptoidea 386, 395
Psychodidae 49, 177
Psychodinae 177
    wing *179*
Psychodopygus, see Lutzomyia
Pteridine for ageing 86
Pterolichoidea 409
Pteropleuron, *see* anepimeron 41, *43*

Pterygota 25, 27–38
Pthirus pubis 363, 367–8
  egg *365*
  Rickettsia prowazekii 519
Ptilinal (frontal) suture *42*
Ptilinum 51
Pulex 326
Pulex irritans 323–4, 33, 336–9 passim
  plague 550
  Rickettsia typhi 522
  Dipylidium caninum in 684
Pulex simulans 324
Pulsating organs 83
Pulvillus 25, *26*, 385, *395*
Pupipara 53
Pycnomonas 613–4
Pyemotes 414
Pyemotes beckeri 423
Pyemotes scolyti 423
Pyemotes tritici 423
  female *424, 425*
Pyemotidae 384
Pyemotoidea 384
Pyroglyphoidea 386
Pyroglyphidae 386, 407
Pyruvate, as an energy source 78

Q fever, *see Coxiella burnetii*

Radfordia affinis 422
Radfordia ensifera 422
Radial sector 44
Radius *44*
Raillietia auris 435–6
  female *435*
Raillietia caprae 435
Raillietiinae 325, 433–6 passim
Rattus
  scrub typhus 526
  sylvatic plague 549–50
Rattus exulans and plague 553
Rattus hawaiiensis and plague 553
Rattus norvegicus and plague 547–8
Rattus rattus and plague
Rectal glands 78
Rectum 77
Reduncinae 623
Reduviidae 344, 350–8
Redwater fever, *see* babesiosis

Reedbuck 623
Reindeer 302
Relapsing fevers, *see Borrelia* spp.
  avian 543
  endemic 541
  epidemic 540–1
  Lyme disease 537–40
  other 542–3
Relict bodies 84
Reoviridae 489, 506
Reproductive system
  female 84, *85*
  male 84, *85*
Resilin 74
Respiratory system 80, *81*
Respiratory trumpet *125*
Rhabdoviridae 489, 505–6,
Rhabdomys pumilio and *Rickettsia conorii* 525
Rhinitis, allergic 407
Rhinoestrus 296
Rhinoestrus latifrons 296
Rhinoestrus purpureus 292, 295–7
  female *297*
Rhinoestrus usbekistanicus 296
Rhinonyssidae 385, 425, 436–7
Rhipicephalus 461, *462*, 595, 603
  Babesia transmission 594–6
  Crimean-Congo haemorrhagic fever 504
Rhipicephalus appendiculatus 471–5, 461–9,
    497
  biology 472–3
    daily activity 472–3
    distribution on host 473
    hosts of 473
    life cycle duration 473, *474*
    seasonal abundance 473–5
  distribution in Africa 472
  vector of
    Nairobi sheep disease 504
    Rickettsia conorii 525
    Theileria parva 603–7
  veterinary importance 475
Rhipicephalus bursa 468
  Babesia ovis transmission 593–603 passim
  Crimean-Congo haemorrhagic fever 504
  Theileria ovis 595
Rhipicephalus duttoni 605
Rhipicephalus evertsi 461, 468, 481
  Babesia bigemina 594
  Babesia ovis 595
  Borrelia theileri 544

Rhipicephalus evertsi — continued
  paralysis caused by 481
  Theileria separata 603
  Theileria ovis 603
Rhipicephalus rossicus and Crimea Congo
    haemorrhagic fever 504
Rhipicephalus sanguineus 461
  male *462*
  vector of
    African horse sickness 509
    Anaplasma 532
    Babesia canis 596, 599
    Ehrlichia canis 529
    Rickettsia conorii 525
    Rickettsia rickettsii 523
    Rickettsia sibirica 524
Rhipicephalus simus 461
Rhipicephalus zambeziensis 472, 605–7
Rhodnius 357
  excretion in 355
Rhodnius ecuadoriensis 354
Rhodnius prolixus 352–8
  life cycle
    egg 352–3
    nymph 353
  vector of
    Trypanosoma cruzi 625–6
    Trypanosoma rangeli 627
Rhombomys opimus (great gerbil) 186, 632–3
Rickettsia 518
Rickettsia akari 525–6
Rickettsia australis 525
Rickettsia conorii 525
Rickettsia prowazekii 518–21
  epidemic typhus 518
  entomological aspects of 520–1
  survival of 520
  transmission of 519–20
Rickettsia rickettsii
Rickettsia sibirica
Rickettsia slovaca
Rickettsia tsutsugamushi
Rickettsia typhi 521–2
Rickettsiaceae 517–30
Rickettsiales 517–33
Rickettsial pox, *see Rickettsia akari*
Rickettsieae 517–29
Rift Valley fever 503
Rochalimaea 528
Rochalimaea quintana 528
Rochalimaea vinsonii 528

Rocky Mountain spotted fever, *see Rickettsia
    rickettsii* 523
Ross River virus 500
Rostrum *61*
Russian spring-summer encephalitis, *see*
    Tick-borne encephalitis 497

Sabethes 114–5
Sabethes chloropterus 133, 141
  and yellow fever 115
Sabethini 114
St Louis encephalitis 496
Saliva, anticoagulants in 58
Salivarium 77
Salivary gland 77
Salmonella 255
Salvinia 117
Sandfly fever virus 503
Sanje group of *Simulium damnosum* 96, *97*
Sarcomastigophora 612
Sarcophaga 270–1, *284*, 286
Sarcophaga melanura and habronemiasis 683
Sarcophagidae 53, 284–6, 268–70
Sarcopromusca arcuata 309, *310*
Sarcoptes scabiei 3, 387–93
  female *387*
  hosts of 388
  life cycle 389–90
  male *388*
  recognition of 390
Sarcoptes scabiei canis 389, 391
Sarcoptes scabiei hominis 391
Sarcoptidae 385–95
Scabies, *see also Sarcoptes scabiei* 391–2
  diagnosis of 390
  treatment of 391–2
Scaptia 217
Sceloporus occidentalis 539
Scepsidinae 214
Schistocerca gregaria 619
Schizophora *42*, 50
Schoengastia 415
  scrub typhus 526
Scirpus americanus 217
Scolopendron morsitans 15
Scorpiones 16, *17*
Scrub typhus, *see Rickettsia tsutsugamushi* 526
Sclerotin 74
Scutellum 41, *43*
Scutum 41, *43*

Selonocosmia javanensis 18
Sergentomyia 177–8, 184
Sergentomyia queenslandi cibarium 184
Setaria 638
Setaria cervi 652
Setaria digitata 682
Shannoniana 114
Sheep (*Ovis aries*)
    arbovirus diseases
        bluetongue 506
        Nairobi sheep disease 504
        Rift Valley fever 503
    arthropod parasites (*see also* myiasis
        below)
        Damalinia ovis 375, 377
        Eutrombicula sarcina 415
        Linognathus spp. 362, 370
        Melophagus ovinus 317
        Psorobia ovis 417
        Psoroptes ovis 396
        Raillietia auris 435
        Sarcoptes scabiei 392
    helminth diseases transmitted by
        arthropods
        Elaeophora schneideri 680
    myiasis caused by
        Calliphorinae 276–9
        Dermatobia hominis 309
        Oestrus ovis 295
    protozoal diseases transmitted by
        arthropods
        babesiasis 595
        theileriasis 603
        trypanosomiasis 620
    rickettsial diseases transmitted by
        arthropods
        Ehrlichia phagocytophila 529
        heartwater 630
        Q fever 529
    bacterial diseases transmitted by
        arthropods tularaemia 554
Shigella 255
Sickle cell anaemia 567
Simuliidae 49, 192–210
    bionomics
        adult feeding 70, 200–3
        breeding sites 201
        dispersal 203–4
        host finding 203
        mating 200
        oviposition 201–2

        survival 204–5
        control 206
        structure
            adult *193, 194*
            antennae *42*
            peritrophic membrane 77–8
            wing *45, 195*
        genera of 193–5
        geographical distribution of *195*
        life cycle
            adult emergence 199
            duration of
                egg 196
                larva *48, 194, 195,* 196–9
                pupa *49,* 199, *200*
        medical and veterinary importance 205
            pest species 205–6
            vectors of disease 205
Simulium 193, 194, 195
    and *Leucocytozoon* 585
    phoresy in 199
Simulium adersi 684
Simulium arcticum 203, 206
Simulium argyreatum 201
Simulium baffinense 204
Simulium bovis and *Onchocerca dukei*
Simulium callidum 201
Simulium columbaschense 206
Simulium damnosum complex 96, *97,* 199–204,
        657–67 passim
    annual biting rate 663
    control of 664–7
    Onchocerca ochengi 678
    Onchocerca volvulus 96, 660–1
    polytene chromosomes 96
    species groups 96–7
        phylogenetic relationships of 97
Simulium damnosum s.s. 97
    Onchocerca volvulus 660–1
Simulium equinum 198
Simulium erythrocephalum 200
Simulium euryadminiculum 203
Simulium exiguum and onchocerciasis 659,
        662
Simulium griseicolle 664
Simulium latipes 202, 204
Simulium luggeri 206
Simulium metallicum 201–2
    onchocerciasis 658–62 passim
Simulium monticola 202, 205
Simulium neavei 202

Simulium neavei – *continued*
  control of 664
  onchocerciasis 658-9, 661, 664
  phoresy 661
  species groups 661
Simulium nyasalandicum 202
Simulium ochraceum 201, 203
Simulium ornatipes 196, 201
Simulium ornatum 198, 200, 203
  animal onchocerciasis 677, 679
Simulium pictipes 201
Simulium piperi 197
Simulium reptans 195, 202
Simulium sanctipauli and onchocerciasis
    660-7
  temephos resistance 667
Simulium simile 200
Simulium sirbanum and onchocerciasis 660-1,
    663, 667
Simulium soubrense and onchocerciasis 660-1,
    667
Simulium squamosum 203
  onchocerciasis 660-1
Simulium tuberosum 201-2
Simulium venustum 203
Simulium vittatum 203
Simulium woodi 202
  onchocerciasis 659-661
Simulium yahense and onchocerciasis
    660-1
Sindbis virus 501
Siphon 119
Siphonaptera 36, *37*, 323-43
  adult behaviour and bionomics
    feeding 333-4
    host finding 332-3
    jumping 337-8
    mating 334-5
    proventriculus, function of 334
    relationship with hosts 336-7
    reproduction 335-6
  life cycle *330*
    adult 332
    duration of 329, 331-2
    egg 328-9
    larva 330-1
    pupa and cocoon 331
  medical and veterinary
    importance 339-40
  mouthparts 66, *67*
  peritrophic membrane 78

structure of adult 325-7, *324, 325, 327,*
    328, 329
  abdomen 326
  combs 327
  head 326
  thorax 326
Siphunculina funicola 52
Siphunculina ceylonica 52
Sitotroga cerealella 422
Sitophilus oryzae 422
Sleeping sickness, *see* trypanosomiasis
Sloth 502
Solenopotes 362, 370-1
Solenopotes capillatus 362, 371
  female *371*
Solenopsis invicta 472
Solifugids 21
Spanish-fly 32
Special Programme (TDR) 10
Species 7
Species complexes 88-99
  Aedes scutellaris 98
  Anopheles barbirostris 98
  Anopheles dirus 98
  Anopheles culicifacies 98
  Anopheles gambiae 93-6
  Anopheles hyrcanus 98
  Anopheles maculatus 98
  Anopheles maculipennis 88-93
  Anopheles minimus 98
  implications of 98-9
  Ornithodoros moubata 542
  Simulium damnosum 96-7, 660-1
Species sanitation 88
Spermatheca 84, *85*
Spermatophore 84
Spermophilus beechyi 551
Spermophilus variegatus 551
Spider 18
Spilopsyllus cuniculi 330-7 passim
  and myxomatosis virus 489
Spiracles 25, 80, *81*
Spirurida 638
Squama *44*
Sternopleuron, *see* katepisternum 41, *43*
Sternum 13
Stipes 66, *67*
Staphylinidae 32
Stegomyia 8
Stegophrynus dammermani 22
Stephanofilaria 638

Stephanofilaria assamensis 681
Stephanofilaria stilesi 681
Sterile insect release 273
Sternostoma tracheacolum 436-7
    female *436*
Stings
    bee 34
    scorpion 17
Stomach bots of horses, *see Gasterophilus* 303
Stomoxyinae 259
    mouthparts of *64, 65*
Stomoxys
    Habronema spp. 261
    Trypanosoma 622
Stomoxys calcitrans 248-52, 259-61
    ageing 86
    biology and bionomics 260
        response to adenosine nucleotides 71
    economic importance 260-1
    pathogens
        Habronema majus 683
        Setaria cervi 261
        Trypanosoma evansi 261
Stomoxys nigra 259, 261
Stomoxys sitiens 259
Streblidae 53
Stylets 68, *69*
Stylet sac 68, *69*
Styloconops, see Leptoconops
Stylosanthes 478
Subcosta *44*
Subgenus 8
Suboesophageal ganglion 83
Subscutellum 43
Subspecies 8
Suncus murinus 551
Superfamily 8
Supraoesophageal ganglion 83
Sweet itch 170
Sylvilagus bachmani 539
Sylvius
Symbionts 80
    rickettsial 517
Symbovines 250
Syncerus caffer, see Buffalo, African
Synonym 7

Tabanidae 4, 38, 50, 211-24
    antennae *42*
    bionomics 217-20

    breeding sites 218
    daily activity 218-20
    dispersal 219-20
    feeding 70
    host finding 219
    mating 218
    oviposition 217
    survival 220
    equine infectious anaemia 221
    hog cholera 221
    life cycle 215-7
        adult 216-7
        duration of 218
        egg 215
        larva *215,* 216
        pupa 216, *217*
    medical and veterinary importance 220-1
    mouthparts *59, 60*
    pathogens
        Anaplasma 532
        trypanosomes 221, 622, 624
        Trypanosoma theileri 627
Tabaninae 214
Tabanid tarsal claws 26
Tabanus 211, 215, 680
    energy source 79
Tabanus americanum 221
Tabanus autumnalis 218, 220
Tabanus biguttatus 216, 220
Tabanus conspicuus 220
Tabanus fraternus and *Anaplasma* 532
Tabanus fulvulus 219
Tabanus fuscicostatus 217, 221
Tabanus iyoensis 219-20
Tabanus molestus 219
Tabanus nigrovittatus 216, 218-20
Tabanus nipponicus 220
Tabanus pallidescens 219
Tabanus paradoxus 217
Tabanus parvicallosus 213
Tabanus punctifer 218
Tabanus sackeni 219
Tabanus septentrionalis 216
Tabanus taeniola 218-9
    Anaplasma 532
Tabanus unilineatus 219
Tachinidae 268
Tamiasciuris 543
Tarsomere 25
Tarsus 25, 385, *395*
Tatera brantsi 551

Tatera indica 552–3
Taurotragus oryx 603
TDR *see* Special Programme for Research and Training in Tropical Diseases 10
Tegmen 29
Teneral adult 75
Tergum 13
Testis 84, *85*
Tetranychidae 414
Thalassaemia 568
Theileria 558, 591, 595, 599, 602–4
   development 603–4
     in tick 604
     in vertebrate 603
   economic importance 602–3
Theileria annulata 602–3, 607–9
Theileria buffeli, see *Theileria orientalis*
Theileria hirci 603
Theileria lestoquardi 603
Theileria mutans 603–4
Theileria orientalis 602, 604
Theileria ovis 603–4
Theileria parva 602–8
   in cattle 602
   epizootiology of 606–7
   in *Rhipicephalus appendiculatus* 605–6
   subspecies 604
Theileria parva bovis 604–5
Theileria parva lawrencei 605
   epizootiology of 606–7
Theileria parva parva 605
Theileria separata 603–4
Theileria sergenti, see *Theileria orientalis*
Theileria velifera 603–4
Theileriidae 558
Theilerioses 602–9
   control in cattle 606
Thelazia 638, 683–4
Thelazia callipaeda 683–4
Thelazia rhodesi 684
Thelyphonus insularis 22
Theobroma cocao pollination 152
Thopha saccata 32
Thysanoptera 29
Thysanura 27, *27*
Tibia 25, 385, *395*
Tick-borne encephalitis 497–8
   subtypes 497
Tick-borne typhus 522–5
TickGARD® 479
Tick paralysis 480

Togaviridae 489, 498–501
Topi 623
Topomyia 114
Toxorhynchites 110, 114
Toxorhynchites amboinensis 126
   virus isolation 490, 500
Toxorhynchites speciosus 115
Toxorhynchitinae 110, 114
Trachea 80, *81*
Tracheoles 80
Tracheomyia macropi 292
Tragelaphinae 623
Tragulus 583
Transovarian transmission
   of arboviruses 490
   of *Babesia* 593
   of *Rickettsia* 522–7
Transverse suture *43*
Triatoma 350, 357
Triatoma braziliensis 358
   Trypanosoma cruzi 625
Triatoma dimidiata 354–8,
   Trypanosoma cruzi 625
Triatoma infestans 352–8
   Trypanosoma cruzi 625–6
   life cycle 352–3
Triatoma protracta 354–6
   Trypanosoma cruzi 626
Triatoma rubida 356
Triatoma rubida uhleri 356
Triatoma rubrofasciata 350, *353*
Triatoma sordida 354
Triatominae 350–7
   bionomics 352–7
     defaecation 355–6
     dispersal 356
     feeding 3354–5
     longevity 357
     resting places 356
     survival 357
   control 357
   head *351*
   life cycle
     egg 352–3
     nymph 353
   medical importance 357–8
   population growth 354
   relationship to humans 384
   wing *352*
Tribolium confusum 685
Trichobothria 414

Trichodectes canis 362, *375*, 377–80
   Dipylidium caninum 684
Trichodectidae 376–8
Trichoprosopon 114, 117
Tripteroides 114, 141
Tritonymph 386
Trochanter 25, 385, *395*
Trogoderma inclusum 33
Trombicula 414
Trombiculidae 414–6
   life cycle 415
Trombidiidae 384–5, 414
Trombidioidea 384, 414
Trypanosoma 612
   cycles in *Glossina* 614–7, *615*
      T. brucei 614–6
      T. congolense 616
      T. vivax 616
   infection rates in *Glossina* 617
   Salivaria 614–24
     species of 613
     subgenera 614
     species, list of 613
   Stercoraria, species of 613, 624–7
   in vertebrate host 617–21
Trypanosoma brucei brucei 613, 615–22 passim
Trypanosoma brucei gambiense 613, 615, 618
Trypanosoma brucei rhodesiense 613–5, 618
Trypanosoma congolense 613, 615–6, 620–2
Trypanosoma cruzi 613
   cycle in triatomine 625
   epidemiology of 626
   geographical distribution of 624
   vectors of 625
Trypanosoma evansi 4, 221, 613–4, 620, 622–4
Trypanosoma grayi 613, 615, 627
Trypanosoma lewisi 613, 627
Trypanosoma melophagium 613, 627
Trypanosoma rangeli 613, 626–7
Trypanosoma simiae 613, 621
Trypanosoma suis 613, 616, 621
Trypanosoma theileri 4, 221, 613, 627
Trypanosoma uniforme 613, 620
Trypanosoma vivax 613, 615–6, 620–2
Trypanosoma vivax viennei 221, 613–4, 622, 624
Trypanosomiasis 5, 612–4
   Chagas' disease 624–6
     control 626
     epidemiology 626
     parasite cycle 625

   vectors 625
   human sleeping sickness 617–9
     gambiense 618–9
     rhodesiense 619
   in African wildlife 621–2, *623*
   in domestic animals 620–1
   treatment 621
Trypanozoon 613–4, 617
Tsetse flies, *see Glossina*
Tularaemia 553–4
   domestic animals 554
   treatment 554
Tunga 325
Tunga penetrans 323–4, 328–30, 335, 339–40
   female 339
Twinnia 198
Typhus
   epidemic 518–21
   murine 521–2
   scrub 526–8
   tick-borne 522–5
Tyrophagus putrescentiae 407–8

Udaya 113
Uranotaenia 114, 117
Uranotaenia lateralis 134
Uranotaeniini 114
Uric acid 79
Uropsylla tasmanica 332
Uropygi 21, *22*

Vagina 85
Varanus 238
Veins on wing 26
Venezuelan equine encephalitis 499–500
Venoms
   cobra 18
   hymenopteran 34
   Latrodectus 19
   Loxosceles 19
   pederin 33
   scorpion 17–8
   Sicarius 20
Vesicula seminalis 84, *85*
Vesicular stomatitis 505
Vespula germanica 35
Viannia 630
Viviparity 84
   Musca bezzii 84

Viviparity—*continued*
    Musca planiceps 84
Vombatus ursinus 435

Warileya 177–8
Warthog 541
Water absorption
    Ixodidae 75
    Xenopsylla cheopis 75
Water buffalo, *see* Zebu
Waterbuck 623
Wax layer in cuticle 73
West Nile virus 495
Western equine encephalitis 499
Wildebeeste 623
Wing structure *44, 45*
    venation *44, 45*
Wohlfahrtia 270
Wohlfahrtia magnifica 285
Wohlfahrtia meigeni 286
Wohlfahrtia nuba 286
Wohlfahrtia opaca, see Wohlfahrtia meigeni
Wohlfahrtia vigil 285–6
Wolbachia 230
Wolbachia pipientis 80
Wolbachieae 517
Wombat 435
Wuchereria bancrofti, see also lymphatic
       filariasis 639–652
    forms of 644–5
Wyeomyia 114

Xenopsylla 326–7
Xenopsylla astia 327, 330, 335
    and plague 548–9
Xenopsylla brasiliensis
    plague 549, 551
Xenopsylla buxtoni and plague 552
Xenopsylla cheopis passim *5, 323–4, 327–37*
    adenosine nucleotides, response to 71
    host finding 333
    structure of
       female abdomen *329*
       male abdomen *328*
       spermatheca *329*
       thorax *327*
       tracheal system *81*
    vector of

Haemobartonella muris 533
Rickettsia typhi 522
Trypanosoma lewisi 627
Yersinia pestis 547–53
    water absorption 75
    water loss 81
Xenopsylla cunicularis 335
Xenopsylla hawaiiensis and plague 553
Xenopsylla philoxera and plague 551, 553

Yaws 52
Yellow fever 490–3
    epidemics since 1958 491
    sylvatic 492–3
Yersinia pestis 5, 517, 544
    characteristics 544
    development in flea 548–9, 548
    distribution 1959–79 546
    in humans 544–5
    in India 547–8
       epidemiology of 547–8
       Plague Commission (early 1900s) 547
       present status, Bombay 548
       present status, Calcutta 548
    pandemics 545
    present status
       Oriental region 542–3
       Palaearctic region 542
       South America 543
       Southern Africa 541
       United States of America 541
    subspecies 544–5
    survival 553
    sylvatic plague 550–1
    vectors 549–50

Zebra 238, 622–3
Zebu (*Bos indicus*)
    Babesia 592
    Dermatobia hominis 311
    Theileria parva 604
    tick resistance 478–9
Zercoseius ometes 21
Zeugnomyia 113
Zoogeographical regions 9
Zoological gardens: myiasis in captive
       animals 274
Zoophily 88